STUDENT'S
SOLUTIONS MANUAL

DAVID DUBRISKE

University of Arkansas for Medical Sciences

CALCULUS
AND ITS APPLICATIONS
ELEVENTH EDITION

Marvin L. Bittinger

Indiana University Purdue University Indianapolis

David J. Ellenbogen

Community College of Vermont

Scott A. Surgent

Arizona State University

PEARSON

Boston   Columbus   Hoboken   Indianapolis   New York   San Francisco
Amsterdam   Cape Town   Dubai   London   Madrid   Milan   Munich   Paris   Montreal   Toronto
Delhi   Mexico City   São Paulo   Sydney   Hong Kong   Seoul   Singapore   Taipei   Tokyo

ISBN-13: 978-0-321-99905-4
ISBN-10: 0-321-99905-3

1 2 3 4 5 6 RRD 18 17 16 15

www.pearsonhighered.com

PEARSON

TABLE OF CONTENTS

Chapter R

Functions, Graphs, and Models

Exercise Set R.1

1. Graph $y = x - 1$.

We choose some x-values and calculate the corresponding y-values to find some ordered pairs that are solutions of the equation. Then we plot the points and connect them with a smooth curve.

x	y	(x, y)
-2	-3	$(-2, -3)$
0	-1	$(0, -1)$
3	2	$(3, 2)$

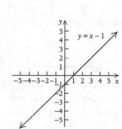

3. Graph $y = -\dfrac{1}{4}x$.

We choose some x-values and calculate the corresponding y-values to find some ordered pairs that are solutions of the equation. Then we plot the points and connect them with a smooth curve.

x	y	(x, y)
-4	1	$(-4, 1)$
0	0	$(0, 0)$
4	-1	$(4, -1)$

5. Graph $y = -\dfrac{5}{3}x + 3$.

We choose some x-values and calculate the corresponding y-values to find some ordered pairs that are solutions of the equation. Then we plot the points and connect them with a smooth curve.

x	y	(x, y)
-3	8	$(-3, 8)$
0	3	$(0, 3)$
3	-2	$(3, -2)$

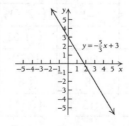

7. Graph $x + y = 5$.

We solve for y first.

$$x + y = 5$$

$$y = 5 - x \qquad \text{subtract } x \text{ from both sides}$$

$$y = -x + 5 \qquad \text{commutative property}$$

Next, we choose some x-values and calculate the corresponding y-values to find some ordered pairs that are solutions of the equation. Then we plot the points and connect them with a smooth curve.

x	y	(x, y)
-1	6	$(-1, 6)$
0	5	$(0, 5)$
2	3	$(2, 3)$

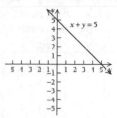

9. Graph $6x + 3y = -9$.

We solve for y first.

$$6x + 3y = -9$$

$$3y = -6x - 9 \qquad \text{subtract } 6x \text{ from both sides}$$

$$y = \frac{-6}{3}x - \frac{9}{3} \qquad \text{divide both sides by 3}$$

$$y = -2x - 3$$

Next, we choose some x-values and calculate the corresponding y-values to find some ordered pairs that are solutions of the equation. Then we plot the points and connect them with a smooth curve.

x	y	(x, y)
-2	1	$(-2, 1)$
0	-3	$(0, -3)$
2	-7	$(2, -7)$

11. Graph $2x + 5y = 10$.

We solve for y first.

$$2x + 5y = 10$$

$$5y = 10 - 2x \qquad \text{subtract } 2x \text{ from both sides}$$

$$y = \frac{1}{5}(10 - 2x) \qquad \text{divide both sides by 5}$$

$$y = 2 - \frac{2}{5}x$$

$$y = -\frac{2}{5}x + 2$$

The solution is continued on the next page.

We choose some x-values and calculate the corresponding y-values to find some ordered pairs that are solutions of the equation. Then we plot the points and connect them with a smooth curve.

x	y	(x, y)
-5	4	$(-5, 4)$
0	2	$(0, 2)$
5	0	$(5, 0)$

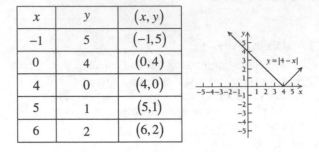

$2x + 5y = 10$

13. Graph $y = x^2 - 5$.

We choose some x-values and calculate the corresponding y-values to find some ordered pairs that are solutions of the equation. Then we plot the points and connect them with a smooth curve.

x	y	(x, y)
-2	-1	$(-2, -1)$
-1	-4	$(-1, -4)$
0	-5	$(0, -5)$
1	-4	$(1, -4)$
2	-1	$(2, -1)$

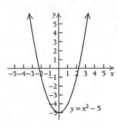

$y = x^2 - 5$

15. Graph $x = y^2 + 2$.

Since x is expressed in terms of y we first choose values for y and then compute x. Then we plot the points that are found and connect them with a smooth curve.

x	y	(x, y)
6	-2	$(-2, -2)$
3	-1	$(1, -1)$
2	0	$(2, 0)$
3	1	$(-1, 1)$
6	2	$(-2, 2)$

$x = y^2 + 2$

17. Graph $y = |4 - x|$.

We choose some x-values and calculate the corresponding y-values to find some ordered pairs that are solutions of the equation. Then we plot the points and connect them with a curve.

x	y	(x, y)
-1	5	$(-1, 5)$
0	4	$(0, 4)$
4	0	$(4, 0)$
5	1	$(5, 1)$
6	2	$(6, 2)$

$y = |4 - x|$

19. Graph $y = 7 - x^2$.

We choose some x-values and calculate the corresponding y-values to find some ordered pairs that are solutions of the equation. Then we plot the points and connect them with a smooth curve.

x	y	(x, y)
-2	3	$(-2, 3)$
-1	6	$(-1, 6)$
0	7	$(0, 7)$
1	6	$(1, 6)$
2	3	$(2, 3)$

$y = 7 - x^2$

21. Graph $y - 7 = x^3$.

First we solve for y.

$$y - 7 = x^3$$
$$y = x^3 + 7$$

Next, we choose some x-values and calculate the corresponding y-values to find some ordered pairs that are solutions of the equation. Then we plot the points and connect them with a smooth curve.

x	y	(x, y)
-2	-1	$(-2, -1)$
-1	6	$(-1, 6)$
0	7	$(0, 7)$
1	8	$(1, 8)$
2	15	$(2, 15)$

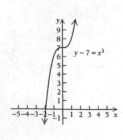

$y - 7 = x^3$

23. $A = 0.5t^4 + 3.45t^3 - 96.65t^2 + 347.7t, \ 0 \le t \le 6$

To determine the amount of ibuprofen in the blood stream that is left after 2 hours, we substitute $t = 2$ into the equation and solve for A.

$A = 0.5(2)^4 + 3.45(2)^3 - 96.65(2)^2 + 347.7(2)$

$= 344.4$

According to the model, approximately 344.4 milligrams of ibuprofen will remain in the blood stream 2 hours after 400 mg have been swallowed.

25. $v(t) = 10.9t$

We substitute 2.5 in for t to get

$v(2.5) = 10.9(2.5)$

$= 27.25$

Shaun White was traveling at 27.25 miles per hour when he reentered the half pipe.

27.
a) Locate 20 on the horizontal axis and go directly up to the graph. Then move left to the vertical axis and read the value there. We estimate the number of hearing-impaired Americans of age 20 is about 1.8 million.

Follow the same process for 40, 50, and 60 to determine the number of hearing-impaired Americans at each of those ages.

We estimate the number of hearing-impaired Americans of age 40 is about 4.1 million.

We estimate the number of hearing-impaired Americans of age 50 is about 5.3 million.

We estimate the number of hearing-impaired Americans of age 60 is about 6.0 million.

b) Locate 4 on the vertical axis and move horizontally across to the graph. There are two x-values that correspond to the y-value of 4. They are 39 and 82, so there are approximately 4 million Americans age 39 who are hearing-impaired and approximately 4 million Americans age 82 who are hearing-impaired.

c) The highest point on the graph appears to correspond to the x-value of 63. Therefore, age 58 appears to be the age at which the greatest number of Americans are hearing-impaired.

d) Visually, we cannot tell precisely which point is the highest point on the graph or which x-value corresponds exactly to that point. The graph is not detailed enough to make that determination.

29.
a) $A = P(1+r)^t$

$A = 100,000(1+0.028)^1$

$= 100,000(1.028)$

$= 102,800$

At the end of 1 year, the investment is worth \$102,800.

b) $A = P\left(1+\dfrac{r}{n}\right)^{nt}$

$A = 100,000\left(1+\dfrac{0.028}{2}\right)^{2 \cdot 1}$

$= 100,000(1+0.014)^2$

$= 100,000(1.014)^2$

$A = 100,000(1.028196)$

$= 102,819.60$

At the end of 1 year, the investment is worth \$102,819.60.

c) $A = P\left(1+\dfrac{r}{n}\right)^{nt}$

$A = 100,000\left(1+\dfrac{0.028}{4}\right)^{4 \cdot 1}$

$= 100,000(1+0.07)^4$

$= 100,000(1.07)^4$

$= 100,000(1.0282953744)$

$= 102,829.537$

$\approx 102,829.54$

At the end of 1 year, the investment is worth \$102,829.54.

d) $A = P\left(1+\dfrac{r}{n}\right)^{nt}$

$A = 100,000\left(1+\dfrac{0.028}{365}\right)^{365 \cdot 1}$

$= 100,000(1+0.00076712329)^{365}$

$= 100,000(1.00076712329)^{365}$

$= 100,000(1.02839458002)$

$= 102,839.458002$

$\approx 102,839.46$

At the end of 1 year, the investment is worth \$102,839.46.

e) There are $24 \cdot 365 = 8760$ hours in one year.

$$A = P\left(1 + \frac{r}{n}\right)^{nt}$$

$$A = 100{,}000\left(1 + \frac{0.028}{8760}\right)^{8760 \cdot 1}$$

$$= 100{,}000\left(1 + 0.000003196347\right)^{8760}$$

$$= 100{,}000\left(1.000003196347\right)^{8760}$$

$$= 100{,}000\left(1.02839563811\right)$$

$$= 102{,}839.563811$$

$$\approx 102{,}839.56$$

At the end of 1 year, the investment is worth $102,839.56.

31. a) $A = P(1 + r)^t$

$$A = 30{,}000(1 + 0.04)^3$$

$$= 30{,}000(1.04)^3$$

$$= 33{,}745.92$$

At the end of 3 years, the investment is worth $33,745.92.

b) $A = P\left(1 + \dfrac{r}{n}\right)^{nt}$

$$A = 30{,}000\left(1 + \frac{0.04}{2}\right)^{2 \cdot 3}$$

$$= 30{,}000(1.02)^6$$

$$= 30{,}000(1.1262)$$

$$= 33{,}784.87$$

At the end of 3 years, the investment is worth $33,784.87.

c) $A = P\left(1 + \dfrac{r}{n}\right)^{nt}$

$$A = 30{,}000\left(1 + \frac{0.04}{4}\right)^{4 \cdot 3}$$

$$= 30{,}000(1.01)^{12}$$

$$= 30{,}000(1.1268)$$

$$= 33{,}804.7509$$

$$\approx 33{,}804.75$$

At the end of 3 years, the investment is worth $33,804.75.

d) $A = P\left(1 + \dfrac{r}{n}\right)^{nt}$

$$A = 30{,}000\left(1 + \frac{0.04}{365}\right)^{365 \cdot 3}$$

$$= 30{,}000(1.000109589)^{1095}$$

$$= 30{,}000(1.1274893877)$$

$$= 33{,}824.68163$$

$$\approx 33{,}824.68$$

At the end of 3 years, the investment is worth $33,824.68.

e) There are $24 \cdot 365 = 8760$ hours in one year.

$$A = P\left(1 + \frac{r}{n}\right)^{nt}$$

$$A = 30{,}000\left(1 + \frac{0.04}{8760}\right)^{8760 \cdot 3}$$

$$= 30{,}000(1.000004566210046)^{26{,}280}$$

$$= 30{,}000(1.127496542677)$$

$$= 33{,}824.89628031334$$

$$\approx 33{,}824.90$$

At the end of 3 years, the investment is worth $33,824.90.

33. Using the formula:

$$M = P\left[\frac{\dfrac{r}{12}\left(1 + \dfrac{r}{12}\right)^n}{\left(1 + \dfrac{r}{12}\right)^n - 1}\right]$$

We substitute 18,000 for P, 0.064 $(6.4\% = 0.064)$ for r, and 36 $(3 \cdot 12 = 36)$ for n. Then we use a calculator to perform the computation.

$$M = 18{,}000\left[\frac{\dfrac{0.064}{12}\left(1 + \dfrac{0.064}{12}\right)^{36}}{\left(1 + \dfrac{0.064}{12}\right)^{36} - 1}\right]$$

$$\approx 550.86$$

The monthly payment on the loan will be approximately $550.86.

35. $W = P\left[\dfrac{(1+r)^n - 1}{r}\right]$

We substitute 3000 for P, 0.0657 $(6.57\% = 0.0657)$ for r, and 18 for n.

$W = 3000\left[\dfrac{(1+0.0657)^{18} - 1}{0.0657}\right]$

$\approx 97,881.97$

Rounded to the nearest cent, Kate's annuity will be worth \$97,881.97 after 18 years.

37. a) Locate 230 on the vertical axis and then think of a horizontal line extending across the graph from this point. The years for which the graph lies above this line are the years for which the condor population was at or above 230. We determine that the condor population is above 230 for the period from 2007 to 2012.

b) Locate 200 on the vertical axis and then think of a horizontal line extending across the graph from this point. The years for which the graph touches this line are the years for which the population was at 200. The condor population was at 200 in 2005.

c) Locate the highest point on the graph and extend a line vertically to the horizontal axis. The year which the condor population was the highest was 2012.

d) Locate the lowest point on the graph and extend a line vertically to the horizontal axis. The year when the condor population was the lowest was 2002.

39. a) Using the formula $W = P\left[\dfrac{(1+r)^n - 1}{r}\right]$ we

substitute 1200 for P, $0.04\,(4\% = 0.04)$ for r and 35 for n.

$W = 1200\left[\dfrac{(1+0.04)^{35} - 1}{0.04}\right]$

$\approx 88,382.67$

Sally will have approximately \$88,382.67 in her account when she retires.

b) Sally invested \$1200 per year for 35 years. Therefore, the total amount of her original payments is: $\$1200 \cdot 35 = \$42,000$. Since the total amount in the account was \$88,382.67, the interest earned over the 35 years is:

$\$88,382.67 - \$42,000 = \$46,382.67$

Therefore, \$42,000 was the total amount of Sally's payments and \$46,382.67 was the total amount of her interest.

41. Substituting the information into the formula for annual yield give us

$Y = \left(1 + \dfrac{0.053}{12}\right)^{12} - 1 = 0.0543.$

Thus, the annual percentage yield is 5.43%

43. Substituting the information into the formula for annual yield give us

$Y = \left(1 + \dfrac{0.0375}{52}\right)^{52} - 1 = 0.0382.$

Thus, the annual percentage yield is 3.82%

45. a) The annual yield for Western Bank is

$Y_{WB} = \left(1 + \dfrac{0.045}{1}\right)^{1} - 1 = 0.045.$

Thus the annual yield for Western Bank is 4.5%.
The annual yield for Commonwealth Savings is

$Y_{CW} = \left(1 + \dfrac{0.0443}{12}\right)^{12} - 1 = 0.0452.$

Thus the annual yield for Commonwealth savings is 4.52%.

b) Commonwealth savings has the higher annual yield.

47. The annual yield for Stockman's Bank is

$Y_{SB} = \left(1 + \dfrac{0.042}{1}\right)^{1} - 1 = 0.042.$

Thus the annual yield for Stockman's Bank is 4.2%.
In order to compete, the annual yield for Mesalands Savings must be 4.2%. We substitute into the annual yield formula and solve for the interest rate on the next page.

Substituting into the annual yield formula, we have:

$$\left(1+\frac{r}{12}\right)^{12}-1=0.042$$

$$\left(1+\frac{r}{12}\right)^{12}=1.042$$

$$1+\frac{r}{12}=(1.042)^{\frac{1}{12}}$$

$$\frac{r}{12}=(1.042)^{\frac{1}{12}}-1$$

$$r=12\left[(1.042)^{\frac{1}{12}}-1\right]$$

$$r=0.0412.$$

Mesalands Savings needs to offer at least 4.12% compounded monthly to be competitive.

49. Graph $y=x-150$

We use the following window.

```
WINDOW
 Xmin=-200
 Xmax=200
 Xscl=50
 Ymin=-200
 Ymax=200
 Yscl=50
↓Xres=1
```

Next, we type the equation into the calculator.

```
Plot1 Plot2 Plot3
\Y1▊X-150
\Y2=
\Y3=
\Y4=
\Y5=
\Y6=
\Y7=
```

The resulting graph is:

$y=x-150$

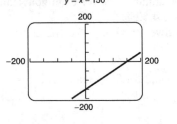

51. Graph $y=x^3+2x^2-4x-13$

We use the following window:

```
WINDOW
 Xmin=-10
 Xmax=10
 Xscl=1
 Ymin=-20
 Ymax=20
 Yscl=5
 Xres=1
```

Next, we type the equation in to the calculator.

```
Plot1 Plot2 Plot3
\Y1▊X^3+2X^2-4X-
13
\Y2=
\Y3=
\Y4=
\Y5=
\Y6=
```

The resulting graph is:

$y=x^3+2x^2-4x-13$

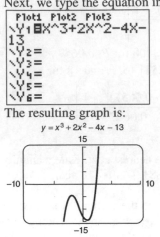

53. Graph $9.6x+4.2y=-100$.

First, we solve for y.
$$9.6x+4.2y=-100$$

$$4.2y=-100-9.6x \qquad \text{subtract } 9.6x \text{ from both sides}$$

$$y=\frac{-9.6x-100}{4.2}$$

Next, we set the window to be:

```
WINDOW
 Xmin=-20
 Xmax=10
 Xscl=5
 Ymin=-40
 Ymax=10
 Yscl=5
 Xres=1
```

Next, we type the equation into the calculator.

```
Plot1 Plot2 Plot3
\Y1▊(-9.6X-100)/
4.2
\Y2=
\Y3=
\Y4=
\Y5=
\Y6=
```

The resulting graph is:

$y=\dfrac{-9.6x-100}{4.2}$

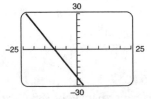

55. Graph $x = 4 + y^2$.

First we solve for y.

$$x = 4 + y^2$$

$$x - 4 = y^2 \qquad \text{subtracting 4}\\ \text{from both sides}$$

$$\pm\sqrt{x-4} = y \qquad \text{taking the square root}\\ \text{of both sides}$$

Next, we set the window to the standard window:

```
WINDOW
 Xmin=-10
 Xmax=10
 Xscl=1
 Ymin=-10
 Ymax=10
 Yscl=1
 Xres=1
```

It is important to remember that we must graph both the positive root and the negative root.
We type both equations into the calculator.

```
Plot1 Plot2 Plot3
\Y1 = -√(X-4)
\Y2 = √(X-4)
\Y3 =
\Y4 =
\Y5 =
\Y6 =
\Y7 =
```

This resulting graph is:

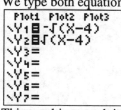
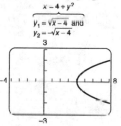

Exercise Set R.2

1. The correspondence is a function because each member of the domain corresponds to only one member of the range.

3. The correspondence is a function because each member of the domain corresponds to only one member of the range.

5. The correspondence is a function because each member of the domain corresponds to only one member of the range, even though two members of the domain, 10 pc. Chicken McNuggets and the Crispy Chicken both correspond to $4.29.

7. The correspondence is a function because each iPod has exactly one amount of memory.

9. The correspondence is a function because each iPod has exactly one number of songs at any given time.

11. The correspondence is a function because any number squared and then increased by 8, corresponds to exactly one number greater than or equal to 8.

12. The correspondence is a function because any number raised to the fourth power corresponds to exactly one nonnegative number.

13. The correspondence is a function because every female has exactly one biological mother.

15. This correspondence is *not* a function, because it is reasonable to assume at least one avenue is intersected by more than one cross street.

17. The correspondence is a function because each shape has exactly one area.

19. a) $f(x) = 4x - 3$
$$f(5.1) = 4(5.1) - 3 = 17.4$$
$$f(5.01) = 4(5.01) - 3 = 17.04$$
$$f(5.001) = 4(5.001) - 3 = 17.004$$
$$f(5) = 4(5) - 3 = 17$$

x	5.1	5.01	5.001	5
$f(x)$	17.4	17.04	17.004	17

b) $f(x) = 4x - 3$
$$f(4) = 4(4) - 3 = 13$$
$$f(3) = 4(3) - 3 = 9$$
$$f(-2) = 4(-2) - 3 = -11$$
$$f(k) = 4(k) - 3 = 4k - 3$$
$$f(1+t) = 4(1+t) - 3 = 4 + 4t - 3 = 4t + 1$$
$$f(x+h) = 4(x+h) - 3 = 4x + 4h - 3$$

21. $g(x) = x^2 - 3$
$$g(-1) = (-1)^2 - 3 = 1 - 3 = -2$$
$$g(0) = (0)^2 - 3 = 0 - 3 = -3$$
$$g(1) = (1)^2 - 3 = 1 - 3 = -2$$
$$g(5) = (5)^2 - 3 = 25 - 3 = 22$$
$$g(u) = (u)^2 - 3 = u^2 - 3$$
$$g(a+h) = (a+h)^2 - 3 = a^2 + 2ah + h^2 - 3$$
$$\frac{g(a+h) - g(a)}{h} = \frac{(a+h)^2 - 3 - \left[(a)^2 - 3\right]}{h}$$
$$= \frac{a^2 + 2ah + h^2 - 3 - \left[a^2 - 3\right]}{h}$$
$$= \frac{2ah + h^2}{h}$$
$$= \frac{h(2a+h)}{h}$$
$$= 2a + h$$

23. $f(x) = \dfrac{1}{(x+3)^2}$

a) $f(4) = \dfrac{1}{((4)+3)^2} = \dfrac{1}{(7)^2} = \dfrac{1}{49}$

$f(-3) = \dfrac{1}{((-3)+3)^2} = \dfrac{1}{(0)^2}$, Output is undefined.

$f(0) = \dfrac{1}{((0)+3)^2} = \dfrac{1}{(3)^2} = \dfrac{1}{9}$

$f(a) = \dfrac{1}{((a)+3)^2} = \dfrac{1}{(a+3)^2}$

The solution is continued on the next page.

$$f(t+4) = \frac{1}{((t+4)+3)^2} = \frac{1}{(t+7)^2}$$

$$f(x+h) = \frac{1}{((x+h)+3)^2} = \frac{1}{(x+h+3)^2}$$

$$\frac{f(x+h) - f(x)}{h}$$

$$= \frac{\dfrac{1}{(x+h+3)^2} - \dfrac{1}{(x+3)^2}}{h}$$

$$= \frac{\dfrac{(x+3)^2}{(x+h+3)^2(x+3)^2} - \dfrac{(x+h+3)^2}{(x+h+3)^2(x+3)^2}}{h}$$

$$= \frac{x^2 + 6x + 9 - \left(x^2 + 2hx + 6x + h^2 + 6h + 9\right)}{h(x+h+3)^2(x+3)^2}$$

$$= \frac{-2hx - h^2 - 6h}{h(x+h+3)^2(x+3)^2}$$

$$= \frac{h(-2x-h-6)}{h(x+h+3)^2(x+3)^2}$$

$$= \frac{-2x-h-6}{(x+h+3)^2(x+3)^2}, \quad h \neq 0$$

b) The function squares the input, then it adds six times the input, then it adds 9 and then it takes the reciprocal of the result.

25. a) A number x multiplied by 4 is $4x$. Then adding 2 to that quantity gives us $4x + 2$. Therefore the function is $f(x) = 4x + 2$.

b) Graph $f(x) = 4x + 2$.

First, we choose some values for x and compute the values for $f(x)$, in order to form the ordered pairs that we will plot on the graph.

$$f(-1) = 4(-1) + 2 = -2$$
$$f(0) = 4(0) + 2 = 2$$
$$f(1) = 4(1) + 2 = 6.$$

x	$f(x)$	$(x, f(x))$
-1	-2	$(-1, -2)$
0	2	$(0, 2)$
1	6	$(1, 6)$

Next we plot the input – output pairs from the table and, in this case, draw the line to complete the graph.

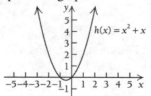

$f(x) = 4x + 2$

27. a) A number x squared is x^2. Then adding x to that quantity gives us $x^2 + x$. Therefore the function is $h(x) = x^2 + x$.

b) Graph $h(x) = x^2 + x$.

First, we choose some values for x and compute the values for $h(x)$, in order to form the ordered pairs that we will plot on the graph.

$$h(-2) = (-2)^2 + (-2) = 2$$
$$h(-1) = (-1)^2 + (-1) = 0$$
$$h(0) = (0)^2 + (0) = 0$$
$$h(1) = (1)^2 + (1) = 2$$
$$h(2) = (2)^2 + (2) = 6.$$

x	$h(x)$	$(x, h(x))$
-2	2	$(-2, 2)$
-1	0	$(-1, 0)$
0	0	$(0, 0)$
1	2	$(1, 2)$
2	6	$(2, 6)$

Next we plot the input – output pairs from the table and, in this case, draw the curve to complete the graph.

$h(x) = x^2 + x$

29. Graph $f(x) = 2x - 5$.

First, we choose some values for x and compute the values for $f(x)$, in order to form the ordered pairs that we will plot on the graph.

$f(-1) = 2(-1) - 5 = -7$

$f(0) = 2(0) - 5 = -5$

$f(1) = 2(1) - 5 = -3$

$f(2) = 2(2) - 5 = -1$

x	$f(x)$	$(x, f(x))$
-1	-7	$(-1, -7)$
0	-5	$(0, -5)$
1	-3	$(1, -3)$
2	-1	$(2, -1)$

Next we plot the input – output pairs from the table and, in this case, draw the line to complete the graph.

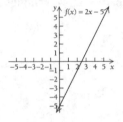

31. Graph $g(x) = -4x$.

First, we choose some values for x and compute the values for $g(x)$, in order to form the ordered pairs that we will plot on the graph.

$g(-1) = -4(-1) = 4$

$g(0) = -4(0) = 0$

$g(1) = -4(1) = -4$.

x	$g(x)$	$(x, g(x))$
-1	4	$(-1, 4)$
0	0	$(0, 0)$
1	-4	$(0, -4)$

Next we plot the input – output pairs from the table and, in this case, draw the line to complete the graph.

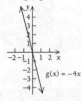

33. Graph $f(x) = x^2 - 2$.

First, we choose some values for x and compute the values for $f(x)$, in order to form the ordered pairs that we will plot on the graph.

$f(-2) = (-2)^2 - 2 = 2$

$f(-1) = (-1)^2 - 2 = -1$

$f(0) = (0)^2 - 2 = -2$

$f(1) = (1)^2 - 2 = -1$

$f(2) = (2)^2 - 2 = 2$

x	$f(x)$	$(x, f(x))$
-2	2	$(-2, 2)$
-1	-1	$(-1, -1)$
0	-2	$(0, -2)$
1	-1	$(1, -1)$
2	2	$(2, 2)$

Next we plot the input – output pairs from the table and, in this case, draw the curve to complete the graph.

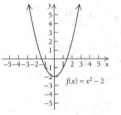

35. Graph $f(x) = 6 - x^2$.

First, we choose some values for x and compute the values for $f(x)$, in order to form the ordered pairs that we will plot on the graph.

$f(-2) = 6 - (-2)^2 = 2$

$f(-1) = 6 - (-1)^2 = 5$

$f(0) = 6 - (0)^2 = 6$

$f(1) = 6 - (1)^2 = 5$

$f(2) = 6 - (2)^2 = 2$

We organize these values into a table on the next page.

Organizing the values on the previous page into a table, we have:

x	$f(x)$	$(x, f(x))$
−2	2	(−2, 2)
−1	5	(−1, 5)
0	6	(0, 6)
1	5	(1, 5)
2	2	(2, 2)

Next we plot the input – output pairs from the table and, in this case, draw the curve to complete the graph.

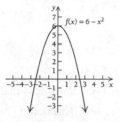

37. Graph $g(x) = x^3$.

First, we choose some values for x and compute the values for $g(x)$, in order to form the ordered pairs that we will plot on the graph.

$g(-2) = (-2)^3 = -8$

$g(-1) = (-1)^3 = -1$

$g(0) = (0)^3 = 0$

$g(1) = (1)^3 = 1$

$g(2) = (2)^3 = 8$

x	$f(x)$	$(x, f(x))$
−2	−8	(−2, 8)
−1	−1	(−1, −1)
0	0	(0, 0)
1	1	(1, 1)
2	8	(2, 8)

Next we plot the input – output pairs from the table and, in this case, draw the curve to complete the graph.

39. The graph is a function, it is impossible to draw a vertical line that intersects the graph more than once.

41. The graph is a function, it is impossible to draw a vertical line that intersects the graph more than once.

43. The graph is not that of a function. A vertical line can intersect the graph more than once.

45. The graph is not that of a function. A vertical line can intersect the graph more than once.

47. The graph is a function, it is impossible to draw a vertical line that intersects the graph more than once.

49. The graph is a function, it is impossible to draw a vertical line that intersects the graph more than once.

51. The graph is not that of a function. A vertical line can intersect the graph more than once.

53. Graph $x = y^2 - 2$.

a) First, we choose some values for y (since x is expressed in terms of y) and compute the values for x, in order to form the ordered pairs that we will plot on the graph.

For $y = -2; x = (-2)^2 - 2 = 2$

For $y = -1; x = (-1)^2 - 2 = -1$

For $y = 0; x = (0)^2 - 2 = -2$

For $y = 1; x = (1)^2 - 2 = -1$

For $y = 2; x = (2)^2 - 2 = 2$

We organize these values into a table on the next page.

Organizing the values on the previous page into a table, we have:

x	y	(x, y)
2	–2	$(2, -2)$
–1	–1	$(-1, -1)$
–2	0	$(-2, 0)$
–1	1	$(-1, 1)$
2	2	$(2, 2)$

Next we plot the input – output pairs from the table and, in this case, draw the curve to complete the graph.

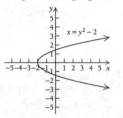

b) The graph is not that of a function. A vertical line can intersect the graph more than once.

55. $f(x) = x^2 - 3x$

$$\frac{f(x+h) - f(x)}{h}$$

$$= \frac{(x+h)^2 - 3(x+h) - \left[x^2 - 3x\right]}{h}$$

$$= \frac{x^2 + 2xh + h^2 - 3x - 3h - \left[x^2 - 3x\right]}{h}$$

$$= \frac{2xh + h^2 - 3h}{h} \quad \text{Combining like terms}$$

$$= \frac{h(2x + h - 3)}{h} \quad \text{Factoring}$$

$$= 2x + h - 3, \quad h \neq 0$$

57. To find $f(-1)$ we need to locate which piece defines the function on the domain that contains $x = -1$. When $x = -1$, the function is defined by $f(x) = -2x + 1$, for $x < 0$; therefore,

$$f(-1) = -2(-1) + 1 = 2 + 1 = 3.$$

To find $f(1)$ we need to locate which piece defines the function on the domain that contains $x = 1$. When $x = 1$, the function is defined by $f(x) = x^2 - 3$, for $0 < x < 4$; therefore,

$$f(1) = (1)^2 - 3 = 1 - 3 = -2.$$

59. To find $f(0)$ we need to locate which piece defines the function on the domain that contains $x = 0$.

When $x = 0$, the function is defined by $f(x) = 17$, for $x = 0$; therefore,

$$f(0) = 17.$$

To find $f(10)$ we need to locate which piece defines the function on the domain that contains $x = 10$. When $x = 10$, the function is defined by $f(x) = \frac{1}{2}x + 1$, for $x \geq 4$; therefore,

$$f(10) = \frac{1}{2}(10) + 1 = 5 + 1 = 6.$$

61. Graph $f(x) = \begin{cases} 1 & \text{for } x < 0 \\ -1 & \text{for } x \geq 0. \end{cases}$

First, we graph $f(x) = 1$ for inputs less than 0. We note for any x-value less than 0, the graph is the horizontal line $y = 1$. Note that for $f(x) = 1$

$$f(-2) = 1$$

$$f(-1) = 1$$

The open circle indicates that $(0, 1)$ is not part of the graph.

Next, we graph $f(x) = -1$ for inputs greater than or equal to 0. We note for any x-value less than 0, the graph is the horizontal line $y = -1$.

Note that for $f(x) = -1$.

$$f(0) = -1$$

$$f(1) = -1$$

$$f(2) = -1$$

The solid dot indicates that $(0, -1)$ is part of the graph.

63. Graph $f(x) = \begin{cases} 6, & \text{for } x = -2 \\ x^2, & \text{for } x \neq -2. \end{cases}$

First, we graph $f(x) = 6$ for $x = -2$.

This graph consists of only one point, $(-2, 6)$.

The solid dot indicates that $(-2, 6)$ is part of the graph.

Next, we graph $f(x) = x^2$ for inputs $x \neq -2$.

Note that for $f(x) = x^2$

$f(-3) = (-3)^2 = 9$

$f(-1) = (-1)^2 = 1$

$f(0) = (0)^2 = 0$

$f(1) = (1)^2 = 1$

$f(2) = (2)^2 = 4$

x	$f(x)$	$(x, f(x))$
-3	9	$(-3, 9)$
-1	1	$(-1, 1)$
0	0	$(0, 0)$
1	1	$(1, 1)$
2	4	$(2, 4)$

Since the input $x = -2$ is not defined on this part of the graph, the point $(-2, 4)$ is not part of the graph. The open circle indicates that $(-2, 4)$ is not part of the graph.

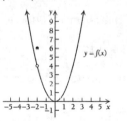

65. Graph $g(x) = \begin{cases} -x, & \text{for } x < 0 \\ 4, & \text{for } x = 0 \\ x + 2, & \text{for } x > 0. \end{cases}$

First, we graph $g(x) = -x$ for inputs $x < 0$.

Creating the input – output table, we have:

x	$g(x)$	$(x, g(x))$
-3	3	$(-3, 3)$
-2	2	$(-2, 2)$
-1	1	$(-1, 1)$

The open circle indicates that $(0, 0)$ is not part of the graph.

Next, we graph $g(x) = 4$ for $x = 0$. This part of the graph consists of a single point. The solid dot indicates that $(0, 4)$ is part of the graph.

Next, we graph $g(x) = x + 2$ for inputs $x > 0$.

Creating the input – output table, we have:

x	$g(x)$	$(x, g(x))$
1	3	$(1, 3)$
2	4	$(2, 4)$
3	5	$(3, 5)$

The open circle indicates that $(0, 2)$ is not part of the graph.

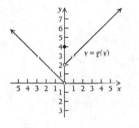

67. Graph $g(x) = \begin{cases} \frac{1}{2}x - 1, & \text{for } x < 2 \\ -4, & \text{for } x = 2 \\ x - 3, & \text{for } x > 2. \end{cases}$

First, we graph $g(x) = \frac{1}{2}x - 1$ for inputs $x < 2$.

Creating the input – output table, we have:

x	$g(x)$	$(x, g(x))$
-2	-2	$(-2, -2)$
0	-1	$(0, -1)$
1	$-\frac{1}{2}$	$\left(1, -\frac{1}{2}\right)$

The open circle indicates that $(2, 0)$ is not part of the graph.

Next, we graph $g(x) = -4$ for $x = 2$. This part of the graph consists of a single point. The solid dot indicates that $(2, -4)$ is part of the graph.

The solution is continued on the next page.

Next, we graph $g(x) = x - 3$ for inputs $x > 2$.
Creating the input – output table, we have:

x	$g(x)$	$(x, g(x))$
3	0	$(3, 0)$
4	1	$(4, 1)$
5	2	$(5, 2)$

The open circle indicates that $(2, -1)$ is not part of the graph.

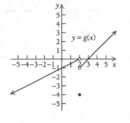

69. Graph $f(x) = \begin{cases} -7, & \text{for } x = 2 \\ x^2 - 3, & \text{for } x \neq 2. \end{cases}$

First, we graph $f(x) = -7$ for $x = 2$.

This graph consists of only one point, $(2, -7)$.

The solid dot indicates that $(2, -7)$ is part of the graph.

Next, we graph $f(x) = x^2 - 3$ for inputs $x \neq -2$.

Note that for $f(x) = x^2 - 3$

$f(-3) = (-3)^2 - 3 = 6$

$f(-1) = (-1)^2 - 3 = -2$

$f(0) = (0)^2 - 3 = -3$

$f(1) = (1)^2 - 3 = -2$

$f(3) = (3)^2 - 3 = 6$

We create the input – output table, we have:

x	$f(x)$	$(x, f(x))$
-3	6	$(-3, 6)$
-1	-2	$(-1, -2)$
0	-3	$(0, -3)$
1	-2	$(1, -2)$
3	6	$(3, 6)$

Since the input $x = 2$ is not defined on this part of the graph, the point $(2, 1)$ is not part of the graph. The open circle indicates that $(2, 1)$ is not part of the graph.

Using the information from the previous column, the graph is:

71. $A(t) = P\left(1 + \dfrac{0.03}{4}\right)^{4t}$

We substitute 500 in for P and 2 in for t:

$A(t) = 500\left(1 + \dfrac{0.03}{4}\right)^{4 \cdot 2}$

$= 500(1.0075)^8$

$= 500(1.061598848)$

$= 530.7994239$

≈ 530.80

The investment will be worth approximately $530.80 after 2 years.

73. $s = \sqrt{\dfrac{hw}{3600}}$

a) We substitute 170 for h and 70 for w.

$s = \sqrt{\dfrac{(170)(70)}{3600}} \approx 1.818$

The patient's approximate surface area is $1.818\,\text{m}^2$

b) We substitute 170 for h and 100 for w.

$s = \sqrt{\dfrac{(170)(100)}{3600}} \approx 2.173$

The patient's approximate surface area is $2.173\,\text{m}^2$

c) We substitute 170 for h and 50 for w.

$s = \sqrt{\dfrac{(170)(50)}{3600}} \approx 1.537$

The patient's approximate surface area is $1.537\,\text{m}^2$

75. a) Yes, the table represents a function. Each event is assigned exactly one scale of impact number.

b) The inputs are the events; the outputs are the scale of impact numbers.

77. First we solve the equation for y.

$2y^2 + 3x = 4x + 5$

$2y^2 = x + 5$ subtract $3x$ from both sides

$y^2 = \dfrac{x+5}{2}$ divide both sides by 2

$y = \pm\sqrt{\dfrac{x+5}{2}}$ use the principal of square roots

We sketch a graph of the equation.

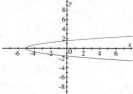

We can see that a vertical line will intersect the graph more than once; therefore, this is not a function.

79. First, we solve the equation for y.

$\left(3y^{3/2}\right)^2 = 72x$

$9y^3 = 72x$

$y^3 = 8x$

$y = \sqrt[3]{8x}$

$y = 2\sqrt[3]{x}$ $[y \geq 0]$

Note: since y must be nonnegative to satisfy the original equation, we only graph the points for which y is nonnegative.

Next, we sketch a graph of the equation:

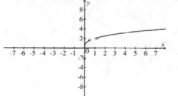

No vertical line meets the graph more than once. Thus, the equation represents a function.

81. ✎ Yes, $x = 4$ is in the domain of f in Exercises 57-60. The function is defined for all values of x.

83. $f(x) = \dfrac{3}{x^2 - 4}$

We begin by setting up the table:

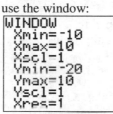

Next, we will type in the equation into the graphing editor.

Now, we are able to look at the table:

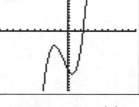

85. Each graph is shown below.

In order to graph $f(x) = x^3 + 2x^2 - 4x - 13$, we use the window:

After entering the function into the graphing editor, we get:

In order to graph $f(x) = \dfrac{3}{x^2 - 4}$, we use the standard window:

The solution is continued on the next page.

After entering the function into the graphing editor, we get:

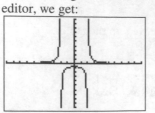

In order to graph $f(x) = |x-2| + |x+1| - 5$, we use the standard window.

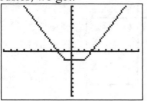

After entering the function into the graphing editor, we get:

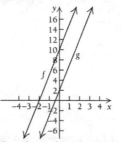

87. a) For *f*, a number *x* adding 2 is $x+2$, then multiplying that quantity by 5 yields $5(x+2)$. So $f(x) = 5(x+2)$.

For *g*, a number *x* multiplied by 5 is $5x$ then adding 2 to that quantity yields $5x + 2$. Therefore, $g(x) = 5x + 2$.

b) Graphing
$f(x) = 5(x+2)$ and $g(x) = 5x + 2$ we have:

c) By observing the graphs, we can deduce these are not the same functions.

89. For *f*, a number *x* multiplied by 3 is $3x$ then adding 6 to that quantity yields $3x + 6$. Therefore, $f(x) = 3x + 6$.

For *g*, a number *x* adding *a* is $x + a$, then multiplying that quantity by 3 yields $3(x + a)$. Therefore, $g(x) = 3(x + a)$.

To determine the value of *a* for when these two functions are equal, we set $f(x)$ equal to $g(x)$ and solve for *a*.

$$f(x) = g(x)$$
$$3x + 6 = 3(x + a)$$
$$3x + 6 = 3x + 3a \qquad \text{distribute the 3}$$
$$3x - 3x + 6 = 3x - 3x + 3a \qquad \text{subtract } 3x \text{ from both sides}$$
$$6 = 3a$$
$$\frac{6}{3} = \frac{3a}{3} \qquad \text{divide both sides by 3}$$
$$2 = a.$$

Therefore the functions are the same when $a = 2$.

Exercise Set R.3

1. $(-1,3)$

3. $(0,5)$

5. $(-9,-5]$

7. $[x,x+h]$

9. $(-\infty,q]$

11. $[-2,2]$

13. $(6,20]$

15. $(-3,\infty)$

17. $(-2,3]$

19. $[12.5,\infty)$

21. a) First, we locate 1 on the horizontal axis and then we look vertically to find the point on the graph for which 1 is the first coordinate. From that point, we look to the vertical axis to find the corresponding y-coordinate, 3. Thus, $f(1)=3$.

b) The domain is the set of all x-values of the points on the graph. The domain is $\{-3,-1,1,3,5\}$.

c) First, we locate 2 on the vertical axis and then we look horizontally to find any points on the graph for which 2 is the second coordinate. One such point exists, $(3,2)$.

Thus the x-value for which $f(x)=2$ is $x=3$.

d) The range is the set of all y-values of the points on the graph. The range is $\{-2,0,2,3,4\}$.

23. a) First, we locate 1 on the horizontal axis and then we look vertically to find the point on the graph for which 1 is the first coordinate. From that point, we look to the vertical axis to find the corresponding y-coordinate, 4. Thus, $f(1)=4$.

b) The domain is the set of all x-values of the points on the graph. The domain is $\{-5,-3,1,2,3,4,5\}$.

c) First, we locate 2 on the vertical axis and then we look horizontally to find any points on the graph for which 2 is the second coordinate. Three such point exists, $(-5,2);(-3,2)$; and $(4,2)$. Thus the x-values for which $f(x)=2$ are $\{-5,-3,4\}$.

d) The range is the set of all y-values of the points on the graph. The range is $\{-3,2,4,5\}$

25. a) First, we locate 1 on the horizontal axis and then we look vertically to find the point on the graph for which 1 is the first coordinate. From that point, we look to the vertical axis to find the corresponding y-coordinate, 1. Thus, $f(1)=-1$.

b) The domain is the set of all x-values of the points on the graph. These extend from -2 to 4. Thus, the domain is $\{x\,|-2\le x\le 4\}$, or in interval notation $[-2,4]$.

c) First, we locate 2 on the vertical axis and then we look horizontally to find any points on the graph for which 2 is the second coordinate. One such point exists, $(3,2)$.

Thus the x-value for which $f(x)=2$ is $x=3$.

d) The range is the set of all y-values of the points on the graph. These extend from -3 to 3. Thus, the range is $\{y\,|-3\le y\le 3\}$, or, in interval notation $[-3,3]$.

27. a) First, we locate 1 on the horizontal axis and then we look vertically to find the point on the graph for which 1 is the first coordinate. From that point, we look to the vertical axis to find the corresponding y-coordinate, -2. Thus, $f(1)=-2$.

b) The domain is the set of all x-values of the points on the graph. These extend from -4 to 2. Thus, the domain is $\{x\,|-4\le x\le 2\}$, or, in interval notation $[-4,2]$.

c) First, we locate 2 on the vertical axis and then we look horizontally to find any points on the graph for which 2 is the second coordinate. One such point exists, $(-2,2)$. Thus the x-value for which $f(x)=2$ are $x=-2$.

d) The range is the set of all y-values of the points on the graph. These extend from -3 to 3. Thus, the range is $\{y\,|-3\le y\le 3\}$, or in interval notation $[-3,3]$.

29. a) First, we locate 1 on the horizontal axis and then we look vertically to find the point on the graph for which 1 is the first coordinate. From that point, we look to the vertical axis to find the corresponding y-coordinate, 3. Thus, $f(1)=3$.

b) The domain is the set of all x-values of the points on the graph. These extend from -3 to 3. Thus, the domain is $\{x\,|-3\le x\le 3\}$, or, in interval notation $[-3,3]$.

c) First, we locate 2 on the vertical axis and then we look horizontally to find any points on the graph for which 2 is the second coordinate. Two such point exists, $(-1.4,2)$ and $(1.4,2)$. Thus the x-values for which $f(x)=2$ are $\{-1.4,1.4\}$.

d) The range is the set of all y-values of the points on the graph. These extend from -5 to 4. Thus, the range is $\{y\,|-5\le y\le 4\}$, or in interval notation $[-5,4]$.

31. a) First, we locate 1 on the horizontal axis and then we look vertically to find the point on the graph for which 1 is the first coordinate. From that point, we look to the vertical axis to find the corresponding y-coordinate, 1. Thus, $f(1)=1$.

b) The domain is the set of all x-values of the points on the graph. These extend from -5 to 5. However, the open circle at the point $(5,2)$ indicates that 5 is not in the domain. Thus, the domain is $\{x\,|-5\le x<5\}$, or in interval notation, $[-5,5)$.

c) First, we locate 2 on the vertical axis and then we look horizontally to find any points on the graph for which 2 is the second coordinate. We notice all the points with x-values in the set $\{x\,|3\le x<5\}$. Thus the x-values for which $f(x)=2$ are $\{x\,|3\le x<5\}$, or $[3,5)$.

d) The range is the set of all y-values of the points on the graph. The range is $\{-2,-1,0,1,2\}$.

33. $f(x)=\dfrac{6}{2-x}$

Since the function value cannot be calculated when the denominator is equal to 0, we solve the following equation to find those real numbers that must be excluded from the domain of f.

$2-x=0$ setting the denominator equal to 0

$2=x$ adding x to both sides

Thus, 2 is not in the domain of f, while all other real numbers are. The domain of f is $\{x\,|\,x$ is a real number and $x\ne 2\}$; or, in interval notation, $(-\infty,2)\cup(2,\infty)$.

35. $f(x)=\sqrt{2x}$

Since the function value cannot be calculated when the radicand is negative, the domain is all real numbers for which $2x\ge 0$. We find them by solving the inequality.

$2x\ge 0$ setting the radicand ≥ 0

$x\ge 0$ dividing both sides by 2

The domain of f is $\{x\,|\,x$ is a real number and $x\ge 0\}$; or, in interval notation, $[0,\infty)$.

37. $f(x)=x^2-2x+3$

We can calculate the function value for all values of x, so the domain is the set of all real numbers $\mathbb{R}$.

39. $f(x) = \dfrac{x-2}{6x-12}$

Since the function value cannot be calculated when the denominator is equal to 0, we solve the following equation to find those real numbers that must be excluded from the domain of f.

$6x - 12 = 0$ setting the denominator equal to 0

$\qquad 6x = 12$ adding 12 to both sides

$\qquad\ x = 2$ dividing both sides by 6

Thus, 2 is not in the domain of f, while all other real numbers are. The domain of f is $\{x \mid x \text{ is a real number and } x \neq 2\}$; or, in interval notation, $(-\infty, 2) \cup (2, \infty)$.

41. $f(x) = |x - 4|$

We can calculate the function value for all values of x, so the domain is the set of all real numbers $\mathbb{R}$.

43. $f(x) = \dfrac{3x-1}{7-2x}$

Since the function value cannot be calculated when the denominator is equal to 0, we solve the following equation to find those real numbers that must be excluded from the domain of f.

$7 - 2x = 0$ setting the denominator equal to 0

$\qquad 7 = 2x$ adding $2x$ to both sides

$\qquad \dfrac{7}{2} = x$ dividing both sides by 2

Thus, $\dfrac{7}{2}$ is not in the domain of f, while all other real numbers are. The domain of f is $\left\{x \mid x \text{ is a real number and } x \neq \dfrac{7}{2}\right\}$; or, in interval notation, $\left(-\infty, \dfrac{7}{2}\right) \cup \left(\dfrac{7}{2}, \infty\right)$.

45. $g(x) = \sqrt{4 + 5x}$

Since the function value cannot be calculated when the radicand is negative, the domain is all real numbers for which $4 + 5x \geq 0$. We find them by solving the inequality.

$4 + 5x \geq 0$ setting the radicand ≥ 0

$\quad 5x \geq -4$ subtracting 4 from both sides

$\quad\ x \geq -\dfrac{4}{5}$ dividing both sides by 5

The domain of g is:

$\left\{x \mid x \text{ is a real number and } x \geq -\dfrac{4}{5}\right\}$; or, in interval notation, $\left[-\dfrac{4}{5}, \infty\right)$.

47. $g(x) = x^2 - 2x + 1$

We can calculate the function value for all values of x, so the domain is the set of all real numbers $\mathbb{R}$.

49. $g(x) = \dfrac{2x}{x^2 - 25}$

Since the function value cannot be calculated when the denominator is equal to 0, we solve the following equation to find those real numbers that must be excluded from the domain of g.

$x^2 - 25 = 0$ setting the denominator equal to 0

$\qquad x^2 = 25$ adding 25 to both sides

$\qquad\ x = \pm\sqrt{25}$ Principal of square roots

$\qquad\ x = \pm 5$

Thus, -5 and 5 are not in the domain of g, while all other real numbers are. The domain of g is $\{x \mid x \text{ is a real number and } x \neq -5,\ x \neq 5\}$; or, in interval notation, $(-\infty, -5) \cup (-5, 5) \cup (5, \infty)$.

51. $g(x) = |x| + 1$

We can calculate the function value for all values of x, so the domain is the set of all real numbers $\mathbb{R}$.

53. $g(x) = \dfrac{2x-6}{x^2 - 6x + 5}$

Since the function value cannot be calculated when the denominator is equal to 0, we solve the following equation to find the real numbers that must be excluded from the domain of g.

$x^2 - 6x + 5 = 0$ setting the denominator equal to 0

$(x-5)(x-1) = 0$ factoring the quadratic equation

$x - 5 = 0 \text{ or } x - 1 = 0$ Using the principle of zero products

$\quad x = 5 \text{ or } \quad x = 1$

Thus, 1 and 5 are not in the domain of g, while all other real numbers are. The domain of g is $\{x \mid x \text{ is a real number and } x \neq 1, x \neq 5\}$; or, in interval notation, $(-\infty, 1) \cup (1, 5) \cup (5, \infty)$.

55. The graph of f lies on or below the x-axis when $f(x) \leq 0$, so we scan the graph from left to right looking for the values of x for which the graph lies on or below the x axis. Those values extend from -1 to 2. So the set of x-values for which $f(x) \leq 0$ is $\{x \,|\, -1 \leq x \leq 2\}$, or, in interval notation, $[-1, 2]$.

57. a) We use the compound interest formula from Theorem 2 in section R.1 and substitute 5000 for P, 2 for n and 0.031 (3.1%) for r. The equation for this function is:

$$A(t) = P\left(1 + \frac{r}{n}\right)^{nt}$$

$$A(t) = 5000\left(1 + \frac{0.031}{2}\right)^{2t}$$

$$A(t) = 5000(1.0155)^{2t}$$

  b) The independent variable t is the time in years the principal has been invested in the account. It would not make sense to have time be a negative number in this case. Therefore, the domain is the set of all non-negative real numbers. $\{t \,|\, 0 \leq t < \infty\}$.

59. a) The graph extends from $x = 25$ to $x = 102$, so the domain, in interval notation, of the function I is $[25, 102]$.

  b) The graph extends from $I(x) = 0$ to approximately $I(x) = 455$. Therefore, the range, in interval notation, of the function I is approximately $[0, 455]$.

  c)

61. a) The graph extends from $t = 0$ to $t = 70$, so the domain, in interval notation, of the function L is $[0, 70]$.

  b) The graph extends from $L(t) = 8$ to $L(t) = 75$, so the range, in interval notation, of the function L is $[8, 75]$.

63.

65. The range in interval notation for each function is:

Exercise 33: $(-\infty, 0) \cup (0, \infty)$

Exercise 35: $[0, \infty)$

Exercise 39: $\left\{\frac{1}{6}\right\}$

Exercise 40: $(-\infty, 0) \cup (0, \infty)$

Exercise 47: $[0, \infty)$

Exercise Set R.4

1. Graph $x = 5$.

The graph consists of all ordered pairs whose first coordinate is 5. This results in a vertical line whose x-intercept is the point $(5,0)$

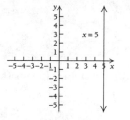

3. Graph $y = -4$.

The graph consists of all ordered pairs whose second coordinate is -4. This results in a horizontal line whose y-intercept is the point $(0,-4)$.

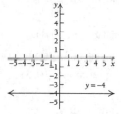

5. Graph $x = -1.5$.

The graph consists of all ordered pairs whose first coordinate is -1.5. This results in a vertical line whose x-intercept is the point $(-1.5,0)$.

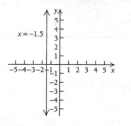

7. Graph $y = 2.25$.

The graph consists of all ordered pairs whose second coordinate is 2.25. This results in a horizontal line whose y-intercept is the point $(0,2.25)$.

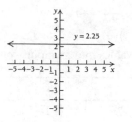

9. Graph $y = -2x$.

Using Theorem 4, The graph of y is the straight line through the origin $(0,0)$ and the point $(1,-2)$. We plot these two points and connect them with a straight line.

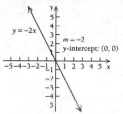

The function $y = -2x$ has slope -2, and y-intercept $(0,0)$.

11. Graph $f(x) = -0.5x$

Using Theorem 4, The graph of $f(x)$ is the straight line through the origin $(0,0)$ and the point $(1,-0.5)$. We plot these two points and connect them with a straight line.

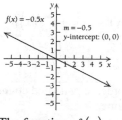

The function $f(x) = -0.5x$ has slope -0.5, and y-intercept $(0,0)$.

13. Graph $y = 3x - 4$

First, we make a table of values. We choose any number for x and then determine y by substitution.

When $x = -1, y = 3(-1) - 4 = -7$.

When $x = 0, y = 3(0) - 4 = -4$.

When $x = 2, y = 3(2) - 4 = 2$.

We organize these values into an input – output table.

x	y	(x, y)
-1	-7	$(-1, -7)$
0	-4	$(0, -4)$
2	2	$(2, 2)$

Next, we plot these ordered pairs and connect them with a straight line.

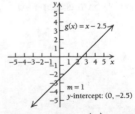

The function $y = 3x - 4$ has slope 3, and y-intercept $(0, -4)$.

15. Graph $g(x) = x - 2.5$.

First, we make a table of values. We choose any number for x and then determine y by substitution.

When $x = 0, g(0) = (0) - 2.5 = -2.5$.

When $x = 1, g(1) = (1) - 2.5 = -1.5$.

We organize these values into an input – output table.

x	$g(x)$	$(x, g(x))$
0	-2.5	$(0, -2.5)$
1	-1.5	$(1, -1.5)$

Next, we plot these ordered pairs and connect them with a straight line.

The function $g(x) = x - 2.5$ has slope 1, and y-intercept $(0, -2.5)$.

17. Graph $y = 7$.

The graph consists of all ordered pairs whose second coordinate is 7. This results in a horizontal line, whose y-intercept is the point $(0, 7)$.

Since the graph is horizontal, the slope is 0 and the y-intercept is $(0, 7)$.

19. First, we solve the equation for y.

$$y - 4x = 1$$
$$y = 4x + 1 \qquad \text{adding } 4x \text{ to both sides}$$

The slope is 4. The y-intercept is $(0, 1)$.

21. First, we solve the equation for y.

$$2x + y - 3 = 0$$
$$2x + y = 3 \qquad \text{adding 3 to both sides}$$
$$y = -2x + 3 \quad \text{subtracting } 2x \text{ from both sides}$$

The slope is -2. The y-intercept is $(0, 3)$.

23. First, we solve the equation for y.

$$3x - 3y + 6 = 0$$
$$3x - 3y = -6 \qquad \text{subtracting 6 from both sides}$$
$$-3y = -3x - 6 \quad \text{subtracting } 3x \text{ from both sides}$$
$$y = \frac{-3x - 6}{-3} \qquad \text{dividing both sides by } -3$$
$$y = x + 2 \qquad \text{simplifying}$$

The slope is 1. The y-intercept is $(0, 2)$.

25. First, we solve the equation for y.

$$x = 3y + 7$$
$$3y + 7 = x \qquad \text{commutative property of equality}$$
$$3y = x - 7 \qquad \text{subtracting 7}$$
$$y = \frac{x - 7}{3} \qquad \text{dividing by 3}$$
$$y = \frac{1}{3}x - \frac{7}{3} \qquad \text{simplifying}$$

The slope is $\dfrac{1}{3}$.

The y-intercept is $\left(0, -\dfrac{7}{3}\right)$.

27. $y - y_1 = m(x - x_1)$

$\quad y - (7) = 7(x - (1))$ substituting

$\quad\quad y - 7 = 7x - 7$ simplifying

$\quad\quad\quad y = 7x$ adding 7 to both sides.

29. $y - y_1 = m(x - x_1)$

$\quad y - (3) = -2(x - (2))$ Substituting

$\quad\quad y - 3 = -2x + 4$

$\quad\quad\quad y = -2x + 7$ Adding 3

31. $y - y_1 = m(x - x_1)$

$\quad y - (0) = -5(x - (5))$ Substituting

$\quad\quad\quad y = -5x + 25$

33. $y = mx + b$

$\quad y = \dfrac{1}{2}x + (-6)$ Substituting

$\quad y = \dfrac{1}{2}x - 6$ Simplifying

35. $y - y_1 = m(x - x_1)$

$\quad y - (8) = 0(x - (4))$ Substituting

$\quad\quad y - 8 = 0$ Simplifying

$\quad\quad\quad y = 8$ Adding 8

37. $m = \dfrac{y_2 - y_1}{x_2 - x_1}$

Substituting, we have:

$m = \dfrac{1 - (-3)}{-2 - 5}$

$\quad = \dfrac{1 + 3}{-2 - 5}$

$\quad = \dfrac{4}{-7}$

$\quad = -\dfrac{4}{7}$

39. $m = \dfrac{y_2 - y_1}{x_2 - x_1}$

Substituting, we have:

$m = \dfrac{-6 - (-5)}{1 - (-3)}$

$\quad = \dfrac{-6 + 5}{1 + 3}$

$\quad = -\dfrac{1}{4}$

41. $m = \dfrac{y_2 - y_1}{x_2 - x_1}$

Substituting, we have:

$m = \dfrac{-9 - (-7)}{3 - 3}$

$\quad = \dfrac{-9 + 7}{3 - 3}$

$\quad = \dfrac{-2}{0}$

Since we cannot divide by 0, the slope is undefined.

43. $m - \dfrac{y_2 - y_1}{x_2 - x_1}$

$m = \dfrac{-\dfrac{3}{4} - \left(-\dfrac{1}{2}\right)}{\dfrac{5}{8} - \left(-\dfrac{3}{16}\right)}$

$\quad = \dfrac{\dfrac{-3}{4} + \dfrac{2}{4}}{\dfrac{10}{16} + \dfrac{3}{16}}$ finding a common denominator

$\quad = \dfrac{-\dfrac{1}{4}}{\dfrac{13}{16}}$ adding fractions

$\quad = -\dfrac{1}{4} \cdot \left(\dfrac{16}{13}\right)$ Multiplying by the reciprical

$\quad = -\dfrac{4}{13}$

45. $m = \dfrac{y_2 - y_1}{x_2 - x_1}$

Substituting, we have:

$m = \dfrac{3 - (3)}{-1 - 2}$

$ = \dfrac{0}{-3}$

$ = 0$

47. $m = \dfrac{y_2 - y_1}{x_2 - x_1}$

Substituting, we have:

$m = \dfrac{4(x+h) - (4x)}{(x+h) - x}$

$ = \dfrac{4x + 4h - 4x}{x + h - x}$

$ = \dfrac{4h}{h}$

$ = 4$

49. $m = \dfrac{y_2 - y_1}{x_2 - x_1}$

Substituting, we have:

$m = \dfrac{[2(x+h)+3] - (2x+3)}{(x+h) - x}$

$ = \dfrac{2x + 2h + 3 - (2x+3)}{x + h - x}$

$ = \dfrac{2h}{h}$

$ = 2$

51. From Exercise 37, we know that the slope is $-\dfrac{4}{7}$. Using the point $(5, -3)$, we substitute into the point-slope equation.

$y - (-3) = -\dfrac{4}{7}(x - 5)$

$y + 3 = -\dfrac{4}{7}x + \dfrac{20}{7}$

$y = -\dfrac{4}{7}x + \dfrac{20}{7} - 3$

$y = -\dfrac{4}{7}x - \dfrac{1}{7}$

Note: We could have found the equation of the line using the point $(-2, 1)$.

$y - (1) = -\dfrac{4}{7}(x - (-2))$

$y - 1 = -\dfrac{4}{7}x - \dfrac{8}{7}$

$y = -\dfrac{4}{7}x - \dfrac{8}{7} + 1$

$y = -\dfrac{4}{7}x - \dfrac{1}{7}$

53. From Exercise 39, we know that the slope is $-\dfrac{1}{4}$. Using the point $(-3, -5)$, we substitute into the point-slope equation.

$y - (-5) = -\dfrac{1}{4}(x - (-3))$

$y + 5 = -\dfrac{1}{4}x - \dfrac{3}{4}$

$y = -\dfrac{1}{4}x - \dfrac{23}{4}$

Alternatively, using the point $(1, -6)$ we get:

$y - (-6) = -\dfrac{1}{4}(x - (1))$

$y + 6 = -\dfrac{1}{4}x + \dfrac{1}{4}$

$y = -\dfrac{1}{4}x - \dfrac{23}{4}$

55. From Exercise 41, we know that the slope is undefined. The graph is a line which contains all ordered pairs whose first coordinate is 3. The equation of the line is $x = 3$

57. From Exercise 43, we know that the slope is $-\dfrac{4}{13}$. Using the point $\left(-\dfrac{3}{16}, -\dfrac{1}{2}\right)$, we substitute into the point-slope equation.

$y - \left(-\dfrac{1}{2}\right) = -\dfrac{4}{13}\left(x - \left(-\dfrac{3}{16}\right)\right)$

$y + \dfrac{1}{2} = -\dfrac{4}{13}x - \dfrac{12}{208}$

$y = -\dfrac{4}{13}x - \dfrac{3}{52} - \dfrac{1}{2}$

$y = -\dfrac{4}{13}x - \dfrac{29}{52}$

The solution is continued on the next page.

Note: We could have found the equation of the line using the point $\left(\dfrac{5}{8}, -\dfrac{3}{4}\right)$.

$$y - \left(-\dfrac{3}{4}\right) = -\dfrac{4}{13}\left(x - \left(\dfrac{5}{8}\right)\right)$$

$$y + \dfrac{3}{4} = -\dfrac{4}{13}x + \dfrac{5}{26}$$

$$y = -\dfrac{4}{13}x - \dfrac{29}{52}$$

59. From Exercise 45, we know that the slope is 0. The line is horizontal, thus the equation is $y = 3$.

61. Slope $= \dfrac{0.4 \text{ ft}}{5 \text{ ft}} = \dfrac{0.4}{5} = \dfrac{2}{25} = 0.08$.
The grade is then 8%.

63. Slope $= \dfrac{43.33 \text{ ft}}{1238 \text{ ft}} = \dfrac{43.33}{1238} = \dfrac{7}{200} = 0.035$.
Expressing the slope as a percentage, we find the head of the river is 3.5%.

65. a) If I, the number of inkjet cartridges required each year, is directly proportional to the number of students, s. Then there exists a constant m such that
$$I = ms$$
To find m, we substitute 16 for I and 2800 for s into $I = ms$ and solve for m.
$$16 = m \cdot 2800$$
$$\dfrac{16}{2800} = m$$
$$0.0057 = m$$
The constant of variation is $m = 0.0057$.
The equation of variation is $I(s) = 0.0057s$.

b) $I(s) = 0.0057s$
$$= 0.0057 \cdot 3100$$
$$= 17.67$$
$$\approx 18$$
The university would need 18 inkjet cartridges if 3100 students were enrolled.

67. a) Total costs = Variables costs + Fixed Costs
To produce x skis, it costs $80 dollars per ski, that is the variable costs are $80x$. In addition to the variable costs the fixed costs are $45,000. The total cost is
$$C(x) = 80x + 45,000$$

b) Revenue = Price times Quantity.
The price of a pair of skis is $450, therefore, the revenue from selling x pairs of skis is given by
$$R(x) = 450x$$

c) Profit = Revenue – Cost.
Using the Cost and Revenue functions found in part (a) and (b) we have:
$$P(x) = R(x) - C(x)$$
$$P(x) = 450x - (80x + 45,000)$$
$$= 370x - 45,000$$
The graph is shown below:

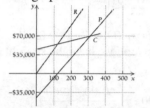

d) Substituting 3000 for x into the profit equation, we have
$$P(3000) = 370(3000) - 45,000$$
$$= 1,065,000.$$
Total profit will be $1,065,000 if they sell the expected 3000 pairs of skis.

e) The break even point occurs when $P(x) = 0$. Therefore, we set the profit function equal to 0 and solve for x:
$$P(x) = 0$$
$$370x - 45,000 = 0$$
$$370x = 45,000$$
$$x = \dfrac{45,000}{370}$$
$$x = 121.62$$
$$x \approx 122$$
They will need to sell 122 pairs of skis in order to break even.

69. a) $V(t) = C - t\left(\dfrac{C-S}{N}\right)$

Substituting 5200 for C, 1100 for S, and 8 for N into the equation we have the straight line depreciation $V(t)$:
$$V(t) = 5200 - t\left(\dfrac{5200 - 1100}{8}\right)$$
$$= 5200 - t\left(\dfrac{4100}{8}\right)$$
$$= 5200 - t(512.5)$$
$$= 5200 - 512.5t$$

b) $V(t) = 5200 - 512.5t$

$V(0) = 5200 - 512.5(0) = \$5200.00$

$V(1) = 5200 - 512.5(1) = \$4687.50$

$V(2) = 5200 - 512.5(2) = \$4175.00$

$V(3) = 5200 - 512.5(3) = \$3662.50$

$V(4) = 5200 - 512.5(4) = \$3150.00$

$V(7) = 5200 - 512.5(7) = \$1612.50$

$V(8) = 5200 - 512.5(8) = \$1100.00$

71. $V(t) = C - t\left(\dfrac{C - S}{N}\right)$

Substituting 60,000 for C, 2000 for S, and 5 for N into the equation we have the straight line depreciation $V(t)$:

$V(t) = 60,000 - t\left(\dfrac{60,000 - 2000}{5}\right)$

$\quad = 60,000 - 11,600t$

$V(3) = 60,000 - 11,600(3) = 25,200$

The computer system will have a book value of $25,200 after 3 years.

73. Slope is rise over run. The maximum riser on a stair is 8.25 inches, and the minimum run for a tread is 9 inches. Therefore:

$\text{Slope} = \dfrac{\text{height of riser}}{\text{length of run}} = \dfrac{8.25 \text{ in}}{9 \text{ in}} = \dfrac{11}{12} = 0.91\overline{6}.$

The maximum grade of stairs in North Carolina is 91%. Don't round up for legal reasons!

75. The average rate of change can be found using the coordinates of any two points on the line. We use the given coordinates $(2005, 4024)$ and $(2013, 5884)$.

$\text{Rate of Change} = \dfrac{5884 - 4024}{2013 - 2005}$

$\quad = \dfrac{1860}{8} = 232.50$

The average rate of change in the annual premium for a single person is $232.50 per year.

77. The average rate of change can be found using the coordinates of any two points on the line. We use the given coordinates $(2001, \$7.4)$ and $(2013, \$29)$.

$\text{Rate of Change} = \dfrac{29 - 7.4}{2013 - 2001}$

$\quad = \dfrac{21.6}{12} \approx 1.8$

The average rate of change of organic food sales in the United States is approximately $1.8 billion per year.

79. The equation of variation is $D = 293t$. The distance the impulse has to travel is 6 ft. So we solve:

$6 = 293t$

$\dfrac{6}{293} = t$

$0.0204778157 = t$

$0.02 \approx t$

It would take an impulse 0.02 seconds to travel from the toes of the person to the brain.

81. a) Since the brain weight B is directly proportional to the person's body weight W, there is a positive constant m such that $B = mW$
To find m, we substitute 3 for B and 120 for W into $B = mW$ solve for m.

$3 = m \cdot 120$

$0.025 = m$

The constant of variation is $m = 0.025$. The equation of variation is $B = 0.025W$.

b) The constant of variation $m = 0.025$ is equivalent to 2.5%, so we have $B = 2.5\%W$. The weight, B, of a human's brain is 2.5% of the persons total body weight, W.

c) Substituting 160 for W into the equation of variation we have:

$B = 0.025(160)$

$\quad = 4$

A person weighing 160 pounds has a brain that weighs 4 pounds.

83. a) $D(r) = \dfrac{11r + 5}{10}$

$D(5) = \dfrac{11(5) + 5}{10} = 6$

When traveling 5 miles per hour, the car's reaction distance is 6 feet.
The solution is continued on the next page.

Substituting 10 for r, we have:

$$D(10) = \frac{11(10)+5}{10} = 11.5$$

When traveling 10 miles per hour, the car's reaction distance is 11.5 feet.

$$D(20) = \frac{11(20)+5}{10} = 22.5$$

When traveling 20 miles per hour, the car's reaction distance is 22.5 feet.

$$D(50) = \frac{11(50)+5}{10} = 55.5$$

When traveling 50 miles per hour, the car's reaction distance is 55.5 feet.

$$D(65) = \frac{11(65)+5}{10} = 72$$

When traveling 65 miles per hour, the car's reaction distance is 72 feet.

b) Plotting the points found in part (a) and connecting the points with a smooth curve we have:

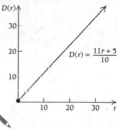

c) ✎

85. a) We find the equation of the line that contains the points $(2006, 49)$ and $(2013, 89)$. First we find the slope of the line.

$$m = \frac{89-49}{2013-2006} = \frac{40}{7} \approx 5.714 \approx 5.7$$

Next, we use the point-slope equation. We will use the point $(2006, 49)$.

$$y - y_1 = m(x - x_1)$$
$$y - 49 = 5.7(x - 2006)$$
$$y - 49 = 5.7x - 11434$$
$$y = 5.7x - 11385$$

Answer may vary based on rounding the calculations.

b) Using the equation and substituting 2014 in for x we have:

$$y = 5.7(2014) - 11385 \approx 94.8$$

In 2014, approximately 94.8% of adults used the internet. *Answer may vary based on rounding the calculations.*

c) Using the equation and substituting 100 in for y we have:

$$100 = 5.7x - 11385$$
$$11485 = 5.7x$$
$$\frac{11485}{5.7} = x$$
$$2014.9 = x$$

According to the equation, 100% of adults will be using the internet in 2015.

d) ✎

87. a) We know that $N = P + 0.02P = 1.02P$. Therefore the equation of variation is $N = 1.02P$.

b) Substituting 200,000 for P we have $N = 1.02(200,000) = 204,000$.

The new population is 204,000 after a growth of 2%.

c) Substituting 367,200 for N we have $367,200 - 1.02P$

$$\frac{367,200}{1.02} = P$$
$$360,000 = P$$

The previous population was 360,000.

89. Consider the three points:

$(2,5), (4,13)$, and $(7,y)$. For these points to lie on the same line, they slope between any two of these points must be the same. First use the known points $(2,5)$ and $(4,13)$ to calculate slope of the line.

$$m = \frac{13-5}{4-2} = \frac{8}{2} = 4.$$

There for the slope of between $(7,y)$ and either of the other two points must be 4 as well. We will substitute into the slope equation and solve for y. Notice using either point will result in the same value for y.

Use $(2,5)$ and $(7,y)$; Use $(4,13)$ and $(7,y)$

$m = 4$	$m = 4$
$\frac{y-5}{7-2} = 4$	$\frac{y-13}{7-4} = 4$
$\frac{y-5}{5} = 4$	$\frac{y-13}{3} = 4$
$y-5 = 20$	$y-13 = 12$
$y = 25$	$y = 25$

Therefore, y must equal 25 in order for the three points to lie on the same line. In other words, the points $(2,5), (4,13)$, and $(7,25)$ all lie on the same line.

91. a) Graph III is appropriate, because it shows the rate before January 1 is approximately $3000 per month, and the rate after January 1 is approximately $2000 per month.
   b) Graph IV is appropriate, because it shows the rate before January 1 is approximately $3000 per month, and the rate after January 1 is approximately –$4000 per month.
   c) Graph I is appropriate, because it shows the rate before January 1 is approximately $1000 per month, and the rate after January 1 is approximately $2000 per month.
   d) Graph II is appropriate, because it shows the rate before January 1 is approximately $4000 per month, and the rate after January 1 is approximately –$2000 per month.

93. Answers may vary. Regions of profit will occur when total revenue is greater than total cost, or when total profit is above the x-axis. Regions of loss will occur when total revenue is less than total cost, or when total profit is below the x-axis.

Exercise Set R.5

1. Graph $y = \frac{1}{4}x^2$ and $y = -\frac{1}{4}x^2$.

   Starting with $y = \frac{1}{4}x^2$, we first find the vertex or the turning point. The x-coordinate of the vertex is

   $$x = -\frac{b}{2a}$$

   $$= -\frac{0}{2\left(\frac{1}{4}\right)}$$

   $$= 0$$

   Substituting 0 for x into the equation, we find the second coordinate of the vertex:

   $$y = \frac{1}{4}(0)^2 = 0.$$

   The vertex is $(0,0)$. The y-axis (The vertical line $x = 0$.) is the axis of symmetry. Next, we choose some x-values on each side of the vertex and compute the y-values.

   When $x = 1$, $y = \frac{1}{4}(1)^2 = \frac{1}{4}\cdot 1 = \frac{1}{4}$

   When $x = 2$, $y = \frac{1}{4}(2)^2 = \frac{1}{4}\cdot 4 = 1$

   When $x = -1$, $y = \frac{1}{4}(-1)^2 = \frac{1}{4}\cdot 1 = \frac{1}{4}$

   When $x = -2$, $y = \frac{1}{4}(-2)^2 = \frac{1}{4}\cdot 4 = 1$

x	y
0	0
1	$\frac{1}{4}$
2	1
-1	$\frac{1}{4}$
-2	1

   We plot these points and connect them with a smooth curve on the axis below.

   Next, we graph $y = -\frac{1}{4}x^2$. First, we find the vertex or the turning point. The x-coordinate of the vertex is

   $$x = -\frac{b}{2a}$$

   $$x = -\frac{0}{2\left(-\frac{1}{4}\right)} = 0$$

   Substituting 0 for x into the equation, we find the second coordinate of the vertex:

   $$y = -\frac{1}{4}(0)^2 = 0.$$

   The vertex is $(0,0)$. The y-axis (The vertical line $x = 0$.) is the axis of symmetry. Next, we choose some x-values on each side of the vertex and compute the y-values.

   When $x = 1$, $y = -\frac{1}{4}(1)^2 = -\frac{1}{4}\cdot 1 = -\frac{1}{4}$

   When $x = 2$, $y = -\frac{1}{4}(2)^2 = -\frac{1}{4}\cdot 4 = -1$

   When $x = -1$, $y = -\frac{1}{4}(-1)^2 = -\frac{1}{4}\cdot 1 = -\frac{1}{4}$

   When $x = -2$, $y = -\frac{1}{4}(-2)^2 = -\frac{1}{4}\cdot 4 = -1$

x	y
0	0
1	$-\frac{1}{4}$
2	-1
-1	$-\frac{1}{4}$
-2	-1

   We plot these points and connect them with a smooth curve on the axis below.

   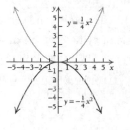

3. Graph $y = x^2$ and $y = x^2 - 1$.

   Starting with $y = x^2$, we first find the vertex or the turning point. The x-coordinate of the vertex is

   $$x = -\frac{b}{2a} = -\frac{0}{2(1)} = 0$$

   Substituting 0 for x into the equation, we find the second coordinate of the vertex:

   $$y = (0)^2 = 0.$$

   The vertex is $(0,0)$. The y-axis (The vertical line $x = 0$.) is the axis of symmetry.

   Next, we choose some x-values on each side of the vertex and compute the y-values on the top of the next page.

Computing the y-values, we have:

When $x =$ 1, $y = (1)^2 = 1$

When $x =$ 2, $y = (2)^2 = 4$

When $x = -1$, $y = (-1)^2 = 1$

When $x = -2$, $y = (-2)^2 = 4$

x	y
0	0
1	1
2	4
−1	1
−2	4

We plot these points and connect them with a smooth curve on the axis below.

Next, we graph $y = x^2 - 1$. First, we find the vertex or the turning point. The x-coordinate of the vertex is

$$x = -\frac{b}{2a}$$

$$= -\frac{0}{2(1)} = 0$$

Substituting 0 for x into the equation, we find the second coordinate of the vertex:

$$y = (0)^2 - 1 = -1.$$

The vertex is $(0, -1)$. The y-axis (The vertical line $x = 0$.) is the axis of symmetry. Next, we choose some x-values on each side of the vertex and compute the y-values.

When $x =$ 1, $y = (1)^2 - 1 = 1 - 1 = 0$

When $x =$ 2, $y = (2)^2 - 1 = 4 - 1 = 3$

When $x = -1$, $y = (-1)^2 - 1 = 1 - 1 = 0$

When $x = -2$, $y = (-2)^2 - 1 = 4 - 1 = 3$

x	y
0	−1
1	0
2	3
−1	0
−2	3

We plot these points and connect them with a smooth curve on the axis below.

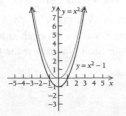

5. Graph $y = -3x^2$ and $y = -3x^2 + 2$.

Starting with $y = -3x^2$, we first find the vertex or the turning point. The x-coordinate of the vertex is:

$$x = -\frac{b}{2a}$$

$$= -\frac{0}{2(-3)} = 0$$

Substituting 0 for x into the equation, we find the second coordinate of the vertex:

$$y = -3(0)^2 = 0.$$

The vertex is $(0, 0)$. The y-axis (The vertical line $x = 0$.) is the axis of symmetry. Next, we choose some x-values on each side of the vertex and compute the y-values.

When $x =$ 1, $y = -3(1)^2 = -3 \cdot 1 = -3$

When $x =$ 2, $y = -3(2)^2 = -3 \cdot 4 = -12$

When $x = -1$, $y = -3(-1)^2 = -3 \cdot 1 = -3$

When $x = -2$, $y = -3(-2)^2 = -3 \cdot 4 = -12$

x	y
0	0
1	−3
2	−12
−1	−3
−2	−12

We plot these points and connect them with a smooth curve on the next page.

Next, we graph $y = -3x^2 + 2$. First, we find the vertex or the turning point. The x-coordinate of the vertex is

$$x = -\frac{b}{2a}$$

$$= -\frac{0}{2(-3)} = 0$$

Substituting 0 for x into the equation, we find the second coordinate of the vertex:

$$y = -3(0)^2 + 2 = 2.$$

The vertex is $(0, 2)$. The y-axis (The vertical line $x = 0$.) is the axis of symmetry. Next, we choose some x-values on each side of the vertex and compute the y-values at the top of the next page.

Computing the y-values, we have:

When $x = 1$, $y = -3(1)^2 + 2 = -3 + 2 = -1$

When $x = 2$, $y = -3(2)^2 + 2 = -12 + 2 = -10$

When $x = -1$, $y = -3(-1)^2 + 2 = -3 + 2 = -1$

When $x = -2$, $y = -3(-2)^2 + 2 = -12 + 2 = -10$

x	y
-2	-10
-1	-1
0	2
1	-1
2	-10

We plot these points and connect them with a smooth curve on the axis below.

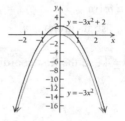

7. Graph $y = |x|$ and $y = |x - 3|$.

Starting with $y = |x|$, we choose some x-values and compute the y values to make a table of points.

When $x = -2$, $y = |-2| = -(-2) = 2$

When $x = -1$, $y = |-1| = -(-1) = 1$

When $x = 0$, $y = |0| = 0$

When $x = 1$, $y = |1| = 1$

When $x = 2$, $y = |2| = 2$

x	y
-2	2
-1	1
0	0
1	1
2	2

We plot these points and connect them with a curve on the axis below.

Next, we graph $y = |x - 3|$. We choose some x-values and compute the y-values to make a table of points at the top of the next column.

When $x = -2$, $y = |-2 - 3| = |-5| = -(-5) = 5$

When $x = 0$, $y = |0 - 3| = |-3| = -(-3) = 3$

When $x = 3$, $y = |3 - 3| = |0| = 0$

When $x = 6$, $y = |6 - 3| = |3| = 3$

When $x = 8$, $y = |8 - 3| = |5| = 5$

x	y
-2	5
0	3
3	0
6	3
8	5

We plot these points and connect them with a curve on the axis below.

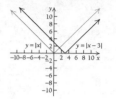

9. Graph $y = x^3$ and $y = x^3 + 1$.

Starting with $y = x^3$, we choose some x-values and compute the y-values to make a table of points.

When $x = -2$, $y = (-2)^3 = -8$

When $x = -1$, $y = (-1)^3 = -1$

When $x = 0$, $y = (0)^3 = 0$

When $x = 1$, $y = (1)^3 = 1$

When $x = 2$, $y = (2)^3 = 8$

x	y
-2	-8
-1	-1
0	0
1	1
2	8

We plot these points and connect them with a smooth curve on the next page.

Next, we graph $y = x^3 + 1$. We choose some x-values and compute the y-values to make a table of points.

When $x = -2$, $y = (-2)^3 + 1 = -8 + 1 = -7$

When $x = -1$, $y = (-1)^3 + 1 = -1 + 1 = 0$

When $x = 0$, $y = (0)^3 + 1 = 0 + 1 = 1$

When $x = 1$, $y = (1)^3 + 1 = 1 + 1 = 2$

When $x = 2$, $y = (2)^3 + 1 = 8 + 1 = 9$

Next, we organize these values into a table at the top of the next page.

Organizing the values into a table, we have:

x	y
-2	-7
-1	0
0	1
1	2
2	9

We plot these points and connect them with a smooth curve.

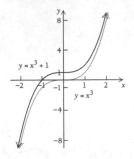

11. Graph $y = \sqrt{x}$ and $y = \sqrt{x-1}$.

Starting with $y = \sqrt{x}$, we choose some x-values and compute the y-values to make a table of points. The domain of the function is the set of all nonnegative real numbers, so we choose x-values that are in the set $[0, \infty)$.

When $x = 0$, $y = \sqrt{0} = 0$

When $x = 1$, $y = \sqrt{1} = 1$

When $x = 4$, $y = \sqrt{4} = 2$

When $x = 9$, $y = \sqrt{9} = 3$

x	y
0	0
1	1
4	2
9	3

We plot these points and connect them with a smooth curve on the axis below.

Next, we graph $y = \sqrt{x-1}$. we choose some x-values and compute the y-values to make a table of points. The domain of the function is the set of all positive real numbers greater than or equal to 1, so we choose x-values that are in the set $[1, \infty)$.

When $x = 1$, $y = \sqrt{1-1} = \sqrt{0} = 0$

When $x = 2$, $y = \sqrt{2-1} = \sqrt{1} = 1$

When $x = 5$, $y = \sqrt{5-1} = \sqrt{4} = 2$

When $x = 10$, $y = \sqrt{10-1} = \sqrt{9} = 3$

Organizing these values into a table, we have:

x	y
1	0
2	1
5	2
10	3

We plot these points and connect them with a smooth curve on the axis below.

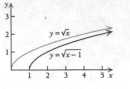

13. $y = x^2 + 4x - 7$

This function is of the form
$y = ax^2 + bx + c$, $a \neq 0$, so its graph is a parabola.
We have $a = 1$ and $b = 4$, so the first coordinate of the vertex is

$$x = -\frac{b}{2a}$$
$$= -\frac{4}{2(1)} = -2.$$

Substituting -2 into the equation, we find the second coordinate of the vertex:

$$y = (-2)^2 + 4(-2) - 7$$
$$= 4 - 8 - 7$$
$$= -11.$$

The vertex is $(-2, -11)$.

15. $y = 2x^4 - 4x^2 - 3$

The function is not of the form
$y = ax^2 + bx + c$, $a \neq 0$, so its graph is not a parabola.

17. Graph $y = x^2 - 6x + 5$.

First, we should recognize that this function is a quadratic function. We find the vertex or the turning point. The x-coordinate of the vertex is

$$x = -\frac{b}{2a}$$
$$= -\frac{-6}{2(1)}$$
$$= 3$$

The solution is continued on the next page.

Substituting 3 for x into the equation, we find the second coordinate of the vertex:

$$y = (3)^2 - 6(3) + 5$$
$$= 9 - 18 + 5 = -4.$$

The vertex is $(3, -4)$. The vertical line $x = 3$ is the axis of symmetry. Next, we choose some x-values on each side of the vertex and compute the y-values.

When $x = -1$, $y = (-1)^2 - 6(-1) + 5 = 12$

When $x = 0$, $y = (0)^2 - 6(0) + 5 = 5$

When $x = 1$, $y = (1)^2 - 6(1) + 5 = 0$

When $x = 2$, $y = (2)^2 - 6(2) + 5 = -3$

When $x = 4$, $y = (4)^2 - 6(4) + 5 = -3$

When $x = 5$, $y = (5)^2 - 6(5) + 5 = 0$

x	y
-1	12
0	5
1	0
2	-3
3	-4
4	-3
5	0

We plot these points and connect them with a smooth curve on the axis below.

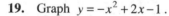

19.   Graph $y = -x^2 + 2x - 1$.

First, we should recognize that this function is a quadratic function. We find the vertex or the turning point. The x-coordinate of the vertex is

$$x = -\frac{b}{2a} = -\frac{2}{2(-1)} = 1$$

Substituting 1 for x into the equation, we find the second coordinate of the vertex:

$$y = -(1)^2 + 2(1) - 1$$
$$= -1 + 2 - 1 = 0.$$

The vertex is $(1, 0)$. The vertical line $x = 1$ is the axis of symmetry. Next, we choose some x-values on each side of the vertex and compute the y-values.

When $x = -1$, $y = -(-1)^2 + 2(-1) - 1 = -4$

When $x = 0$, $y = -(0)^2 + 2(0) - 1 = -1$

When $x = 2$, $y = -(2)^2 + 2(2) - 1 = -1$

When $x = 3$, $y = -(3)^2 + 2(3) - 1 = -4$

Organizing the values into a table, we have:

x	y
-1	-4
0	-1
1	0
2	-1
3	-4

We plot these points and connect them with a smooth curve on the axis below.

21.   Graph $f(x) = 3x^2 - 6x + 4$.

First, we should recognize that this function is a quadratic function. We find the vertex or the turning point. The x-coordinate of the vertex is

$$x = -\frac{b}{2a} = -\frac{-6}{2(3)} = 1$$

Substituting 1 for x into the equation, we find the second coordinate of the vertex:

$$f(1) = 3(1)^2 - 6(1) + 4$$
$$= 3(1) - 6 + 4 = 1.$$

The vertex is $(1, 1)$. The vertical line $x = 1$ is the axis of symmetry.

Next, we choose some x-values on each side of the vertex and compute the y-values.

When $x = -1$, $y = 3(-1)^2 - 6(-1) + 4 = 13$

When $x = 0$, $y = 3(0)^2 - 6(0) + 4 = 4$

When $x = 2$, $y = 3(2)^2 - 6(2) + 4 = 4$

When $x = 3$, $y = 3(3)^2 - 6(3) + 4 = 13$

x	y
-1	13
0	4
1	1
2	4
3	13

We plot these points and connect them with a smooth curve on the axis below.

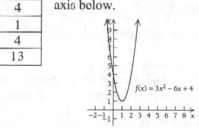

23.   Graph $g(x) = -3x^2 - 4x + 5$.

First, we should recognize that this function is a quadratic function. We find the vertex or the turning point. The x-coordinate of the vertex is

$$x = -\frac{b}{2a} = -\frac{-4}{2(-3)} = -\frac{2}{3}$$

Substituting $-\frac{2}{3}$ for x into the equation, we find the second coordinate of the vertex on the next page.

Substituting in to the function, we find the second coordinate of the vertex.

$$f\left(-\frac{2}{3}\right) = -3\left(-\frac{2}{3}\right)^2 - 4\left(-\frac{2}{3}\right) + 5$$

$$= -3\left(\frac{4}{9}\right) + \frac{8}{3} + 5 =$$

$$= -\frac{4}{3} + \frac{8}{3} + 5 = \frac{19}{3}.$$

The vertex is $\left(-\frac{2}{3}, \frac{19}{3}\right)$. The vertical line

$x = -\frac{2}{3}$ is the axis of symmetry. Next, we choose some x-values on each side of the vertex and compute the y-values.

When $x = -2$, $y = -3(-2)^2 - 4(-2) + 5 = 1$

When $x = -1$, $y = -3(-1)^2 - 4(-1) + 5 = 6$

When $x = 0$, $y = -3(0)^2 - 4(0) + 5 = 5$

When $x = 1$, $y = -3(1)^2 - 4(1) + 5 = -2$

x	y
-2	1
-1	6
$-\frac{2}{3}$	$\frac{19}{3}$
0	5
1	-2

We plot these points and connect them with a smooth curve on the axis below.

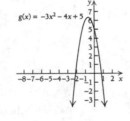

25. Graph $y = \frac{3}{x}$.

First we determine the domain. The domain is all real numbers except for 0, since substituting 0 for x would result in division by 0. Now, to find y-values, we substitute any value for x other than 0, and compute the value for y.

When $x = -6$, $y = \frac{3}{-6} = -\frac{1}{2}$

When $x = -3$, $y = \frac{3}{-3} = -1$

When $x = -1$, $y = \frac{3}{-1} = -3$

When $x = 1$, $y = \frac{3}{1} = 3$

When $x = 3$, $y = \frac{3}{3} = 1$

When $x = 6$, $y = \frac{3}{6} = \frac{1}{2}$

x	y
-6	$-\frac{1}{2}$
-3	-1
-1	-2
1	2
3	1
6	$\frac{1}{2}$

We plot these points and connect them with a smooth curve on the axis below.

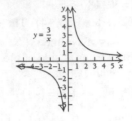

27. Graph $y = -\frac{2}{x}$.

First we determine the domain. The domain is all real numbers except for 0, since substituting 0 for x would result in division by 0. Now, to find y-values, we substitute any value for x other than 0, and compute the value for y.

When $x = -4$, $y = -\frac{2}{-4} = \frac{1}{2}$

When $x = -2$, $y = -\frac{2}{-2} = 1$

When $x = -1$, $y = -\frac{2}{-1} = 2$

When $x = 1$, $y = -\frac{2}{1} = -2$

When $x = 2$, $y = -\frac{2}{2} = -1$

When $x = 4$, $y = -\frac{2}{4} = -\frac{1}{2}$

x	y
-4	$\frac{1}{2}$
-2	1
-1	2
1	-2
2	-1
4	$-\frac{1}{2}$

We plot these points and connect them with a smooth curve on the axis below.

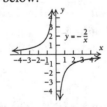

29. Graph $y = \frac{1}{x-1}$.

First we determine the domain. The domain is all real numbers except for 1, since substituting 1 for x would result in division by 0. Now, to find y-values, we substitute any value for x other than 1, and compute the value for y at the top of the next page.

Computing the values for y, we have:

When $x = -1$, $y = \dfrac{1}{(-1)-1} = -\dfrac{1}{2}$

When $x = 0$, $y = \dfrac{1}{(0)-1} = \dfrac{1}{-1} = -1$

When $x = \dfrac{1}{2}$, $y = \dfrac{1}{\left(\frac{1}{2}\right)-1} = \dfrac{1}{-\frac{1}{2}} = -2$

When $x = \dfrac{3}{2}$, $y = \dfrac{1}{\left(\frac{3}{2}\right)-1} = \dfrac{1}{\frac{1}{2}} = 2$

When $x = 2$, $y = \dfrac{1}{(2)-1} = \dfrac{1}{1} = 1$

When $x = 3$, $y = \dfrac{1}{(3)-1} = \dfrac{1}{2}$

We plot these points and connect them with a smooth curve on the axis below.

x	y
-1	$-\dfrac{1}{2}$
0	-1
$\dfrac{1}{2}$	-2
$\dfrac{3}{2}$	2
2	1
3	$\dfrac{1}{2}$

31. Graph $y = \sqrt[3]{x}$.

First we determine the domain of the function. Since the index of the radicand is odd (3) the domain is all real numbers. We are free to choose any number for x and compute the value for y.

When $x = -8$, $y = \sqrt[3]{-8} = -2$

When $x = -1$, $y = \sqrt[3]{-1} = -1$

When $x = 0$, $y = \sqrt[3]{0} = 0$

When $x = 1$, $y = \sqrt[3]{1} = 1$

When $x = 8$, $y = \sqrt[3]{8} = 2$

x	y
-8	-2
-1	-1
0	0
1	1
8	2

We plot these points and connect them with a smooth curve on the axis below.

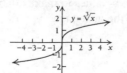

33. Graph $g(x) = \dfrac{x^2 + 7x + 10}{x+2}$.

First, we simplify the function, by factoring the numerator and removing a factor of 1 as follows:

$$f(x) = \dfrac{x^2 + 7x + 10}{x+2}$$

$$= \dfrac{(x+5)(x+2)}{x+2}$$

$$f(x) = \dfrac{x+2}{x+2} \cdot \dfrac{x+5}{1}$$

$$= x + 5, \quad x \neq -2$$

The simplification assumes that x is not -2. The number -2 is not in the domain of the original function because it would result in division by zero. Thus we can express the function as follows:

$$y = f(x) = x + 5, \quad x \neq -2.$$

To find function values, we substitute any value for x other than -2 and calculate the y-values.

x	y
-5	0
-4	1
-3	2
-1	4
0	5
1	6
2	7

We plot these points and draw the graph. The open circle at the point $(2, 3)$ indicates that it is not part of the graph.

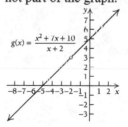

35. Graph $f(x) = \dfrac{x^2 - 1}{x - 1}$.

First, we simplify the function, by factoring the numerator and removing a factor of 1 as follows:

$$f(x) = \dfrac{x^2 - 1}{x - 1}$$

$$= \dfrac{(x-1)(x+1)}{x-1}$$

$$f(x) = \dfrac{x-1}{x-1} \cdot \dfrac{x+1}{1}$$

$$= x + 1, \quad x \neq 1$$

The simplification assumes that x is not 1. The number 1 is not in the domain of the original function because it would result in division by zero. Thus we can express the function as follows:

$$y = f(x) = x + 1, \quad x \neq 1.$$

To find function values, we substitute any value for x other than 1 and calculate the y-values.

x	y
−2	−1
−1	0
0	1
2	3
3	4
4	5

We plot these points and draw the graph. The open circle at the point $(1, 2)$ indicates that it is not part of the graph.

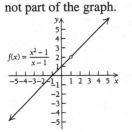

37. Graph $f(x) = (1.25)^x$.

First we recognize this is an exponential function, we choose some x-values and compute the y-values to make a table of points.

When $x = -2, y = (1.25)^{-2} = 0.64$

When $x = -1, y = y = (1.25)^{-1} = 0.80$

When $x = \ 0, y = y = (1.25)^{0} = 1$

When $x = \ 1, y = y = (1.25)^{1} = 1.25$

When $x = \ 2, y = y = (1.25)^{2} = 1.5625$

Organizing the values into a table, we have:

x	y
−2	0.64
−1	0.80
0	1
1	1.25
2	1.5625

We plot these points and connect them with a smooth curve on the axis below.

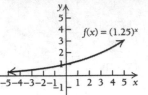

39. Graph $f(x) = (0.4)^x$.

First we recognize this is an exponential function, we choose some x-values and compute the y-values to make a table of points.

When $x = -2, y = (0.4)^{-2} = 6.25$

When $x = -1, y = y = (0.4)^{-1} = 2.5$

When $x = \ 0, y = y = (0.4)^{0} = 1$

When $x = \ 1, y = y = (0.4)^{1} = 0.4$

When $x = \ 2, y = y = (0.4)^{2} = 0.16$

x	y
−2	6.25
−1	2.5
0	1
1	0.4
2	0.16

We plot these points and connect them with a smooth curve on the axis below.

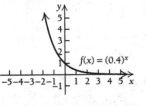

41. Graph $f(x) = 3 \cdot (1.1)^x$.

First we recognize this is an exponential function, we choose some x-values and compute the y-values to make a table of points at the top of the next page.

Computing y-values, we have:

When $x = -2$, $y = 3 \cdot (1.1)^{-2} = 2.47934$

When $x = -1$, $y = y = 3 \cdot (1.1)^{-1} = 2.727273$

When $x = 0$, $y = y = 3 \cdot (1.1)^{0} = 3$

When $x = 1$, $y = y = 3 \cdot (1.1)^{1} = 3.3$

When $x = 2$, $y = y = 3 \cdot (1.1)^{2} = 3.63$

x	y
-2	2.479
-1	2.727
0	3
1	3.3
2	3.63

We plot these points and connect them with a smooth curve on the axis below.

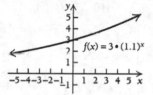

43. Graph $f(x) = 4 \cdot (0.8)^{x}$.

First we recognize this is an exponential function, we choose some x-values and compute the y-values to make a table of points.

When $x = -2$, $y = 4 \cdot (0.8)^{-2} = 10.9375$

When $x = -1$, $y = y = 4 \cdot (0.8)^{-1} = 8.75$

When $x = 0$, $y = y = 4 \cdot (0.8)^{0} = 4$

When $x = 1$, $y = y = 4 \cdot (0.8)^{1} = 5.6$

When $x = 2$, $y = y = 4 \cdot (0.8)^{2} = 4.48$

x	y
-2	10.9375
-1	8.75
0	4
1	5.6
2	4.48

We plot these points and connect them with a smooth curve on the axis below.

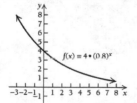

45. Solve $x^2 - 2x = 2$.

First, we put the equation in standard form.

$x^2 - 2x - 2 = 0$ subtract 2 from both sides

The equation is in standard form with

$a = 1$, $b = -2$, $c = -2$

Next, we apply the quadratic formula:

$$x = \frac{-b \pm \sqrt{b^2 - 4ac}}{2a}$$

Substituting the values for a, b, and c, we get:

$$x = \frac{-(-2) \pm \sqrt{(-2)^2 - 4(1)(-2)}}{2(1)}$$

$$= \frac{2 \pm \sqrt{4+8}}{2} = \frac{2 \pm \sqrt{12}}{2}$$

$$= \frac{2 \pm 2\sqrt{3}}{2} \qquad \left(\sqrt{12} = \sqrt{4 \cdot 3} = 2\sqrt{3}\right)$$

$$= \frac{2\left(1 \pm \sqrt{3}\right)}{2 \cdot 1} = 1 \pm \sqrt{3}$$

The solutions are $1 + \sqrt{3}$ and $1 - \sqrt{3}$.

47. Solve $x^2 + 6x = 1$.

First, we put the equation in standard form.

$x^2 + 6x - 1 = 0$ subtract 1 from both sides

The equation is in standard form with

$a = 1$, $b = 6$, $c = -1$

Next, we apply the quadratic formula.

$$x = \frac{-b \pm \sqrt{b^2 - 4ac}}{2a}$$

Substituting the values for a, b, and c, we get:

$$x = \frac{-(6) \pm \sqrt{(6)^2 - 4(1)(-1)}}{2(1)}$$

$$= \frac{-6 \pm \sqrt{36+4}}{2} = \frac{-6 \pm \sqrt{40}}{2}$$

$$= \frac{-6 \pm 2\sqrt{10}}{2} \qquad \left(\sqrt{40} = \sqrt{4 \cdot 10} = 2\sqrt{10}\right)$$

$$= \frac{2\left(-3 \pm \sqrt{10}\right)}{2 \cdot 1} = -3 \pm \sqrt{10}$$

The solutions are $-3 + \sqrt{10}$ and $-3 - \sqrt{10}$.

49. Solve $4x^2 = 4x + 1$.

First, we put the equation in standard form.

$4x^2 - 4x - 1 = 0$ subtract $4x$ and 1 from both sides

The equation is in standard form with

$a = 4$, $b = -4$, $c = -1$

The solution is continued on the next page.

Next, we apply the quadratic formula.

$$x = \frac{-b \pm \sqrt{b^2 - 4ac}}{2a}$$

Substituting the values for a, b, and c, we get:

$$x = \frac{-(-4) \pm \sqrt{(-4)^2 - 4(4)(-1)}}{2(4)}$$

$$= \frac{4 \pm \sqrt{16 + 16}}{8}$$

$$= \frac{4 \pm \sqrt{32}}{8}$$

$$x = \frac{4 \pm 4\sqrt{2}}{8} \qquad \left(\sqrt{32} = \sqrt{16 \cdot 2} = 4\sqrt{2}\right)$$

$$= \frac{4\left(1 \pm \sqrt{2}\right)}{4 \cdot 2} = \frac{1 \pm \sqrt{2}}{2}$$

The solutions are $\dfrac{1 + \sqrt{2}}{2}$ and $\dfrac{1 - \sqrt{2}}{2}$.

51. Solve $3y^2 + 8y + 2 = 0$.

The equation is in standard form with
$a = 3$, $b = 8$, $c = 2$

Next, we apply the quadratic formula.

$$y = \frac{-b \pm \sqrt{b^2 - 4ac}}{2a}$$

Substituting the values for a, b, and c, we get:

$$y = \frac{-(8) \pm \sqrt{(8)^2 - 4(3)(2)}}{2(3)}$$

$$= \frac{-8 \pm \sqrt{64 - 24}}{6} = \frac{-8 \pm \sqrt{40}}{6}$$

$$= \frac{-8 \pm 2\sqrt{10}}{6} \qquad \left(\sqrt{40} = \sqrt{4 \cdot 10} = 2\sqrt{10}\right)$$

$$= \frac{2\left(-4 \pm \sqrt{10}\right)}{2 \cdot 3} = \frac{-4 \pm \sqrt{10}}{3}$$

The solutions are $\dfrac{-4 + \sqrt{10}}{3}$ and $\dfrac{-4 - \sqrt{10}}{3}$.

53. Solve $x + 7 + \dfrac{9}{x} = 0$.

Multiplying both sides by x, we get:

$$x \cdot \left(x + 7 + \frac{9}{x}\right) = 0 \cdot x$$

$$x^2 + 7x + 9 = 0.$$

This is a quadratic equation in standard form
with $a = 1, b = 7$, and $c = 9$.

Next, we apply the quadratic formula.

$$x = \frac{-b \pm \sqrt{b^2 - 4ac}}{2a}$$

Substituting the values for a, b, and c, we get:

$$x = \frac{-(7) \pm \sqrt{(7)^2 - 4(1)(9)}}{2(1)}$$

$$= \frac{-7 \pm \sqrt{49 - 36}}{2} = \frac{-7 \pm \sqrt{13}}{2}$$

The solutions are $\dfrac{-7 + \sqrt{13}}{2}$ and $\dfrac{-7 - \sqrt{13}}{2}$.

55. $x^{1/5} = \sqrt[5]{x^1} = \sqrt[5]{x} \qquad \left(a^{m/n} = \sqrt[n]{a^m}\right)$

57. $y^{2/3} = \sqrt[3]{y^2} \qquad \left(a^{m/n} = \sqrt[n]{a^m}\right)$

59. $t^{-2/5} = \dfrac{1}{t^{2/5}} \qquad \left(a^{-n} = \dfrac{1}{a^n}\right)$

$\phantom{t^{-2/5}} = \dfrac{1}{\sqrt[5]{t^2}} \qquad \left(a^{m/n} = \sqrt[n]{a^m}\right)$

61. $b^{-1/3} = \dfrac{1}{b^{1/3}} \qquad \left(a^{-n} = \dfrac{1}{a^n}\right)$

$\phantom{b^{-1/3}} = \dfrac{1}{\sqrt[3]{b}} \qquad \left(a^{m/n} = \sqrt[n]{a^m}\right)$

63. $\left(x^2 - 3\right)^{-1/2} = \dfrac{1}{\left(x^2 - 3\right)^{1/2}} \qquad \left(a^{-n} = \dfrac{1}{a^n}\right)$

$ = \dfrac{1}{\sqrt{x^2 - 3}} \qquad \left(a^{m/n} = \sqrt[n]{a^m}\right)$

65. $\sqrt{x^3} = x^{3/2} \qquad \left(\text{The index is 2; } \sqrt[n]{a^m} = a^{m/n}\right)$

67. $\sqrt[5]{a^3} = a^{3/5} \qquad \left(\sqrt[n]{a^m} = a^{m/n}\right)$

69. $\sqrt[4]{x^{12}} = x^{12/4} = x^3$

71. $\dfrac{1}{\sqrt{t^5}} = \dfrac{1}{t^{5/2}} \qquad \left(\sqrt[n]{a^m} = a^{m/n}\right)$

$\phantom{\dfrac{1}{\sqrt{t^5}}} = t^{-5/2} \qquad \left(\dfrac{1}{a^n} = a^{-n}\right)$

73. $\dfrac{1}{\sqrt{x^2+7}} = \dfrac{1}{\left(x^2+7\right)^{1/2}}$ $\left(\sqrt[n]{a^m} = a^{m/n}\right)$

$= \left(x^2+7\right)^{-1/2}$ $\left(\dfrac{1}{a^n} = a^{-n}\right)$

75. $9^{3/2}$

$= \left(9^{1/2}\right)^3$ $\left(\tfrac{3}{2} = \tfrac{1}{2}\cdot 3;\ a^{m\cdot n} = \left(a^m\right)^n\right)$

$= \left(\sqrt{9}\right)^3$ $\left(a^{1/n} = \sqrt[n]{a}\right)$

$= (3)^3 = 27$

77. $64^{2/3}$

$= \left(64^{1/3}\right)^2$ $\left(\tfrac{2}{3} = \tfrac{1}{3}\cdot 2;\ a^{m\cdot n} = \left(a^m\right)^n\right)$

$= \left(\sqrt[3]{64}\right)^2$ $\left(a^{1/n} = \sqrt[n]{a}\right)$

$= (4)^2$ $\left(\sqrt[3]{64} = 4\right)$

$= 16$

79. The domain of a rational function is restricted to those input values that do not result in division by 0. To determine the domain of

$f(x) = \dfrac{x^2-25}{x-5}$

we set the denominator equal to zero and solve:

$x-5=0$

$x=5$.

Therefore, 5 is not in the domain. The domain of f consists of all real numbers except 5.

81. The domain of a rational function is restricted to those input values that do not result in division by 0.

To determine the domain of

$f(x) = \dfrac{x^3}{x^2-5x+6}$

we set the denominator equal to zero and solve:

$x^2-5x+6=0$

$(x-2)(x-3)=0$ factoring

$x-2=0$ or $x-3=0$ Principle of Zero Products

$x=2$ or $x=3$.

Therefore, 2 and 3 are not in the domain. The domain of f consists of all real numbers except 2 and 3.

83. The domain of the radical function $f(x) = \sqrt{5x+4}$ is restricted to those input values that result in the value of the radicand being greater than or equal to 0. In other words, the domain will be the set of real numbers that satisfy the inequality $5x+4\ge 0$.

To find the domain, we solve the inequality:

$5x+4\ge 0$

$5x\ge -4$

$x\ge -\dfrac{4}{5}$

Therefore, the domain of f consists of all real numbers greater than or equal to $-\dfrac{4}{5}$, or in interval notation $\left[-\dfrac{4}{5},\infty\right)$.

85. The domain of the radical function $f(x) = \sqrt[4]{7-x}$ is restricted to those input values that result in the value of the radicand being greater than or equal to 0. In other words, the domain will be the set of real numbers that satisfy the inequality $7-x\ge 0$.

To find the domain, we solve the inequality:

$7-x\ge 0$

$7\ge x$

Therefore, the domain of f consists of all real numbers less than or equal to 7, or in interval notation $(-\infty,7]$.

87. We set the demand equation equal to the supply equation and solve for x.

$1000-10x = 250+5x$

$750 = 15x$

$50 = x$

Thus, the equilibrium price is $50. To find the equilibrium quantity, we substitute 50 for x into either the demand equation or supply equation. We use the demand equation.

$q = 1000-10(50)$

$q = 1000-500$

$q = 500$

The equilibrium quantity is 500 units. The equilibrium point is $(50,500)$.

89. We set the demand equation equal to the supply equation and solve for x.

$$\frac{5}{x} = \frac{x}{5}$$

$25 = x^2$ multiply both sides by $5x$

$\sqrt{25} = \sqrt{x^2}$ take the square root of both sides

$\pm 5 = x$

Since it is not appropriate to have a negative price, the equilibrium price is 5 hundred dollars or $500. To find the equilibrium quantity, we substitute 5 for x into either the demand equation or supply equation. We use the demand equation.

$$q = \frac{5}{(5)} = 1$$

The equilibrium quantity is 1 thousand units or 1000 units. The equilibrium point is $(5,1)$.

91. We set the demand equation equal to the supply equation and solve for x.

$$(x-3)^2 = x^2 + 2x + 1$$

$$x^2 - 6x + 9 = x^2 + 2x + 1$$

$-6x + 9 = 2x + 1$ subtracting x^2 from both sides

$8 = 8x$

$1 = x$

The equilibrium price is $1.
To find the equilibrium quantity, we substitute 1 for x into either the demand equation or supply equation. We use the demand equation.

$$q = (1-3)^2$$

$$= (-2)^2 = 4$$

The equilibrium quantity is 4 hundred units or 400 units. The equilibrium point is $(1,4)$.

93. We set the demand equation equal to the supply equation and solve for x.

$$5 - x = \sqrt{x+7} \qquad 0 \le x \le 5$$

$(5-x)^2 = \left(\sqrt{x+7}\right)^2$ Squaring both sides

$$25 - 10x + x^2 = x + 7$$

$$18 - 11x + x^2 = 0$$

$(9-x)(2-x) = 0$ Factoring the quadratic

$9 - x = 0$ or $2 - x = 0$ Principle of Zero Products

$9 = x$ or $2 = x$

Since 9 is not in the domain of the demand function, the equilibrium price is 2 thousand dollars or $2000.

To find the equilibrium quantity, we substitute 2 for x into either the demand equation or supply equation. We use the demand equation.

$$q = 5 - (2) = 3$$

The equilibrium quantity is 3 thousand units or 3000 units. The equilibrium point is $(2,3)$.

95. If the number of tickets sold S is inversely proportional to price of the ticket p, then we have:

$$S = \frac{k}{p}.$$

We find the constant of variation by substituting 175 for S and 20 for p.

$$175 = \frac{k}{20}$$

$$3500 = k$$

The equation of variation is $S = \dfrac{3500}{p}$.

If the price of a ticket was $25, we find S by substituting 25 in for p.

$$S = \frac{3500}{25}$$

$$= 140$$

They would sell 140 tickets at $25 per ticket.

97. a) $R(x) = 11.74x^{0.25}$

$$R(40,000) = 11.74(40,000)^{0.25}$$

$$= 11.74(14.14213562)$$

$$= 166.0286722$$

$$\approx 166$$

The maximum range will be approximately 166 miles when the peak power is 40,000 watts.

$$R(50,000) = 11.74(50,000)^{0.25}$$

$$= 11.74(14.95348781)$$

$$= 175.5539469$$

$$\approx 176$$

The maximum range will be approximately 176 miles when the peak power is 50,000 watts.

$$R(60,000) = 11.74(60,000)^{0.25}$$

$$= 11.74(15.6508458)$$

$$= 183.7409297$$

$$\approx 184$$

The maximum range will be approximately 184 miles when the peak power is 60,000 watts.

b) Plotting the points found in part (a) and connecting them with a smooth curve we see:

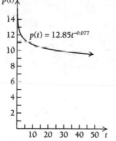

99. $P(t) = 12.85t^{-0.077}$; $2000 \Rightarrow t = 0$.

a) In 2015, $t = 2015 - 2000 = 15$.

$$P = 12.85(15)^{-0.077} \approx 10.43$$

In 2015, average pollution was Approximately 10.43 micrograms per cubic meter.
In 2020, $t = 2020 - 2000 = 20$.

$$P = 12.85(20)^{-0.077} \approx 10.20$$

In 2020, average pollution was Approximately 10.20 micrograms per cubic meter.
In 2025, $t = 2025 - 2000 = 25$.

$$P = 12.85(25)^{-0.077} \approx 10.03$$

In 2025, average pollution was Approximately 10.03 micrograms per cubic meter.

b) Plot the points above and others, if necessary, and draw the graph.

101. The number of cities N with a population greater than S is inversely proportional to S.

$$N = \frac{k}{S}$$

$$285 = \frac{k}{100,000} \quad \text{Substituting}$$

$$28,500,000 = k$$

The equation of variation is

$$N = \frac{28,500,000}{S}.$$

We find N when S is 350,000.

$$N = \frac{28,500,000}{350,000} = 81.42 \approx 81.$$

We find N when S is 500,000.

$$N = \frac{28,500,000}{500,000} = 57.$$

Using the fact that there were 81 cities with a population greater than 350,000 and 57 cities with a population of 500,000 or greater, we estimate that there are $81 - 57 = 24$ cities with a population between 350,000 and 500,000.
To estimate the number of cities with a population between 300,000 and 600,000 we find N when S is 300,000 and when S is 600,000.

$$N = \frac{28,500,000}{300,000} = 95$$

$$N = \frac{28,500,000}{600,000} = 47.5 \approx 48$$

There are 95 cities with a population greater than 300,000 and 48 cities with a population greater than 600,000, so there are $95 - 48 = 47$ cities with a population between 300,000 and 600,000.

103.

105. $f(x) = 2x^3 - x^2 - 14x - 10$

Enter the function into your calculator.

```
Plot1  Plot2  Plot3
\Y1■2X^3-X^2-14X
-10
\Y2=
\Y3=
\Y4=
\Y5=
\Y6=
```

Using the window:

```
WINDOW
 Xmin=-10
 Xmax=10
 Xscl=1
 Ymin=-30
 Ymax=10
 Yscl=10
 Xres=1
```

We see the graph:

Now using the ZERO feature on the calculator, we approximate the zeros. The zeros are -1.831, -0.856, 3.188.

107. $f(x) = x^4 + 4x^3 - 36x^2 - 160x + 300$

Enter the function into your calculator.

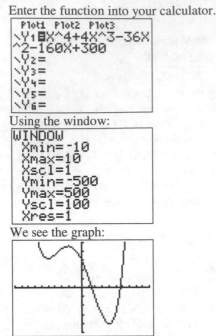

Using the window:

WINDOW
Xmin=-10
Xmax=10
Xscl=1
Ymin=-500
Ymax=500
Yscl=100
Xres=1

We see the graph:

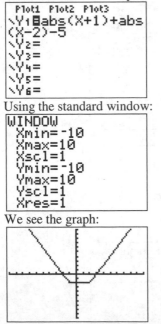

Now using the ZERO feature on the calculator, we approximate the zeros. The zeros are 1.489 and 5.673.

109. $f(x) = |x+1| + |x-2| - 5$

Enter the function into your calculator.

Plot1 Plot2 Plot3
\Y1◼abs(X+1)+abs
(X-2)-5
\Y2=
\Y3=
\Y4=
\Y5=
\Y6=

Using the standard window:

WINDOW
Xmin=-10
Xmax=10
Xscl=1
Ymin=-10
Ymax=10
Yscl=1
Xres=1

We see the graph:

Now using the ZERO feature on the calculator, we approximate the zeros. The zeros are –2 and 3.

111. $f(x) = |x+1| + |x-2| - 3$

Enter the function into your calculator.

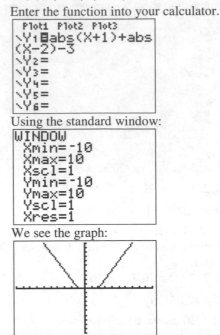

Using the standard window:

WINDOW
Xmin=-10
Xmax=10
Xscl=1
Ymin=-10
Ymax=10
Yscl=1
Xres=1

We see the graph:

We see that the graph intersects the x-axis between $-1 \le x \le 2$. The zeros of this function are all real numbers in $[-1, 2]$

113. We enter the demand and supply equations into the graphing editor on the calculator.

Plot1 Plot2 Plot3
\Y1◼15(0.95)^X
\Y2◼2.5(1.13)^X
\Y3=
\Y4=
\Y5=
\Y6=

Using the window:

WINDOW
Xmin=0
Xmax=30
Xscl=10
Ymin=0
Ymax=25
Yscl=5
↓Xres=1

Using the intersect feature on the calculator we find the intersection to be:

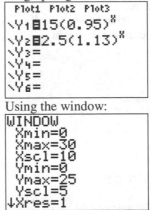

Intersection
X=10.326494 Y=8.8319014

The equilibrium point for this market is $(10.33, 8.83)$. In other words, 8830 units will be sold at a price of $10.33.

Exercise Set R.6

1. The data is decreasing at a constant rate, so a linear function $f(x) = mx + b$ could be used to model the data.

3. The data rises first and then falls over the domain, so a quadratic function $f(x) = ax^2 + bx + c$, $a < 0$ could be used to model the data.

5. The data rises first and then falls over the domain, so a quadratic function $f(x) = ax^2 + bx + c$, $a < 0$ could be used to model the data.

7. The data rises first and then falls over the domain, so a quadratic function $f(x) = ax^2 + bx + c$, $a < 0$ could be used to model the data.

9. a) We find the linear function $f(x) = mx + b$ that contains the data points $(0, 65.4)$ and $(6, 56.4)$. Note the year 2006 corresponds to $x = 0$.
We find the slope containing the two points
$$m = \frac{56.4 - 65.4}{6 - 0} = \frac{-9}{6} = -\frac{3}{2} = -1.5$$
Next, we can choose the y-intercept $(0, 65.4)$, and we can substitute into the slope-intercept equation.
$$y = mx + b$$
$$y = -1.5x + 65.4$$
The linear function that models the data is:
$$f(x) = -1.5x + 65.4.$$
Alternatively, we could have used the slope and one of the points to substitute into the point-slope equation to find the equation of the line.
$$y - y_1 = m(x - x_1)$$
$$y - 56.4 = -1.5(x - 6)$$
$$y - 56.4 = -1.5x + 9$$
$$y = -1.5x + 65.4$$
We notice that the linear function that models the data remains:
$$f(x) = -1.5x + 65.4.$$

One other alternative method, we substitute in to the equation $y = mx + b$ to obtain a system of equations.
$$65.4 = m \cdot 0 + b \qquad (1)$$
$$56.4 = m \cdot 6 + b \qquad (2)$$
Subtracting each side of Equation (1) from each side of Equation (2) we get:
$$-9 = 6m$$
$$\frac{-9}{6} = m$$
$$-1.5 = m$$
Now substitute $m = -1.5$ in for either Equation (1) or (2) and solve for b. We use Equation (1).
$$65.4 = (-1.5)(0) + b$$
$$65.4 = b$$
Once again, we notice that the linear function that models the data remains:
$$f(x) = -1.5x + 65.4.$$

b) In June of 2018, $x = 12$.
$$f(12) = -1.5(12) + 65.4 = 47.4$$
The number of cable video customers in 2018 is approximately 47.4 million.

11. a) Consider the general quadratic function
$$y = ax^2 + bx + c$$
Using the points
$(0,0), (2,200),$ and $(3,167)$, we substitute each point into the general quadratic function to obtain the system of equations
$$0 = a \cdot 0^2 + b \cdot 0 + c$$
$$200 = a \cdot 2^2 + b \cdot 2 + c$$
$$167 = a \cdot 3^2 + b \cdot 3 + c$$
Which gives us:
$$0 = c$$
$$200 = 4a + 2b + c$$
$$167 = 9a + 3b + c$$
From the first equation we see that $c = 0$. We substitute this value for c into the other two equations and our system is reduced to:
$$200 = 4a + 2b$$
$$167 = 9a + 3b$$
Solving this system of equations, we get:
$$a = -\frac{133}{3}, b = \frac{566}{3}, c = 0$$

The solution is continued on the next page.

Using the information from the previous page, we have the quadratic function:

$$f(x) = -\frac{133}{3}x^2 + \frac{566}{3}x + 0$$

$$= -\frac{133}{3}x^2 + \frac{566}{3}x$$

Writing the equation with proper fractions we have:

$$f(x) = -44\frac{1}{3}x^2 + 188\frac{2}{3}x.$$

b) $f(4) = -44\frac{1}{3}(4)^2 + 188\frac{2}{3}(4) \approx 45.3$.

Approximately 45.3 mg of albuterol will be in the bloodstream after 4 hours.

c) ✏ \

13. a) Consider the general quadratic function

$y = ax^2 + bx + c$, where y is the braking distance, in feet, and x is the speed, in miles per hour.

Using the points

$(20, 25), (40, 105)$, and $(60, 300)$, we substitute each point into the general quadratic function to get the system of equations:

$$25 = a \cdot 20^2 + b \cdot 20 + c$$

$$105 = a \cdot 40^2 + b \cdot 40 + c$$

$$300 = a \cdot 60^2 + b \cdot 60 + c$$

or,

$$25 = 400a + 20b + c$$

$$105 = 1600a + 40b + c$$

$$300 = 3600a + 60b + c$$

Solving the system of equations, we get:
$a = 0.14375, b = -4.625,$ and $c = 60$.

Therefore, the function is

$$y = 0.144x^2 - 4.63x + 60.$$

b) We substitute 50 for x and compute the value of y.

$$y = 0.144(50)^2 - 4.63(50) + 60 = 188.5$$

The breaking distance of a car traveling at 50 mph is about 188.5 ft.

c) ✏

15. a) Answers will vary depending on which points are used to find the function.

We will use the points $(30, 1.4)$ and $(70, 53.0)$. We substitute in to the equation

$y = mx + b$ to obtain a system of equations.

$$1.4 = m \cdot 30 + b \qquad (1)$$

$$53.0 = m \cdot 70 + b \qquad (2)$$

Subtracting each side of Equation (1) from each side of Equation (2) we get:

$$51.6 = 40m$$

$$1.29 = m$$

Now substitute $m = 1.29$ in for either Equation (1) or (2) and solve for b. We use Equation (1).

$$1.4 = (1.29)(30) + b$$

$$1.4 = 38.7 + b$$

$$-37.3 = b$$

Therefore, the linear function that fits the data is $y = 1.29x - 37.3$.

b) We plot the points in the table in the text and then graph the function found in part (a) on the same set of axes.

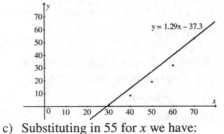

c) Substituting in 55 for x we have:

$$y = 1.29(55) - 37.3 = 33.65.$$

Approximately 33.65% of 55-yr-old women have high blood pressure.

17. a) First, we enter the data into the statistic editor on the calculator, letting L_1 be the values for x and L_2 be the values for y. Remember x is the year since 1970.

L1	L2	L3	3
0	11.2	▆▆▆▆	
10	14.2		
20	17		
30	20.9		
43	26.4		
------	------		

L3(1)=

Using the exponential regression feature we get:

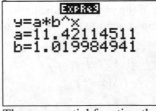

ExpReg
y=a*b^x
a=11.42114511
b=1.019984941

The exponential function that best fits the following data is: $y = 11.421(1.020)^x$.

b) Graphing the scatter plot of the data and the function we have:

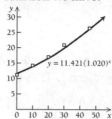

c) Since $x = 0$ corresponds to 1970, in 2020 $x = 2020 - 1970 = 50$. Substituting in 50 for x into the function found in part a) we have

$y = 11.421(1.020)^{50} = 30.740 \approx 30.7$. In 2020, the population of Texas will be approximately 30.7 million.

19. a) First, we enter the data into the statistic editor on the calculator, letting L_1 be the values for x and L_2 be the values for y. Remember x is the year since 1980.

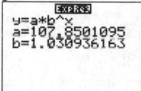

Using the exponential regression feature we get:

```
y=a*b^x
a=107.8501095
b=1.030936163
```

The exponential function that best fits the following data is: $y = 107.850(1.031)^x$.

b) Graphing the scatter plot of the data and the function we have:

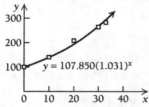

c) Since $x = 0$ corresponds to 1980, in 2020 $x = 2020 - 1980 = 40$. Substituting 40 for x into the function found in part a) we have

$y = 107.850(1.031)^{40} = 364.83$. In 2020, you will need \$364.83 to have the same buying power \$100 had in 1980.

d) ✎

21. ✎

23. ✎

25. a) First, we enter the data into the statistic editor on the calculator, letting L_1 be the values for x and L_2 be the values for y. Remember x is the year since 2006.

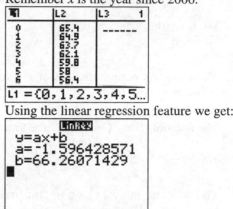

Using the linear regression feature we get:

```
LinReg
y=ax+b
a=-1.596428571
b=66.26071429
```

The linear function that fits the data is $y = -1.5964x + 66.2607$.

b) In 2018, $x = 2018 - 2006 = 12$. Substituting 12 in for x we get:

$y = -1.5964(12) + 66.2607$

$\approx 47.1039 \approx 47.10$

The number of cable video customers in 2018 will be 47.10 million.

c) The answers are very close, giving essentially the same information.

d) Using the exponential regression feature on the calculator results in:

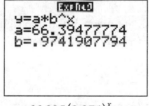

```
ExpReg
y=a*b^x
a=66.39477774
b=.9741907794
```

$y = 66.395(0.974)^x$

Substituting in 12 for x we get:

$y = 66.395(0.974)^{12}$

$= 66.395(0.728966) \approx 48.4$

The exponential function with coefficients rounded to 3 decimal places estimates 48.4 million subscribers in 2018. Note: if you do not round the coefficients and base of the exponential function the model will estimate 49 million subscribers in 2018.

e) ✎

Chapter 1

Differentiation

Exercise Set 1.1

---

1. We solve the equation:
$$-3x = 6$$
$$x = -2$$
Therefore, As x approaches -2, the value of $-3x$ approaches 6.

3. The notation $\lim\limits_{x \to 4} f(x)$ is read "the limit, as x approaches 4, of $f(x)$."

5. The notation $\lim\limits_{x \to 5^-} F(x)$ is read "the limit, as x approaches 5 from the left, of $F(x)$."

7. The notation $\lim\limits_{x \to 2^+}$ is read "the limit, as x approaches 2 from the right."

9. The notation $\lim\limits_{x \to 5}$ is read "the limit, as x approaches 5."

11. a) As inputs x approach 3 from the left, outputs $f(x)$ approach 1. That is,
$$\lim_{x \to 3^-} f(x) = 1.$$
b) As inputs x approach 3 from the right, outputs $f(x)$ approach 2. Thus the limit from the right is 2. That is,
$$\lim_{x \to 3^+} f(x) = 2.$$
c) From part (a) and part(b) we know that $\lim\limits_{x \to 3^-} f(x) = 1$ and $\lim\limits_{x \to 3^+} f(x) = 2$. Since the limit from the left, 1, is not the same as the limit from the right, 2, $\lim\limits_{x \to 3} f(x)$ *does not exist.*

13. a) As inputs x approach -2 from the left, outputs $g(x)$ approach 4. Thus the limit from the left is 4. That is, $\lim\limits_{x \to -2^-} g(x) = 4$.

b) As inputs x approach -2 from the right, outputs $g(x)$ approach 2. Thus the limit from the right is 2. That is, $\lim\limits_{x \to -2^+} g(x) = 2$.

c) From part (a) and part (b) we know that $\lim\limits_{x \to -2^-} g(x) = 4$ and $\lim\limits_{x \to -2^+} g(x) = 2$. Since the limit from the left, 4, is not the same as the limit from the right, 2, $\lim\limits_{x \to -2} g(x)$ *does not exist.*

15. As inputs x approach 2 from the left, outputs $F(x)$ approach 4. Thus the limit from the left is 4. That is, $\lim\limits_{x \to 2^-} F(x) = 4$. As inputs x approach 2 from the right, outputs $F(x)$ approach 4. Thus the limit from the right is 4. That is, $\lim\limits_{x \to 2^+} F(x) = 4$.
Since the limit from the left, 4, is the same as the limit from the right, 4, we have
$$\lim_{x \to 2} F(x) = 4.$$

17. As inputs x approach -5 from the left, outputs $F(x)$ approach 0. Thus the limit from the left is 0. That is, $\lim\limits_{x \to -5^-} F(x) = 0$.
As inputs x approach -5 from the right, outputs $F(x)$ approach 0. Thus the limit from the right is 0. That is, $\lim\limits_{x \to -5^+} F(x) = 0$.
Since the limit from the left, 0, is the same as the limit from the right, 0, we have
$$\lim_{x \to -5} F(x) = 0.$$

19. As inputs x approach 6 from the left, outputs $F(x)$ approach 0. Thus the limit from the left is 0. That is, $\lim\limits_{x \to 6^-} F(x) = 0$.
As inputs x approach 6 from the right, outputs $F(x)$ approach 0. Thus the limit from the right is 0. That is, $\lim\limits_{x \to 6^+} F(x) = 0$.
Since the limit from the left, 0, is the same as the limit from the right, 0, we have
$$\lim_{x \to 6} F(x) = 0.$$

21. As inputs x approach -2 from the left, outputs $F(x)$ approach 4. Thus the limit from the left is 4. That is, $\lim\limits_{x \to -2^-} F(x) = 4$.

23. As inputs x approach 0 from the left, outputs $G(x)$ approach 3. Thus the limit from the left is 3. That is, $\lim\limits_{x \to 0^-} G(x) = 3$.

    As inputs x approach 0 from the right, outputs $G(x)$ approach 3. Thus the limit from the right is 3. That is $\lim\limits_{x \to 0^+} G(x) = 3$.

    Since the limit from the left, 3, is the same as the limit from the right, 3, we have $\lim\limits_{x \to 0} G(x) = 3$.

25. As inputs x approach 1 from the right, outputs $G(x)$ approach -1. Thus the limit from the right is -1. That is, $\lim\limits_{x \to 1^+} G(x) = -1$

27. As inputs x approach 1 from the left, outputs $G(x)$ approach 4. Thus the limit from the left is 4. That is, $\lim\limits_{x \to 1^-} G(x) = 4$.

    As inputs x approach 1 from the right, outputs $G(x)$ approach -1. Thus the limit from the right is -1. That is, $\lim\limits_{x \to 1^+} G(x) = -1$.

    Since the limit from the left, 4, is not the same as the limit from the right, -1, we have $\lim\limits_{x \to 1} G(x)$ *does not exist.*

29. As inputs x approach 3 from the left, outputs $G(x)$ approach 0. Thus the limit from the left is 0. That is, $\lim\limits_{x \to 3^-} G(x) = 0$.

    As inputs x approach 3 from the right, outputs $G(x)$ approach 0. Thus the limit from the right is 0. That is, $\lim\limits_{x \to 3^+} G(x) = 0$.

    Since the limit from the left, 0, is the same as the limit from the right, 0, we have $\lim\limits_{x \to 3} G(x) = 0$.

31. As inputs x approach 2 from the left, outputs $H(x)$ approach 1. Thus the limit from the left is 1. That is, $\lim\limits_{x \to -2^-} H(x) = 1$.

33. As inputs x approach 2 from the left, outputs $H(x)$ approach 1. Thus the limit from the left is 1. That is, $\lim\limits_{x \to -2^-} H(x) = 1$.

    As inputs x approach 2 from the right, outputs $H(x)$ approach 1. Thus the limit from the right is 1. That is, $\lim\limits_{x \to -2^+} H(x) = 1$.

    Since the limit from the left, 1, is the same as the limit from the right, 1, we have $\lim\limits_{x \to -2} H(x) = 1$.

35. As inputs x approach 1 from the right, outputs $H(x)$ approach 2. Thus the limit from the right is 2. That is, $\lim\limits_{x \to -1^+} H(x) = 2$.

37. As inputs x approach 1 from the left, outputs $H(x)$ approach 4. Thus the limit from the left is 4. That is, $\lim\limits_{x \to 1^-} H(x) = 4$.

    As inputs x approach 1 from the right, outputs $H(x)$ approach 2. Thus the limit from the right is 2. That is, $\lim\limits_{x \to 1^+} H(x) = 2$.

    Since the limit from the left, 4, is not the same as the limit from the right, 2, we have $\lim\limits_{x \to 1} H(x)$ *does not exist.*

39. As inputs x approach 3 from the left, outputs $H(x)$ approach 1. Thus the limit from the left is 1. That is, $\lim\limits_{x \to 3^-} H(x) = 1$.

    As inputs x approach 3 from the right, outputs $H(x)$ approach 1. Thus the limit from the right is 1. That is, $\lim\limits_{x \to 3^+} H(x) = 1$.

    Since the limit from the left, 1, is the same as the limit from the right, 1, we have $\lim\limits_{x \to 3} H(x) = 1$.

41. As inputs x approach 2 from the left, outputs $f(x)$ approach -1. Thus the limit from the left is -1. That is, $\lim\limits_{x \to 2^-} f(x) = -1$.

As inputs x approach 2 from the right, outputs $f(x)$ approach -1. Thus the limit from the right is -1. That is, $\lim\limits_{x \to 2^+} f(x) = -1$.

Since the limit from the left, -1, is the same as the limit from the right, -1, we have $\lim\limits_{x \to 2} f(x) = -1$.

43. As inputs x approach 0 from the left, outputs $f(x)$ approach 2. Thus the limit from the left is 2. That is $\lim\limits_{x \to 0^-} f(x) = 2$.

As inputs x approach 0 from the right, outputs $f(x)$ approach 2. Thus the limit from the right is 2. That is, $\lim\limits_{x \to 0^+} f(x) = 2$.

Since the limit from the left, 2, is the same as the limit from the right, 2, we have $\lim\limits_{x \to 0} f(x) = 2$.

45. As inputs x approach 1 from the left, outputs $f(x)$ increase without bound. We say that the limit from the left is infinity. That is $\lim\limits_{x \to 1^-} f(x) = \infty$.

As inputs x approach 1 from the right, outputs $f(x)$ decrease without bound. We say that limit from the right is negative infinity. That is, $\lim\limits_{x \to 1^+} f(x) = -\infty$.

Since the function values as $x \to 1$ from the left increase without bound, and the function values as $x \to 1$ from the right decrease without bound, the limit does not exist. We have, $\lim\limits_{x \to 1} f(x)$ *does not exist*.

47. As inputs x approach 2 from the left, outputs $f(x)$ approach 0. Thus the limit from the left is 0. That is $\lim\limits_{x \to -2^-} f(x) = 0$.

As inputs x approach 2 from the right, outputs $f(x)$ approach 0. Thus the limit from the right is 0. That is, $\lim\limits_{x \to -2^+} f(x) = 0$.

Since the limit from the left, 0, is the same as the limit from the right, 0, we have $\lim\limits_{x \to -2} f(x) = 0$.

49. As inputs x get more and more negative, output $f(x)$ get closer and closer to 2. $\lim\limits_{x \to -\infty} f(x) = 2$.

51. Defining $f(x) = |x|$ as a piecewise defined function we have:

$$f(x) = \begin{cases} -x, & x < 0 \\ x, & x \geq 0. \end{cases}$$

We graph the function by creating an input-output table.

x	-2	-1	0	1	2
$f(x)$	2	1	0	1	2

Next, we plot the points and draw the graph.

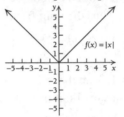

Find $\lim\limits_{x \to 0} f(x)$.

As inputs x approach 0 from the left, outputs $f(x)$ approach 0. We have, $\lim\limits_{x \to 0^-} f(x) = 0$.

As inputs x approach 0 from the right, outputs $f(x)$ approach 0. We have, $\lim\limits_{x \to 0^+} f(x) = 0$.

Since the limit from the left is the same as the limit from the right, we have $\lim\limits_{x \to 0} f(x) = 0$.

Find $\lim\limits_{x \to -2} f(x)$.

As inputs x approach -2 from the left, outputs $f(x)$ approach 2. We have, $\lim\limits_{x \to -2^-} f(x) = 2$.

As inputs x approach -2 from the right, outputs $f(x)$ approach 2. We have, $\lim\limits_{x \to -2^+} f(x) = 2$.

Since the limit from the left is the same as the limit from the right, we have $\lim\limits_{x \to -2} f(x) = 2$.

53. $g(x) = x^2 - 5$

We graph the function by creating an input-output table.

x	-2	-1	0	1	2
$g(x)$	-1	-4	-5	-4	-1

Next, we plot the points from the table and draw the graph.

Find $\lim\limits_{x \to 0} g(x)$.

As inputs x approach 0 from the left, outputs $g(x)$ approach -5. We have,

$\lim\limits_{x \to 0^-} g(x) = -5$.

As inputs x approach 0 from the right, outputs $g(x)$ approach -5. We have,

$\lim\limits_{x \to 0^+} g(x) = -5$

Since the limit from the left is the same as the limit from the right, we have

$\lim\limits_{x \to 0} g(x) = -5$.

Find $\lim\limits_{x \to -1} g(x)$.

As inputs x approach -1 from the left, outputs $g(x)$ approach -4. We have,

$\lim\limits_{x \to -1^-} g(x) = -4$.

As inputs x approach -1 from the right, outputs $g(x)$ approach -4. We have,

$\lim\limits_{x \to -1^+} g(x) = -4$

Since the limit from the left is the same as the limit from the right, we have

$\lim\limits_{x \to -1} g(x) = -4$.

55. $G(x) = \dfrac{1}{x+2}$

Since $x = -2$ makes the denominator zero, we exclude the value -2 from the domain. Creating an input-output table we have

x	-4	-3	-2.1	-1.9	-1	0
$G(x)$	$-\frac{1}{2}$	-1	-10	10	1	$\frac{1}{2}$

Next we plot the points and draw the graph.

Find $\lim\limits_{x \to -1} G(x)$.

As inputs x approach -1 from the left, outputs $G(x)$ approach 1. We have, $\lim\limits_{x \to -1^-} G(x) = 1$.

As inputs x approach -1 from the right, outputs $G(x)$ approach 1. We have,

$\lim\limits_{x \to -1^+} G(x) = 1$. Since the limit from the left is the same as the limit from the right, we have

$\lim\limits_{x \to -1} G(x) = 1$.

Find $\lim\limits_{x \to -2} G(x)$

As inputs x approach -2 from the left, outputs $G(x)$ decrease without bound. We have,

$\lim\limits_{x \to -2^-} G(x) = -\infty$

As inputs x approach -2 from the right, outputs $G(x)$ increase without bound. We have,

$\lim\limits_{x \to -2^+} G(x) = \infty$.

Since the function values as $x \to -2$ from the left decrease without bound, and the function values as $x \to -2$ from the right increase without bound, the limit does not exist. We have, $\lim\limits_{x \to -2} G(x)$ *does not exist.*

57. $f(x) = \dfrac{1}{x} - 2$

Since $x = 0$ makes the denominator zero, we exclude the value 0 from the domain. Creating an input-output table we have

x	-1	-0.5	-0.1	0.1	0.5	1
$f(x)$	-3	-4	-12	8	0	-1

The solution is continued on the next page.

Next we plot the points from the table on the previous page and draw the graph.

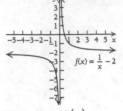

Find $\lim\limits_{x\to\infty} f(x)$.

As inputs x get larger and larger, outputs $f(x)$ get closer and closer to -2. We have

$\lim\limits_{x\to\infty} f(x) = -2$.

Find $\lim\limits_{x\to 0} f(x)$.

As inputs x approach 0 from the left, outputs $f(x)$ decrease without bound. We have,

$\lim\limits_{x\to 0^-} f(x) = -\infty.$

As inputs x approach 0 from the right, outputs $f(x)$ increase without bound. We have,

$\lim\limits_{x\to 0^+} f(x) = \infty.$

Since the function values as $x\to 0$ from the left decrease without bound, and the function values as $x\to 0$ from the right increase without bound, the limit does not exist. We have,

$\lim\limits_{x\to 0} f(x)\, does\ not\ exist.$

59. $g(x) = \dfrac{1}{x-3} + 2$

Since $x = 3$ makes the denominator zero, we exclude the value 3 from the domain. Creating an input-output table we have

x	1	2	2.5	2.9	3.1	3.5	4	5
$g(x)$	$\frac{3}{2}$	1	0	-8	12	4	3	$\frac{5}{2}$

Next we plot the points and draw the graph.

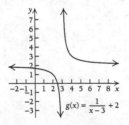

Find $\lim\limits_{x\to\infty} g(x)$.

As inputs x get larger and larger, outputs $g(x)$ get closer and closer to 2. We have

$\lim\limits_{x\to\infty} g(x) = 2$.

Find $\lim\limits_{x\to 3} g(x)$.

As inputs x approach 3 from the left, outputs $g(x)$ decrease without bound. We have,

$\lim\limits_{x\to 3^-} g(x) = -\infty$.

As inputs x approach 3 from the right, outputs $g(x)$ increase without bound. We have,

$\lim\limits_{x\to 3^+} g(x) = \infty$.

Since the function values as $x\to 3$ from the left decrease without bound, and the function values as $x\to 3$ from the right increase without bound, the limit does not exist. We have,

$\lim\limits_{x\to 3} g(x)\, does\ not\ exist.$

61. $F(x) = \begin{cases} 2x+1, & \text{for } x < 1 \\ x, & \text{for } x \geq 1. \end{cases}$

We create an input-output table for each piece of the function.

For $x < 1$

x	-1	0	0.9
$F(x)$	-1	1	2.8

We plot the points and draw the graph. Notice we draw an open circle at the point $(1,3)$ to indicate that the point is not part of the graph.

For $x \geq 1$

x	1	2	3
$F(x)$	1	2	3

We plot the points and draw the graph. Notice we draw a solid circle at the point $(1,1)$ to indicate that the point is part of the graph.

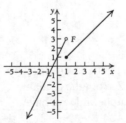

The solution is continued on the next page.

Find $\lim\limits_{x\to 1^-} F(x)$.

As inputs x approach 1 from the left, outputs $F(x)$ approach 3. That is,

$\lim\limits_{x\to 1^-} F(x) = 3$.

Find $\lim\limits_{x\to 1^+} F(x)$.

As inputs x approach 1 from the right, outputs $F(x)$ approach 1. That is,

$\lim\limits_{x\to 1^+} F(x) = 1$.

Find $\lim\limits_{x\to 1} F(x)$.

Since the limit from the left, 3, is not the same as the limit from the right, 1, we have

$\lim\limits_{x\to 1} F(x)$ *does not exist.*

63. $g(x) = \begin{cases} -x+4, & \text{for } x < 3 \\ x-3, & \text{for } x > 3. \end{cases}$

We create an input-output table for each piece of the function.

For $x < 3$

x	0	1	2	2.9
$g(x)$	4	3	2	1.1

We plot the points and draw the graph. Notice we draw an open circle at the point $(3,1)$ to indicate that the point is not part of the graph.
For $x > 3$

x	3.1	4	5	6
$g(x)$	0.1	1	2	3

We plot the points and draw the graph. Notice we draw an open circle at the point $(3,0)$ to indicate that the point is not part of the graph.

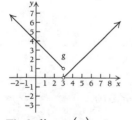

Find $\lim\limits_{x\to 3^-} g(x)$.

As inputs x approach 3 from the left, outputs $g(x)$ approach 1. That is,

$\lim\limits_{x\to 3^-} g(x) = 1$.

Find $\lim\limits_{x\to 3^+} g(x)$.

As inputs x approach 3 from the right, outputs $g(x)$ approach 0. That is,

$\lim\limits_{x\to 3^+} g(x) = 0$.

Find $\lim\limits_{x\to 3} g(x)$

Since the limit from the left, 1, is not the same as the limit from the right, 0, we have

$\lim\limits_{x\to 3} g(x)$ *does not exist.*

65. $F(x) = \begin{cases} -2x-3, & \text{for } x < -1 \\ x^3, & \text{for } x > -1. \end{cases}$

We create an input-output table for each piece of the function.
For $x < -1$

x	-3	-2	-1.1
$F(x)$	3	1	-0.8

We plot the points and draw the graph. Notice we draw an open circle at the point $(-1,-1)$ to indicate that the point is not part of the graph.
For $x > -1$

x	-0.9	0	1
$F(x)$	-0.729	0	1

We plot the points and draw the graph. Notice we draw an open circle at the point $(-1,-1)$ to indicate that the point is not part of the graph.

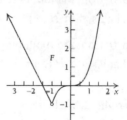

As inputs x approach -1 from the left, outputs $F(x)$ approach -1. We have,

$\lim\limits_{x\to -1^-} F(x) = -1$.

As inputs x approach -1 from the right, outputs $F(x)$ approach -1. We have,

$\lim\limits_{x\to -1^+} F(x) = -1$.

Since the limit from the left is the same as the limit from the right, we have

$\lim\limits_{x\to -1} F(x) = -1$.

67. $H(x) = \begin{cases} x+1, & \text{for } x < 0 \\ 2, & \text{for } 0 \le x < 1 \\ 3-x, & \text{for } x \ge 1. \end{cases}$

We create an input-output table for each piece of the function.

For $x < 0$

x	-1	-0.5	-0.1
$H(x)$	0	0.5	0.9

We plot the points and draw the graph. Notice we draw an open circle at the point $(0,1)$ to indicate that the point is not part of the graph. For $0 \le x < 1$, the function has value of 2. We draw a solid circle at the point $(0,2)$ to indicate the point is part of the graph and we draw an open circle at $(1,2)$ to indicate that the point is not part of the graph.

For $x \ge 1$

x	1	2	3
$H(x)$	2	1	0

We plot the points and draw the graph. Notice we draw a solid circle at the point $(1,2)$ to indicate that the point is part of the graph.

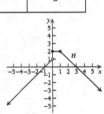

Find $\lim\limits_{x \to 0} H(x)$

As inputs x approach 0 from the left, outputs $H(x)$ approach 1. That is, $\lim\limits_{x \to 0^-} H(x) = 1$.

As inputs x approach 0 from the right, outputs $H(x)$ approach 2. That is, $\lim\limits_{x \to 0^+} H(x) = 2$.

Since the limit from the left, 1, is not the same as the limit from the right, 2, we have

$\lim\limits_{x \to 0} H(x)$ *does not exist.*

Find $\lim\limits_{x \to 1} H(x)$

As inputs x approach 1 from the left, outputs $H(x)$ approach 2. That is, $\lim\limits_{x \to 1^-} H(x) = 2$.

As inputs x approach 1 from the right, outputs $H(x)$ approach 2. That is, $\lim\limits_{x \to 1^+} H(x) = 2$.

Since the limit from the left, 2, is the same as the limit from the right, 2, we have

$\lim\limits_{x \to 0} H(x) = 2$.

69. $\lim\limits_{x \to 0.25^-} C(x) = \$3.50$

$\lim\limits_{x \to 0.25^+} C(x) = \$3.50$

$\lim\limits_{x \to 0.25} C(x) = \$3.50$

71. $\lim\limits_{x \to 0.6^-} C(x) = \$4.00.$

$\lim\limits_{x \to 0.6^+} C(x) = \$4.50.$

$\lim\limits_{x \to 0.6} C(x)$ *does not exist.*

73. $\lim\limits_{x \to 2^-} p(x) = \$1.19.$

$\lim\limits_{x \to 2^+} p(x) = \$1.40.$

$\lim\limits_{x \to 2} p(x)$ *does not exist.*

75. $\lim\limits_{x \to 3^-} p(x) = \$1.40.$

$\lim\limits_{x \to 3^+} p(x) = \$1.61.$

$\lim\limits_{x \to 3} p(x)$ *does not exist.*

77. $\lim\limits_{x \to 8925^-} r(x) = 10\%.$

$\lim\limits_{x \to 8925^+} r(x) = 15\%.$

$\lim\limits_{x \to 8925} r(x)$ *does not exist.*

79. $\lim\limits_{x \to 50,000^-} r(x) = 25\%.$

$\lim\limits_{x \to 50,000^+} r(x) = 25\%.$

$\lim\limits_{x \to 50,000} r(x) = 25\%.$

$\lim\limits_{x \to 87,850^-} r(x) = 25\%.$

$\lim\limits_{x \to 87,850^+} r(x) = 28\%.$

$\lim\limits_{x \to 87,850} r(x)$ *does not exist.*

81. $\lim\limits_{x \to 48,600^-} r(x) = 15\%.$

$\lim\limits_{x \to 48,600^+} r(x) = 25\%.$

$\lim\limits_{x \to 48,600} r(x)$ *does not exist.*

83. As inputs x approach 2 from the right, outputs $f(x)$ approach 4. We have,

$\lim\limits_{x \to 2^+} f(x) = 4$. In order for $\lim\limits_{x \to 2} f(x)$ to exist

we need $\lim\limits_{x \to 2^-} f(x) = 4$. We will use the letter c

for the unknown in the equation; therefore,

$\lim\limits_{x \to 2^-} \frac{1}{2}(x) + c = 4$.

Substitute 2 in for x to get the equation:

$\frac{1}{2}(2) + c = 4$ and solving for c we get

$1 + c = 4$

$c = 3$.

Therefore, in order for the limit to exist as x approaches 2, the function must be:

$f(x) = \begin{cases} \frac{1}{2}x + \underline{3} & \text{for } x < 2 \\ -x + 6 & \text{for } x > 2. \end{cases}$

85. As inputs x approach 2 from the left, outputs $f(x)$ approach -5. We have,

$\lim\limits_{x \to 2^-} f(x) = -5$. In order for $\lim\limits_{x \to 2} f(x)$ to exist

we need $\lim\limits_{x \to 2^+} f(x) = -5$. We will use the letter

c for the unknown in the equation and this gives us

$\lim\limits_{x \to 2^+} \left(-x^2 + c\right) = -5$

Substitute 2 in for x to get the equation:

$-(2)^2 + c = -5$ and solving for c we get

$-(4) + c = -5$

$c = -1$.

Therefore, in order for the limit to exist as x approaches 2, the function must be:

$f(x) = \begin{cases} x^2 - 9 & \text{for } x < 2 \\ -x^2 + \underline{-1} & \text{for } x > 2. \end{cases}$

87. Graph $f(x) = \begin{cases} x^2 - 2, & \text{for } x < 0 \\ 2 - x^2, & \text{for } x \geq 0. \end{cases}$

Using the calculator we enter the function into the graphing editor as follows:

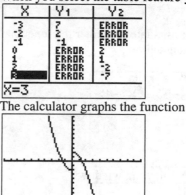

When you select the table feature you get:

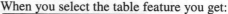

The calculator graphs the function

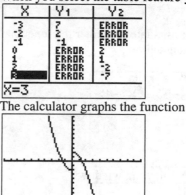

Using the trace feature, we find the limits.

Find $\lim\limits_{x \to 0} f(x)$.

As inputs x approach 0 from the left, outputs $f(x)$ approach -2. We have,

$\lim\limits_{x \to 0^-} f(x) = -2$.

As inputs x approach 0 from the right, outputs $f(x)$ approach 2. We have,

$\lim\limits_{x \to 0^+} f(x) = 2$

Since the limit from the left, -2, is not the same as the limit from the right, 2, we have

$\lim\limits_{x \to 0} f(x)$ *does not exist.*

Find $\lim\limits_{x \to -2} f(x)$.

As inputs x approach -2 from the left, outputs $f(x)$ approach 2. We have,

$\lim\limits_{x \to -2^-} f(x) = 2$.

As inputs x approach -2 from the right, outputs $f(x)$ approach 2. We have,

$\lim\limits_{x \to -2^+} f(x) = 2$

Since the limit from the left is the same as the limit from the right, we have

$\lim\limits_{x \to -2} f(x) = 2$.

89. Graph $f(x) = \dfrac{x-5}{x^2 - 4x - 5}$

Using the calculator we enter the function into the graphing editor.

```
Plot1 Plot2 Plot3
\Y1◻(X-5)/(X²-4)▶
\Y2=
\Y3=
\Y4=
\Y5=
\Y6=
```

Using the following window:

```
WINDOW
 Xmin=-5
 Xmax=10
 Xscl=1
 Ymin=-5
 Ymax=5
 Yscl=1
 Xres=1
```

The calculator graphs the function:

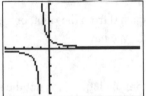

Using the trace feature on the calculator we find the limits.

Find $\lim\limits_{x \to -1} f(x)$.

As inputs x approach -1 from the left, outputs $f(x)$ increase without bound. We have,

$$\lim_{x \to -1^-} f(x) = -\infty.$$

As inputs x approach -1 from the right, outputs $f(x)$ decrease without bound. We have,

$$\lim_{x \to -1^+} f(x) = \infty$$

Since the function values as $x \to -1$ from the left increase without bound, and the function values as $x \to -1$ from the right decrease without bound, the limit does not exist.

$\lim\limits_{x \to -1} f(x)$ *does not exist.*

Find $\lim\limits_{x \to 5} f(x)$.

As inputs x approach 5 from the left, outputs $f(x)$ approach $\dfrac{1}{6}$. We have,

$$\lim_{x \to 5^-} f(x) = \frac{1}{6}.$$

As inputs x approach 5 from the right, outputs $f(x)$ approach $\dfrac{1}{6}$. We have,

$$\lim_{x \to 5^+} f(x) = \frac{1}{6}.$$

Since the limit from the left is the same as the limit from the right, we have

$$\lim_{x \to 5} f(x) = \frac{1}{6}.$$

Exercise Set 1.2

1. By limit property L1, $\lim\limits_{x \to 3} 7 = 7$.

Therefore, the statement $\lim\limits_{x \to 3} 7 = 3$ is a false.

3. By limit property L2,

$$\lim_{x \to 1}\left[g(x)\right]^2 = \left[\lim_{x \to 1} g(x)\right]^2 = \left[5\right]^2 = 25.$$

Therefore, the statement is true.

5. The statement is false. $g(3)$ could exist and if $\lim\limits_{x \to 3} g(x)$ *does not exist*, then the function would still be discontinuous at $x = 3$. .

7. This statement is false. If $\lim\limits_{x \to 4} F(x)$ exists but is not equal to $F(4)$, then F is not continuous.

9. It follows from the Theorem on Limits of Rational Functions that we can find the limit by substitution:

$$\lim_{x \to 2} (4x - 5) = 4(2) - 5 = 3.$$

11. It follows from the Theorem on Limits of Rational Functions that we can find the limit by substitution:

$$\lim_{x \to -1}\left(x^2 - 4\right) = (-1)^2 - 4$$
$$= 1 - 4 = -3.$$

13. It follows from the Theorem on Limits of Rational Functions that we can find the limit by substitution:

$$\lim_{x \to 5}\left(x^2 - 6x + 9\right) = (5)^2 - 6(5) + 9$$
$$= 25 - 30 + 9$$
$$= 4.$$

15. It follows from the Theorem on Limits of Rational Functions that we can find the limit by substitution:

$$\lim_{x \to 2}\left(2x^4 - 3x^3 + 4x - 1\right)$$
$$= 2(2)^4 - 3(2)^3 + 4(2) - 1$$
$$= 2(16) - 3(8) + 8 - 1$$
$$= 32 - 24 + 8 - 1$$
$$= 15.$$

17. It follows from the Theorem on Limits of Rational Functions that we can find the limit by substitution:

$$\lim_{x \to 3} \frac{x^2 - 25}{x^2 - 5} = \frac{(3)^2 - 25}{(3)^2 - 5}$$
$$= \frac{9 - 25}{9 - 5}$$
$$= \frac{-16}{4}$$
$$= -4.$$

19. We verify the expression yields an indeterminate form by substitution:

$$\lim_{x \to 3} \frac{x^2 - 9}{x - 3} = \frac{(3)^2 - 9}{(3) - 3}$$
$$= \frac{0}{0}.$$

This is an indeterminate form. In order to find the limit we will simplify the function by factoring the numerator and canceling common factors. Then we will apply the Theorem on Limits of Rational Functions to the simplified function.

$$\lim_{x \to 3} \frac{x^2 - 9}{x - 3} = \lim_{x \to 3} \frac{(x - 3)(x + 3)}{x - 3}$$
$$= \lim_{x \to 3}(x + 3) \qquad \text{simplifying, assuming } x \neq 3$$
$$= 3 + 3 \qquad \text{substitution}$$
$$= 6.$$

21. We verify the expression yields an indeterminate form by substitution.

$$\lim_{x \to -2} \frac{x^2 - 2x - 8}{x^2 - 4} = \frac{(-2)^2 - 2(-2) - 8}{(-2)^2 - 4}$$
$$= \frac{0}{0}.$$

This is an indeterminate form. In order to find the limit we will simplify the function by factoring the numerator and denominator and canceling common factors. Then we will apply the Theorem on Limits of Rational Functions to the simplified function.

The solution is continued on the next page.

We determine the limit as follows:

$$\lim_{x \to -2} \frac{x^2 - 2x - 8}{x^2 - 4} = \lim_{x \to -2} \frac{(x-4)(x+2)}{(x-2)(x+2)}$$

$$= \lim_{x \to -2} \left(\frac{x-4}{x-2} \right) \quad \text{simplifying, assuming } x \neq -2$$

$$= \frac{-2-4}{-2-2} \quad \text{substitution}$$

$$= \frac{-6}{-4}$$

$$= \frac{3}{2}.$$

23.   We verify the expression yields an indeterminate form by substitution:

$$\lim_{x \to 2} \frac{3x^2 + x - 14}{x^2 - 4} = \frac{3(2)^2 + (2) - 14}{(2)^2 - 4} = \frac{0}{0}.$$

This is an indeterminate form. In order to find the limit we will simplify the function by factoring the numerator and denominator and canceling common factors. Then we will apply the Theorem on Limits of Rational Functions to the simplified function.

$$\lim_{x \to 2} \frac{3x^2 + x - 14}{x^2 - 4} = \lim_{x \to 2} \frac{(3x+7)(x-2)}{(x+2)(x-2)}$$

$$= \lim_{x \to 2} \left(\frac{3x+7}{x+2} \right) \quad \text{simplifying, assuming } x \neq 2$$

$$= \frac{3(2)+7}{2+2} \quad \text{substitution}$$

$$= \frac{13}{4}.$$

25.   We verify the expression yields an indeterminate form by substitution:

$$\lim_{x \to 2} \frac{x^3 - 8}{2 - x} = \frac{(2)^3 - 8}{2 - 2} = \frac{0}{0}.$$

This is an indeterminate form. In order to find the limit we will simplify the function by factoring the numerator and canceling common factors. Then we will apply the Theorem on Limits of Rational Functions to the simplified function.

$$\lim_{x \to 2} \frac{x^3 - 8}{2 - x} = \lim_{x \to 2} \frac{(x-2)(x^2 + 2x + 4)}{-(x-2)}$$

$$= \lim_{x \to 2} \left[-\left(x^2 + 2x + 4 \right) \right] \quad \text{simplifying, assuming } x \neq 2$$

$$= -\left((2)^2 + 2(2) + 4 \right) \quad \text{substitution}$$

$$= -12.$$

27.   We verify the expression yields an indeterminate form by substitution:

$$\lim_{x \to 25} \frac{\sqrt{x} - 5}{x - 25} = \frac{\sqrt{25} - 5}{(25) - 25} = \frac{0}{0}.$$

This is an indeterminate form. In order to find the limit we will simplify the function by factoring the denominator and canceling common factors. Then we will apply the Theorem on Limits of Rational Functions to the simplified function.

$$\lim_{x \to 25} \frac{\sqrt{x} - 5}{x - 25} = \lim_{x \to 25} \frac{\sqrt{x} - 5}{\left(\sqrt{x} - 5 \right)\left(\sqrt{x} + 5 \right)}$$

$$= \lim_{x \to 25} \frac{1}{\left(\sqrt{x} + 5 \right)} \quad \text{simplifying, assuming } x \neq 25$$

$$= \frac{1}{\sqrt{25} + 5} \quad \text{substitution}$$

$$= \frac{1}{10}.$$

29.   We verify the expression yields an indeterminate form by substitution:

$$\lim_{x \to 2} \frac{x^2 + 3x - 10}{x^2 - 4x + 4} = \frac{(2)^2 + 3(2) - 10}{(2)^2 - 4(2) + 4}$$

$$= \frac{0}{0}.$$

This is an indeterminate form. In order to find the limit we will simplify the function by factoring the numerator and the denominator then canceling common factors. Then we will apply the Theorem on Limits of Rational Functions to the simplified function.

$$\lim_{x \to 2} \frac{x^2 + 3x - 10}{x^2 - 4x + 4} = \lim_{x \to 2} \frac{(x-2)(x+5)}{(x-2)(x-2)}$$

$$= \lim_{x \to 2} \left(\frac{x+5}{x-2} \right) \quad \text{simplifying, assuming } x \neq 2$$

$$= \frac{2+5}{2-2} \quad \text{substitution}$$

$$= \frac{7}{0}.$$

Substitution yields division by zero. Therefore,

$$\lim_{x \to 2} \frac{x^2 + 3x - 10}{x^2 - 4x + 4} \ does\ not\ exist.$$

31. $\lim\limits_{x \to 5} \sqrt{x^2 - 16}$

By limit property L2,

$\lim\limits_{x \to 5} \sqrt{x^2 - 16} = \sqrt{\lim\limits_{x \to 5} \left(x^2 - 16\right)}$

$\qquad = \sqrt{\lim\limits_{x \to 5} x^2 - \lim\limits_{x \to 5} 16}$ By L3

$\qquad = \sqrt{\left(\lim\limits_{x \to 5} x\right)^2 - 16}$ By L2 and L1

$\qquad = \sqrt{(5)^2 - 16}$

$\qquad = \sqrt{25 - 16}$

$\qquad = \sqrt{9}$

$\qquad = 3.$

33. $\lim\limits_{x \to 2} \sqrt{x^2 - 9}$

By limit property L2,

$\lim\limits_{x \to 2} \sqrt{x^2 - 9} = \sqrt{\lim\limits_{x \to 2} \left(x^2 - 9\right)}$

$\qquad = \sqrt{\lim\limits_{x \to 2} x^2 - \lim\limits_{x \to 2} 9}$ By L3

$\qquad = \sqrt{\left(\lim\limits_{x \to 2} x\right)^2 - 9}$ By L2 and L1

$\qquad = \sqrt{(2)^2 - 9}$

$\qquad = \sqrt{4 - 9}$

$\qquad = \sqrt{-5}.$

Therefore, $\lim\limits_{x \to 2} \sqrt{x^2 - 9}$ *does not exist.*

35. $\lim\limits_{x \to -4} \sqrt{x^2 - 16}$

By limit property L2,

$\lim\limits_{x \to -4^-} \sqrt{x^2 - 16} = \sqrt{\lim\limits_{x \to -4^-} \left(x^2 - 16\right)}$

$\qquad = \sqrt{\lim\limits_{x \to -4^-} x^2 - \lim\limits_{x \to -4^-} 16}$ By L3

$\qquad = \sqrt{\left(\lim\limits_{x \to -4^-} x\right)^2 - 16}$ By L2 and L1

$\qquad = \sqrt{(-4)^2 - 16}$

$\qquad = \sqrt{16 - 16}$

$\qquad = \sqrt{0}$

$\qquad = 0.$

37. The function is not continuous over the interval, because $g(x)$ is not continuous at $x = -2$. As x approaches -2 from the left, $g(x)$ approaches 4. However, as x approaches -2 from the right $f(x)$ approaches -3. Therefore, $g(x)$ is not continuous at -2.

39. The function is continuous over the interval.

41. The function is not continuous over the interval, because it is not continuous at $x = -2$. We see that $\lim\limits_{x \to -2} t(x) = \infty$, therefore, the limit does not exist as x approaches -2; furthermore, the function is not defined at $x = -2$, in other words $t(-2)$ does not exist. Therefore the function is not continuous at $x = -2$.

43. a) As inputs x approach 1 from the right, outputs $f(x)$ approach -1. Thus, the limit from the right is -1. $\lim\limits_{x \to 1^+} f(x) = -1$.
As inputs x approach 1 from the left, outputs $f(x)$ approach 2. Thus, the limit from the left is 2. $\lim\limits_{x \to 1} f(x) = 2$.
Since the limit from the left, -1, is not the same as the limit from the right, 2, the $\lim\limits_{x \to 1} f(x)$ *does not exist*

b) When the input is 1, the output $f(1)$ is -1. That is $f(1) = -1$

c) Since the limit of $f(x)$ as x approaches 1 does not exist, the function is not continuous at $x = 1$.

d) As inputs x approach -2 from the right, outputs $f(x)$ approach 3. Thus, the limit from the right is 3. $\lim\limits_{x \to -2^+} f(x) = 3$.
As inputs x approach -2 from the left, outputs $f(x)$ approach 3. Thus, the limit from the left is 3. $\lim\limits_{x \to -2^-} f(x) = 3$.
Since the limit from the left, 3, is the same as the limit from the right, 3, we say $\lim\limits_{x \to -2} f(x) = 3..$

e) When the input is -2, the output $f(-2)$ is 3. That is $f(-2) = 3$

f) The function $f(x)$ is continuous at $x = -2$, because
   1) $f(-2)$ exists, $f(-2) = 3$,
   2) $\lim\limits_{x \to -2} f(x)$ exists, $\lim\limits_{x \to -2} f(x) = 3$, and
   3) $\lim\limits_{x \to -2} f(x) = 3 = f(-2)$.

45. a) As inputs x approach -1 from the right, outputs $k(x)$ approach 2. Thus, the limit from the right is 2. $\lim\limits_{x \to -1^+} k(x) = 2$.
   As inputs x approach -1 from the left, outputs $k(x)$ approach 2. Thus, the limit from the left is 2. $\lim\limits_{x \to -1^-} k(x) = 2$.
   Since the limit from the left, 2, is the same as the limit from the right, 2, we have $\lim\limits_{x \to -1} k(x) = 2$.

   b) $k(-1)$ is not defined, or does not exist.

   c) Since the value of $k(-1)$ does not exist, the function is not continuous at $x = -1$.

   d) As inputs x approach 3 from the right, outputs $k(x)$ approach -2. Thus, the limit from the right is -2. $\lim\limits_{x \to 3^+} k(x) = -2$.
   As inputs x approach 3 from the left, outputs $k(x)$ approach -2. Thus, the limit from the left is -2. $\lim\limits_{x \to 3^-} k(x) = -2$.
   Since the limit from the left, -2, is the same as the limit from the right, -2, we have $\lim\limits_{x \to 3} k(x) = -2$.

   e) $k(3) = -2$

   f) The function $k(x)$ is continuous at $x = 3$, because
   1) $k(3)$ exists,
   2) $\lim\limits_{x \to 3} k(x)$ exists, and
   3) $\lim\limits_{x \to 3} k(x) = -2 = k(3)$.

47. a) As inputs x approach 3 from the right, outputs $G(x)$ approach 3. Thus,
   $\lim\limits_{x \to 3^+} G(x) = 3$.

   b) As inputs x approach 3 from the left, outputs $G(x)$ approach 1. Thus,
   $\lim\limits_{x \to 3^-} G(x) = 1$.

   c) Since the limit from the left, 1, is not the same as the limit from the right, 3, the limit does not exist. $\lim\limits_{x \to 3} G(x)$ does not exist.

   d) $G(3) = 1$

   e) The function $G(x)$ is not continuous at $x = 3$ because the limit does not exist as x approaches 3.

   f) The function $G(x)$ is continuous at $x = 0$, because
   1) $G(0)$ exists,
   2) $\lim\limits_{x \to 0} G(x)$ exists, and
   3) $\lim\limits_{x \to 0} G(x) = G(0)$.

   g) The function $G(x)$ is continuous at $x = 2.9$, because
   1) $G(2.9)$ exists,
   2) $\lim\limits_{x \to 2.9} G(x)$ exists, and
   3) $\lim\limits_{x \to 2.9} G(x) = G(2.9)$.

49. First we find the function value when $x = 4$.
   $g(4) = (4)^2 - 3(4) = 4$. Hence, $g(4)$ exists.
   Next, we find the limit as x approaches 4. It follows from the Theorem on Limits of Rational Functions that we can find the limit by substitution: $\lim\limits_{x \to 4} g(x) = (4)^2 - 3(4) = 4$.
   Therefore, $\lim\limits_{x \to 4} g(x) = 4 = g(4)$ and the function is continuous at $x = 4$.

51. The function $G(x) = \dfrac{1}{x}$ is not continuous at $x = 0$ because $G(0) = \dfrac{1}{0}$ is undefined.

53. First we find the function value when $x = 4$.
$f(4) = -(4) + 7 = 3$. Hence, $f(4)$ exists. Next we find the limit as x approaches 4. As the inputs x approach 4 from the left, the outputs $f(x)$ approach 3, that is,

$$\lim_{x \to 4^-} f(x) = \frac{1}{2}(4) + 1 = 3.$$

As the inputs x approach 4 from the right, the outputs $f(x)$ approach 3, that is,

$$\lim_{x \to 4^+} f(x) = -(4) + 7 = 3.$$

Since the limit from the left, 3, is the same as the limit from the right, 3. The limit exists. We have:

$$\lim_{x \to 4} f(x) = 3.$$

Therefore, we have $\lim_{x \to 4} f(x) = 3 = f(4)$.

Thus the function is continuous at $x = 4$.

55. The function is not continuous at $x = 3$ because the limit does not exist as x approaches 3. To verify this we take the limit as x approaches 3 from the left and the limit as x approaches 3 from the right.

As x approaches 3 from the left we have

$$\lim_{x \to 3^-} F(x) = \frac{1}{3}(3) + 4 = 5.$$

As x approaches 3 from the right we have

$$\lim_{x \to 3^+} F(x) = 2(3) - 5 = 1.$$

Since the limit from the left, 5, is not the same as the limit from the right, 1, the limit does not exist. $\lim_{x \to 3} F(x)$ does not exist.

57. The function is not continuous at $x = 4$ because the function is not defined at $x = 4$. Therefore, $g(4)$ does not exist.

59. The function is not continuous at $x = 2$. To verify this, we take the limit as x approaches 2. Using the Theorem on Limits of Rational Functions, we simplify the function near 2 by factoring the numerator and canceling common factors.

$$\lim_{x \to 2} G(x) = \lim_{x \to 2} \frac{x^2 - 4}{x - 2}$$

$$= \lim_{x \to 2} \frac{(x - 2)(x + 2)}{x - 2}$$

$$= \lim_{x \to 2} (x + 2)$$

$$= 2 + 2 = 4$$

Therefore,

$$\lim_{x \to 2} G(x) = 4.$$

However, when $x = 2$, the output $G(2)$ is defined to be 5. That is, $G(2) = 5$. Therefore, $\lim_{x \to 2} G(x) = 4 \neq 5 = G(2)$. Thus the function is not continuous at $x = 2$.

61. First we find the function value when $x = 4$.
$G(4) = 2(4) - 3 = 5$, $G(4)$ exists.

Next we find the limit as x approaches 4. To find the limit as x approaches 4 from the left, we first simplify the rational function by factoring the numerator and canceling common factors.

$$\lim_{x \to 4^-} G(x) = \lim_{x \to 4^-} \frac{x^2 - 3x - 4}{x - 4}$$

$$= \lim_{x \to 4^-} \frac{(x - 4)(x + 1)}{x - 4}$$

$$= \lim_{x \to 4^-} (x + 1)$$

$$= 4 + 1$$

$$= 5$$

To find the limit as x approaches 4 from the right, we can use substitution.

$$\lim_{x \to 4^+} G(x) = \lim_{x \to 4^+} (2x - 3)$$

$$= 2(4) - 3$$

$$= 5$$

Therefore, the limit exists.

$$\lim_{x \to 4} G(x) = 5.$$

Thus we have,

$$\lim_{x \to 4} G(x) = 5 = G(4).$$

Therefore, the function in continuous at $x = 4$.

63. The function is not continuous at $x = 5$ because $g(5)$ does not exist.

$$g(5) = \frac{1}{(5)^2 - 7(5) + 10}$$

$$= \frac{1}{25 - 35 + 10}$$

$$= \frac{1}{0}.$$

65. The function is not continuous at $x = 2$, because $G(2)$ does not exist.

$$G(2) = \frac{1}{(2)^2 - 6(2) + 8}$$

$$= \frac{1}{0}$$

67. Yes, the function is continuous over the interval $(-4, 4)$. Since the function is defined for every value in the interval, the Theorem on Limits of Rational Functions tells us $\lim_{x \to a} g(x) = g(a)$ for all values a in the interval. Thus $g(x)$ is continuous over the interval.

69. No, the function is not continuous over the interval $(0, \infty)$ because the function does not exist at $x = 1$. $G(1) = \frac{1}{1-1} = \frac{1}{0}$, which is undefined.

71. Yes, the function is continuous on $\mathbb{R}$. The function is defined for all real numbers, so by the Theorem on Limits of Rational Functions, $\lim_{x \to a} g(x) = g(a)$ for all a in $\mathbb{R}$.

73. The limit as x approaches 20 from the left is found in this case using Limit Property L3 by substituting on the piece of the function that is defined for values less than 20. The limit from the left of the function is:
$$\lim_{x \to 20^-} (1.5x) = 1.50(20) = 30.$$
The limit as x approaches 20 from the right is found in this case using Limit Property L3 by substituting on the piece of the function that is defined for values greater than 20. The limit from the right of the function is:
$$\lim_{x \to 20^+} (1.25x) = 1.25(20) = 25.$$
Since the limit from the left, 30, is not the same as the limit from the right, 25, the limit does not exist. $\lim_{x \to 20} p(x)$ *does not exist.*

75. The limit as t approaches 60 from the left is found in this case using Limit Property L3 by substituting on the piece of the function that is defined for values less than 60. The limit from the left of the function is:
$$\lim_{t \to 60^-} (2t) = 2(60) = 120.$$
The limit as t approaches 20 from the right is found in this case using Limit Property L3 by substituting on the piece of the function that is defined for values greater than 20. The limit from the right of the function is:
$$\lim_{t \to 60^+} (300 - 3t) = 300 - 3(60) = 120.$$
Since the limit from the left, 120, is the same as the limit from the right, 120, the limit is $\lim_{t \to 60} T(t) = 120$.

77. In exercise 74, we found that $\lim_{x \to 100^-} = 0.08(100) = 8$. In order for $p(x)$ to be continuous, the limit from the right must equal 8. Therefore,
$$\lim_{x \to 100^+} 0.06x + k = 8$$
$$0.06(100) + k = 8 \qquad \text{By Limit Property L3}$$
$$6 + k = 8$$
$$k = 2$$
The constant k must equal 2 in order for the function to be continuous at $x = 100$.

79. 0.5, or $\frac{1}{2}$

81. 0.5, or $\frac{1}{2}$

83. 0.378, or $\frac{1}{\sqrt{7}}$

85. 0

Exercise Set 1.3

1. a) $f(x) = 5x^2$

    so,

    $f(x+h) = 5(x+h)^2$ substituting $x+h$ for x

    $\qquad = 5(x^2 + 2xh + h^2)$

    $\qquad = 5x^2 + 10xh + 5h^2$

    Then

    $\dfrac{f(x+h) - f(x)}{h}$ Difference quotient

    $= \dfrac{(5x^2 + 10xh + 5h^2) - 5x^2}{h}$ Substituting

    $= \dfrac{10xh + 5h^2}{h}$

    $= \dfrac{h(10x + 5h)}{h}$ Factoring the numerator

    $= \dfrac{h}{h} \cdot (10x + 5h)$ Removing a factor = 1.

    $= 10x + 5h,$ Simplified difference quotient

    or $5(2x + h)$

    b) The difference quotient column in the table can be completed using the simplified difference quotient.

    $10x + 5h$ Simplified difference quotient

    $= 10(5) + 5(2) = 60$

    substituting 5 for x and 2 for h;

    $= 10(5) + 5(1) = 55$

    substituting 5 for x and 1 for h;

    $= 10(5) + 5(0.1) = 50.5$

    substituting 5 for x and 0.1 for h;

    $= 10(5) + 5(0.01) = 50.05$

    substituting 5 for x and 0.01 for h.

    The completed table is:

x	h	$\dfrac{f(x+h) - f(x)}{h}$
5	2	60
5	1	55
5	0.1	50.5
5	0.01	50.05

3. a) $f(x) = -5x^2$

    so,

    $f(x+h) = -5(x+h)^2$ substituting $x+h$ for x

    $\qquad = -5(x^2 + 2xh + h^2)$

    $\qquad = -5x^2 - 10xh - 5h^2$

    Then

    $\dfrac{f(x+h) - f(x)}{h}$ Difference quotient

    $= \dfrac{(-5x^2 - 10xh - 5h^2) - (-4x^2)}{h}$ Substituting

    $= \dfrac{-10xh - 5h^2}{h}$

    $= \dfrac{h(-10x - 5h)}{h}$ Factoring the numerator

    $= \dfrac{h}{h} \cdot (-10x - 5h)$ Removing a factor = 1.

    $= -10x - 5h,$ Simplified difference quotient

    or $-5(2x + h)$

    b) The difference quotient column in the table can be completed using the simplified difference quotient.

    $-10x - 5h$ Simplified difference quotient

    $= -10(5) - 5(2) = -60$

    substituting 5 for x and 2 for h;

    $= -10(5) - 5(1) = -55$

    substituting 5 for x and 1 for h;

    $= -10(5) - 5(0.1) = -50.5$

    substituting 5 for x and 0.1 for h;

    $= -10(5) - 5(0.01) = -50.05$

    substituting 5 for x and 0.01 for h.

    The completed table is:

x	h	$\dfrac{f(x+h) - f(x)}{h}$
5	2	-60
5	1	-55
5	0.1	-50.5
5	0.01	-50.05

5. a) $f(x) = x^2 - x$

We substitute $x+h$ for x

$f(x+h) = (x+h)^2 - (x+h)$

$\qquad = (x^2 + 2xh + h^2) - x - h$

$\qquad = x^2 + 2xh + h^2 - x - h$

Then

$\dfrac{f(x+h) - f(x)}{h}$ Difference quotient

$= \dfrac{(x^2 + 2xh + h^2 - x - h) - (x^2 - x)}{h}$

$= \dfrac{2xh + h^2 - h}{h}$

$= \dfrac{h(2x + h - 1)}{h}$ Factoring the numerator

$= \dfrac{h}{h} \cdot (2x + h - 1)$ Removing a factor $= 1$.

$= 2x + h - 1$ Simplified difference quotient

b) The difference quotient column in the table can be completed using the simplified difference quotient.

$2x + h - 1$ Simplified difference quotient

$= 2(5) + (2) - 1 = 11$

     substituting 5 for x and 2 for h;

$= 2(5) + (1) - 1 = 10$

     substituting 5 for x and 1 for h;

$= 2(5) + (0.1) - 1 = 9.1$

     substituting 5 for x and 0.1 for h;

$= 2(5) + (0.01) - 1 = 9.01$

     substituting 5 for x and 0.01 for h.

The completed table is:

x	h	$\dfrac{f(x+h) - f(x)}{h}$
5	2	11
5	1	10
5	0.1	9.1
5	0.01	9.01

7. a) $f(x) = \dfrac{9}{x}$

We substitute $x+h$ for x

$f(x+h) = \dfrac{9}{x+h}$

Then

$\dfrac{f(x+h) - f(x)}{h}$ Difference quotient

$= \dfrac{\left(\dfrac{9}{x+h}\right) - \left(\dfrac{9}{x}\right)}{h}$

$= \dfrac{\left(\dfrac{9}{x+h} \cdot \dfrac{x}{x}\right) - \left(\dfrac{9}{x} \cdot \dfrac{(x+h)}{(x+h)}\right)}{h}$ multiplying by 1

$= \dfrac{\left(\dfrac{9x}{x(x+h)}\right) - \left(\dfrac{9(x+h)}{x(x+h)}\right)}{h}$

$= \dfrac{\dfrac{-9h}{x(x+h)}}{h}$ adding fractions

$= \dfrac{\dfrac{-9h}{x(x+h)}}{\dfrac{h}{1}}$ $h = \dfrac{h}{1}$

$= \dfrac{-9h}{x(x+h)} \cdot \dfrac{1}{h}$ multiplying by the reciprocal

$= \dfrac{h}{h} \cdot \left(\dfrac{-9}{x(x+h)}\right)$ Removing a factor $= 1$.

$= \dfrac{-9}{x(x+h)}$ Simplified difference quotient

b) The difference quotient column in the table can be completed using the simplified difference quotient.

$-\dfrac{9}{x(x+h)}$ Simplified difference quotient

$= -\dfrac{9}{5(5+2)} = -\dfrac{9}{35}$

     substituting 5 for x and 2 for h;

$= -\dfrac{9}{5(5+1)} = -\dfrac{9}{30} = -\dfrac{3}{10}$

     substituting 5 for x and 1 for h;

$= -\dfrac{9}{5(5+0.1)} = -\dfrac{9}{25.5} = -\dfrac{6}{17}$

     substituting 5 for x and 0.1 for h;

$= -\dfrac{9}{5(5+0.01)} = -\dfrac{9}{25.05} = -\dfrac{60}{167}$

     substituting 5 for x and 0.01 for h.

We organize the values into a table on the next page.

The completed table is:

x	h	$\dfrac{f(x+h)-f(x)}{h}$
5	2	$-\dfrac{9}{35}$
5	1	$-\dfrac{3}{10}$
5	0.1	$-\dfrac{6}{17}$
5	0.01	$-\dfrac{60}{167}$

9. a) $f(x)=2x+3$

We substitute $x+h$ for x

$f(x+h)=2(x+h)+3$

$\qquad = 2x+2h+3$

Then

$\dfrac{f(x+h)-f(x)}{h}$ Difference quotient

$= \dfrac{(2x+2h+3)-(2x+3)}{h}$

$= \dfrac{2h}{h}$

$= 2$ Simplified difference quotient

b) The difference quotient is 2 for all values of x and h. Therefore, the completed table is:

x	h	$\dfrac{f(x+h)-f(x)}{h}$
5	2	2
5	1	2
5	0.1	2
5	0.01	2

11. a) $f(x)=12x^3$

so,

$f(x+h)=12(x+h)^3$ substituting $x+h$ for x

$\qquad = 12\left(x^3+3x^2h+3xh^2+h^3\right)$

Then

$\dfrac{f(x+h)-f(x)}{h}$ Difference quotient

$= \dfrac{12\left(x^3+3x^2h+3xh^2+h^3\right)-\left(12x^3\right)}{h}$

$= \dfrac{36x^2h+36xh^2+12h^3}{h}$

$= \dfrac{h\left(36x^2+36xh+12h^2\right)}{h}$

   Factoring the numerator.

$= \dfrac{h}{h}\cdot\left(36x^2+36xh+12h^2\right)$

   Removing a factor $= 1$.

$= 36x^2+36xh+12h^2$

   Simplified difference quotient

b) The difference quotient column in the table can be completed using the simplified difference quotient.

$36x^2+36xh+12h^2$

$= 36(5)^2+36(5)(2)+12(2)^2 = 1308$

   substituting 5 for x and 2 for h;

$= 36(5)^2+36(5)(1)+12(1)^2 = 1092$

   substituting 5 for x and 1 for h;

$= 36(5)^2+36(5)(0.1)+12(0.1)^2 = 918.12$

   substituting 5 for x and 0.1 for h;

$= 36(5)^2+36(5)(0.01)+12(0.01)^2 = 901.8012$

   substituting 5 for x and 0.01 for h.

The completed table is:

x	h	$\dfrac{f(x+h)-f(x)}{h}$
5	2	1308
5	1	1092
5	0.1	918.12
5	0.01	901.8012

13. a) $f(x)=x^2-4x$

We substitute $x+h$ for x

$f(x+h)=(x+h)^2-4(x+h)$

$\qquad = \left(x^2+2xh+h^2\right)-4x-4h$

$\qquad = x^2+2xh+h^2-4x-4h$

The solution is continued on the next page.

Then

$\dfrac{f(x+h)-f(x)}{h}$ Difference quotient

$= \dfrac{\left((x+h)^2 - 4(x+h)\right) - \left(x^2 - 4x\right)}{h}$

$= \dfrac{\left(x^2 + 2xh + h^2 - 4x - 4h\right) - \left(x^2 - 4x\right)}{h}$

$= \dfrac{2xh + h^2 - 4h}{h}$

$= \dfrac{h(2x + h - 4)}{h}$ Factoring the numerator.

$= \dfrac{h}{h} \cdot (2x + h - 4)$ Removing a factor = 1.

$= 2x + h - 4$ Simplified difference quotient

b) The difference quotient column in the table can be completed using the simplified difference quotient.

    $2x + h - 4$ Simplified difference quotient

    $= 2(5) + (2) - 4 = 8$

           substituting 5 for x and 2 for h;

    $= 2(5) + (1) - 4 = 7$

           substituting 5 for x and 1 for h;

    $= 2(5) + (0.1) - 4 = 6.1$

           substituting 5 for x and 0.1 for h;

    $= 2(5) + (0.01) - 4 = 6.01$

           substituting 5 for x and 0.01 for h.

The completed table is:

x	h	$\dfrac{f(x+h)-f(x)}{h}$
5	2	8
5	1	7
5	0.1	6.1
5	0.01	6.01

15. a) $f(x) = x^2 - 3x + 5$

We substitute $x + h$ for x

$f(x+h) = (x+h)^2 - 3(x+h) - 5$

         $= \left(x^2 + 2xh + h^2\right) - 3x - 3h + 5$

         $= x^2 + 2xh + h^2 - 3x - 3h + 5$

We substitute to find the simplified difference quotient at the top of the next column.

$\dfrac{f(x+h)-f(x)}{h}$ Difference quotient

$= \dfrac{\left(x^2 + 2xh + h^2 - 3x - 3h + 5\right) - \left(x^2 - 3x + 5\right)}{h}$

$= \dfrac{2xh + h^2 - 3h}{h}$

$= \dfrac{h(2x + h - 3)}{h}$ Factoring the numerator.

$= \dfrac{h}{h} \cdot (2x + h - 3)$ Removing a factor = 1.

$= 2x + h - 3$ Simplified difference quotient

b) The difference quotient column in the table can be completed using the simplified difference quotient.

    $2x + h - 3$ Simplified difference quotient

    $= 2(5) + (2) - 3 = 9$

           substituting 5 for x and 2 for h;

    $= 2(5) + (1) - 3 = 8$

           substituting 5 for x and 1 for h;

    $= 2(5) + (0.1) - 3 = 7.1$

           substituting 5 for x and 0.1 for h;

    $= 2(5) + (0.01) - 3 = 7.01$

           substituting 5 for x and 0.01 for h.

The completed table is:

x	h	$\dfrac{f(x+h)-f(x)}{h}$
5	2	9
5	1	8
5	0.1	7.1
5	0.01	7.01

17. To find the average rate of change from 2003 to 2007, we locate the corresponding points $(2003, 0)$ and $(2007, 1.1)$. Using these two points we calculate the average rate of change

$\dfrac{1.1 - 0}{2007 - 2003} = \dfrac{1.1}{4} = 0.275 \approx 0.28$.

The average rate of change of total employment from 2003 to 2007 increased approximately 0.28% per year.

To find the average rate of change from 2007 to 2012, we locate the corresponding points $(2007, 1.1)$ and $(2012, 1.9)$.

The solution is continued on the next page.

Using the two points on the previous page we calculate the average rate of change

$$\frac{1.9-1.1}{2012-2007} = \frac{0.8}{5} = 0.16.$$

The average rate of change of total employment from 2007 to 2012 increased approximately 0.16% per year.

To find the average rate of change from 2003 to 2012, we locate the corresponding points $(2003,0)$ and $(2012,1.9)$. Using these two points we calculate the average rate of change

$$\frac{1.9-0}{2012-2003} = \frac{1.9}{9} = 0.211\overline{1} \approx 0.21.$$

The average rate of change of total employment from 2003 to 2012 increased approximately 0.21% per year.

19. To find the average rate of change from 2003 to 2007, we locate the corresponding points $(2003,0)$ and $(2007,1.4)$. Using these two points we calculate the average rate of change

$$\frac{1.4-0}{2007-2003} = \frac{1.4}{4} = 0.35.$$

The average rate of change of professional services occupations employment from 2003 to 2007 increased approximately 0.35% per year.
To find the average rate of change from 2007 to 2012, we locate the corresponding points $(2007,1.4)$ and $(2012,2.7)$. Using these two points we calculate the average rate of change

$$\frac{2.7-1.4}{2012-2007} = \frac{1.3}{5} = 0.26.$$

The average rate of change of professional services occupations employment from 2007 to 2012 increased approximately 0.16% per year.

To find the average rate of change from 2003 to 2012, we locate the corresponding points $(2003,0)$ and $(2012,2.7)$. Using these two points we calculate the average rate of change

$$\frac{2.7-0}{2012-2003} = \frac{2.7}{9} = 0.3.$$

The average rate of change of professional services occupations employment from 2003 to 2012 increased approximately 0.3% per year.

21. To find the average rate of change from 2003 to 2007, we locate the corresponding points $(2003,0)$ and $(2007,-0.1)$.

Using the two points in the previous column, we calculate the average rate of change

$$\frac{0-(-0.1)}{2007-2003} = \frac{-0.1}{4} = -0.025 \approx -0.03.$$

The average rate of change of education employment from 2003 to 2007 increased approximately −0.03% per year.

To find the average rate of change from 2007 to 2012, we locate the corresponding points $(2007,-0.1)$ and $(2012,-0.7)$. Using these two points we calculate the average rate of change

$$\frac{-0.7-(-0.1)}{2012-2007} = \frac{-0.6}{5} = -0.12.$$

The average rate of change of education employment from 2007 to 2012 increased approximately −0.12% per year.

To find the average rate of change from 2003 to 2012, we locate the corresponding points $(2003,0)$ and $(2012,-0.7)$. Using these two points we calculate the average rate of change

$$\frac{-0.7-0}{2012-2003} = \frac{-0.7}{9} = -0.077 \approx -0.08.$$

The average rate of change of education employment from 2003 to 2012 increased approximately −0.08% per year.

23. To find the average rate of change from 2003 to 2007, we locate the corresponding points $(2003,0)$ and $(2007,0.2)$. Using these two points we calculate the average rate of change

$$\frac{0.2-0}{2007-2003} = \frac{0.2}{4} = 0.05.$$

The average rate of change of mining and logging employment from 2003 to 2007 increased approximately 0.05% per year.

To find the average rate of change from 2007 to 2012, we locate the corresponding points $(2007,0.2)$ and $(2012,3.8)$. Using these two points we calculate the average rate of change

$$\frac{3.8-0.2}{2012-2007} = \frac{3.6}{5} = 0.72.$$

The average rate of change of mining and logging employment from 2007 to 2012 increased approximately 0.72% per year.

To find the average rate of change from 2003 to 2012, we locate the corresponding points $(2003,0)$ and $(2012,3.8)$. Using these two points we calculate the average rate of change

$$\frac{3.8-0}{2012-2003} = \frac{3.8}{9} = 0.422\overline{2} \approx 0.42$$

The average rate of change of mining and logging employment from 2003 to 2012 increased approximately 0.42% per year.

25. In order to find the average rate of change from 1982 to 1992, we use the data points $(1982,73.10)$ and $(1992,85.78)$. The average rate of change is

$$\frac{85.78-73.10}{1992-1982} = \frac{12.68}{10} = 1.268 .$$

Between the years 1982 to 1992 the average rate of change in U.S. energy consumption was about 1.268 quadrillion BTUs per year.

In order to find the average rate of change from 1992 to 2002, we use the data points $(1992,85.78)$ and $(2002,97.65)$. The average rate of change is

$$\frac{97.65-85.78}{2002-1992} = \frac{11.87}{10} = 1.187 .$$

Between the years 1992 to 2002 the average rate of change in U.S. energy consumption was about 1.187 quadrillion BTUs per year.

In order to find the average rate of change from 2002 to 2012, we use the data points $(2002,97.65)$ and $(2012,95.10)$. The average rate of change is

$$\frac{95.10-97.65}{2012-2002} = \frac{-2.55}{10} = -0.255 .$$

Between the years 2002 to 2012 the average rate of change in U.S. energy consumption was about -0.225 quadrillion BTUs per year.

27. a) From 0 units to 1 unit the average rate of change is

$$\frac{70-0}{1-0} = 70 \text{ pleasure units per unit.}$$

From 1 unit to 2 units the average rate of change is

$$\frac{109-70}{2-1} = \frac{39}{1} = 39 \text{ pleasure units per unit.}$$

From 2 units to 3 units the average rate of change is

$$\frac{138-109}{3-2} = \frac{29}{1} = 29 \text{ pleasure units per unit.}$$

From 3 units to 4 units the average rate of change is

$$\frac{161-138}{4-3} = \frac{23}{1} = 23 \text{ pleasure units per unit.}$$

b)

29. $p(x) = 0.06x^3 - 0.5x^2 + 1.64x + 24.76$

a) $p(4) = 0.06(4)^3 - 0.5(4)^2 + 1.64(4) + 24.76$

$p(4) = 27.16$

b) $p(6) = 0.06(6)^3 - 0.5(6)^2 + 1.64(6) + 24.76$

$p(6) = 29.56$

c) $P(6) - p(4) = 29.56 - 27.16 = 2.40$

d) $\dfrac{p(6) - p(4)}{6-4} = \dfrac{2.40}{2} = 1.20$

This result implies that the average price of a ticket between 2012 ($x = 4$) and 2014 ($x = 6$) grew at an average rate of $1.20 per year.

31. $P(t) = 5400(0.975)^t$

$P(8) = 5400(0.975)^8 = 4409.919 \approx 4410$

$P(5) = 5400(0.975)^5 = 4757.916 \approx 4758$

$\dfrac{P(8) - P(5)}{8-5} = \dfrac{4410 - 4758}{8-5} = -116$

The population of Payton county lost on average 116 people per year between the 5th and 8th years after the last census.

33. $C(x) = -0.05x^2 + 50x$

First substitute 305 for x.

$C(305) = -0.05(305)^2 + 50(305)$

$\qquad = -4651.25 + 15,250 = 10,598.75$

The total cost of producing 305 units is $10,598.75.

Next substitute 300 for x.

$C(300) = -0.05(300)^2 + 50(300)$

$\qquad = -4500 + 15,000 = 10,500.00$

The total cost of producing 300 units is $10,500.00.

$\dfrac{C(305) - C(300)}{305 - 300} = \dfrac{10,598.75 - 10,500}{305 - 300}$

$\qquad\qquad = \dfrac{98.75}{5}$

$\qquad\qquad = 19.75$

The average cost of production between the 300th unit and 305th unit is 19.75 per unit.

35. Note: Answers will vary according to the values estimated from the graph.

   a) Locate the points $(0, 8)$ and $(12, 20)$ on the girls growth median weight chart. Using these points we calculate the average growth rate.
   $$\frac{20 - 8}{12 - 0} = \frac{12}{12} = 1.$$
   The average growth rate of a girl during her first 12 months is 1 pound per month.

   b) Locate the points $(12, 20)$ and $(24, 26.5)$ on the girls growth median weight chart. Using these points we calculate the average growth rate.
   $$\frac{26.5 - 20}{24 - 12} = \frac{6.5}{12} = 0.5416 \approx 0.54.$$
   The average growth rate of a girl during her second 12 months is approximately 0.54 pounds per month.

   c) Locate the points $(0, 8)$ and $(24, 26.5)$ on the girls growth median weight chart. Using these points we calculate the average growth rate.
   $$\frac{26.5 - 8}{24 - 0} = \frac{18.5}{24} = 0.7708\overline{3} \approx 0.77.$$
   The average growth rate of a girl during her first 24 months is approximately 0.77 pounds per month.

   d) We estimate the growth rate of a 12 month old girl to be approximately 0.67 pounds per month. This answer will vary depending upon your tangent line.

   e) The graph indicates that the growth rate is fastest during the first 3 months.

37. $H(w) = 0.11 w^{1.36}$

   a) First we substitute 500 and 700 in for w to find the home range at the respective weights at the top of the next column.
   $$H(500) = 0.11(500)^{1.36}$$
   $$= 0.11(4683.809314)$$
   $$= 515.2190246$$
   $$\approx 515.22$$
   $$H(700) = 0.11(700)^{1.36}$$
   $$= 0.11(7401.731628)$$
   $$= 814.1904791$$
   $$\approx 814.19$$

Next we use the function values to find the average rate at which the mammal's home range will increase
$$\frac{H(700) - H(500)}{700 - 500} = \frac{814.19 - 515.22}{700 - 500}$$
$$= \frac{298.97}{200}$$
$$\approx 1.49485$$
The average rate at which a carnivorous mammal's home range increases as the animal's weight grows from 500 g to 700 g is approximately 1.49 hectares per gram.

   b) First we substitute 200 and 300 in for w to find the home range at the respective weights.
   $$H(200) = 0.11(200)^{1.36}$$
   $$= 0.11(1347.102971)$$
   $$= 148.1813269$$
   $$\approx 148.18$$
   $$H(300) = 0.11(300)^{1.36}$$
   $$= 0.11(2338.217499)$$
   $$= 257.2039249$$
   $$\approx 257.20$$
Next we use the function values to find the average rate at which the mammal's home range will increase
$$\frac{H(300) - H(200)}{300 - 200} = \frac{257.20 - 148.18}{300 - 200}$$
$$= \frac{109.02}{100}$$
$$\approx 1.0902$$
The average rate at which a carnivorous mammal's home range increases as the animal's weight grows from 200 g to 300 g is approximately 1.09 hectares per gram.

39. a) We locate the points $(0, 0)$ and $(8, 10)$ on the graph and use them to calculate the average rate of change.
   $$\frac{10 - 0}{8 - 0} = \frac{10}{8} = \frac{5}{4} = 1.25$$
   The average rate of change is 1.25 words per minute.

   The solution is continued on the next page.

We locate the points $(8,10)$ and $(16,20)$ on the graph and use them to calculate the average rate of change.

$$\frac{20-10}{16-8}=\frac{10}{8}=\frac{5}{4}=1.25$$

The average rate of change is 1.25 words per minute.

We locate the points $(16,20)$ and $(24,25)$ on the graph and use them to calculate the average rate of change.

$$\frac{25-20}{24-16}=\frac{5}{8}=0.625$$

The average rate of change is 0.625 words per minute.

We locate the points $(24,25)$ and $(32,25)$ on the graph and use them to calculate the average rate of change.

$$\frac{25-25}{32-24}=\frac{0}{8}=0$$

The average rate of change is 0 words per minute.

We locate the points $(32,25)$ and $(36,25)$ on the graph and use them to calculate the average rate of change.

$$\frac{25-25}{36-32}=\frac{0}{4}=0$$

The average rate of change is 0 words per minute.

b) ✎

41. $s(t)=16t^2$

a) First, we find the function values by substituting 3 and 5 in for t respectively.

$$s(3)=16(3)^2=16(9)=144$$

$$s(5)=16(5)^2=16(25)=400$$

Next we subtract the function values.

$$s(5)-s(3)=400-144=256$$

The object will fall 256 feet in the two second time period between $t=3$ and $t=5$.

b) The average rate of change is calculated as

$$\frac{s(5)-s(3)}{5-3}=\frac{400-144}{5-3}$$

$$=\frac{256}{2}=128$$

The average velocity of the object during the two second time period from $t=3$ to $t=5$ is 128 feet per second.

43. a) For each curve, as t changes from 0 to 4, $P(t)$ changes from 0 to 500. Thus, the average growth rate for each country is

$$\frac{500-0}{4-0}=\frac{500}{4}=125.$$

The average growth rate for each country is approximately 125 million people per year.

b) ✎

c) For Country A:

As t changes from 0 to 1, $P(t)$ changes from 0 to 290. Thus the average growth rate is

$$\frac{290-0}{1-0}=290 \text{ million people per year.}$$

As t changes from 1 to 2, $P(t)$ changes from 290 to 250. Thus the average growth rate is

$$\frac{250-290}{2-1}=-40 \text{ million people per year.}$$

As t changes from 2 to 3, $P(t)$ changes from 250 to 200. Thus the average growth rate is

$$\frac{200-250}{3-2}=-50 \text{ million people per year.}$$

As t changes from 3 to 4, $P(t)$ changes from 200 to 500. Thus the average growth rate is

$$\frac{500-200}{4-3}=300 \text{ million people per year.}$$

For Country B we calculate the average growth rates at the top of the next column.

As t changes from 0 to 1, $P(t)$ changes from 0 to 125. Thus the average growth rate is

$$\frac{125-0}{1-0}=125 \text{ million people per year.}$$

As t changes from 1 to 2, $P(t)$ changes from 125 to 250. Thus the average growth rate is

$$\frac{250-125}{2-1}=125 \text{ million people per year.}$$

The solution is continued on the next page.

As t changes from 2 to 3, $P(t)$ changes from 250 to 375. Thus the average growth rate is

$$\frac{375-250}{3-2}=125 \text{ million people per year.}$$

As t changes from 3 to 4, $P(t)$ changes from 375 to 500. Thus the average growth rate is

$$\frac{500-375}{4-3}=125 \text{ million people per year.}$$

d) ✎

45.
a) Tracing along the 4-year private school graph, we see that the largest increase in costs occurred in the 2006-07 year.

b) Tracing along the 4-year public school graph, we see that the largest increases in costs occurred during the 2003-04 year and the 2008-09 year.

c) For the 4-year public school, the cost in year 2010 dollars is approximately \$10,711. To find out what the cost was in 2000 dollars assuming a 3% inflation rate over the 10 years we create the following equation using the simple compound interest formula from section R1. $C(1+r)^t = A$.

$$C(1+.03)^{10} = 10,711$$

$$C(1.03)^{10} = 10,711$$

$$C = \frac{10,711}{(1.03)^{10}}$$

$$C = 7969.989 \approx 7970$$

The cost of attending a 4 year public school in 2000 was \$7970 in year 2000 dollars. Likewise, for the 4-year private school, the cost in year 2010 dollars is approximately \$27,054

To find out what the cost was in year 2000 dollars assuming a 3% inflation rate over the 10 years. Using the simple compound interest formula from section R1, we have:

$$C(1+.03)^{10} = 27,054$$

$$C(1.03)^{10} = 27,054$$

$$C = \frac{27,054}{(1.03)^{10}}$$

$$C = 20,130.71 \approx 20,130$$

The cost of attending a 4 year private school in 2000 was \$20,130 in year 2000 dollars.

47. $f(x) = ax^2 + bx + c$

Substituting $x+h$ for x we have,

$$f(x+h) = a(x+h)^2 + b(x+h) + c$$
$$= a\left(x^2 + 2xh + h^2\right) + bx + bh + c$$
$$= ax^2 + 2axh + ah^2 + bx + bh + c$$

Thus,

$$\frac{f(x+h) - f(x)}{h} \quad \text{Difference quotient}$$

$$= \frac{\left(ax^2 + 2axh + ah^2 + bx + bh + c\right) - \left(ax^2 + bx + c\right)}{h}$$

$$= \frac{2axh + ah^2 + bh}{h}$$

$$= \frac{h(2ax + ah + b)}{h} \quad \text{Factoring the numerator}$$

$$= 2ax + ah + b \quad \text{Simplified difference quotient}$$

49. $f(x) = x^4$

Substituting $x+h$ for x we have,

$$f(x+h) = (x+h)^4$$
$$= x^4 + 4x^3h + 6x^2h^2 + 4xh^3 + h^4$$

Thus,

$$\frac{f(x+h) - f(x)}{h} \quad \text{Difference quotient}$$

$$= \frac{\left(x^4 + 4x^3h + 6x^2h^2 + 4xh^3 + h^4\right) - \left(x^4\right)}{h}$$

$$= \frac{4x^3h + 6x^2h^2 + 4xh^3 + h^4}{h}$$

$$= \frac{h\left(4x^3 + 6x^2h + 4xh^2 + h^3\right)}{h} \quad \text{Factoring the numerator}$$

$$= 4x^3 + 6x^2h + 4xh^2 + h^3 \quad \text{Simplified difference quotient}$$

51. $f(x) = ax^5 + bx^4$

Substituting $x + h$ for x we have,

$$f(x+h) = a(x+h)^5 + b(x+h)^4$$
$$= ax^5 + 5ax^4h + 10ax^3h^2 + 10ax^2h^3 + 5axh^4 + ah^5 + bx^4 + 4bx^3h + 6bx^2h^2 + 4bxh^3 + bh^4$$

Thus,

$\dfrac{f(x+h) - f(x)}{h}$ Difference quotient

$$= \frac{ax^5 + 5ax^4h + 10ax^3h^2 + 10ax^2h^3 + 5axh^4 + ah^5 + bx^4 + 4bx^3h + 6bx^2h^2 + 4bxh^3 + bh^4 - \left(ax^5 + bx^4\right)}{h}$$

$$= \frac{5ax^4h + 10ax^3h^2 + 10ax^2h^3 + 5axh^4 + ah^5 + 4bx^3h + 6bx^2h^2 + 4bxh^3 + bh^4}{h}$$

$$= \frac{h\left(5ax^4 + 10ax^3h + 10ax^2h^2 + 5axh^3 + ah^4 + 4bx^3 + 6bx^2h + 4bxh^2 + bh^3\right)}{h}$$ Factoring the numerator

$$= 5ax^4 + 10ax^3h + 10ax^2h^2 + 5axh^3 + ah^4 + 4bx^3 + 6bx^2h + 4bxh^2 + bh^3$$ Simplified difference quotient

53. $f(x) = \dfrac{1}{1-x}$

$\dfrac{f(x+h) - f(x)}{h}$

$$= \frac{\dfrac{1}{1-(x+h)} - \left(\dfrac{1}{1-x}\right)}{h}$$

$$= \frac{\dfrac{1}{1-x-h} \cdot \dfrac{1-x}{1-x} - \left(\dfrac{1}{1-x} \cdot \dfrac{1-x-h}{1-x-h}\right)}{h}$$

$$= \frac{\dfrac{1-x}{(1-x-h)(1-x)} - \dfrac{1-x-h}{(1-x-h)(1-x)}}{h}$$

$$= \frac{\dfrac{h}{(1-x-h)(1-x)}}{h}$$

$$= \frac{h}{(1-x-h)(1-x)} \cdot \frac{1}{h}$$

$$= \frac{1}{(1-x-h)(1-x)}$$

55. $f(x) = \sqrt{2x+1}$

Substituting $x + h$ for x we have,

$$f(x+h) = \sqrt{2(x+h)+1}$$

Thus,

$\dfrac{f(x+h) - f(x)}{h}$ Difference quotient

$$= \frac{\sqrt{2(x+h)+1} - \sqrt{2x+1}}{h}$$

Next we rationalize the numerator and simplify.

$$= \frac{\sqrt{2(x+h)+1} - \sqrt{2x+1}}{h} \cdot \frac{\sqrt{2(x+h)+1} + \sqrt{2x+1}}{\sqrt{2(x+h)+1} + \sqrt{2x+1}}$$

$$= \frac{2(x+h)+1 - (2x+1)}{h\left(\sqrt{2(x+h)+1} + \sqrt{2x+1}\right)}$$

$$= \frac{2x - 2h + 1 - 2x - 1}{h\left(\sqrt{2(x+h)+1} + \sqrt{2x+1}\right)}$$

$$= \frac{-2h}{h\left(\sqrt{2(x+h)+1} + \sqrt{2x+1}\right)}$$

$$= \frac{-2}{\sqrt{2(x+h)+1} + \sqrt{2x+1}}$$

56. $f(x) = \dfrac{1}{\sqrt{x}}$

Substituting $x+h$ for x we have,

$f(x+h) = \dfrac{1}{\sqrt{x+h}}$

Thus,

$\dfrac{f(x+h)-f(x)}{h}$ Difference quotient

$= \dfrac{\left(\dfrac{1}{\sqrt{x+h}}\right) - \left(\dfrac{1}{\sqrt{x}}\right)}{h}$

Find a common denominator in the numerator.

$= \dfrac{\left(\dfrac{1}{\sqrt{x+h}} \cdot \dfrac{\sqrt{x}}{\sqrt{x}}\right) \left(\dfrac{1}{\sqrt{x}} \cdot \dfrac{\sqrt{x+h}}{\sqrt{x+h}}\right)}{h}$

$= \dfrac{\dfrac{\sqrt{x}}{\sqrt{x}\sqrt{x+h}} - \dfrac{\sqrt{x+h}}{\sqrt{x}\sqrt{x+h}}}{h}$

$= \dfrac{\dfrac{\sqrt{x}-\sqrt{x+h}}{\sqrt{x}\sqrt{x+h}}}{h}$

$= \dfrac{\sqrt{x}-\sqrt{x+h}}{\sqrt{x}\sqrt{x+h}} \cdot \dfrac{1}{h}$ Simplifying the complex fraction

$\dfrac{f(x+h)-f(x)}{h} = \dfrac{\sqrt{x}-\sqrt{x+h}}{h\sqrt{x}\sqrt{x+h}}$

Rationalizing the numerator, we have:

$\dfrac{f(x+h)-f(x)}{h} = \dfrac{\sqrt{x}-\sqrt{x+h}}{h\sqrt{x}\sqrt{x+h}} \cdot \dfrac{\sqrt{x}+\sqrt{x+h}}{\sqrt{x}+\sqrt{x+h}}$

$= \dfrac{x-(x+h)}{h\sqrt{x}\sqrt{x+h}\left(\sqrt{x}+\sqrt{x+h}\right)}$

$= \dfrac{-h}{h\sqrt{x}\sqrt{x+h}\left(\sqrt{x}+\sqrt{x+h}\right)}$

$= \dfrac{-1}{\sqrt{x}\sqrt{x+h}\left(\sqrt{x}+\sqrt{x+h}\right)}$

Exercise Set 1.4

1. $f(x) = \dfrac{1}{2}x^2$

a), b)

x-axis is tangent to curve at (0, 0).

c) Find the simplified difference quotient first.

$$\frac{f(x+h)-f(x)}{h}$$

$$=\frac{\frac{1}{2}(x+h)^2-\frac{1}{2}x^2}{h}$$

$$=\frac{\frac{1}{2}(x^2+2xh+h^2)-\frac{1}{2}x^2}{h}$$

$$=\frac{\frac{1}{2}x^2+xh+\frac{1}{2}h^2-\frac{1}{2}x^2}{h}$$

$$=\frac{xh+\frac{1}{2}h^2}{h}$$

$$=\frac{h\left(x+\frac{1}{2}h\right)}{h}$$

$$=x+\frac{1}{2}h \quad \text{Simplified difference quotient}$$

Now we will find the limit of the difference quotient as $h \to 0$ using the simplified difference quotient.

$$\lim_{h\to 0}\frac{f(x+h)-f(x)}{h}=\lim_{h\to 0}\left(x+\frac{1}{2}h\right)$$
$$=x$$

Thus, $f'(x)=x$.

d) Find the values of the derivative by making the appropriate substitutions.

$f'(-2)=(-2)=-2$ Substituting -2 for x

$f'(0)=(0)=0$ Substituting 0 for x

$f'(1)=(1)=1$ Substituting 1 for x

3. $f(x)=-2x^2$

a), b)

x-axis is tangent to curve at (0, 0).

c) Find the simplified difference quotient first.

$$\frac{f(x+h)-f(x)}{h}$$

$$=\frac{-2(x+h)^2-\left(-2x^2\right)}{h}$$

$$=\frac{-2\left(x^2+2xh+h^2\right)-\left(-2x^2\right)}{h}$$

$$=\frac{-2x^2-4xh-2h^2+2x^2}{h}$$

$$=\frac{-4xh-2h^2}{h}$$

$$=\frac{h(-4x-2h)}{h}$$

$$=-4x-2h \quad \text{Simplified difference quotient}$$

Now we will find the limit of the difference quotient as $h \to 0$ using the simplified difference quotient.

$$\lim_{h\to 0}\frac{f(x+h)-f(x)}{h}=\lim_{h\to 0}(-4x-2h)$$
$$=-4x$$

Thus, $f'(x)=-4x$.

d) Find the values of the derivative by making the appropriate substitutions.

$f'(-2)=-4(-2)=8$ Substituting -2 for x

$f'(0)=-4(0)=0$ Substituting 0 for x

$f'(1)=-4(1)=-4$ Substituting 1 for x

5. $f(x) = -x^3$

a), b)

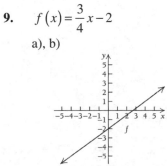

(–2, 8)

x-axis is tangent to curve at (0, 0).

(1, –1)

c) Find the simplified difference quotient first.

$$\frac{f(x+h)-f(x)}{h}$$

$$= \frac{-(x+h)^3 - (-x^3)}{h}$$

$$= \frac{-(x^3 + 3x^2h + 3xh^2 + h^3) - (-x^3)}{h}$$

$$= \frac{-x^3 - 3x^2h - 3xh^2 - h^3 + x^3}{h}$$

$$= \frac{-3x^2h - 3xh^2 - h^3}{h}$$

$$= \frac{h(-3x^2 - 3xh - h^2)}{h}$$

$$= -3x^2 - 3xh - h^2 \quad \text{Simplified difference quotient}$$

Now we will find the limit of the difference quotient as $h \to 0$ using the simplified difference quotient.

$$\lim_{h \to 0} \frac{f(x+h)-f(x)}{h} = \lim_{h \to 0}\left(-3x^2 - 3xh - h^2\right)$$

$$= -3x^2$$

Thus, $f'(x) = -3x^2$.

d) Find the values of the derivative by making the appropriate substitutions.

$$f'(-2) = -3(-2)^2 = -12 \quad \text{Substituting } -2 \text{ for } x$$

$$f'(0) = -3(0)^2 = 0 \quad \text{Substituting } 0 \text{ for } x$$

$$f'(1) = -3(1)^2 = -3 \quad \text{Substituting } 1 \text{ for } x$$

7. $f(x) = 2x + 3$

a), b)

Note: for linear functions the tangent line is the line itself.

c) Find the simplified difference quotient first.

$$\frac{f(x+h)-f(x)}{h}$$

$$= \frac{2(x+h)+3-(2x+3)}{h}$$

$$= \frac{2x+2h+3-2x-3}{h}$$

$$= \frac{2h}{h}$$

$$= 2 \quad \text{Simplified difference quotient}$$

Now we will find the limit of the difference quotient as $h \to 0$ using the simplified difference quotient.

$$\lim_{h \to 0} \frac{f(x+h)-f(x)}{h} = \lim_{h \to 0}(2)$$

$$= 2$$

Thus, $f'(x) = 2$.

d) Since the derivative is a constant, the value of the derivative will be 2 regardless of the value of x.

$$f'(-2) = 2 \quad \text{Substituting } -2 \text{ for } x$$

$$f'(0) = 2 \quad \text{Substituting } 0 \text{ for } x$$

$$f'(1) = 2 \quad \text{Substituting } 1 \text{ for } x$$

9. $f(x) = \frac{3}{4}x - 2$

a), b)

Note: for linear functions the tangent line is the line itself.

c) Find the simplified difference quotient first.

$$\frac{f(x+h)-f(x)}{h}$$

$$=\frac{\frac{3}{4}(x+h)-2-\left(\frac{3}{4}x-2\right)}{h}$$

$$=\frac{\frac{3}{4}x+\frac{3}{4}h-2-\frac{3}{4}x+2}{h}$$

$$=\frac{\frac{3}{4}h}{h}$$

$$=\frac{3}{4} \quad \text{Simplified difference quotient}$$

Now we will find the limit of the difference quotient as $h \to 0$ using the simplified difference quotient.

$$\lim_{h\to 0}\frac{f(x+h)-f(x)}{h}=\lim_{h\to 0}\left(\frac{3}{4}\right)$$

$$=\frac{3}{4}$$

Thus, $f'(x)=\frac{3}{4}$.

d) Since the derivative is a constant, the value of the derivative will be $\frac{3}{4}$ regardless of the value of x.

$$f'(-2)=\frac{3}{4} \quad \text{Substituting } -2 \text{ for } x$$

$$f'(0)=\frac{3}{4} \quad \text{Substituting } 0 \text{ for } x$$

$$f'(1)=\frac{3}{4} \quad \text{Substituting } 1 \text{ for } x$$

11. $f(x)=x^2+x$

a), b)

c) Find the simplified difference quotient first.

$$\frac{f(x+h)-f(x)}{h}$$

$$=\frac{(x+h)^2+(x+h)-\left(x^2+x\right)}{h}$$

$$=\frac{x^2+2xh+h^2+x+h-x^2-x}{h}$$

$$=\frac{2xh+h^2+h}{h}$$

$$=\frac{h(2x+h+1)}{h}$$

$$=2x+h+1 \quad \text{Simplified difference quotient}$$

Now we will find the limit of the difference quotient as $h \to 0$ using the simplified difference quotient.

$$\lim_{h\to 0}\frac{f(x+h)-f(x)}{h}=\lim_{h\to 0}(2x+h+1)$$

$$=2x+1$$

Thus, $f'(x)=2x+1$.

d) Find the values of the derivative by making the appropriate substitutions.

$$f'(-2)=2(-2)+1=-3$$

$$f'(0)=2(0)+1=1$$

$$f'(1)=2(1)+1=3$$

13. $f(x)=-5x^2-2x+7$

a), b)

c) Find the simplified difference quotient first.

$$\frac{f(x+h)-f(x)}{h}$$

$$=\frac{\left(-5(x+h)^2-2(x+h)+7\right)-\left(-5x^2-2x+7\right)}{h}$$

$$=\frac{-5\left(x^2+2xh+h^2\right)-2x-2h+7+5x^2+2x-7}{h}$$

$$=\frac{-5x^2-10xh-5h^2-2x-2h+7+5x^2+2x-7}{h}$$

$$=\frac{-10xh-5h^2-2h}{h}$$

$$=\frac{h(-10x-5h-2)}{h}$$

$$=-10x-5h-2 \quad \text{Simplified difference quotient}$$

Now we will find the limit of the difference quotient as $h \to 0$ using the simplified difference quotient.

$$\lim_{h\to 0}\frac{f(x+h)-f(x)}{h}=\lim_{h\to 0}(-10x-5h-2)$$

$$=-10x-2$$

Thus, $f'(x)=-10x-2$.

d) Find the values of the derivative by making the appropriate substitutions.

$$f'(-2)=-10(-2)-2=18$$

$$f'(0)=-10(0)-2=-2$$

$$f'(1)=-10(1)-2=-12$$

15. $f(x)=\dfrac{1}{x}$

a), b)

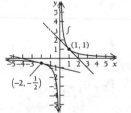

There is no tangent line for $x=0$

c) Find the simplified difference quotient first.

$$\frac{f(x+h)-f(x)}{h}$$

$$=\frac{\left(\dfrac{1}{x+h}\right)-\left(\dfrac{1}{x}\right)}{h}$$

$$=\frac{\left(\dfrac{1}{x+h}\cdot\dfrac{x}{x}\right)-\left(\dfrac{1}{x}\cdot\dfrac{(x+h)}{(x+h)}\right)}{h}$$

$$=\frac{\left(\dfrac{x}{x(x+h)}\right)-\left(\dfrac{(x+h)}{x(x+h)}\right)}{h}$$

$$=\frac{\dfrac{-h}{x(x+h)}}{h}$$

$$=\frac{-h}{x(x+h)}\cdot\frac{1}{h}$$

$$=\frac{-1}{x(x+h)} \quad \text{Simplified difference quotient}$$

Now we will find the limit of the difference quotient as $h \to 0$ using the simplified difference quotient.

$$\lim_{h\to 0}\frac{f(x+h)-f(x)}{h}=\lim_{h\to 0}\left(\frac{-1}{x(x+h)}\right)$$

$$=\frac{-1}{x(x+0)}$$

$$=\frac{-1}{x^2}$$

Thus, $f'(x)=\dfrac{-1}{x^2}$.

d) Find the values of the derivative by making the appropriate substitutions.

$$f'(-2)=\frac{-1}{(-2)^2}=-\frac{1}{4}$$

$$f'(0)=\frac{-1}{(0)^2}\,; \text{Thus, } f'(0) \text{ does not exist.}$$

$$f'(1)=\frac{-1}{(1)^2}=-1$$

17. From Example 2 we know that $f'(x) = 3x^2$.

a) $f'(-2) = 3(-2)^2 = 12$, so the slope of the line tangent to the curve at $(-2, -8)$ is 12. We substitute the point and the slope into the point-slope equation to find the equation of the tangent line.

$$y - y_1 = m(x - x_1)$$
$$y - (-8) = 12(x - (-2))$$
$$y + 8 = 12x + 24$$
$$y = 12x + 16$$

b) $f'(0) = 3(0)^2 = 0$, so the slope of the line tangent to the curve at $(0, 0)$ is 0. We substitute the point and the slope into the point-slope equation to find the equation of the tangent line.

$$y - y_1 = m(x - x_1)$$
$$y - (0) = 0(x - (0))$$
$$y = 0$$

c) $f'(4) = 3(4)^2 = 48$, so the slope of the line tangent to the curve at $(4, 64)$ is 48. We substitute the point and the slope into the point-slope equation to find the equation of the tangent line.

$$y - y_1 = m(x - x_1)$$
$$y - 64 = 48(x - 4)$$
$$y - 64 = 48x - 192$$
$$y = 48x - 128$$

19. From Exercise 16 we know that $f'(x) = \dfrac{-2}{x^2}$.

a) $f'(1) = \dfrac{-2}{(1)^2} = -2$, so the slope of the line tangent to the curve at $(1, 2)$ is -2. We substitute the point and the slope into the point-slope equation to find the equation of the tangent line.

$$y - y_1 = m(x - x_1)$$
$$y - 2 = -2(x - 1)$$
$$y - 2 = -2x + 2$$
$$y = -2x + 4$$

b) $f'(-1) = \dfrac{-2}{(-1)^2} = -2$, so the slope of the line tangent to the curve at $(-1, -2)$ is -2. We substitute the point and the slope into the point-slope equation to find the equation of the tangent line.

$$y - y_1 = m(x - x_1)$$
$$y - (-2) = -2(x - (-1))$$
$$y + 2 = -2x - 2$$
$$y = -2x - 4$$

c) $f'(100) = \dfrac{-2}{(100)^2} = -0.0002$, so the slope of the line tangent to the curve at $(100, 0.02)$ is -0.0002. We substitute the point and the slope into the point-slope equation to find the equation of the tangent line.

$$y - y_1 = m(x - x_1)$$
$$y - 0.02 = -0.0002(x - 100)$$
$$y - 0.02 = -0.0002x + 0.02$$
$$y = -0.0002x + 0.04$$

21. First, we will find $f'(x)$

$$\frac{f(x+h) - f(x)}{h}$$

$$= \frac{\left((x+h)^2 - 2(x+h)\right) - \left(x^2 - 2x\right)}{h}$$

$$= \frac{x^2 + 2xh + h^2 - 2x - 2h - x^2 + 2x}{h}$$

$$= \frac{2xh + h^2 - 2h}{h}$$

$$= \frac{h(2x + h - 2)}{h}$$

$$= 2x + h - 2 \quad \text{Simplified difference quotient}$$

$$f'(x) = \lim_{h \to 0} \frac{f(x+h) - f(x)}{h}$$
$$= \lim_{h \to 0} (2x + h - 2)$$
$$= 2x - 2$$

Thus, $f'(x) = 2x - 2$.

The solution is continued on the next page.

a) $f'(-2) = 2(-2) - 2 = -6$, so the slope of the line tangent to the curve at $(-2, 8)$ is -6. We substitute the point and the slope into the point-slope equation to find the equation of the tangent line.
$$y - y_1 = m(x - x_1)$$
$$y - 8 = -6(x - (-2))$$
$$y - 8 = -6x - 12$$
$$y = -6x - 4$$

b) $f'(1) = 2(1) - 2 = 0$, so the slope of the line tangent to the curve at $(1, -1)$ is 0. We substitute the point and the slope into the point-slope equation to find the equation of the tangent line.
$$y - y_1 = m(x - x_1)$$
$$y - (-1) = 0(x - 1)$$
$$y + 1 = 0$$
$$y = -1$$

c) $f'(4) = 2(4) - 2 = 6$, so the slope of the line tangent to the curve at $(4, 8)$ is 6. We substitute the point and the slope into the point-slope equation to find the equation of the tangent line.
$$y - y_1 = m(x - x_1)$$
$$y - 8 = 6(x - 4)$$
$$y - 8 = 6x - 24$$
$$y = 6x - 16$$

23. Find the simplified difference quotient for $f(x) = mx + b$ first.
$$\frac{f(x+h) - f(x)}{h}$$
$$= \frac{m(x+h) + b - (mx + b)}{h}$$
$$= \frac{mx + mh + b - mx - b}{h}$$
$$= \frac{mh}{h}$$
$$= m \qquad \text{Simplified difference quotient}$$
Now we will find the limit of the difference quotient as $h \to 0$ using the simplified difference quotient.
$$\lim_{h \to 0} \frac{f(x+h) - f(x)}{h} = \lim_{h \to 0}(m) = m$$
Thus, $f'(x) = m$.

25. If a function has a "corner," it will not be differentiable at that point. Thus, the function is not differentiable at x_3, x_4, x_6. The function has a vertical tangent at x_{12}. Vertical lines have undefined slope, hence the function is not differentiable at x_{12}. Also, if a function is discontinuous at some point a, then it is not differentiable at a. The function is discontinuous at the point x_0, thus it is not differentiable at x_0.
Therefore, the graph is not differentiable at the points $x_0, x_3, x_4, x_6, x_{12}$.

27. If a function has a "corner," it will not be differentiable at that point. Thus, the function is not differentiable at x_3. The function has a vertical tangent at x_1. Vertical lines have undefined slope, hence the function is not differentiable at x_1. Also, if a function is discontinuous at some point a, then it is not differentiable at a. The function is discontinuous at the points x_1, x_2, x_4, thus it is not differentiable at x_1, x_2, x_4.
Therefore, the graph is not differentiable at the points x_1, x_2, x_3, x_4.

29. The following graph is continuous but not differentiable, at $x = 3$.

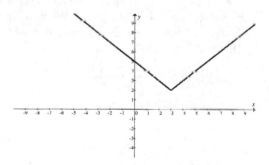

31. The following graph has a horizontal tangent line at $x = 5$.

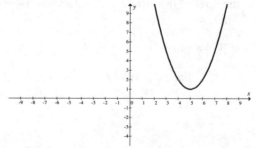

33. The following graph has horizontal tangent lines at $x = 2$ and $x = 5$ and is continuous but not differentiable at $x = 3$.

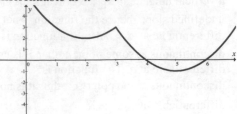

35. The postage function does not have any "corners" nor does it have any vertical tangents. However it is discontinuous at all natural numbers. Therefore the postage function is not differentiable for 1, 2, 3, 4, … 12.

37. The largest increase occurs on December 13th. On this day rate of increase was 16 points per day. The greatest decrease occurred on December 10th. On this day rate of decrease was 53 points per day.

39. ✎ The lines L_2, L_3, L_4, L_6 appear to be tangent lines. The slopes appear to be the same as the instantaneous rate of change of the function at the indicated points.

41. $f(x) = x^4$

We found the simplified difference quotient in Exercise 49 of Exercise Set 1.3. We now find the limit of the difference quotient as $h \to 0$. Thus,

$$f'(x) = \lim_{h \to 0} \frac{f(x+h) - f(x)}{h}$$
$$= \lim_{h \to 0} 4x^3 + 6x^2 h + 4xh^2 + h^3$$
$$= 4x^3$$

43. $f(x) = x^5$

We found the simplified difference quotient in Exercise 50 of Exercise Set 1.3. We now find the limit of the difference quotient as $h \to 0$. Thus,

$$f'(x) = \lim_{h \to 0} \frac{f(x+h) - f(x)}{h}$$
$$= \lim_{h \to 0} 5x^4 + 10x^3 h + 10x^2 h^2 + 5xh^3 + h^4$$
$$= 5x^4$$

45. $f(x) = \sqrt{x}$

We found the simplified difference quotient in Example 8 of Section 1.3. We now find the limit of the difference quotient as $h \to 0$.

$$\lim_{h \to 0} \frac{f(x+h) - f(x)}{h} = \lim_{h \to 0} \frac{1}{\sqrt{x+h} + \sqrt{x}}$$
$$= \frac{1}{\sqrt{x+0} + \sqrt{x}}$$
$$= \frac{1}{2\sqrt{x}}$$

Thus, $f'(x) = \dfrac{1}{2\sqrt{x}}$

47. $f(x) = \dfrac{1}{\sqrt{x}}$

We found the simplified difference quotient in Exercise 56 of Exercise Set 1.3. We now find the limit of the difference quotient as $h \to 0$.

$$\lim_{h \to 0} \frac{f(x+h) - f(x)}{h}$$
$$= \lim_{h \to 0} \frac{-1}{\sqrt{x}\sqrt{x+h}\left(\sqrt{x} + \sqrt{x+h}\right)}$$
$$= \frac{-1}{\sqrt{x}\sqrt{x+0}\left(\sqrt{x} + \sqrt{x+0}\right)}$$
$$= \frac{-1}{x\left(2\sqrt{x}\right)}$$
$$= \frac{-1}{2x\sqrt{x}}$$

Thus, $f'(x) = \dfrac{-1}{2x\sqrt{x}}$.

49. a) The domain of the rational function is restricted to those input values that do not result in division by 0. The domain for

$f(x) = \dfrac{x^2 - 9}{x + 3}$ consists of all real numbers

except -3. Since $f(-3)$ does not exist, the function is not continuous at -3. Thus, the function is not differentiable at $x = -3$.

b) Answers will vary. The simplest way of finding $f'(4)$ is to use the nDeriv function on your calculator. However, without using advanced technology the easiest way is to approximate the difference quotient using a very small value of h, and allowing the calculator to perform the basic computations as shown on the next page.

We illustrate using $h = 0.0001$. The difference quotient will be

$$\frac{f(x+h) - f(x)}{h} = \frac{f(4+0.0001) - f(4)}{0.0001}$$

Using the calculator to evaluate the function we have:

$$f(4.0001) = 1.0001$$

$$f(4) = 1$$

Plugging in these values we have:

$$\frac{f(4+0.0001) - f(4)}{0.0001} = \frac{1.0001 - 1}{0.0001} = 1$$

Therefore, $f'(4) = 1$.

51. a) Looking at the graph of the function, we see there is a "corner" when $x = 3$. Therefore, $k(x) = |x-3| + 2$ is not differentiable at $x = 3$.

b) Using the piecewise definition of

$$k(x) = |x-3| + 2$$

$$= \begin{cases} -(x-3)+2, & \text{for } x < 3 \\ (x-3)+2, & \text{for } x \geq 3 \end{cases}$$

We notice that:

$$k'(x) = \begin{cases} -1, & \text{for } x < 3 \\ 1, & \text{for } x > 3 \end{cases}$$

Therefore,

$$k'(0) = -1; \ k'(1) = -1;$$

$$k'(4) = 1; \ k'(10) = 1.$$

The "shortcut" is noticing that this function is a linear function with slope $m = -1$ for $x < 3$ and slope $m = 1$ for $x > 3$.

53. The error was made when the student did not determine the implied domain of the function.

$$f(x) = \frac{x^2 + 4x + 3}{x+1}$$

is undefined at $x = -1$. Therefore, $f(x)$ is not differentiable at $x = -1$. Once the domain is properly defined, the student can find the derivative of the function.

55. a) The function $F(x)$ is continuous at $x = 2$.

1) $F(2)$ exists, $F(2) = 5$

2) $\lim\limits_{x \to 2^-} F(x) = 5$ and $\lim\limits_{x \to 2^+} F(x) = 5$,

Therefore, $\lim\limits_{x \to 2} F(x) = 5$.

3) $\lim\limits_{x \to 2} F(x) = 5 = F(2)$.

b) The function $F(x)$ is not differentiable at $x = 2$ because there is a "corner" at $x = 2$.

57. In order for $H(x)$ to be differentiable at $x = 3$. $H(x)$ must be continuous at $x = 3$. It must also be "smooth" at $x = 3$ which means the slope as x approaches 3 from the left must equal the slope as x approaches 3 from the right.

For $x \leq 3$

We find the derivative by differentiating the piece of the function that is defined on the interval $x \leq 3$. Therefore,

$$H'(x) = \frac{d}{dx}\left(2x^2 - x\right)$$

$$= 4x - 1.$$

Therefore, when $x = 3$

$$H'(3) = 4(3) - 1 = 11.$$

For $x > 3$ the derivative is given by

$$H'(x) = \frac{d}{dx}(mx + b)$$

$$= m.$$

Using this information, we know in order for the slope as x approaches from the right to equal the slope as x approaches from the left, we must have: $m = 11$.

We also know:

$\lim\limits_{x \to 3^-} H(x) = 15$, since $H(3) = 15$. In order for $H(x)$ to be continuous, we must have

$$\lim\limits_{x \to 3^+} H(x) = 15.$$

Using the above information and substituting $m = 11$ we have:

$$\lim\limits_{x \to 3^+} 11(x) + b = 15$$

$$11(3) + b = 15$$

$$33 + b = 15$$

$$b = -18$$

Therefore, the values $m = 11$ and $b = -18$ will make $H(x)$ differentiable at $x = 3$.

58-63. Left to the student.

65. There is a vertical tangent at $x = 5$, therefore, $f'(x)$ does not exist at $x = 5$.

Exercise Set 1.5

1. $y = x^8$

$$\frac{dy}{dx} = \frac{d}{dx}x^8$$

$$= 8x^{8-1} \qquad \text{Theorem 1}$$

$$= 8x^7$$

3. $y = -0.5x$

$$\frac{dy}{dx} = \frac{d}{dx}(-0.5x)$$

$$= -0.5\frac{d}{dx}x \qquad \text{Theorem 3}$$

$$= -0.5\left(1x^{1-1}\right) \qquad \text{Theorem 1}$$

$$= -0.5\left(x^0\right)$$

$$= -0.5 \qquad\qquad \left[a^0 = 1\right]$$

5. $y = 7$ Constant function

$$\frac{dy}{dx} = \frac{d}{dx}7$$

$$= 0 \qquad\qquad \text{Theorem 2}$$

7. $y = 3x^{10}$

$$\frac{dy}{dx} = \frac{d}{dx}\left(3x^{10}\right)$$

$$= 3\frac{d}{dx}\left(x^{10}\right) \qquad \text{Theorem 3}$$

$$= 3\left(10x^{10-1}\right) \qquad \text{Theorem 1}$$

$$= 30x^9$$

9. $y = x^{-8}$

$$\frac{dy}{dx} = \frac{d}{dx}x^{-8}$$

$$= -8x^{-8-1} \qquad \text{Theorem 1}$$

$$= -8x^{-9}$$

11. $y = 3x^{-5}$

$$\frac{dy}{dx} = \frac{d}{dx}\left(3x^{-5}\right)$$

$$= 3\frac{d}{dx}\left(x^{-5}\right) \qquad \text{Theorem 3}$$

$$= 3\left(-5x^{-5-1}\right) \qquad \text{Theorem 1}$$

$$= -15x^{-6}$$

13. $y = x^4 - 7x$

$$\frac{dy}{dx} = \frac{d}{dx}\left(x^4 - 7x\right)$$

$$= \frac{d}{dx}x^4 - \frac{d}{dx}7x \quad \text{Theorem 4}$$

$$= \frac{d}{dx}x^4 - 7\frac{d}{dx}x \quad \text{Theorem 3}$$

$$= 4x^{4-1} - 7\left(1x^{1-1}\right) \quad \text{Theorem 1}$$

$$= 4x^3 - 7x^0$$

$$= 4x^3 - 7 \qquad\qquad \left[a^0 = 1\right]$$

15. $y = 4\sqrt{x} = 4x^{\frac{1}{2}}$

$$\frac{dy}{dx} = \frac{d}{dx}4x^{\frac{1}{2}}$$

$$= 4\frac{d}{dx}x^{\frac{1}{2}} \qquad \text{Theorem 3}$$

$$\frac{dy}{dx} = 4\left(\frac{1}{2}x^{\frac{1}{2}-1}\right) \qquad \text{Theorem 1}$$

$$= 2x^{-\frac{1}{2}}$$

$$= \frac{2}{x^{\frac{1}{2}}} = \frac{2}{\sqrt{x}} \qquad \text{Properties of exponents}$$

17. $y = x^{0.7}$

$$\frac{dy}{dx} = \frac{d}{dx}x^{0.7}$$

$$= 0.7x^{0.7-1} \qquad \text{Theorem 1}$$

$$= 0.7x^{-0.3}$$

19. $y = -4.8x^{\frac{1}{3}}$

$\dfrac{dy}{dx} = \dfrac{d}{dx}\left(-4.8x^{\frac{1}{3}}\right)$

$\quad = -4.8\dfrac{d}{dx}\left(x^{\frac{1}{3}}\right)$ 　　Theorem 3

$\quad = -4.8\left(\dfrac{1}{3}x^{\frac{1}{3}-1}\right)$ 　　Theorem 1

$\quad = -1.6x^{-\frac{2}{3}}$

21. $y = \dfrac{6}{x^4} = 6x^{-4}$

$\dfrac{dy}{dx} = \dfrac{d}{dx}\left(6x^{-4}\right)$

$\quad = 6\dfrac{d}{dx}\left(x^{-4}\right)$ 　　Theorem 3

$\quad = 6\left(-4x^{-4-1}\right)$ 　　Theorem 1

$\quad = -24x^{-5}$

$\quad = -\dfrac{24}{x^5}$ 　　Properties of exponents

23. $y = \dfrac{3x}{4} = \dfrac{3}{4}x$

$\dfrac{dy}{dx} = \dfrac{d}{dx}\left(\dfrac{3}{4}x\right)$

$\quad = \dfrac{3}{4}\cdot\dfrac{d}{dx}(x)$ 　　Theorem 3

$\quad = \dfrac{3}{4}\cdot\left(1x^{1-1}\right)$ 　　Theorem 1

$\quad = \dfrac{3}{4}$ 　　Properties of exponents

25. $\dfrac{d}{dx}\left(\sqrt[4]{x} - \dfrac{3}{x}\right)$

$\quad = \dfrac{d}{dx}\sqrt[4]{x} - \dfrac{d}{dx}\dfrac{3}{x}$ 　　Theorem 4

$\quad = \dfrac{d}{dx}x^{\frac{1}{4}} - \dfrac{d}{dx}3x^{-1}$ 　Properties of exponents

$\quad = \dfrac{d}{dx}x^{\frac{1}{4}} - 3\dfrac{d}{dx}x^{-1}$ 　　Theorem 3

$\quad = \dfrac{1}{4}x^{\frac{1}{4}-1} - 3\left(-1x^{-1-1}\right)$ 　Theorem 1

$\quad = \dfrac{1}{4}x^{-\frac{3}{4}} + 3x^{-2}$

$\quad = \dfrac{1}{4x^{\frac{3}{4}}} + \dfrac{3}{x^2}$

$\quad = \dfrac{1}{4\sqrt[4]{x^3}} + \dfrac{3}{x^2}$

27. $\dfrac{d}{dx}\left(-2\sqrt[3]{x^5}\right)$

$\quad = -2\dfrac{d}{dx}\left(\sqrt[3]{x^5}\right)$ 　　Theorem 3

$\quad = -2\dfrac{d}{dx}\left(x^{\frac{5}{3}}\right)$

$\quad = -2\left(\dfrac{5}{3}x^{\frac{5}{3}-1}\right)$ 　　Theorem 1

$\quad = -\dfrac{10}{3}x^{\frac{2}{3}} = -\dfrac{10\sqrt[3]{x^2}}{3}$

29. $\dfrac{d}{dx}\left(5x^2 - 7x + 3\right)$

$\quad = \dfrac{d}{dx}5x^2 - \dfrac{d}{dx}7x + \dfrac{d}{dx}3$ 　　Theorem 4

$\quad = 5\dfrac{d}{dx}x^2 - 7\dfrac{d}{dx}x + \dfrac{d}{dx}3$ 　　Theorem 3

$\quad = 5\left(2x^{2-1}\right) - 7\left(1x^{1-1}\right) + 0$ 　Theorems 1 and 2

$\quad = 10x - 7$

31. $f(x) = 0.3x^{1.2}$

$f'(x) = \dfrac{d}{dx}0.3x^{1.2}$

$\quad = 0.3\dfrac{d}{dx}x^{1.2}$ 　　Theorem 3

$\quad = 0.3\left(1.2x^{1.2-1}\right)$ 　　Theorem 1

$\quad = 0.36x^{0.2}$

33. $f(x) = \dfrac{3x}{4} = \dfrac{3}{4}x$

$f'(x) = \dfrac{d}{dx}\left(\dfrac{3}{4}x\right)$

$\qquad = \dfrac{3}{4}\dfrac{d}{dx}(x)$

$\qquad = \dfrac{3}{4}\left(1x^{1-1}\right)$

$\qquad = \dfrac{3}{4}$

35. $f(x) = \dfrac{2}{5x^6} = \dfrac{2}{5}x^{-6}$

$f'(x) = \dfrac{d}{dx}\left(\dfrac{2}{5}x^{-6}\right)$

$\qquad = \dfrac{2}{5}\dfrac{d}{dx}\left(x^{-6}\right)$

$\qquad = \dfrac{2}{5}\left(-6x^{-6-1}\right)$

$\qquad = \dfrac{-12}{5}x^{-7}$

$\qquad = -\dfrac{12}{5x^7}$

37. $f(x) = \dfrac{4}{x} - x^{3/5} = 4x^{-1} - x^{3/5}$

$f'(x) = \dfrac{d}{dx}\left(4x^{-1} - x^{3/5}\right)$

$\qquad = \dfrac{d}{dx}\left(4x^{-1}\right) - \dfrac{d}{dx}\left(x^{3/5}\right)$

$\qquad = 4\dfrac{d}{dx}\left(x^{-1}\right) - \dfrac{d}{dx}\left(x^{3/5}\right)$

$\qquad = 4\left(-1x^{-1-1}\right) - \left(\dfrac{3}{5}x^{3/5-1}\right)$

$\qquad = -4x^{-2} - \dfrac{3}{5}x^{-2/5}$

$\qquad = -\dfrac{4}{x^2} - \dfrac{3}{5}x^{-2/5}$

39. $f(x) = 7x - 14$

$f'(x) = \dfrac{d}{dx}(7x - 14)$

$\qquad = \dfrac{d}{dx}(7x) - \dfrac{d}{dx}(14)$

$f'(x) = 7\dfrac{d}{dx}(x) - \dfrac{d}{dx}(14)$

$\qquad = 7\left(1x^{1-1}\right) - 0$

$\qquad = 7$

41. $f(x) = \dfrac{x^{3/2}}{3} = \dfrac{1}{3}x^{3/2}$

$f'(x) = \dfrac{d}{dx}\left(\dfrac{1}{3}x^{3/2}\right)$

$\qquad = \dfrac{1}{3}\dfrac{d}{dx}\left(x^{3/2}\right)$

$\qquad = \dfrac{1}{3}\left(\dfrac{3}{2}x^{3/2-1}\right)$

$\qquad = \dfrac{1}{2}x^{1/2}$, or $\dfrac{\sqrt{x}}{2}$

43. $f(x) = -0.01x^2 + 0.4x + 50$

$f'(x) = \dfrac{d}{dx}\left(-0.01x^2 + 0.4x + 50\right)$

$\qquad = \dfrac{d}{dx}\left(-0.01x^2\right) + \dfrac{d}{dx}(0.4x) + \dfrac{d}{dx}(50)$

$\qquad = -0.01\dfrac{d}{dx}\left(x^2\right) + 0.4\dfrac{d}{dx}(x) + \dfrac{d}{dx}(50)$

$\qquad = -0.01\left(2x^{2-1}\right) + 0.4\left(1x^{1-1}\right) + 0$

$\qquad = -0.02x + 0.4$

45. $y = x^{-3/4} - 3x^{2/3} + x^{5/4} + \dfrac{2}{x^4}$

$y = x^{-3/4} - 3x^{2/3} + x^{5/4} + 2x^{-4}$

$y' = \dfrac{d}{dx}\left(x^{-3/4}\right) - 3\dfrac{d}{dx}\left(x^{2/3}\right) + \dfrac{d}{dx}\left(x^{5/4}\right) + 2\dfrac{d}{dx}\left(x^{-4}\right)$

$\quad = \left(\dfrac{-3}{4}x^{-3/4-1}\right) - 3\left(\dfrac{2}{3}x^{2/3-1}\right) +$

$\qquad\qquad\qquad \left(\dfrac{5}{4}x^{5/4-1}\right) + 2\left(-4x^{-4-1}\right)$

$\quad = \dfrac{-3}{4}x^{-7/4} - 2x^{-1/3} + \dfrac{5}{4}x^{1/4} - 8x^{-5}$

$\quad = \dfrac{-3}{4}x^{-7/4} - 2x^{-1/3} + \dfrac{5}{4}x^{1/4} - \dfrac{8}{x^5}$

47. $y = \dfrac{x}{7} + \dfrac{7}{x} = \dfrac{1}{7}x + 7x^{-1}$

$y' = \dfrac{d}{dx}\left(\dfrac{1}{7}x + 7x^{-1}\right)$

$\quad = \dfrac{d}{dx}\left(\dfrac{1}{7}x\right) + \dfrac{d}{dx}\left(7x^{-1}\right)$

$\quad = \dfrac{1}{7}\dfrac{d}{dx}\left(x^{1}\right) + 7\dfrac{d}{dx}\left(x^{-1}\right)$

$\quad = \dfrac{1}{7}\left(1x^{1-1}\right) + 7\left(-1x^{-1-1}\right)$

$\quad = \dfrac{1}{7} - 7x^{-2}$

$\quad = \dfrac{1}{7} - \dfrac{7}{x^{2}}$

49. $f(x) = \sqrt{x} = x^{\frac{1}{2}}$

First, we find $f'(x)$

$f'(x) = \dfrac{d}{dx}\left(x^{\frac{1}{2}}\right)$

$\quad = \dfrac{1}{2}x^{\frac{1}{2}-1}$

$\quad = \dfrac{1}{2}x^{-\frac{1}{2}}$

$\quad = \dfrac{1}{2x^{\frac{1}{2}}} = \dfrac{1}{2\sqrt{x}}$

Therefore,

$f'(4) = \dfrac{1}{2\sqrt{4}} = \dfrac{1}{2\cdot 2} = \dfrac{1}{4}$

51. $y = x + \dfrac{2}{x^{3}} = x + 2x^{-3}$

Find $\dfrac{dy}{dx}$ first.

$\dfrac{dy}{dx} = \dfrac{d}{dx}\left(x + 2x^{-3}\right)$

$\quad = \dfrac{d}{dx}(x) + 2\dfrac{d}{dx}\left(x^{-3}\right)$

$\quad = 1x^{1-1} + 2\left(-3x^{-3-1}\right)$

$\quad = 1 - 6x^{-4}$

$\quad = 1 - \dfrac{6}{x^{4}}$

Therefore,

$\dfrac{dy}{dx}\bigg|_{x=1} = 1 - \dfrac{6}{(1)^{4}}$

$\quad\quad = 1 - 6$

$\quad\quad = -5$

53. $y = \sqrt[3]{x} + \sqrt{x} = x^{\frac{1}{3}} + x^{\frac{1}{2}}$

Find $\dfrac{dy}{dx}$ first.

$\dfrac{dy}{dx} = \dfrac{d}{dx}\left(x^{\frac{1}{3}} + x^{\frac{1}{2}}\right)$

$\quad = \dfrac{d}{dx}\left(x^{\frac{1}{3}}\right) + \dfrac{d}{dx}\left(x^{\frac{1}{2}}\right)$

$\quad = \dfrac{1}{3}x^{\frac{1}{3}-1} + \dfrac{1}{2}x^{\frac{1}{2}-1}$

$\quad = \dfrac{1}{3\sqrt[3]{x^{2}}} + \dfrac{1}{2\sqrt{x}}$

Therefore,

$\dfrac{dy}{dx}\bigg|_{x=64} = \dfrac{1}{3\sqrt[3]{(64)^{2}}} + \dfrac{1}{2\sqrt{64}}$

$\quad\quad = \dfrac{1}{48} + \dfrac{1}{16}$

$\quad\quad = \dfrac{1}{12}$

55. $y = \dfrac{2}{5x^{3}} = \dfrac{2}{5}x^{-3}$

Find $\dfrac{dy}{dx}$ first.

$\dfrac{dy}{dx} = \dfrac{d}{dx}\left(\dfrac{2}{5}x^{-3}\right)$

$\quad = \dfrac{2}{5}\dfrac{d}{dx}\left(x^{-3}\right)$

$\quad = \dfrac{2}{5}\left(-3x^{-3-1}\right)$

$\quad = -\dfrac{6}{5}x^{-4}$

$\quad = -\dfrac{6}{5x^{4}}$

Therefore,

$\dfrac{dy}{dx}\bigg|_{x=4} = -\dfrac{6}{5(4)^{4}} = -\dfrac{3}{640}$

57. We will need the derivative to find the slope of the tangent line at each of the indicated points. We find the derivative first.

$$f(x) = x^2 - \sqrt{x} = x^2 - x^{1/2}$$

$$f'(x) = \frac{d}{dx}\left(x^2 - x^{1/2}\right)$$

$$= \frac{d}{dx}\left(x^2\right) - \frac{d}{dx}\left(x^{1/2}\right)$$

$$= 2x^{2-1} - \frac{1}{2}x^{1/2-1}$$

$$= 2x - \frac{1}{2x^{1/2}}$$

$$= 2x - \frac{1}{2\sqrt{x}}$$

a) Using the derivative, we find the slope of the line tangent to the curve at point $(1,0)$ by evaluating the derivative at $x = 1$.

$$f'(1) = 2(1) - \frac{1}{2\sqrt{1}} = 2 - \frac{1}{2} = \frac{3}{2} \text{. Therefore}$$

the slope of the tangent line is $\frac{3}{2}$. We use the point-slope equation to find the equation of the tangent line.

$$y - y_1 = m(x - x_1)$$

$$y - 0 = \frac{3}{2}(x - 1)$$

$$y = \frac{3}{2}x - \frac{3}{2}$$

b) Using the derivative, we find the slope of the line tangent to the curve at point $(4,14)$ by evaluating the derivative at $x = 4$.

$$f'(4) = 2(4) - \frac{1}{2\sqrt{4}} = 8 - \frac{1}{4} = \frac{31}{4}.$$

Therefore the slope of the tangent line is $\frac{31}{4}$. We use the point-slope equation to find the equation of the tangent line.

$$y - y_1 = m(x - x_1)$$

$$y - 14 = \frac{31}{4}(x - 4)$$

$$y - 14 = \frac{31}{4}x - 31$$

$$y = \frac{31}{4}x - 17$$

c) Using the derivative, we find the slope of the line tangent to the curve at point $(9, 78)$ by evaluating the derivative at $x = 9$.

$$f'(9) = 2(9) - \frac{1}{2\sqrt{9}} = 18 - \frac{1}{6} = \frac{107}{6}.$$

Therefore the slope of the tangent line is $\frac{107}{6}$. We use the point-slope equation to find the equation of the tangent line.

$$y - y_1 = m(x - x_1)$$

$$y - 78 = \frac{107}{6}(x - 9)$$

$$y - 78 = \frac{107}{6}x - \frac{321}{2}$$

$$y = \frac{107}{6}x - \frac{165}{2}$$

59. We will need the derivative to find the slope of the tangent line at each of the indicated points. We find the derivative first.

$$g(x) = \sqrt[3]{x^2} = x^{\frac{2}{3}}$$

$$g'(x) = \frac{d}{dx}\left(x^{\frac{2}{3}}\right)$$

$$= \frac{2}{3}x^{\frac{2}{3}-1}$$

$$= \frac{2}{3\sqrt[3]{x}}$$

a) Using the derivative, we find the slope of the line tangent to the curve at point $(-1,1)$ by evaluating the derivative at $x = -1$.

$$g'(-1) = \frac{2}{3\sqrt[3]{-1}} = -\frac{2}{3} \text{. Therefore the slope}$$

of the tangent line is $-\frac{2}{3}$. We use the point-slope equation to find the equation of the tangent line.

$$y - y_1 = m(x - x_1)$$

$$y - 1 = -\frac{2}{3}(x + 1)$$

$$y - 1 = -\frac{2}{3}x - \frac{2}{3}$$

$$y = -\frac{2}{3}x + \frac{1}{3}$$

b) Using the derivative, we find the slope of the line tangent to the curve at point $(1,1)$ by evaluating the derivative at $x=1$.

$$g'(1)=\frac{2}{3\sqrt[3]{1}}=\frac{2}{3}.$$

Therefore the slope of the tangent line is $\frac{2}{3}$.
We use the point-slope equation to find the equation of the tangent line.

$$y-y_1=m(x-x_1)$$
$$y-1=\frac{2}{3}(x-1)$$
$$y-1=\frac{2}{3}x-\frac{2}{3}$$
$$y=\frac{2}{3}x+\frac{1}{3}$$

c) Using the derivative, we find the slope of the line tangent to the curve at point $(8,4)$ by evaluating the derivative at $x=8$.

$$g'(8)=\frac{2}{3\sqrt[3]{8}}=\frac{1}{3}.$$ Therefore the slope of the tangent line is $\frac{1}{3}$.

We use the point-slope equation to find the equation of the tangent line.

$$y-y_1=m(x-x_1)$$
$$y-4=\frac{1}{3}(x-8)$$
$$y-4=\frac{1}{3}x-\frac{8}{3}$$
$$y=\frac{1}{3}x+\frac{4}{3}$$

61. $y=-x^2+4$

A horizontal tangent line has slope equal to 0, so we first find the values of x that make

$$\frac{dy}{dx}=0.$$
First, we find the derivative.
$$\frac{dy}{dx}=\frac{d}{dx}\left(-x^2+4\right)$$
$$\frac{dy}{dx}=-\frac{d}{dx}x^2+\frac{d}{dx}4$$
$$=-2x+0$$
$$=-2x$$

Next, we set the derivative equal to zero and solve for x.

$$\frac{dy}{dx}=0$$
$$-2x=0$$
$$x=0$$

So the horizontal tangent will occur when $x=0$. Next we find the point on the graph. For $x=0$, $y=-(0)^2+4=4$, so there is a horizontal tangent at the point $(0,4)$.

63. $y=x^3-2$

A horizontal tangent line has slope equal to 0, so we first find the values of x that make

$$\frac{dy}{dx}=0.$$
First, we find the derivative.
$$\frac{dy}{dx}=\frac{d}{dx}\left(x^3-2\right)$$
$$=\frac{d}{dx}x^3-\frac{d}{dx}2$$
$$=3x^2-0$$
$$=3x^2$$
Next, we set the derivative equal to zero and solve for x.
$$\frac{dy}{dx}=0$$
$$3x^2=0$$
$$x=0$$
So the horizontal tangent will occur when $x=0$. Next we find the point on the graph. For $x=0$, $y=(0)^3-2=-2$, so there is a horizontal tangent at the point $(0,-2)$.

65. $y=5x^2-3x+8$

A horizontal tangent line has slope equal to 0, so we need to find the values of x that make

$$\frac{dy}{dx}=0.$$
First, we find the derivative.
$$\frac{dy}{dx}=\frac{d}{dx}\left(5x^2-3x+8\right)$$
$$=\frac{d}{dx}5x^2-\frac{d}{dx}3x+\frac{d}{dx}8$$
$$=5\left(2x^{2-1}\right)-3\left(1x^{1-1}\right)+0$$
$$=10x-3$$
The solution is continued on the next page.

Next, we set the derivative on the previous page equal to zero and solve for x.

$$\frac{dy}{dx} = 0$$

$$10x - 3 = 0$$

$$10x = 3$$

$$x = \frac{3}{10}$$

So the horizontal tangent will occur when $x = \frac{3}{10}$. Next we find the point on the graph.

For $x = \frac{3}{10}$,

$$y = 5\left(\frac{3}{10}\right)^2 - 3\left(\frac{3}{10}\right) + 8$$

$$= 5\left(\frac{9}{100}\right) - \frac{9}{10} + 8$$

$$= \frac{45}{100} - \frac{9}{10} + 8$$

$$= \frac{45}{100} - \frac{9}{10} \cdot \frac{10}{10} + \frac{8}{1} \cdot \frac{100}{100}$$

$$= \frac{45}{100} - \frac{90}{100} + \frac{800}{100}$$

$$= \frac{45 - 90 + 800}{100}$$

$$= \frac{755}{100} = \frac{151}{20}$$

Therefore, there is a horizontal tangent at the point $\left(\frac{3}{10}, \frac{151}{20}\right)$, or $(0.3, 7.55)$.

67. $y = -0.01x^2 + 0.4x + 50$

A horizontal tangent line has slope equal to 0, so we need to find the values of x that make $\frac{dy}{dx} = 0$.

First, we find the derivative.

$$\frac{dy}{dx} = \frac{d}{dx}\left(-0.01x^2 + 0.4x + 50\right)$$

$$= -0.02x + 0.4 \qquad \text{See Exercise 46}$$

Next, we set the derivative equal to zero and solve for x.

$$\frac{dy}{dx} = 0$$

$$-0.02x + 0.4 = 0$$

$$-0.02x = -0.4$$

$$x = \frac{-0.4}{-0.02}$$

$$x = 20$$

So the horizontal tangent will occur when $x = 20$. Next we find the point on the graph. For $x = 20$,

$$y = -0.01(20)^2 + 0.4(20) + 50$$

$$= -0.01(400) + 8 + 50$$

$$= -4 + 8 + 50$$

$$= 54$$

Therefore, there is a horizontal tangent at the point $(20, 54)$.

69. $y = -2x + 5$ Linear function

$$\frac{dy}{dx} = -2 \qquad \text{Slope is } -2$$

There are no values of x for which $\frac{dy}{dx} = 0$, so there are no points on the graph at which there is a horizontal tangent.

71. $y = -3$ Constant Function

$$\frac{dy}{dx} = 0 \qquad \text{Theorem 2}$$

$\frac{dy}{dx} = 0$ for all values of x, so the tangent line is horizontal for all points on the graph.

73. $y = -\frac{1}{3}x^3 + 6x^2 - 11x - 50$

A horizontal tangent line has slope equal to 0, so we need to find the values of x that make $\frac{dy}{dx} = 0$.

First, we find the derivative.

$$\frac{dy}{dx} = \frac{d}{dx}\left(-\frac{1}{3}x^3 + 6x^2 - 11x - 50\right)$$

$$= \frac{d}{dx}\left(-\frac{1}{3}x^3\right) + \frac{d}{dx}\left(6x^2\right) - \frac{d}{dx}(11x) - \frac{d}{dx}(50)$$

$$= -x^2 + 12x - 11$$

Next, we set the derivative equal to zero and solve for x at the top of the next page.

From the previous page, we solve:

$$\frac{dy}{dx} = 0$$

$$-x^2 + 12x - 11 = 0$$

$$x^2 - 12x + 11 = 0$$

$$(x-1)(x-11) = 0$$

$$x - 1 = 0 \quad \text{or} \quad x - 11 = 0$$

$$x = 1 \quad \text{or} \quad x = 11$$

There are two horizontal tangents. One at $x = 1$ and one at $x = 11$.

Next we find the points on the graph where the horizontal tangents occur.

For $x = 1$,

$$y = -\frac{1}{3}(1)^3 + 6(1)^2 - 11(1) - 50$$

$$= -\frac{1}{3} + 6 - 11 - 50$$

$$= -\frac{166}{3} = -55\frac{1}{3}$$

For $x = 11$,

$$y = -\frac{1}{3}(11)^3 + 6(11)^2 - 11(11) - 50$$

$$= -\frac{1331}{3} + 726 - 121 - 50$$

$$= \frac{334}{3} = 111\frac{1}{3}$$

Therefore, there are horizontal tangents at the points $\left(1, -55\frac{1}{3}\right)$ and $\left(11, 111\frac{1}{3}\right)$.

75. $y = x^3 - 6x + 1$

A horizontal tangent line has slope equal to 0, so we need to find the values of x that make

$$\frac{dy}{dx} = 0.$$

First, we find the derivative.

$$\frac{dy}{dx} = \frac{d}{dx}\left(x^3 - 6x + 1\right)$$

$$= \frac{d}{dx}\left(x^3\right) - \frac{d}{dx}(6x) + \frac{d}{dx}(1)$$

$$= 3x^2 - 6$$

Next, we set the derivative equal to zero and solve for x.

$$\frac{dy}{dx} = 0$$

$$3x^2 - 6 = 0$$

$$x^2 = 2$$

$$x = \pm\sqrt{2}$$

There are two horizontal tangents. One at $x = -\sqrt{2}$ and one at $x = \sqrt{2}$. Next we find the points on the graph where the horizontal tangents occur.

For $x = -\sqrt{2}$,

$$y = \left(-\sqrt{2}\right)^3 - 6\left(-\sqrt{2}\right) + 1$$

$$= -2\sqrt{2} + 6\sqrt{2} + 1$$

$$= 1 + 4\sqrt{2}$$

For $x = \sqrt{2}$,

$$y = \left(\sqrt{2}\right)^3 - 6\left(\sqrt{2}\right) + 1$$

$$= 2\sqrt{2} - 6\sqrt{2} + 1$$

$$= 1 - 4\sqrt{2}$$

Therefore, there are horizontal tangents at the points $\left(-\sqrt{2}, 1 + 4\sqrt{2}\right)$ and $\left(\sqrt{2}, 1 - 4\sqrt{2}\right)$.

77. $y = \frac{1}{3}x^3 - 3x^2 + 9x - 9$

A horizontal tangent line has slope equal to 0, so we need to find the values of x that make

$$\frac{dy}{dx} = 0.$$

First, we find the derivative.

$$\frac{dy}{dx} = \frac{d}{dx}\left(\frac{1}{3}x^3 - 3x^2 + 9x - 9\right)$$

$$= \frac{d}{dx}\left(\frac{1}{3}x^3\right) - \frac{d}{dx}\left(3x^2\right) + \frac{d}{dx}(9x) - \frac{d}{dx}(9)$$

$$= x^2 - 6x + 9$$

Next, we set the derivative equal to zero and solve for x.

$$\frac{dy}{dx} = 0$$

$$x^2 - 6x + 9 = 0$$

$$(x - 3)^2 = 0$$

$$x - 3 = 0$$

$$x = 3$$

There is one horizontal tangent at $x = 3$.

The solution is continued on the next page.

Using the information from the previous page, we find the point on the graph where the horizontal tangent occurs.

For $x = 3$,

$$y = \frac{1}{3}(3)^3 - 3(3)^2 + 9(3) - 9$$

$$= 9 - 27 + 27 - 9 = 0$$

Therefore, there is horizontal tangent at the point $(3, 0)$.

79. $y = 6x - x^2$

To find the tangent line that has slope equal to 1, so we need to find the values of x that make

$$\frac{dy}{dx} = 1.$$

First, we find the derivative.

$$\frac{dy}{dx} = \frac{d}{dx}\left(6x - x^2\right)$$

$$= \frac{d}{dx}6x - \frac{d}{dx}x^2$$

$$= 6 - 2x$$

Next, we set the derivative equal to 1 and solve for x.

$$\frac{dy}{dx} = 1$$

$$6 - 2x = 1$$

$$-2x = 1 - 6$$

$$-2x = -5$$

$$x = \frac{5}{2}$$

So the tangent will occur when $x = \frac{5}{2}$.

Next we find the point on the graph.

For $x = \frac{5}{2}$,

$$y = 6\left(\frac{5}{2}\right) - \left(\frac{5}{2}\right)^2$$

$$= 15 - \frac{25}{4} = \frac{35}{4}$$

The tangent line has slope 1 at $\left(\frac{5}{2}, \frac{35}{4}\right)$ or $(2.5, 8.75)$.

81. $y = -0.01x^2 + 2x$

To find the tangent line that has slope equal to 1, we need to find the values of x that make

$$\frac{dy}{dx} = 1.$$

First, we find the derivative.

$$\frac{dy}{dx} = \frac{d}{dx}\left(-0.01x^2 + 2x\right)$$

$$= \frac{d}{dx}\left(-0.01x^2\right) + \frac{d}{dx}(2x)$$

$$= -0.01\left(2x^{2-1}\right) + 2\left(x^{1-1}\right)$$

$$= -0.02x + 2$$

Next, we set the derivative equal to 1 and solve for x.

$$\frac{dy}{dx} = 1$$

$$-0.02x + 2 = 1$$

$$-0.02x = -1$$

$$x = \frac{-1}{-0.02}$$

$$x = 50$$

So the tangent will occur when $x = 50$. Next we find the point on the graph.

For $x = 50$,

$$y = -0.01(50)^2 + 2(50)$$

$$= -0.01(2500) + 100$$

$$= -25 + 100$$

$$= 75$$

The tangent line has slope 1 at $(50, 75)$.

83. $y = \frac{1}{3}x^3 - x^2 - 4x + 1$

To find the tangent line that has slope equal to 1, we need to find the values of x that make

$$\frac{dy}{dx} = 1.$$

First, we find the derivative.

$$\frac{dy}{dx} = \frac{d}{dx}\left(\frac{1}{3}x^3 - x^2 - 4x + 1\right)$$

$$= \frac{d}{dx}\left(\frac{1}{3}x^3\right) - \frac{d}{dx}x^2 - \frac{d}{dx}4x + \frac{d}{dx}1$$

$$= x^2 - 2x - 4$$

The solution is continued on the next page.

Next, we set the derivative from the previous page equal to 1 and solve for x.

$$\frac{dy}{dx} = 1$$

$$x^2 - 2x - 4 = 1$$

$$x^2 - 2x - 5 = 0$$

This is a quadratic equation, not readily factorable, so we use the quadratic formula where $a = 1, b = 4,$ and, $c = 1$.

$$x = \frac{-(-2) \pm \sqrt{(-2)^2 - 4(1)(-5)}}{2(1)}$$

$$= \frac{2 \pm \sqrt{4 + 20}}{2}$$

$$= \frac{2 \pm \sqrt{24}}{2}$$

$$= \frac{2 \pm 2\sqrt{6}}{2}$$

$$= 1 \pm \sqrt{6}$$

There are two tangent lines that have slope equal to 1. The first one occurs at $x = 1 + \sqrt{6}$ and the second one occurs at $x = 1 - \sqrt{6}$. We use the original equation to find the points on the graph.

For $x = 1 + \sqrt{6}$,

$$y = \frac{1}{3}(1 + \sqrt{6})^3 - (1 + \sqrt{6})^2 - 4(1 + \sqrt{6}) + 1$$

$$= \frac{1}{3}(19 + 9\sqrt{6}) - (7 + 2\sqrt{6}) - 4 - 4\sqrt{6} + 1$$

$$= \frac{19}{3} + 3\sqrt{6} - 7 - 2\sqrt{6} - 4 - 4\sqrt{6} + 1$$

$$= -\frac{11}{3} - 3\sqrt{6}$$

For $x = 1 - \sqrt{6}$,

$$y = \frac{1}{3}(1 - \sqrt{6})^3 - (1 - \sqrt{6})^2 - 4(1 - \sqrt{6}) + 1$$

$$= \frac{1}{3}(19 - 9\sqrt{6}) - (7 - 2\sqrt{6}) - 4 + 4\sqrt{6} + 1$$

$$= \frac{19}{3} - 3\sqrt{6} - 7 + 2\sqrt{6} - 4 + 4\sqrt{6} + 1$$

$$= -\frac{11}{3} + 3\sqrt{6}$$

Therefore, the tangent line has slope 1 at the points

$$\left(1 + \sqrt{6}, -\frac{11}{3} - 3\sqrt{6}\right) \text{ and } \left(1 - \sqrt{6}, -\frac{11}{3} + 3\sqrt{6}\right).$$

85. a) In order to find the rate of change of the circumference with respect to the radius, we must find the derivative of the function with respect to r.

$$C(r) = 6.28r$$

$$C'(r) = 6.28$$

b) $C'(4) = 6.28$

c) ✏

87. $w(t) = 8.15 + 1.82t - 0.0596t^2 + 0.000758t^3$

a) In order to find the rate of change of weight with respect to time, we take the derivative of the function with respect to t.

$$w'(t)$$

$$= \frac{d}{dt}(8.15 + 1.82t - 0.0596t^2 + 0.000758t^3)$$

$$= 0 + 1.82 - 0.0596(2t) + 0.000758(3t^2)$$

$$= 1.82 - 0.1192t + 0.002274t^2$$

Therefore, the rate of change of weight with respect to time is given by:

$$w'(t) = 1.82 - 0.1192t + 0.002274t^2$$

b) The weight of the baby at age 10 months can be found by evaluating the function when $t = 10$.

$$w(10) = 8.15 + 1.82(10) - 0.0596(10)^2$$
$$+ 0.000758(10)^3$$

$$\approx 21.148 \qquad \text{Using a calculator}$$

Therefore, a 10 month old boy weighs approximately 21.148 pounds.

c) The rate of change of the baby's weight with respect to time at age of 10 months can be found by evaluating the derivative when $t = 10$.

$$w'(10)$$

$$= 1.82 - 0.1192(10) + 0.002274(10)^2$$

$$\approx 0.8554$$

A 10 month old boys weight will be increasing at a rate of 0.86 pounds per month.

89. $R(v) = \dfrac{6000}{v} = 6000v^{-1}$

   a) Using the power rule, we take the derivative of R with respect to v.

$$R'(v) = 6000\left(-1v^{-1-1}\right)$$

$$= -6000v^{-2}$$

$$= -\dfrac{6000}{v^2}$$

   The rate of change of heart rate with respect to the output per beat is

$$R'(v) = -\dfrac{6000}{v^2}.$$

   b) To find the heart rate at $v = 80$ mL per beat, we evaluate the function $R(v)$ when $v = 80$.

$$R(80) = \dfrac{6000}{80} = 75.$$

   The heart rate is 75 beats per minute when the output per beat is 80 mL per beat.

   c) To find the rate of change of the heart beat at $v = 80$ mL per beat, we evaluate the derivative $R'(v)$ at $v = 80$.

$$R'(80) = -\dfrac{6000}{80^2}$$

$$R'(80) = -\dfrac{15}{16}$$

$$= -0.9375$$

   The heart rate is decreasing at a rate of 0.94 beats per minute when the output per beat is 80 mL per beat.

91. a) Using the power rule, we find the growth rate $\dfrac{dP}{dt}$.

$$\dfrac{dP}{dt} = \dfrac{d}{dt}\left(100{,}000 + 2000t^2\right)$$

$$= 0 + 2000(2t)$$

$$= 4000t$$

   b) Evaluate the function P when $t = 10$.

$$P(10) = 100{,}000 + 2000(10)^2$$

$$= 100{,}000 + 2000(100)$$

$$= 300{,}000$$

   The population of the city will be 300,000 people after 10 years.

   c) Evaluate the derivative $P'(t)$ when $t = 10$.

$$\left.\dfrac{dP}{dt}\right|_{t=10} = P'(10) = 4000(10) = 40{,}000$$

   The population's growth rate after 10 years is 40,000 people per year.

   d) ✎

93. $V = 1.22\sqrt{h} = 1.22h^{\frac{1}{2}}$

   a) Using the power rule,

$$\dfrac{dV}{dh} = \dfrac{d}{dh}\left(1.22h^{\frac{1}{2}}\right)$$

$$= 1.22\left(\dfrac{1}{2}h^{\frac{1}{2}-1}\right)$$

$$= 0.61h^{-\frac{1}{2}}$$

$$= \dfrac{0.61}{h^{\frac{1}{2}}} = \dfrac{0.61}{\sqrt{h}}$$

   b) Evaluate the function V when $h = 40{,}000$.

$$V = 1.22\sqrt{40{,}000}$$

$$= 244$$

   A person would be able to see 244 miles to the horizon from a height of 40,000 feet.

   c) Evaluate the derivative $\dfrac{dV}{dh}$ when $h = 40{,}000$.

$$\left.\dfrac{dV}{dh}\right|_{h=40{,}000} = \dfrac{0.61}{\sqrt{40{,}000}}$$

$$= \dfrac{0.61}{200}$$

$$= 0.00305$$

   The rate of change at $h = 40{,}000$ is 0.0031 miles per foot.

   d) ✎

95. $f(x) = x^2 - 4x + 1$

   The derivative is positive when $f'(x) > 0$.

   Find $f'(x)$.

$$f'(x) = \dfrac{d}{dx}\left(x^2 - 4x + 1\right)$$

$$= 2x - 4$$

   Next, we solve the inequality

$$f'(x) > 0$$

$$2x - 4 > 0$$

$$2x > 4$$

$$x > 2$$

   Therefore, the interval for which $f'(x)$ is positive is $(2, \infty)$.

97. $y = x^4 - \frac{4}{3}x^2 - 4$

A horizontal tangent line has slope equal to 0, so we need to find the values of x that make $\frac{dy}{dx} = 0$.

First, we find the derivative.

$$\frac{dy}{dx} = \frac{d}{dx}\left(x^4 - \frac{4}{3}x^2 - 4\right)$$

$$= \frac{d}{dx}x^4 - \frac{d}{dx}\left(\frac{4}{3}x^2\right) - \frac{d}{dx}4$$

$$= 4x^{4-1} - \frac{4}{3}\left(2x^{2-1}\right) - 0$$

$$= 4x^3 - \frac{8}{3}x$$

Next, we set the derivative equal to zero and solve for x.

$$\frac{dy}{dx} = 0$$

$$4x^3 - \frac{8}{3}x = 0$$

$$4x\left(x^2 - \frac{2}{3}\right) = 0$$

$$4x = 0 \quad \text{or} \quad x^2 - \frac{2}{3} = 0$$

$$x = 0 \quad \text{or} \quad x^2 = \frac{2}{3}$$

$$x = 0 \quad \text{or} \quad x = \pm\sqrt{\frac{2}{3}}$$

So the horizontal tangent will occur when $x = 0$, $x = \sqrt{\frac{2}{3}}$, and $x = -\sqrt{\frac{2}{3}}$. Next we find the points on the graph.

For $x = 0$,

$$y = (0)^4 - \frac{4}{3}(0)^2 - 4 = -4.$$

For $x = \sqrt{\frac{2}{3}}$

$$y = \left(\sqrt{\frac{2}{3}}\right)^4 - \frac{4}{3}\left(\sqrt{\frac{2}{3}}\right)^2 - 4$$

$$= -\frac{40}{9}$$

For $x = -\sqrt{\frac{2}{3}}$

$$y = \left(-\sqrt{\frac{2}{3}}\right)^4 - \frac{4}{3}\left(-\sqrt{\frac{2}{3}}\right)^2 - 4$$

$$= -\frac{40}{9}$$

There are three points on the graph for which the tangent line is horizontal.

$$(0,-4), \quad \left(\sqrt{\frac{2}{3}}, -\frac{40}{9}\right), \quad \text{and,} \quad \left(-\sqrt{\frac{2}{3}}, -\frac{40}{9}\right).$$

99. $f(x) = x^5 + x^3$

Taking the derivative we have:

$$f'(x) = \frac{d}{dx}\left(x^5 + x^3\right)$$

$$= \frac{d}{dx}x^5 + \frac{d}{dx}x^3$$

$$= 5x^4 + 3x^2$$

Notice that $f'(x) \geq 0$ for all values of x.

Therefore, $f(x)$ is always increasing.

101. $k(x) = \frac{1}{x^2}, \quad x \neq 0$

Taking the derivative we have:

$$k'(x) = \frac{d}{dx}\left(\frac{1}{x^2}\right)$$

$$= \frac{d}{dx}x^{-2}$$

$$= -2x^{-2-1}$$

$$= -2x^{-3} = -\frac{2}{x^3}$$

Notice $k'(x) < 0$ for all values of x over the interval $(0,\infty)$. Therefore, $k(x)$ is always decreasing over the interval $(0,\infty)$.

103. $y = (x+3)(x-2)$

First, we multiply the two binomials.

$y = (x+3)(x-2) = x^2 + x - 6$

Therefore,

$\dfrac{dy}{dx} = \dfrac{d}{dx}(x^2 + x - 6)$

$\quad = \dfrac{d}{dx}(x^2) + \dfrac{d}{dx}(x) - \dfrac{d}{dx}(6)$

$\quad = 2x + 1$

105. $y = \dfrac{x^5 - x^3}{x^2}$

First, we separate the fraction.

$y = \dfrac{x^5}{x^2} - \dfrac{x^3}{x^2}$

$\quad = x^{5-2} - x^{3-2} \qquad \left[\dfrac{a^m}{a^n} = a^{m-n}\right]$

$\quad = x^3 - x^1$

Therefore,

$\dfrac{dy}{dx} = \dfrac{d}{dx}(x^3 - x)$

$\quad = \dfrac{d}{dx}x^3 - \dfrac{d}{dx}x$

$\quad = 3x^2 - 1$

107. $y = \sqrt{7x}$

First, we simplify the radical.

$y = \sqrt{7 \cdot x}$

$\quad = \sqrt{7}\sqrt{x} \qquad \left[\sqrt{m \cdot n} = \sqrt{m}\sqrt{n}\right]$

$\quad = \sqrt{7}\left(x^{\frac{1}{2}}\right) \qquad \left[\sqrt[m]{a} = a^{\frac{1}{m}}; m = 2\right]$

Therefore,

$\dfrac{dy}{dx} = \dfrac{d}{dx}\left(\sqrt{7}(x)^{\frac{1}{2}}\right)$

$\quad = \sqrt{7}\dfrac{d}{dx}\left(x^{\frac{1}{2}}\right)$

$\quad = \sqrt{7}\left(\dfrac{1}{2}x^{\frac{1}{2}-1}\right)$

$\quad = \dfrac{\sqrt{7}}{2}x^{-\frac{1}{2}}$

$\quad = \dfrac{\sqrt{7}}{2x^{\frac{1}{2}}} = \dfrac{\sqrt{7}}{2\sqrt{x}}$

109. $y = \left(\sqrt{x} - \dfrac{1}{\sqrt{x}}\right)^2$

$y = \left(x^{\frac{1}{2}} - x^{-\frac{1}{2}}\right)^2$

$\quad = \left(x^{\frac{1}{2}} - x^{-\frac{1}{2}}\right)\left(x^{\frac{1}{2}} - x^{-\frac{1}{2}}\right)$

$\quad = x - 2x^0 + x^{-1}$

$\quad = x + x^{-1} - 2$

$\dfrac{dy}{dx} = \dfrac{d}{dx}\left(x + x^{-1} - 2\right)$

$\quad = 1 - x^{-2}$

$\quad = 1 - \dfrac{1}{x^2}$

111. ✎

113. $f(x) = \dfrac{5x^2 + 8x - 3}{3x^2 + 2}$

First we enter the equation into the graphing editor on the calculator.

```
Plot1 Plot2 Plot3
\Y1冃(5X^2+8X-3)/
(3X^2+2)
\Y2=
\Y3=
\Y4=
\Y5=
\Y6=
```

The solution is continued on the next page.

Using the window:

```
WINDOW
 Xmin=-4
 Xmax=4
 Xscl=1
 Ymin=-3
 Ymax=3
 Yscl=1
 Xres=1
```

We get the graph:

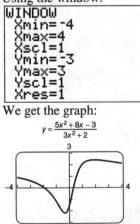

The horizontal tangents occur at the turning points of this function. Using the trace feature, or the minimum/maximum feature on the calculator, we find the turning points. We estimate the x-values at which the tangent lines are horizontal are

$x = -0.346$ and $x = 1.929$.

115. $f(x) = x^4 - x^3$

Using the calculator, we graph the function and the derivative in the same window. We can use the nDeriv feature to graph the derivative without actually calculating the derivative.

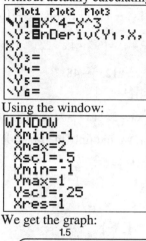

Using the window:

We get the graph:

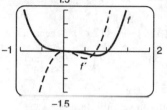

Note, the function $f(x)$ is the solid graph.

Using the calculator, we can find the derivative of the function when $x = 1$.

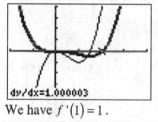

We have $f'(1) = 1$.

117. $f(x) = \dfrac{5x^2 + 8x - 3}{3x^2 + 2}$

Using the calculator, we graph the function and the derivative in the same window. We can use the nDeriv feature to graph the derivative without actually calculating the derivative.

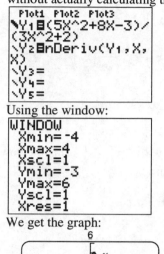

Using the window:

We get the graph:

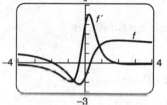

Note, the function $f(x)$ is the thicker graph.

Using the calculator, we can find the derivative of the function when $x - 1$.

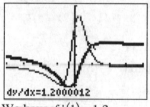

We have $f'(1) = 1.2$.

Exercise Set 1.6

1. Differentiate $y = x^9 \cdot x^4$ using the Product Rule (Theorem 5).

$$\frac{dy}{dx} = \frac{d}{dx}\left(x^9 \cdot x^4\right)$$

$$= x^9 \cdot \frac{d}{dx}\left(x^4\right) + x^4 \cdot \frac{d}{dx}\left(x^9\right)$$

$$= x^9 \cdot 4x^3 + x^4 \cdot 9x^8$$

$$= 4x^{12} + 9x^{12}$$

$$= 13x^{12}$$

Differentiate $y = x^9 \cdot x^4 = x^{13}$ using the Power Rule (Theorem 1).

$$\frac{dy}{dx} = \frac{d}{dx}\left(x^{13}\right)$$

$$= 13x^{13-1}$$

$$= 13x^{12}$$

The two results are equivalent.

3. Differentiate $f(x) = (2x+5)(3x-4)$ using the Product Rule (Theorem 5).

$$f'(x) = \frac{d}{dx}\left[(2x+5)(3x-4)\right]$$

$$= (2x+5) \cdot \frac{d}{dx}(3x-4) +$$

$$\qquad (3x-4) \cdot \frac{d}{dx}(2x+5)$$

$$= (2x+5) \cdot 3 + (3x-4) \cdot 2$$

$$= 6x + 15 + 6x - 8$$

$$= 12x + 7$$

Differentiate $f(x) = (2x+5)(3x-4)$ using the Power Rule (Theorem 1). First, we multiply the binomial terms in the function.

$$f(x) = (2x+5)(3x-4)$$

$$= 6x^2 + 7x - 20$$

Therefore, by Theorem 1, Theorem 2, and Theorem 4 we have:

$$f'(x) = \frac{d}{dx}\left(6x^2 + 7x - 20\right)$$

$$= \frac{d}{dx}\left(6x^2\right) + \frac{d}{dx}(7x) - \frac{d}{dx}(20) \quad \text{Theorem 4}$$

$$= 12x + 7 \qquad\qquad \text{Theorem 1 and 2}$$

The two results are equivalent.

5. Differentiate $F(x) = 3x^4\left(x^2 - 4x\right)$ using the Product Rule.

$$F'(x) = \frac{d}{dx}\left[3x^4\left(x^2 - 4x\right)\right]$$

$$= 3x^4 \cdot \frac{d}{dx}\left(x^2 - 4x\right) + \left(x^2 - 4x\right) \cdot \frac{d}{dx}\left(3x^4\right)$$

$$= 3x^4 \cdot (2x-4) + \left(x^2 - 4x\right) \cdot 12x^3$$

$$= 6x^5 - 12x^4 + 12x^5 - 48x^4$$

$$= 18x^5 - 60x^4$$

Differentiate $F(x) = 3x^4\left(x^2 - 4x\right)$ using the Power Rule. First, we multiply the function.

$$F(x) = 3x^4\left(x^2 - 4x\right)$$

$$= 3x^6 - 12x^5$$

Therefore, we have:

$$F'(x) = \frac{d}{dx}\left(3x^6 - 12x^5\right)$$

$$= \frac{d}{dx}\left(3x^6\right) - \frac{d}{dx}\left(12x^5\right) \quad \text{Theorem 4}$$

$$= 3\left(6x^5\right) - 12\left(5x^4\right) \quad \begin{array}{l}\text{Theorem 1}\\\text{Theorem 3}\end{array}$$

$$= 18x^5 - 60x^4$$

The two results are equivalent.

7. Differentiate $y = \left(3\sqrt{x} + 2\right)x^2$ using the Product Rule.

$$\frac{dy}{dx} = \frac{d}{dx}\left[\left(3\sqrt{x} + 2\right)x^2\right]$$

$$= \left(3\sqrt{x} + 2\right) \cdot \frac{d}{dx}\left(x^2\right) + x^2 \cdot \frac{d}{dx}\left(3\sqrt{x} + 2\right)$$

$$= \left(3x^{\frac{1}{2}} + 2\right) \cdot \frac{d}{dx}\left(x^2\right) + x^2 \cdot \frac{d}{dx}\left(3x^{\frac{1}{2}} + 2\right)$$

$$= \left(3x^{\frac{1}{2}} + 2\right) \cdot 2x + x^2 \cdot \frac{3}{2}x^{-\frac{1}{2}} \quad \begin{array}{l}\text{Theorems 1}\\\text{2 and 4.}\end{array}$$

$$= 6x^{\frac{3}{2}} + 4x + \frac{3}{2}x^{\frac{3}{2}}$$

$$= \frac{15}{2}x^{\frac{3}{2}} + 4x$$

Differentiate $y = \left(3\sqrt{x} + 2\right)x^2$ using the Power Rule. First, we multiply the function.

$$y = \left(3\sqrt{x} + 2\right)x^2$$

$$= 3x^{\frac{1}{2}+2} + 2x^2$$

$$= 3x^{\frac{5}{2}} + 2x^2$$

The solution is continued on the next page.

Therefore, we have:

$$\frac{dy}{dx} = \frac{d}{dx}\left(3x^{5/2} + 2x^2\right)$$

$$= \frac{d}{dx}\left(3x^{5/2}\right) + \frac{d}{dx}\left(2x^2\right) \quad \text{Theorem 4}$$

$$= 3\left(\frac{5}{2}x^{5/2-1}\right) + 2\left(2x^1\right) \quad \begin{array}{l}\text{Theorem 1}\\\text{Theorem 3}\end{array}$$

$$= \frac{15}{2}x^{3/2} + 4x$$

The two results are equivalent.

9. Differentiate $f(x) = (2x+5)\left(3x^2 - 4x + 1\right)$
using the Product Rule.

$$f'(x) = \frac{d}{dx}\left[(2x+5)\left(3x^2 - 4x + 1\right)\right]$$

$$= (2x+5)\cdot\frac{d}{dx}\left(3x^2 - 4x + 1\right) +$$

$$\qquad\qquad \left(3x^2 - 4x + 1\right)\cdot\frac{d}{dx}(2x+5)$$

$$= (2x+5)\cdot(6x-4) + \left(3x^2 - 4x + 1\right)\cdot 2$$

$$= 12x^2 + 22x - 20 + 6x^2 - 8x + 2$$

$$= 18x^2 + 14x - 18$$

Differentiate $f(x) = (2x+5)\left(3x^2 - 4x + 1\right)$
using the Power Rule. First, we multiply the
terms in the function.

$$f(x) = \left[(2x+5)\left(3x^2 - 4x + 1\right)\right]$$

$$= 6x^3 + 7x^2 - 18x + 5$$

Thus, we have:

$$f'(x) = \frac{d}{dx}\left(6x^3 + 7x^2 - 18x + 5\right)$$

$$= \frac{d}{dx}\left(6x^3\right) + \frac{d}{dx}\left(7x^2\right) -$$

$$\qquad\qquad \frac{d}{dx}(18x) + \frac{d}{dx}(5)$$

$$= 18x^2 + 14x - 18$$

The two results are equivalent.

11. Differentiate $F(t) = \left(\sqrt{t} + 2\right)\left(3t - 4\sqrt{t} + 7\right)$
using the Product Rule.

$$F'(t) = \frac{d}{dt}\left[\left(\sqrt{t} + 2\right)\left(3t - 4\sqrt{t} + 7\right)\right]$$

$$= \left(t^{1/2} + 2\right)\cdot\frac{d}{dt}\left(3t - 4t^{1/2} + 7\right) +$$

$$\qquad \left(3t - 4t^{1/2} + 7\right)\cdot\frac{d}{dt}\left(t^{1/2} + 2\right) \quad \left[\sqrt{t} = t^{1/2}\right]$$

$$= \left(t^{1/2} + 2\right)\cdot\left(3 - 4\left(\frac{1}{2}t^{-1/2}\right)\right) +$$

$$\qquad \left(3t - 4t^{1/2} + 7\right)\cdot\left(\frac{1}{2}t^{-1/2}\right)$$

$$= \left(t^{1/2} + 2\right)\cdot\left(3 - 2t^{-1/2}\right) +$$

$$\qquad \left(3t - 4t^{1/2} + 7\right)\cdot\left(\frac{1}{2}t^{-1/2}\right)$$

$$= 3t^{1/2} - 2 + 6 - 4t^{-1/2} + \frac{3}{2}t^{1/2} - 2 + \frac{7}{2}t^{-1/2}$$

$$= \frac{9}{2}t^{1/2} - \frac{1}{2}t^{-1/2} + 2$$

$$= \frac{9\sqrt{t}}{2} - \frac{1}{2\sqrt{t}} + 2$$

Differentiate $F(t) = \left(\sqrt{t} + 2\right)\left(3t - 4\sqrt{t} + 7\right)$
using the Power Rule

$$F(t) = \left(\sqrt{t} + 2\right)\left(3t - 4\sqrt{t} + 7\right)$$

$$= 3t^{3/2} - 4t + 7t^{1/2} + 6t - 8t^{1/2} + 14$$

$$= 3t^{3/2} - t^{1/2} + 2t + 14$$

Therefore, we have:

$$F'(t) = \frac{d}{dt}\left(3t^{3/2} - t^{1/2} + 2t + 14\right)$$

$$= \frac{d}{dt}\left(3t^{3/2}\right) - \frac{d}{dt}\left(t^{1/2}\right) + \frac{d}{dt}(2t) + \frac{d}{dt}(14)$$

$$= \frac{9t^{1/2}}{2} - \frac{1}{2}t^{-1/2} + 2$$

$$= \frac{9\sqrt{t}}{2} - \frac{1}{2\sqrt{t}} + 2$$

The two results are equivalent.

13. Differentiate $y = \dfrac{x^6}{x^4}$ using the Quotient Rule (Theorem 6).

$$\frac{dy}{dx} = \frac{d}{dx}\left(\frac{x^6}{x^4}\right)$$

$$= \frac{x^4 \dfrac{d}{dx}\left(x^6\right) - x^6 \dfrac{d}{dx}\left(x^4\right)}{\left(x^4\right)^2}$$

$$= \frac{x^4\left(6x^5\right) - x^6\left(4x^3\right)}{x^8}$$

$$\frac{dy}{dx} = \frac{6x^9 - 4x^9}{x^8}$$

$$= \frac{2x^9}{x^8}$$

$$= 2x, \qquad \text{for } x \neq 0$$

Differentiate $y = \dfrac{x^6}{x^4} = x^2$ using the Power Rule.

$$\frac{dy}{dx} = \frac{d}{dx}x^2$$

$$= 2x^{2-1}$$

$$= 2x, \qquad \text{for } x \neq 0$$

The two results are equivalent.

15. Differentiate $g(x) = \dfrac{3x^7 - x^3}{x}$ using the Quotient Rule.

$$g'(x) = \frac{d}{dx}\left(\frac{3x^7 - x^3}{x}\right)$$

$$= \frac{x\dfrac{d}{dx}\left(3x^7 - x^3\right) - \left(3x^7 - x^3\right)\dfrac{d}{dx}(x)}{(x)^2}$$

$$= \frac{x\left(21x^6 - 3x^2\right) - \left(3x^7 - x^3\right)(1)}{x^2}$$

$$= \frac{18x^7 - 2x^3}{x^2}$$

$$= \frac{x^2\left(18x^5 - 2x\right)}{x^2}$$

$$= 18x^5 - 2x, \qquad \text{for } x \neq 0$$

Differentiate $g(x) = \dfrac{3x^7 - x^3}{x}$ using the Power Rule. First, factor the numerator and divide the common factors.

$$g(x) = \frac{3x^7 - x^3}{x}$$

$$= \frac{x\left(3x^6 - x^2\right)}{x}$$

$$= 3x^6 - x^2$$

$$g'(x) = \frac{d}{dx}\left(3x^6 - x^2\right)$$

$$= 18x^5 - 2x, \qquad \text{for } x \neq 0$$

The two results are equivalent.

17. Differentiate $G(x) = \dfrac{8x^3 - 1}{2x - 1}$ using the Quotient Rule.

$$G'(x) = \frac{d}{dx}\left(\frac{8x^3 - 1}{2x - 1}\right)$$

$$= \frac{(2x-1)\dfrac{d}{dx}\left(8x^3 - 1\right) - \left(8x^3 - 1\right)\dfrac{d}{dx}(2x-1)}{(2x-1)^2}$$

$$= \frac{(2x-1)\left(24x^2\right) - \left(8x^3 - 1\right)(2)}{(2x-1)^2}$$

$$= \frac{(2x-1)\left(24x^2\right) - (2x-1)\left(4x^2 + 2x + 1\right)(2)}{(2x-1)^2}$$

$$= \frac{(2x-1)\left[\left(24x^2\right) - \left(4x^2 + 2x + 1\right)(2)\right]}{(2x-1)^2}$$

$$= \frac{(2x-1)\left[\left(24x^2\right) - \left(8x^2 + 4x + 2\right)\right]}{(2x-1)^2}$$

$$= \frac{\left[16x^2 - 4x - 2\right]}{(2x-1)}$$

$$= \frac{(2x-1)(8x+2)}{(2x-1)}$$

$$= 8x + 2; \qquad x \neq \frac{1}{2}$$

The solution is continued on the next page.

Differentiate $G(x) = \dfrac{8x^3 - 1}{2x - 1}$ using the Power Rule.

First, factor the numerator and divide the common factors.

$$G(x) = \frac{8x^3 - 1}{2x - 1}$$

$$= \frac{(2x - 1)(4x^2 + 2x + 1)}{2x - 1} \quad \text{Difference of cubes}$$

$$= 4x^2 + 2x + 1$$

$$G'(x) = \frac{d}{dx}(4x^2 + 2x + 1)$$

$$= 8x + 2; \qquad x \neq \tfrac{1}{2}$$

The two results are equivalent.

19. Differentiate $y = \dfrac{t^2 - 16}{t + 4}$ using the Quotient Rule.

$$\frac{dy}{dt} = \frac{d}{dt}\left(\frac{t^2 - 16}{t + 4}\right)$$

$$= \frac{(t + 4)\frac{d}{dt}(t^2 - 16) - (t^2 - 16)\frac{d}{dt}(t + 4)}{(t + 4)^2}$$

$$= \frac{(t + 4)(2t) - (t^2 - 16)(1)}{(t + 4)^2}$$

$$= \frac{2t^2 + 8t - t^2 + 16}{(t + 4)^2}$$

$$= \frac{t^2 + 8t + 16}{(t + 4)^2}$$

The derivative is simplified as follows:

$$\frac{dy}{dt} = \frac{(t + 4)^2}{(t + 4)^2}$$

$$= 1; \qquad t \neq -4$$

Differentiate $y = \dfrac{t^2 - 16}{t + 4}$ using the Power Rule.

First, factor the numerator and divide the common factors.

$$y = \frac{t^2 - 16}{t + 4}$$

$$= \frac{(t + 4)(t - 4)}{t + 4} \quad \text{Difference of squares}$$

$$= t - 4$$

Therefore,

$$\frac{dy}{dt} = \frac{d}{dt}(x - 4)$$

$$= 1, \qquad \text{for } t \neq -4$$

The two results are equivalent.

21. $g(x) = (5x^2 + 4x - 3)(2x^2 - 3x + 1)$

Using the Product Rule, we have:

$$g'(x) = \frac{d}{dx}\left[(5x^2 + 4x - 3)(2x^2 - 3x + 1)\right]$$

$$= (5x^2 + 4x - 3) \cdot \frac{d}{dx}(2x^2 - 3x + 1) +$$

$$(2x^2 - 3x + 1) \cdot \frac{d}{dx}(5x^2 + 4x - 3)$$

$$= (5x^2 + 4x - 3) \cdot (4x - 3) +$$

$$(2x^2 - 3x + 1) \cdot (10x + 4)$$

Simplifying we get

$$= (20x^3 + x^2 - 24x + 9) +$$

$$(20x^3 - 22x^2 - 2x + 4)$$

$$= 40x^3 - 21x^2 - 26x + 13$$

23. $y = \dfrac{5x^2 - 1}{2x^3 + 3}$

Using the Quotient Rule.

$$\frac{dy}{dx} = \frac{d}{dx}\left(\frac{5x^2 - 1}{2x^3 + 3}\right)$$

$$= \frac{(2x^3 + 3)\frac{d}{dx}(5x^2 - 1) - (5x^2 - 1)\frac{d}{dx}(2x^3 + 3)}{(2x^3 + 3)^2}$$

$$= \frac{(2x^3 + 3)(10x) - (5x^2 - 1)(6x^2)}{(2x^3 + 3)^2}$$

$$= \frac{20x^4 + 30x - 30x^4 + 6x^2}{(2x^3 + 3)^2}$$

$$= \frac{-10x^4 + 6x^2 + 30x}{(2x^3 + 3)^2}$$

$$= \frac{-2x(5x^3 - 3x - 15)}{(2x^3 + 3)^2}$$

25. $F(x) = (-3x^2 + 4x)(7\sqrt{x} + 1)$

$F(x) = (-3x^2 + 4x)(7x^{\frac{1}{2}} + 1) \quad \left[\sqrt{x} = x^{\frac{1}{2}}\right]$

Using the Product Rule, we have:

$F(x) = \dfrac{d}{dx}\left[(-3x^2 + 4x)(7x^{\frac{1}{2}} + 1)\right]$

$\quad = (-3x^2 + 4x) \cdot \dfrac{d}{dx}(7x^{\frac{1}{2}} + 1) +$

$\qquad (7x^{\frac{1}{2}} + 1) \cdot \dfrac{d}{dx}(-3x^2 + 4x)$

$\quad = (-3x^2 + 4x) \cdot \left(\dfrac{7}{2}x^{-\frac{1}{2}}\right) +$

$\qquad (7x^{\frac{1}{2}} + 1) \cdot (-6x + 4)$

Simplifying we get

$\quad = \left(-\dfrac{21}{2}x^{\frac{3}{2}} + 14x^{\frac{1}{2}}\right) +$

$\qquad \left(-42x^{\frac{3}{2}} - 6x + 28x^{\frac{1}{2}} + 4\right)$

$\quad = -\dfrac{105}{2}x^{\frac{3}{2}} - 6x + 42x^{\frac{1}{2}} + 4$

27. $g(t) = \dfrac{t}{3 - t} + 5t^3$

Differentiating we have:

$g'(t) = \dfrac{d}{dt}\left(\dfrac{t}{3 - t} + 5t^3\right)$

$\quad = \dfrac{d}{dt}\left(\dfrac{t}{3 - t}\right) + \dfrac{d}{dt}\left(5t^3\right)$

Using the derivative, we will apply the Quotient Rule to the first term, and the Power Rule to the second term.

$g'(t) = \underbrace{\dfrac{(3 - t) \cdot \dfrac{d}{dt}(t) - t \cdot \dfrac{d}{dt}(3 - t)}{(3 - t)^2}}_{\text{Quotient Rule}} + 15t^2$

$\quad = \dfrac{(3 - t)(1) - t(-1)}{(3 - t)^2} + 15t^2$

$\quad = \dfrac{3}{(3 - t)^2} + 15t^2$

29. $G(x) = (5x - 4)^2 = (5x - 4)(5x - 4)$

Using the Product Rule, we have

$G'(x) = \dfrac{d}{dx}\left[(5x - 4)(5x - 4)\right]$

$\quad = (5x - 4) \cdot \dfrac{d}{dx}(5x - 4) +$

$\qquad (5x - 4) \cdot \dfrac{d}{dx}(5x - 4)$

$\quad = (5x - 4) \cdot \dfrac{d}{dx}(5x - 4) +$

$\qquad (5x - 4) \cdot \dfrac{d}{dx}(5x - 4)$

$\quad = (5x - 4) \cdot (5) + (5x - 4) \cdot (5)$

$\quad = 50x - 40$

31. $y = (x^3 - 4x)^2 = (x^3 - 4x)(x^3 - 4x)$

Using the Product Rule, we have

$\dfrac{dy}{dx} = \dfrac{d}{dx}\left[(x^3 - 4x)(x^3 - 4x)\right]$

$\quad = (x^3 - 4x) \cdot \dfrac{d}{dx}(x^3 - 4x) +$

$\qquad (x^3 - 4x) \cdot \dfrac{d}{dx}(x^3 - 4x)$

$\quad = (x^3 - 4x) \cdot (3x^2 - 4) +$

$\qquad (x^3 - 4x) \cdot (3x^2 - 4)$

Simplifying, we get

$\quad = 2(x^3 - 4x)(3x^2 - 4)$

$\quad = 2x(x^2 - 4)(3x^2 - 4)$

33. $f(x) = 6x^{-4}(6x^3 + 10x^2 - 8x + 3)$

Using the Product Rule:

$f'(x) = \left(6x^{-4}\right)\dfrac{d}{dx}\left(6x^3 + 10x^2 - 8x + 3\right) +$

$\qquad \left(6x^3 + 10x^2 - 8x + 3\right)\dfrac{d}{dx}\left(6x^{-4}\right)$

$\quad = \left(6x^{-4}\right)\left(18x^2 + 20x - 8\right) +$

$\qquad \left(6x^3 + 10x^2 - 8x + 3\right)\left(-24x^{-5}\right)$

Simplifying, we get

$\quad = 108x^{-2} + 120x^{-3} - 48x^{-4} -$

$\qquad 144x^{-2} - 240x^{-3} + 192x^{-4} - 72x^{-5}$

$\quad = -36x^{-2} - 120x^{-3} + 144x^{-4} - 72x^{-5}$

35. $F(t) = \left(t + \dfrac{2}{t}\right)(t^2 - 3) = (t + 2t^{-1})(t^2 - 3)$

Using the Product Rule, we have:

$F'(t) = \dfrac{d}{dt}\left[(t + 2t^{-1})(t^2 - 3)\right]$

$\quad = (t + 2t^{-1}) \cdot \dfrac{d}{dt}(t^2 - 3) +$

$\quad\quad (t^2 - 3) \cdot \dfrac{d}{dt}(t + 2t^{-1})$

$\quad = (t + 2t^{-1}) \cdot (2t) + (t^2 - 3) \cdot (1 - 2t^{-2})$

Simplifying, we get

$\quad = 2t^2 + 4 + (t^2 - 2 - 3 + 6t^{-2})$

$\quad = 3t^2 - 1 + 6t^{-2}$

$\quad = 3t^2 - 1 + \dfrac{6}{t^2}$

37. $y = \dfrac{x^3 - 1}{x^2 + 1} + 4x^3$

Differentiating we have:

$\dfrac{dy}{dx} = \dfrac{d}{dx}\left(\dfrac{x^3 - 1}{x^2 + 1} + 4x^3\right)$

$\quad = \dfrac{d}{dx}\left(\dfrac{x^3 - 1}{x^2 + 1}\right) + \dfrac{d}{dx}(4x^3)$

We will apply the Quotient Rule to the first term, and the Power Rule to the second term.

$\dfrac{dy}{dx} = \underbrace{\dfrac{(x^2 + 1)\cdot(3x^2) - (x^3 - 1)\cdot(2x)}{(x^2 + 1)^2}}_{\text{Quotient Rule}} + 12x$

Simplifying, we get

$\quad = \dfrac{(3x^4 + 3x^2) - (2x^4 - 2x)}{(x^2 + 1)^2} + 12x^2$

$\quad = \dfrac{x^4 + 3x^2 + 2x}{(x^2 + 1)^2} + 12x^2$

39. $y = \dfrac{\sqrt[3]{x} - 7}{\sqrt{x} + 3} = \dfrac{x^{1/3} - 7}{x^{1/2} + 3}$

Using the Quotient Rule.

$\dfrac{dy}{dx} = \dfrac{d}{dx}\left(\dfrac{x^{1/3} - 7}{x^{1/2} + 3}\right)$

$\quad = \dfrac{(x^{1/2} + 3)\dfrac{d}{dx}(x^{1/3} - 7) - (x^{1/3} - 7)\dfrac{d}{dx}(x^{1/2} + 3)}{(x^{1/2} + 3)^2}$

$\quad = \dfrac{(x^{1/2} + 3)\left(\dfrac{1}{3}x^{-2/3}\right) - (x^{1/3} - 7)\left(\dfrac{1}{2}x^{-1/2}\right)}{(x^{1/2} + 3)^2}$

Note, the previous derivative can be simplified as follows

$\dfrac{dy}{dx} = \dfrac{(x^{1/2} + 3)\left(\dfrac{1}{3}x^{-2/3}\right) - (x^{1/3} - 7)\left(\dfrac{1}{2}x^{-1/2}\right)}{(x^{1/2} + 3)^2}$

$\quad = \dfrac{\dfrac{1}{3}x^{-1/6} + x^{-2/3} - \dfrac{1}{2}x^{-1/6} + \dfrac{7}{2}x^{-1/2}}{(x^{1/2} + 3)^2}$

$\quad = \dfrac{x^{-2/3}}{6} \cdot \dfrac{\dfrac{1}{6}x^{-1/6} + \dfrac{7}{2}x^{-1/2}}{(x^{1/2} + 3)^2} \cdot \dfrac{6x^{2/3}}{6x^{2/3}}$

$\quad = \dfrac{6 - \sqrt{x} + 21x^{1/6}}{6x^{2/3}(\sqrt{x} + 3)^2}$

41. $f(x) = \dfrac{x^{-1}}{x + x^{-1}}$

Using the Quotient Rule, we have

$f'(x) = \dfrac{d}{dx}\left(\dfrac{x^{-1}}{x + x^{-1}}\right)$

$\quad = \dfrac{(x + x^{-1})(-1x^{-2}) - (x^{-1})(1 - x^{-2})}{(x + x^{-1})^2}$

$\quad = \dfrac{-x^{-1} - x^{-3} - x^{-1} + x^{-3}}{(x^{-1} + x)^2}$

$\quad = \dfrac{-2x^{-1}}{(x^{-1} + x)^2}, \qquad \text{for } x \ne 0$

The solution is continued on the next page.

Simplify the derivative as follows:

$$f'(x) = \frac{-2x^{-1}}{\left(x^{-1} + x\right)^2}$$

$$= \frac{-\frac{2}{x}}{\left(\frac{1}{x} + x\right)^2}$$

$$= \frac{-\frac{2}{x}}{\left(\frac{1+x^2}{x}\right)^2}$$

$$= \frac{-2}{x} \cdot \frac{x^2}{\left(1+x^2\right)^2}$$

$$= \frac{-2x}{\left(x^2+1\right)^2}, \qquad \text{for } x \neq 0$$

43. $F(t) = \dfrac{1}{t-4}$

Using the Quotient Rule, we have

$$F'(t) = \frac{d}{dt}\left(\frac{1}{t-4}\right)$$

$$= \frac{(t-4)\frac{d}{dt}(1) - (1)\frac{d}{dt}(t-4)}{(t-4)^2}$$

$$= \frac{(t-4)(0) - (1)(1)}{(t-4)^2}$$

$$= \frac{-1}{(t-4)^2}$$

45. $f(x) = \dfrac{3x^2 - 5x}{x^2 - 1}$

Using the Quotient Rule, we have

$$f'(x) = \frac{d}{dx}\left(\frac{3x^2 - 5x}{x^2 - 1}\right)$$

$$= \frac{(x^2-1)\frac{d}{dx}(3x^2-5x) - (3x^2-5x)\frac{d}{dx}(x^2-1)}{(x^2-1)^2}$$

$$= \frac{(x^2-1)(6x-5) - (3x^2-5x)(2x)}{(x^2-1)^2}$$

$$= \frac{6x^3 - 5x^2 - 6x + 5 - (6x^3 - 10x^2)}{(x^2-1)^2}$$

$$= \frac{5x^2 - 6x + 5}{(x^2-1)^2}$$

47. $g(t) = \dfrac{-t^2 + 3t + 5}{t^2 - 2t + 4}$

Using the Quotient Rule, we have

$$g'(t) = \frac{d}{dt}\left(\frac{-t^2 + 3t + 5}{t^2 - 2t + 4}\right)$$

$$= \frac{(t^2 - 2t + 4)\frac{d}{dt}(-t^2 + 3t + 5)}{(t^2 - 2t + 4)^2} -$$

$$\frac{(-t^2 + 3t + 5)\frac{d}{dt}(t^2 - 2t + 4)}{(t^2 - 2t + 4)^2}$$

$$= \frac{(t^2 - 2t + 4)(-2t + 3) - (-t^2 + 3t + 5)(2t - 2)}{(t^2 - 2t + 4)^2}$$

Note, the previous derivative could be simplified as follows:

$$g'(t) = \frac{-2t^3 + 7t^2 - 14t + 12 - \left(-2t^3 + 8t^2 + 4t - 10\right)}{(t^2 - 2t + 4)^2}$$

$$= \frac{-t^2 - 18t + 22}{(t^2 - 2t + 4)^2}$$

49. $y = \dfrac{8}{x^2 + 4}$

$$\frac{dy}{dx} = \frac{(x^2 + 4)(0) - 8(2x)}{(x^2 + 4)^2}$$

$$\frac{dy}{dx} = \frac{-16x}{(x^2 + 4)^2}$$

a) When $x = 0$, $\dfrac{dy}{dx} = \dfrac{-16(0)}{(0^2 + 4)^2} = 0$, so the

slope of the tangent line at $(0, 2)$ is 0. The equation of the horizontal line passing through $(0, 2)$ is $y = 2$.

b) When $x = -2$,

$$\frac{dy}{dx} = \frac{-16(-2)}{\left((-2)^2 + 4\right)^2} = \frac{32}{64} = \frac{1}{2}, \text{ so the slope}$$

of the tangent line at $(-2,1)$ is $\frac{1}{2}$. Using

the point-slope equation, we have:

$$y - y_1 = m(x - x_1)$$

$$y - 1 = \frac{1}{2}\left(x - (-2)\right)$$

$$y - 1 = \frac{1}{2}x + 1$$

$$y = \frac{1}{2}x + 2$$

51. $y = x^2 + \dfrac{3}{x-1}$

$$\frac{dy}{dx} = 2x + \frac{(x-1)(0) - 3(1)}{(x-1)^2}$$

$$= 2x - \frac{3}{(x-1)^2}$$

a) When $x = 2$, $y = (2)^2 + \dfrac{3}{2-1} = 4 + 3 = 7$,

and $\dfrac{dy}{dx} = 2(2) - \dfrac{3}{(2-1)^2} = 4 - 3 = 1$.

Therefore, the slope of the tangent line at $(2,7)$ is 1.

Using the point-slope equation, we have:

$$y - y_1 = m(x - x_1)$$

$$y - 7 = 1(x - 2)$$

$$y - 7 = x - 2$$

$$y = x + 5$$

b) When $x = 3$, $y = (3)^2 + \dfrac{3}{3-1} = 9 + \dfrac{3}{2} = \dfrac{21}{2}$,

and $\dfrac{dy}{dx} = 2(3) - \dfrac{3}{(3-1)^2} = 6 - \dfrac{3}{4} = \dfrac{21}{4}$.

Therefore, the slope of the tangent line at

$\left(3, \dfrac{21}{2}\right)$ is $\dfrac{21}{4}$.

Using the point-slope equation, we have:

$$y - y_1 = m(x - x_1)$$

$$y - \frac{21}{2} = \frac{21}{4}(x - 3)$$

$$y - \frac{21}{2} = \frac{21}{4}x - \frac{63}{4}$$

$$y = \frac{21}{4}x - \frac{21}{4}$$

53. The average cost of producing x items is

$A_C(x) = \dfrac{C(x)}{x}$. Therefore,

$$A_C(x) = \frac{750 + 34x - 0.068x^2}{x}.$$

Next, we take the derivative using the Quotient Rule to find the rate at which average cost is changing.

$$A_C'(x) = \frac{d}{dx}\left(\frac{750 + 34x - 0.068x^2}{x}\right)$$

$$A_C'(x) = \frac{x(34 - 0.136x) - \left(750 + 34x - 0.068x^2\right)(1)}{(x)^2}$$

$$= \frac{34x - 0.136x^2 - \left(750 + 34x - 0.068x^2\right)}{x^2}$$

$$= \frac{-0.068x^2 - 750}{x^2}$$

Substituting 175 for x, we have

$$A_C'(175) = \frac{-0.068(175)^2 - 750}{(175)^2}$$

$$= \frac{-2082.5 - 750}{30,625}$$

$$\approx -0.09249$$

Therefore, when 175 belts have been produced, average cost is changing at a rate of -0.09249 dollars per belt.

55. The average revenue of producing x items is

$A_R(x) = \dfrac{R(x)}{x}$. Therefore,

$$A_R(x) = \frac{45x^{9/10}}{x} = \frac{45}{x^{1/10}}.$$

The solution is continued on the next page.

Next, we take the derivative of the function on the previous page using the Quotient Rule to find the rate at which average revenue is changing.

$$A_R'(x) = \frac{d}{dx}\left(\frac{45}{x^{1/10}}\right)$$

$$= \frac{x^{1/10}(0) - (45)\left(\frac{1}{10}x^{-9/10}\right)}{\left(x^{1/10}\right)^2}$$

$$= \frac{-\frac{45}{10}x^{-9/10}}{x^{2/10}}$$

$$= -\frac{9}{2x^{11/10}}$$

Substituting 175 for x, we have

$$A_R'(175) = -\frac{9}{2(175)^{11/10}}$$

$$= -\frac{9}{586.64023}$$

$$= -0.0153$$

Therefore, when 175 belts have been produced, average revenue is changing at a rate of -0.0153 dollars per belt.

57. $A_P(x) = \dfrac{P(x)}{x} = \dfrac{R(x) - C(x)}{x}$

From Exercises 53 and 55, we know that

$$A_P(x) = \frac{45x^{9/10} - \left(750 + 34x - 0.068x^2\right)}{x}$$

$$= \frac{0.068x^2 - 34x - 750 + 45x^{9/10}}{x}$$

Using the Quotient Rule to take the derivative, we have:

$$A_P'(x) = \frac{x\left(0.136x - 34 + 40.5x^{-1/10}\right)}{(x)^2} -$$

$$\frac{\left(0.068x^2 - 34x - 750 + 45x^{9/10}\right)(1)}{(x)^2}$$

$$= \frac{0.068x^2 + 750 - 4.5x^{9/10}}{x^2}$$

Substituting 175 for x, we have:

$$A_P'(175) = \frac{0.068(175)^2 + 750 - 4.5(175)^{9/10}}{(175)^2}$$

$$= \frac{2082.5 + 750 - 469.8365}{30,625}$$

$$= \frac{2362.6635}{30,625}$$

$$\approx 0.0772$$

When 175 belts have been produced and sold, the average profit is changing at a rate of 0.0772 dollars per belt.

Alternatively, we could have used the information in Exercises 53 and 55 to find the rate of change of average profit when 175 belts are produced and sold. Notice that

$$A_P'(x) = A_R'(x) - A_C'(x)$$

$$= -0.0153 - (-0.0925)$$

$$= 0.0772$$

59. The average profit of producing x items is

$$A_P(x) = \frac{R(x) - C(x)}{x}. \text{ Therefore,}$$

$$A_P(x) = \frac{65x^{0.9} - \left(4300 + 2.1x^{0.6}\right)}{x}$$

$$= \frac{65x^{0.9} - 2.1x^{0.6} - 4300}{x}.$$

Using the Quotient Rule to take the derivative, we have

$$A_P'(x)$$

$$= \frac{x\left(65\left(0.9x^{-0.1}\right) - 2.1\left(0.6x^{-0.4}\right)\right) - \left(65x^{0.9} - 2.1x^{0.6} - 4300\right)(1)}{(x)^2}$$

$$= \frac{58.5x^{0.9} - 1.26x^{0.6} - \left(65x^{0.9} - 2.1x^{0.6} - 4300\right)}{x^2}$$

$$= \frac{-6.5x^{0.9} + 0.84x^{0.6} + 4300}{x^2}$$

Substituting 50 for x, we have

$$A_P'(50) = \frac{-6.5(50)^{0.9} + 0.84(50)^{0.6} + 4300}{(50)^2}$$

$$= \frac{4089.00428745}{2500}$$

$$= 1.63560171$$

$$\approx 1.64$$

Therefore, when 50 vases have been produced and sold, the average profit is changing at rate of 1.64 dollars per vase.

61. $P(t) = 567 + t\left(36t^{0.6} - 104\right)$

a) Using the Product Rule and remembering that the derivative of a constant is 0, we have

$$P'(t) = 0 + t\left(36\left(0.6t^{-0.4}\right)\right) + \left(36t^{0.6} - 104\right)(1)$$

$$= 21.6t^{0.6} + 36t^{0.6} - 104$$

$$= 57.6t^{0.6} - 104$$

b) Substituting 45 for t, we have:

$$P'(45) = 57.6\left(45\right)^{0.6} - 104$$

$$= 565.39243291 - 104$$

$$= 461.39243291$$

c) ✎

63. $T(t) = \dfrac{4t}{t^2 + 1} + 98.6$

a) $T'(t) = \dfrac{\left(t^2 + 1\right)(4) - (4t)(2t)}{\left(t^2 + 1\right)^2} + 0$

$$= \dfrac{4t^2 + 4 - 8t^2}{\left(t^2 + 1\right)^2}$$

$$= \dfrac{-4t^2 + 4}{\left(t^2 + 1\right)^2}$$

$$= \dfrac{-4\left(t^2 - 1\right)}{\left(t^2 + 1\right)^2}$$

b) $T(2) = \dfrac{4(2)}{(2)^2 + 1} + 98.6 = \dfrac{8}{5} + 98.6 = 100.2$

After 2 hours, Gloria's temperature is approximately 100.2 degrees Fahrenheit.

c) $T'(2) = \dfrac{-4\left((2)^2 - 1\right)}{\left((2)^2 + 1\right)^2} = \dfrac{-12}{25} = -0.48$

After 2 hours, Gloria's temperature is changing at rate of -0.48 degrees per hour.

65. $y(t) = 5t(t - 1)(2t + 3)$

First, group the factors of $y(t)$ in order to apply the product rule.

$$y(t) = \left[5t(t - 1)\right] \cdot (2t + 3)$$

Notice that when we take the derivative of the first term, $\left[5t(t - 1)\right]$ we will have to apply the Product Rule again.

$$y'(t) = \left[5t(t - 1)\right](2) + (2t + 3)\underbrace{\left[(5t)(1) + (t - 1)(5)\right]}_{\text{Product Rule for } [5t(t-1)]}$$

$$= 10t(t - 1) + (2t + 3)[5t + 5t - 5]$$

$$= 10t(t - 1) + (2t + 3)(10t - 5)$$

The previous derivative can be simplified as follows:

$$y'(t) = 10t^2 - 10t + 20t^2 + 30t - 10t - 15$$

$$= 30t^2 + 10t - 15$$

67. $g(x) = \left(x^3 - 8\right) \cdot \dfrac{x^2 + 1}{x^2 - 1}$

We will begin by applying the Product Rule.

$$g'(x) = \left(x^3 - 8\right)\dfrac{d}{dx}\left(\dfrac{x^2 + 1}{x^2 - 1}\right) + \dfrac{x^2 + 1}{x^2 - 1} \cdot \dfrac{d}{dx}\left(x^3 - 8\right)$$

Notice, that we will have to apply the Quotient Rule to take the derivative of $\dfrac{x^2 + 1}{x^2 - 1}$.

$$g'(x) = \left(x^3 - 8\right)\dfrac{\left(x^2 - 1\right)(2x) - \left(x^2 + 1\right)(2x)}{\left(x^2 - 1\right)^2} +$$

$$\left(\dfrac{x^2 + 1}{x^2 - 1}\right) \cdot \left(3x^2\right)$$

$$= \left(x^3 - 8\right)\dfrac{-4x}{\left(x^2 - 1\right)^2} + \left(\dfrac{x^2 + 1}{x^2 - 1}\right) \cdot \left(3x^2\right)$$

The solution is continued on the next page.

The derivative on the previous page can be simplified as follows.

$$g'(x) = \frac{-4x(x^3-8)}{(x^2-1)^2} + \frac{3x^2(x^2+1)}{x^2-1}$$

$$= \frac{-4x^4+32x}{(x^2-1)^2} + \frac{3x^4+3x^2}{x^2-1} \cdot \frac{x^2-1}{x^2-1}$$

$$= \frac{-4x^4+32x+3x^6-3x^2}{(x^2-1)^2}$$

$$= \frac{3x^6-4x^4-3x^2+32x}{(x^2-1)^2}$$

69. $f(x) = \dfrac{(x-1)(x^2+x+1)}{x^4-3x^3-5}$

First we will group the numerator, to apply the Quotient Rule. Remember that we will have to apply the Product Rule when taking the derivative of the numerator.

$$f(x) = \frac{\left[(x-1)(x^2+x+1)\right]}{x^4-3x^3-5}$$

Calculating the derivative, we have:

$f'(x)$

$$= \frac{(x^4-3x^3-5)\left[(x-1)(2x+1)+(x^2+x+1)(1)\right]}{(x^4-3x^3-5)^2} -$$

$$\frac{\left[(x-1)(x^2+x+1)\right](4x^3-9x^2)}{(x^4-3x^3-5)^2}$$

$$= \frac{(x^4-3x^3-5)\left[2x^2-x-1+x^2+x+1\right]}{(x^4-3x^3-5)^2} -$$

$$\frac{\left[x^3-1\right](4x^3-9x^2)}{(x^4-3x^3-5)^2}$$

$$= \frac{(x^4-3x^3-5)\left[3x^2\right]-\left[x^3-1\right](4x^3-9x^2)}{(x^4-3x^3-5)^2}$$

$$= \frac{3x^6-9x^5-15x^2-4x^6+9x^5+4x^3-9x^2}{(x^4-3x^3-5)^2}$$

$$= \frac{-x^6+4x^3-24x^2}{(x^4-3x^3-5)^2}$$

71. $f(x) = \dfrac{x^2}{x^2-1}$ and $g(x) = \dfrac{1}{x^2-1}$

a) $f'(x) = \dfrac{(x^2-1)(2x)-x^2(2x)}{(x^2-1)^2} = \dfrac{-2x}{(x^2-1)^2}$

b) $g'(x) = \dfrac{(x^2-1)(0)-1(2x)}{(x^2-1)^2} = \dfrac{-2x}{(x^2-1)^2}$

c) ✎

73. ✎

75. a) Definition of the derivative.
b) Adding and subtracting the same quantity is the same as adding 0.
c) The limit of a sum is the sum of the limits.
d) Factoring common factors.
e) The limit of a product is the product of the limits and $\lim\limits_{h\to 0} f(x+h) = f(x)$.
f) Definition of the derivative.
g) Using Leibniz's notation.

77. The break-even point occurs when $P(x) = 0$.

$$P(x) = R(x) - C(x)$$

$$= 45x^{9/10} - (750+34x-0.068x^2)$$

$$= 0.068x^2 - 34x - 750 + 45x^{9/10}$$

Using the window:

```
WINDOW
 Xmin=0
 Xmax=500
 Xscl=100
 Ymin=-500
 Ymax=500
 Yscl=100
 Xres=1
```

We graph the profit function on the calculator.

The solution is continued on the next page.

The break-even point will be the zero of the function. Using the zero finder on the calculator we have:

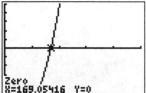

We see that the break-even point occurs at $x = 170$ belts.

The profit is changing at rate of

$$P(x) = 0.068x^2 - 34x - 750 + 45x^{\frac{9}{10}}$$

$$P'(x) = 0.136x - 34 + 40.5x^{-\frac{1}{10}}$$

Substituting 170 for x we have:

$$P'(170) = 0.136(170) - 34 + 40.5(170)^{-\frac{1}{10}}$$

$$= 13.353$$

$$\approx 13.35$$

Therefore, at the break-even point, profit is increasing at a rate of 13.35 dollars per belt. From Exercise 57 we know that:

$$A_P'(x) = \frac{0.068x^2 + 750 - 4.5x^{\frac{9}{10}}}{x^2}$$

Substituting 170 for x we get:

$$A_P'(170) = \frac{0.068(170)^2 + 750 - 4.5(170)^{\frac{9}{10}}}{(170)^2}$$

$$\approx 0.07811 \approx 0.078$$

At the break-even point, average profit is changing at a rate of 0.078 dollars per belt.

79. $f(x) = \left(x + \dfrac{2}{x}\right)(x^2 - 3)$

Using the calculator, we graph the function and the derivative in the same window. We can use the nDeriv feature to graph the derivative without actually calculating the derivative.

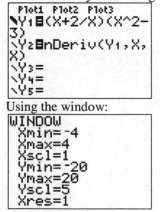

Using the window:

```
WINDOW
 Xmin=-4
 Xmax=4
 Xscl=1
 Ymin=-20
 Ymax=20
 Yscl=5
 Xres=1
```

We get the graph:

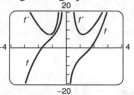

Note, the function $f(x)$ is the thicker graph.

The horizontal tangents occur at the turning points of this function, or at the x-intercepts of the derivative. We can see that the derivative never intersects the x-axis, therefore, there are no points at which the tangent line is horizontal.

81. $f(x) = \dfrac{0.01x^2}{x^4 + 0.0256}$

Using the calculator, we graph the function and the derivative in the same window. We can use the nDeriv feature to graph the derivative without actually calculating the derivative.

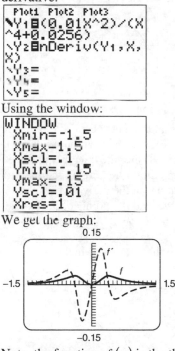

Using the window:

```
WINDOW
 Xmin=-1.5
 Xmax=1.5
 Xscl=.1
 Ymin=-.15
 Ymax=.15
 Yscl=.01
 Xres=1
```

We get the graph:

Note, the function $f(x)$ is the thicker graph.

The horizontal tangents occur at the turning points of this function, or at the x-intercepts of the derivative. Using the trace feature, the minimum/maximum feature on the function, or the zero feature on the derivative on the calculator, we find the points of horizontal tangency.

We estimate the points at which the tangent lines are horizontal are

$(-0.4, 0.03125)$, $(0,0)$, and $(0.4, 0.03125)$.

83. $f(x) = \dfrac{4x}{x^2 + 1}$

Using the calculator, we graph the function and the derivative in the same window. We can use the nDeriv feature to graph the derivative without actually calculating the derivative.

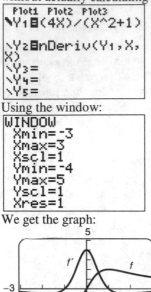

Using the window:

```
WINDOW
 Xmin=-3
 Xmax=3
 Xscl=1
 Ymin=-4
 Ymax=5
 Yscl=1
 Xres=1
```

We get the graph:

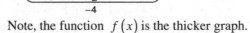

Note, the function $f(x)$ is the thicker graph.

The horizontal tangents occur at the turning points of this function, or at the *x*-intercepts of the derivative. Using the trace feature, the minimum/maximum feature on the function, or the zero feature on the derivative on the calculator, we find the points of horizontal tangency.

We estimate the points at which the tangent lines are horizontal are $(-1, -2)$ and $(1, 2)$.

Exercise Set 1.7

1. $y = (3-2x)^2$

Using the Extended Power Rule:

$\dfrac{dy}{dx} = \dfrac{d}{dx}\left[(3-2x)^2\right]$

$\quad = 2(3-2x)^{2-1} \cdot \dfrac{d}{dx}(3-2x)$

$\quad = 2(3-2x)(-2)$

$\quad = 8x - 12$

Simplifying the function first, we have:

$y = (3-2x)^2$

$\quad = (3-2x)(3-2x)$

$\quad = 4x^2 - 12x + 9$

We take the derivative using the Power Rule.

$\dfrac{dy}{dx} = \dfrac{d}{dx}\left(4x^2 - 12x + 9\right)$

$\quad = \dfrac{d}{dx}\left(4x^2\right) - \dfrac{d}{dx}(12x) + \dfrac{d}{dx}(9)$

$\quad = 8x - 12$

The results are the same.

3. $y = (7-x)^{55}$

Using the Extended Power Rule:

$\dfrac{dy}{dx} = \dfrac{d}{dx}\left[(7-x)^{55}\right]$

$\quad = 55(7-x)^{55-1} \cdot \dfrac{d}{dx}(7-x)$

$\dfrac{dy}{dx} = 55(7-x)^{54}(-1)$

$\quad = -55(7-x)^{54}$

5. $y = \sqrt{1-x} = (1-x)^{1/2}$

Using the Extended Power Rule

$\dfrac{dy}{dx} = \dfrac{d}{dx}(1-x)^{1/2}$

$\quad = \dfrac{1}{2}(1-x)^{1/2-1} \cdot \dfrac{d}{dx}(1-x)$

$\quad = \dfrac{1}{2}(1-x)^{-1/2} \cdot (-1)$

$\quad = \dfrac{-1}{2(1-x)^{1/2}}$

$\quad = \dfrac{-1}{2\sqrt{1-x}}$

7. $y = \sqrt{3x^2-4} = \left(3x^2-4\right)^{1/2}$

Using the Extended Power Rule

$\dfrac{dy}{dx} = \dfrac{d}{dx}\left[\left(3x^2-4\right)^{1/2}\right]$

$\quad = \dfrac{1}{2}\left(3x^2-4\right)^{-1/2} \cdot \dfrac{d}{dx}\left(3x^2-4\right)$

$\quad = \dfrac{1}{2\left(3x^2-4\right)^{1/2}} \cdot (6x)$

$\quad = \dfrac{3x}{\sqrt{3x^2-4}}$

9. $y = \left(4x^2+1\right)^{-50}$

Using the Extended Power Rule

$\dfrac{dy}{dx} = \dfrac{d}{dx}\left[\left(4x^2+1\right)^{-50}\right]$

$\quad = -50\left(4x^2+1\right)^{-50-1} \cdot \dfrac{d}{dx}\left(4x^2+1\right)$

$\quad = -50\left(4x^2+1\right)^{-51} \cdot (8x)$

$\quad = -400x\left(4x^2+1\right)^{-51}$

$\quad = \dfrac{-400x}{\left(4x^2+1\right)^{51}}$

11. $y = (x-4)^8 (2x+3)^6$

Using the Product Rule, we have

$\dfrac{dy}{dx} = \dfrac{d}{dx}\left[(x-4)^8 (2x+3)^6\right]$

$\quad = (x-4)^8 \dfrac{d}{dx}(2x+3)^6 + (2x+3)^6 \dfrac{d}{dx}(x-4)^8$

Next, we will apply the Extended Power Rule.

$\dfrac{dy}{dx} = (x-4)^8\left[6(2x+3)^{6-1} \cdot \dfrac{d}{dx}(2x+3)\right] +$

$\qquad (2x+3)^6\left[8(x-4)^{8-1} \cdot \dfrac{d}{dx}(x-4)\right]$

$\quad = (x-4)^8\left[6(2x+3)^5 (2)\right] +$

$\qquad (2x+3)^6\left[8(x-4)^7 (1)\right]$

$\quad = 12(x-4)^8 (2x+3)^5 + 8(2x+3)^6 (x-4)^7$

The solution is continued on the next page.

Factoring out common factors, we have:

$$\frac{dy}{dx} = 4(x-4)^7 (2x+3)^5 \left[3(x-4)+2(2x+3)\right]$$

$$= 4(x-4)^7 (2x+3)^5 \left[3x-12+4x+6\right]$$

$$= 4(x-4)^7 (2x+3)^5 (7x-6)$$

13. $y = \dfrac{1}{(4x+5)^2} = (4x+5)^{-2}$

Using the Extended Power Rule

$$\frac{dy}{dx} = \frac{d}{dx}\left[(4x+5)^{-2}\right]$$

$$= -2(4x+5)^{-2-1} \cdot \frac{d}{dx}(4x+5)$$

$$= -2(4x+5)^{-3} \cdot (4)$$

$$= -8(4x+5)^{-3}$$

$$= \frac{-8}{(4x+5)^3}$$

15. $y = \dfrac{4x^2}{(7-5x)^3}$

First, we use the Quotient Rule.

$$\frac{dy}{dx} = \frac{d}{dx}\left[\frac{4x^2}{(7-5x)^3}\right]$$

$$= \frac{(7-5x)^3 \frac{d}{dx}(4x^2) - 4x^2 \frac{d}{dx}(7-5x)^3}{\left((7-5x)^3\right)^2}$$

Next, using the Extended Power Rule, we have:

$$\frac{dy}{dx} = \frac{(7-5x)^3 (8x) - 4x^2 \left[3(7-5x)^2 (-5)\right]}{(7-5x)^6}$$

$$= \frac{8x(7-5x)^3 + 60x^2 (7-5x)^2}{(7-5x)^6}$$

$$= \frac{(7-5x)^2 \left[8x(7-5x)+60x^2\right]}{(7-5x)^6} \quad \text{Factoring}$$

$$= \frac{56x - 40x^2 + 60x^2}{(7-5x)^4} \quad \begin{array}{l}\text{Dividing}\\\text{common factors}\end{array}$$

$$= \frac{20x^2 + 56x}{(7-5x)^4}$$

$$= \frac{4x(5x+14)}{(7-5x)^4}$$

17. $f(x) = (3+x^3)^5 - (1+x^7)^4$

Using the Difference Rule and then the Extended Power Rule we have:

$$f'(x) = \frac{d}{dx}\left[(3+x^3)^5 - (1+x^7)^4\right]$$

$$= \frac{d}{dx}(3+x^3)^5 - \frac{d}{dx}(1+x^7)^4$$

$$= 5(3+x^3)^{5-1}\left(\frac{d}{dx}(3+x^3)\right) -$$

$$\qquad 4(1+x^7)^{4-1}\left(\frac{d}{dx}(1+x^7)\right)$$

$$= 5(3+x^3)^4 (3x^2) - 4(1+x^7)^3 (7x^6)$$

$$= 15x^2 (3+x^3)^4 - 28x^6 (1+x^7)^3$$

19. $f(x) = x^2 + (200-x)^2$

Using the Sum Rule and the Extended Power Rule, we have:

$$f'(x) = \frac{d}{dx}\left[x^2 + (200-x)^2\right]$$

$$= \frac{d}{dx}(x^2) + \frac{d}{dx}(200-x)^2$$

$$= 2x + 2(200-x)^{2-1}\left[\frac{d}{dx}(200-x)\right]$$

$$= 2x + 2(200-x)(-1)$$

$$= 2x + 2x - 400$$

$$= 4x - 400$$

21. $G(x) = \sqrt[3]{2x-1} + (4-x)^2 = (2x-1)^{1/3} + (4-x)^2$

Using the Sum Rule, we have:

$$G'(x) = \frac{d}{dx}\left[(2x-1)^{1/3} + (4-x)^2\right]$$

$$= \frac{d}{dx}(2x-1)^{1/3} + \frac{d}{dx}(4-x)^2$$

Next we apply the Extended Power Rule on the next page to complete the derivative.

Using the Extended power Rule, we have:

$$G'(x) = \frac{1}{3}(2x-1)^{\frac{1}{3}-1}\left(\frac{d}{dx}(2x-1)\right) +$$
$$2(4-x)^{2-1}\left(\frac{d}{dx}(4-x)\right)$$

$$= \frac{1}{3}(2x-1)^{-\frac{2}{3}}(2) + 2(4-x)^{1}(-1)$$

$$= \frac{2}{3}(2x-1)^{-\frac{2}{3}} - 2(4-x)$$

$$= \frac{2}{3(2x-1)^{\frac{2}{3}}} - 8 + 2x$$

$$= \frac{2}{3\sqrt[3]{(2x-1)^{2}}} - 8 + 2x$$

23. $f(x) = -5x(2x-3)^{4}$

Using the Product Rule, we have

$$f'(x) = \frac{d}{dx}\left[-5x(2x-3)^{4}\right]$$

$$= -5x\frac{d}{dx}\left[(2x-3)^{4}\right] + (2x-3)^{4}\frac{d}{dx}(-5x)$$

Using the Extended Power Rule, we have

$$f'(x) = -5x\left[4(2x-3)^{3}\left(\frac{d}{dx}(2x-3)\right)\right] +$$
$$(2x-3)^{4}(-5)$$

$$= -5x\left[4(2x-3)^{3}(2)\right] + (2x-3)^{4}(-5)$$

$$= -40x(2x-3)^{3} - 5(2x-3)^{4}$$

$$= -5(2x-3)^{3}\left[8x + (2x-3)\right] \quad \text{Factoring}$$

$$= -5(2x-3)^{3}(10x-3)$$

25. $F(x) = (5x+2)^{4}(2x-3)^{8}$

Using the Product Rule, we have:

$$F'(x) = \frac{d}{dx}\left[(5x+2)^{4}(2x-3)^{8}\right]$$

$$= (5x+2)^{4}\frac{d}{dx}(2x-3)^{8} + (2x-3)^{8}\frac{d}{dx}(5x+2)^{4}$$

We complete the derivative by applying the Extended Power Rule at the top of the next column.

Using the Extended Power Rule, we have:

$$F'(x) = (5x+2)^{4}\left[8(2x-3)^{7}(2)\right] +$$
$$(2x-3)^{8}\left[4(5x+2)^{3}(5)\right]$$

$$= 16(5x+2)^{4}(2x-3)^{7} + 20(2x-3)^{8}(5x+2)^{3}$$

$$= 4(5x+2)^{3}(2x-3)^{7}\left[4(5x+2) + 5(2x-3)\right]$$

$$= 4(5x+2)^{3}(2x-3)^{7}(30x-7)$$

27. $f(x) = x^{2}\sqrt{4x-1} = x^{2}(4x-1)^{\frac{1}{2}}$

Using the Product Rule and the Extended Power Rule, we have

$$f'(x) = \frac{d}{dx}\left[x^{2}(4x-1)^{\frac{1}{2}}\right]$$

$$= x^{2}\left[\frac{1}{2}(4x-1)^{-\frac{1}{2}}(4)\right] + (4x-1)^{\frac{1}{2}}(2x)$$

$$f'(x) = \frac{2x^{2}}{(4x-1)^{\frac{1}{2}}} + 2x(4x-1)^{\frac{1}{2}}$$

$$= \frac{2x^{2}}{\sqrt{(4x-1)}} + 2x\sqrt{(4x-1)}$$

The derivative can be simplified as follows:

$$f'(x) = \frac{2x^{2}}{\sqrt{4x-1}} + \frac{2x\sqrt{4x-1}}{1} \cdot \frac{\sqrt{4x-1}}{\sqrt{4x-1}}$$

$$= \frac{2x^{2}}{\sqrt{4x-1}} + \frac{2x(4x-1)}{\sqrt{4x-1}}$$

$$= \frac{2x^{2}}{\sqrt{4x-1}} + \frac{8x^{2}-2x}{\sqrt{4x-1}}$$

$$= \frac{10x^{2}-2x}{\sqrt{4x-1}}$$

$$= \frac{2x(5x-1)}{\sqrt{4x-1}}$$

29. $F(x) = \sqrt[4]{x^{2}-5x+2} = (x^{2}-5x+2)^{\frac{1}{4}}$

Using the Extended Power Rule, we have

$$F'(x) = \frac{d}{dx}(x^{2}-5x+2)^{\frac{1}{4}}$$

$$= \frac{1}{4}(x^{2}-5x+2)^{\frac{1}{4}-1} \cdot \frac{d}{dx}(x^{2}-5x+2)$$

$$= \frac{1}{4}(x^{2}-5x+2)^{-\frac{3}{4}}(2x-5)$$

$$= \frac{2x-5}{4(x^{2}-5x+2)^{\frac{3}{4}}}$$

31. $f(x) = \left(\dfrac{3x-1}{5x+2}\right)^4$

Using the Extended Power Rule, we have

$$f'(x) = 4\left(\dfrac{3x-1}{5x+2}\right)^3 \dfrac{d}{dx}\left[\dfrac{3x-1}{5x+2}\right]$$

Using the Quotient Rule, we have

$$f'(x) = 4\left(\dfrac{3x-1}{5x+2}\right)^3\left[\dfrac{(5x+2)(3)-(3x-1)(5)}{(5x+2)^2}\right]$$

$$= 4\left(\dfrac{3x-1}{5x+2}\right)^3\left[\dfrac{15x+6-15x+5}{(5x+2)^2}\right]$$

$$= 4\left(\dfrac{3x-1}{5x+2}\right)^3\left[\dfrac{11}{(5x+2)^2}\right]$$

$$= \dfrac{44(3x-1)^3}{(5x+2)^5}$$

33. $g(x) = \sqrt{\dfrac{3+2x}{5-x}} = \left(\dfrac{3+2x}{5-x}\right)^{\frac{1}{2}}$

Using the Extended Power Rule, we have

$$g'(x) = \dfrac{d}{dx}\left(\left(\dfrac{3+2x}{5-x}\right)^{\frac{1}{2}}\right)$$

$$= \dfrac{1}{2}\left(\dfrac{3+2x}{5-x}\right)^{\frac{1}{2}-1}\cdot\dfrac{d}{dx}\left(\dfrac{3+2x}{5-x}\right)$$

Using the Quotient Rule, we have

$$= \dfrac{1}{2}\left(\dfrac{3+2x}{5-x}\right)^{-\frac{1}{2}}\left[\dfrac{(5-x)(2)-(3+2x)(-1)}{(5-x)^2}\right]$$

$$= \dfrac{1}{2}\left(\dfrac{5-x}{3+2x}\right)^{\frac{1}{2}}\left[\dfrac{10-2x+3+2x}{(5-x)^2}\right]$$

$$= \dfrac{1}{2}\left(\dfrac{5-x}{3+2x}\right)^{\frac{1}{2}}\left[\dfrac{13}{(5-x)^2}\right]$$

$$= \dfrac{13}{2(5-x)^{\frac{3}{2}}(3+2x)^{\frac{1}{2}}}$$

$$= \dfrac{13}{2\sqrt{(5-x)^3}\cdot\sqrt{(3+2x)}}$$

35. $f(x) = \left(2x^3-3x^2+4x+1\right)^{100}$

Using the Extended Power Rule, we have

$$f'(x) = 100\left(2x^3-3x^2+4x+1\right)^{99}\left(6x^2-6x+4\right)$$

$$= 200\left(2x^3-3x^2+4x+1\right)^{99}\left(3x^2-3x+2\right)$$

37. $h(x) = \left(\dfrac{1-3x}{2-7x}\right)^{-5} = \left(\dfrac{2-7x}{1-3x}\right)^5$

Using the Extended Power Rule, we have

$$h'(x) = \dfrac{d}{dx}\left[\left(\dfrac{2-7x}{1-3x}\right)^5\right]$$

$$= 5\left(\dfrac{2-7x}{1-3x}\right)^{5-1}\left[\dfrac{d}{dx}\left(\dfrac{2-7x}{1-3x}\right)\right]$$

Next, using the Quotient Rule, we have

$h'(x)$

$$= 5\left(\dfrac{2-7x}{1-3x}\right)^4\left(\dfrac{(1-3x)(-7)-(2-7x)(-3)}{(1-3x)^2}\right)$$

$$= 5\left(\dfrac{2-7x}{1-3x}\right)^4\left(\dfrac{-7+21x+6-21x}{(1-3x)^2}\right)$$

$$= 5\left(\dfrac{2-7x}{1-3x}\right)^4\left(\dfrac{-1}{(1-3x)^2}\right)$$

$$= \dfrac{-5(2-7x)^4}{(1-3x)^6}$$

39. $f(x) = \sqrt{\dfrac{x^2+x}{x^2-x}} = \left(\dfrac{x^2+x}{x^2-x}\right)^{\frac{1}{2}}$

Using the Extended Power Rule, we have

$$f'(x) = \dfrac{1}{2}\left(\dfrac{x^2+x}{x^2-x}\right)^{\frac{1}{2}-1}\cdot\dfrac{d}{dx}\left[\dfrac{x^2+x}{x^2-x}\right]$$

Using the Quotient Rule, we have

$f'(x)$

$$= \dfrac{1}{2}\left(\dfrac{x^2+x}{x^2-x}\right)^{-\frac{1}{2}}\left[\dfrac{(x^2-x)(2x+1)-(x^2+x)(2x-1)}{(x^2-x)^2}\right]$$

$$= \dfrac{1}{2}\left(\dfrac{x^2+x}{x^2-x}\right)^{-\frac{1}{2}}\left[\dfrac{2x^3-x^2-x-2x^3-x^2+x}{(x^2-x)^2}\right]$$

Continued at the top of the next page.

Simplifying the derivative from the previous page, we have:

$$f'(x) = \frac{1}{2}\left(\frac{x^2-x}{x^2+x}\right)^{\frac{1}{2}}\left(\frac{-2x^2}{\left(x^2-x\right)^2}\right)$$

$$= \frac{-x^2}{\left(x^2-x\right)^{\frac{3}{2}}\left(x^2+x\right)^{\frac{1}{2}}}$$

The previous derivative can be further simplified as follows:

$$f'(x) = \frac{-x^2}{\left(x^2-x\right)^{\frac{3}{2}}\left(x^2+x\right)^{\frac{1}{2}}}$$

$$= \frac{-x^2}{x^{\frac{3}{2}}\left(x-1\right)^{\frac{3}{2}}x^{\frac{1}{2}}\left(x+1\right)^{\frac{1}{2}}} \quad \text{Factoring}$$

$$= \frac{-x^2}{x^2\left(x-1\right)^{\frac{3}{2}}\left(x+1\right)^{\frac{1}{2}}}$$

$$= \frac{-1}{\left(x-1\right)^{\frac{3}{2}}\left(x+1\right)^{\frac{1}{2}}}$$

41. $f(x) = \dfrac{(5x-4)^7}{(6x+1)^3}$

Using the Quotient Rule and the Extended Power Rule, we have:

$$f'(x) = \frac{d}{dx}\left[\frac{(5x-4)^7}{(6x+1)^3}\right]$$

$$= \frac{(6x+1)^3\left[7(5x-4)^6(5)\right]-(5x-4)^7\left[3(6x+1)^2(6)\right]}{\left((6x+1)^3\right)^2}$$

Simplifying the expression, we have

$$f'(x) = \frac{35(5x-4)^6(6x+1)^3-18(5x-4)^7(6x+1)^2}{(6x+1)^6}$$

$$= \frac{(5x-4)^6(6x+1)^2\left[35(6x+1)-18(5x-4)\right]}{(6x+1)^6}$$

$$= \frac{(5x-4)^6\left[210x+35-90x+72\right]}{(6x+1)^4}$$

$$= \frac{(5x-4)^6\left[120x+107\right]}{(6x+1)^4}$$

Therefore,

$$f'(x) = \frac{(5x-4)^6(120x+107)}{(6x+1)^4}$$

43. $f(x) = 12(2x+1)^{\frac{2}{3}}(3x-4)^{\frac{5}{4}}$

Using the Product Rule and the Extended Power Rule, we have:

$$f'(x) = 12\frac{d}{dx}\left[(2x+1)^{\frac{2}{3}}(3x-4)^{\frac{5}{4}}\right]$$

$$= 12\left[(2x+1)^{\frac{2}{3}}\left[\frac{5}{4}(3x-4)^{\frac{1}{4}}(3)\right]\right]+$$

$$\qquad 12\left[(3x-4)^{\frac{5}{4}}\left[\frac{2}{3}(2x+1)^{-\frac{1}{3}}(2)\right]\right]$$

$$= 12\left[\frac{15}{4}(2x+1)^{\frac{2}{3}}(3x-4)^{\frac{1}{4}}\right]+$$

$$\qquad 12\left[\frac{4}{3}(3x-4)^{\frac{5}{4}}(2x+1)^{-\frac{1}{3}}\right]$$

$$= 45(2x+1)^{\frac{2}{3}}(3x-4)^{\frac{1}{4}}+\frac{16(3x-4)^{\frac{5}{4}}}{(2x+1)^{\frac{1}{3}}}$$

We simplify the derivative as follows:

$f'(x)$

$$= \frac{45(2x+1)^{\frac{2}{3}}(3x-4)^{\frac{1}{4}}}{1}\cdot\frac{(2x+1)^{\frac{1}{3}}}{(2x+1)^{\frac{1}{3}}}+\frac{16(3x-4)^{\frac{5}{4}}}{(2x+1)^{\frac{1}{3}}}$$

$$= \frac{45(2x+1)(3x-4)^{\frac{1}{4}}}{(2x+1)^{\frac{1}{3}}}+\frac{16(3x-4)^{\frac{5}{4}}}{(2x+1)^{\frac{1}{3}}}$$

$$= \frac{45(2x+1)(3x-4)^{\frac{1}{4}}+16(3x-4)^{\frac{5}{4}}}{(2x+1)^{\frac{1}{3}}}$$

$$= \frac{(3x-4)^{\frac{1}{4}}\left[45(2x+1)+16(3x-4)^{\frac{4}{4}}\right]}{(2x+1)^{\frac{1}{3}}}$$

$$= \frac{(3x-4)^{\frac{1}{4}}\left[45(2x+1)+16(3x-4)\right]}{(2x+1)^{\frac{1}{3}}}$$

$$= \frac{(3x-4)^{\frac{1}{4}}\left[90x+45+48x-64\right]}{(2x+1)^{\frac{1}{3}}}$$

$$= \frac{(3x-4)^{\frac{1}{4}}\left[138x-19\right]}{(2x+1)^{\frac{1}{3}}}$$

Therefore,

$$f'(x) = \frac{(3x-4)^{\frac{1}{4}}(138x-19)}{(2x+1)^{\frac{1}{3}}}$$

45. $y = \dfrac{15}{u^3} = 15u^{-3}$ and $u = 2x+1$

$$\frac{dy}{du} = 15\left(-3u^{-3-1}\right) = -45u^{-4} = \frac{-45}{u^4}$$

$$\frac{du}{dx} = 2$$

Applying the Chain Rule, we have:

$$\frac{dy}{dx} = \frac{dy}{du} \cdot \frac{du}{dx}$$

$$= \frac{-45}{u^4} \cdot 2$$

$$= \frac{-45}{(2x+1)^4} \cdot 2 \quad \text{Substituting } 2x+1 \text{ for } u$$

$$= \frac{-90}{(2x+1)^4} \quad \text{Simplifying}$$

47. $y = u^{50}$ and $u = 4x^3 - 2x^2$

$$\frac{dy}{du} = 50u^{50-1} = 50u^{49}$$

$$\frac{du}{dx} = 4\left(3x^{3-1}\right) - 2\left(2x^{2-1}\right) = 12x^2 - 4x$$

$$\frac{dy}{dx} = \frac{dy}{du} \cdot \frac{du}{dx}$$

$$= 50u^{49} \cdot \left(12x^2 - 4x\right)$$

$$\text{Substituting } 4x^3 - 2x^2 \text{ for } u$$

$$= 50\left(4x^3 - 2x^2\right)^{49} \cdot \left(12x^2 - 4x\right)$$

$$= 200x(3x-1)\left(4x^3 - 2x^2\right)^{49} \quad \text{Simplifying}$$

49. $y = (u+1)(u-1)$ and $u = x^3 + 1$

$$\frac{dy}{du} = (u+1)(1) + (u-1)(1) \quad \text{Product Rule}$$

$$= 2u$$

$$\frac{du}{dx} = 3x^2$$

$$\frac{dy}{dx} = \frac{dy}{du} \cdot \frac{du}{dx}$$

$$= (2u) \cdot \left(3x^2\right) \quad \text{Substituting } x^3 + 1 \text{ for } u$$

$$= \left(2\left(x^3 + 1\right)\right) \cdot \left(3x^2\right)$$

$$= 6x^2\left(x^3 + 1\right) \quad \text{Simplifying}$$

51. $y = 5u^2 + 3u$ where $u = x^3 + 1$

$$\frac{dy}{du} = 10u + 3$$

$$\frac{du}{dx} = 3x^2$$

Applying the Chain Rule, we have:

$$\frac{dy}{dx} = \frac{dy}{du} \cdot \frac{du}{dx}$$

$$= (10u + 3) \cdot \left(3x^2\right)$$

$$= \left(10\left(x^3 + 1\right) + 3\right) \cdot \left(3x^2\right) \quad \text{Substituting for } u$$

$$= 3x^2\left(\left(10x^3 + 10\right) + 3\right)$$

$$= 3x^2\left(10x^3 + 13\right)$$

53. $y = \sqrt{7 - 3u} = (7 - 3u)^{1/2}$ where $u = x^2 - 9$

$$\frac{dy}{du} = \frac{1}{2}(7 - 3u)^{-1/2}(-3) \quad \text{Extended Power Rule}$$

$$= \frac{-3}{2(7 - 3u)^{1/2}}$$

$$\frac{du}{dx} = 2x$$

We apply the Chain Rule.

$$\frac{dy}{dx} = \frac{dy}{du} \cdot \frac{du}{dx}$$

$$= \left(\frac{-3}{2(7 - 3u)^{1/2}}\right) \cdot (2x)$$

$$= \left(\frac{-3}{2\left(7 - 3\left(x^2 - 9\right)\right)^{1/2}}\right) \cdot (2x) \quad \text{Substituting}$$

$$= \frac{-3x}{\left(34 - 3x^2\right)^{1/2}}$$

$$= \frac{-3x}{\sqrt{34 - 3x^2}}$$

55. $y = \dfrac{1}{u^2+u}$ and $u = 5+3t$

$\dfrac{dy}{du} = \dfrac{\left(u^2+u\right)(0)-(1)(2u+1)}{\left(u^2+u\right)^2}$ Quotient Rule

$= \dfrac{-(2u+1)}{\left(u^2+u\right)^2}$

$\dfrac{du}{dt} = 3$

We apply the Chain Rule.

$\dfrac{dy}{dt} = \dfrac{dy}{du} \cdot \dfrac{du}{dt}$

$= \left(\dfrac{-(2u+1)}{\left(u^2+u\right)^2}\right)\cdot(3)$

$= \left(\dfrac{-(2(5+3t)+1)}{\left((5+3t)^2+(5+3t)\right)^2}\right)\cdot(3)$ Substituting

$= \dfrac{-3(10+6t+1)}{\left((5+3t)^2+(5+3t)\right)^2}$

$= \dfrac{-3(6t+11)}{(5+3t)^2\left((5+3t)+1\right)^2}$ Factoring

$= \dfrac{-3(6t+11)}{(5+3t)^2(6+3t)^2}$

57. $y = \left(x^3-4x\right)^{10}$

First, we find the derivative using the Extended Power Rule.

$\dfrac{dy}{dx} = 10\left(x^3-4x\right)^9\left(3x^2-4\right)$

When $x = 2$,

$\dfrac{dy}{dx} = 10\left((2)^3-4(2)\right)^9\left(3(2)^2-4\right)$

$= 10(0)^9(8)$

$= 0$

Thus, the slope of the tangent line at the point, $(2,0)$ is 0. The equation of the horizontal line passing through the point $(2,0)$ is $y = 0$.

59. $y = x\sqrt{2x+3} = x(2x+3)^{\frac{1}{2}}$

First, we find the derivative using the Product Rule and the Extended Power Rule.

$\dfrac{dy}{dx} = x\left(\dfrac{1}{2}(2x+3)^{\frac{1}{2}-1}(2)\right)+(2x+3)^{\frac{1}{2}}(1)$

$= \dfrac{x}{\sqrt{2x+3}}+\sqrt{2x+3}$

When $x = 3$,

$\dfrac{dy}{dx} = \dfrac{(3)}{\sqrt{2(3)+3}}+\sqrt{2(3)+3}$

$= \dfrac{3}{\sqrt{9}}+\sqrt{9}$

$= \dfrac{3}{3}+3$

$= 1+3 = 4$

Thus, the slope of the tangent line at $(3,9)$ is 4.

Using the point-slope equation, we find the equation of the tangent line.

$y-y_1 = m(x-x_1)$

$y-9 = 4(x-3)$

$y-9 = 4x-12$

$y = 4x-3$

61. $g(x) = \left(\dfrac{6x+1}{2x-5}\right)^2$

a) Using Extended Power Rule and the Quotient Rule, we have

$g'(x) = 2\left(\dfrac{6x+1}{2x-5}\right)^{2-1}\cdot\dfrac{d}{dx}\left(\dfrac{6x+1}{2x-5}\right)$

$= 2\left(\dfrac{6x+1}{2x-5}\right)\left[\dfrac{(2x-5)(6)-(6x+1)(2)}{(2x-5)^2}\right]$

$= \dfrac{2(6x+1)}{2x-5}\left[\dfrac{12x-30-12x-2}{(2x-5)^2}\right]$

$= \dfrac{2(6x+1)}{2x-5}\left[\dfrac{-32}{(2x-5)^2}\right]$

$= \dfrac{-64(6x+1)}{(2x-5)^3}$

b) Using the Quotient Rule on

$$g(x) = \frac{36x^2 + 12x + 1}{4x^2 - 20x + 25}, \text{ we have}$$

$$g'(x) = \frac{\left(4x^2 - 20x + 25\right)\left(72x + 12\right)}{\left(4x^2 - 20x + 25\right)^2} -$$

$$\frac{\left(36x^2 + 12x + 1\right)\left(8x - 20\right)}{\left(4x^2 - 20x + 25\right)^2}$$

$$= \frac{288x^3 - 1392x^2 + 1560x + 300}{\left(4x^2 - 20x + 25\right)^2} -$$

$$\frac{288x^3 - 624x^2 - 232x - 20}{\left(4x^2 - 20x + 25\right)^2}$$

$$= \frac{-768x^2 + 1792x + 320}{\left(4x^2 - 20x + 25\right)^2}$$

$$= \frac{-64(2x - 5)(6x + 1)}{(2x - 5)^4}$$

$$= \frac{-64(6x + 1)}{(2x - 5)^3}$$

c) ✎ The results are the same. Which method is easier depends on the student. We believe that the Extended Power rule offers us a more efficient approach. It takes too much time to expand the function, and then factor it back to binomials.

63. Using the Chain Rule:

$$f(u) = u^3, \; g(x) = u = 2x^4 + 1$$

First find $f'(u)$ and $g'(x)$.

$$f'(u) = 3u^2$$

$$f'(g(x)) = 3\left(2x^4 + 1\right)^2 \quad \text{Substituting } g(x) \text{ for } u$$

$$g'(x) = 8x^3$$

The Chain Rule states

$$(f \circ g)'(x) = f'(g(x)) \cdot g'(x)$$

Substituting, we have:

$$(f \circ g)'(x) = 3\left(2x^4 + 1\right)^2 \cdot \left(8x^3\right)$$

$$= 24x^3 \left(2x^4 + 1\right)^2$$

Therefore,

$$(f \circ g)'(-1) = 24(-1)^3 \left(2(-1)^4 + 1\right)^2$$

$$= -24(2 + 1)^2$$

$$= -24 \cdot 9 = -216$$

Alternatively, finding $f(g(x))$ first, we have:

$$f(g(x)) = f\left(2x^4 + 1\right) = \left(2x^4 + 1\right)^3$$

By the Extended Power Rule:

$$f'(g(x)) = 3\left(2x^4 + 1\right)^2 \left(8x^3\right)$$

$$= 24x^3 \left(2x^4 + 1\right)^2$$

Therefore, $f'(g(-1)) = -216$ as above.

65. Using the Chain Rule:

$$f(u) = \sqrt[3]{u} = u^{1/3}, \; g(x) = u = 1 + 3x^2$$

First find $f'(u)$ and $g'(x)$.

$$f'(u) = \frac{1}{3}u^{-2/3} = \frac{1}{3 \cdot \sqrt[3]{u^2}}$$

$$f'(g(x)) = \frac{1}{3 \cdot \sqrt[3]{\left(1 + 3x^2\right)^2}} \quad \begin{array}{l}\text{Substituting } g(x) \\ \text{for } u\end{array}$$

$$g'(x) = 6x$$

The Chain Rule states

$$(f \circ g)'(x) = f'(g(x)) \cdot g'(x)$$

Substituting, we have:

$$(f \circ g)'(x) = \frac{1}{3 \cdot \sqrt[3]{\left(1 + 3x^2\right)^2}} \cdot (6x)$$

$$= \frac{2x}{\sqrt[3]{\left(1 + 3x^2\right)^2}}$$

Therefore,

$$(f \circ g)'(2) = \frac{2(2)}{\sqrt[3]{\left(1 + 3(2)^2\right)^2}}$$

$$= \frac{4}{\sqrt[3]{(13)^2}}$$

$$\approx 0.72348760$$

The solution is continued on the next page.

Alternatively, finding $f(g(x))$ first, we have:

$$f(g(x)) = f(1+3x^2) = (1+3x^2)^{\frac{1}{3}}$$

By the Extended Power Rule:

$$f'(g(x)) = \frac{1}{3}(1+3x^2)^{-\frac{2}{3}}(6x)$$

$$= \frac{2x}{(1+3x^2)^{\frac{2}{3}}}$$

Therefore $(f \circ g)'(2) = \dfrac{4}{\sqrt[3]{(13)^2}} \approx 0.72348760$.

67. $f(x) = \left(2x^3 + (4x-5)^2\right)^6$

Letting $u = 2x^3 + (4x-5)^2$ and applying the Chain Rule, we have:

$$f'(x)$$

$$= 6\left(2x^3 + (4x-5)^2\right)^{6-1} \cdot \frac{d}{dx}\left(2x^3 + (4x-5)^2\right)$$

We will have to apply the chain rule again to find $\dfrac{d}{dx}\left(2x^3 + (4x-5)^2\right)$.

Applying the chain rule again, we have:

$$\frac{d}{dx}\left(2x^3 + (4x-5)^2\right)$$

$$= 2\frac{d}{dx}x^3 + \frac{d}{dx}(4x-5)^2$$

$$= 2(3x^2) + 2(4x-5)^{2-1} \cdot \frac{d}{dx}(4x-5)$$

$$= 6x^2 + 2(4x-5) \cdot 4$$

$$= 6x^2 + 32x - 40.$$

Therefore, the derivative is:

$$f'(x) = 6\left(2x^3 + (4x-5)^2\right)^5 \left(6x^2 + 32x - 40\right).$$

69. $f(x) = \sqrt{x^2 + \sqrt{1-3x}} = \left(x^2 + (1-3x)^{\frac{1}{2}}\right)^{\frac{1}{2}}$

Applying the Chain Rule, we have

$$f'(x)$$

$$= \frac{1}{2}\left(x^2 + (1-3x)^{\frac{1}{2}}\right)^{\frac{1}{2}-1} \cdot \frac{d}{dx}\left(x^2 + (1-3x)^{\frac{1}{2}}\right)$$

$$= \frac{1}{2}\left(x^2 + (1-3x)^{\frac{1}{2}}\right)^{-\frac{1}{2}} \cdot \frac{d}{dx}\left(x^2 + (1-3x)^{\frac{1}{2}}\right)$$

The solution is continued at the top of the next column.

Applying the chain rule again, we have

$$\frac{d}{dx}\left(x^2 + (1-3x)^{\frac{1}{2}}\right)$$

$$= \frac{d}{dx}x^2 + \frac{d}{dx}(1-3x)^{\frac{1}{2}}$$

$$= 2x^{2-1} + \frac{1}{2}(1-3x)^{\frac{1}{2}-1} \cdot \frac{d}{dx}(1-3x)$$

$$= 2x + \frac{1}{2\sqrt{1-3x}} \cdot (-3)$$

$$= 2x - \frac{3}{2\sqrt{1-3x}}.$$

Therefore, the derivative is:

$$f'(x)$$

$$\frac{1}{2}\left(x^2 + (1-3x)^{\frac{1}{2}}\right)^{-\frac{1}{2}}\left(2x - \frac{3}{2\sqrt{1-3x}}\right)$$

$$= \frac{1}{2\sqrt{x^2 + \sqrt{1-3x}}}\left(2x - \frac{3}{2\sqrt{1-3x}}\right).$$

71. $R(x) = 1000\sqrt{x^2 - 0.1x} = 1000\left(x^2 - 0.1x\right)^{\frac{1}{2}}$

Using the Extended Power Rule, we have

$$R'(x) = 1000\left[\frac{1}{2}\left(x^2 - 0.1x\right)^{-\frac{1}{2}}(2x - 0.1)\right]$$

$$= 500\left[\frac{2x - 0.1}{\left(x^2 - 0.1x\right)^{\frac{1}{2}}}\right]$$

$$= \frac{500(2x - 0.1)}{\sqrt{x^2 - 0.1x}}$$

Substituting 20 for x, we have

$$R'(20) = \frac{500(2(20) - 0.1)}{\sqrt{(20)^2 - 0.1(20)}}$$

$$= 1000.00314070$$

$$\approx 1000$$

When 20 planes have been sold, revenue is changing at a rate of 1,000 thousand of dollars per plane, or 1,000,000 dollars per plane.

73. $P(x) = R(x) - C(x)$ and
$$P'(x) = R'(x) - C'(x)$$

Since we are trying to find the rate at which total profit is changing as a function of x, we can use the derivatives found in Exercise 71 and 72 to find the derivative of the profit function. There is no need to find the Profit function first and then take the derivative.

$$P'(x) = R'(x) - C'(x)$$
$$= \frac{500(2x - 0.1)}{\sqrt{x^2 - 0.1x}} - \frac{4000x}{3(x^2 + 2)^{2/3}}$$

75. Let x be the number of years since 2008.
$$C(x) = 9.26x^4 - 85.27x^3 + 287.24x^2 -$$
$$309.12x + 2651.4$$

a) Taking the derivative with respect to x, we have
$$\frac{dC}{dx} = 9.26(4x^3) - 85.27(3x^2) +$$
$$287.24(2x) - 309.12$$
$$\frac{dC}{dx} = 37.04x^3 - 255.81x^2 + 574.48x - 309.12$$

b) ✎

77. $A = 1000\left(1 + \dfrac{r}{4}\right)^{20}$

a) Using the Extended Power Rule, we have
$$\frac{dA}{dr} = 1000(20)\left(1 + \frac{r}{4}\right)^{19}\left(\frac{1}{4}\right)$$
$$= 5000\left(1 + \frac{r}{4}\right)^{19}$$

b) ✎

79. $P(x) = 0.08x^2 + 80x$ and $x = 5t + 1$

a) Substituting $5t + 1$ for x, we have
$$P(t) = 0.08(5t + 1)^2 + 80(5t + 1)$$
$$= 0.08(25t^2 + 10t + 1) + 400t + 80$$
$$= 2t^2 + 400.8t + 80.08$$

b) First we find the derivative with respect to t.
$$P'(t) = \frac{dP}{dt}\left(2t^2 + 400.8t + 80.08\right)$$
$$= 4t + 400.8$$

Substituting 48 in for t, we have
$$P'(48) = 4(48) + 400.8 = 592.80.$$
After 48 months, profit is increasing at rate of 592.80 dollars per month.

81. $D = 0.85A(c + 25)$ and $c = (140 - y)\dfrac{w}{72x}$

a) Substituting 5 for A we have:
$$D(c) = 0.85(5)(c + 25)$$
$$= 4.25(c + 25)$$
$$= 4.25c + 106.25$$
Substituting 0.6 for x, 45 for y, we have:
$$c(w) = (140 - 45)\frac{w}{72(0.6)}$$
$$= 95\frac{w}{43.2}$$
$$= \frac{95w}{43.2} \approx 2.199w$$

b) $\dfrac{dD}{dc} = \dfrac{d}{dc}(4.25c + 106.25) = 4.25$
The dosage changes at a rate of 4.25 mg per unit of creatine clearance.

c) $\dfrac{dc}{dw} = \dfrac{d}{dw}(2.199w) = 2.199$
The creatine clearance changes at a rate of 2.199 unit of creatine clearance per kilogram.

d) By the Chain Rule:
$$\frac{dD}{dw} = \frac{dD}{dc} \cdot \frac{dc}{dw} = (4.25)(2.199) \approx 9.346$$
The dosage changes at a rate of 9.35 milligrams per kilogram.

e) ✎

83. $f(x) = x + \sqrt{x}$

Note that $f'(x) = 1 + \dfrac{1}{2\sqrt{x}}$ and
$$f'(f(x)) = 1 + \frac{1}{2\sqrt{x + \sqrt{x}}}.$$

The solution is continued on the next page.

Applying the Chain Rule to the iterated function, we have

$$\frac{d}{dx}\left[(f \circ f)(x)\right]$$

$$= \frac{d}{dx}\left[f\left(f(x)\right)\right]$$

$$= f'\left(f(x)\right)\cdot f'(x)$$

$$= \left[1 + \frac{1}{2\sqrt{x+\sqrt{x}}}\right]\cdot\left[1 + \frac{1}{2\sqrt{x}}\right]$$

$$= 1 + \frac{1}{2\sqrt{x}} + \left(\frac{1}{2\sqrt{x+\sqrt{x}}}\right)\left(1 + \frac{1}{2\sqrt{x}}\right)$$

85. $f(x) = \sqrt[3]{x} = x^{\frac{1}{3}}$

Note that $f'(x) = \frac{1}{3}x^{-\frac{2}{3}}$,

$f'\left(f(x)\right) = \frac{1}{3}x^{-\frac{2}{9}}$, and

$f'\left(f\left(f(x)\right)\right) = \frac{1}{3}x^{-\frac{2}{27}}$

Applying the Chain Rule to the iterated function, we have

$$\frac{d}{dx}\left[(f \circ f \circ f)(x)\right]$$

$$= \frac{d}{dx}\left[f\left(f\left(f(x)\right)\right)\right]$$

$$= f'\left(f\left(f(x)\right)\right)\cdot f'\left(f(x)\right)\cdot f'(x)$$

$$= \frac{1}{3}x^{-\frac{2}{27}}\cdot\frac{1}{3}\left(x^{-\frac{2}{9}}\right)\cdot\frac{1}{3}\left(x^{-\frac{2}{3}}\right)$$

$$= \frac{1}{27}x^{-\frac{2}{27}-\frac{2}{9}-\frac{2}{3}}$$

$$= \frac{1}{27}x^{-\frac{26}{27}}.$$

87. $y = \sqrt[3]{x^3+6x+1}\cdot x^5 = \left(x^3+6x+1\right)^{\frac{1}{3}}\cdot x^5$

Using the Product Rule and the Extended Power Rule, we have

$$\frac{dy}{dx} = \left(x^3+6x+1\right)^{\frac{1}{3}}\cdot\left(5x^4\right) +$$

$$x^5\left[\frac{1}{3}\left(x^3+6x+1\right)^{-\frac{2}{3}}\left(3x^2+6\right)\right]$$

$$= 5x^4\left(x^3+6x+1\right)^{\frac{1}{3}} + \frac{3x^5\left(x^2+2\right)}{3\left(x^3+6x+1\right)^{\frac{2}{3}}}$$

The derivative can be further simplified by finding a common denominator and combining the fractions.

$$\frac{dy}{dx} = \frac{\left(x^3+6x+1\right)^{\frac{2}{3}}}{\left(x^3+6x+1\right)^{\frac{2}{3}}}\cdot\frac{5x^4\left(x^3+6x+1\right)^{\frac{1}{3}}}{1} +$$

$$\frac{x^5\left(x^2+2\right)}{\left(x^3+6x+1\right)^{\frac{2}{3}}}$$

Therefore,

$$\frac{dy}{dx} = \frac{5x^4\left(x^3+6x+1\right)}{\left(x^3+6x+1\right)^{\frac{2}{3}}} + \frac{x^5\left(x^2+2\right)}{\left(x^3+6x+1\right)^{\frac{2}{3}}}$$

$$= \frac{5x^7+30x^5+5x^4+x^7+2x^5}{\left(x^3+6x+1\right)^{\frac{2}{3}}}$$

$$= \frac{6x^7+32x^5+5x^4}{\left(x^3+6x+1\right)^{\frac{2}{3}}}$$

89. $y = \left(x\sqrt{1+x^2}\right)^3 = \left(x\left(1+x^2\right)^{\frac{1}{2}}\right)^3$

Using the Extended Power Rule and the Product Rule, we find the derivative:

$$\frac{dy}{dx} = \frac{d}{dx}\left(x\sqrt{1+x^2}\right)^3$$

$$= 3\left(x\left(1+x^2\right)^{\frac{1}{2}}\right)^2\cdot$$

$$\left(x\left(\frac{1}{2}\left(1+x^2\right)^{-\frac{1}{2}}\cdot 2x\right) + \left(1+x^2\right)^{\frac{1}{2}}\cdot 1\right)$$

The derivative can be simplified as follows:

$$\frac{dy}{dx} = 3x^2\left(1+x^2\right)\cdot\left(\frac{x^2}{\left(1+x^2\right)^{\frac{1}{2}}} + \left(1+x^2\right)^{\frac{1}{2}}\right)$$

$$= 3x^2\left(1+x^2\right)\cdot\left(\frac{x^2}{\left(1+x^2\right)^{\frac{1}{2}}} + \frac{1+x^2}{\left(1+x^2\right)^{\frac{1}{2}}}\right)$$

$$= 3x^2\left(1+x^2\right)\cdot\left(\frac{1+2x^2}{\left(1+x^2\right)^{\frac{1}{2}}}\right)$$

$$= \left(3x^2+6x^4\right)\left(1+x^2\right)^{\frac{1}{2}}$$

$$= \left(3x^2+6x^4\right)\sqrt{1+x^2}$$

91.　$y = \left(\dfrac{x^2 - x - 1}{x^2 + 1} \right)^3$

Using the Extended Power Rule and the Quotient Rule, we have

$\dfrac{dy}{dx}$

$= 3\left(\dfrac{x^2 - x - 1}{x^2 + 1} \right)^2 \left[\dfrac{(x^2 + 1)(2x - 1) - (x^2 - x - 1)(2x)}{(x^2 + 1)^2} \right]$

$= 3\left(\dfrac{x^2 - x - 1}{x^2 + 1} \right)^2 \left[\dfrac{x^2 + 4x - 1}{(x^2 + 1)^2} \right]$

$= \dfrac{3(x^2 - x - 1)^2 (x^2 + 4x - 1)}{(x^2 + 1)^4}$

93.　$f(t) = \sqrt{3t + \sqrt{t}} = \left(3t + t^{1/2} \right)^{1/2}$

Using the Extended Power Rule, we have

$f'(t) = \dfrac{1}{2}\left(3t + t^{1/2} \right)^{-1/2}\left(3 + \dfrac{1}{2}t^{-1/2} \right)$

$= \dfrac{3 + \dfrac{1}{2\sqrt{t}}}{2\sqrt{3t + \sqrt{t}}}$

$= \dfrac{\dfrac{3}{1} \cdot \dfrac{2\sqrt{t}}{2\sqrt{t}} + \dfrac{1}{2\sqrt{t}}}{2\sqrt{3t + \sqrt{t}}}$

Simplifying, we have:

$f'(t) = \dfrac{\dfrac{6\sqrt{t} + 1}{2\sqrt{t}}}{2\sqrt{3t + \sqrt{t}}}$

$= \dfrac{6\sqrt{t} + 1}{4\sqrt{t}\sqrt{3t + \sqrt{t}}}$

95.　a) Applying the product rule, we have:

$\dfrac{d}{dx}\left[f(x) \right]^3$

$= \dfrac{d}{dx}\left[\left[f(x) \right]^2 \cdot \left[f(x) \right] \right]$

$= \left[f(x) \right]^2 \dfrac{d}{dx}\left(f(x) \right) + \dfrac{d}{dx}\left[f(x) \right]^2 f(x)$

$= \left[f(x) \right]^2 f'(x) + \dfrac{d}{dx}\left[f(x) \right]^2 f(x)$

We apply the product rule to $\dfrac{d}{dx}\left[f(x) \right]^2$

$\dfrac{d}{dx}\left[f(x) \right]^2 = f(x)f'(x) + f'(x)f(x)$

$= 2f(x)f'(x).$

Therefore, we now have:

$\dfrac{d}{dx}\left[f(x) \right]^3 = \left[f(x) \right]^2 f'(x) + 2\left[f(x) \right]^2 f'(x)$

$= 3\left[f(x) \right]^2 f'(x)$

b)　Applying the product rule we have:

$\dfrac{d}{dx}\left[f(x) \right]^4$

$= \dfrac{d}{dx}\left[\left[f(x) \right]^3 \cdot \left[f(x) \right] \right]$

$= \left[f(x) \right]^3 \dfrac{d}{dx}\left(f(x) \right) + \dfrac{d}{dx}\left[f(x) \right]^3 f(x)$

$= \left[f(x) \right]^3 f'(x) + \dfrac{d}{dx}\left[f(x) \right]^3 f(x).$

From part (a) we know that:

$\dfrac{d}{dx}\left[f(x) \right]^3 = 3\left[f(x) \right]^2 f'(x)$

Therefore, substituting into the derivative we have:

$\dfrac{d}{dx}\left[f(x) \right]^4$

$= \left[f(x) \right]^3 f'(x) + 3\left[f(x) \right]^2 f'(x)f(x)$

$= 4\left[f(x) \right]^3 f'(x)$

97.　$f(x) = 1.68x\sqrt{9.2 - x^2}$;　$[-3, 3]$

Using the calculator, we graph the function and the derivative in the same window. We can use the nDeriv feature to graph the derivative without actually calculating the derivative.

```
Plot1  Plot2  Plot3
\Y1⊟1.68X√(9.2-X
^2)
\Y2⊟nDeriv(Y1,X,
X)
\Y3■
\Y4=
\Y5=
```

Using the window:

```
WINDOW
 Xmin=-3
 Xmax=3
 Xscl=1
 Ymin=-8
 Ymax=8
 Yscl=1
 Xres=1
```

The solution is continued on the next page.

The graph is:

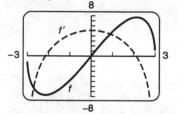

Note, the function $f(x)$ is the solid graph. The horizontal tangents occur at the turning points of this function, or at the x-intercepts of the derivative. Using the trace feature, the minimum/maximum feature on the function, or the zero feature on the derivative on the calculator, we find the points of horizontal tangency. We estimate the points at which the tangent lines are horizontal are $(-2.14476, -7.728)$ and $(2.14476, 7.728)$.

99. $f(x) = x\sqrt{4-x^2} = x(4-x^2)^{\frac{1}{2}}$

$$f'(x) = x\left[\frac{1}{2}(4-x^2)^{-\frac{1}{2}}(-2x)\right] + (4-x^2)^{\frac{1}{2}}(1)$$

$$= \frac{-x^2}{(4-x^2)^{\frac{1}{2}}} + (4-x^2)^{\frac{1}{2}}$$

$$= \frac{-x^2}{(4-x^2)^{\frac{1}{2}}} + \frac{(4-x^2)}{(4-x^2)^{\frac{1}{2}}}$$

$$= \frac{4-2x^2}{(4-x^2)^{\frac{1}{2}}}$$

$$= \frac{4-2x^2}{\sqrt{4-x^2}}$$

Using the window:

```
WINDOW
 Xmin=-2
 Xmax=2
 Xscl=1
 Ymin=-5
 Ymax=5
 Yscl=1
 Xres=1
```

The graph of the function and the derivative are shown below.

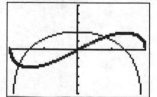

Note: The graph of the function is the thicker graph.

Exercise Set 1.8

1. $y = x^4 - 7$

$\dfrac{dy}{dx} = 4x^{4-1} = 4x^3$ First Derivative

$\dfrac{d^2 y}{dx^2} = 4\left(3x^{3-1}\right) = 12x^2$ Second Derivative

3. $y = 2x^4 - 5x$

$\dfrac{dy}{dx} = 2\left(4x^3\right) - 5 =$

$\quad = 8x^3 - 5$ First Derivative

$\dfrac{d^2 y}{dx^2} = 8\left(3x^2\right)$

$\quad = 24x^2$ Second Derivative

5. $y = 4x^2 - 5x + 7$

$\dfrac{dy}{dx} = 4\left(2x^{2-1}\right) - 5$

$\quad = 8x - 5$ First Derivative

$\dfrac{d^2 y}{dx^2} = 8$ Second Derivative

7. $y = 7x + 2$

$\dfrac{dy}{dx} = 7$ First Derivative

$\dfrac{d^2 y}{dx^2} = 0$ Second Derivative

9. $y = \dfrac{1}{x^3} = x^{-3}$

$\dfrac{dy}{dx} = -3x^{-3-1}$

$\quad = -3x^{-4}$

$\quad = \dfrac{-3}{x^4}$ First Derivative

$\dfrac{d^2 y}{dx^2} = -3\left(-4x^{-4-1}\right)$

$\quad = 12x^{-5}$

$\quad = \dfrac{12}{x^5}$ Second Derivative

11. $y = \sqrt{x} = x^{\frac{1}{2}}$

$\dfrac{dy}{dx} = \dfrac{1}{2} x^{\frac{1}{2}-1}$

$\quad = \dfrac{1}{2} x^{-\frac{1}{2}}$

$\quad = \dfrac{1}{2x^{\frac{1}{2}}} = \dfrac{1}{2\sqrt{x}}$ First Derivative

$\dfrac{d^2 y}{dx^2} = \dfrac{1}{2} \cdot \left(-\dfrac{1}{2} x^{-\frac{1}{2}-1}\right)$

$\quad = -\dfrac{1}{4} x^{-\frac{3}{2}}$

$\quad = -\dfrac{1}{4x^{\frac{3}{2}}} = -\dfrac{1}{4\sqrt{x^3}}$ Second Derivative

13. $f(x) = x^3 - \dfrac{5}{x} = x^3 - 5x^{-1}$

$f'(x) = 3x^{3-1} + 5\left(-1x^{-1-1}\right)$

$\quad = 3x^2 + 5x^{-2}$

$\quad = 3x^2 + \dfrac{5}{x^2}$ First Derivative

$f''(x) = 3\left(2x^{2-1}\right) + 5\left(-2x^{-2-1}\right)$

$\quad = 6x - 10x^{-3}$

$\quad = 6x - \dfrac{10}{x^3}$ Second Derivative

15. $f(x) = x^{\frac{1}{5}}$

$f'(x) = \dfrac{1}{5} x^{\frac{1}{5}-1}$

$\quad = \dfrac{1}{5} x^{-\frac{4}{5}}$

$\quad = \dfrac{1}{5x^{\frac{4}{5}}}$ First Derivative

$f''(x) = \dfrac{1}{5}\left(-\dfrac{4}{5}\right) x^{-\frac{4}{5}-1}$

$\quad = -\dfrac{4}{25} x^{-\frac{9}{5}}$

$\quad = -\dfrac{4}{25x^{\frac{9}{5}}}$ Second Derivative

17. $f(x) = 2x^{-2}$

$f'(x) = 2\left(-2x^{-2-1}\right)$

$\qquad = -4x^{-3}$

$\qquad = -\dfrac{4}{x^3}$ First Derivative

$f''(x) = -4\left(-3x^{-3-1}\right)$

$\qquad = 12x^{-4}$

$\qquad = \dfrac{12}{x^4}$ Second Derivative

19. $f(x) = \left(x^2 + 3x\right)^7$

$f'(x) = 7\left(x^2 + 3x\right)^{7-1}(2x+3)$ Theorem 7

$\qquad = 7(2x+3)\left(x^2+3x\right)^6$ First Derivative

$f''(x) = 7(2x+3)\left[6\left(x^2+3x\right)^{6-1}(2x+3)\right]+$

$\qquad\quad 7\left(x^2+6x\right)^6(2)$ Theorem 5

$\qquad = 42(2x+3)^2\left(x^2+3x\right)^5 + 14\left(x^2+3x\right)^6$

We can simplify the second derivative by factoring out common factors.

$f''(x)$

$= 14\left(x^2+3x\right)^5\left[3(2x+3)^2 + \left(x^2+3x\right)\right]$

$= 14\left(x^2+3x\right)^5\left[3\left(4x^2+12x+9\right) + \left(x^2+3x\right)\right]$

$= 14\left(x^2+3x\right)^5\left[12x^2+36x+27 + x^2+3x\right]$

$= 14\left(x^2+3x\right)^5\left(13x^2+39x+27\right)$ $\substack{\text{Second}\\\text{Derivative}}$

21. $f(x) = \left(3x^2+2x+1\right)^5$

$f'(x) = 5\left(3x^2+2x+1\right)^{5-1}(6x+2)$ Theorem 7

$\qquad = 5\left(3x^2+2x+1\right)^4(6x+2)$ $\substack{\text{First}\\\text{Derivative}}$

The second derivative is calculated at the top of the next column.

Taking the second derivative, we have:

$f''(x) = 5\left(3x^2+2x+1\right)^4(6) +$ Theorem 5

$\qquad\quad 5(6x+2)\cdot 4\left(3x^2+2x+1\right)^3(6x+2)$

$\qquad = 30\left(3x^2+2x+1\right)^4 +$

$\qquad\quad 20(6x+2)^2\left(3x^2+2x+1\right)^3$

$\qquad = 10\left(3x^2+2x+1\right)^3\left[3\left(3x^2+2x+1\right) + \right.$

$\qquad\qquad\qquad\qquad\qquad \left. 2\left(36x^2+24x+4\right)\right]$

$\qquad = 10\left(3x^2+2x+1\right)^3\left[9x^2+6x+3 + \right.$

$\qquad\qquad\qquad\qquad\qquad \left. 72x^2+48x+8\right]$

$\qquad = 10\left(3x^2+2x+1\right)^3\left(81x^2+54x+11\right)$

          Second Derivative

23. $f(x) = \sqrt[4]{\left(x^2+1\right)^3} = \left(x^2+1\right)^{3/4}$

$f'(x) = \dfrac{3}{4}\left(x^2+1\right)^{-1/4}(2x)$ Theorem 7

$\qquad = \dfrac{3}{2}x\left(x^2+1\right)^{-1/4}$ First Derivative

$f''(x) = \dfrac{3}{2}x\cdot\dfrac{-1}{4}\left(x^2+1\right)^{-5/4}(2x) +$

$\qquad\quad \dfrac{3}{2}\left(x^2+1\right)^{-1/4}(1)$ Theorem 5

$\qquad = -\dfrac{3}{4}x^2\left(x^2+1\right)^{-5/4} + \dfrac{3}{2}\left(x^2+1\right)^{-1/4}$

$\qquad = \dfrac{-3x^2}{4\left(x^2+1\right)^{5/4}} + \dfrac{3}{2\left(x^2+1\right)^{1/4}}$

We simplify the second derivative as follows:

$f''(x) = \dfrac{-3x^2}{4\left(x^2+1\right)^{5/4}} + \dfrac{3}{2\left(x^2+1\right)^{1/4}}\cdot\dfrac{2\left(x^2+1\right)}{2\left(x^2+1\right)}$

$\qquad = \dfrac{-3x^2}{4\left(x^2+1\right)^{5/4}} + \dfrac{6\left(x^2+1\right)}{4\left(x^2+1\right)^{5/4}}$

$\qquad = \dfrac{3x^2+6}{4\left(x^2+1\right)^{5/4}}$

$\qquad = \dfrac{3\left(x^2+2\right)}{4\left(x^2+1\right)^{5/4}}$

25.
$$y = x^{3/2} - 5x$$
$$y' = \frac{3}{2}x^{3/2-1} - 5$$
$$= \frac{3}{2}x^{1/2} - 5 \qquad \text{First Derivative}$$
$$y'' = \frac{3}{2}\left(\frac{1}{2}x^{1/2-1}\right)$$
$$= \frac{3}{4}x^{-1/2}$$
$$= \frac{3}{4\sqrt{x}} \qquad \text{Second Derivative}$$

27.
$$y = \left(x^3 - x\right)^{3/4}$$
$$y' = \frac{3}{4}\left(x^3 - x\right)^{3/4-1}\left(3x^2 - 1\right) \qquad \text{Theorem 7}$$
$$= \frac{3}{4}\left(x^3 - x\right)^{-1/4}\left(3x^2 - 1\right) \qquad \text{First Derivative}$$
$$y'' = \frac{3}{4}\left(x^3 - x\right)^{-1/4}(6x) + \qquad \text{Theorem 5}$$
$$\frac{3}{4}\left(3x^2 - 1\right) \cdot \frac{-1}{4}\left(x^3 - x\right)^{-1/4-1}\left(3x^2 - 1\right)$$
$$= \frac{9}{2}x\left(x^3 - x\right)^{-1/4} +$$
$$\frac{-3}{16}\left(3x^2 - 1\right)^2\left(x^3 - x\right)^{-5/4}$$
$$= \frac{9x}{2\left(x^3 - x\right)^{1/4}} - \frac{3\left(3x^2 - 1\right)^2}{16\left(x^3 - x\right)^{5/4}}$$

The second derivative can be simplified by finding a common denominator and combining the fractions.
$$y'' = \frac{9x}{2\left(x^3 - x\right)^{1/4}} \cdot \frac{8\left(x^3 - x\right)}{8\left(x^3 - x\right)} - \frac{3\left(3x^2 - 1\right)^2}{16\left(x^3 - x\right)^{5/4}}$$
$$= \frac{72x^4 - 72x^2}{16\left(x^3 - x\right)^{5/4}} - \frac{3\left(9x^4 - 6x^2 + 1\right)}{16\left(x^3 - x\right)^{5/4}}$$
$$= \frac{72x^4 - 72x^2 - 27x^4 + 18x^2 - 3}{16\left(x^3 - x\right)^{5/4}}$$
$$= \frac{45x^4 - 54x^2 - 3}{16\left(x^3 - x\right)^{5/4}} \qquad \text{Second Derivative}$$

29.
$$y = 3x^{4/3} - x^{1/2}$$
$$y' = 3 \cdot \frac{4}{3}x^{4/3-1} - \frac{1}{2}x^{1/2-1}$$
$$= 4x^{1/3} - \frac{1}{2}x^{-1/2} \qquad \text{First Derivative}$$
$$y'' = 4 \cdot \frac{1}{3}x^{1/3-1} - \frac{1}{2} \cdot \frac{-1}{2}x^{-1/2-1}$$
$$= \frac{4}{3}x^{-2/3} + \frac{1}{4}x^{-3/2}$$
$$= \frac{4}{3x^{2/3}} + \frac{1}{4x^{3/2}} \qquad \text{Second Derivative}$$

31.
$$y = \frac{2}{x^3} + \frac{1}{x^2} = 2x^{-3} + x^{-2}$$
$$y' = 2\left(-3x^{-3-1}\right) + \left(-2x^{-2-1}\right)$$
$$= -6x^{-4} - 2x^{-3} \qquad \text{First Derivative}$$
$$y'' = -6\left(-4x^{-4-1}\right) - 2\left(-3x^{-3-1}\right)$$
$$= 24x^{-5} + 6x^{-4}$$
$$= \frac{24}{x^5} + \frac{6}{x^4} \qquad \text{Second Derivative}$$

33.
$$y = \left(x^3 - 2\right)(5x + 1)$$
$$y' = \left(x^3 - 2\right)(5) + (5x + 1)\left(3x^2\right) \qquad \text{Theorem 5}$$
$$= 5x^3 - 10 + 15x^3 + 3x^2$$
$$= 20x^3 + 3x^2 - 10 \qquad \text{First Derivative}$$
$$y'' = 20\left(3x^{3-1}\right) + 3\left(2x^{2-1}\right)$$
$$= 60x^2 + 6x \qquad \text{Second Derivative}$$

35.
$$y = \frac{3x + 1}{2x - 3}$$
$$y' = \frac{(2x - 3)(3) - (3x + 1)(2)}{(2x - 3)^2} \qquad \text{Theorem 6}$$
$$= \frac{6x - 9 - 6x - 2}{(2x - 3)^2}$$
$$= \frac{-11}{(2x - 3)^2} \qquad \text{First Derivative}$$
The solution is continued on the next page.

Differentiating the derivative on the previous page, we find the second derivative.

$$y'' = \frac{(2x-3)^2(0)-(-11)\left(2(2x-3)^{2-1}(2)\right)}{\left((2x-3)^2\right)^2}$$

Theorem 6 and Theorem 7

$$= \frac{44(2x-3)}{(2x-3)^4}$$

$$= \frac{44}{(2x-3)^3} \qquad \text{Second Derivative}$$

37. $y = x^5$

$$\frac{dy}{dx} = 5x^{5-1} = 5x^4 \qquad \text{First Derivative}$$

$$\frac{d^2y}{dx^2} = 5\left(4x^{4-1}\right) = 20x^3 \qquad \text{Second Derivative}$$

$$\frac{d^3y}{dx^3} = 20\left(3x^{3-1}\right) = 60x^2 \qquad \text{Third Derivative}$$

$$\frac{d^4y}{dx^4} = 60\left(2x^{2-1}\right) = 120x \qquad \text{Fourth Derivative}$$

39. $y = x^6 - x^3 + 2x$

$$\frac{dy}{dx} = 6x^{6-1} - 3x^{3-1} + 2$$

$$= 6x^5 - 3x^2 + 2 \qquad \text{First Derivative}$$

$$\frac{d^2y}{dx^2} = 6\left(5x^{5-1}\right) - 3\left(2x^{2-1}\right)$$

$$= 30x^4 - 6x \qquad \text{Second Derivative}$$

$$\frac{d^3y}{dx^3} = 30\left(4x^{4-1}\right) - 6$$

$$= 120x^3 - 6 \qquad \text{Third Derivative}$$

$$\frac{d^4y}{dx^4} = 120\left(3x^{3-1}\right)$$

$$= 360x^2 \qquad \text{Fourth Derivative}$$

$$\frac{d^5y}{dx^5} = 360\left(2x^{2-1}\right)$$

$$= 720x \qquad \text{Fifth Derivative}$$

41. $f(x) = x^{-3} + 2x^{\frac{1}{3}}$

$$f'(x) = -3x^{-3-1} + 2\left(\frac{1}{3}x^{\frac{1}{3}-1}\right)$$

$$= -3x^{-4} + \frac{2}{3}x^{-\frac{2}{3}} \qquad \text{First Derivative}$$

$$f''(x) = -3\left(-4x^{-4-1}\right) + \frac{2}{3}\cdot\frac{-2}{3}x^{-\frac{2}{3}-1}$$

$$= 12x^{-5} - \frac{4}{9}x^{-\frac{5}{3}} \qquad \text{Second Derivative}$$

$$f'''(x) = 12\left(-5x^{-5-1}\right) - \frac{4}{9}\cdot\frac{-5}{3}x^{-\frac{5}{3}-1}$$

$$= -60x^{-6} + \frac{20}{27}x^{-\frac{8}{3}} \qquad \text{Third Derivative}$$

$$f^{(4)}(x) = -60\left(-6x^{-6-1}\right) + \frac{20}{27}\cdot\frac{-8}{3}x^{-\frac{8}{3}-1}$$

$$= 360x^{-7} - \frac{160}{81}x^{-\frac{11}{3}} \qquad \text{Fourth Derivative}$$

$$f^{(5)}(x) = 360\left(-7x^{-7-1}\right) - \frac{160}{81}\cdot\frac{-11}{3}x^{-\frac{11}{3}-1}$$

$$= -2520x^{-8} + \frac{1760}{243}x^{-\frac{14}{3}} \qquad \text{Fifth Derivative}$$

43. $g(x) = x^4 - 3x^3 - 7x^2 - 6x + 9$

$$g'(x) = 4x^{4-1} - 3\left(3x^{3-1}\right) - 7\left(2x^{2-1}\right) - 6$$

$$= 4x^3 - 9x^2 - 14x - 6 \qquad \text{First Derivative}$$

$$g''(x) = 4\left(3x^{3-1}\right) - 9\left(2x^{2-1}\right) - 14$$

$$= 12x^2 - 18x - 14 \qquad \text{Second Derivative}$$

$$g'''(x) = 12\left(2x^{2-1}\right) - 18$$

$$= 24x - 18 \qquad \text{Third Derivative}$$

$$g^{(4)}(x) = 24 \qquad \text{Fourth Derivative}$$

$$g^{(5)}(x) = 0 \qquad \text{Fifth Derivative}$$

$$g^{(6)}(x) = 0 \qquad \text{Sixth Derivative}$$

45. $s(t) = -10t^2 + 2t + 5$

a) $v(t) = s'(t) = -20t + 2$

b) $a(t) = v'(t) = s''(t) = -20$

c) When $t = 1$,
$$v(1) = -20(1) + 2 = -18 \frac{m}{\sec}$$
$$a(1) = -20 \frac{m}{\sec^2}$$
After 1 second, the velocity is -18 meters per second, and the acceleration is -20 meters per second squared.

47. $s(t) = 3t + 10$

a) $v(t) = s'(t) = 3$

b) $a(t) = v'(t) = s''(t) = 0$

c) When $t = 2$
$$v(2) = 3$$
$$a(2) = 0$$
After 2 hours, the velocity is 3 miles per hour, and the acceleration is 0 miles per hour squared.

d) ✎

49. $s(t) = 16t^2$

a) When $t = 3$, $s(3) = 16(3)^2 = 144$.
The hammer falls 144 feet in 3 seconds.

b) $v(t) = s'(t) = 32t$
When $t = 3$, $v(3) = 32(3) = 96$
the hammer is falling at 96 feet per second after 3 seconds.

c) $a(t) = v'(t) = s''(t) = 32$
When $t = 3$, $a(3) = 32$
the hammer is accelerating at 32 feet per second squared after 3 seconds.

51. $s(t) = 4.905t^2$

The velocity and acceleration are given by:
$$v(t) = s'(t) = 9.81t$$
$$a(t) = v'(t) = s''(t) = 9.81$$

After 2 seconds, we have
$$v(2) = 9.81(2) = 19.62$$
The stone is falling at 19.62 meters per second.
$$a(2) = 9.81$$
The stone is accelerating at 9.81 meters per second squared.

53. a) The bicyclist's velocity is the greatest at time $t = 0$. The tangent line at $t = 0$ has the greatest slope.

b) The bicyclist's acceleration is negative, since the slopes of the tangent line are decreasing with time.

55. a) $f'(t) = 0$ indicates that the rate of change is zero so the graph will be horizontal. The graph appears to be horizontal over the interval $7 < t < 11$ or $(7,11)$.

b) $f''(t) = 0$ indicates where the rate of change of the graph is changing at a constant rate. This appears to be the intervals $2 < t < 4$ or $(2,4)$; $7 < t < 11$ or $(7,11)$; $13 < t < 15$ or $(13,15)$.

c) $f''(t) > 0$ indicates that the rate of change of the graph is increasing. This occurs over the intervals $0 < t < 2$ or $(0,2)$; $11 < t < 13$ or $(11,13)$.

d) $f''(t) < 0$ indicates that the rate of change of the graph is decreasing. This occurs over the interval $4 < t < 7$ or $(4,7)$.

d) ✎

57. $S(t) = 2t^3 - 40t^2 + 220t + 160$

a) $S'(t) = 6t^2 - 80t + 220$
When $t = 1$,
$$S'(1) = 6(1)^2 - 80(1) + 220 = 146$$
After 1 month, sales are increasing at 146 thousand $(146,000)$ dollars per month.
When $t = 2$,
$$S'(2) = 6(2)^2 - 80(2) + 220 = 84$$
After 2 month, sales are increasing at 84 thousand $(84,000)$ dollars per month.
When $t = 4$,
$$S'(4) = 6(4)^2 - 80(4) + 220 = -4$$
After 4 months, sales are changing at a rate of -4 thousand (-4000) dollars per month.

b) $S''(t) = 12t - 80$
When $t = 1$, $S''(1) = 12(1) - 80 = -68$
After 1 month, the rate of change of sales are changing at a rate of -68 thousand $(-68,000)$ dollars per month squared.

The solution is continued on the next page.

When $t = 2$,
$$S''(2) = 12(2) - 80 = -56$$

After 2 months, the rate of change of sales are changing at a rate of -56 thousand $(-56,000)$ dollars per month squared.

When $t = 4$, $S''(4) = 12(4) - 80 = -32$

After 4 months, the rate of change of sales are changing at a rate of -32 thousand $(-32,000)$ dollars per month squared.

c) ✎

59. a) $p(t) = \dfrac{2000t}{4t + 75}$

First find the derivative of the population function. Using the quotient rule, we have:

$$p'(t)$$
$$= \frac{d}{dt}\left(\frac{2000t}{4t + 75} \right)$$
$$= \frac{(4t + 75)\frac{d}{dt}(2000t) - (2000t)\frac{d}{dt}(4t + 75)}{(4t + 75)^2}$$
$$= \frac{(4t + 75)(2000) - (2000t)(4)}{(4t + 75)^2}$$
$$= \frac{8000t + 150,000 - 8000t}{(4t + 75)^2}$$
$$- \frac{150,000}{(4t + 75)^2}$$

Next we substitute the appropriate values in for t.

$$p'(10) = \frac{150,000}{(4(10) + 75)^2}$$
$$= \frac{150,000}{13,225}$$
$$\approx 11.34 .$$

$$p'(50) = \frac{150,000}{(4(50) + 75)^2}$$
$$= \frac{150,000}{75,625}$$
$$\approx 1.98 .$$

$$p'(100) = \frac{150,000}{(4(100) + 75)^2}$$
$$= \frac{150,000}{225,625}$$
$$\approx 0.665 .$$

b) First find the second derivative.

$$p''(t) = \frac{d}{dt}\left(\frac{150,000}{(4t + 75)^2} \right)$$
$$= \frac{d}{dt}\left(150,000(4t + 75)^{-2} \right)$$
$$= -300,000(4t + 75)^{-3} \cdot 4$$
$$= -\frac{1,200,000}{(4t + 75)^3}$$

Next we substitute values for t:

$$p''(10) = -\frac{1,200,000}{(4(10) + 75)^3}$$
$$= -\frac{1,200,000}{1,520,875}$$
$$\approx -0.789 .$$

$$p''(50) = -\frac{1,200,000}{(4(50) + 75)^3}$$
$$= -\frac{1,200,000}{20,796,875}$$
$$\approx -0.0577 .$$

$$p''(100) = -\frac{1,200,000}{(4(100) + 75)^3}$$
$$= -\frac{1,200,000}{107,171,875}$$
$$\approx -0.0112 .$$

c) ✎

61. $y = \dfrac{1}{(1 - x)} = (1 - x)^{-1}$

$$y' = -1(1 - x)^{-1-1}(-1) = 1(1 - x)^{-2}$$
$$y'' = -2(1 - x)^{-2-1}(-1) = 2(1 - x)^{-3}$$
$$y''' = 2(-3)(1 - x)^{-3-1}(-1) = 6(1 - x)^{-4}$$

Therefore,

$$y''' = \frac{6}{(1 - x)^4}$$

63. $y = \dfrac{\sqrt{x}+1}{\sqrt{x}-1} = \dfrac{x^{1/2}+1}{x^{1/2}-1}$

$y' = \dfrac{\left(x^{1/2}-1\right)\left(\dfrac{1}{2}x^{-1/2}\right)-\left(x^{1/2}+1\right)\left(\dfrac{1}{2}x^{-1/2}\right)}{\left(x^{1/2}-1\right)^2}$

$= \dfrac{\dfrac{1}{2}-\dfrac{1}{2}x^{-1/2}-\dfrac{1}{2}-\dfrac{1}{2}x^{-1/2}}{\left(x^{1/2}-1\right)^2}$

$= \dfrac{-x^{-1/2}}{\left(x^{1/2}-1\right)^2}$

We find the second derivative.

$y'' = \dfrac{\left(x^{1/2}-1\right)^2\left(\dfrac{1}{2}x^{-3/2}\right)-\left(-x^{-1/2}\right)\left[2\left(x^{1/2}-1\right)\dfrac{1}{2}x^{-1/2}\right]}{\left(\left(x^{1/2}-1\right)^2\right)^2}$

$= \dfrac{\dfrac{1}{2}x^{-3/2}\left(x^{1/2}-1\right)^2+x^{-1}\left(x^{1/2}-1\right)}{\left(x^{1/2}-1\right)^4}$

$= \dfrac{\left(x^{1/2}-1\right)\left(\dfrac{1}{2}x^{-3/2}\left(x^{1/2}-1\right)+x^{-1}\right)}{\left(x^{1/2}-1\right)^4}$

$= \dfrac{\dfrac{1}{2}x^{-1}-\dfrac{1}{2}x^{-3/2}+x^{-1}}{\left(x^{1/2}-1\right)^3}$

$= \dfrac{\left(\dfrac{3}{2}x^{-1}-\dfrac{1}{2}x^{-3/2}\right)}{\left(x^{1/2}-1\right)^3}$

$= \dfrac{\dfrac{3}{2x}-\dfrac{1}{2x^{3/2}}}{\left(x^{1/2}-1\right)^3}$

$= \dfrac{\dfrac{3x^{1/2}}{2x^{3/2}}-\dfrac{1}{2x^{3/2}}}{\left(x^{1/2}-1\right)^3}$

$= \dfrac{3x^{1/2}-1}{2x^{3/2}\left(x^{1/2}-1\right)^3}$

65. $y = x^k$

$\dfrac{dy}{dx} = kx^{k-1}$

$\dfrac{d^2y}{dx^2} = k(k-1)x^{k-2}$

$\dfrac{d^3y}{dx^3} = k(k-1)(k-2)x^{k-3}$

$\dfrac{d^4y}{dx^4} = k(k-1)(k-2)(k-3)x^{k-4}$

$\dfrac{d^5y}{dx^5} = k(k-1)(k-2)(k-3)(k-4)x^{k-5}$

67. $f(x) = \dfrac{x-1}{x+2}$

$f'(x) = \dfrac{(x+2)(1)-(x-1)(1)}{(x+2)^2}$

$= \dfrac{3}{(x+2)^2}$

Notice, $f'(x) = \dfrac{3}{(x+2)^2} = 3(x+2)^{-2}$ so,

$f''(x) = 3(-2)(x+2)^{-2-1}\cdot(1)$

$= -6(x+2)^{-3}$

$= -\dfrac{6}{(x+2)^3}$

Notice, $f''(x) = -\dfrac{6}{(x+2)^3} = -6(x+2)^{-3}$ so,

$f'''(x) = -6(-3)(x+2)^{-3-1}(1)$

$= 18(x+2)^{-4}$

$= \dfrac{18}{(x+2)^4}$

Notice, $f'''(x) = \dfrac{18}{(x+2)^4} = 18(x+2)^{-4}$ so,

$f^{(4)}(x) = 18(-4)(x+2)^{-4-1}(1)$

$= -72(x+2)^{-5}$

$= -\dfrac{72}{(x+2)^5}$

69. Consider the function $s(t) = 4.905t^2$. We find the first and second derivatives.

$$s'(t) = 4.905\left(2t^{2-1}\right) = 9.81t$$

$$s''(t) = 9.81$$

The second derivative, $s''(t) = 9.81$, represents acceleration due to gravity on Earth which is 9.81 meters per second squared.

71. ✎

73. a) Graph I matches the description.
b) Graph IV matches the description.
c) Graph II matches the description.
d) Graph III matches the description.

75. We verify the expression yields an indeterminate form by substitution:

$$\lim_{x\to 5}\frac{x^2-25}{2x-10} = \frac{(5)^2-25}{2(5)-10}$$

$$= \frac{0}{0}.$$

This is an indeterminate form.
We now apply L'Hôpital's Rule to find the limit. By taking the derivative of the numerator and denominator separately we have.

$$\lim_{x\to 5}\frac{x^2-25}{2x-10} = \lim_{x\to 5}\left(\frac{\frac{d}{dx}\left(x^2-25\right)}{\frac{d}{dx}\left(2x-10\right)}\right)$$

$$= \lim_{x\to 5}\left(\frac{2x}{2}\right)$$

$$= \lim_{x\to 5}(x)$$

$$= 5. \qquad \text{Substitution}$$

77. We verify the expression yields an indeterminate form by substitution:

$$\lim_{x\to 1}\frac{x^3+2x-3}{x^2-1} = \frac{(1)^3+2(1)-3}{(1)^2-1}$$

$$= \frac{0}{0}.$$

This is an indeterminate form.

We now apply L'Hôpital's Rule to find the limit. By taking the derivative of the numerator and denominator separately we have:

$$\lim_{x\to 1}\frac{x^3+2x-3}{x^2-1} = \lim_{x\to 1}\left(\frac{\frac{d}{dx}\left(x^3+2x-3\right)}{\frac{d}{dx}\left(x^2-1\right)}\right)$$

$$= \lim_{x\to 1}\left(\frac{3x^2+2}{2x}\right)$$

$$= \frac{3(1)^2+2}{2(1)} \qquad \text{Substitution}$$

$$= \frac{5}{2}.$$

79. We verify the expression yields an indeterminate form by substitution:

$$\lim_{x\to -3}\frac{x^2-9}{x+3} = \frac{(-3)^2-9}{(-3)+3}$$

$$= \frac{0}{0}.$$

This is an indeterminate form.
We now apply L'Hôpital's Rule to find the limit. By taking the derivative of the numerator and denominator separately we have:

$$\lim_{x\to -3}\frac{x^2-9}{x+3} = \lim_{x\to -3}\left(\frac{\frac{d}{dx}\left(x^2-9\right)}{\frac{d}{dx}\left(x+3\right)}\right)$$

$$= \lim_{x\to -3}\left(\frac{2x}{1}\right)$$

$$= \lim_{x\to -3}(2x)$$

$$= 2(-3) \qquad \text{Substitution}$$

$$= -6.$$

81. We verify the expression yields an indeterminate form by substitution:

$$\lim_{x\to 2}\frac{x^3+5x-18}{2x^2-8} = \frac{(2)^3+5(2)-18}{2(2)^2-8}$$

$$= \frac{0}{0}.$$

This is an indeterminate form.

The solution is continued on the next page.

We now apply L'Hôpital's Rule to find the limit. By taking the derivative of the numerator and denominator separately we have:

$$\lim_{x \to 2} \frac{x^3 + 5x - 18}{2x^2 - 8} = \lim_{x \to 2}\left(\frac{\frac{d}{dx}\left(x^3 + 5x - 18\right)}{\frac{d}{dx}\left(2x^2 - 8\right)} \right)$$

$$= \lim_{x \to 2}\left(\frac{3x^2 + 5}{4x} \right)$$

$$= \frac{3(2)^2 + 5}{4(2)} \qquad \text{Substitution}$$

$$= \frac{17}{8}.$$

83. We verify the expression yields an indeterminate form by substitution:

$$\lim_{x \to \infty} \frac{4x^2 + x - 3}{2x^2 + 1} = \frac{\infty}{\infty}.$$

This is an indeterminate form.

We now apply L'Hôpital's Rule to find the limit. By taking the derivative of the numerator and denominator separately we have:

$$\lim_{x \to \infty} \frac{4x^2 + x - 3}{2x^2 + 1} = \lim_{x \to \infty}\left(\frac{\frac{d}{dx}\left(4x^2 + x - 3\right)}{\frac{d}{dx}\left(2x^2 + 1\right)} \right)$$

$$= \lim_{x \to \infty}\left(\frac{8x + 1}{4x} \right)$$

$$= \frac{\infty}{\infty} \qquad \text{Substitution}$$

This is still indeterminate so we apply L'Hôpital's Rule again.

$$\lim_{x \to \infty} \frac{4x^2 + x - 3}{2x^2 + 1} = \lim_{x \to \infty}\left(\frac{8x + 1}{4x} \right)$$

$$= \lim_{x \to \infty}\left(\frac{\frac{d}{dx}\left(8x + 1\right)}{\frac{d}{dx}\left(4x\right)} \right)$$

$$= \lim_{x \to \infty}\left(\frac{8}{4} \right)$$

$$= 2.$$

85. $s(t) = 0.1t^4 - t^2 + 0.4;$ $\qquad [-5, 5]$

From the graph we see that $v(t)$ switches at $t = -1.29$ and $t = 1.29$.

87. $s(t) = t^4 + t^3 - 4t^2 - 2t + 4;$ $\qquad [-3, 3]$

From the graph we see that $v(t)$ switches at $t = 0.604$ and $t = -1.104$.

Chapter 2

Applications of Differentiation

Exercise Set 2.1

1. $f(x) = x^2 + 6x - 3$

First, find the critical points.

$f'(x) = 2x + 6$

$f'(x)$ exists for all real numbers. We solve

$f'(x) = 0$

$2x + 6 = 0$

$2x = -6$

$x = -3$

The only critical value is -3. We use -3 to divide the real number line into two intervals,

A: $(-\infty, -3)$ and B: $(-3, \infty)$.

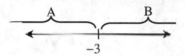

We use a test value in each interval to determine the sign of the derivative in each interval.

A: Test -4, $f'(-4) = 2(-4) + 6 = -2 < 0$

B: Test 0, $f'(0) = 2(0) + 6 = 6 > 0$

We see that $f(x)$ is decreasing on $(-\infty, -3)$ and increasing on $(-3, \infty)$, and the change from decreasing to increasing indicates that a relative minimum occurs at $x = -3$. We substitute into the original equation to find $f(-3)$:

$f(-3) = (-3)^2 + 6(-3) - 3 = -12$

Thus, there is a relative minimum at $(-3, -12)$.

We use the information obtained to sketch the graph. Other function values are listed below.

x	$f(x)$
-6	-3
-5	-8
-4	-11
-3	-12
-2	-11
-1	-8
0	-3

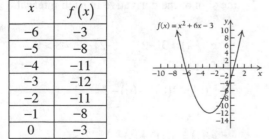

3. $f(x) = 2 - 3x - 2x^2$

First, find the critical points.

$f'(x) = -3 - 4x$

$f'(x)$ exists for all real numbers. We solve

$f'(x) = 0$

$-3 - 4x = 0$

$x = -\dfrac{3}{4}$

The only critical value is $-\dfrac{3}{4}$. We use $-\dfrac{3}{4}$ to divide the real number line into two intervals,

A: $\left(-\infty, -\dfrac{3}{4}\right)$ and B: $\left(-\dfrac{3}{4}, \infty\right)$.

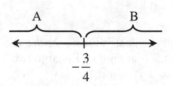

We use a test value in each interval to determine the sign of the derivative in each interval.

A: Test -1, $f'(-1) = -3 - 4(-1) = 1 > 0$

B: Test 0, $f'(0) = -3 - 4(0) = -3 < 0$

We see that $f(x)$ is increasing on $\left(-\infty, -\dfrac{3}{4}\right)$ and decreasing on $\left(-\dfrac{3}{4}, \infty\right)$, and the change from increasing to decreasing indicates that a relative maximum occurs at $x = -\dfrac{3}{4}$. We substitute into the original equation to find $f\left(-\dfrac{3}{4}\right)$:

$f\left(-\dfrac{3}{4}\right) = 2 - 3\left(-\dfrac{3}{4}\right) - 2\left(-\dfrac{3}{4}\right)^2 = \dfrac{25}{8}$

Thus, there is a relative maximum at $\left(-\dfrac{3}{4}, \dfrac{25}{8}\right)$.

The solution is continued on the next page.

We use the information obtained on the previous page to sketch the graph. Other function values are listed below.

x	$f(x)$
-3	-7
-2	0
-1	3
$-\frac{3}{4}$	$\frac{25}{8}$
0	2
1	-3
2	-12

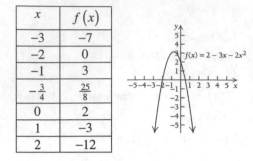

We use the information obtained to sketch the graph. Other function values are listed below.

x	$F(x)$
-5	$-\frac{17}{2}$
-4	-11
-3	$-\frac{25}{2}$
-2	-13
-1	$-\frac{25}{2}$
0	-11
1	$-\frac{17}{2}$

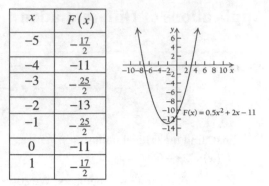

5. $F(x) = 0.5x^2 + 2x - 11$

First, find the critical points.

$F'(x) = x + 2$

$F'(x)$ exists for all real numbers. We solve:

$F'(x) = 0$

$x + 2 = 0$

$x = -2$

The only critical value is -2. We use -2 to divide the real number line into two intervals,

A: $(-\infty, -2)$ and B:$(-2, \infty)$.

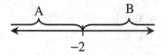

We use a test value in each interval to determine the sign of the derivative in each interval.

A: Test -3, $F'(-3) = (-3) + 2 = -1 < 0$

B: Test 0, $F'(0) = (0) + 2 = 2 > 0$

We see that $F(x)$ is decreasing on $(-\infty, -2)$ and increasing on $(-2, \infty)$, and the change from decreasing to increasing indicates that a relative minimum occurs at $x = -2$. We substitute into the original equation to find $F(-2)$:

$F(-2) = 0.5(-2)^2 + 2(-2) - 11 = -13$

Thus, there is a relative minimum at $(-2, -13)$.

7. $g(x) = x^3 + \dfrac{1}{2}x^2 - 2x + 5$

First, find the critical points.

$g'(x) = 3x^2 + x - 2$

$g'(x)$ exists for all real numbers. We solve

$$g'(x) = 0$$

$$3x^2 + x - 2 = 0$$

$$(3x - 2)(x + 1) = 0$$

$3x - 2 = 0$ $\qquad$ or $\qquad$ $x + 1 = 0$

$x = \dfrac{2}{3}$ $\qquad$ or $\qquad$ $x = -1$

The critical values are -1 and $\dfrac{2}{3}$. We use them to divide the real number line into three intervals,

A: $(-\infty, -1)$, B: $\left(-1, \dfrac{2}{3}\right)$, and C:$\left(\dfrac{2}{3}, \infty\right)$.

We use a test value in each interval to determine the sign of the derivative in each interval.

A: Test -2,

$g'(-2) = 3(-2)^2 + (-2) - 2 = 8 > 0$

B: Test 0,

$g'(0) = 3(0)^2 + (0) - 2 = -2 < 0$

C: Test 1,

$g'(1) = 3(1)^2 + (1) - 2 = 2 > 0$

The solution is continued on the next page.

We see that $g(x)$ is increasing on $(-\infty, -1)$,

decreasing on $\left(-1, \dfrac{2}{3}\right)$, and increasing on

$\left(\dfrac{2}{3}, \infty\right)$. So there is a relative maximum at

$x = -1$ and a relative minimum at $x = \dfrac{2}{3}$.

We find $g(-1)$:

$$g(-1) = (-1)^3 + \dfrac{1}{2}(-1)^2 - 2(-1) + 5$$

$$= -1 + \dfrac{1}{2} + 2 + 5 = \dfrac{13}{2}$$

Then we find $g\left(\dfrac{2}{3}\right)$:

$$g\left(\dfrac{2}{3}\right) = \left(\dfrac{2}{3}\right)^3 + \dfrac{1}{2}\left(\dfrac{2}{3}\right)^2 - 2\left(\dfrac{2}{3}\right) + 5$$

$$-\dfrac{8}{27} + \dfrac{2}{9} - \dfrac{4}{3} + 5 = \dfrac{113}{27}$$

There is a relative maximum at $\left(-1, \dfrac{13}{2}\right)$, and

there is a relative minimum at $\left(\dfrac{2}{3}, \dfrac{113}{27}\right)$.

We use the information obtained to sketch the graph. Other function values are listed

x	$g(x)$
-2	3
0	5
1	$\dfrac{9}{2}$
2	11

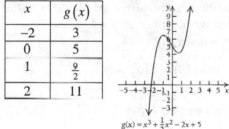

$g(x) = x^3 + \frac{1}{2}x^2 - 2x + 5$

9. $f(x) = x^3 - 3x^2$

First, find the critical points.

$f'(x) = 3x^2 - 6x$

$f'(x)$ exists for all real numbers. We solve

$$f'(x) = 0:$$

$$3x^2 - 6x = 0$$

$$3x(x - 2) = 0$$

$$x = 0 \quad \text{or} \quad x = 2$$

The critical values are 0 and 2.

We use them to divide the real number line into three intervals,

A: $(-\infty, 0)$, B: $(0, 2)$, and C: $(2, \infty)$.

![number line with intervals A, B, C divided at 0 and 2]

We use a test value in each interval to determine the sign of the derivative in each interval.

A: Test -1, $f'(-1) = 3(-1)^2 - 6(-1) = 9 > 0$

B: Test 1, $\quad f'(1) = 3(1)^2 - 6(1) = -3 < 0$

C: Test 3, $\quad f'(3) = 3(3)^2 - 6(3) = 9 > 0$

We see that $f(x)$ is increasing on $(-\infty, 0)$,

decreasing on $(0, 2)$, and increasing on $(2, \infty)$.

So there is a relative maximum at $x = 0$ and a relative minimum at $x = 2$.

We find $f(0)$:

$$f(0) = (0)^3 - 3(0)^2 - 0.$$

Then we find $f(2)$:

$$f(2) = (2)^3 - 3(2)^2 = -4.$$

There is a relative maximum at $(0, 0)$, and there is a relative minimum at $(2, -4)$. We use the information obtained to sketch the graph. Other function values are listed below.

x	$f(x)$
-2	-20
-1	-4
1	-2
3	0
4	16

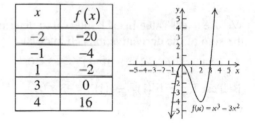

$f(x) = x^3 - 3x^2$

11. $f(x) = x^3 + 3x$

First, find the critical points.

$f'(x) = 3x^2 + 3$

$f'(x)$ exists for all real numbers. We solve

$$f'(x) = 0$$

$$3x^2 + 3 = 0$$

$$x^2 = -1$$

There are no real solutions to this equation. Therefore, the function does not have any critical values.

The solution is continued on the next page.

We test a point

$$f'(0) = 3(0)^2 + 3 = 3 > 0$$

We see that $f(x)$ is increasing on $(-\infty, \infty)$, and that there are no relative extrema. We use the information obtained to sketch the graph. Other function values are listed below.

x	$f(x)$
-2	-14
-1	-4
0	0
1	4
2	14

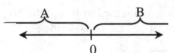

13. $F(x) = 1 - x^3$

First, find the critical points.

$$F'(x) = -3x^2$$

$F'(x)$ exists for all real numbers. We solve

$$F'(x) = 0$$

$$-3x^2 = 0$$

$$x = 0$$

The only critical value is 0. We use 0 to divide the real number line into two intervals, A: $(-\infty, 0)$, and B: $(0, \infty)$.

We use a test value in each interval to determine the sign of the derivative in each interval.

A: Test -1, $F'(-1) = -3(-1)^2 = -3 < 0$

B: Test 1, $F'(1) = -3(1)^2 = -3 < 0$

We see that $F(x)$ is decreasing on $(-\infty, 0)$ and decreasing on $(0, \infty)$, so the function has no relative extrema. We use the information obtained and determing other function values are listed below, we sketch the graph.

x	$F(x)$
-2	9
-1	2
0	1
1	0
2	-7

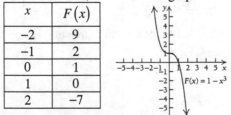

15. $G(x) = x^3 - 6x^2 + 10$

First, find the critical points.

$$G'(x) = 3x^2 - 12x$$

$G'(x)$ exists for all real numbers. We solve

$$G'(x) = 0$$

$$x^2 - 4x = 0 \qquad \text{Dividing by 3}$$

$$x(x-4) = 0$$

$$x = 0 \qquad \text{or} \qquad x - 4 = 0$$

$$x = 0 \qquad \text{or} \qquad x = 4$$

The critical values are 0 and 4. We use them to divide the real number line into three intervals, A: $(-\infty, 0)$, B: $(0, 4)$, and C: $(4, \infty)$.

We use a test value in each interval to determine the sign of the derivative in each interval.

A: Test -1, $G'(-1) = 3(-1)^2 - 12(-1) = 15 > 0$

B: Test 1, $G'(1) = 3(1)^2 - 12(1) = -9 < 0$

C: Test 5, $G'(5) = 3(5)^2 - 12(5) = 15 > 0$

We see that $G(x)$ is increasing on $(-\infty, 0)$, decreasing on $(0, 4)$, and increasing on $(4, \infty)$. So there is a relative maximum at $x = 0$ and a relative minimum at $x = 4$.

We find $G(0)$:

$$G(0) = (0)^3 - 6(0)^2 + 10$$

$$= 10$$

Then we find $G(4)$:

$$G(4) = (4)^3 - 6(4)^2 + 10$$

$$= 64 - 96 + 10$$

$$= -22$$

There is a relative maximum at $(0, 10)$, and there is a relative minimum at $(4, -22)$. We use the information obtained to sketch the graph. Other function values are listed below.

x	$G(x)$
-2	-22
-1	3
1	5
2	-6
3	-17

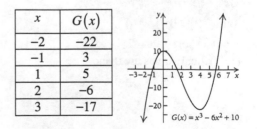

17. $g(x) = x^3 - x^4$

First, find the critical points.

$g'(x) = 3x^2 - 4x^3$

$g'(x)$ exists for all real numbers. We solve

$g'(x) = 0$

$3x^2 - 4x^3 = 0$

$x^2(3 - 4x) = 0$

$x^2 = 0 \qquad \text{or} \qquad 3 - 4x = 0$

$x = 0 \qquad \text{or} \qquad -4x = -3$

$x = 0 \qquad \text{or} \qquad x = \dfrac{3}{4}$

The critical values are 0 and $\dfrac{3}{4}$.

We use the critical values to divide the real number line into three intervals,

A: $(-\infty, 0)$, B: $\left(0, \dfrac{3}{4}\right)$, and C: $\left(\dfrac{3}{4}, \infty\right)$.

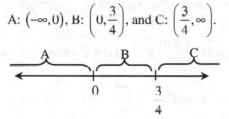

We use a test value in each interval to determine the sign of the derivative in each interval.

A: Test -1, $g'(-1) = 3(-1)^2 - 4(-1)^3 = 7 > 0$

B: Test $\dfrac{1}{2}$, $g'\left(\dfrac{1}{2}\right) = 3\left(\dfrac{1}{2}\right)^2 - 4\left(\dfrac{1}{2}\right)^3$

$\qquad = 3\left(\dfrac{1}{4}\right) - 4\left(\dfrac{1}{8}\right) = \dfrac{1}{4} > 0$

C: Test 1, $g'(1) = 3(1)^2 - 4(1)^3 = -1 < 0$

We see that $g(x)$ is increasing on $(-\infty, 0)$ and $\left(0, \dfrac{3}{4}\right)$, and is decreasing on $\left(\dfrac{3}{4}, \infty\right)$. So there is no relative extrema at $x = 0$ but there is a relative maximum at $x = \dfrac{3}{4}$.

We find $g\left(\dfrac{3}{4}\right)$:

$g\left(\dfrac{3}{4}\right) = \left(\dfrac{3}{4}\right)^3 - \left(\dfrac{3}{4}\right)^4 = \dfrac{27}{64} - \dfrac{81}{256} = \dfrac{27}{256}$

From the previous column, we determine there is a relative maximum at $\left(\dfrac{3}{4}, \dfrac{27}{256}\right)$. We use the information obtained to sketch the graph. Other function values are listed below.

x	$g(x)$
-2	-24
-1	-2
0	0
$\dfrac{1}{2}$	$\dfrac{1}{16}$
1	0
2	-8

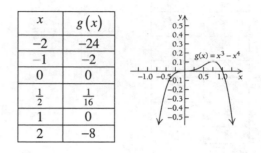

19. $f(x) = \dfrac{1}{3}x^3 - 2x^2 + 4x - 1$

First, find the critical points.

$f'(x) = x^2 - 4x + 4$

$f'(x)$ exists for all real numbers. We solve

$f'(x) = 0$

$x^2 - 4x + 4 = 0$

$(x - 2)^2 = 0$

$x = 2$

The only critical value is 2.

We divide the real number line into two intervals,

A: $(-\infty, 2)$ and B: $(2, \infty)$.

We use a test value in each interval to determine the sign of the derivative in each interval.

A: Test 0, $f'(0) = (0)^2 - 4(0) + 4 = 4 > 0$

B: Test 3, $f'(3) = (3)^2 - 4(3) + 4 = 1 > 0$

We see that $f(x)$ is increasing on both $(-\infty, 2)$ and $(2, \infty)$. Therefore, there are no relative extrema.

The solution is continued on the next page.

We use the information obtained on the previous page to sketch the graph. Other function values are listed below.

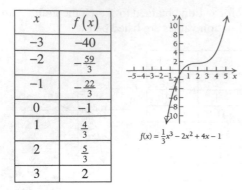

x	$f(x)$
-3	-40
-2	$-\dfrac{59}{3}$
-1	$-\dfrac{22}{3}$
0	-1
1	$\dfrac{4}{3}$
2	$\dfrac{5}{3}$
3	2

$$f(x) = \tfrac{1}{3}x^3 - 2x^2 + 4x - 1$$

21. $f(x) = 3x^4 - 15x^2 + 12$

First, find the critical points.

$$f'(x) = 12x^3 - 30x$$

$f'(x)$ exists for all real numbers. We solve

$$f'(x) = 0$$

$$12x^3 - 30x = 0$$

$$6x(2x^2 - 5) = 0$$

$$6x = 0 \qquad \text{or} \qquad 2x^2 - 5 = 0$$

$$x = 0 \qquad \text{or} \qquad x^2 = \frac{5}{2}$$

$$x = 0 \qquad \text{or} \qquad x = \pm\frac{\sqrt{10}}{2}$$

The critical values are 0, $\dfrac{\sqrt{10}}{2}$ and $-\dfrac{\sqrt{10}}{2}$. We use them to divide the real number line into four intervals,

A: $\left(-\infty, -\dfrac{\sqrt{10}}{2}\right)$, B: $\left(-\dfrac{\sqrt{10}}{2}, 0\right)$,

C: $\left(0, \dfrac{\sqrt{10}}{2}\right)$, and D: $\left(\dfrac{\sqrt{10}}{2}, \infty\right)$.

We use a test value in each interval to determine the sign of the derivative in each interval.

A: Test -2,

$$f'(-2) = 12(-2)^3 - 30(-2) = -36 < 0$$

B: Test -1,

$$f'(-1) = 12(-1)^3 - 30(-1) = 18 > 0$$

C: Test 1,

$$f'(1) = 12(1)^3 - 30(1) = -18 < 0$$

D: Test 2,

$$f'(2) = 12(2)^3 - 30(2) = 36 > 0$$

We see that $f(x)$ is decreasing on

$\left(-\infty, -\dfrac{\sqrt{10}}{2}\right)$, increasing on $\left(-\dfrac{\sqrt{10}}{2}, 0\right)$,

decreasing again on $\left(0, \dfrac{\sqrt{10}}{2}\right)$, and increasing

again on $\left(\dfrac{\sqrt{10}}{2}, \infty\right)$. Thus, there is a relative

minimum at $x = -\dfrac{\sqrt{10}}{2}$, a relative maximum at

$x = 0$, and another relative minimum at

$x = \dfrac{\sqrt{10}}{2}$.

We find $f\left(-\dfrac{\sqrt{10}}{2}\right)$:

$$f\left(-\frac{\sqrt{10}}{2}\right) = 3\left(-\frac{\sqrt{10}}{2}\right)^4 - 15\left(-\frac{\sqrt{10}}{2}\right)^2 + 12$$

$$= -\frac{27}{4}$$

Then we find $f(0)$:

$$f(0) = 3(0)^4 - 15(0)^2 + 12 = 12$$

Then we find $f\left(\dfrac{\sqrt{10}}{2}\right)$:

$$f\left(\frac{\sqrt{10}}{2}\right) = 3\left(\frac{\sqrt{10}}{2}\right)^4 - 15\left(\frac{\sqrt{10}}{2}\right)^2 + 12$$

$$= -\frac{27}{4}$$

The solution is continued on the next page.

There are relative minima at $\left(-\dfrac{\sqrt{10}}{2},-\dfrac{27}{4}\right)$ and

$\left(\dfrac{\sqrt{10}}{2},-\dfrac{27}{4}\right)$.

There is a relative maximum at $(0,12)$.

We use the information obtained above to sketch the graph. Other function values are listed below.

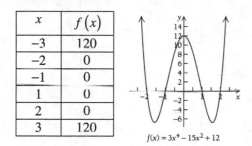

x	$f(x)$
-3	120
-2	0
-1	0
1	0
2	0
3	120

$f(x) = 3x^4 - 15x^2 + 12$

23. $G(x) = \sqrt[3]{x+2} = (x+2)^{1/3}$

First, find the critical points.

$G'(x) = \dfrac{1}{3}(x+2)^{-2/3}(1)$

$\quad\quad \dfrac{1}{3(x+2)^{2/3}}$

$G'(x)$ does not exist when $x = -2$. The equation $G'(x) = 0$ has no solution, therefore, the only critical value is $x = -2$.

We use -2 to divide the real number line into two intervals,

A: $(-\infty,-2)$ and B: $(-2,\infty)$:

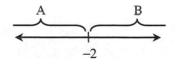

We use a test value in each interval to determine the sign of the derivative in each interval.

A: Test -3, $G'(-3) = \dfrac{1}{3(-3+2)^{2/3}} = \dfrac{1}{3} > 0$

B: Test -1, $G'(-1) = \dfrac{1}{3(-1+2)^{2/3}} = \dfrac{1}{3} > 0$

We see that $G(x)$ is increasing on both $(-\infty,-2)$ and $(-2,\infty)$. Thus, there are no relative extrema for $G(x)$.

We use the information obtained to sketch the graph. Other function values are listed below.

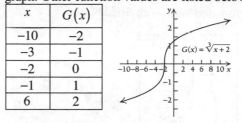

x	$G(x)$
-10	-2
-3	-1
-2	0
-1	1
6	2

$G(x) = \sqrt[3]{x+2}$

25. $f(x) = 1 - x^{2/3}$

First, find the critical points.

$f'(x) = \dfrac{-2}{3}x^{-1/3}$

$\quad\quad = \dfrac{-2}{3\sqrt[3]{x}}$

$f'(x)$ does not exist when

$3\sqrt[3]{x} = 0$, which means that $f'(x)$ does not exist when $x = 0$. The equation $f'(x) = 0$ has no solution, therefore, the only critical value is $x = 0$.

We use 0 to divide the real number line into two intervals,

A: $(-\infty,0)$ and B: $(0,\infty)$:

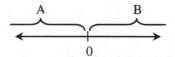

We use a test value in each interval to determine the sign of the derivative in each interval.

A: Test -1, $f'(-1) = -\dfrac{2}{3\sqrt[3]{-1}} = \dfrac{2}{3} > 0$

B: Test 1, $f'(1) = -\dfrac{2}{3\sqrt[3]{1}} = -\dfrac{2}{3} < 0$

We see that $f(x)$ is increasing on $(-\infty,0)$ and decreasing on $(0,\infty)$. Thus, there is a relative maximum at $x = 0$.

We find $f(0)$:

$f(0) = 1 - (0)^{2/3} = 1$.

Therefore, there is a relative maximum at $(0,1)$.

The solution is continued on the next page.

We use the information obtained on the previous page to sketch the graph. Other function values are listed below.

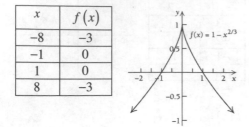

x	$f(x)$
-8	-3
-1	0
1	0
8	-3

27. $G(x) = \dfrac{-8}{x^2+1} = -8(x^2+1)^{-1}$

First, find the critical points.

$G'(x) = -8(-1)(x^2+1)^{-2}(2x)$

$ = \dfrac{16x}{(x^2+1)^2}$

$G'(x)$ exists for all real numbers. Setting the derivative equal to zero, we have:

$G'(x) = 0$

$\dfrac{16x}{(x^2+1)^2} = 0$

$16x = 0$

$x = 0$

The only critical value is 0.
We use 0 to divide the real number line into two intervals,

  A: $(-\infty, 0)$ and B: $(0, \infty)$:

```
        A           B
     ⏞‾‾‾‾   ⏞‾‾‾
    ←——————|——————→
            0
```

We use a test value in each interval to determine the sign of the derivative in each interval.

A: Test -1, $G'(-1) = \dfrac{16(-1)}{((-1)^2+1)^2} = \dfrac{-16}{4} = -4 < 0$

B: Test 1, $G'(1) = \dfrac{16(1)}{((1)^2+1)^2} = \dfrac{16}{4} = 4 > 0$

We see that $G(x)$ is decreasing on $(-\infty, 0)$ and increasing on $(0, \infty)$. Thus, a relative minimum occurs at $x = 0$.

We find $G(0)$:

$G(0) = \dfrac{-8}{(0)^2+1} = -8$

Thus, there is a relative minimum at $(0, -8)$.

We use the information obtained to sketch the graph. Other function values are listed below.

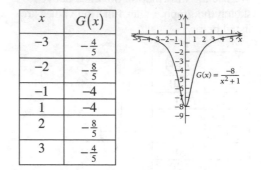

x	$G(x)$
-3	$-\dfrac{4}{5}$
-2	$-\dfrac{8}{5}$
-1	-4
1	-4
2	$-\dfrac{8}{5}$
3	$-\dfrac{4}{5}$

29. $g(x) = \dfrac{4x}{x^2+1}$

First, find the critical points.

$g'(x) = \dfrac{(x^2+1)(4) - 4x(2x)}{(x^2+1)^2}$ Quotient Rule

$ = \dfrac{4x^2+4-8x^2}{(x^2+1)^2}$

$ = \dfrac{4-4x^2}{(x^2+1)^2}$

$g'(x)$ exists for all real numbers. We solve

$g'(x) = 0$

$\dfrac{4-4x^2}{(x^2+1)^2} = 0$

$4 - 4x^2 = 0$ Multiplying by $(x^2+1)^2$

$x^2 - 1 = 0$ Dividing by -4

$x^2 = 1$

$x = \pm\sqrt{1}$

$x = \pm 1$

The critical values are -1 and 1.

The solution is continued on the next page.

We use the critical values found on the previous page to divide the real number line into three intervals, A: $(-\infty,-1)$, B: $(-1,1)$, and C: $(1,\infty)$.

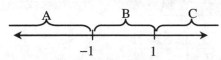

We use a test value in each interval to determine the sign of the derivative in each interval.

A: Test -2, $g'(-2)=\dfrac{4-4(-2)^2}{\left((-2)^2+1\right)^2}=-\dfrac{12}{25}<0$

B: Test 0, $\quad g'(0)=\dfrac{4-4(0)^2}{\left((0)^2+1\right)^2}=4>0$

C: Test 2, $\quad g'(2)=\dfrac{4-4(2)^2}{\left((2)^2+1\right)^2}=-\dfrac{12}{25}<0$

We see that $g(x)$ is decreasing on $(-\infty,\ 1)$, increasing on $(-1,1)$, and decreasing again on $(1,\infty)$. So there is a relative minimum at $x=-1$ and a relative maximum at $x=1$. We find $g(-1)$:

$$g(-1)=\dfrac{4(-1)}{(-1)^2+1}=\dfrac{-4}{2}=-2$$

Then we find $g(1)$:

$$g(1)=\dfrac{4(1)}{(1)^2+1}=\dfrac{4}{2}=2$$

There is a relative minimum at $(-1,-2)$, and there is a relative maximum at $(1,2)$. We use the information obtained to sketch the graph. Other function values are listed below.

x	$g(x)$
-3	$-\dfrac{6}{5}$
-2	$-\dfrac{8}{5}$
0	0
2	$\dfrac{8}{5}$
3	$\dfrac{6}{5}$

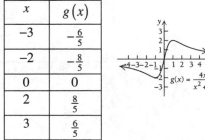

31. $f(x)=\sqrt[3]{x}=(x)^{\frac{1}{3}}$

First, find the critical points.

$$f'(x)=\frac{1}{3}(x)^{-\frac{2}{3}}$$

$$=\frac{1}{3(x)^{\frac{2}{3}}}=\frac{1}{3\cdot\sqrt[3]{x^2}}$$

$f'(x)$ does not exist when $x=0$. The equation $f'(x)=0$ has no solution, therefore, the only critical value is $x=0$.

We use 0 to divide the real number line into two intervals,

A: $(-\infty,0)$ and B: $(0,\infty)$:

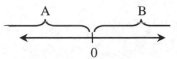

We use a test value in each interval to determine the sign of the derivative in each interval.

A: Test -1, $f'(-1)=\dfrac{1}{3\sqrt[3]{(-1)^2}}=\dfrac{1}{3}>0$

B: Test 1, $f'(1)=\dfrac{1}{3\left(\sqrt[3]{(1)^2}\right)}=\dfrac{1}{3}>0$

We see that $f(x)$ is increasing on both $(-\infty,0)$ and $(0,\infty)$. Thus, there are no relative extrema for $f(x)$. We use the information obtained to sketch the graph. Other function values are listed below.

x	$f(x)$
-8	-2
-1	-1
0	0
1	1
8	2

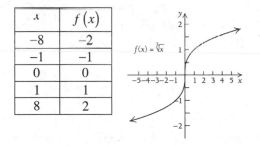

33. $g(x) = \sqrt{x^2 + 2x + 5} = \left(x^2 + 2x + 5\right)^{\frac{1}{2}}$

First, find the critical points.

$g'(x) = \dfrac{1}{2}\left(x^2 + 2x + 5\right)^{-\frac{1}{2}}(2x + 2)$

$= \dfrac{2(x+1)}{2\left(x^2 + 2x + 5\right)^{\frac{1}{2}}}$

$= \dfrac{x+1}{\sqrt{x^2 + 2x + 5}}$

The equation $x^2 + 2x + 5 = 0$ has no real-number solution, so $g'(x)$ exists for all real numbers. Next we find out where the derivative is zero. We solve

$g'(x) = 0$

$\dfrac{x+1}{\sqrt{x^2 + 2x + 5}} = 0$

$x + 1 = 0$

$x = -1$

The only critical value is -1.

We use -1 to divide the real number line into two intervals,

  A: $(-\infty, -1)$ and B: $(-1, \infty)$:

$$A \qquad\qquad B$$
$$\xleftarrow{\qquad\qquad}\underset{-1}{\,|\,}\xrightarrow{\qquad\qquad}$$

We use a test value in each interval to determine the sign of the derivative in each interval.
A: Test -2,

$g'(-2) = \dfrac{(-2)+1}{\sqrt{(-2)^2 + 2(-2) + 5}} = \dfrac{-1}{\sqrt{5}} < 0$

B: Test 0,

$g'(0) = \dfrac{(0)+1}{\sqrt{(0)^2 + 2(0) + 5}} = \dfrac{1}{\sqrt{5}} > 0$

We see that $g(x)$ is decreasing on $(-\infty, -1)$ and increasing on $(-1, \infty)$, and the change from decreasing to increasing indicates that a relative minimum occurs at $x = -1$. We substitute into the original equation to find $g(-1)$:

$g(-1) = \sqrt{(-1)^2 + 2(-1) + 5} = \sqrt{4} = 2$

Thus, there is a relative minimum at $(-1, 2)$.

We use the information obtained to sketch the graph. Other function values are listed below.

x	$g(x)$
-4	3.61
-2	2.24
0	2.24
1	2.83
3	4.47

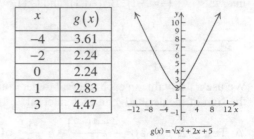

$g(x) = \sqrt{x^2 + 2x + 5}$

35. – 68.   Left to the student.

69. Answers may vary, one such graph is:

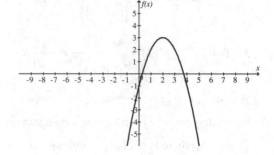

71. Answers may vary, one such graph is:

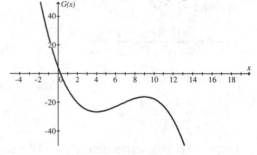

73. Answers may vary, one such graph is:

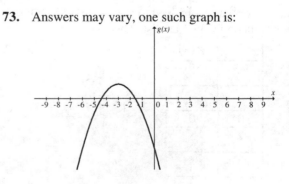

75. Answers may vary, one such graph is:

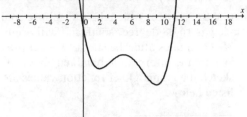

77. Answers may vary, one such graph is:

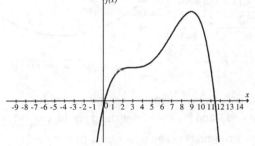

79. Answers may vary, one such graph is:

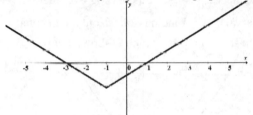

81. Answers may vary, one such graph is:

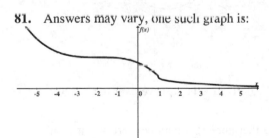

83. Answers may vary, one such graph is:

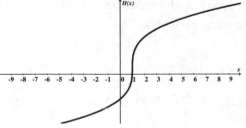

85.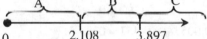

87. Letting t be years since 2000 and E be thousand of employees, we have the function:

$$E(t) = 107.833t^3 - 971.369t^2 + 2657.917t + 50347.83$$

First, we find the critical points.

$$E'(t) = 323.499t^2 - 1942.738t + 2657.917$$

$E'(t)$ exists for all real numbers. Solve

$$E'(t) = 0$$

$323.499t^2 - 1942.738t + 2657.917 = 0$

Using the quadratic formula, we have:

$$t = \frac{1942.738 \pm \sqrt{(-1942.738)^2 - 4(323.499)(2657.917)}}{2(323.499)}$$

$$= \frac{1942.738 \pm \sqrt{334,896.970312}}{646.998}$$

$t \approx 2.108 \qquad$ or $\qquad t \approx 3.897$

There are two critical values.

We use them to divide the interval $[0, \infty)$ into three intervals:

A: $[0, 2.108)$ B: $(2.08, 3.897)$, and C: (3.897∞)

Next, we test a point in each interval to determine the sign of the derivative.

A: Test 1,

$$E'(1) = 323.499(1)^2 - 1942.738(1) + 2657.917$$
$$= 1038.678 > 0$$

B: Test 3,

$$E'(3) = 323.499(3)^2 - 1942.738(3) + 2657.917$$
$$= -258.806 < 0$$

C: Test 4,

$$E'(4) = 323.499(4)^2 - 1942.738(4) + 2657.917$$
$$= 62.949 > 0$$

Since, $E(t)$ is increasing on $[0, 2.108)$ and decreasing on $(2.108, 3.897)$ and there is a relative maximum at $t = 2.108$.

$$E(2.108) = 107.833(2.108)^3 - 971.369(2.108)^2 + 2657.917(2.108) + 50347.83$$

$\approx 52,644.383$

There is a relative minimum at $(2.108,\ 52,644.383)$. The solution is continued.

Since, $E(t)$ is decreasing on $(2.108, 3.897)$ and increasing $[3.897, \infty)$ on and there is a relative minimum at $t = 3.897$.

$$E(3.897) = 107.833(3.897)^3 - 971.369(3.897)^2 + $$
$$2657.917(3.897) + 50347.83$$

$$\approx 52,335.73$$

There is a relative maxnimum at $(3.897,\ 52,335.73)$.

We sketch the graph.

t	$T(t)$
0	50,347
1	52,142
3	52,491
5	52,832
8	64,654

$E(t)$, $E(t) = 107.833t^3 - 971.369t^2 + 2657.917t + 50347.833$

60,000

50,000

40,000

$1\ 2\ 3\ 4\ 5\ 6\ 7\ t$

89. $f(t) = 0.00259t^2 - 0.457t + 36.237$

First, we find the critical points.

$f'(t) = 0.00518t - 0.457$

$f'(x)$ exists everywhere, so we solve

$$f'(t) = 0$$

$$0.00518t - 0.457 = 0$$

$$t = 88.22$$

The only critical value is about 88.22 we use it to break up the interval $(0, \infty)$ into two intervals

A: $(0, 88.28)$ and B: $(88.22, \infty)$.

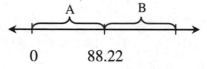

$$0 \qquad 88.22$$

Testing a point in each interval, we have:

A: Test 20,

$$f'(20) = 0.00518(20) - 0.457 = -0.3534 < 0$$

B: Test 100,

$$f'(100) = 0.00518(100) - 0.457 = 0.061 > 0$$

We see that $f(t)$ is decreasing on $(0, 88.22)$ and increasing on $(88.22, \infty)$, so there is a relative minimum at $t = 88.22$.

$f(88.22)$

$$= 0.00259(88.22)^2 - 0.457(88.22) + 36.237$$

$$\approx 16.09$$

There is a relative minimum at about $(88.22, 16.08)$. Thus, the latitude that is closest to the equator at which the full eclipse could be view is 16.08 degrees south and will occur 88.22 minutes after the start of the eclipse. We use the information obtained above to sketch the graph. Other function values are listed below.

x	$f(x)$
20	28.133
30	24.858
40	22.101
60	18.141
80	16.253

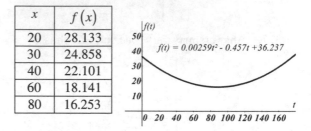

91. The derivative is negative over the interval $(-\infty, -1)$ and positive over the interval $(-1, \infty)$. Furthermore, it is equal to zero when $x = -1$. This means that the function is decreasing over the interval $(-\infty, -1)$, increasing over the interval $(-1, \infty)$ and has a horizontal tangent at $x = -1$. A possible graph is shown below.

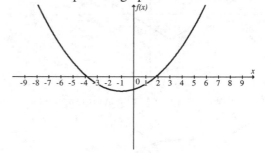

93. The derivative is positive over the interval $(-\infty, 1)$ and negative over the interval $(1, \infty)$. Furthermore, it is equal to zero when $x = 1$. This means that the function is increasing over the interval $(-\infty, 1)$, decreasing over the interval $(1, \infty)$ and has a horizontal tangent at $x = 1$. A possible graph is shown below.

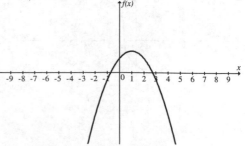

95. The derivative is positive over the interval $(-4, 2)$ and negative over the intervals $(-\infty, -4)$ and $(2, \infty)$. Furthermore, it is equal to zero when $x = -4$ and $x = 2$. This means that the function is decreasing over the interval $(-\infty, -4)$, then increasing over the interval $(-4, 2)$, and then decreasing again over the interval $(2, \infty)$. The function has horizontal tangents at $x = -4$ and $x = 2$. A possible graph is shown below.

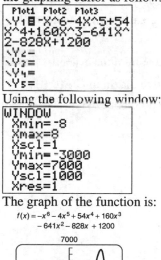

97. $f(x) = -x^6 - 4x^5 + 54x^4 + 160x^3 - 641x^2$
$$-828x + 1200$$

Using the calculator we enter the function into the graphing editor as follows:

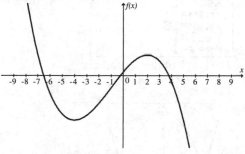

Using the following window:

```
WINDOW
 Xmin=-8
 Xmax=8
 Xscl=1
 Ymin=-3000
 Ymax=7000
 Yscl=1000
 Xres=1
```

The graph of the function is:

$f(x) = -x^6 - 4x^5 + 54x^4 + 160x^3$
$- 641x^2 - 828x + 1200$

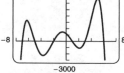

We find the relative extrema using the minimum/maximum feature on the calculator. There are relative minima at $(-3.683, -2288.03)$ and $(2.116, -1083.08)$.

There are relative maxima at $(-6.262, 3213.8)$, $(-0.559, 1440.06)$, and $(5.054, 6674.12)$.

99. $f(x) = \sqrt[3]{\left| 4 - x^2 \right|} + 1$

Using the calculator we enter the function into the graphing editor as follows:

```
Plot1 Plot2 Plot3
\Y1■³√(abs(4-X^2
))+1
\Y2=
\Y3=
\Y4=
\Y5=
\Y6=
```

Using the following window:

```
WINDOW
 Xmin=-10
 Xmax=10
 Xscl=1
 Ymin=0
 Ymax=6
 Yscl=.5
 Xres=1
```

The graph of the function is:

$f(x) = \sqrt[3]{|4 - x^2|} + 1$

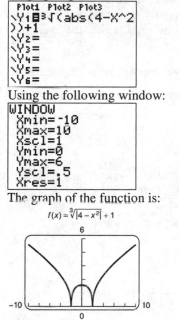

We find the relative extrema using the minimum/maximum feature on the calculator. There are relative minima at $(-2, 1)$ and $(2, 1)$.

There is a relative maximum at $(0, 2.587)$.

101. $f(x) = \left| x - 2 \right|$

Using the calculator we enter the function into the graphing editor as follows:

```
Plot1 Plot2 Plot3
\Y1■abs(X-2)■
\Y2=
\Y3=
\Y4=
\Y5=
\Y6=
\Y7=
```

Using the following window:

```
WINDOW
 Xmin=-10
 Xmax=10
 Xscl=1
 Ymin=-10
 Ymax=10
 Yscl=1
 Xres=1
```

The solution is continued on the enxt page.

The graph of the function is:

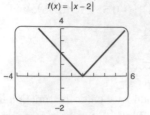

We find the relative extrema using the minimum/maximum feature on the calculator. The graph is decreasing over the interval $(-\infty, 2)$.

The graph is increasing over the interval $(2, \infty)$.

There is a relative minimum at $(2, 0)$.

The derivative does not exist at $x = 2$

103. $f(x) = \left| x^2 - 1 \right|$

Using the calculator we enter the function into the graphing editor as follows:

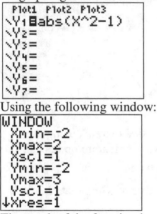

Using the following window:

The graph of the function is:

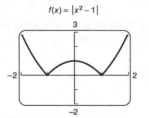

We find the relative extrema using the minimum/maximum feature on the calculator. The graph is decreasing over the interval $(-\infty, -1)$ and $(0, 1)$.

The graph is increasing over the interval $(-1, 0)$ and $(2, \infty)$.

There are relative mimima at $(-1, 0)$ and $(1, 0)$.

There is a relative maximum at $(0, 1)$.

The derivative does not exist at $x = -1$ and $x = 1$.

105. $f(x) = \left| 9 - x^2 \right|$

Using the calculator we enter the function into the graphing editor as follows:

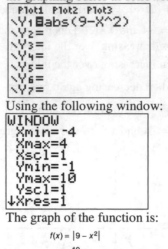

Using the following window:

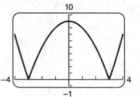

The graph of the function is:

We find the relative extrema using the minimum/maximum feature on the calculator. The graph is decreasing over the interval $(-\infty, -3)$ and $(0, 3)$.

The graph is increasing over the interval $(-3, 0)$ and $(3, \infty)$.

There are relative mimima at $(-3, 0)$ and $(3, 0)$.

There is a relative maximum at $(0, 9)$.

The derivative does not exist at $x = -3$ and $x = 3$.

107. $f(x) = |x^3 - 1|$

Using the calculator we enter the function into the graphing editor as follows:

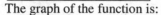

Using the following window:

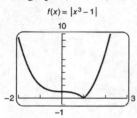

The graph of the function is:

$f(x) = |x^3 - 1|$

We find the relative extrema using the minimum/maximum feature on the calculator. The graph is decreasing over the interval $(-\infty, -1)$.

The graph is increasing over the interval $(1, \infty)$.

There is a relative minimum at $(1, 0)$.

The derivative does not exist at $x - 1$.

109.

111.

Exercise Set 2.2

1. $f(x) = 4 - x^2$

First, find $f'(x)$ and $f''(x)$.

$f'(x) = -2x$

$f''(x) = -2$

Next, find the critical points of $f(x)$. Since $f'(x)$ exists for all real numbers x, the only critical points occur when $f'(x) = 0$.

$f'(x) = 0$

$-2x = 0$

$x = 0$

We find the function value at $x = 0$.

$f(0) = 4 - (0)^2 = 4$.

The critical point is $(0, 4)$.

Next, we apply the Second Derivative test.

$f''(x) = -2$

$f''(0) = -2 < 0$

Therefore, $f(0) = 4$ is a relative maximum.

3. $f(x) = x^2 + x - 1$

First, find $f'(x)$ and $f''(x)$.

$f'(x) = 2x + 1$

$f''(x) = 2$

Next, find the critical points of $f(x)$. Since $f'(x)$ exists for all real numbers x, the only critical points occur when $f'(x) = 0$.

$2x + 1 = 0$

$2x = -1$

$x = -\dfrac{1}{2}$

We find the function value at $x = -\dfrac{1}{2}$.

$f\left(-\dfrac{1}{2}\right) = \left(-\dfrac{1}{2}\right)^2 + \left(-\dfrac{1}{2}\right) - 1 = \dfrac{1}{4} - \dfrac{1}{2} - 1 = -\dfrac{5}{4}$.

The critical point is $\left(-\dfrac{1}{2}, -\dfrac{5}{4}\right)$.

Next, we apply the Second Derivative test.

$f''(x) = 2$

$f''\left(-\dfrac{1}{2}\right) = 2 > 0$

Therefore, $f\left(-\dfrac{1}{2}\right) = -\dfrac{5}{4}$ is a relative minimum.

5. $f(x) = -4x^2 + 3x - 1$

First, find $f'(x)$ and $f''(x)$.

$f'(x) = -8x + 3$

$f''(x) = -8$

Next, find the critical points of $f(x)$. Since $f'(x)$ exists for all real numbers x, the only critical points occur when $f'(x) = 0$.

$f'(x) = 0$

$-8x + 3 = 0$

$-8x = -3$

$x = \dfrac{3}{8}$

We find the function value at $x = \dfrac{3}{8}$.

$f\left(\dfrac{3}{8}\right) = -4\left(\dfrac{3}{8}\right)^2 + 3\left(\dfrac{3}{8}\right) - 1$

$= -4\left(\dfrac{9}{64}\right) + \dfrac{9}{8} - 1$

$= -\dfrac{9}{16} + \dfrac{18}{16} - \dfrac{16}{16}$

$= -\dfrac{7}{16}$

The critical point is $\left(\dfrac{3}{8}, -\dfrac{7}{16}\right)$.

Next, we apply the Second Derivative test.

$f''(x) = -8$

$f''\left(\dfrac{3}{8}\right) = -8 < 0$

Therefore, $f\left(\dfrac{3}{8}\right) = -\dfrac{7}{16}$ is a relative maximum.

7. $f(x) = x^3 - 12x - 1$

First, find $f'(x)$ and $f''(x)$.

$f'(x) = 3x^2 - 12$

$f''(x) = 6x$

The solution is continued on the next page.

Next, find the critical points of $f(x)$. Since $f'(x)$ exists for all real numbers x, the only critical points occur when $f'(x) = 0$.

$$f'(x) = 0$$
$$3x^2 - 12 = 0$$
$$3x^2 = 12$$
$$x^2 = 4$$
$$x = \pm\sqrt{4}$$
$$x = \pm 2$$

There are two critical values.
First, we find the function value at $x = -2$.
$$f(-2) = (-2)^3 - 12(-2) - 1$$
$$= -8 + 24 - 1$$
$$= 15$$

The critical point is $(-2, 15)$.
Next, we apply the Second Derivative test.
$$f''(x) = 6x$$
$$f''(-2) = 6(-2) = -12 < 0$$

Therefore, $f(-2) = 15$ is a relative maximum.
Next, we find the function value at $x = 2$.
$$f(2) = (2)^3 - 12(2) - 1$$
$$= 8 - 24 - 1$$
$$= -17$$

The critical point is $(2, -17)$.
Next, we apply the Second Derivative test.
$$f''(x) = 6x$$
$$f''(2) = 6(2) = 12 > 0$$

Therefore, $f(2) = -17$ is a relative minimum.

9. $f(x) = x^3 - 27x$

a) Find $f'(x)$ and $f''(x)$.
$$f'(x) = 3x^2 - 27$$
$$f''(x) = 6x$$
The domain of f is $\mathbb{R}$.

b) Find the critical points of $f(x)$. Since $f'(x)$ exists for all real numbers x, the only critical points occur when $f'(x) = 0$.
$$3x^2 - 27 = 0$$
$$x^2 = 9$$
$$x = \pm 3$$
There are two critical values $x = -3$ and $x = 3$.
$$f(-3) = (-3)^3 - 27(-3) = 54.$$
The critical point on the graph is $(-3, 54)$.
$$f(3) = (3)^3 - 27(3) = -54.$$
The critical point on the graph is $(3, -54)$.

c) We apply the Second Derivative test to the critical points.
For $x = -3$
$$f''(x) = 6x$$
$$f''(-3) = 6(-3) = -18 < 0$$
The critical point $(-3, 54)$ is a relative maximum.
For $x = 3$
$$f''(x) = 6x$$
$$f''(3) = 6(3) = 18 > 0$$
The critical point $(3, -54)$ is a relative minimum.
If we use the critical values $x = -3$ and $x = 3$ to divide the real line into three intervals, $(-\infty, -3), (-3, 3)$, and $(3, \infty)$, we know from the extrema above, that $f(x)$ is increasing over the interval $(-\infty, -3)$, decreasing over the interval $(-3, 3)$ and then increasing again over the interval $(3, \infty)$.

d) Find the points of inflection. $f''(x)$ exists for all real numbers, so we solve the equation
$$f''(x) = 0$$
$$6x = 0$$
$$x = 0$$
Therefore, a possible inflection point occurs at $x = 0$.
$$f(0) = (0)^3 - 27(0) = 0.$$
A possible inflection point on the graph is the point $(0, 0)$.

e) To determine concavity, we use the possible inflection point to divide the real number line into two intervals
$A : (-\infty, 0)$ and $B : (0, \infty)$. We test a point in each interval.

A: Test -1: $f''(-1) = 6(-1) = -6 < 0$

B: Test 1: $f''(1) = 6(1) = 6 > 0$

Then, $f(x)$ is concave down on the interval $(-\infty, 0)$ and concave up on the interval $(0, \infty)$, so $(0, 0)$ is an inflection point.

f) Finally, we use the preceding information to sketch the graph of the function. Additional function values can also be calculated as needed.

x	$f(x)$
-4	44
-1	26
1	-26
4	-44

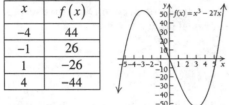

11. $f(x) = 2x^3 - 3x^2 - 36x + 28$

a) First, find $f'(x)$ and $f''(x)$.

$f'(x) = 6x^2 - 6x - 36$

$f''(x) = 12x - 6$

The domain of f is $\mathbb{R}$.

b) Find the critical points of $f(x)$. Since $f'(x)$ exists for all real numbers x, the only critical points occur when $f'(x) = 0$.

$$f'(x) = 0$$
$$6x^2 - 6x - 36 = 0$$
$$x^2 - x - 6 = 0$$
$$(x - 3)(x + 2) = 0$$
$$x - 3 = 0 \ \text{ or } \ x + 2 = 0$$
$$x = 3 \ \text{ or } \ \ \ \ x = -2$$

There are two critical values $x = -2$ and $x = 3$.

We find the function value at $x = -2$

$f(-2) = 2(-2)^3 - 3(-2)^2 - 36(-2) + 28$

$\ \ \ \ \ \ \ = -16 - 12 + 72 + 28$

$\ \ \ \ \ \ \ = 72$

The critical point on the graph is $(-2, 72)$.

Next, we find the function value at $x = 3$.

$f(3) = 2(3)^3 - 3(3)^2 - 36(3) + 28$

$\ \ \ \ \ \ = 54 - 27 - 108 + 28$

$\ \ \ \ \ \ = -53$

The critical point on the graph is $(3, -53)$.

c) Apply the Second Derivative test to the critical points.

For $x = -2$

$\ \ \ f''(x) = 12x - 6$

$\ \ \ f''(-2) = 12(-2) - 6 = -30 < 0$

The critical point $(-2, 72)$ is a relative maximum.

For $x = 3$

$\ \ \ f''(x) = 12x - 6$

$\ \ \ f''(3) = 12(3) - 6 = 30 > 0$

The critical point $(3, -53)$ is a relative minimum.

We use the critical values $x = -2$ and $x = 3$ to divide the real line into three intervals, $A : (-\infty, -2)$, B: $(-2, 3)$, and C: $(3, \infty)$, we know from the extrema above, that $f(x)$ is increasing over the interval $(-\infty, -2)$, decreasing over the interval $(-2, 3)$ and then increasing again over the interval $(3, \infty)$.

d) Find the points of inflection. $f''(x)$ exists for all real numbers, so we solve the equation

$$f''(x) = 0$$
$$12x - 6 = 0$$
$$x = \frac{1}{2}$$

Therefore, a possible inflection point occurs at $x = \frac{1}{2}$.

$$f\left(\frac{1}{2}\right) = 2\left(\frac{1}{2}\right)^3 - 3\left(\frac{1}{2}\right)^2 - 36\left(\frac{1}{2}\right) + 28$$

$$= \frac{19}{2}.$$

A possible inflection point on the graph is the point $\left(\frac{1}{2}, \frac{19}{2}\right)$.

e) To determine concavity, we use the possible inflection point to divide the real number line into two intervals

$A: \left(-\infty, \frac{1}{2}\right)$ and $B: \left(\frac{1}{2}, \infty\right)$. We test a point in each interval

A: Test $0: f''(0) = 12(0) - 6 = -6 < 0$

B: Test $1: f''(1) = 12(1) - 6 = 6 > 0$

Then, $f(x)$ is concave down on the interval

$\left(-\infty, \frac{1}{2}\right)$ and concave up on the interval

$\left(\frac{1}{2}, \infty\right)$, so $\left(\frac{1}{2}, \frac{19}{2}\right)$ is an inflection point.

f) We use the preceding information to sketch the graph of the function. Additional function values can also be calculated as needed.

x	$f(x)$
-3	55
-1	59
0	28
1	-9
4	-36

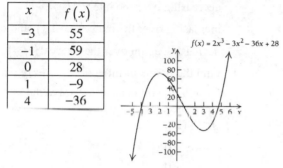

$f(x) = 2x^3 - 3x^2 - 36x + 28$

13. $f(x) = 80 - 9x^2 - x^3$

a) First, find $f'(x)$ and $f''(x)$.

$f'(x) = -18x - 3x^2$

$f''(x) = -18 - 6x$

The domain of f is $\mathbb{R}$.

b) Next, find the critical points of $f(x)$.

Since $f'(x)$ exists for all real numbers x, the only critical points occur when $f'(x) = 0$.

$f'(x) = 0$

$-18x - 3x^2 = 0$

$-3x(x + 6) = 0$

$-3x = 0$ or $x + 6 = 0$

$x = 0$ or $x = -6$

There are two critical values $x = -6$ and $x = 0$.

We find the function value at $x = -6$.

$f(-6) = 80 - 9(-6)^2 - (-6)^3$

$= 80 - 324 + 216$

$= -28$

The critical point on the graph is $(-6, -28)$.

Next, we find the function value at $x = 0$.

$f(0) = 80 - 9(0)^2 - (0)^3 = 80$.

The critical point on the graph is $(0, 80)$.

c) We Apply the Second Derivative test to the critical points.

For $x = -6$

$f''(x) = -18 - 6x$

$f''(-6) = -18 - 6(-6) = 18 > 0$

The critical point $(-6, -28)$ is a relative minimum.

For $x = 0$

$f''(x) = -18 - 6x$

$f''(0) = -18 - 6(0) = -18 < 0$

The critical point $(0, 80)$ is a relative maximum.

We use the critical values $x = -6$, and $x = 0$ to divide the real line into three intervals,

$A: (-\infty, -6)$, $B: (-6, 0)$, and $C: (0, \infty)$, we know from the extrema above, that $f(x)$ is decreasing over the interval $(-\infty, -6)$, increasing over the interval $(-6, 0)$ and then decreasing again over the interval $(0, \infty)$.

d) Find the points of inflection. $f''(x)$ exists for all real numbers, so we solve the equation

$f''(x) = 0$

$-18 - 6x = 0$

$x = -3$

Therefore, a possible inflection point occurs at $x = -3$.

$f(-3) = 80 - 9(-3)^2 - (-3)^3 = 26$.

A possible inflection point on the graph is the point $(-3, 26)$.

e) To determine concavity, we use the possible inflection point to divide the real number line into two intervals

$A:(-\infty,-3)$ and $B:(-3,\infty)$. We test a point in each interval

A: Test -4: $f''(-4)=-18-6(-4)=6>0$

B: Test 0: $f''(0)=-18-6(0)=-18<0$

Then, $f(x)$ is concave up on the interval $(-\infty,-3)$ and concave down on the interval $(-3,\infty,)$, so $(-3,26)$ is an inflection point.

f) We use the preceding information to sketch the graph of the function. Additional function values are also calculated.

x	$f(x)$
-9	80
-4	0
-2	52
0	80
2	36
3	-28

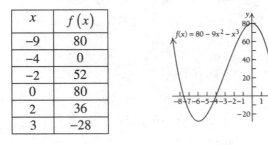

15. $f(x)=-x^3+3x-2$

a) First, find $f'(x)$ and $f''(x)$.

$f'(x)=-3x^2+3$

$f''(x)=-6x$

The domain of f is $\mathbb{R}$.

b) Find the critical points of $f(x)$. Since $f'(x)$ exists for all real numbers x, the only critical points occur when $f'(x)=0$.

$f'(x)=0$

$-3x^2+3=0$

$-3(x^2-1)=0$

$x^2-1=0$

$x=\pm1$

There are two critical values $x=-1$ and $x=1$.

We find the function value at $x=-1$.

$f(-1)=-(-1)^3+3(-1)-2=-4$.

The critical point on the graph is $(-1,-4)$.

We find the function value at $x=1$.

$f(1)=-(1)^3+3(1)-2=0$.

The critical point on the graph is $(1,0)$.

c) We apply the Second Derivative test to the critical points.

For $x=-1$

$f''(x)=-6x$

$f''(-1)=-6(-1)=6>0$

The critical point $(-1,-4)$ is a relative minimum.

For $x=1$

$f''(x)=-6x$

$f''(1)=-6(1)=-6<0$

The critical point $(1,0)$ is a relative maximum.

We use the critical values $x=-1$ and $x=1$ to divide the real line into three intervals, $A:(-\infty,-1)$, $B:(-1,1)$, and $C:(1,\infty)$, we know from the extrema above, that $f(x)$ is decreasing over the interval $(-\infty,-1)$, increasing over the interval $(-1,1)$ and then decreasing again over the interval $(1,\infty)$.

d) Find the points of inflection. $f''(x)$ exists for all real numbers, so we solve the equation

$f''(x)=0$

$-6x=0$

$x=0$

Therefore, a possible inflection point occurs at $x=0$.

$f(0)=-(0)^3+3(0)-2=-2$.

A possible inflection point on the graph is the point $(0,-2)$.

f) To determine concavity, we use the possible inflection point to divide the real number line into two intervals

$A:(-\infty,0)$ and $B:(0,\infty)$.

We test a point in each interval

A: Test -1: $f''(-1)=-6(-1)=6>0$

B: Test 1: $f''(1)=-6(1)=-6<0$

Then, $f(x)$ is concave up on the interval $(-\infty,0)$ and concave down on the interval $(0,\infty)$, so $(0,-2)$ is an inflection point.

e) We use the preceding information to sketch the graph of the function. Additional function values can also be calculated as needed.

x	$f(x)$
-3	16
-2	0
2	-4
3	-20

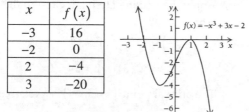

17. $f(x) = 3x^4 - 16x^3 + 18x^2$

a) First, find $f'(x)$ and $f''(x)$.

$$f'(x) = 12x^3 - 48x^2 + 36x$$

$$f''(x) = 36x^2 - 96x + 36$$

The domain of f is $\mathbb{R}$.

b) Next, find the critical points of $f(x)$.

Since $f'(x)$ exists for all real numbers x, the only critical points occur when $f'(x) = 0$.

$$f'(x) = 0$$

$$12x^3 - 48x^2 + 36x = 0$$

$$12x(x^2 - 4x + 3) = 0$$

$$12x(x-1)(x-3) = 0$$

$$12x = 0 \quad \text{or} \quad x - 1 = 0 \quad \text{or} \quad x - 3 = 0$$

$$x = 0 \quad \text{or} \quad x = 1 \quad \text{or} \quad x = 3$$

There are three critical values $x = 0$, $x = 1$, and $x = 3$.

Then

$$f(0) = 3(0)^4 - 16(0)^3 + 18(0)^2 = 0$$

$$f(1) = 3(1)^4 - 16(1)^3 + 18(1)^2 = 5$$

$$f(3) = 3(3)^4 - 16(3)^3 + 18(3)^2 = -27$$

Thus, the critical points $(0,0)$, $(1,5)$, and $(3,-27)$ are on the graph.

c) Apply the Second Derivative test to the critical points.

$$f''(0) = 36(0)^2 - 96(0) + 36 = 36 > 0$$

The critical point $(0,0)$ is a relative minimum.

$$f''(1) = 36(1)^2 - 96(1) + 36 = -24 < 0$$

The critical point $(1,5)$ is a relative maximum.

Testing $x = 3$.

$$f''(3) = 36(3)^2 - 96(3) + 36 = 72 > 0$$

The critical point $(3,-27)$ is a relative minimum.

We use the critical values $0, 1,$ and 3 to divide the real line into four intervals, $A : (-\infty, 0)$, $B : (0,1)$, $C : (1,3)$ and $D : (3,\infty)$, we know from the extrema above, that $f(x)$ is decreasing over the intervals $(-\infty, 0)$ and $(1,3)$ and $f(x)$ increasing over the intervals $(0,1)$ and $(3,\infty)$.

d) Find the points of inflection. $f''(x)$ exists for all real numbers, so we solve the equation $f''(x) = 0$.

$$f''(x) = 0$$

$$36x^2 - 96x + 36 = 0$$

$$12(3x^2 - 8x + 3) = 0$$

$$3x^2 - 8x + 3 = 0$$

Using the quadratic formula, we find that

$$x = \frac{4 \pm \sqrt{7}}{3}, \text{ so } x \approx 0.451 \text{ or } x \approx 2.215 \text{ are}$$

possible inflection points.

$$f(0.451) \approx 2.321$$

$$f(2.215) \approx -13.358$$

So, $(0.451, 2.321)$ and $(2.215, -13.358)$ are two more points on the graph.

e) To determine concavity, we use the possible inflection point to divide the real number line into three intervals $A : (-\infty, 0.451)$, $B : (0.451, 2.215)$, and $C : (2.215, \infty)$.

We test a point in each interval to determine the sign of the second derivative.

A: Test 0:

$$f''(0) = 36(0)^2 - 96(0) + 36 = 36 > 0$$

B: Test 1:

$$f''(1) = 36(1)^2 - 96(1) + 36 = -24 < 0$$

C: Test 3:

$$f''(3) = 36(3)^2 - 96(3) + 36 = 72 > 0$$

The solution is continued on the next page.

From the previous page, we determine $f(x)$ is concave up on the interval $(-\infty, 0.451)$ and concave down on the interval $(0.451, 2.215)$ and concave up on the interval $(2.215, \infty)$, so $(0.451, 2.321)$ and $(2.215, -13.358)$ are inflection points.

f) We use the preceding information to sketch the graph of the function. Additional function values can also be calculated as needed.

x	$f(x)$
-1	37
2	-8
4	32

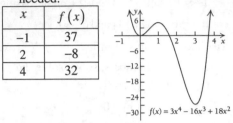

$f(x) = 3x^4 - 16x^3 + 18x^2$

19. $f(x) = x^4 - 6x^2$

a) First, find $f'(x)$ and $f''(x)$.

$$f'(x) = 4x^3 - 12x$$

$$f''(x) = 12x^2 - 12$$

The domain of f is $\mathbb{R}$.

b) Find the critical points of $f(x)$. Since $f'(x)$ exists for all real numbers x, the only critical points occur when $f'(x) = 0$.

$$f'(x) = 0$$
$$4x^3 - 12x = 0$$
$$4x(x^2 - 3) = 0$$
$$4x = 0 \quad \text{or} \quad x^2 - 3 = 0$$
$$x = 0 \quad \text{or} \quad x = \pm\sqrt{3}$$

There are three critical values $-\sqrt{3}, 0,$ and $\sqrt{3}$.

Then

$$f(-\sqrt{3}) = (-\sqrt{3})^4 - 6(-\sqrt{3})^2$$
$$= 9 - 6(3)$$
$$= -9$$
$$f(0) = (0)^4 - 6(0)^2 = 0$$
$$f(\sqrt{3}) = (\sqrt{3})^4 - 6(\sqrt{3})^2$$
$$= 9 - 6(3)$$
$$= -9$$

Thus, the critical points $(-\sqrt{3}, -9), (0,0),$ $(\sqrt{3}, -9)$ and are on the graph.

c) Apply the Second Derivative test to the critical points.

$$f''(-\sqrt{3}) = 12(-\sqrt{3})^2 - 12$$
$$= 12(3) - 12 = 24 > 0$$

The critical point $(-\sqrt{3}, -9)$ is a relative minimum.

$$f''(0) = 12(0)^2 - 12 = -12 < 0$$

The critical point $(0,0)$ is a relative maximum.

$$f''(\sqrt{3}) = 12(\sqrt{3})^2 - 12$$
$$= 12(3) - 12 = 24 > 0$$

The critical point $(\sqrt{3}, -9)$ is a relative minimum.

If we use the critical values $-\sqrt{3}, 0,$ and $\sqrt{3}$ to divide the real line into four intervals,

A: $(-\infty, -\sqrt{3})$, B: $(-\sqrt{3}, 0)$, C: $(0, \sqrt{3})$, and D: $(\sqrt{3}, \infty)$

Then $f(x)$ is decreasing over the intervals $(-\infty, -\sqrt{3})$ and $(0, \sqrt{3})$, and $f(x)$ increasing over the intervals $(-\sqrt{3}, 0)$ and $(\sqrt{3}, \infty)$.

d) Find the points of inflection. $f''(x)$ exists for all real numbers, so we solve the equation

$$f''(x) = 0$$
$$12x^2 - 12 = 0$$
$$x^2 - 1 = 0$$
$$x^2 = 1$$
$$x = \pm 1$$

So $x = -1$ or $x = 1$ are possible inflection points.

$$f(-1) = (-1)^4 - 6(-1)^2 = 1 - 6 = -5$$
$$f(1) = (1)^4 - 6(1)^2 = 1 - 6 = -5$$

So, $(-1, -5)$ and $(1, -5)$ are two more points on the graph.

e) To determine concavity, we use the possible inflection point to divide the real number line into three intervals $A: (-\infty, -1)$,

B: $(-1,1)$, and C: $(1, \infty)$. We test a point in each interval

A: Test -2:
$$f''(-2) = 12(-2)^2 - 12 = 36 > 0$$
B: Test 0:
$$f''(0) = 12(0)^2 - 12 = -12 < 0$$
C: Test 2:
$$f''(2) = 12(2)^2 - 12 = 36 > 0$$

Then, $f(x)$ is concave up on the intervals $(-\infty, -1)$ and $(1, \infty)$ and concave down on the interval $(-1,1)$, so $(-1,-5)$ and $(1,-5)$ are inflection points.

f) We use the preceding information to sketch the graph of the function. Additional function values can also be calculated as needed.

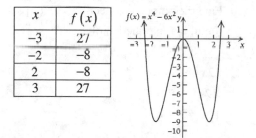

x	$f(x)$
-3	27
-2	-8
2	-8
3	27

21. $f(x) = x^3 - 6x^2 + 9x + 1$

a) First, find $f'(x)$ and $f''(x)$.
$$f'(x) = 3x^2 - 12x + 9$$
$$f''(x) = 6x - 12$$
The domain of f is $\mathbb{R}$.

b) $f'(x)$ exists for all values of x, so the only critical points of f are where $f'(x) = 0$.
$$f'(x) = 0$$
$$3x^2 - 12x + 9 = 0$$
$$3\left(x^2 - 4x + 3\right) = 0$$
$$3(x-1)(x-3) = 0$$
$$x - 1 = 0 \quad \text{or} \quad x - 3 = 0$$
$$x = 1 \quad \text{or} \quad x = 3$$
The critical values are 1 and 3.

Then,
$$f(1) = (1)^3 - 6(1)^2 + 9(1) + 1 = 5$$
$$f(3) = (3)^3 - 6(3)^2 + 9(3) + 1 = 1$$
So, $(1,5)$ and $(3,1)$ are on the graph.

c) Applying the Second Derivative Test, we have:
$$f''(1) = 6(1) - 12 = -6 < 0$$
$$f''(3) = 6(3) - 12 = 6 > 0$$
Therefore, $(1,5)$ is a relative maximum and $(3,1)$ is a relative minimum.

The solution is continued on the next page. If we use the points 1 and 3 to divide the real number line into three intervals, $(-\infty, 1)$, $(1,3)$, and $(3, \infty)$. Using the extrema, we know $f(x)$ is increasing on the intervals $(-\infty, 1)$ and $(3, \infty)$, and $f(x)$ is decreasing on the interval $(1,3)$.

d) Find the points of inflection. $f''(x)$ exists for all real numbers. Solve
$$f''(x) = 0$$
$$6x - 12 = 0$$
$$6x = 12$$
$$x = 2$$
There is a possible inflection point at $x = 2$.
$$f(2) = (2)^3 - 6(2)^2 + 9(2) + 1 = 3$$
Thus, the point $(2,3)$ on the graph is a possible inflection point.

e) To determine concavity, we use 2 to divide the real number line into two intervals, A: $(-\infty, 2)$ and B: $(2, \infty)$. Then test a point in each interval.
A: Test 0, $f''(0) = 6(0) - 12 = -12 < 0$
B: Test 3, $f''(3) = 6(3) - 12 = 6 > 0$
Thus, $f(x)$ is concave down on the interval $(-\infty, 2)$ and concave up on the interval $(2, \infty)$ and $(2,3)$ is an inflection point.

f) We sketch the graph. Additional points may be found as necessary.

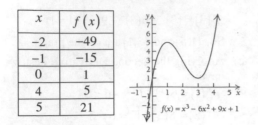

x	$f(x)$
-2	-49
-1	-15
0	1
4	5
5	21

$f(x) = x^3 - 6x^2 + 9x + 1$

23. $f(x) = x^4 - 2x^3$

a) First, find $f'(x)$ and $f''(x)$.

$$f'(x) = 4x^3 - 6x^2$$
$$f''(x) = 12x^2 - 12x$$

The domain of f is $\mathbb{R}$.

b) $f'(x)$ exists for all values of x, so the only critical points of f are where $f'(x) = 0$.

$$f'(x) = 0$$
$$4x^3 - 6x^2 = 0$$
$$2x^2(2x - 3) = 0$$
$$2x^2 = 0 \quad \text{or} \quad 2x - 3 = 0$$
$$x = 0 \quad \text{or} \quad x = \frac{3}{2}$$

The critical values are 0 and $\frac{3}{2}$.

$$f(0) = (0)^4 - 2(0)^3 = 0$$
$$f\left(\frac{3}{2}\right) = \left(\frac{3}{2}\right)^4 - 2\left(\frac{3}{2}\right)^3 = -\frac{27}{16}$$

So, $(0,0)$ and $\left(\frac{3}{2}, -\frac{27}{16}\right)$ are on the graph.

c) Applying the Second Derivative Test, we have:

$$f''(0) = 12(0)^2 - 12(0) = 0$$

The Second Derivative Test fails, we will have to use the First Derivative Test for $x = 0$. Divide $\left(-\infty, \frac{3}{2}\right)$ into two intervals, A: $(-\infty, 0)$ and B: $\left(0, \frac{3}{2}\right)$, and test a point in each interval at the top of the next column.

Testing a point in each interval, we have:
A: Test -1,

$$f'(-1) = 4(-1)^3 - 6(-1)^2 = -10 < 0$$

B: Test 1,

$$f'(1) = 4(1)^3 - 6(1)^2 = -2 < 0$$

Since, f is decreasing on both intervals, $(0,0)$ is not a relative extremum.

We use the Second Derivative Test for $x = \frac{3}{2}$.

$$f''\left(\frac{3}{2}\right) = 12\left(\frac{3}{2}\right)^2 - 12\left(\frac{3}{2}\right) = 9 > 0$$

Therefore, $\left(\frac{3}{2}, -\frac{27}{16}\right)$ is a relative minimum.

Thus, $f(x)$ is increasing on the interval $\left(\frac{3}{2}, \infty\right)$, and $f(x)$ is decreasing on the intervals $(-\infty, 0)$ and $\left(0, \frac{3}{2}\right)$.

d) Find the points of inflection. $f''(x)$ exists for all real numbers. Solve

$$f''(x) = 0$$
$$12x^2 - 12x = 0$$
$$12x(x - 1) = 0$$
$$12x = 0 \quad \text{or} \quad x - 1 = 0$$
$$x = 0 \quad \text{or} \quad x = 1$$

There are a possible inflection points at $x = 0$ and $x = 1$.

$f(0) = 0$ found earlier

$$f(1) = (1)^4 - 2(1)^3 = -1$$

Thus, the points $(0,0)$ and $(1,-1)$ on the graph are possible inflection points.

e) Use 0 and 1 to divide the real number line into two intervals, A: $(-\infty, 0)$, B: $(0,1)$, and C: $(1, \infty)$. Then test a point in each interval:

A: Test -1,

$$f''(-1) = 12(-1)^2 - 12(-1) = 24 > 0$$

B: Test $\frac{1}{2}$,

$$f''\left(\frac{1}{2}\right) = 12\left(\frac{1}{2}\right)^2 - 12\left(\frac{1}{2}\right) = -3 < 0$$

The solution is continued on the next page.

Continued from the previous page.
C: Test 2,

$$f''(2) = 12(2)^2 - 12(2) = 24 > 0$$

Thus, $f(x)$ is concave up on the intervals $(-\infty, 0)$ and $(1, \infty)$ and concave down on the interval $(0,1)$. Also, the points $(0,0)$ and $(1,-1)$ are inflection points.

f) Using the preceding information, we sketch the graph. Additional points may be found as necessary.

x	$f(x)$
-2	32
-1	3
2	0
3	27

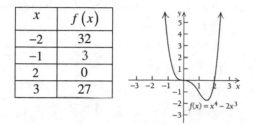

25. $f(x) = x^3 - 6x^2 - 135x$

a) $f'(x) = 3x^2 - 12x - 135$
    $f''(x) = 6x - 12$
    The domain of f is $\mathbb{R}$.

b) $f'(x)$ exists for all values of x, so the only critical points of f are where $f'(x) = 0$.

$$3x^2 - 12x - 135 = 0$$
$$x^2 - 4x - 45 = 0$$
$$(x-9)(x+5) = 0$$
$$x - 9 = 0 \quad \text{or} \quad x+5 = 0$$
$$x = 9 \quad \text{or} \quad x = -5$$

The critical values are -5 and 9.
The function values are

$$f(-5) = (-5)^3 - 6(-5)^2 - 135(-5)$$
$$= -125 - 150 + 675$$
$$= 400$$

$$f(9) = (9)^3 - 6(9)^2 - 135(9)$$
$$= 729 - 486 - 1215$$
$$= -972$$

The critical points $(-5, 400)$ and $(9, -972)$ are on the graph.

c) Applying the Second Derivative Test, we have:

$$f''(-5) = 6(-5) - 12 = -30 - 12$$
$$= -42 < 0$$

The critical point $(-5, 400)$ is a relative maximum.

$$f''(9) = 6(9) - 12 = 54 - 12$$
$$= 42 > 0$$

The critical point $(9, -972)$ is a relative minimum.
If we use the points -5 and 9 to divide the real number line into three intervals $(-\infty, 5)$, $(-5, 9)$, and $(9, \infty)$ we see that $f(x)$ is increasing on the intervals $(-\infty, -5)$ and $(9, \infty)$ and $f(x)$ is decreasing on the interval $(-5, 9)$.

d) Find the points of inflection. $f''(x)$ exists for all values of x, so the only possible inflection points occur when $f''(x) = 0$. We set the second derivative equal to zero and find possible inflection points.

$$6x - 12 = 0$$
$$6x = 12$$
$$x = 2$$

The only possible inflection point is 2.

$$f(2) = (2)^3 - 6(2)^2 - 135(2)$$
$$= 8 - 24 - 270$$
$$= -286$$

The point $(2, -286)$ is a possible inflection point on the graph.

e) To determine concavity we use 2 to divide the real number line into two intervals, A: $(-\infty, 2)$ and B: $(2, \infty)$, and we test a point in each interval.

A: Test 0, $f''(0) = 6(0) - 12 = -12 < 0$

B: Test 3, $f''(3) = 6(3) - 12 = 6 > 0$

We see that f is concave down on the interval $(-\infty, 2)$ and concave up on the interval $(2, \infty)$. Therefore $(2, -286)$ is an inflection point.

f) We sketch the graph using the preceding information. Additional function values may also be calculated as necessary.

x	$f(x)$
-11	-572
-10	-250
-3	324
0	0
2	-286
5	-700
15	0
16	400

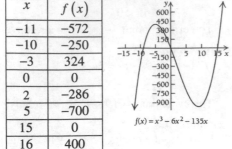

$f(x) = x^3 - 6x^2 - 135x$

27. $f(x) = x^4 - 4x^3 + 10$

a) $f'(x) = 4x^3 - 12x^2$

   $f''(x) = 12x^2 - 24x$

   The domain of f is $\mathbb{R}$.

b) $f'(x)$ exists for all values of x, so the only critical points of f are where $f'(x) = 0$.

   $4x^3 - 12x^2 = 0$

   $4x^2(x - 3) = 0$

   $4x^2 = 0$ or $x - 3 = 0$

   $x = 0$ or $x = 3$

   The critical values are 0 and 3.

   $f(0) = (0)^4 - 4(0)^3 + 10 = 10$

   $f(3) = (3)^4 - 4(3)^3 + 10 = -17$

   The critical points $(0, 10)$ and $(3, -17)$ are on the graph.

c) Applying the Second Derivative Test, we have:

   $f''(0) = 12(0)^2 - 24(0) = 0$

   The test fails, we will use the First Derivative Test.

   Divide $(-\infty, 3)$ into two intervals,

   A: $(-\infty, 0)$ and B: $(0, 3)$, and test a point in each interval.

   A: Test -1,

   $f'(-1) = 4(-1)^3 - 12(-1)^2 = -16 < 0$

   B: Test 1, $f'(1) = 4(1)^3 - 12(1)^2 = -8 < 0$

   Since, f is decreasing on both intervals, $(0, 10)$ is not a relative extremum.

We use the Second Derivative Test for $x = 3$.

$f''(3) = 12(3)^2 - 24(3) = 36 > 0$

The critical point $(3, -17)$ is a relative minimum.

When we applied the First Derivative Test, we saw that $f(x)$ was decreasing on the intervals $(-\infty, 0)$ and $(0, 3)$. Since $(3, -17)$ is a relative minimum, we know that $f(x)$ is increasing on $(3, \infty)$.

d) Find the points of inflection. $f''(x)$ exists for all values of x, so the only possible inflection points occur when $f''(x) = 0$.

   $12x^2 - 24x = 0$

   $12x(x - 2) = 0$

   $12x = 0$ or $x - 2 = 0$

   $x = 0$ or $x = 2$

   Possible points of inflection occur at $x = 0$ and $x = 2$.

   $f(0) = (0)^4 - 4(0)^3 + 10 = 10$

   $f(2) = (2)^4 - 4(2)^3 + 10 = -6$

   The points $(0, 10)$ and $(2, -6)$ are possible inflection points on the graph.

e) To determine concavity we use 0 and 2 to divide the real number line into three intervals,

   A: $(-\infty, 0)$, B: $(0, 2)$, and C: $(2, \infty)$, Then test a point in each interval.

   A: Test -1,

   $f''(-1) = 12(-1)^2 - 24(-1) = 36 > 0$

   B: Test 1,

   $f''(1) = 12(1)^2 - 24(1) = -12 < 0$

   C: Test 3,

   $f''(3) = 12(3)^2 - 24(3) = 36 > 0$

   We see that f is concave up on the intervals $(-\infty, 0)$ and $(2, \infty)$ and concave down on the interval $(0, 2)$. Therefore both $(0, 10)$ and $(2, -6)$ are inflection points.

f) We sketch the graph using the preceding information. Additional function values may also be calculated as necessary.

x	$f(x)$
-2	58
-1	15
1	7
4	10
5	135

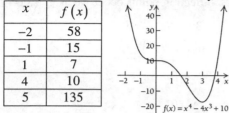

$f(x) = x^4 - 4x^3 + 10$

29. $f(x) = x^3 - 6x^2 + 12x - 6$

a) $f'(x) = 3x^2 - 12x + 12$

$f''(x) = 6x - 12$

The domain of f is $\mathbb{R}$.

b) $f'(x)$ exists for all values of x, so the only critical points of f are where $f'(x) = 0$.

$3x^2 - 12x + 12 = 0$

$x^2 - 4x + 4 = 0$ Dividing by 3

$(x - 2)^2 = 0$

$x - 2 = 0$

$x = 2$

The critical value is 2.

$f(2) = (2)^3 - 6(2)^2 + 12(2) - 6 = 2$

The critical point $(2,2)$ is on the graph.

c) Applying the Second Derivative Test, we have:

$f''(2) = 6(2) - 12 = 0$

The test fails, we will use the First Derivative Test.

Divide the real line into two intervals,

A: $(-\infty, 2)$ and B: $(2, \infty)$, and test a point in each interval.

A: Test 0,

$f'(0) = 3(0)^2 - 12(0) + 12 = 12 > 0$

B: Test 3,

$f'(3) = 3(3)^2 - 12(3) + 12 = 3 > 0$

Since, f is increasing on both intervals, $(2,2)$ is not a relative extremum.

When we applied the First Derivative Test, we saw that $f(x)$ was increasing on the intervals $(-\infty, 2)$ and $(2, \infty)$.

d) Find the points of inflection. $f''(x)$ exists for all values of x, so the only possible inflection points occur when $f''(x) = 0$.

$6x - 12 = 0$

$6x = 12$

$x = 2$

We have already seen that $f(2) = 2$, so the point $(2,2)$ is a possible inflection point on the graph.

e) To determine concavity we use 2 to divide the real number line into two intervals, A: $(-\infty, 2)$ and B: $(2, \infty)$, Then test a point in each interval.

A: Test 0, $f''(0) = 6(0) - 12 = -12 < 0$

B: Test 3, $f''(3) = 6(3) - 12 = 6 > 0$

We see that $f(x)$ is concave down on the interval $(-\infty, 2)$ and concave up on the interval $(2, \infty)$. Therefore, the point $(2,2)$ is an inflection point.

f) We sketch the graph using the preceding information. Additional function values may also be calculated as necessary.

x	$f(x)$
-1	-25
0	-6
1	1
3	3
4	10

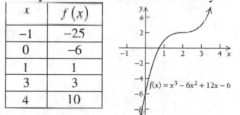

$f(x) = x^3 - 6x^2 + 12x - 6$

31. $f(x) = 5x^3 - 3x^5$

a) $f'(x) = 15x^2 - 15x^4$

$f''(x) = 30x - 60x^3$

The domain of f is $\mathbb{R}$.

b) $f'(x)$ exists for all values of x, so the only critical points of f are where $f'(x) = 0$.

$15x^2 - 15x^4 = 0$

$15x^2(1 - x^2) = 0$

$15x^2 = 0$ or $1 - x^2 = 0$

$x = 0$ or $x = \pm 1$

The critical values are -1, 0, and 1.

We evaluate the function at the critical values at the top of the next page.

Using the critical values from the previous page, we have:

$f(-1) = 5(-1)^3 - 3(-1)^5 = -2$

$f(0) = 5(0)^3 - 3(0)^5 = 0$

$f(1) = 5(1)^3 - 3(1)^5 = 2$

The critical points $(-1, -2)$, $(0, 0)$ and $(1, 2)$ are on the graph.

c) Applying the Second Derivative Test, we have:

$f''(-1) = 30(-1) - 60(-1)^3 = 30 > 0$

So, the critical point $(-1, -2)$ is a relative minimum.

$f''(0) = 30(0) - 60(0)^3 = 0$

The test fails, we will use the First Derivative Test.

Divide $(-1, 1)$ into two intervals, A: $(-1, 0)$ and B: $(0, 1)$, and test a point in each interval.

A: Test $-\dfrac{1}{2}$,

$f'\left(-\dfrac{1}{2}\right) = 15\left(-\dfrac{1}{2}\right)^2 - 15\left(-\dfrac{1}{2}\right)^4$

$= \dfrac{45}{16} > 0$

B: Test $\dfrac{1}{2}$,

$f'\left(\dfrac{1}{2}\right) = 15\left(\dfrac{1}{2}\right)^2 - 15\left(\dfrac{1}{2}\right)^4$

$= \dfrac{45}{16} > 0$

Since, f is increasing on both intervals, $(0, 0)$ is not a relative extremum.

We use the Second Derivative Test for $x = 1$.

$f''(1) = 30(1) - 60(1)^3 = -30 < 0$

The critical point $(1, 2)$ is a relative maximum.

When we applied the First Derivative Test, we saw that $f(x)$ was increasing on the intervals $(-1, 0)$ and $(0, 1)$. Since $(-1, -2)$ is a relative minimum, we know that $f(x)$ is decreasing on $(-\infty, -1)$. Since $(1, 2)$ is a relative maximum, we know that $f(x)$ is decreasing on $(1, \infty)$.

d) Find the points of inflection. $f''(x)$ exists for all values of x, so the only possible inflection points occur when $f''(x) = 0$.

$30x - 60x^3 = 0$

$30x(1 - 2x^2) = 0$

$30x = 0$ or $1 - 2x^2 = 0$

$x = 0$ or $x^2 = \dfrac{1}{2}$

$x = 0$ or $x = \pm\sqrt{\dfrac{1}{2}} = \pm\dfrac{1}{\sqrt{2}}$

$f\left(-\dfrac{1}{\sqrt{2}}\right) = 5\left(-\dfrac{1}{\sqrt{2}}\right)^3 - 3\left(-\dfrac{1}{\sqrt{2}}\right)^5$

$= -1.237$

$f(0) = 5(0)^3 - 3(0)^5 = 0$

$f\left(\dfrac{1}{\sqrt{2}}\right) = 5\left(\dfrac{1}{\sqrt{2}}\right)^3 - 3\left(\dfrac{1}{\sqrt{2}}\right)^5$

$= 1.237$

The points $\left(-\dfrac{1}{\sqrt{2}}, -1.237\right)$, $(0, 0)$ and

$\left(\dfrac{1}{\sqrt{2}}, 1.237\right)$ are possible inflection points on the graph.

e) To determine concavity we use

$-\dfrac{1}{\sqrt{2}}$, 0, and $\dfrac{1}{\sqrt{2}}$ to divide the real number

line into four intervals, A: $\left(-\infty, -\dfrac{1}{\sqrt{2}}\right)$,

B: $\left(-\dfrac{1}{\sqrt{2}}, 0\right)$, C: $\left(0, \dfrac{1}{\sqrt{2}}\right)$, and

D: $\left(\dfrac{1}{\sqrt{2}}, \infty\right)$.

Testing the point in each interval, we have

A: Test -1, $f''(-1) = 30(-1) - 60(-1)^3$

$= 30 > 0$

B: Test $-\dfrac{1}{2}$,

$f''\left(-\dfrac{1}{2}\right) = 30\left(-\dfrac{1}{2}\right) - 60\left(-\dfrac{1}{2}\right)^3$

$= -\dfrac{15}{2} < 0$

The solution is continued on the next page.

Continued from the previous page.

C: Test $\dfrac{1}{2}$,

$$f''\left(\frac{1}{2}\right) = 30\left(\frac{1}{2}\right) - 60\left(\frac{1}{2}\right)^3$$

$$= \frac{15}{2} > 0$$

D: Test 1,

$$f''(1) = 30(1) - 60(1)^3$$

$$= -30 < 0$$

We see that f is concave up on the intervals

$\left(-\infty, -\dfrac{1}{\sqrt{2}}\right)$ and $\left(0, \dfrac{1}{\sqrt{2}}\right)$ and concave

down on the intervals

$\left(-\dfrac{1}{\sqrt{2}}, 0\right)$ and $\left(\dfrac{1}{\sqrt{2}}, \infty\right)$. Therefore, the

points $\left(-\dfrac{1}{\sqrt{2}}, -1.237\right)$, $(0,0)$ and

$\left(\dfrac{1}{\sqrt{2}}, 1.237\right)$ are inflection points.

f) We sketch the graph using the preceding information. Additional function values may also be calculated as necessary.

x	$f(x)$
-2	56
$-\dfrac{1}{2}$	$-\dfrac{17}{32}$
$\dfrac{1}{2}$	$\dfrac{17}{32}$
2	-56

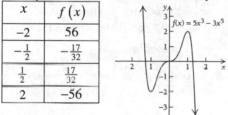

33. $f(x) = x^2(1-x)^2 = x^2 - 2x^3 + x^4$

a) $f'(x) = 2x - 6x^2 + 4x^3$

$f''(x) = 2 - 12x + 12x^2$

b) $f'(x)$ exists for all values of x, so the only critical points of f are where $f'(x) = 0$.

$$2x - 6x^2 + 4x^3 = 0$$

$$2x(1 - 3x + 2x^2) = 0$$

$$2x(1-x)(1-2x) = 0$$

$2x = 0$ or $1 - 2x = 0$ or $1 - x = 0$

$x = 0$ or $x = \dfrac{1}{2}$ or $x = 1$

The critical values are 0, $\dfrac{1}{2}$, and 1.

Evaluating the function at the critical values, we have:

$$f(0) = (0)^2(1-(0))^2 = 0$$

$$f\left(\frac{1}{2}\right) = \left(\frac{1}{2}\right)^2\left(1-\left(\frac{1}{2}\right)\right)^2 = \frac{1}{16}$$

$$f(1) = (1)^2(1-(1))^2 = 0$$

The critical points $(0,0)$, $\left(\dfrac{1}{2}, \dfrac{1}{16}\right)$, and

$(1,0)$ are on the graph.

c) Applying the Second Derivative Test, we have:

$$f''(0) = 2 - 12(0) + 12(0)^2 = 2 > 0$$

So, the critical point $(0,0)$ is a relative minimum.

$$f''\left(\frac{1}{2}\right) = 2 - 12\left(\frac{1}{2}\right) + 12\left(\frac{1}{2}\right)^2 = -1 < 0$$

So, the critical point $\left(\dfrac{1}{2}, \dfrac{1}{16}\right)$ is a relative

maximum.

$$f''(1) = 2 - 12(1) + 12(1)^2 = 2 > 0$$

So, the critical point $(1,0)$ is a relative minimum. The solution is continued.

We use the points 0, $\dfrac{1}{2}$, and 1 to divide the

real number line into four intervals,

$(-\infty, 0)$, $\left(0, \dfrac{1}{2}\right)$, $\left(\dfrac{1}{2}, 1\right)$, and $(1, \infty)$, we

know that $f(x)$ is decreasing on the

intervals $(-\infty, 0)$ and $\left(\dfrac{1}{2}, 1\right)$, and $f(x)$ is

increasing on the intervals

$\left(0, \dfrac{1}{2}\right)$ and $(1, \infty)$.

d) Find the points of inflection. $f''(x)$ exists for all values of x, so the only possible inflection points occur when $f''(x) = 0$.

$$2 - 12x + 12x^2 = 0$$

$$1 - 6x + 6x^2 = 0 \qquad \text{Dividing by 2}$$

Using the quadratic formula we have:

$$x = \frac{3 \pm \sqrt{3}}{6}$$

$x \approx 0.211$ or $x \approx 0.789$

The solution is continued on the next page.

Evaluating the function at the values found on the previous page, we have:

$f(0.211) \approx 0.028$

$f(0.789) \approx 0.028$

The points, $(0.211, 0.028)$ and

$(0.789, 0.028)$ are possible inflection points on the graph.

e) To determine concavity we use 0.211 and 0.789 to divide the real number line into three intervals,

$A:(-\infty, 0.211)$, $B: (0.211, 0.789)$,

and C: $(0.789, \infty)$

Then test a point in each interval.

A: Test 0, $\quad f''(0) = 2 - 12(0) + 12(0)^2$

$\qquad = 2 > 0$

B: Test $\dfrac{1}{2}$, $f''\left(\dfrac{1}{2}\right) = 2 - 12\left(\dfrac{1}{2}\right) + 12\left(\dfrac{1}{2}\right)^2$

$\qquad = -1 < 0$

C: Test 1, $\quad f''(1) = 2 - 12(1) + 12(1)^2$

$\qquad = 2 > 0$

We see that f is concave up on the intervals $(-\infty, 0.211)$ and $(0.789, \infty)$ and concave down on the interval $(0.211, 0.789)$.

Therefore, the points $(0.211, 0.028)$ and

$(0.789, 0.028)$ are inflection points.

f) We sketch the graph using the preceding information. Additional function values may also be calculated as necessary.

x	$f(x)$
-2	36
-1	4
2	4
3	36

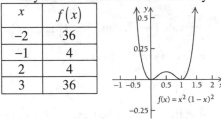

$f(x) = x^2 (1-x)^2$

35. $f(x) = (x-1)^{\frac{2}{3}}$

a) $f'(x) = \dfrac{2}{3}(x-1)^{-\frac{1}{3}} = \dfrac{2}{3(x-1)^{\frac{1}{3}}}$

$f''(x) = -\dfrac{2}{9}(x-1)^{-\frac{4}{3}} = -\dfrac{2}{9(x-1)^{\frac{4}{3}}}$

The domain of f is $\mathbb{R}$.

b) $f'(x)$ does not exist for $x = 1$. The equation $f'(x) = 0$ has no solution, therefore, $x = 1$ is the only critical point.

$f(1) = (1-1)^{\frac{2}{3}} = 0$.

So, the critical point, $(1, 0)$ is on the graph.

c) We apply the First Derivative Test. We use 1 to divide the real number line into two intervals $A:(-\infty, 1)$ and B: $(1, \infty)$.

Testing a point in each interval, we have

A: Test 0, $f'(0) = \dfrac{2}{3((0)-1)^{\frac{1}{3}}} = -\dfrac{2}{3} < 0$

B: Test 2, $f'(2) = \dfrac{2}{3((2)-1)^{\frac{1}{3}}} = \dfrac{2}{3} > 0$

$(1, 0)$ is a relative minimum. $f(x)$ is decreasing on the interval $(-\infty, 1)$ and increasing on the interval $(1, \infty)$.

d) Find the points of inflection. $f''(x)$ does not exist when $x = 1$. The equation $f''(x) = 0$ has no solution, so $x = 1$ is the only possible inflection point. We know that $f(1) = 0$.

e) To determine concavity, we divide the real number line into two intervals,

$A:(-\infty, 1)$ and B: $(1, \infty)$ and then we test a point in each interval.

A: Test 0, $f''(0) = -\dfrac{2}{9((0)-1)^{\frac{4}{3}}} = -\dfrac{2}{9} < 0$

B: Test 2, $f''(2) = -\dfrac{2}{9((2)-1)^{\frac{4}{3}}} = -\dfrac{2}{9} < 0$

Thus, $f(x)$ is concave down on the intervals $(-\infty, 1)$ and $(1, \infty)$. Therefore, the point $(1, 0)$ is not an inflection point.

f) We sketch the graph using the preceding information. Additional function values may also be calculated as necessary.

x	$f(x)$
-7	4
0	1
2	1
9	4

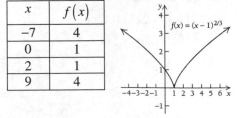

$f(x) = (x-1)^{2/3}$

37. $f(x) = (x-3)^{\frac{1}{3}} - 1$

a) $f'(x) = \frac{1}{3}(x-3)^{-\frac{2}{3}} = \frac{1}{3(x-3)^{\frac{2}{3}}}$

$f''(x) = -\frac{2}{9}(x-3)^{-\frac{5}{3}} = -\frac{2}{9(x-3)^{\frac{5}{3}}}$

The domain of f is $\mathbb{R}$.

b) $f'(x)$ does not exist for $x = 3$. The equation $f'(x) = 0$ has no solution, therefore, $x = 3$ is the only critical value.

$f(3) = ((3)-3)^{\frac{1}{3}} - 1 = -1$.

So, the critical point, $(3, -1)$ is on the graph.

c) We apply the First Derivative Test. We use 3 to divide the real number line into two intervals $A : (-\infty, 3)$ and $B : (3, \infty)$ and then we test a point in each interval.

A: Test 2, $f'(2) = \frac{1}{3((2)-3)^{\frac{2}{3}}} = \frac{1}{3} > 0$

B: Test 4, $f'(4) = \frac{1}{3((4)-3)^{\frac{2}{3}}} = \frac{1}{3} > 0$

$f(x)$ is increasing on both intervals $(-\infty, 3)$ and $(3, \infty)$, therefore $(3, -1)$ is not a relative extremum.

d) Find the points of inflection. $f''(x)$ does not exist when $x = 3$. The equation $f''(x) = 0$ has no solution, so at $x = 3$ is the only possible inflection point. We know that $f(3) = -1$.

e) To determine concavity, we divide the real number line into two intervals, $A : (-\infty, 3)$ and $B : (3, \infty)$ and then we test a point in each interval.

A: Test 2, $f''(2) = -\frac{2}{9((2)-3)^{\frac{5}{3}}} = \frac{2}{9} > 0$

B: Test 4, $f''(4) = -\frac{2}{9((4)-3)^{\frac{5}{3}}} = -\frac{2}{9} < 0$

Thus, $f(x)$ is concave up on the interval $(-\infty, 3)$ and $f(x)$ is concave down on the interval $(3, \infty)$. Therefore, the point $(3, -1)$ is an inflection point.

f) We sketch the graph using the preceding information. Additional function values may also be calculated as necessary.

x	$f(x)$
-5	-3
2	-2
4	0
11	1

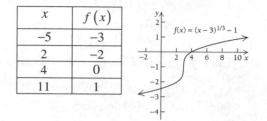

39. $f(x) = -2(x-4)^{\frac{2}{3}} + 5$

a) $f'(x) = -\frac{4}{3}(x-4)^{-\frac{1}{3}} = -\frac{4}{3(x-4)^{\frac{1}{3}}}$

$f''(x) = \frac{4}{9}(x-4)^{-\frac{4}{3}} = \frac{4}{9(x-4)^{\frac{4}{3}}}$

The domain of f is $\mathbb{R}$.

b) $f'(x)$ does not exist for $x = 4$. The equation $f'(x) = 0$ has no solution, therefore, $x = 4$ is the only critical point.

$f(4) = -2((4)-4)^{\frac{2}{3}} + 5 = 5$.

So, the critical point $(4, 5)$, is on the graph.

c) We apply the First Derivative Test. We use 4 to divide the real number line into two intervals $A : (-\infty, 4)$ and $B : (4, \infty)$ and then we test a point in each interval at the top of the next column:

A: Test 3, $f'(3) = -\frac{4}{3((3)-4)^{\frac{1}{3}}} = \frac{4}{3} > 0$

B: Test 5, $f'(5) = -\frac{4}{3((5)-4)^{\frac{1}{3}}} = -\frac{4}{3} < 0$

Thus, $(4, 5)$ is a relative maximum. We also know that $f(x)$ is increasing on the interval $(-\infty, 4)$ and decreasing on the interval $(4, \infty)$.

d) Find the points of inflection. $f''(x)$ does not exist when $x = 4$. The equation $f''(x) = 0$ has no solution, so $x = 4$ is the only possible inflection point. We know that $f(4) = 5$.

e) To determine concavity, we divide the real number line into two intervals, $A : (-\infty, 4)$ and $B : (4, \infty)$ and then we test a point in each interval at the top of the next page.

Testing a point in each interval, we have:

A: Test 3, $f''(3) = \dfrac{4}{9\left((3)-4\right)^{4/3}} = \dfrac{4}{9} > 0$

B: Test 5, $f''(5) = \dfrac{4}{9\left((5)-4\right)^{4/3}} = \dfrac{4}{9} > 0$

Thus, $f(x)$ is concave up on both intervals $(-\infty, 4)$ and $(4, \infty)$ Therefore, the point $(4,5)$ is not an inflection point.

f) We sketch the graph using the preceding information. Additional function values may also be calculated as necessary.

x	$f(x)$
−4	−3
3	3
5	3
12	−3

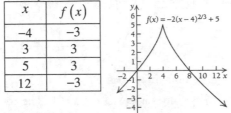

$f(x) = -2(x-4)^{2/3} + 5$

41. $f(x) = -x\sqrt{1-x^2} = -x\left(1-x^2\right)^{1/2}$

a) $f'(x) = -x \cdot \dfrac{1}{2}\left(1-x^2\right)^{-1/2}(-2x)$

$\qquad + \left(1-x^2\right)^{1/2} \cdot (-1)$

$\qquad = \dfrac{2x^2-1}{\left(1-x^2\right)^{1/2}} = \dfrac{2x^2-1}{\sqrt{1-x^2}}$

$f''(x) = \left(2x^2-1\right)\left(-\dfrac{1}{2}\right)\left(1-x^2\right)^{-3/2}(-2x) +$

$\qquad \left(1-x^2\right)^{-1/2}(4x)$

$\qquad = \dfrac{-2x^3+3x}{\left(1-x^2\right)^{3/2}}$

The domain of $f(x)$ is $[-1,1]$.

b) $f'(x)$ does not exist when $x = \pm 1$.

However, the domain of $f(x)$ is $[-1,1]$. Therefore, relative extrema cannot occur at $x = -1$ or $x = 1$ because there is not an open interval containing −1 or 1 on which the function is defined.

The other critical points occur where $f'(x) = 0$.

$\dfrac{2x^2-1}{\sqrt{1-x^2}} = 0$

$2x^2 - 1 = 0$

$x^2 = \dfrac{1}{2}$

$x = \pm\dfrac{1}{\sqrt{2}}$

The critical values are $-\dfrac{1}{\sqrt{2}}$ and $\dfrac{1}{\sqrt{2}}$.

$f\left(-\dfrac{1}{\sqrt{2}}\right) = -\left(-\dfrac{1}{\sqrt{2}}\right)\sqrt{1-\left(-\dfrac{1}{\sqrt{2}}\right)^2}$

$\qquad = \left(\dfrac{1}{\sqrt{2}}\right)\sqrt{1-\dfrac{1}{2}}$

$\qquad = \left(\dfrac{1}{\sqrt{2}}\right)\dfrac{1}{\sqrt{2}}$

$\qquad = \dfrac{1}{2}$

Evaluating the second critical value, we have:

$f\left(\dfrac{1}{\sqrt{2}}\right) = -\left(\dfrac{1}{\sqrt{2}}\right)\sqrt{1-\left(\dfrac{1}{\sqrt{2}}\right)^2}$

$\qquad = \left(-\dfrac{1}{\sqrt{2}}\right)\sqrt{1-\dfrac{1}{2}}$

$\qquad = \left(\dfrac{1}{\sqrt{2}}\right)\dfrac{1}{\sqrt{2}}$

$\qquad = -\dfrac{1}{2}$

Therefore, $\left(-\dfrac{1}{\sqrt{2}}, \dfrac{1}{2}\right)$ and $\left(\dfrac{1}{\sqrt{2}}, -\dfrac{1}{2}\right)$ are critical points on the graph.

c) We use the Second Derivative Test.

$f''\left(-\dfrac{1}{\sqrt{2}}\right) = -4 < 0$

The critical point $\left(-\dfrac{1}{\sqrt{2}}, \dfrac{1}{2}\right)$ is a relative maximum.

$f''\left(\dfrac{1}{\sqrt{2}}\right) = 4 > 0$

The critical point $\left(\dfrac{1}{\sqrt{2}}, -\dfrac{1}{2}\right)$ is a relative minimum.

The solution is continued on the next page.

If we use the points $-\dfrac{1}{\sqrt{2}}$ and $\dfrac{1}{\sqrt{2}}$ to divide

the interval $[-1,1]$ into three intervals

$$\left[-1,-\dfrac{1}{\sqrt{2}}\right),\ \left(-\dfrac{1}{\sqrt{2}},\dfrac{1}{\sqrt{2}}\right),\ \text{and}\ \left(\dfrac{1}{\sqrt{2}},1\right],$$

we see that $f(x)$ is increasing on the

intervals $\left(-1,-\dfrac{1}{\sqrt{2}}\right)$ and $\left(\dfrac{1}{\sqrt{2}},1\right)$ and

$f(x)$ is decreasing on the interval

$$\left(-\dfrac{1}{\sqrt{2}},\dfrac{1}{\sqrt{2}}\right).$$

d) Find the points of inflection. $f''(x)$ does not exist when $x=-1$ and $x=1$. However, inflection points cannot occur at those values because the domain of the function is $[-1,1]$. The remaining possible inflection points occur when $f''(x)=0$. We set the second derivative equal to zero and solve for the possible inflection points.

$$f''(x)=0$$

$$\dfrac{-2x^3+3x}{\left(1-x^2\right)^{5/2}}=0$$

$$-2x^3+3x=0$$

$$x\left(-2x^2+3\right)=0$$

$$x=0\quad\text{or}\quad 2x^2-3=0$$

$$x=0\quad\text{or}\quad x^2=\dfrac{3}{2}$$

$$x=0\quad\text{or}\quad x=\dfrac{\pm\sqrt{6}}{2}$$

Note that $f(x)$ is not defined for $x=\pm\dfrac{\sqrt{6}}{2}$.

Therefore, the only possible inflection point is $x=0$. Evaluating the function we have

$$f(0)=0\sqrt{1-(0)^2}=0.$$

Therefore, $(0,0)$ is a possible inflection point on the graph.

e) To determine concavity, we use 0 to divide the interval $(-1,1)$ into two intervals, A: $(-1,0)$ and B: $(0,1)$ and then we test a point in each interval.

Testing a point in each interval, we have:

A: Test $-\dfrac{1}{2},\ f''\left(-\dfrac{1}{2}\right)=\dfrac{-10}{3^{5/2}}<0$

B: Test $\dfrac{1}{2},\quad f''\left(\dfrac{1}{2}\right)=\dfrac{10}{3^{5/2}}>0$

Thus, $f(x)$ is concave down on the interval $(-1,0)$ and $f(x)$ is concave up on the interval $(0,1)$. Therefore, the point $(0,0)$ is an inflection point.

f) We sketch the graph using the preceding information. Additional function values may also be calculated as necessary.

x	$f(x)$
-1	0
$-\dfrac{1}{2}$	$\dfrac{\sqrt{3}}{4}$
$\dfrac{1}{2}$	$-\dfrac{\sqrt{3}}{4}$
1	0

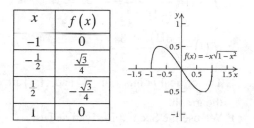

43. $f(x)=\dfrac{8x}{x^2+1}$

a) $f'(x)=\dfrac{\left(x^2+1\right)(8)-(2x)(8x)}{\left(x^2+1\right)^2}$ Quotient Rule

$$=\dfrac{8x^2+8-16x^2}{\left(x^2+1\right)^2}$$

$$=\dfrac{8-8x^2}{\left(x^2+1\right)^2}$$

Next we find the second derivative

$$f''(x)=\dfrac{\left(x^2+1\right)^2(-16x)-\left(8-8x^2\right)\left[2\left(x^2+1\right)^1(2x)\right]}{\left(\left(x^2+1\right)^2\right)^2}$$

$$=\dfrac{\left(x^2+1\right)\left[-16x\left(x^2+1\right)-4x\left(8-8x^2\right)\right]}{\left(x^2+1\right)^4}$$

$$=\dfrac{-16x^3-16x-32x+32x^3}{\left(x^2+1\right)^3}$$

$$=\dfrac{16x^3-48x}{\left(x^2+1\right)^3}$$

The domain of f is $\mathbb{R}$.

b) $f'(x)$ exists for all real numbers. Solve:

$$f'(x)=0$$

$$\frac{8-8x^2}{\left(x^2+1\right)^2}=0$$

$$8-8x^2=0$$

$$x^2=1$$

$$x=\pm1$$

The two critical values are $x=-1$ and $x=1$.
Evaluating the function at the critical values, we have:

$$f(-1)=\frac{8(-1)}{(-1)^2+1}=-\frac{8}{2}=-4$$

$$f(1)=\frac{8(1)}{(1)^2+1}=\frac{8}{2}=4$$

The critical points $(-1,-4)$ and $(1,4)$ are on the graph.

c) We use the Second Derivative Test.

$$f''(-1)=4>0$$

So the point $(-1,-4)$ is a relative minimum.

$$f''(1)=-4<0$$

So the point $(1,4)$ is a relative maximum.

$f(x)$ is decreasing on the intervals $(-\infty,1)$ and $(1,\infty)$, and $f(x)$ is increasing on the interval $(-1,1)$.

d) Find the points of inflection. $f''(x)$ exists for all real numbers, so the only possible points of inflection occur when $f''(x)=0$.

$$\frac{16x^3-48x}{\left(x^2+1\right)^3}=0$$

$$16x^3-48x=0$$

$$16x\left(x^2-3\right)=0$$

$$16x=0 \quad \text{or} \quad x^2-3=0$$

$$x=0 \quad \text{or} \quad x^2=3$$

$$x=0 \quad \text{or} \quad x=\pm\sqrt{3}$$

There are three possible inflection points $x=-\sqrt{3},0,$ and $\sqrt{3}$.

$$f\left(-\sqrt{3}\right)=-2\sqrt{3}$$

$$f(0)=0$$

$$f\left(\sqrt{3}\right)=2\sqrt{3}$$

The points $\left(-\sqrt{3},-2\sqrt{3}\right)$, $(0,0)$, and $\left(\sqrt{3},2\sqrt{3}\right)$ are three possible inflection points on the graph.

e) To determine concavity we use $-\sqrt{3},0,$ and $\sqrt{3}$ to divide the real number line into four intervals, A: $\left(-\infty,-\sqrt{3}\right)$, B: $\left(-\sqrt{3},0\right)$, C: $\left(0,\sqrt{3}\right)$, and D: $\left(\sqrt{3},\infty\right)$

We will test a point in each interval.

A: Test $-2, f''(-2)=-\frac{32}{125}<0$

B: Test $-1, f''(-1)=4>0$

C: Test $1, f''(1)=-4<0$

D: Test $2, f''(2)=\frac{32}{125}>0$

We see that f is concave down on the intervals $\left(-\infty,-\sqrt{3}\right)$ and $\left(0,\sqrt{3}\right)$ and concave up on the intervals $\left(-\sqrt{3},0\right)$ and $\left(\sqrt{3},\infty\right)$. Therefore the points $\left(-\sqrt{3},-2\sqrt{3}\right)$, $(0,0)$, and $\left(\sqrt{3},2\sqrt{3}\right)$ are inflection points.

f) We sketch the graph using the preceding information. Additional function values may also be calculated as necessary.

x	$f(x)$
-3	$-\frac{12}{5}$
-2	$-\frac{16}{5}$
2	$\frac{16}{5}$
3	$\frac{12}{5}$

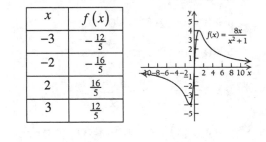

45. $f(x)=\dfrac{-4}{x^2+1}=-4\left(x^2+1\right)^{-1}$

a) $f'(x)=-4(-1)\left(x^2+1\right)^{-2}(2x)$ Extended Power Rule

$$f'(x)=8x\left(x^2+1\right)^{-2}$$

$$=\frac{8x}{\left(x^2+1\right)^2}$$

The solution is continued on the next page.

Next, we find the second derivative.

$f''(x)$

$$= \frac{\left(x^2+1\right)^2 (8) - (8x)\left(2\left(x^2+1\right)(2x)\right)}{\left(\left(x^2+1\right)^2\right)^2}$$

$$= \frac{\left(x^2+1\right)\left[\left(x^2+1\right)(8) - (8x)(2)(2x)\right]}{\left(x^2+1\right)^4}$$

$$= \frac{8x^2+8-32x^2}{\left(x^2+1\right)^3}$$

$$= \frac{8-24x^2}{\left(x^2+1\right)^3}$$

The domain of f is $\mathbb{R}$.

b) $f'(x)$ exists for all real numbers Solve:

$f'(x) = 0$

$\dfrac{8x}{\left(x^2+1\right)^2} = 0$ Multiplying by $\left(x^2+1\right)^2$

$8x = 0$

$x = 0$

The critical value is $x = 0$.

$f(0) = \dfrac{-4}{(0)^2+1} = -4$

The critical point $(0, -4)$ is on the graph.

c) We use the Second Derivative Test.

$f''(0) = 8 > 0$

So the point $(0, \ 4)$ is a relative minimum.

$f(x)$ is decreasing on the interval $(-\infty, 0)$,

and $f(x)$ is increasing on the interval $(0, \infty)$.

d) $f''(x)$ exists for all real numbers. Solve $f''(x) = 0$.

$\dfrac{8-24x^2}{\left(x^2+1\right)^3} = 0$

$8-24x^2 = 0$ Multiplying by $\left(x^2+1\right)^3$

$x^2 = \dfrac{1}{3}$

$x = \pm\dfrac{1}{\sqrt{3}}$

There are two possible inflection points

$-\dfrac{1}{\sqrt{3}}$ and $\dfrac{1}{\sqrt{3}}$.

$f\left(\pm\dfrac{1}{\sqrt{3}}\right) = \dfrac{-4}{\left(\pm\dfrac{1}{\sqrt{3}}\right)^2+1} = \dfrac{-4}{\dfrac{1}{3}+1} = \dfrac{-4}{\dfrac{4}{3}} = -3$

The points $\left(-\dfrac{1}{\sqrt{3}}, -3\right)$ and $\left(\dfrac{1}{\sqrt{3}}, -3\right)$ are possible inflection points on the graph.

e) To determine concavity we use $-\dfrac{1}{\sqrt{3}}$ and $\dfrac{1}{\sqrt{3}}$ to divide the real number line into three intervals, $A: \left(-\infty, -\dfrac{1}{\sqrt{3}}\right)$, $B: \left(-\dfrac{1}{\sqrt{3}}, \dfrac{1}{\sqrt{3}}\right)$, and $C: \left(\dfrac{1}{\sqrt{3}}, \infty\right)$.

Then test a point in each interval.

A: Test -1, $f''(-1) = -2 < 0$

B: Test 0, $f''(0) = 8 > 0$

C: Test 1, $f''(1) = -2 < 0$

We see that f is concave down on the intervals $\left(-\infty, -\dfrac{1}{\sqrt{3}}\right)$ and $\left(\dfrac{1}{\sqrt{3}}, \infty\right)$ and concave up on the interval $\left(-\dfrac{1}{\sqrt{3}}, \dfrac{1}{\sqrt{3}}\right)$.

Therefore the points $\left(-\dfrac{1}{\sqrt{3}}, -3\right)$ and $\left(\dfrac{1}{\sqrt{3}}, -3\right)$ are inflection points.

f) We sketch the graph using the preceding information. Additional function values may also be calculated as necessary.

x	$f(x)$
-4	$-\frac{4}{17}$
-2	$-\frac{4}{5}$
-1	-2
1	-2
2	$-\frac{4}{5}$
4	$-\frac{4}{17}$

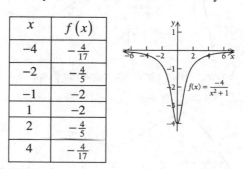

47. Answers may vary, one possible graph is:

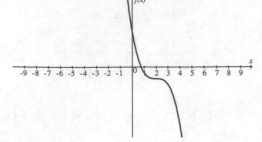

49. Answers may vary, one possible graph is:

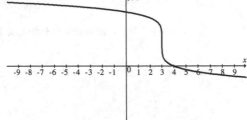

51. Answers may vary, one possible graph is:

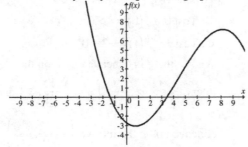

53. Answers may vary, one possible graph is:

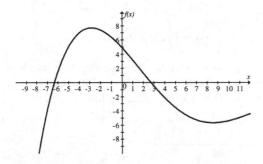

55. Answers may vary, one possible graph is:

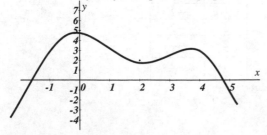

57. $R(x) = 50x - 0.5x^2$

$C(x) = 4x + 10$

$P(x) = R(x) - C(x)$

$ = (50x - 0.5x^2) - (4x + 10)$

$ = -0.5x^2 + 46x - 10$

We will restrict the domains of all three functions to $x \geq 0$ since a negative number of units cannot be produced and sold.

First graph $R(x) = 50x - 0.5x^2$

$R'(x) = 50 - x$

$R''(x) = -1$

Since $R'(x)$ exists for all $x \geq 0$, the only critical points are where $R'(x) = 0$.

$50 - x = 0$

$50 = x \qquad$ Critical Value

Find the function value at $x = 50$.

$R(50) = 50(50) - 0.5(50)^2$

$ = 2500 - 1250$

$ = 1250$

This critical point $(50, 1250)$ is on the graph.

We use the Second Derivative Test:

$R''(50) = -1 < 0$

The point $(50, 1250)$ is a relative maximum.

We use 50 to divide the interval $[0, \infty)$ into two intervals, $[0, 50)$ and $(50, \infty)$. We know that R is increasing on $(0, 50)$ and decreasing on $(50, \infty)$.

Next, find the inflection points. Since $R''(x)$ exists for all $x \geq 0$, and $R''(x) = -1$, there are no possible inflection points.

Furthermore, since $R''(x) < 0$ for all $x \geq 0$, R is concave down over the interval $(0, \infty)$.

Sketch the graph using the preceding information. The x-intercepts of R are found by solving $R(x) = 0$.

$50x - 0.5x^2 = 0$

$0.5x(100 - x) = 0$

$0.5x = 0 \text{ or } 100 - x = 0$

$x = 0 \text{ or } 100 = x$

The x-intercepts are $(0, 0)$ and $(100, 0)$.

The solution is continued on the next page.

Next, we graph $C(x) = 4x + 10$. This is a linear function with slope 4 and y-intercept $(0,10)$.

$C(x)$ is increasing over the entire domain $x \geq 0$ and has no relative extrema or points of inflection.

Finally, we graph $P(x) = -0.5x^2 + 46x - 10$

$P'(x) = -x + 46$

$P''(x) = -1$

Since $P'(x)$ exists for all $x \geq 0$, the only critical points occur when $P'(x) = 0$.

$-x + 46 = 0$

$\qquad 46 = x \qquad$ Critical Value

Find the function value at $x = 46$.

$P(46) = -0.5(46)^2 + 46(46) - 10$

$\qquad = -1058 + 2116 - 10$

$\qquad = 1048$

The critical point $(46, 1048)$ is on the graph.

We use the Second Derivative Test:

$P''(46) = -1 < 0$

The point $(46, 1048)$ is a relative maximum.

We use 46 to divide the interval $[0, \infty)$ into two intervals, $[0, 46)$ and $(46, \infty)$, we know that P is increasing on $(0, 46)$ and decreasing on $(46, \infty)$.

Next, find the inflection points. Since $P''(x)$ exists for all $x \geq 0$, and $P''(x) = -1$, there are no possible inflection points. Furthermore, since $P''(x) < 0$ for all $x \geq 0$, P is concave down over the interval $(0, \infty)$.

Sketch the graph using the preceding information.

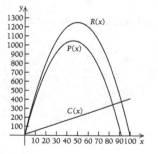

59. $p(x) = \dfrac{13x^3 - 240x^2 - 2460x + 585,000}{75,000}$

$p'(x) = \dfrac{39x^2 - 480x - 2460}{75,000}$

$p''(x) = \dfrac{78x - 480}{75,000}$

Since $p'(x)$ exists for all real numbers, the only critical points are where $p'(x) = 0$.

$\dfrac{39x^2 - 480x - 2460}{75,000} = 0$

$39x^2 - 480x - 2460 = 0$

Using the quadratic formula, we have:

$x = \dfrac{-b \pm \sqrt{b^2 - 4ac}}{2a}$

$\quad = \dfrac{-(-480) \pm \sqrt{(-480)^2 - 4(39)(-2460)}}{2(39)}$

$\quad = \dfrac{480 \pm \sqrt{614,160}}{78}$

$x \approx -3.89$ or $x \approx 16.20 \qquad$ Critical values

Since the domain of the function is $0 \leq x \leq 40$, we consider only $x \approx 16.20$

$p(16.20)$

$= \dfrac{13(16.20)^3 - 240(16.20)^2 - 2460(16.20) - 585,000}{75,000}$

≈ 7.17

The critical point $(16.20, 7.17)$ is on the graph.

We use the Second Derivative Test:

$p''(x) = \dfrac{78(16.20) - 480}{75,000} \approx 0.01 > 0$

The point $(16.20, 7.17)$ is a relative minimum.

If we use the point 16.20 to divide the domain into two intervals, $[0, 16.20)$ and $(16.20, 40]$, we know that p is decreasing on $(0, 16.20)$ and increasing on $(16.20, 40)$.

Next, we find the inflection points. $p''(x)$ exists for all real numbers, so the only possible inflection points are where $p''(x) = 0$. We solve this equation at the top of the next page.

$$p''(x) = 0$$

$$\frac{78x - 480}{75,000} = 0$$

$$78x - 480 = 0$$

$$78x = 480$$

$$x \approx 6.15$$

$$p(6.15)$$

$$= \frac{13(6.15)^3 - 240(6.15)^2 - 2460(6.15) - 585,000}{75,000}$$

$$\approx 7.52$$

The point $(6.15, 7.52)$ is a possible inflection point.

To determine concavity, we use 6.15 to divide the domain into two intervals
A: $[0, 6.15)$ and B: $(6.15, 40]$ and test a point in each interval.

A: Test 1, $p''(1) = \dfrac{78(1) - 480}{75,000} = -0.005 < 0$

B: Test 7, $p''(7) = \dfrac{78(7) - 480}{75,000} = 0.00088 > 0$

Then p is concave down on $(0, 6.15)$ and concave up on $(6.15, 40)$ and the point $(6.15, 7.52)$ is a point of inflection.

Sketch the graph for $0 \le x \le 40$ using the preceding information. Additional function values may be calculated if necessary.

x	$p(x)$
0	7.8
8	7.42
12	7.25
20	7.25
24	7.57
32	9.15
40	12.46

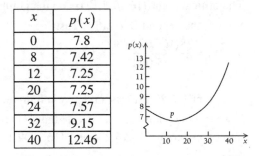

61. The monthly rainfall is approximated by

$$R(t) = -0.006t^4 + 0.213t^3 - 1.702t^2 + 0.615t + 27.745$$

To find the inflection points we find the first and second derivatives.

$$R'(t) = -0.024t^3 + 0.639t^2 - 3.404t + 0.615$$

$$R''(t) = -0.072t^2 + 1.278t - 3.404$$

Since $R''(t)$ exists for all real values, we solve $R''(t) = 0$ By the quadratic formula.

$$t = \frac{-1.278 \pm \sqrt{(1.278)^2 - 4(-0.072)(-3.404)}}{2(-0.072)}$$

$$t = 3.26 \quad \text{or} \quad t = 14.48$$

These are two possible points of inflection.

$$R(3.26) \approx 18.36$$

$$R(14.48) \approx 62.69$$

The two points of inflection are
$(3.26, 18.36)$ and $(14.48, 62.69)$. However, $t = 14.48$ is not in the domain of this function. The left most inflection point $(3.26, 18.36)$ implies that rate of change of the amount of rainfall is decreasing the fastest at this point.

63. ✎

65. $f(x) = ax^2 + bx + c, \qquad a \ne 0$

$$f'(x) = 2ax + b$$

$$f''(x) = 2a$$

Since $f'(x)$ exists for all real numbers, the only critical points occur when $f'(x) = 0$. We solve:

$$2ax + b = 0$$

$$2ax = -b$$

$$x = \frac{-b}{2a}$$

So the critical value will occur at $x = \dfrac{-b}{2a}$.

Applying the second derivative test, we see that

$$f''(x) = 2a > 0, \quad \text{for } a > 0$$

$$f''(x) = 2a < 0, \quad \text{for } a < 0$$

Therefore, a relative maximum occurs at $x = \dfrac{-b}{2a}$ when $a < 0$ and a relative minimum occurs at $x = \dfrac{-b}{2a}$ when $a > 0$.

67. True.

69. True.

71. True.

73. True.

75. $f(x) = 4x - 6x^{\frac{2}{3}}$

Graphing the function on the calculator we have:

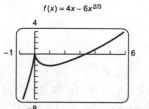

$f(x) = 4x - 6x^{2/3}$

Using the minimum/maximum feature on the calculator, we estimate a relative maximum at $(0,0)$ and a relative minimum at $(1,-2)$.

77. $f(x) = x^2 (1-x)^3$

Graphing the function on the calculator we have:

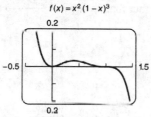

$f(x) = x^2 (1-x)^3$

Using the minimum/maximum feature on the calculator, we estimate a relative maximum at $(0.4, 0.035)$ and a relative minimum at $(0,0)$.

79. $f(x) = (x-1)^{\frac{2}{3}} - (x+1)^{\frac{2}{3}}$

Graphing the function on the calculator we have:

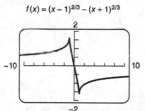

$f(x) = (x-1)^{2/3} - (x+1)^{2/3}$

Using the minimum/maximum feature on the calculator, we estimate a relative maximum at $(-1, 1.587)$ and a relative minimum at $(1, -1.587)$.

81.

Exercise Set 2.3

1. $f(x) = \dfrac{x+4}{x-2}$

The expression is in simplified form. We set the denominator equal to zero and solve.

$x - 2 = 0$

$x = 2$

The vertical asymptote is the line $x = 2$.

3. $f(x) = \dfrac{5x}{x^2 - 25}$

First, we write the function in simplified form.

$f(x) = \dfrac{5x}{(x-5)(x+5)}$

Once the expression is in simplified form, we set the denominator equal to zero and solve.

$(x-5)(x+5) = 0$

$x - 5 = 0 \quad \text{or} \quad x + 5 = 0$

$x = 5 \quad \text{or} \quad x = -5$

The vertical asymptotes are the lines $x = -5$ and $x = 5$.

5. $f(x) = \dfrac{x+3}{x^3 - x}$

First, we write the function in simplified form.

$f(x) = \dfrac{x+3}{x(x^2 - 1)}$

$= \dfrac{x+3}{x(x-1)(x+1)}$

Once the expression is in simplified form, we set the denominator equal to zero and solve.

$x(x+1)(x-1) = 0$

$x = 0 \quad \text{or} \quad x + 1 = 0 \quad \text{or} \quad x - 1 = 0$

$x = 0 \quad \text{or} \quad x = -1 \quad \text{or} \quad x = 1$

The vertical asymptotes are the lines $x = 0$, $x = -1$, and $x = 1$.

7. $f(x) = \dfrac{x+2}{x^2 + 6x + 8}$

First, we write the function in simplified form.

$f(x) = = \dfrac{x+2}{(x+2)(x+4)}$

$= \dfrac{1}{x+4} \quad , \; x \neq -2$ Dividing common factors

Once the expression is in simplified form, we set the denominator equal to zero and solve.

$x + 4 = 0$

$x = -4$

The vertical asymptote is the line $x = -4$.

9. $f(x) = \dfrac{7}{x^2 + 49}$

The function is in simplified form. The equation $x^2 + 49 = 0$ has no real solution; therefore, the function does not have any vertical asymptotes.

11. $f(x) = \dfrac{6x}{8x + 3}$

To find the horizontal asymptote, we consider $\lim\limits_{x \to \infty} f(x)$. To find the limit, we will use some

algebra and the fact that as $x \to \infty$, $\dfrac{b}{ax^n} \to 0$ for

any positive integer n.

$\lim\limits_{x \to \infty} f(x) = \lim\limits_{x \to \infty} \dfrac{6x}{8x + 3}$

$= \lim\limits_{x \to \infty} \dfrac{6x}{8x + 3} \cdot \dfrac{\frac{1}{x}}{\frac{1}{x}}$ Multiplying by a form of 1

$= \lim\limits_{x \to \infty} \dfrac{\frac{6x}{x}}{\frac{8x}{x} + \frac{3}{x}}$

$= \lim\limits_{x \to \infty} \dfrac{6}{8 + \frac{3}{x}}$

$= \dfrac{6}{8 + 0}$ $\left[\text{as } x \to \infty, \dfrac{b}{ax^n} \to 0 \right]$

$= \dfrac{6}{8} = \dfrac{3}{4}$.

In a similar manner, it can be shown that

$\lim\limits_{x \to -\infty} f(x) = \dfrac{3}{4}$.

The horizontal asymptote is the line $y = \dfrac{3}{4}$.

13. $f(x) = \dfrac{4x}{x^2 - 3x}$

To find the horizontal asymptote, we consider $\lim\limits_{x \to \infty} f(x)$. To find the limit on the next page,

we will use some algebra and the fact that as

$x \to \infty$, $\dfrac{b}{ax^n} \to 0$ for any positive integer n.

Determining the limit of the function on the previous page, we have:

$$\lim_{x\to\infty} f(x) = \lim_{x\to\infty} \frac{4x}{x^2 - 3x}$$

$$= \lim_{x\to\infty} \frac{4x}{x^2 - 3x} \cdot \frac{\frac{1}{x^2}}{\frac{1}{x^2}} \quad \text{Multiplying by a form of 1}$$

$$= \lim_{x\to\infty} \frac{\frac{4x}{x^2}}{\frac{x^2}{x^2} - \frac{3}{x^2}}$$

$$= \lim_{x\to\infty} \frac{\frac{4}{x}}{1 + \frac{3}{x^2}}$$

$$= \frac{0}{1 + 0} \quad \left[\text{as } x\to\infty, \frac{b}{ax^n} \to 0 \right]$$

$$= 0.$$

In a similar manner, it can be shown that $\lim_{x\to-\infty} f(x) = 0$.

The horizontal asymptote is the line $y = 0$.

15. $f(x) = 4 + \dfrac{2}{x}$

To find the horizontal asymptote, we consider $\lim_{x\to\infty} f(x)$. To find the limit, we will use the fact that as $x\to\infty, \dfrac{b}{ax^n} \to 0$ for any positive integer n.

$$\lim_{x\to\infty} f(x) = \lim_{x\to\infty} 4 + \frac{2}{x}$$

$$= 4 + 0 \quad \left[\text{as } x\to\infty, \frac{b}{ax^n} \to 0 \right].$$

$$= 4$$

In a similar manner, it can be shown that $\lim_{x\to-\infty} f(x) = 4$.

The horizontal asymptote is the line $y = 4$.

17. $f(x) = \dfrac{6x^3 + 4x}{3x^2 - x}$

To find the horizontal asymptote, we consider $\lim_{x\to\infty} f(x)$. To find the limit, we will use some algebra and the fact that as $x\to\infty, \dfrac{b}{ax^n} \to 0$ for any positive integer .

$$\lim_{x\to\infty} f(x) = \lim_{x\to\infty} \frac{6x^3 + 4x}{3x^2 - x}$$

$$= \lim_{x\to\infty} \frac{6x^3 + 4x}{3x^2 - x} \cdot \frac{\frac{1}{x^2}}{\frac{1}{x^2}} \quad \text{Multiplying by a form of 1}$$

$$= \lim_{x\to\infty} \frac{6x + \frac{4}{x}}{3 - \frac{1}{x}}$$

$$= \frac{\lim_{x\to\infty} 6x + \frac{4}{x}}{3 - 0} = \infty$$

In a similar manner, it can be shown that $\lim_{x\to-\infty} f(x) = -\infty$.

The function increases without bound as $x\to\infty$ and decreases without bound as $x\to-\infty$. Therefore, the function does not have a horizontal asymptote.

19. $f(x) = \dfrac{4x^3 - 3x + 2}{x^3 + 2x - 4}$

To find the horizontal asymptote, we consider $\lim_{x\to\infty} f(x)$. To find the limit, we will use some algebra and the fact that as $x\to\infty, \dfrac{b}{ax^n} \to 0$ for any positive integer n.

$$\lim_{x\to\infty} f(x) = \lim_{x\to\infty} \frac{4x^3 - 3x + 2}{x^3 + 2x - 4}$$

$$= \lim_{x\to\infty} \frac{4x^3 - 3x + 2}{x^3 + 2x - 4} \cdot \frac{\frac{1}{x^3}}{\frac{1}{x^3}}$$

$$= \lim_{x\to\infty} \frac{4 - \frac{3}{x^2} + \frac{2}{x^3}}{1 + \frac{2}{x^2} - \frac{4}{x^3}}$$

$$= \frac{4}{1} \quad \left[\text{as } x\to\infty, \frac{b}{ax^n} \to 0 \right]$$

$$= 4$$

In a similar manner, it can be shown that $\lim_{x\to-\infty} f(x) = 4$.

The horizontal asymptote is the line $y = 4$.

21. $f(x) = \dfrac{2x^3 - 4x + 1}{4x^3 + 2x - 3}$

To find the horizontal asymptote, we consider $\lim\limits_{x \to \infty} f(x)$. To find the limit, we will use some

algebra and the fact that as $x \to \infty$, $\dfrac{b}{ax^n} \to 0$ for

any positive integer n.

$\lim\limits_{x \to \infty} f(x) = \lim\limits_{x \to \infty} \dfrac{2x^3 - 4x + 1}{4x^3 + 2x - 3}$

$= \lim\limits_{x \to \infty} \dfrac{2x^3 - 4x + 1}{4x^3 + 2x - 3} \cdot \dfrac{\frac{1}{x^3}}{\frac{1}{x^3}}$

$= \lim\limits_{x \to \infty} \dfrac{2 - \dfrac{4}{x^2} + \dfrac{1}{x^3}}{4 + \dfrac{2}{x^2} - \dfrac{3}{x^3}}$

$= \dfrac{2 - 0 + 0}{4 + 0 - 0} \quad \left[\text{as } x \to \infty, \dfrac{b}{ax^n} \to 0 \right]$

$= \dfrac{2}{4} = \dfrac{1}{2}$

In a similar manner, it can be shown that

$\lim\limits_{x \to -\infty} f(x) = \dfrac{1}{2}.$

The horizontal asymptote is the line $y = \dfrac{1}{2}$.

23. $f(x) = \dfrac{4}{x} = 4x^{-1}$

a) *Intercepts.* Since the numerator is the constant 4, there are no x-intercepts. The number 0 is not in the domain of the function, so there are no y-intercepts.

b) *Asymptotes.*
Vertical. The denominator is 0 for $x = 0$, so the line $x = 0$ is a vertical asymptote.
Horizontal. The degree of the numerator is less than the degree of the denominator, so $y = 0$ is the horizontal asymptote.
Slant. There is no slant asymptote since the degree of the numerator is not one more than the degree of the denominator.

c) *Derivatives and Domain.*

$f'(x) = -4x^{-2} = -\dfrac{4}{x^2}$

$f''(x) = 8x^{-3} = \dfrac{8}{x^3}$

The domain of f is $(-\infty, 0) \cup (0, \infty)$ as determined in step (b).

d) *Critical Points.* $f'(x)$ exists for all values of x except 0, but 0 is not in the domain of the function, so $x = 0$ is not a critical value. The equation $f'(x) = 0$ has no solution, so there are no critical points.

e) *Increasing, decreasing, relative extrema.* We use 0 to divide the real number line into two intervals A: $(-\infty, 0)$ and B: $(0, \infty)$, and we test a point in each interval.

A: Test -1, $f'(-1) = -\dfrac{4}{(-1)^2} = -4 < 0$

B: Test 1, $f'(1) = -\dfrac{4}{(1)^2} = -4 < 0$

Then $f(x)$ is decreasing on both intervals. Since there are no critical points, there are no relative extrema.

f) *Inflection points.* $f''(x)$ does not exist at 0, but because 0 is not in the domain of the function, there cannot be an inflection point at 0. The equation $f''(x) = 0$ has no solution; therefore, there are no inflection points.

g) *Concavity.* We use 0 to divide the real number line into two intervals A: $(-\infty, 0)$ and B: $(0, \infty)$, and we test a point in each interval.

A: Test -1, $f''(-1) = \dfrac{8}{(-1)^3} = -8 < 0$

B: Test 1, $f''(1) = \dfrac{8}{(1)^3} = 8 > 0$

Therefore, $f(x)$ is concave down on $(-\infty, 0)$ and concave up on $(0, \infty)$.

h) *Sketch.*

25. $f(x) = \dfrac{-2}{x-5} = -2(x-5)^{-1}$

a) *Intercepts.* Since the numerator is the constant -2, there are no x-intercepts. To find the y-intercepts we compute $f(0)$.

$$f(0) = \frac{-2}{(0)-5} = \frac{2}{5}$$

The point $\left(0, \dfrac{2}{5}\right)$ is the y-intercept.

b) *Asymptotes.*
Vertical. The denominator is 0 for $x = 5$, so the line $x = 5$ is a vertical asymptote.
Horizontal. The degree of the numerator is less than the degree of the denominator, so $y = 0$ is the horizontal asymptote.
Slant. There is no slant asymptote since the degree of the numerator is not one more than the degree of the denominator.

c) *Derivatives and Domain.*

$$f'(x) = 2(x-5)^{-2} = \frac{2}{(x-5)^2}$$

$$f''(x) = -4(x-5)^{-3} = \frac{-4}{(x-5)^3}$$

The domain of f is $(-\infty,5) \cup (5,\infty)$ as determined in step (b).

d) *Critical Points.* $f'(x)$ exists for all values of x except 5, but 5 is not in the domain of the function, so $x = 5$ is not a critical value. The equation $f'(x) = 0$ has no solution, so there are no critical points.

e) *Increasing, decreasing, relative extrema.* We use 5 to divide the real number line into two intervals A: $(-\infty,5)$ and B: $(5,\infty)$, and we test a point in each interval.

A: Test 4, $f'(4) = \dfrac{2}{(4-5)^2} = \dfrac{2}{1} = 2 > 0$

B: Test 6, $f'(6) = \dfrac{2}{(6-5)^2} = \dfrac{2}{1} = 2 > 0$

Then $f(x)$ is increasing on both intervals. Since there are no critical points, there are no relative extrema.

f) *Inflection points.* $f''(x)$ does not exist at 5, but because 5 is not in the domain of the function, there cannot be an inflection point at 5. The equation $f''(x) = 0$ has no solution; therefore, there are no inflection points.

g) *Concavity.* We use 5 to divide the real number line into two intervals A: $(-\infty,5)$ and B: $(5,\infty)$, and we test a point in each interval.

A: Test 4, $f''(4) = \dfrac{-4}{(4-5)^3} = \dfrac{-4}{-1} = 4 > 0$

B: Test 6, $f''(6) = \dfrac{-4}{(6-5)^3} = \dfrac{-4}{1} = -4 < 0$

Therefore, $f(x)$ is concave up on $(-\infty,5)$ and concave down on $(5,\infty)$.

h) *Sketch.*

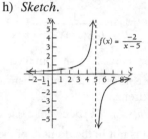

27. $f(x) = \dfrac{1}{x-3} = (x-3)^{-1}$

a) *Intercepts.* Since the numerator is the constant 1, there are no x-intercepts. To find the y-intercepts we compute $f(0)$.

$$f(0) = \frac{1}{(0)-3} = -\frac{1}{3}$$

The point $\left(0, -\dfrac{1}{3}\right)$ is the y-intercept.

b) *Asymptotes.*
Vertical. The denominator is 0 for $x = 3$, so the line $x = 3$ is a vertical asymptote.
Horizontal. The degree of the numerator is less than the degree of the denominator, so $y = 0$ is the horizontal asymptote.
Slant. There is no slant asymptote since the degree of the numerator is not one more than the degree of the denominator.

c) *Derivatives and Domain.*

$$f'(x) = -(x-3)^{-2} = \frac{-1}{(x-3)^2}$$

$$f''(x) = 2(x-3)^{-3} = \frac{2}{(x-3)^3}$$

The domain of f is $(-\infty,3) \cup (3,\infty)$ as determined in step (b).

d) *Critical Points.* $f'(x)$ exists for all values of x except 3, but 3 is not in the domain of the function, so $x = 3$ is not a critical value. The equation $f'(x) = 0$ has no solution, so there are no critical points.

e) *Increasing, decreasing, relative extrema.* We use 3 to divide the real number line into two intervals A: $(-\infty, 3)$ and B: $(3, \infty)$, and we test a point in each interval.

A: Test 2, $f'(2) = \dfrac{-1}{\left((2) - 3\right)^2} = -1 < 0$

B: Test 4, $f'(4) = \dfrac{-1}{\left((4) - 3\right)^2} = -1 < 0$

Then $f(x)$ is decreasing on both intervals. Since there are no critical points, there are no relative extrema.

f) *Inflection points.* $f''(x)$ does not exist at 3, but because 3 is not in the domain of the function, there cannot be an inflection point at 3. The equation $f''(x) = 0$ has no solution; therefore, there are no inflection points.

g) *Concavity.* We use 3 to divide the real number line into two intervals A: $(-\infty, 3)$ and B: $(3, \infty)$, and we test a point in each interval.

A: Test 2, $f''(2) = \dfrac{2}{\left((2) - 3\right)^3} = -2 < 0$

B: Test 4, $f''(4) = \dfrac{2}{\left((4) - 3\right)^3} = 2 > 0$

Therefore, $f(x)$ is concave down on $(-\infty, 3)$ and concave up on $(3, \infty)$.

h) *Sketch.*

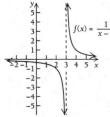

29. $f(x) = \dfrac{-2}{x+5} = -2(x+5)^{-1}$

a) *Intercepts.* Since the numerator is the constant -2, there are no x-intercepts. To find the y-intercepts we compute $f(0)$.

$$f(0) = \frac{-2}{(0) + 5} = \frac{-2}{5} = -\frac{2}{5}$$

The point $\left(0, -\dfrac{2}{5}\right)$ is the y-intercept.

b) *Asymptotes.*
Vertical. The denominator is 0 for $x = -5$, so the line $x = -5$ is a vertical asymptote.
Horizontal. The degree of the numerator is less than the degree of the denominator, so $y = 0$ is the horizontal asymptote.

Slant. There is no slant asymptote since the degree of the numerator is not one more than the degree of the denominator.

c) *Derivatives and Domain.*

$$f'(x) = 2(x+5)^{-2} = \frac{2}{(x+5)^2}$$

$$f''(x) = -4(x+5)^{-3} = \frac{-4}{(x+5)^3}$$

The domain of f is $(-\infty, -5) \cup (-5, \infty)$ as determined in step (b).

d) *Critical Points.* $f'(x)$ exists for all values of x except -5, but -5 is not in the domain of the function, so $x = -5$ is not a critical value. The equation $f'(x) = 0$ has no solution, so there are no critical points.

e) *Increasing, decreasing, relative extrema.* We use -5 to divide the real number line into two intervals A: $(-\infty, -5)$ and B: $(-5, \infty)$, and we test a point in each interval.

A: Test -6, $f'(-6) = \dfrac{2}{\left((-6) + 5\right)^2} = 2 > 0$

B: Test -4, $f'(-4) = \dfrac{2}{\left((-4) + 5\right)^2} = 2 > 0$

Then $f(x)$ is increasing on both intervals. Since there are no critical points, there are no relative extrema.

f) *Inflection points.* $f''(x)$ does not exist at -5, but because -5 is not in the domain of the function, there cannot be an inflection point at -5. The equation $f''(x) = 0$ has no solution; therefore, there are no inflection points.

g) *Concavity.* We use -5 to divide the real number line into two intervals

A: $(-\infty, -5)$ and B: $(-5, \infty)$, and we test a point in each interval.

A: Test -6, $f''(-6) = \dfrac{-4}{\left((-6)+5\right)^3} = 4 > 0$

B: Test -4, $f''(-4) = \dfrac{-4}{\left((-4)+5\right)^3} = -4 < 0$

Therefore, $f(x)$ is concave up on $(-\infty, -5)$ and concave down on $(-5, \infty)$.

h) *Sketch.*

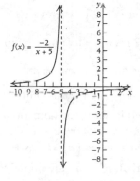

$f(x) = \dfrac{-2}{x+5}$

31. $f(x) = \dfrac{2x+1}{x}$

a) *Intercepts.* To find the x-intercepts, solve $f(x) = 0$.

$$\frac{2x+1}{x} = 0$$

$$2x+1 = 0$$

$$x = -\frac{1}{2}$$

This value does not make the denominator 0; therefore, the x-intercept is $\left(-\dfrac{1}{2}, 0\right)$.

The number 0 is not in the domain of $f(x)$ so there are no y-intercepts.

b) *Asymptotes.*
Vertical. The denominator is 0 for $x = 0$, so the line $x = 0$ is a vertical asymptote.

Horizontal. The numerator and the denominator have the same degree, so

$y = \dfrac{2}{1}$, or $y = 2$ is the horizontal asymptote.

Slant. There is no slant asymptote since the degree of the numerator is not one more than the degree of the denominator.

c) *Derivatives and Domain.*

$$f'(x) = -\frac{1}{x^2}$$

$$f''(x) = 2x^{-3} = \frac{2}{x^3}$$

The domain of f is $(-\infty, 0) \cup (0, \infty)$ as determined in step (b).

d) *Critical Points.* $f'(x)$ exists for all values of x except 0, but 0 is not in the domain of the function, so $x = 0$ is not a critical value. The equation $f'(x) = 0$ has no solution, so there are no critical points.

e) *Increasing, decreasing, relative extrema.* We use 0 to divide the real number line into two intervals A: $(-\infty, 0)$ and B: $(0, \infty)$, and we test a point in each interval.

A: Test -1, $f'(-1) = -\dfrac{1}{(-1)^2} = -1 < 0$

B: Test 1, $f'(1) = -\dfrac{1}{(1)^2} = -1 < 0$

Then $f(x)$ is decreasing on both intervals. Since there are no critical points, there are no relative extrema.

f) *Inflection points.* $f''(x)$ does not exist at 0, but because 0 is not in the domain of the function, there cannot be an inflection point at 0. The equation $f''(x) = 0$ has no solution; therefore, there are no inflection points.

g) *Concavity.* We use 0 to divide the real number line into two intervals A: $(-\infty, 0)$ and B: $(0, \infty)$, and we test a point in each interval.

A: Test -1, $f''(-1) = \dfrac{2}{(-1)^3} = -2 < 0$

B: Test 1, $f''(1) = \dfrac{2}{(1)^3} = 2 > 0$

Therefore, $f(x)$ is concave down on $(-\infty, 0)$ and concave up on $(0, \infty)$.

h) *Sketch.*

33. $f(x) = x + \dfrac{2}{x} = \dfrac{x^2 + 2}{x}$

a) *Intercepts.* The equation $f(x) = 0$ has no real solutions, so there are no x-intercepts. The number 0 is not in the domain of $f(x)$ so there are no y-intercepts.

b) *Asymptotes.*
Vertical. The denominator is 0 for $x = 0$, so the line $x = 0$ is a vertical asymptote.
Horizontal. The degree of the numerator is greater than the degree of the denominator, so there are no horizontal asymptotes.
Slant. The degree of the numerator is exactly one greater than the degree of the denominator. As $|x|$ approaches ∞,

$f(x) = x + \dfrac{2}{x}$ approaches x. Therefore,

$y = x$ is the slant asymptote.

c) *Derivatives and Domain.*

$f'(x) = 1 - 2x^{-2} = 1 - \dfrac{2}{x^2}$

$f''(x) = 4x^{-3} = \dfrac{4}{x^3}$

The domain of f is $(-\infty, 0) \cup (0, \infty)$ as determined in step (b).

d) *Critical Points.* $f'(x)$ exists for all values of x except 0, but 0 is not in the domain of the function, so $x = 0$ is not a critical value. The critical points will occur when $f'(x) = 0$.

$1 - \dfrac{2}{x^2} = 0$

$1 = \dfrac{2}{x^2}$

$x^2 = 2$

$x = \pm\sqrt{2}$

Thus, $-\sqrt{2}$ and $\sqrt{2}$ are critical values.
$f\left(-\sqrt{2}\right) = -2\sqrt{2}$ and $f\left(\sqrt{2}\right) = 2\sqrt{2}$, so the critical points $\left(-\sqrt{2}, -2\sqrt{2}\right)$ and $\left(\sqrt{2}, 2\sqrt{2}\right)$ are on the graph.

e) *Increasing, decreasing, relative extrema.*
We use $-\sqrt{2}, 0,$ and $\sqrt{2}$ to divide the real number line into four intervals
A: $\left(-\infty, -\sqrt{2}\right)$ B: $\left(-\sqrt{2}, 0\right)$, C: $\left(0, \sqrt{2}\right)$, and D: $\left(\sqrt{2}, \infty\right)$.

A: Test -2, $f'(-2) = 1 - \dfrac{2}{(-2)^2} = \dfrac{1}{2} > 0$

B: Test -1, $f'(-1) = 1 - \dfrac{2}{(-1)^2} = -1 < 0$

C: Test 1, $f'(1) = 1 - \dfrac{2}{(1)^2} = -1 < 0$

D: Test 2, $f'(2) = 1 - \dfrac{2}{(2)^2} = \dfrac{1}{2} > 0$

Then $f(x)$ is increasing on

$\left(-\infty, -\sqrt{2}\right)$ and $\left(\sqrt{2}, \infty\right)$ and is decreasing on

$\left(-\sqrt{2}, 0\right)$ and $\left(0, \sqrt{2}\right)$. Therefore,

$\left(-\sqrt{2}, -2\sqrt{2}\right)$ is a relative maximum, and

$\left(\sqrt{2}, 2\sqrt{2}\right)$ is a relative minimum.

f) *Inflection points.* $f''(x)$ does not exist at 0, but because 0 is not in the domain of the function, there cannot be an inflection point at 0. The equation $f''(x) = 0$ has no solution; therefore, there are no inflection points.

g) *Concavity.* We use 0 to divide the real number line into two intervals
A: $(-\infty, 0)$ and B: $(0, \infty)$, and we test a point in each interval.

A: Test -1, $f''(-1) = \dfrac{4}{(-1)^3} = -4 < 0$

B: Test 1, $f''(1) = \dfrac{4}{(1)^3} = 4 > 0$

Therefore, $f(x)$ is concave down on

$(-\infty, 0)$ and concave up on $(0, \infty)$.

h) *Sketch.* Use the preceding information to sketch the graph. Compute additional function values as needed.

35. $f(x) = \dfrac{-1}{x^2} = -x^{-2}$

a) *Intercepts.* Since the numerator is the constant -1, there are no x-intercepts. The number 0 is not in the domain of the function, so there are no y-intercepts.

b) *Asymptotes.*
Vertical. The denominator is 0 for $x = 0$, so the line $x = 0$ is a vertical asymptote.
Horizontal. The degree of the numerator is less than the degree of the denominator, so $y = 0$ is the horizontal asymptote.
Slant. There is no slant asymptote since the degree of the numerator is not one more than the degree of the denominator.

c) *Derivatives and Domain.*

$$f'(x) = 2x^{-3} = \dfrac{2}{x^3}$$

$$f''(x) = -6x^{-4} = -\dfrac{6}{x^4}$$

The domain of f is $(-\infty, 0) \cup (0, \infty)$ as determined in step (b).

d) *Critical Points.* $f'(x)$ exists for all values of x except 0, but 0 is not in the domain of the function, so $x = 0$ is not a critical value. The equation $f'(x) = 0$ has no solution, so there are no critical points.

e) *Increasing, decreasing, relative extrema.* We use 0 to divide the real number line into two intervals A: $(-\infty, 0)$ and B: $(0, \infty)$, and we test a point in each interval.

A: Test -1, $f'(-1) = \dfrac{2}{(-1)^3} = -2 < 0$

B: Test 1, $f'(1) = \dfrac{2}{(1)^3} = 2 > 0$

Then $f(x)$ is decreasing on $(-\infty, 0)$ and is increasing on $(0, \infty)$. Since there are no critical points, there are no relative extrema.

f) *Inflection points.* $f''(x)$ does not exist at 0, but because 0 is not in the domain of the function, there cannot be an inflection point at 0. The equation $f''(x) = 0$ has no solution; therefore, there are no inflection points.

g) *Concavity.* We use 0 to divide the real number line into two intervals A: $(-\infty, 0)$ and B: $(0, \infty)$, and we test a point in each interval.

A: Test -1, $f''(-1) = -\dfrac{6}{(-1)^4} = -6 < 0$

B: Test 1, $f''(1) = -\dfrac{6}{(1)^4} = -6 < 0$

Therefore, $f(x)$ is concave down on both intervals.

h) *Sketch.* Use the preceding information to sketch the graph. Compute additional function values as needed.

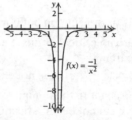

$f(x) = \dfrac{-1}{x^2}$

37. $f(x) = \dfrac{x}{x-3}$

a) *Intercepts.* To find the x-intercepts, solve $f(x) = 0$.

$$\dfrac{x}{x-3} = 0$$

$$x = 0$$

Since $x = 0$ does not make the denominator 0, the x-intercept is $(0,0)$. $f(0) = 0$, so the y-intercept is $(0,0)$ also.

b) *Asymptotes.*
Vertical. The denominator is 0 for $x = 3$, so the line $x = 3$ is a vertical asymptote.
Horizontal. The numerator and the denominator have the same degree, so $y = \dfrac{1}{1}$, or $y = 1$ is the horizontal asymptote.
Slant. There is no slant asymptote since the degree of the numerator is not one more than the degree of the denominator.

c) *Derivatives and Domain.*

$$f'(x) = \dfrac{-3}{(x-3)^2}$$

$$f''(x) = 6(x-3)^{-3} = \dfrac{6}{(x-3)^3}$$

The domain of f is $(-\infty, 3) \cup (3, \infty)$ as determined in step (b).

d) *Critical Points.* $f'(x)$ exists for all values of x except 3, but 3 is not in the domain of the function, so $x = 3$ is not a critical value. The equation $f'(x) = 0$ has no solution, so there are no critical points.

e) *Increasing, decreasing, relative extrema.* We use 3 to divide the real number line into two intervals A: $(-\infty, 3)$ and B: $(3, \infty)$, and we test a point in each interval.

A: Test 2, $f'(2) = \dfrac{-3}{((2)-3)^2} = -3 < 0$

B: Test 4, $f'(4) = \dfrac{-3}{((4)-3)^2} = -3 < 0$

Then $f(x)$ is decreasing on both intervals. Since there are no critical points, there are no relative extrema.

f) *Inflection points.* $f''(x)$ does not exist at 3, but because 3 is not in the domain of the function, there cannot be an inflection point at 3. The equation $f''(x) = 0$ has no solution; therefore, there are no inflection points.

g) *Concavity.* We use 3 to divide the real number line into two intervals A: $(-\infty, 3)$ and B: $(3, \infty)$, and we test a point in each interval.

A: Test 2, $f''(2) = \dfrac{6}{((2)-3)^3} = -6 < 0$

B: Test 4, $f''(4) = \dfrac{6}{((4)-3)^3} = 6 > 0$

Therefore, $f(x)$ is concave down on $(-\infty, 3)$ and concave up on $(3, \infty)$.

h) *Sketch.* Use the preceding information to sketch the graph. Compute additional function values as needed.

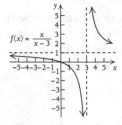

39. $f(x) = \dfrac{1}{x^2 + 3} = \left(x^2 + 3\right)^{-1}$

a) *Intercepts.* Since the numerator is the constant 1, there are no x-intercepts.

$f(0) = \dfrac{1}{(0)^2 + 3} = \dfrac{1}{3}$, so the y-intercept is $\left(0, \dfrac{1}{3}\right)$.

b) *Asymptotes.*
Vertical. $x^2 + 3 = 0$ has no real solution, so there are no vertical asymptotes.
Horizontal. The degree of the numerator is less than the degree of the denominator, so $y = 0$ is the horizontal asymptote.
Slant. There is no slant asymptote since the degree of the numerator is not one more than the degree of the denominator.

c) *Derivatives and Domain.*

$f'(x) = -2x\left(x^2 + 3\right)^{-2} = -\dfrac{2x}{\left(x^2 + 3\right)^2}$

$f''(x) = \dfrac{6x^2 - 6}{\left(x^2 + 3\right)^3}$

The domain of f is $\mathbb{R}$ as determined in step (b).

d) *Critical Points.* $f'(x)$ exists for all real numbers. Solve $f'(x) = 0$

$-\dfrac{2x}{\left(x^2 + 3\right)^2} = 0$

$2x = 0$

$x = 0$

The critical value is 0. From step (a) we found $\left(0, \dfrac{1}{3}\right)$ is on the graph.

e) *Increasing, decreasing, relative extrema.* We use 0 to divide the real number line into two intervals A: $(-\infty, 0)$ and B: $(0, \infty)$, and we test a point in each interval.

A: Test -1, $f'(-1) = \dfrac{1}{8} > 0$

B: Test 1, $f'(1) = -\dfrac{1}{8} < 0$

Then $f(x)$ is increasing on $(-\infty, 0)$ and is decreasing on $(0, \infty)$. Thus $\left(0, \dfrac{1}{3}\right)$ is a relative maximum.

f) *Inflection points.* $f''(x)$ exists for all real numbers. Solve $f''(x) = 0$.

$$\frac{6x^2 - 6}{\left(x^2 + 2\right)^3} = 0$$

$$6x^2 - 6 = 0$$

$$6x^2 = 6$$

$$x^2 = 1$$

$$x = \pm 1$$

$f(-1) = \dfrac{1}{4}$ and $f(1) = \dfrac{1}{4}$

So, $\left(-1, \dfrac{1}{4}\right)$ and $\left(1, \dfrac{1}{4}\right)$ are possible points of inflection.

g) *Concavity.* We use -1 and 1 to divide the real number line into three intervals

A: $(-\infty, -1)$ B: $(-1, 1)$, and C: $(1, \infty)$

A: Test -2, $f''(-2) = \dfrac{18}{343} > 0$

B: Test $\quad 0$, $f''(0) = -\dfrac{2}{9} < 0$

C: Test $\quad 2$, $f''(2) = \dfrac{18}{343} > 0$

Therefore, $f(x)$ is concave up on $(-\infty, -1)$ and $(1, \infty)$, and concave down on $(-1, 1)$.

Thus the points $\left(-1, \dfrac{1}{4}\right)$ and $\left(1, \dfrac{1}{4}\right)$ are points of inflection.

h) *Sketch.*

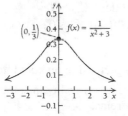

41. $f(x) = \dfrac{x+3}{x^2 - 9} = \dfrac{x+3}{(x+3)(x-3)} = \dfrac{1}{x-3}$, $x \ne \pm 3$

We write the expression in simplified form noting that the domain is restricted to all real numbers except for $x = \pm 3$.

a) *Intercepts.* $f(x) = 0$ has no solution. $x = -3$ is not in the domain of the function. Therefore, there are no x-intercepts.

To find the y-intercepts we compute $f(0)$.

$$f(0) = \frac{1}{(0) - 3} = -\frac{1}{3}$$

The point $\left(0, -\dfrac{1}{3}\right)$ is the y-intercept.

b) *Asymptotes.*
Vertical. In the original function, the denominator is 0 for $x = -3$ or $x = 3$, however, $x = -3$ also made the numerator equal to 0. We look at the limits to determine if there are vertical asymptotes at these points.

$$\lim_{x \to -3} \frac{x+3}{x^2 - 9} = \lim_{x \to -3} \frac{1}{x-3} = \frac{1}{-3-3} = -\frac{1}{6}.$$

Because the limit exists, the line $x = -3$ is not a vertical asymptote. Instead, we have a removable discontinuity, or a "hole" at the point $\left(-3, -\dfrac{1}{6}\right)$.

An open circle is drawn at $\left(-3, -\dfrac{1}{6}\right)$ to show that it is not part of the graph.
The denominator is 0 for $x = 3$ and the numerator is not 0 at this value, so the line $x = 3$ is a vertical asymptote.
Horizontal. The degree of the numerator is less than the degree of the denominator, so $y = 0$ is the horizontal asymptote.
Slant. There is no slant asymptote since the degree of the numerator is not one more than the degree of the denominator.

c) *Derivatives and Domain.*

$$f'(x) = -(x-3)^{-2} = \frac{-1}{(x-3)^2}$$

$$f''(x) = 2(x-3)^{-3} = \frac{2}{(x-3)^3}$$

The domain of f as determined in step (b) is $(-\infty, -3) \cup (-3, 3) \cup (3, \infty)$.

d) *Critical Points.* $f'(x)$ exists for all values of x except 3, but 3 is not in the domain of the function, so $x = 3$ is not a critical value. The equation $f'(x) = 0$ has no solution, so there are no critical points.

e) *Increasing, decreasing, relative extrema.* We use -3 and 3 to divide the real number line into three intervals

A: $(-\infty, -3)$ B: $(-3, 3)$ and C: $(3, \infty)$.

The solution is continued on the next page.

We notice on the previous page that $f'(x) < 0$ for all real numbers, $f(x)$ is decreasing on all three intervals $(-\infty, -3)$, $(-3, 3)$, and $(3, \infty)$. Since there are no critical points, there are no relative extrema.

f) *Inflection points.* $f''(x)$ does not exist at 3, but because 3 is not in the domain of the function, there cannot be an inflection point at 3. The equation $f''(x) = 0$ has no solution; therefore, there are no inflection points.

g) *Concavity.* We use -3 and 3 to divide the real number line into three intervals
A: $(-\infty, -3)$ B: $(-3, 3)$ and C: $(3, \infty)$ and we test a point in each interval.

A: Test -4, $f''(-4) = \dfrac{2}{((-4)-3)^3} = -\dfrac{2}{343} < 0$

B: Test 2, $f''(2) = \dfrac{2}{((2)-3)^3} = -2 < 0$

C: Test 4, $f''(4) = \dfrac{2}{((4)-3)^3} = 2 > 0$

Therefore, $f(x)$ is concave down on $(-\infty, -3)$ and $(-3, 3)$ and concave up on $(3, \infty)$.

h) *Sketch.* Use the preceding information to sketch the graph. Compute additional function values as needed.

43. $f(x) = \dfrac{x-1}{x+2}$

a) *Intercepts.* To find the x-intercepts, solve $f(x) = 0$.

$\dfrac{x-1}{x+2} = 0$

$x = 1$

Since $x = 1$ does not make the denominator 0, the x-intercept is $(1, 0)$.

We evaluate the function at $x = 0$.

$f(0) = \dfrac{0-1}{0+2} = -\dfrac{1}{2}$, so the y-intercept is $\left(0, -\dfrac{1}{2}\right)$.

b) *Asymptotes.*
Vertical. The denominator is 0 for $x = -2$, so the line $x = -2$ is a vertical asymptote.
Horizontal. The numerator and the denominator have the same degree, so $y = \dfrac{1}{1}$, or $y = 1$ is the horizontal asymptote.
Slant. There is no slant asymptote since the degree of the numerator is not one more than the degree of the denominator.

c) *Derivatives and Domain.*

$f'(x) = \dfrac{3}{(x+2)^2}$

$f''(x) = -\dfrac{6}{(x+2)^3}$

The domain of f is $(-\infty, -2) \cup (-2, \infty)$ as determined in part (b).

d) *Critical Points.* $f'(x)$ exists for all values of x except -2, but -2 is not in the domain of the function, so $x = -2$ is not a critical value. The equation $f'(x) = 0$ has no solution, so there are no critical points.

e) *Increasing, decreasing, relative extrema.* We use -2 to divide the real number line into two intervals
A: $(-\infty, -2)$ and B: $(-2, \infty)$, and we test a point in each interval.

A: Test -3, $f'(-3) = 3 > 0$

B: Test -1, $f'(-1) = 3 > 0$

Then $f(x)$ is increasing on both intervals. Since there are no critical points, there are no relative extrema.

f) *Inflection points.* $f''(x)$ does not exist at -2, but because -2 is not in the domain of the function, there cannot be an inflection point at -2. The equation $f''(x) = 0$ has no solution; therefore, there are no inflection points.

g) *Concavity.* We use −2 to divide the real number line into two intervals
   A: $(-\infty,-2)$ and B: $(-2,\infty)$, and we test a point in each interval.
   A: Test − 3, $f''(-3) = 6 > 0$
   B: Test − 1, $f''(-1) = -6 < 0$
   Therefore, $f(x)$ is concave up on $(-\infty,-2)$ and concave down on $(-2,\infty)$.

h) *Sketch.* Use the preceding information to sketch the graph. Compute additional function values as needed.

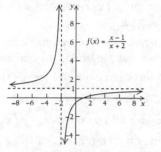

45. $f(x) = \dfrac{x^2 - 9}{x+1}$

a) *Intercepts.* To find the x intercepts, solve $f(x) = 0$.
$$\frac{x^2 - 9}{x+1} = 0$$
$$x^2 - 9 = 0$$
$$x = \pm 3$$
Neither of these values make the denominator 0, so the x-intercepts are $(-3,0)$ and $(3,0)$.

$f(0) = -9$, so the y-intercept is $(0,-9)$.

b) *Asymptotes.*
   Vertical. The denominator is 0 for $x = -1$, so the line $x = -1$ is a vertical asymptote.
   Horizontal. The degree of the numerator is greater than the degree of the denominator, so there are no horizontal asymptotes.
   Slant. Divide the numerator by the denominator.

$$\begin{array}{r} x \quad\quad -1 \\ x+1\overline{)x^2 \quad\quad -9} \\ \underline{x^2 + x} \\ -x - 9 \\ \underline{-x - 1} \\ -8 \end{array}$$

By dividing, we get
$$f(x) = x - 1 - \frac{8}{x+1}$$
As $|x|$ approaches ∞, $f(x)$ approaches $x - 1$, so $y = x - 1$ is the slant asymptote.

c) *Derivatives and Domain.*
$$f'(x) = \frac{x^2 + 2x + 9}{(x+1)^2}$$
$$f''(x) = -\frac{16}{(x+1)^3}$$
The domain of f is $(-\infty,-1) \cup (-1,\infty)$ as determined in part (b).

d) *Critical Points.* $f'(x)$ exists for all values of x except −1, but −1 is not in the domain of the function, so $x = -1$ is not a critical value.
   $f'(x) = 0$ has no real solution, so there are no critical points.

e) *Increasing, decreasing, relative extrema.* We use −1 to divide the real number line into two intervals
   A: $(-\infty,-1)$ and B: $(-1,\infty)$
   We test a point in each interval.
   A: Test − 2, $f'(-2) = 9 > 0$
   B: Test   0, $f'(0) = 9 > 0$
   Then $f(x)$ is increasing on both intervals.
   Since there are no critical points, there are no relative extrema.

f) *Inflection points.* $f''(x)$ does not exist at 1, but because −1 is not in the domain of the function, there cannot be an inflection point at −1. The equation $f''(x) = 0$ has no solution; therefore, there are no inflection points.

g) *Concavity.* We use −1 to divide the real number line into two intervals
   A: $(-\infty,-1)$ and B: $(-1,\infty)$, and we test a point in each interval.
   A: Test − 2, $f''(-2) = 16 > 0$
   B: Test   0, $f''(0) = -16 < 0$
   Therefore, $f(x)$ is concave up on $(-\infty,-1)$ and concave down on $(-1,\infty)$.

h) *Sketch.*

$f(x) = \frac{x^2 - 9}{x + 1}$

47. $f(x) = \dfrac{x-3}{x^2+2x-15} = \dfrac{x-3}{(x-3)(x+5)} = \dfrac{1}{x+5}$,

$x \neq 3$.

We write the expression in simplified form noting that the domain is restricted to all real numbers except for $x = -5$ and $x = 3$.

a) *Intercepts.* $f(x) = 0$ has no solution. $x = 3$ is not in the domain of the function. Therefore, there are no x-intercepts. To find the y-intercepts we compute $f(0)$:

$$f(0) = \frac{0-3}{(0)^2 + 2(0) - 15} = \frac{1}{5}$$

The point $\left(0, \dfrac{1}{5}\right)$ is the y-intercept.

b) *Asymptotes.*
Vertical. In the original function, the denominator is 0 for $x = -5$ or $x = 3$; however, $x = 3$ also made the numerator equal to 0.
We look at the limits to determine if there are vertical asymptotes at these points.

$$\lim_{x \to 3} \frac{x-3}{x^2+2x-15} = \lim_{x \to 3} \frac{1}{x+5} = \frac{1}{8}$$

Because the limit exists, the line $x = 3$ is not a vertical asymptote. Instead, we have a removable discontinuity, or a "hole" at the

point $\left(3, \dfrac{1}{8}\right)$. An open circle is drawn at this

point to show that it is not part of the graph. The denominator is 0 for $x = -5$ and the numerator is not 0 at this value, so the line $x = -5$ is a vertical asymptote.
Horizontal. The degree of the numerator is less than the degree of the denominator, so $y = 0$ is the horizontal asymptote.

Slant. There is no slant asymptote since the degree of the numerator is not one more than the degree of the denominator.

c) *Derivatives and Domain.*

$$f'(x) = -(x+5)^{-2} = \frac{-1}{(x+5)^2}$$

$$f''(x) = 2(x+5)^{-3} = \frac{2}{(x+5)^3}$$

The domain of f as determined in part (b) is $(-\infty, -5) \cup (-5, 3) \cup (3, \infty)$.

d) *Critical Points.* $f'(x)$ exists for all values of x except -5, but -5 is not in the domain of the function, so $x = -5$ is not a critical value. The equation $f'(x) = 0$ has no solution, so there are no critical points.

e) *Increasing, decreasing, relative extrema.* We use -5 and 3 to divide the real number line into three intervals A: $(-\infty, -5)$, B: $(-5, 3)$, and C: $(3, \infty)$. We notice that $f'(x) < 0$ for all real numbers, $f(x)$ is decreasing on all three intervals $(-\infty, -5)$, $(-5, 3)$, and $(3, \infty)$. Since there are no critical points, there are no relative extrema.

f) *Inflection points.* $f''(x)$ does not exist at -5, but because -5 is not in the domain of the function, there cannot be an inflection point at -5. The equation $f''(x) = 0$ has no solution; therefore, there are no inflection points.

g) *Concavity.* We use -5 and 3 to divide the real number line into three intervals A: $(-\infty, -5)$ B: $(-5, 3)$ and C: $(3, \infty)$, and we test a point in each interval.
A: Test -6, $f''(-6) = -2 < 0$
B: Test -4, $f''(-4) = 2 > 0$
C: Test 4, $f''(4) = \dfrac{2}{729} > 0$

Therefore, $f(x)$ is concave down on $(-\infty, -5)$ and concave up on $(-5, 3)$ and $(3, \infty)$.

h) *Sketch.* Use the preceding information to sketch the graph. Compute additional function values as needed.

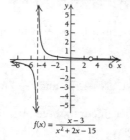

$$f(x) = \frac{x-3}{x^2 + 2x - 15}$$

49. $f(x) = \dfrac{2x^2}{x^2 - 16}$

a) *Intercepts.* The numerator is 0 for $x = 0$ and this value does not make the denominator 0, so the x-intercept is $(0,0)$.

$f(0) = 0$, so the y-intercept is $(0,0)$ also.

b) *Asymptotes.*
Vertical. The denominator is 0 when

$$x^2 - 16 = 0$$
$$x^2 = 16$$
$$x = \pm 4$$

So the lines $x = -4$ and $x = 4$ are vertical asymptotes.
Horizontal. The numerator and the denominator have the same degree, so

$y = \dfrac{2}{1}$, or $y = 2$ is the horizontal asymptote.

Slant. There is no slant asymptote since the degree of the numerator is not one more than the degree of the denominator.

c) *Derivatives and Domain.*

$$f'(x) = -\frac{64x}{\left(x^2 - 16\right)^2}$$

$$f''(x) = \frac{192x^2 + 1024}{\left(x^2 - 16\right)^3}$$

The domain of f is

$(-\infty, -4) \cup (-4, 4) \cup (4, \infty)$ as determined in part (b).

d) *Critical Points.* $f'(x)$ exists for all values of x except $x = -4$ and $x = 4$, but -4 and 4 are not in the domain of the function, so $x = -4$ and $x = 4$ are not critical values.
$f'(x) = 0$ for $x = 0$, so $(0,0)$ is the only critical point.

e) *Increasing, decreasing, relative extrema.* We use -4, 0, and 4 to divide the real number line into four intervals
A: $(-\infty, -4)$ B: $(-4, 0)$, C: $(0, 4)$, and D: $(4, \infty)$.
We test a point in each interval.

A: Test -5, $f'(-5) = \dfrac{320}{81} > 0$

B: Test -1, $f'(-1) = \dfrac{64}{225} > 0$

C: Test 1, $f'(1) = -\dfrac{64}{225} < 0$

D: Test 5, $f'(5) = -\dfrac{320}{81} < 0$

Then $f(x)$ is increasing on the intervals $(-\infty, -4)$ and $(-4, 0)$, and is decreasing on the intervals $(0, 4)$ and $(4, \infty)$. Thus, there is a relative maximum at $(0, 0)$.

f) *Inflection points.* $f''(x)$ does not exist at -4 and 4, but because -4 and 4 are not in the domain of the function, there cannot be an inflection point at -4 or 4. The equation $f''(x) = 0$ has no real solution; therefore, there are no inflection points.

g) *Concavity.* We use -4 and 4 to divide the real number line into three intervals
A: $(-\infty, -4)$ B: $(-4, 4)$ and C: $(4, \infty)$, and we test a point in each interval.

A: Test -5, $f''(-5) = \dfrac{5824}{729} > 0$

B: Test 0, $f''(0) = -\dfrac{1}{4} < 0$

C: Test 5, $f''(5) = \dfrac{5824}{729} > 0$

Therefore, $f(x)$ is concave up on the intervals $(-\infty, -4)$ and $(4, \infty)$ and concave down on the interval $(-4, 4)$.

h) *Sketch.* Use the preceding information to sketch the graph.

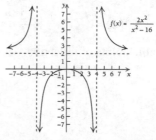

$$f(x) = \frac{2x^2}{x^2 - 16}$$

51. $f(x) = \dfrac{10}{x^2 + 4}$

a) *Intercepts.* Since the numerator is the constant 10, there are no x-intercepts.

$f(0) = \dfrac{5}{2}$, so the y-intercept is $\left(0, \dfrac{5}{2}\right)$.

b) *Asymptotes.*

Vertical. $x^2 + 4 = 0$ has no real solution, so there are no vertical asymptotes.

Horizontal. The degree of the numerator is less than the degree of the denominator, so $y = 0$ is the horizontal asymptote.

Slant. There is no slant asymptote since the degree of the numerator is not one more than the degree of the denominator.

c) *Derivatives and Domain.*

$$f'(x) = -20x\left(x^2 + 4\right)^{-2} = -\dfrac{20x}{\left(x^2 + 4\right)^2}$$

$$f''(x) = \dfrac{20\left(3x^2 - 4\right)}{\left(x^2 + 4\right)^3}$$

The domain of f is $\mathbb{R}$.

d) *Critical Points.* $f'(x)$ exists for all real numbers. $f'(x) = 0$ for $x = 0$, so 0 is a critical value. From step (a) we already know $\left(0, \dfrac{5}{2}\right)$ is on the graph.

e) *Increasing, decreasing, relative extrema.* We use 0 to divide the real number line into two intervals A: $(-\infty, 0)$ and B: $(0, \infty)$, and we test a point in each interval.

A: Test -1, $f'(-1) = \dfrac{4}{5} > 0$

B: Test 1, $f'(1) = -\dfrac{4}{5} < 0$

Then $f(x)$ is increasing on $(-\infty, 0)$ and is decreasing on $(0, \infty)$. Thus $\left(0, \dfrac{5}{2}\right)$ is a relative maximum.

f) *Inflection points.* $f''(x)$ exists for all real numbers. $f''(x) = 0$ for $x = \pm\dfrac{2}{\sqrt{3}}$, which are possible points of inflection.

We have:

$$f\left(-\dfrac{2}{\sqrt{3}}\right) = \dfrac{15}{8} \text{ and } f\left(\dfrac{2}{\sqrt{3}}\right) = \dfrac{15}{8}.$$

So, $\left(-\dfrac{2}{\sqrt{3}}, \dfrac{15}{8}\right)$ and $\left(\dfrac{2}{\sqrt{3}}, \dfrac{15}{8}\right)$ are inflection points.

g) *Concavity.* We use $-\dfrac{2}{\sqrt{3}}$ and $\dfrac{2}{\sqrt{3}}$ to divide the real number line into three intervals

A: $\left(-\infty, -\dfrac{2}{\sqrt{3}}\right)$ B: $\left(-\dfrac{2}{\sqrt{3}}, \dfrac{2}{\sqrt{3}}\right)$, and

C: $\left(\dfrac{2}{\sqrt{3}}, \infty\right)$.

A: Test -2, $f''(-2) = \dfrac{5}{16} > 0$

B: Test 0, $f''(0) = -\dfrac{5}{4} < 0$

C: Test 2, $f''(2) = \dfrac{5}{16} > 0$

Therefore, $f(x)$ is concave up on

$\left(-\infty, -\dfrac{2}{\sqrt{3}}\right)$ and $\left(\dfrac{2}{\sqrt{3}}, \infty\right)$ and concave

down on $\left(-\dfrac{2}{\sqrt{3}}, \dfrac{2}{\sqrt{3}}\right)$. Thus, the points

$\left(-\dfrac{2}{\sqrt{3}}, \dfrac{15}{8}\right)$ and $\left(\dfrac{2}{\sqrt{3}}, \dfrac{15}{8}\right)$ are points of inflection.

h) *Sketch.* Use the preceding information to sketch the graph. Compute additional function values as needed.

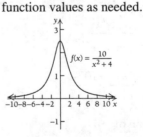

53. $f(x) = \dfrac{x^2 + 1}{x}$

a) *Intercepts.* The equation $f(x) = 0$ has no real solutions, so there are no x-intercepts. The number 0 is not in the domain of $f(x)$ so there are no y-intercepts.

b) *Asymptotes.*
Vertical. The denominator is 0 for $x = 0$, so the line $x = 0$ is a vertical asymptote.
Horizontal. The degree of the numerator is greater than the degree of the denominator, so there are no horizontal asymptotes.
Slant. The degree of the numerator is exactly one greater than the degree of the denominator. When we divide the numerator by the denominator we have

$$f(x) = \frac{x^2+1}{x} = x + \frac{1}{x}.$$ As $|x|$ approaches

∞, $f(x) = x + \dfrac{1}{x}$ approaches x. Therefore,

$y = x$ is the slant asymptote.

c) *Derivatives and Domain.*

$$f'(x) = \frac{x^2-1}{x^2}$$

$$f''(x) = \frac{2}{x^3}$$

The domain of f is $(-\infty, -0) \cup (0, \infty)$ as determined in part (b).

d) *Critical Points.* $f'(x)$ exists for all values of x except 0, but 0 is not in the domain of the function, so $x = 0$ is not a critical value. The critical points will occur when $f'(x) = 0$.

$$\frac{x^2-1}{x^2} = 0$$

$$x^2 - 1 = 0$$

$$x^2 = 1$$

$$x = \pm 1$$

So, −1 and 1 are critical values, thus $f(-1) = -2$ and $f(1) = 2$, so the critical points $(-1, -2)$ and $(1, 2)$ are on the graph.

e) *Increasing, decreasing, relative extrema.* We use −1, 0, and 1 to divide the real number line into four intervals
A: $(-\infty, -1)$ B: $(-1, 0)$, C: $(0, 1)$, and
D: $(1, \infty)$. We test a point in each interval.

A: Test −2, $f'(-2) = \dfrac{3}{4} > 0$

B: Test $-\dfrac{1}{2}$, $f'\left(-\dfrac{1}{2}\right) = -3 < 0$

C: Test $\dfrac{1}{2}$, $f'\left(\dfrac{1}{2}\right) = -3 < 0$

D: Test 2, $f'(2) = \dfrac{3}{4} > 0$

Then $f(x)$ is increasing on $(-\infty, -1)$ and $(1, \infty)$ and is decreasing on $(-1, 0)$ and $(0, 1)$. Therefore, $(-1, -2)$ is a relative maximum, and $(1, 2)$ is a relative minimum.

f) *Inflection points.* $f''(x)$ does not exist at 0, but because 0 is not in the domain of the function, there cannot be an inflection point at 0. The equation $f''(x) = 0$ has no solution; therefore, there are no inflection points.

g) *Concavity.* We use 0 to divide the real number line into two intervals
A: $(-\infty, 0)$ and B: $(0, \infty)$, and we test a point in each interval.
A: Test −1, $f''(-1) = -2 < 0$
B: Test 1, $f''(1) = 2 > 0$
Therefore, $f(x)$ is concave down on $(-\infty, 0)$ and concave up on $(0, \infty)$.

h) *Sketch.* Use the preceding information to sketch the graph. Compute additional function values as needed.

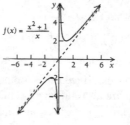

55. $f(x) = \dfrac{x^2-16}{x+4} = \dfrac{(x-4)(x+4)}{x+4} = x-4,\ x \neq -4$
Notice that $f(x) = x - 4$ for all values of x except $x = -4$, where it is undefined. The graph of $f(x)$ will be the graph of $y = x - 4$ except at the point $x = -4$.

a) *Intercepts.* $f(x) = 0$ when $x = 4$, so the x-intercept is $(4, 0)$.
$f(0) = -4$.
The point $(0, -4)$ is the y-intercept.

b) *Asymptotes.*

In simplified form $f(x) = x - 4$, a linear function everywhere except $x = -4$. So there are no asymptotes of any kind.

In the original function, the denominator is 0 for $x = -4$; however, $x = -4$ also made the numerator equal to 0. We look at the limits to determine if there are vertical asymptotes at these points.

$\lim\limits_{x \to -4} \dfrac{x^2 - 16}{x + 4} = \lim\limits_{x \to -4} (x - 4) = -8$. Because the limit exists, the line $x = -4$ is not a vertical asymptote. Instead, we have a removable discontinuity, or a "hole" at the point $(-4, -8)$. An open circle is drawn at $(-4, -8)$ to show that it is not part of the graph.

c) *Derivatives and Domain.*

$f'(x) = 1, \ x \neq -4$

$f''(x) = 0, \ x \neq -4$

d) *Critical Points.* There are no critical points.

e) *Increasing, decreasing, relative extrema.*

We use -4 to divide the real number line into two intervals

A: $(-\infty, -4)$ and B: $(-4, \infty)$.

We notice that $f'(x) > 0$ for all real numbers in the domain, $f(x)$ is increasing on both intervals. Since there are no critical points, there are no relative extrema.

f) *Inflection points.* $f''(x)$ is constant; therefore, there are no points of inflection.

g) *Concavity.* $f''(x)$ is 0; therefore, there is no concavity.

h) *Sketch.* Use the preceding information to sketch the graph.

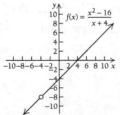

Note: In the preceding problem, we could have noticed that the graph of

$f(x) = \dfrac{x^2 - 16}{x + 4}$ is the graph of $f(x) = x - 4$

with the exception of the point $(-4, -8)$ which is a removable discontinuity.

We simply need to graph $f(x) = x - 4$ with a hole at the point $(-4, -8)$ and determine all other aspects of the graph of $f(x)$ from the linear graph.

57. $C(x) = 3x^2 + 80$

a) $A(x) = \dfrac{C(x)}{x} = \dfrac{3x^2 + 80}{x} = 3x + \dfrac{80}{x}$

b) Using the techniques of this section we find the following information. We will only consider the values of x is $(0, \infty)$.

Intercepts. None.

Asymptotes. $x = 0$ is the vertical asymptote. There is no horizontal asymptote. As $|x|$

approaches ∞, $A(x) = 3x + \dfrac{80}{x}$ approaches

$3x$. Therefore, $y = 3x$ is the slant asymptote.

Increasing, decreasing, relative extrema.

$A'(x) = 3 - \dfrac{80}{x^2}$. $A'(x)$ is not defined for

$x = 0$, however that value is outside the domain of the function. $A'(x) = 0$ when

$x = \sqrt{\dfrac{80}{3}}$, and $A\left(\sqrt{\dfrac{80}{3}}\right) = 2\sqrt{240}$.

Using $x = \sqrt{\dfrac{80}{3}}$ to divide the interval $(0, \infty)$

into two intervals, $\left(0, \sqrt{\dfrac{80}{3}}\right)$ and $\left(\sqrt{\dfrac{80}{3}}, \infty\right)$,

and testing a point in each interval, we find

that $A(x)$ is decreasing on $\left(0, \sqrt{\dfrac{80}{3}}\right)$ and

increasing on $\left(\sqrt{\dfrac{80}{3}}, \infty\right)$. Therefore, the

point $\left(\sqrt{\dfrac{80}{3}}, 2\sqrt{240}\right)$ is a relative minimum.

Inflection points, concavity.

$A''(x) = \dfrac{160}{x^3}$ exists for all values of t in

$(0, \infty)$. The equation $A''(x) = 0$ has no real solution, so there are no possible points of inflection. Furthermore, $A''(x) > 0$ for all x in the domain, so $A(x)$ is concave up on $(0, \infty)$. The solution is continued.

We use the information on the previous page to sketch the graph. Additional values may be computed as necessary.

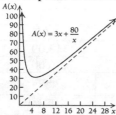

c) The degree of the numerator is exactly one greater than the degree of the denominator. When we divide the numerator by the denominator we have

$$A(x) = \frac{3x^2 + 80}{x} = 3x + \frac{80}{x}.$$ As $|x|$

approaches ∞, $A(x) = 3x + \frac{80}{x}$ approaches

$3x$. Therefore, $y = 3x$ is the slant asymptote. This means that when a large number of pairs of rocket skates are produced, the average cost can be estimated by multiplying the number of pairs produced by 3 thousand dollars.

59. $C(x) = 5000 + 600x$

$$R(x) = -\frac{1}{2}x^2 + 1000x$$

a) Profit is revenue minus cost, therefore, the total profit function is:

$$P(x) = R(x) - C(x)$$

$$= -\frac{1}{2}x^2 + 1000x - (5000 + 600x)$$

$$= -\frac{1}{2}x^2 + 400x - 5000$$

b) $A(x) = \dfrac{P(x)}{x}$

$$= \frac{-\frac{1}{2}x^2 + 400x - 5000}{x}$$

$$= -\frac{1}{2}x + 400 - \frac{5000}{x}$$

c) As $|x|$ approaches ∞, $A(x)$ approaches

$-\frac{1}{2}x + 400$. Therefore, $y = -\frac{1}{2}x + 400$ is

the slant asymptote. This represents the average profit for x items, when x is a large number of items.

d) Using the techniques of this section we find the following additional information.

Intercepts. The x-intercepts are $(12.70, 0)$ and $(787.30, 0)$. There is no P-intercept.

Asymptotes.   Vertical: $x = 0$
                 Horizonta: None

$$\text{Slant:} \qquad y = -\frac{1}{2}x + 400$$

Increasing, decreasing, relative extrema. $A(x)$ is increasing over the interval $(0, 100]$ and decreasing over the interval $[100, \infty)$.

The point $(100, 300)$ is a relative maximum.

Inflection points, concavity. $A(x)$ is concave down on the interval $(0, \infty)$. There are no inflection points. We use this information and compute other function values as necessary to sketch the graph.

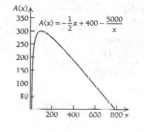

61. $P(x) = \dfrac{1}{1 + 0.0362x}$

a)

$$P(5) = \frac{1}{1 + 0.0362(5)} = 0.84674$$

In 1995, the purchasing power of a dollar was $0.85.

$$P(10) = \frac{1}{1 + 0.0362(10)} = 0.73421439$$

In 2000, the purchasing power of a dollar was $0.73.

$$P(25) = \frac{1}{1 + 0.0362(25)} = 0.5249343$$

In 2015, the purchasing power of a dollar was $0.52.

b) Solve $P(x) = 0.50$

$$\frac{1}{1+0.0362x} = 0.50$$

$$1 = 0.50(1+0.0362x)$$

$$1 = 0.50 + 0.0181x$$

$$0.50 = 0.0181x$$

$$27.624 = x$$

27.6 years after 1990, or in 2017, the purchasing power of a dollar will be $0.50.

c) Find $\lim\limits_{x \to \infty} P(x)$.

$$\lim_{x \to \infty} P(x) = \lim_{x \to \infty} \frac{1}{1+0.0362x}$$

$$= \lim_{x \to \infty} \frac{1}{1+0.0362x} \cdot \frac{\frac{1}{x}}{\frac{1}{x}}$$

$$= \lim_{x \to \infty} \frac{\frac{1}{x}}{\frac{1}{x} + 0.0362}$$

$$= \frac{0}{0 + 0.0362} = 0$$

$$\lim_{x \to \infty} P(x) = 0.$$

63. $E(n) = 9 \cdot \dfrac{4}{n}$

a) Calculate each value for the given n.

$$E(9) = 9 \cdot \frac{4}{9} = 4.00$$

$$E(6) = 9 \cdot \frac{4}{6} = 6.00$$

$$E(3) = 9 \cdot \frac{4}{3} = 12.00$$

$$E(1) = 9 \cdot \frac{4}{1} = 36.00$$

$$E\left(\frac{2}{3}\right) = 9 \cdot \frac{4}{\frac{2}{3}} = 9\left(4 \cdot \frac{3}{2}\right) = 54.00$$

$$E\left(\frac{1}{3}\right) = 9 \cdot \frac{4}{\frac{1}{3}} = 9\left(4 \cdot \frac{3}{1}\right) = 108.00$$

We complete the table.

Innings Pitched (n)	Earned-Run average (E)
9	4.00
6	6.00
3	12.00
1	36.00
$\frac{2}{3}$	54.00
$\frac{1}{3}$	108.00

b) $\lim\limits_{n \to 0} E(n) = \lim\limits_{n \to 0} 9 \cdot \dfrac{4}{n} = \lim\limits_{n \to 0} \dfrac{36}{n} = \infty$

If the pitcher gives up one or more runs but gets no one out, the pitcher would be credited with zero innings pitched.

65. ✎

67. $\lim\limits_{x \to 0} \dfrac{|x|}{x}$

Using $|x| = \begin{cases} -x, & \text{for } x < 0 \\ x, & \text{for } x \geq 0 \end{cases}$, we have

$$\lim_{x \to 0^-} \frac{|x|}{x} = -1 \text{ and } \lim_{x \to 0^+} \frac{|x|}{x} = 1.$$

Therefore, $\lim\limits_{x \to 0} \dfrac{|x|}{x}$ does not exist.

69. $\lim\limits_{x \to -\infty} \dfrac{-6x^3 + 7x}{2x^2 - 3x - 10} = \lim\limits_{x \to -\infty} \dfrac{-6x + \dfrac{7}{x}}{2 - \dfrac{3}{x} - \dfrac{10}{x^2}}$

$$= \frac{\lim\limits_{x \to -\infty}(-6x) + 0}{2 - 0 - 0}$$

$$= \infty$$

71. $f(x) = x^2 + \dfrac{1}{x^2}$

$f(x) = x^2 + \frac{1}{x^2}$

73. $f(x) = \dfrac{x^3 + 4x^2 + x - 6}{x^2 - x - 2}$

$f(x) = \frac{x^3 + 4x^2 + x - 6}{x^2 - x - 2}$

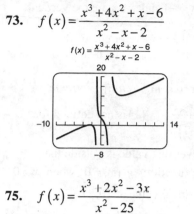

75. $f(x) = \dfrac{x^3 + 2x^2 - 3x}{x^2 - 25}$

$f(x) = \frac{x^3 + 2x^2 - 3x}{x^2 - 25}$

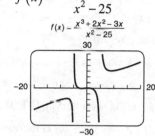

77. $f(x) = \dfrac{x^2 - 3}{2x - 4}$

$f(x) = \frac{x^2 - 3}{2x - 4}$

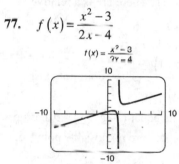

a) Solve $f(x) = 0$.

$$\frac{x^2 - 3}{2x - 4} = 0$$

$$x^2 - 3 = 0 \qquad x \neq \frac{1}{2}$$

$$x^2 = 3$$

$$x = \pm\sqrt{3} \approx \pm 1.732$$

The x-intercepts are:
$(-1.732, 0)$ and $(1.732, 0)$.

b) Evaluate at $x = 0$.

$$f(0) = \frac{(0)^2 - 3}{2(0) - 4} = \frac{3}{4}$$

The y-intercept is $(0, 0.75)$.

c) *Vertical.* $\qquad x = 2$
Horizontal. $\qquad$ None
Slant. Dividing the numerator by the denominator we have

$$f(x) = \frac{x^2 - 3}{2x - 4}$$

$$= \frac{1}{2}x + 1 + \frac{1}{2x - 4}$$

As $|x|$ approaches ∞,

$$f(x) \text{ approaches } \frac{x}{2} + 1 = 0.5x + 1.$$

The slant asymptote is
$y = 0.5x + 1$

79. $f(x) = \dfrac{x^5 + x - 9}{x^3 + 6x}$

Using long division we have:

$$\begin{array}{r} x^2 \qquad\quad -6 \\ x^3 + 6x \overline{\smash{\big)}\, x^5 \qquad\qquad + x - 9} \\ \underline{x^5 + 6x^3} \qquad\qquad\quad \\ -6x^3 + 0x^2 + x \\ \underline{-6x^3 \qquad\quad +36x} \\ 37x - 9 \end{array}$$

Therefore,

$$f(x) = \frac{x^5 + x - 9}{x^3 + 6x} = x^2 - 6 + \frac{37x - 9}{x^3 + 6x}.$$

As $|x|$ gets large, $f(x)$ approaches $x^2 - 6$.
Therefore, the nonlinear asymptote is
$y = x^2 - 6$.

Using a calculator, we graph the function, f, and the asymptote, y, below.

$f(x) = \frac{x^5 + x - 9}{x^3 + 6x}$ and $y = x^2 - 6$

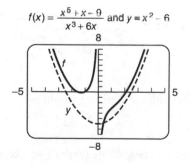

81. One possible rational function would be

$$f(x) = \frac{-2x}{x-2}.$$

Using the techniques in this section, we sketch the graph.

Intercepts. $f(x) = 0$ when $x = 0$. It turns out the x-intercept and the y-intercept is $(0,0)$.

Asymptotes. $x = 2$ is the vertical asymptote. The degree of the numerator equals that of the denominator, the line $y = -2$ is the horizontal asymptote.

Increasing, decreasing, relative extrema.

$$f'(x) = \frac{4}{(x-2)^2}.$$ $f'(x)$ is not defined for

$x = 2$, however that value is outside the domain of the function. $f'(x) > 0$ so $f(x)$ is increasing on the intervals $(-\infty, 2)$ and $(2, \infty)$. There are no relative extrema.

Inflection points, concavity.

$$f''(x) = \frac{-8}{(x-2)^3}$$ does not exist when $x = 2$.

The equation $f''(x) = 0$ has no real solution, so there are no possible points of inflection. Furthermore, $f''(x) > 0$ for all x in $(-\infty, 2)$, so $f(x)$ is concave up on $(-\infty, 2)$, and, $f''(x) < 0$ for all x in $(2, \infty)$, so $f(x)$ is concave down on $(2, \infty)$.

We use this information to sketch the graph. Additional values may be computed as necessary.

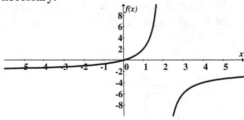

83. One possible rational function would be

$$g(x) = \frac{x^2 - 2}{x^2 - 1}.$$

Using the techniques in this section, we sketch the graph.

Intercepts. $g(x) = 0$ when $x = \pm\sqrt{2}$. The x-intercepts are $\left(-\sqrt{2}, 0\right)$ and $\left(\sqrt{2}, 0\right)$ When

$x = 0$, $g(x) = 2$, so the y-intercept is $(0, 2)$.

Asymptotes. $x = -1$ and $x = 1$ are the vertical asymptote. The degree of the numerator equals that of the denominator, the line $y = 1$ is the horizontal asymptote.

Increasing, decreasing, relative extrema.

$$g'(x) = \frac{2x}{\left(x^2 - 1\right)^2}.$$ $g'(x)$ is not defined for

$x = \pm 1$, however that value is outside the domain of the function. $g'(x) = 0$, when $x = 0$, so there is a critical value at $(0, 2)$.

We notice that $g'(x) < 0$ when $x < -1$ so $g(x)$ is decreasing on the interval $(-\infty, -1)$. $g'(x) > 0$ when $1 < x$ so $g(x)$ is increasing on the interval $(1, \infty)$. $g'(x) < 0$ when $-1 < x < 0$ so $g(x)$ is decreasing on the interval $(-1, 0)$. $g'(x) > 0$ when $0 < x < 1$ so $g(x)$ is increasing on the interval $(0, 1)$. There are is a relative minimum at $(0, 2)$.

Inflection points, concavity.

$$g''(x) = \frac{-2\left(3x^2 + 1\right)}{\left(x^2 - 1\right)^3}.$$ $g''(x)$ does not exist

when $x = \pm 1$. The equation $g''(x) = 0$ has no real solution, so there are no possible points of inflection.

Furthermore, $g''(x) > 0$ for all x in $(-1, 1)$, so $g(x)$ is concave up on $(-1, 1)$, and, $g''(x) < 0$ for all x in $(-\infty, -1)$ and $(1, \infty)$, so $g(x)$ is concave down on $(-\infty, -1)$ and $(1, \infty)$.

We use this information to sketch the graph. Additional values may be computed as necessary.

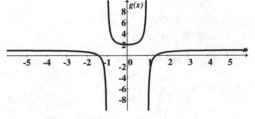

85. One possible rational function would be

$$h(x) = \frac{-8}{x^2 + x - 6}.$$

Using the techniques in this section, we sketch the graph.

Intercepts. $h(x) = 0$ has no real solution. The are no x-intercepts. When $x = 0$, $h(x) = \frac{4}{3}$, so

the y-intercept is $\left(0, \frac{4}{3}\right)$.

Asymptotes. $x = -3$ and $x = 2$ are the vertical asymptotes.

The degree of the numerator is less than that of the denominator, the line $y = 0$ is the horizontal asymptote.

Increasing, decreasing, relative extrema.

$h'(x) = \frac{8(2x+1)}{(x-2)^2(x+3)^2}$. $h'(x)$ is not defined

for $x = -3$ or $x = 2$, however those values are outside the domain of the function. $h'(x) = 0$,

when $x = \frac{-1}{2}$ so there is a critical value at

$\left(\frac{-1}{2}, \frac{32}{25}\right)$.

$h'(x) < 0$ when $-3 < x < \frac{-1}{2}$ so $h(x)$ is

decreasing on the interval $\left(-3, \frac{-1}{2}\right)$.

We find $h'(x) > 0$ when $\frac{-1}{2} < x < 2$ so $h(x)$ is

increasing on the interval $\left(\frac{-1}{2}, 2\right)$. There is a

relative minimum at $\left(\frac{-1}{2}, \frac{32}{25}\right)$. Furthermore,

$h'(x) < 0$ when $x < -3$ so $h(x)$ is decreasing on the interval $(-\infty, -3)$. $h'(x) > 0$ when $2 < x$ so $h(x)$ is increasing on the interval $(2, \infty)$.

Inflection points, concavity.

$h''(x) = \frac{-16(3x^2 + 3x + 7)}{(x-2)^3(x+3)^3}$. $h''(x)$ does not

exist when $x = -2$ and $x = 0$. The equation $h''(x) = 0$ has no real solution. Therefore $h(x)$ has no points of inflection. $h''(x) < 0$ when $x < -3$ and $x > 2$.

Therefore, $h(x)$ is concave down on the intervals $(-\infty, -3)$ and $(2, \infty)$. $h''(x) > 0$ when $-3 < x < 2$, so $h(x)$ is concave up on $(-3, 2)$.

We use this information to sketch the graph. Additional values may be computed as necessary.

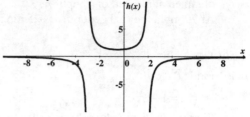

Exercise Set 2.4

1. a) The absolute maximum gasoline mileage is obtained at a speed of 55 mph.
b) The absolute minimum gasoline mileage is obtained at a speed of 5 mph.
c) At 70 mph, the fuel economy is 25 mpg.

3. $f(x) = 5 + x - x^2$; $[0,2]$

a) Find $f'(x)$

$$f'(x) = 1 - 2x$$

b) Find the Critical values. The derivative exists for all real numbers. Thus, we solve

$$f'(x) = 0$$
$$1 - 2x = 0$$
$$1 = 2x$$
$$\frac{1}{2} = x$$

c) List the critical values and endpoints. These values are 0, $\frac{1}{2}$, and 2.

d) Evaluate $f(x)$ at each value in step (c).

$$f(0) = 5 + (0) - (0)^2 = 5$$
$$f\left(\frac{1}{2}\right) = 5 + \left(\frac{1}{2}\right) - \left(\frac{1}{2}\right)^2 = \frac{21}{4} = 5.25$$
$$f(2) = 5 + (2) - (2)^2 = 3$$

The largest of these values, $\frac{21}{4}$, is the absolute maximum, it occurs at $x = \frac{1}{2}$. The smallest of these values, 3, is the absolute minimum, it occurs at $x = 2$.

5. $f(x) = x^3 + \frac{1}{2}x^2 - 2x + 5$; $[-2,1]$

a) Find $f'(x)$

$$f'(x) = 3x^2 + x - 2$$

b) Find the critical values. The derivative exists for all real numbers. Thus, we solve

$$f'(x) = 0$$
$$3x^2 + x - 2 = 0$$
$$(3x - 2)(x + 1) = 0$$
$$3x - 2 = 0 \text{ or } x + 1 = 0$$
$$x = \frac{2}{3} \text{ or } \quad x = -1$$

c) List the critical values and endpoints. These values are are: -2, -1, $\frac{2}{3}$, and 1.

d) Evaluate $f(x)$ for each value in step (c).

$$f(-2) = (-2)^3 + \frac{1}{2}(-2)^2 - 2(-2) + 5 = 3$$
$$f(-1) = (-1)^3 + \frac{1}{2}(-1)^2 - 2(-1) + 5 = \frac{13}{2} = 6.5$$
$$f\left(\frac{2}{3}\right) = \left(\frac{2}{3}\right)^3 + \frac{1}{2}\left(\frac{2}{3}\right)^2 - 2\left(\frac{2}{3}\right) + 5 = \frac{113}{27} \approx 4.2$$
$$f(1) = (1)^3 + \frac{1}{2}(1)^2 - 2(1) + 5 = \frac{9}{2} = 4.5$$

On the interval $[-2,1]$, the absolute maximum is $\frac{13}{2}$, which occurs at $x = -1$. The absolute minimum is 3, which occurs at $x = -2$.

7. $f(x) = x^3 + \frac{1}{2}x^2 - 2x + 4$; $[-2,0]$

a) Find $f'(x)$

$$f'(x) = 3x^2 + x - 2$$

b) Find the critical values. The derivative exists for all real numbers. Thus, we solve

$$3x^2 + x - 2 = 0$$
$$(3x - 2)(x + 1) = 0$$
$$3x - 2 = 0 \text{ or } x + 1 = 0$$
$$x = \frac{2}{3} \text{ or } \quad x = -1$$

c) The critical value $x = \frac{2}{3}$ is not in the interval, so we exclude it. We will test the values: -2, -1, and 0.

d) Evaluate $f(x)$ for each value in step (c).

$$f(-2) = (-2)^3 + \frac{1}{2}(-2)^2 - 2(-2) + 4 = 2$$

$$f(-1) = (-1)^3 + \frac{1}{2}(-1)^2 - 2(-1) + 4 = \frac{11}{2} = 5.5$$

$$f(0) = (0)^3 + \frac{1}{2}(0)^2 - 2(0) + 4 = 4$$

On the interval $[-2, 0]$, the absolute maximum is 5.5, which occurs at $x = -1$. The absolute minimum is 2, which occurs at $x = -2$.

9. $f(x) = 2x + 4$; $[-1, 1]$

a) Find $f'(x)$

$$f'(x) = 2$$

b) and c)

The derivative exists and $f'(x) = 2$ for all real numbers. Note that the derivative is never 0. Thus, there are no critical values for $f(x)$, and the absolute maximum and absolute minimum will occur at the endpoints of the interval.

d) Evaluate $f(x)$ at the endpoints.

$$f(-1) = 2(-1) + 4 = 2$$

$$f(1) = 2(1) + 4 = 6$$

On the interval $[-1, 1]$, the absolute maximum is 6, which occurs at $x = 1$. The absolute minimum is 2, which occurs at $x = -1$.

11. $f(x) = 7 - 4x$; $[-2, 5]$

a) Find $f'(x)$

$$f'(x) = -4$$

b) and c)

The derivative exists and is -4 for all real numbers. Note that the derivative is never 0. Thus, there are no critical values for $f(x)$, and the absolute maximum and absolute minimum will occur at the endpoints of the interval.

d) Evaluate $f(x)$ at the endpoints.

$$f(-2) = 7 - 4(-2) = 15$$

$$f(5) = 7 - 4(5) = -13$$

On the interval $[-2, 5]$, the absolute maximum is 15, which occurs at $x = -2$. The absolute minimum is -13, which occurs at $x = 5$.

13. $f(x) = -5$; $[-1, 1]$

Note for all values of x, $f(x) = -5$. Thus, the absolute maximum is -5 for $-1 \le x \le 1$ and the absolute minimum is -5 for $-1 \le x \le 1$.

15. $g(x) = 24$; $[4, 13]$

Note for all values of x, $g(x) = 24$. Thus, the absolute maximum is 24 for $4 \le x \le 13$ and the absolute minimum is 24 for $4 \le x \le 13$.

17. $f(x) = x^2 - 4x + 5$; $[-1, 3]$

a) $f'(x) = 2x - 4$

b) $f'(x)$ exists for all real numbers. Solve:

$$2x - 4 = 0$$

$$2x = 4$$

$$x = 2$$

c) The critical value and the endpoints are -1, 2, and 3.

d) Evaluate $f(x)$ for each value in step (c).

$$f(-1) = (-1)^2 - 4(-1) + 5 = 10$$

$$f(2) = (2)^2 - 4(2) + 5 = 1$$

$$f(3) = (3)^2 - 4(3) + 5 = 2$$

On the interval $[-1, 3]$, the absolute maximum is 10, which occurs at $x = -1$. The absolute minimum is 1, which occurs at $x = 2$.

19. $f(x) = 1 + 6x - 3x^2$; $[0, 4]$

a) $f'(x) = 6 - 6x$

b) $f'(x)$ exists for all real numbers. Solve:

$$6 - 6x = 0$$

$$x = 1$$

c) The critical value and the endpoints are 0, 1, and 4. Note, since 0 is an endpoint of the interval, $x = 0$ is included in this list as an endpoint, not as a critical value.

d) Evaluate $f(x)$ for each value in step (c).

$$f(0) = 1 + 6(0) - 3(0)^2 = 1$$

$$f(1) = 1 + 6(1) - 3(1)^2 = 4$$

$$f(4) = 1 + 6(4) - 3(4)^2 = -23$$

On the interval $[0,4]$, the absolute maximum is 4, which occurs at $x = 1$. The absolute minimum is -23, which occurs at $x = 4$.

21. $f(x) = x^3 - 3x;$ $[-5,1]$

a) $f'(x) = 3x^2 - 3$

b) $f'(x)$ exists for all real numbers. Solve:

$$3x^2 - 3 = 0$$

$$x^2 - 1 = 0$$

$$x = \pm 1$$

c) The critical value and the endpoints are -5, -1, and 1. Note, since 1 is an endpoint of the interval, $x = 1$ is included in this list as an endpoint, not a critical value.

d) Evaluate $f(x)$ for each value in step (c).

$$f(-5) = (-5)^3 - 3(-5) = -110$$

$$f(-1) = (-1)^3 - 3(-1) = 2$$

$$f(1) = (1)^3 - 3(1) = -2$$

On the interval $[-5,1]$, the absolute maximum is 2, which occurs at $x = -1$. The absolute minimum is -110, which occurs at $x = -5$.

23. $f(x) = 1 - x^3;$ $[-8,8]$

a) $f'(x) = -3x^2$

b) $f'(x)$ exists for all real numbers. Solve:

$$-3x^2 = 0$$

$$x = 0$$

c) The critical value and the endpoints are -8, 0, and 8.

d) Evaluate $f(x)$ for each value in step (c).

$$f(-8) = 1 - (-8)^3 = 513$$

$$f(0) = 1 - (0)^3 = 1$$

$$f(8) = 1 - (8)^3 = -511$$

On the interval $[-8,8]$, the absolute maximum is 513, which occurs at $x = -8$. The absolute minimum is -511, which occurs at $x = 8$.

25. $f(x) = x^3 - 6x^2 + 10;$ $[0,4]$

a) $f'(x) = 3x^2 - 12x$

b) $f'(x)$ exists for all real numbers. Solve:

$$3x^2 - 12x = 0$$

$$3x(x - 4) = 0$$

$$3x = 0 \quad \text{or} \quad x - 4 = 0$$

$$x = 0 \quad \text{or} \quad x = 4$$

c) The critical values and the endpoints are 0 and 4. Note, since the possible critical values are the endpoints of the interval, they are included in this list as endpoints, not as critical values.

d) Evaluate $f(x)$ for each value in step (c).

$$f(0) = (0)^3 - 6(0)^2 + 10 = 10$$

$$f(4) = (4)^3 - 6(4)^2 + 10 = -22$$

On the interval $[0,4]$, the absolute maximum is 10, which occurs at $x = 0$. The absolute minimum is -22, which occurs at $x = 4$.

27. $f(x) = x^3 - x^4;$ $[-1,1]$

a) $f'(x) = 3x^2 - 4x^3$

b) $f'(x)$ exists for all real numbers. Solve:

$$3x^2 - 4x^3 = 0$$

$$x^2(3 - 4x) = 0$$

$$x^2 = 0 \quad \text{or} \quad 3 - 4x = 0$$

$$x = 0 \quad \text{or} \quad x = \frac{3}{4}$$

c) The critical values and the endpoints are -1, 0, $\frac{3}{4}$, and 1.

d) Evaluate $f(x)$ for each value in step (c).

$$f(-1) = (-1)^3 - (-1)^4 = -2$$

$$f(0) = (0)^3 - (0)^4 = 0$$

$$f\left(\frac{3}{4}\right) = \left(\frac{3}{4}\right)^3 - \left(\frac{3}{4}\right)^4 = \frac{27}{256} \approx 0.105$$

$$f(1) = (1)^3 - (1)^4 = 0$$

e) On the interval $[-1,1]$, the absolute maximum is $\dfrac{27}{256}$, which occurs at $x=\dfrac{3}{4}$. The absolute minimum is -2, which occurs at $x=-1$.

29. $f(x)=x^4-2x^2+5;$ $[-2,2]$

a) $f'(x)=4x^3-4x$

b) $f'(x)$ exists for all real numbers. Solve:

$$4x^3-4x=0$$
$$4x(x^2-1)=0$$
$$4x=0 \quad \text{or} \quad x^2-1=0$$
$$x=0 \quad \text{or} \quad x=\pm 1$$

c) The critical values and the endpoints are $-2,\ -1,\ 0,\ 1,$ and 2.

d) Evaluate $f(x)$ for each value in step (c).

$$f(-2)=(-2)^4-2(-2)^2+5=13$$
$$f(-1)=(-1)^4-2(-1)^2+5=4$$
$$f(0)=(0)^4-2(0)^2+5=5$$
$$f(1)=(1)^4-2(1)^2+5=4$$
$$f(2)=(2)^4-2(2)^2+5=13$$

On the interval $[-2,2]$, the absolute maximum is 13, which occurs at $x=-2$ and $x=2$. The absolute minimum is 4, which occurs at $x=-1$ and $x=1$.

31. $f(x)=1-x^{2/3};$ $[-8,8]$

a) $f'(x)=-\dfrac{2}{3}x^{-1/3}=-\dfrac{2}{3x^{1/3}}$

b) $f'(x)$ does not exist for $x=0$. The equation $f'(x)=0$ has no solution, so $x=0$ is the only critical value.

c) The critical values and the endpoints are $-8,\ 0,$ and 8.

d) Evaluate $f(x)$ for each value in step (c).

$$f(-8)=1-(-8)^{2/3}=-3$$
$$f(0)=1-(0)^{2/3}=1$$
$$f(8)=1-(8)^{2/3}=-3$$

On the interval $[-8,8]$, the absolute maximum is 1, which occurs at $x=0$. The absolute minimum is -3, which occurs at $x=-8$ and $x=8$.

33. $f(x)=x+\dfrac{4}{x};$ $[-8,-1]$

a) $f'(x)=1-4x^{-2}=1-\dfrac{4}{x^2}$

b) $f'(x)$ does not exist for $x=0$. However, $x=0$ is not in the interval. Solve $f'(x)=0$.

$$1-\dfrac{4}{x^2}=0$$
$$1=\dfrac{4}{x^2}$$
$$x^2=4$$
$$x=\pm 2$$

The only critical value in the interval is at $x=-2$.

c) The critical values and the endpoints are $-8,\ -2$ and -1.

d) Evaluate $f(x)$ for each value in step (c).

$$f(-8)=(-8)+\dfrac{4}{(-8)}=-\dfrac{17}{2}-8.5$$
$$f(-2)=(-2)+\dfrac{4}{(-2)}=-4$$
$$f(-1)=(-1)+\dfrac{4}{(-1)}=-5$$

On the interval $[-8,-1]$, the absolute maximum is -4, which occurs at $x=-2$. The absolute minimum is $-\dfrac{17}{2}$, which occurs at $x=-8$.

35. $f(x)=\dfrac{x^2}{x^2+1};$ $[-2,2]$

a) $f'(x)=\dfrac{(x^2+1)(2x)-x^2(2x)}{(x^2+1)^2}$ Quotient Rule

$$=\dfrac{2x^3+2x-2x^3}{(x^2+1)^2}$$

$$=\dfrac{2x}{(x^2+1)^2}$$

b) $f'(x)$ exists for all real numbers. Solve:

$$f'(x) = 0$$

$$\frac{2x}{\left(x^2+1\right)^2} = 0$$

$$2x = 0$$

$$x = 0$$

c) The critical values and the endpoints are -2, 0, and 2.

d) Evaluate $f(x)$ for each value in step (c).

$$f(-2) = \frac{(-2)^2}{(-2)^2+1} = \frac{4}{5}$$

$$f(0) = \frac{(0)^2}{(0)^2+1} = 0$$

$$f(2) = \frac{(2)^2}{(2)^2+1} = \frac{4}{5}$$

On the interval $[-2,2]$, the absolute maximum is $\frac{4}{5}$, which occurs at $x = -2$ and $x = 2$. The absolute minimum is 0, which occurs at $x = 0$.

37. $f(x) = (x+1)^{\frac{1}{3}}$; $[-2, 26]$

a) $f'(x) = \frac{1}{3}(x+1)^{-\frac{2}{3}} = \frac{1}{3(x+1)^{\frac{2}{3}}}$

b) $f'(x)$ does not exist for $x = -1$. The equation $f'(x) = 0$ has no solution, so $x = -1$ is the only critical value.

c) The critical values and the endpoints are -2, -1, and 26.

d) Evaluate $f(x)$ for each value in step (c).

$$f(-2) = ((-2)+1)^{\frac{1}{3}} = -1$$

$$f(-1) = ((-1)+1)^{\frac{1}{3}} = 0$$

$$f(26) = ((26)+1)^{\frac{1}{3}} = 3$$

On the interval $[-2, 26]$, the absolute minimum is -1, which occurs at $x = -2$. The absolute maximum is 3, which occurs at $x = 26$.

39. – 48.   Left to the student.

49. $f(x) = 30x - x^2$

When no interval is specified, we use the real line $(-\infty, \infty)$.

a) Find $f'(x)$

$$f'(x) = 30 - 2x$$

b) Find the critical values. The derivative exists for all real numbers. Thus, we solve $f'(x) = 0$.

$$f'(x) = 0$$

$$30 - 2x = 0$$

$$-2x = -30$$

$$x = 15$$

The only critical value is $x = 15$.

c) Since there is only one critical value, we can apply Max-Min Principle 2. First we find $f''(x)$.

$$f''(x) = -2$$

The second derivative is constant, so $f''(15) = -2$. Since the second derivative is negative at 15, we have a maximum at $x = 15$. Next, we find the function value at $x = 15$.

$$f(15) = 30(15) - (15)^2 = 225$$

Therefore, the absolute maximum is 225, which occurs at $x = 15$. There is no minimum value.

51. $f(x) = 2x^2 - 40x + 270$

When no interval is specified, we use the real line $(-\infty, \infty)$.

a) Find $f'(x)$

$$f'(x) = 4x - 40$$

b) Find the critical values. The derivative exists for all real numbers. Thus, we solve $f'(x) = 0$.

$$f'(x) = 0$$

$$4x - 40 = 0$$

$$4x = 40$$

$$x = 10$$

The only critical value is $x = 10$.

c) Since there is only one critical value, we can apply Max-Min Principle 2. First we find $f''(x)$.

$$f''(x) = 4.$$

The second derivative is constant, so $f''(10) = 4$.

Since the second derivative is positive at 10, we have a minimum at $x = 10$. Next, we find the function value at $x = 10$.

$$f(10) = 2(10)^2 - 40(10) + 270 = 70$$

Therefore, the absolute minimum is 70, which occurs at $x = 10$. The function has no maximum value.

53. $f(x) = 16x - \dfrac{4}{3}x^3;$ $\qquad (0, \infty)$

a) Find $f'(x)$

$$f'(x) = 16 - 4x^2$$

b) Find the critical values. The derivative exists for all real numbers. Thus, we solve

$$f'(x) = 0$$

$$16 - 4x^2 = 0$$

$$4x^2 = 16$$

$$x^2 = 4$$

$$x = \pm 2$$

There are two critical values; however, $x = 2$ is the only critical value on the interval $(0, \infty)$.

c) Since there is only one critical value in the interval, we can apply Max-Min Principle 2. First we find $f''(x)$.

$$f''(x) = -8x.$$

Next evaluate the second derivative at $x = 2$.

$$f''(2) = -16 < 0$$

Since the second derivative is negative at 2, we have a maximum at $x = 2$.

$$f(2) = 16(2) - \frac{4}{3}(2)^3 = \frac{64}{3}$$

Therefore, the absolute maximum is $\dfrac{64}{3}$, which occurs at $x = 2$. There is no minimum value.

55. $f(x) = x(60 - x) = 60x - x^2$

When no interval is specified, we use the real line $(-\infty, \infty)$.

a) Find $f'(x)$.

$$f'(x) = 60 - 2x.$$

b) Find the critical values. The derivative exists for all real numbers.

Thus, we solve $f'(x) = 0$.

$$f'(x) = 0$$

$$60 - 2x = 0$$

$$60 = 2x$$

$$30 = x$$

The only critical value is $x = 30$.

c) Since there is only one critical value, we can apply Max-Min Principle 2. First we find $f''(x)$.

$$f''(x) = -2.$$

The second derivative is constant, so $f''(30) = -2$. Since the second derivative is negative at 30, we have a maximum at $x = 30$. Next, we find the function value at $x = 30$.

$$f(30) = 30(60 - 30) = 900$$

Therefore, the absolute maximum is 900, which occurs at $x = 30$. The function has no minimum value.

57. $f(x) = \dfrac{1}{3}x^3 - 5x;$ $\qquad [-3, 3]$

a) Find $f'(x)$.

$$f'(x) = x^2 - 5.$$

b) Find the critical values. The derivative exists for all real numbers. Thus, we solve $f'(x) = 0$.

$$x^2 - 5 = 0$$

$$x^2 = 5$$

$$x = \pm\sqrt{5} \approx \pm 2.236$$

Both critical values are in the interval $[-3, 3]$.

c) The interval is closed and there is more than one critical value, so we use Max-Min Principle 1.

The critical points and the endpoints are -3, $-\sqrt{5}$, $\sqrt{5}$, and 3.

The solution is continued on the next page.

Next, we find the function values at these points.

$$f(-3) = \frac{1}{3}(-3)^3 - 5(-3) = 6$$

$$f(-\sqrt{5}) = \frac{1}{3}(-\sqrt{5})^3 - 5(-\sqrt{5}) = \frac{10\sqrt{5}}{3} \approx 7.454$$

$$f(\sqrt{5}) = \frac{1}{3}(\sqrt{5})^3 - 5(\sqrt{5}) = -\frac{10\sqrt{5}}{3} \approx -7.454$$

$$f(3) = \frac{1}{3}(3)^3 - 5(3) = -6$$

Thus, the absolute maximum over the interval $[-3,3]$, is $\frac{10\sqrt{5}}{3}$, which occurs at $x = -\sqrt{5}$, and the absolute minimum over $[-3,3]$ is $-\frac{10\sqrt{5}}{3}$, which occurs at $x = \sqrt{5}$.

59. $f(x) = -0.001x^2 + 4.8x - 60$

When no interval is specified , we use the real line $(-\infty, \infty)$.

a) Find $f'(x)$

$$f'(x) = -0.002x + 4.8$$

b) Find the critical values. The derivative exists for all real numbers. Thus, we solve $f'(x) = 0$.

$$-0.002x + 4.8 = 0$$
$$-0.002x = -4.8$$
$$x = 2400$$

The only critical value is $x = 2400$.

c) Since there is only one critical value, we can apply Max-Min Principle 2. First we find $f''(x)$.

$$f''(x) = -0.002 .$$

The second derivative is constant, so $f''(2400) = -0.002$.

Since the second derivative is negative at 2400, we have a maximum at $x = 2400$. Next, we find the function value at $x = 2400$.

$$f(2400) = -0.001(2400)^2 + 4.8(2400) - 60$$
$$= 5700$$

Therefore, the absolute maximum is 5700, which occurs at $x = 2400$. The function has no minimum value.

61. $f(x) = -x^3 + x^2 + 5x - 1;$ $(0, \infty)$

a) Find $f'(x)$.

$$f'(x) = -3x^2 + 2x + 5 .$$

b) Find the critical values. The derivative exists for all real numbers. Thus, we solve $f'(x) = 0$.

$$-3x^2 + 2x + 5 = 0$$
$$-(3x - 5)(x + 1) = 0$$
$$3x - 5 = 0 \quad \text{or} \quad x + 1 = 0$$
$$x = \frac{5}{3} \quad \text{or} \qquad x = -1$$

$x = \frac{5}{3}$ is the only critical value on the interval $(0, \infty)$.

c) The interval $(0, \infty)$ is not closed. The only critical value in the interval is $x = \frac{5}{3}$. Therefore, we can apply Max-Min Principle 2. First, we find the second derivative.

$$f''(x) = -6x + 2$$

$$f''\left(\frac{5}{3}\right) = -6\left(\frac{5}{3}\right) + 2 = -8 < 0$$

Since the second derivative is negative when $x = \frac{5}{3}$, there is a maximum at $x = \frac{5}{3}$.

Next, find the function value:

$$f\left(\frac{5}{3}\right) = -\left(\frac{5}{3}\right)^3 + \left(\frac{5}{3}\right)^2 + 5\left(\frac{5}{3}\right) - 1 = \frac{148}{27}$$

Thus, the absolute maximum over the interval $(0, \infty)$ is $\frac{148}{27}$, which occurs at $x = \frac{5}{3}$. The function has no minimum value.

63. $f(x) = 15x^2 - \frac{1}{2}x^3;$ $[0, 30]$

a) Find $f'(x)$.

$$f'(x) = 30x - \frac{3}{2}x^2 .$$

b) Find the critical values. The derivative exists for all real numbers. Thus, we solve $f'(x) = 0$.

$$30x - \frac{3}{2}x^2 = 0$$

$$60x - 3x^2 = 0$$

$$3x(20 - x) = 0$$

$$3x = 0 \quad \text{or} \quad 20 - x = 0$$

$$x = 0 \quad \text{or} \quad x = 20$$

Both critical values are in the interval $[0, 30]$.

c) Since the interval is closed and there is more than one critical value, we apply the Max-Min Principle 1.

The critical values and the endpoints are 0, 20, and 30.

Next, we find the function values.

$$f(0) = 15(0)^2 - \frac{1}{2}(0)^3 = 0$$

$$f(20) = 15(20)^2 - \frac{1}{2}(20)^3 = 2000$$

$$f(30) = 15(30)^2 - \frac{1}{2}(30)^3 = 0$$

The largest of these values, 2000, is the maximum. It occurs at $x = 20$. The smallest of these values, 0, is the minimum. It occurs at $x = 0$ and $x = 30$.

Thus, the absolute maximum over the interval $[0, 30]$, is 2000, which occurs at $x = 20$, and the absolute minimum over $[0, 30]$ is 0, which occurs at $x = 0$ and $x = 30$.

b) Find the critical values. $f'(x)$ does not exist for $x = 0$; however, 0 is not in the interval $(0, \infty)$. Therefore, we solve

$$f'(x) = 0$$

$$2 - \frac{72}{x^2} = 0$$

$$2 = \frac{72}{x^2}$$

$$2x^2 = 72 \quad \text{Multiplying by } x^2, \text{ since } x \neq 0.$$

$$x^2 = 36$$

$$x = \pm 6$$

c) The interval $(0, \infty)$ is not closed. The only critical value in the interval is $x = 6$. Therefore, we can apply Max-Min Principle 2. First, we find the second derivative.

$$f''(x) = 144x^{-3} = \frac{144}{x^3}$$

Evaluating the second derivative at $x = 6$, we have:

$$f''(6) = \frac{144}{(6)^3} = \frac{2}{3} > 0.$$

Since the second derivative is positive when $x = 6$, there is a minimum at $x = 6$.

Next, find the function value at $x = 6$.

$$f(6) = 2(6) + \frac{72}{6} = 24$$

Thus, the absolute minimum over the interval $(0, \infty)$ is 24, which occurs at $x = 6$. The function has no maximum value over the interval $(0, \infty)$.

65. $f(x) = 2x + \frac{72}{x}; \qquad (0, \infty)$

$$f(x) = 2x + 72x^{-1}$$

a) Find $f'(x)$.

$$f'(x) = 2 - 72x^{-2} = 2 - \frac{72}{x^2}.$$

67. $f(x) = x^2 + \frac{432}{x}; \qquad (0, \infty)$

$$f(x) = x^2 + 432x^{-1}$$

a) Find $f'(x)$.

$$f'(x) = 2x - 432x^{-2} = 2x - \frac{432}{x^2}.$$

b) Find the critical values. $f'(x)$ does not exist for $x = 0$; however, 0 is not in the interval $(0, \infty)$. Therefore, we solve

$$f'(x) = 0$$

$$2x - \frac{432}{x^2} = 0$$

$$2x = \frac{432}{x^2}$$

$$2x^3 = 432 \qquad \text{Multiplying by } x^2, \text{ since } x \neq 0.$$

$$x^3 = 216$$

$$x = 6$$

c) The interval $(0, \infty)$ is not closed. The only critical value in the interval is $x = 6$. Therefore, we can apply Max-Min Principle 2.
First, we find the second derivative.

$$f''(x) = 2 + 864x^{-3} = 2 + \frac{864}{x^3}$$

Evaluating the second derivative at $x = 6$, we have:

$$f''(6) = 2 + \frac{864}{(6)^3} = 6 > 0.$$

Since the second derivative is positive when $x = 6$, there is a minimum at $x = 6$. Next, find the function value at $x = 6$.

$$f(6) = (6)^2 + \frac{432}{6} = 108$$

Thus, the absolute minimum over the interval $(0, \infty)$ is 108, which occurs at $x = 6$. The function has no maximum value over the interval $(0, \infty)$.

69. $f(x) = 2x^4 + x; \qquad [-1, 1]$

a) Find $f'(x)$.

$$f'(x) = 8x^3 + 1.$$

b) Find the critical values. The derivative exists for all real numbers. Thus, we solve $f'(x) = 0$.

$$8x^3 + 1 = 0$$

$$8x^3 = -1$$

$$x^3 = -\frac{1}{8}$$

$$x = -\frac{1}{2}$$

The only critical value $x = -\frac{1}{2}$ is in the interval $[-1, 1]$.

c) The interval is closed, and we are looking for both the absolute maximum and absolute minimum values, so we use Max-Min Principle 1.
The critical points and the endpoints are

$$-1, \ -\frac{1}{2}, \text{ and } 1.$$

Next, we find the function values at these points.

$$f(-1) = 2(-1)^4 + (-1) = 1$$

$$f\left(-\frac{1}{2}\right) = 2\left(-\frac{1}{2}\right)^4 + \left(-\frac{1}{2}\right) = -\frac{3}{8}$$

$$f(1) = 2(1)^4 + (1) = 3$$

The largest of these values, 3, is the maximum. It occurs at $x = 1$. The smallest of these values, $-\frac{3}{8}$, is the minimum. It occurs at $x = -\frac{1}{2}$.
Thus, the absolute maximum over the interval $[-1, 1]$, is 3, which occurs at $x = 1$, and the absolute minimum over $[-1, 1]$ is $-\frac{3}{8}$, which occurs at $x = -\frac{1}{2}$.

71. $f(x) = \sqrt[3]{x} = x^{\frac{1}{3}}; \qquad [0, 8]$

a) Find $f'(x)$

$$f'(x) = \frac{1}{3}x^{-\frac{2}{3}} = \frac{1}{3 \cdot \sqrt[3]{x^2}}$$

b) Find the critical values. $f'(x)$ does not exist for $x = 0$. The equation $f'(x) = 0$ has no solution, so the only critical value is 0, which is also an endpoint.

c) The interval is closed, and we are looking for both the absolute maximum and absolute minimum values, so we use Max-Min Principle 1.
The only critical value is an endpoint. The endpoints are 0 and 8.

The solution is continued on the next page.

Next, we find the function values at these points.

$$f(0) = \sqrt[3]{0} = 0$$

$$f(8) = \sqrt[3]{8} = 2$$

The largest of these values, 2, is the maximum. It occurs at $x = 8$. The smallest of these values, 0, is the minimum. It occurs at $x = 0$.

Thus, the absolute maximum over the interval $[0,8]$, is 2, which occurs at $x = 8$, and the absolute minimum over $[0,8]$ is 0, which occurs at $x = 0$.

73. $f(x) = (x-1)^3$

When no interval is specified , we use the real line $(-\infty, \infty)$.

a) Find $f'(x)$,

$$f'(x) = 3(x-1)^2.$$

b) Find the critical values. The derivative exists for all real numbers. Thus, we solve $f'(x) = 0$,

$$3(x-1)^2 = 0$$
$$x-1 = 0$$
$$x = 1$$

The only critical value is $x = 1$.

c) Since there is only one critical value, we can apply Max-Min Principle 2. First we find $f''(x)$.

$$f''(x) = 6(x-1).$$

Now,

$$f''(1) = 6((1)-1) = 0 \text{, so the Max-Min}$$

Principle 2 fails. We cannot use Max-Min Principle 1, because there are no endpoints.

We note that $f'(x) = 3(x-1)^2$ is never negative. Thus, $f(x)$ is increasing everywhere except at $x = 1$. Therefore, the function has no maximum or minimum over the interval $(-\infty, \infty)$.

Notice

$$f''(0) = -6 < 0$$
$$f''(2) = 6 > 0$$
and
$$f(1) = 0$$

Therefore, there is a point of inflection at $(1,0)$.

75. $f(x) = 2x - 3; \quad [-1,1]$

a) Find $f'(x)$.

$$f'(x) = 2.$$

b) and c)

The derivative exists and is 2 for all real numbers. Therefore, $f'(x)$ is never 0. Thus, there are no critical values. We apply the Max-Min Principle 1. The endpoints are -1 and 1. We find the function values at the endpoints.

$$f(-1) = 2(-1) - 3 = -5$$
$$f(1) = 2(1) - 3 = -1$$

Therefore, the absolute maximum over the interval $[-1,1]$ is -1, which occurs at $x = 1$, and the absolute minimum over the interval $[-1,1]$ is -5, which occurs at $x = -1$.

77. $f(x) = 2x - 3; \quad [-1,5)$

a) Find $f'(x)$

$$f'(x) = 2$$

b) and c)

The derivative exists and is 2 for all real numbers. Therefore, $f'(x)$ is never 0. Thus, there are no critical values. We apply the Max-Min Principle 1. There is only one endpoint, $x = -1$. We find the function value at the endpoint.

$$f(-1) = 2(-1) - 3 = -5$$

We know $f'(x) > 0$ over the interval, so the function is increasing over the interval $(-1,5)$. Therefore, the minimum value will be the left hand endpoint. The absolute minimum over the interval $[-1,5)$ is -5, which occurs at $x = -1$. Since the right endpoint is not included in the interval, the function has no maximum value over the interval $[-1,5)$.

79. $f(x) = 9 - 5x; \quad [-2,3)$

a) Find $f'(x)$.

$$f'(x) = -5.$$

b) and c)
    The derivative exists for all real numbers
    and is never 0. There are no critical values.
    The only endpoint is the left endpoint -2.
    $f'(x)<0$ over the interval, so the function

    is decreasing and a maximum occurs at
    $x=-2$. We find the function value at
    $x=-2$.
    $$f(-2)=9-5(-2)=19.$$
    The absolute maximum over the interval
    $[-2,3)$ is 19, which occurs at $x=-2$. The
    function has no minimum value over the
    interval $[-2,3)$.

81. $g(x)=x^{\frac{2}{3}}$

When no interval is specified , we use the real
line $(-\infty,\infty)$.

a) Find $g'(x)$.
    $$g'(x)=\frac{2}{3}x^{-\frac{1}{3}}=\frac{2}{3\cdot\sqrt[3]{x}}.$$

b) Find the critical values. $g'(x)$ does not exist
    for $x=0$. The equation $g'(x)=0$ has no
    solution, so the only critical value is 0.

c) We apply the Max-Min Principle 2.
    $$g''(x)=-\frac{2}{9x^{\frac{4}{3}}}$$
    $g''(0)$ does not exist.

    Note that $g'(x)<0$ for $x<0$ and $g'(x)>0$
    for $x>0$, so $g(x)$ is decreasing on $(-\infty,0)$
    and increasing on $(0,\infty)$.
    Therefore, the absolute minimum over the
    interval $(-\infty,\infty)$ is 0, which occurs at $x=0$.
    The function has no maximum value.

83. $f(x)=\frac{1}{3}x^3-\frac{1}{2}x^2-2x+1$

When no interval is specified , we use the real
line $(-\infty,\infty)$.

a) Find $f'(x)$
    $$f'(x)=x^2-x-2$$

b) Find the critical values. The derivative
    exists for all real numbers. Thus, we solve
    $f'(x)=0$.
    $$x^2-x-2=0$$
    $$(x-2)(x+1)=0$$
    $$x=2 \quad \text{or} \quad x=-1$$
    There are two critical values -1 and 2.

c) The interval $(-\infty,\infty)$ is not closed, so the
    Max-Min Principle 1 does not apply. Since
    there is more than one critical value, the
    Max-Min Principle 2 does not apply.
    A quick sketch of the graph will help us
    determine absolute or relative extrema occur
    at the critical values.

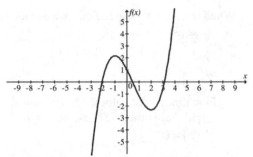

We determine that the function has no
absolute extrema over the interval $(-\infty,\infty)$.

85. $t(x)=x^4-2x^2$

When no interval is specified , we use the real
line $(-\infty,\infty)$.

a) Find $t'(x)$
    $$t'(x)=4x^3-4x$$

b) Find the critical values. The derivative exists
    for all real numbers. Thus, we solve
    $t'(x)=0$.
    $$4x^3-4x=0$$
    $$4x(x^2-1)=0$$
    $$x=0 \quad \text{or} \quad x^2-1=0$$
    $$x=0 \quad \text{or} \quad x=\pm1$$
    There are three critical values -1, 0, and 1.

c) The interval $(-\infty,\infty)$ is not closed, so the
    Max-Min Principle 1 does not apply. Since
    there is more than one critical value, the
    Max-Min Principle 2 does not apply.

    The solution is continued on the next page.

A quick sketch of the graph will help us determine whether absolute or relative extrema occur at the critical values.

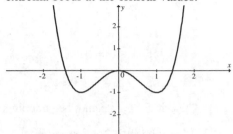

We determine that the function has no absolute maximum over the interval $(-\infty,\infty)$. The function's absolute minimum is -1, which occurs at $x=-1$ and $x-1$.

87. – 96.   Left to the student.

97. $M(t)=-2t^2+100t+180, \quad 0 \le t \le 40$

a) $M'(t)=-4t+100$

b) $M'(t)$ exists for all real numbers. We solve
$M'(t)=0$,
$-4t+100=0$
$\qquad 4t=100$
$\qquad\quad t=25$

c)   Since there is only one critical value, we apply the Max-Min Principle 2. First, we find the second derivative.

$M''(t)=-4$. The second derivative is negative for all values of t in the interval, therefore, a maximum occurs at $t=25$.

$M(25)=-2(25)^2+100(25)+180=1430$

The maximum productivity for $0 \le t \le 40$ is 1430 units per month, which occurs after $t=25$ years of employment.

99. $p(x)=-0.039x^3+0.594x^2-1.967x+7.555$

We restrict our attention to the years 2003 to 2013. That is, we will look at the x-values $0 \le x \le 10$.

a)   Find $p'(x)$.

$p'(x)=-0.117x^2+1.188x-1.967$.

b)   Find the critical values. $p'(x)$ exists for all real numbers. Therefore, we solve $p'(x)=0$.

$-0.117x^2+1.188x-1.967=0$
Using the quadratic formula, we have:

$$x=\frac{-(1.188)\pm\sqrt{(1.188)^2-4(-0.117)(-1.967)}}{2(-0.117)}$$

$$=\frac{-1.188\pm\sqrt{0.490788}}{-0.234}$$

$x \approx 2.08 \quad$ or $\quad x \approx 8.07$
The critical values are $x \approx 2.08$ and $x \approx 8.07$ is in the interval $[0,10]$.

c)   The critical values and the endpoints are 0, 2.08, 8.07, and 10.

d)   Using a calculator, we find the function values.

$p(0) \approx 7.555$

$p(2.08) \approx 5.683$

$p(8.07) \approx 9.869$

$p(10) \approx 8.285$

The absolute maximum occurs when $x=8.07$.
According to this model, the maximum percentage of unemployed workers in service occupations occurred in 2011.

101. We use the model
$$P(t)=2.69t^4-63.941t^3+459.895t^2-$$
$$688.692t+24{,}150.217$$

We consider the interval $[0,\infty)$, where $t=0$ corresponds to the year 2000.

a)   Find $P'(t)$.

$P'(t)=10.76t^3-191.823t^2+919.79t-688.692$

b) $P'(t)$ exists for all real numbers. Solve:
$$P'(t)=0$$

$10.76t^3-191.823t^2+919.79t-688.692=0$
Using a graphing calculator, we approximate the zeros of $P'(t)$. We find the solutions:

$t \approx 0.914$

$t \approx 7.232$

$t \approx 9.681$

c)   The critical values and the endpoints are: 0, 0.914, 7.232, and 9.681.

d) Find the function values.
$$P(0) = 24,150$$
$$P(0.914) \approx 23,858$$
$$P(7.232) \approx 26,396$$
$$P(9.681) \approx 39,533$$
The absolute minimum production of world wide oil was 23,858,000 barrels. The world achieved this production 0.914 years after 2000, or approximately the year 2001.

103. $C(x) = 5000 + 600x$

$$R(x) = -\frac{1}{2}x^2 + 1000x, \ 0 \le x \le 600$$

a) $P(x) = R(x) - C(x)$
$$= -\frac{1}{2}x^2 + 1000x - (5000 + 600x)$$
$$= -\frac{1}{2}x^2 + 400x - 5000$$

b) First, we find the critical values.
$$P'(x) = -x + 400$$

$P'(x)$ exists for all real numbers. Solve:
$$P'(x) = 0$$
$$-x + 400 = 0$$
$$x = 400$$
The critical value is 400 and the endpoints are 0 and 600. Using the Max-Min Principle 1. We evaluate the function at the endpoints and critical values:

$$P(0) = -\frac{1}{2}(0)^2 + 400(0) - 5000$$
$$= -5000$$

$$P(400) = -\frac{1}{2}(400)^2 + 400(400) - 5000$$
$$= 75,000$$

$$P(600) = -\frac{1}{2}(600)^2 + 400(600) - 5000$$
$$= 55,000$$
The total profit is maximized when 400 items are produced.

105. $B(x) = 305x^2 - 1830x^3, \quad 0 \le x \le 0.16$

a) $B'(x) = 610x - 5490x^2$

b) $B'(x)$ exists for all real numbers. Solve:
$$B'(x) = 0$$
$$610x - 5490x^2 = 0$$
$$610x(1 - 9x) = 0$$
$$x = 0 \quad \text{or} \quad 1 - 9x = 0$$
$$x = 0 \quad \text{or} \quad x = \frac{1}{9} \approx 0.11$$

c) The critical points and the endpoints are $0, \frac{1}{9}$, and 0.16.

d) We find the function values.
$$B(0) = 305(0)^2 - 1830(0)^3 = 0$$
$$B\left(\frac{1}{9}\right) = 305\left(\frac{1}{9}\right)^2 - 1830\left(\frac{1}{9}\right)^3$$
$$= \frac{305}{243} \approx 1.255$$
$$B(0.16) = 305(0.16)^2 - 1830(0.16)^3$$
$$\approx 0.312$$
The maximum blood pressure is approximately 1.255, which occurs at a dose of $x = \frac{1}{9}$ cc, or about 0.11 cc of the drug.

107. We look at the derivative on each piece of the function to determine any critical values. For $-3 < x < 1$, $f'(x) = 2$ so there are no critical values for this part of the function. For $1 < x \le 2$, $f'(x) = -2x$. $f'(x) = 0$ when $x = 0$, which is outside the domain of this piece of the function. Therefore, there are no critical values of $f(x)$. The absolute extrema will occur at one of the endpoints. The function values are:
$$f(-3) = 2(-3) + 1 = -5$$
$$f(1) = 2(1) + 1 = 3$$
$$f(2) = 4 - (2)^2 = 0$$
On the interval $[-3, 2]$, the absolute minimum is -5, which occurs at $x = -3$. The absolute maximum is 3, which occurs at $x = 1$. We sketch a graph of the function.

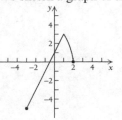

109. We look at the derivative on each piece of the function to determine any critical values. For $-4 < x < 0$, $h'(x) = -2x$. $h'(x) = 0$ when $x = 0$, which is also the endpoint of this part of the domain. For $0 < x < 1$, $h'(x) = -1$.

Therefore, there are no critical values of $h(x)$ on this part of the domain. For $1 \le x \le 2$, $h'(x) = 1$. Therefore, there are no critical values of $h(x)$ on this part of the domain. The absolute extrema will occur at one of the endpoints. We determine the function values.

$$h(-4) = 1 - (-4)^2 = -15$$
$$h(0) = 1 - (0) = 1$$
$$h(1) = (1) - 1 = 0$$
$$h(2) = (2) - 1 = 1$$

On the interval $[-4, 2]$, the absolute minimum is -15, which occurs at $x = -4$. The absolute maximum is 1, which occurs at $x = 0$ and $x = 2$. A sketch of the graph is shown below.

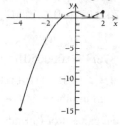

111. a) The sketch of the graph is shown below:

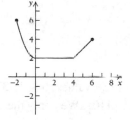

b) From the graph, the absolute maximum is 6 and occurs at $x = -2$.

c) The absolute minimum value for this function is 2. This value occurs over the range $0 \le x \le 4$.

113. $g(x) = x\sqrt{x+3}$; $\quad [-3, 3]$

a) Find $g'(x)$.

$$g'(x) = x\left[\frac{1}{2}(x+3)^{-\frac{1}{2}}(1)\right] + (1)(x+3)^{\frac{1}{2}}$$

$$= \frac{x}{2(x+3)^{\frac{1}{2}}} + (x+3)^{\frac{1}{2}}$$

$$= \frac{x}{2(x+3)^{\frac{1}{2}}} + \frac{(x+3)^{\frac{1}{2}}}{1} \cdot \frac{2(x+3)^{\frac{1}{2}}}{2(x+3)^{\frac{1}{2}}}$$

Multiplying by a form of 1

$$= \frac{x}{2(x+3)^{\frac{1}{2}}} + \frac{2(x+3)}{2(x+3)^{\frac{1}{2}}}$$

$$= \frac{3x+6}{2(x+3)^{\frac{1}{2}}}, \text{ or } \frac{3x+6}{2\sqrt{x+3}}$$

b) Find the critical values. $g'(x)$ exists for all values in $[-3, 3]$ except -3. This is a critical value as well as an endpoint. To find the other critical values, we solve

$$g'(x) = 0$$
$$\frac{3x+6}{2\sqrt{x+3}} = 0$$
$$3x+6 = 0$$
$$3x = -6$$
$$x = -2$$

The second critical value on the interval is -2.

c) On a closed interval, the Max-Min Principle 1 can always be used. The critical values and the endpoints are -3, -2, and 3.

d) Find the function value at each value in step (c)..

$$g(-3) = (-3)\sqrt{(-3)+3} = 0$$
$$g(-2) = (-2)\sqrt{(-2)+3} = -2$$
$$g(3) = (3)\sqrt{(3)+3} = 3\sqrt{6}$$

Thus, the absolute maximum over the interval $[-3, 3]$ is $3\sqrt{6}$, which occurs at $x = 3$, and the absolute minimum is -2, which occurs at $x = -2$.

115. $C(x) = (2x+4) + \left(\dfrac{2}{x-6}\right), \quad x > 6$

$= 2x + 4 + 2(x-6)^{-1}$

a) Find $C'(x)$.

$C'(x) = 2 - 2(x-6)^{-2}(1) = 2 - \dfrac{2}{(x-6)^2}$

b) Find the critical values.

$C'(x)$ does not exist for $x = 6$; however, this value is not in the domain interval, so it is not a critical value. Solve $C'(x) = 0$.

$2 - \dfrac{2}{(x-6)^2} = 0$

$2 = \dfrac{2}{(x-6)^2}$

$2(x-6)^2 = 2 \qquad \begin{array}{l}\text{Multiplying by } (x-6)^2 \\ \text{Since } x \neq 6.\end{array}$

$(x-6)^2 = 1$

$x - 6 = \pm 1 \qquad \begin{array}{l}\text{Taking the square root} \\ \text{of both sides.}\end{array}$

$x = 6 \pm 1$

$x = 5 \quad \text{or} \quad x = 7$

The only critical value in $(6, \infty)$ is 7.

c) Since there is only one critical value, we apply the Max-Min Principle 2.

$C''(x) = 4(x-6)^{-3} = \dfrac{4}{(x-6)^3}$

$C''(7) = \dfrac{4}{(7-6)^3} = 4 > 0$

Therefore, since $C''(7) > 0$, there is a minimum at $x = 7$.

The Katie's Clocks should use 7 "quality units" to minimize its total cost of service.

117. From exercise 101, we know that the first derivative is

$R(t) = P'(t) = 10.76t^3 - 191.823t^2 + 919.79t - 688.692$

a) We find the maximum rate of change, by finding the critical values of the derivative. $(0,8)$. Taking the derivative we have:

$P''(t) = 32.28t^2 - 383.646t + 919.79$

b) $P''(t)$ exists for all real numbers. Solve:

$P''(t) = 0$

$32.28t^2 - 383.646t + 919.79 = 0$

Using the quadratic formula, we find the zeros of $P''(t)$.

$t = \dfrac{-(-383.646) \pm \sqrt{(-383.646)^2 - 4(32.28)(919.79)}}{2(32.28)}$

The two solutions are $t \approx 3.33$ and $t \approx 8.554$. Only one of the critical values, $t = 3.33$, is in the interval.

c) Since there is only one critical value, we apply the Max-Min Principle 2.

$P'''(t) = 64.56t - 383.646$

$P'''(3.33) = 64.56(3.33) - 383.646 = -168.6 < 0$

Therefore, the absolute maximum over the interval $(0,8)$ occurs at $t \approx 3.33$.

$P'(3.33) = 644.427 \approx 640.$

In the year 2003 – 2004, worldwide oil production was increasing most rapidly. It was increasing at a rate of approximately 640,000 barrels per year.

119. $P(t) = 0.0000000219t^4 - 0.0000167t^3 + 0.00155t^2 + 0.002t + 0.22, \quad 0 \leq t \leq 110$

a) Find $P'(t)$.

$P'(t) = 0.0000000876t^3 - 0.0000501t^2 + 0.0031t + 0.002$

$P'(t)$ exists for all real numbers. Solve $P'(t) = 0$. We use a calculator to find the zeros of $P'(t)$. We estimate the solutions to be:

$x \approx -0.639$

$x \approx 71.333$

$x \approx 501.223$

Only one of the critical values, $x \approx 71.333$, is in the interval $[0,110]$. We apply the Max-Min Principle 1, to find the absolute maximum.

The critical values and the endpoints are 0, 71.333, and 110.

The function values at these points are

$P(0) = 0.22$

$P(71.333) \approx 2.755$

$P(110) \approx 0.174$

Thus, the absolute maximum oil production for the U.S. after 1910 was 2.755 billion barrels per year. This production level occurred 71.333 year after 1910, or in 1981.

b) In 2010, $t = 2010 - 1910 = 100$. We plug this value into the first derivative to obtain:

$P'(100)$

$$= 0.0000000876(100)^3 - 0.0000501(100)^2 +$$
$$0.0031(100) + 0.002$$

≈ -0.1014.

The rate of oil was declining at approximately 0.1014 billion of barrels per year.

In 2015, $t = 2015 - 1910 = 105$. We plug this value into the first derivative to obtain:

$P'(105)$

$$= 0.0000000876(105)^3 - 0.0000501(105)^2 +$$
$$0.0031(105) + 0.002$$

≈ -0.1234.

The rate of oil was declining at approximately 0.1234 billion of barrels per year.

121. $f(x) = \dfrac{3}{4}(x^2 - 1)^{\frac{2}{3}}$; $\left[\frac{1}{2}, \infty\right)$

Using a calculator, we enter the equation into the graphing editor:

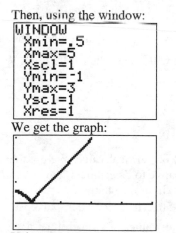

Then, using the window:

```
WINDOW
 Xmin=.5
 Xmax=5
 Xscl=1
 Ymin=-1
 Ymax=3
 Yscl=1
 Xres=1
```

We get the graph:

Using the table feature, we locate the extrema. We estimate the absolute minimum to be 0, which occurs at $x = 1$. There is no absolute maximum.

123. a) Using a graphing calculator, we fit the linear equation $y = x + 8.857$. This corresponds to the model $P(t) = t + 8.857$. Where P is the pressure of the contractions and t is the time in minutes. We substitute 7 for t to find the pressure at 7 minutes.

$P(7) = 7 + 8.857 = 15.857$.

The pressure at 7 minutes is 15.857 mm of Hg.

b) Rounding the coefficients to 3 decimal places, we find the quartic regression

$$y = 0.117x^4 - 1.520x^3 + 6.193x^2 -$$
$$7.018x + 10.009$$

Changing the variables we get the model

$$P(t) = 0.117t^4 - 1.520t^3 + 6.193t^2 -$$
$$7.018t + 10.009$$

Using the table feature, when $x = 7$, $y = 24.857$. So the pressure at 7 minutes is 24.86 mm of mercury. (If we use the rounded coefficients above, we get 23.897 mm of Hg.)

Using the trace feature, we estimate the smallest contraction on the interval $[0, 10]$ was about 7.62 mm of Hg. This occurred when $x \approx 0.765$ min.

Exercise Set 2.5

1. Express $Q = xy$ as a function of one variable.
First, we solve $x + y = 70$ for y.

$x + y = 70$

$y = 70 - x$

Next, we substitute $70 - x$ for y in $Q = xy$.

$Q = xy$

$Q = x(70 - x)$ Substituting

$\quad = 70x - x^2$

Now that Q is a function of one variable we can find the maximum. First, we find the critical values.

$Q'(x) = 70 - 2x$. Since $Q'(x)$ exists for all real numbers, the only critical value will occur when $Q'(x) = 0$. We solve:

$70 - 2x = 0$

$\quad 70 = 2x$

$\quad 35 = x$

There is only one critical value. We use the second derivative to determine if the critical value is a maximum. Note that:

$Q''(x) = -2 < 0$. The second derivative is negative for all values of x. Therefore, a maximum occurs at $x = 35$.

Now, $Q(35) = 70(35) - (35)^2 = 1225$

Therefore, the maximum product is 1225, which occurs when $x = 35$. If $x = 35$, then $y = 70 - 35 = 35$. The two numbers are 35 and 35.

3. Let x be one number and y be the other number. Since the difference of the two numbers must be 6, we have $x - y = 6$.

The product, Q, of the two numbers is given by $Q = xy$, so our task is to minimize $Q = xy$, where $x - y = 6$.

First, we express $Q = xy$ as a function of one variable.

Solving $x - y = 6$ for y, we have:

$x - y = 6$

$-y = 6 - x$

$\quad y = x - 6$

Next, we substitute $x - 6$ for y in $Q = xy$.

$Q(x) = x(x - 6) = x^2 - 6x$.

Finding the deriviative, we have:

$Q'(x) = 2x - 6$

The derivative exists for all values of x; thus, the only critical values are where $Q'(x) = 0$.

$2x - 6 = 0$

$\quad 2x = 6$

$\quad x = 3$

There is only one critical value. We can use the second derivative to determine whether we have a minimum.

$Q''(x) = 2 > 0$ for all values of x. Therefore, a minimum occurs at $x = 3$.

$Q(3) = (3)^2 - 6(3) = -9$

Thus, the minimum product is -9 when $x = 3$, and $y = 3 - 6 = -3$.

5. Maximize $Q = xy^2$, where x and y are positive numbers such that $x + y^2 = 4$.

Express $Q = xy^2$ as a function of one variable

First, we solve $x + y^2 = 4$ for y^2.

$x + y^2 = 4$

$\quad y^2 = 4 - x$

Next, we substitute $4 - x$ for y^2 in $Q = xy^2$.

$Q = xy^2$

$Q = x(4 - x)$

$\quad = 4x - x^2$

Now that Q is a function of one variable we can find the maximum. First, we find the critical values.

$Q'(x) = 4 - 2x$.

Since $Q'(x)$ exists for all real numbers, the only critical value will occur when $Q'(x) = 0$. We set $Q'(x) = 0$ and solve for x:

$4 - 2x = 0$

$-2x = -4$

$\quad x = 2$.

There is only one critical value. We use the second derivative to determine if the critical value is a maximum. Note that:

$Q''(x) = -2 < 0$. The second derivative is negative for all values of x. Therefore, a maximum occurs at $x = 2$.

Now,

$Q(2) = (4)(2) - (2)^2 = 4$

The solution is continued on the next page.

Substitute 2 in for x in $x + y^2 = 4$ and solve for y.

$$2 + y^2 = 4$$
$$y^2 = 4 - 2$$
$$y^2 = 2$$
$$y = \pm\sqrt{2}$$
$$y = \sqrt{2} \quad x \text{ and } y \text{ are positive}$$

Then Q is a maximum when $x = 2$ and $y = \sqrt{2}$.

7. Minimize $Q = x^2 + 2y^2$, where $x + y = 3$.
Express Q as a function of one variable. First, solve $x + y - 3$ for y.

$$x + y = 3$$
$$y = 3 - x$$

Then substitute $3 - x$ for y in $Q = x^2 + 2y^2$.

$$Q = x^2 + 2(3 - x)^2$$
$$= x^2 + 2(9 - 6x + x^2)$$
$$= 3x^2 - 12x + 18$$

Find $Q'(x)$, where $Q(x) = 3x^2 - 12x + 18$.

$$Q'(x) = 6x - 12$$

This derivative exists for all values of x; thus the only critical values are where

$$Q'(x) = 0$$
$$6x - 12 = 0$$
$$6x = 12$$
$$x - 2$$

Since there is only one critical value, we can use the second derivative to determine whether we have a minimum. Note that:

$Q''(x) = 6$, which is positive for all real numbers. Thus $Q''(2) > 0$, so a minimum occurs when $x = 2$. The value of Q is

$$Q(2) = 2^2 + 2(3 - 2)^2$$
$$= 4 + 2$$
$$= 6.$$

Substitute 2 for x in $y = 3 - x$ to find y.

$$y = 3 - x$$
$$y = 3 - 2$$
$$y = 1$$

Thus, the minimum value of Q is 6 when $x = 2$ and $y = 1$.

9. Maximize $Q = xy$, where x and y are positive numbers such that $x + \dfrac{4}{3}y^2 = 1$.

Express Q as a function of one variable. First, solve $x + \dfrac{4}{3}y^2 = 1$ for x.

$$x + \frac{4}{3}y^2 = 1$$
$$x = 1 - \frac{4}{3}y^2$$

Then substitute $1 - \dfrac{4}{3}y^2$ for x in $Q = xy$.

$$Q = xy = \left(1 - \frac{4}{3}y^2\right)y$$
$$= y - \frac{4}{3}y^3$$

Find $Q'(y)$, where $Q(y) = y - \dfrac{4}{3}y^3$.

$$Q'(y) = 1 - 4y^2$$

This derivative exists for all values of y; thus the only critical values are where

$$Q'(y) = 0$$
$$1 - 4y^2 = 0$$
$$-4y^2 = -1$$
$$y^2 = \frac{1}{4}$$
$$y = \pm\sqrt{\frac{1}{4}}$$
$$y = \pm\frac{1}{2}$$
$$y = \frac{1}{2} \quad y \text{ must be positive}$$

Since there is only one critical value, we can use the second derivative to determine whether we have a maximum.
Note that:

$$Q''(y) = -8y$$

and

$$Q''\left(\frac{1}{2}\right) = -8\left(\frac{1}{2}\right) = -4 < 0.$$

Since $Q''\left(\dfrac{1}{2}\right)$ is negative, a maximum occurs at

$$y = \frac{1}{2}.$$

The solution is continued on the next page.

Evaluating the function at $y = \dfrac{1}{2}$ we have:

$$Q\left(\dfrac{1}{2}\right) = \dfrac{1}{2} - \dfrac{4}{3}\left(\dfrac{1}{2}\right)^3$$

$$= \dfrac{1}{2} - \dfrac{4}{3} \cdot \dfrac{1}{8}$$

$$= \dfrac{1}{2} - \dfrac{1}{6}$$

$$= \dfrac{1}{3}.$$

Substitute $\dfrac{1}{2}$ for y in $x = 1 - \dfrac{4}{3}y^2$ to find x.

$$x = 1 - \dfrac{4}{3}y^2$$

$$x = 1 - \dfrac{4}{3}\left(\dfrac{1}{2}\right)^2$$

$$x = 1 - \dfrac{1}{3}$$

$$x = \dfrac{2}{3}$$

Thus, the maximum value of Q is $\dfrac{1}{3}$ when

$$x = \dfrac{2}{3} \text{ and } y = \dfrac{1}{2}.$$

11. Let x represent the width and y represent the length of the area. It is helpful to draw a picture.

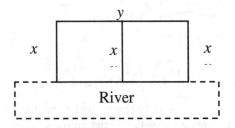

Since the rancher has 240 yards of fence, the perimeter is $3x + y = 240$. Solving this equation for y, we have $y = 240 - 3x$. Since x and y must be positive, we are restricted to the interval $0 < x < 80$.

The objective is to maximize area, which is given by

$$A = l \cdot w$$

Substituting $y = 240 - 3x$ for the width and x for the length, we have:

$$A(x) = x(240 - 3x) = 240x - 3x^2.$$

We will maximize the area over the restricted interval by first finding the derivative of the area function.

$$A(x) = 240x - 3x^2$$

$$A'(x) = 240 - 6x$$

The derivative exists for all values of x in the interval $(0, 80)$. The only critical values are where:

$$A'(x) = 0$$

$$240 - 6x = 0$$

$$-6x = -240$$

$$x = 40$$

Since there is only one critical value in the interval, we use the second derivative to determine whether we have a maximum. Note, $A''(x) = -6 < 0$ for all values of x, so there is a maximum at $x = 40$.

Next we find the dimensions and the area. When $x = 40$,

$$y = 240 - 3x = 240 - 3(40) = 120$$

and

$$A(40) = 240(40) - 3(40)^2 = 4800.$$

Therefore, the maximum area is 4800 yd^2 when the overall dimensions are 40 yd by 120 yd.

13. Let x represent the width and y represent the length of the area. It is helpful to draw a picture.

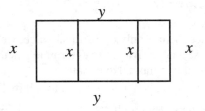

Since 1200 yards of fence is available, the perimeter is $4x + 2y = 1200$. Solving this equation for y, we have $y = 600 - 2x$. Since x and y must be positive, we are restricted to the interval $0 < x < 300$.

The objective is to maximize area, which is given by

$$A = l \cdot w$$

Substituting $y = 600 - 2x$ for the width and x for the length, we have:

$$A(x) = xy = x(600 - 2x) = 600x - 2x^2.$$

The solution is continued on the next page.

We will maximize the area over the restricted interval by first finding the derivative of the area function.

$$A(x) = 600x - 2x^2$$

$$A'(x) = 600 - 4x$$

The derivative exists for all values of x in the interval $(0, 300)$. The only critical values are where:

$$A'(x) = 0$$

$$600 - 4x = 0$$

$$-4x = -600$$

$$x = 150$$

Since there is only one critical value in the interval, we use the second derivative to determine whether we have a maximum. Note, $A''(x) = -4 < 0$ for all values of x, so there is a maximum at $x = 150$.

Next we find the dimensions and the area. When $x = 150$,

$$y = 600 - 2x = 600 - 2(150) = 300$$

and

$$A(150) = 600(150) - 2(150)^2$$

$$= 45,000.$$

Therefore, the maximum area is $45,000 \text{ yd}^2$ when the overall dimensions are 150 yd by 300 yd.

15. Let x represent the length and y represent the width. It is helpful to draw a picture.

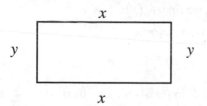

The perimeter is found by adding up the length of the sides. Since it is fixed at 42 feet, the equation of the perimeter is $2x + 2y = 42$.

The area is given by $A = xy$.

First, we solve the perimeter equation for y.

$$2x + 2y = 42$$

$$2y = 42 - 2x$$

$$y = 21 - x$$

Then we substitute for y into the area formula.

$$A = xy$$

$$= x(21 - x) = 21x - x^2$$

We want to maximize the area on the interval $(0, 21)$. We consider this interval because x is the length of the rectangle and cannot be negative.

Since the perimeter cannot exceed 42 feet, x cannot be greater than 21, also if x is 21 feet, the width of the rectangle would be 0 feet. We begin by finding $A'(x)$.

$$A'(x) = 21 - 2x.$$

This derivative exists for all values of x in $(0, 21)$. Thus, the only critical values occur where $A'(x) = 0$. Solving $A'(x) = 0$, we have:

$$21 - 2x = 0$$

$$-2x = -21$$

$$x = \frac{21}{2} = 10.5$$

Since there is only one critical value in the interval, we can use the second derivative to determine whether we have a maximum. Note that

$$A''(x) = -2 < 0 \text{ for all values of } x. \text{ Thus,}$$

$A''(10.5) < 0$, so a maximum occurs at $x = 10.5$.

Now,

$$A(x) = 21x - x^2$$

$$A(10.5) = 21(10.5) - (10.5)^2$$

$$= 110.25$$

The maximum area is 110.25 ft^2.

Note: when $x = 10.5$, $y = 21 - 10.5 = 10.5$, so the overall dimensions that will achieve the maximum area are 10.5 ft by 10.5 ft.

17. When squares of length h on a side are cut out of the corners, we are left with a square base of length x. A picture will help.

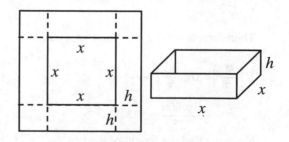

The resulting volume of the box is

$$V = lwh = x \cdot x \cdot h = x^2 h.$$

We want to express V in terms of one variable. The solution is continued on the next page.

Note that the overall length of a side of the cardboard is 20 in. We see from the drawing, that $h + x + h = 20,$ *or* $x + 2h = 20.$ Solving for h we get:

$$2h = 20 - x$$

$$h = \frac{1}{2}(20 - x) = 10 - \frac{1}{2}x.$$

Substituting h into the volume equation, we have:

$$V = x^2\left(10 - \frac{1}{2}x\right) = 10x^2 - \frac{1}{2}x^3.$$ The objective

is to maximize $V(x)$ on the interval $(0, 20)$.

First, we find the derivative.

$$V'(x) = 20x - \frac{3}{2}x^2$$

This derivative exists for all x in the interval $(0, 20)$. We set the derivative equal to zero and solve for the critical values

$$V'(x) = 0$$

$$20x - \frac{3}{2}x^2 = 0$$

$$x\left(20 - \frac{3}{2}x\right) = 0$$

$$x = 0 \quad \text{or} \quad 20 - \frac{3}{2}x = 0$$

$$x = 0 \quad \text{or} \quad -\frac{3}{2}x = -20$$

$$x = 0 \quad \text{or} \quad x = \frac{40}{3} = 13\tfrac{1}{3}$$

The only critical value in $(0, 20)$ is $\dfrac{40}{3}$ or about 13.33. Therefore, we can use the second derivative $V''(x) = 20 - 3x$ to determine if we have a maximum. We have

$$V''\left(\frac{40}{3}\right) = 20 - 3\left(\frac{40}{3}\right) = -20 < 0.$$

Therefore, there is a maximum at $\dfrac{40}{3}$.

$$V\left(\frac{40}{3}\right) = 10\left(\frac{40}{3}\right)^2 - \frac{1}{2}\left(\frac{40}{3}\right)^3$$

$$= \frac{16,000}{27} = 592\tfrac{16}{27}$$

Now, we find the height of the box.

$$h = 10 - \frac{1}{2}\left(\frac{40}{3}\right) = \frac{10}{3} = 3\tfrac{1}{3}.$$

Therefore, a box with dimensions $13\tfrac{1}{3}$ in. by $13\tfrac{1}{3}$ in. by $3\tfrac{1}{3}$ in. will yield a maximum volume of $592\tfrac{16}{27}$ in^3.

19. First, we make a drawing.

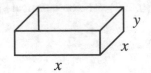

The surface area of the open-top, square-based, rectangular tank is found by adding the area of the base and the four sides. x^2 is the area of the base, xy is the area of one of the sides and there are four sides, therefore the surface area is given by $S = x^2 + 4xy$.

The volume must by 32 cubic feet, and is given by $V = l \cdot w \cdot h = x^2 y = 32$.

To express S in terms of one variable, we solve $x^2 y = 32$ for y:

$$y = \frac{32}{x^2}.$$

Substituting, we have:

$$S(x) = x^2 + 4x\left(\frac{32}{x^2}\right)$$

$$= x^2 + \frac{128}{x} = x^2 + 128x^{-1}$$

Now S is defined only for positive numbers, so we minimize S on the interval $(0, \infty)$.

First, we find $S'(x)$.

$$S'(x) = 2x - 128x^{-2}$$

$$= 2x - \frac{128}{x^2}$$

Since $S'(x)$ exists for all x in $(0, \infty)$, the only critical values are where $S'(x) = 0$. We solve the following equation:

$$S'(x) = 0$$

$$2x - \frac{128}{x^2} = 0$$

$$x^3 = 64$$

$$x = 4$$

Since there is only one critical value, we use the second derivative to determine whether we have a minimum.

The solution is continued on the next page.

Note that $S''(x) = 2 + 256x^{-3} = 2 + \dfrac{256}{x^3}$.

$S''(4) = 2 + \dfrac{256}{4^3} = 6 > 0$. Since the second

derivative is positive, we have a minimum at $x = 4$. We find y when $x = 4$.

$y = \dfrac{32}{x^2}$

$= \dfrac{32}{4^2}$

$= 2$

The surface area is minimized when $x = 4$ ft and $y = 2$ ft. We find the minimum surface area by substituting these values into the surface area equation.

$S = x^2 + 4xy$

$= (4)^2 + 4(4)(2)$

$= 16 + 32$

$= 48$

$S(4) = 4^2 + 4 \cdot 4 \cdot 2 = 48.$

Therefore, when the dimensions are 4 ft by 4 ft by 2 ft, the minimum surface area will be 48 ft².

21. First, we make a drawing.

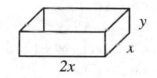

The surface area of the open-top, rectangular dumpster is found by adding the area of the base and the four sides. $2x^2$ is the area of the base, xy is the area of two of the sides, while $2xy$ is the area of the other two sides. Therefore the surface area is given by

$S = 2x^2 + 2xy + 2(2xy) = 2x^2 + 6xy$.

The volume must by 12 cubic yards, and is given by $V = l \cdot w \cdot h = 2x \cdot x \cdot y = 2x^2 y = 12$.

To express S in terms of one variable, we solve $2x^2 y = 12$ for y:

$y = \dfrac{6}{x^2}$

Then

$S(x) = 2x^2 + 6x\left(\dfrac{6}{x^2}\right)$

$= 2x^2 + \dfrac{36}{x} = 2x^2 + 36x^{-1}$

Now S is defined only for positive numbers, so we minimize S on the interval $(0, \infty)$.

First, we find $S'(x)$.

$S'(x) = 4x - 36x^{-2}$

$= 4x - \dfrac{36}{x^2}$

Since $S'(x)$ exists for all x in $(0, \infty)$, the only critical values are where $S'(x) = 0$. We solve the following equation:

$4x - \dfrac{36}{x^2} = 0$

$4x = \dfrac{36}{x^2}$

$x^3 = 9$

$x = \sqrt[3]{9} \approx 2.08$

Since there is only one critical value, we use the second derivative to determine whether we have a minimum.

Note that $S''(x) = 4 + 72x^{-3} = 4 + \dfrac{72}{x^3}$.

$S''\left(\sqrt[3]{9}\right) = 4 + \dfrac{72}{\left(\sqrt[3]{9}\right)^3} = 12 > 0$. Since the second

derivative is positive, we have a minimum at $x = \sqrt[3]{9} \approx 2.08$. The width is 2.08 yd.; therefore, the length is $2(2.08) \approx 4.16$. We find the height y

$y = \dfrac{6}{x^2}$

$= \dfrac{6}{(2.08)^2}$

≈ 1.387

The overall dimensions of the dumpster that will minimize surface area are 2.08 yd by 4.16 yd by 1.387 yd.

23. $R(x) = 50x - 0.5x^2$; $C(x) = 4x + 10$

Profit is equal to revenue minus cost.

$$P(x) = R(x) - C(x)$$
$$= 50x - 0.5x^2 - (4x + 10)$$
$$= -0.5x^2 + 46x - 10$$

Because x is the number of units produced and sold, we are only concerned with the nonnegative values of x. Therefore, we will find the maximum of $P(x)$ on the interval $[0, \infty)$.

First, we find $P'(x)$.

$$P'(x) = -x + 46$$

The derivative exists for all values of x in $[0, \infty)$. Thus, we solve $P'(x) = 0$.

$$-x + 46 = 0$$
$$-x = -46$$
$$x = 46$$

There is only one critical value. We can use the second derivative to determine whether we have a maximum.

$$P''(x) = -1 < 0$$

The second derivative is less than zero for all values of x. Thus, a maximum occurs at $x = 46$.

$$P(46) = -0.5(46)^2 + 46(46) - 10$$
$$= -1058 + 2116 - 10$$
$$= 1048$$

The maximum profit is $1048 when 46 units are produced and sold.

25. $R(x) = 2x$; $C(x) = 0.01x^2 + 0.6x + 30$

Profit is equal to revenue minus cost.

$$P(x) = R(x) - C(x)$$
$$= 2x - (0.01x^2 + 0.6x + 30)$$
$$= -0.01x^2 + 1.4x - 30$$

Because x is the number of units produced and sold, we are only concerned with the nonnegative values of x. Therefore, we will find the maximum of $P(x)$ on the interval $[0, \infty)$.

First, we find $P'(x)$.

$$P'(x) = -0.02x + 1.4$$

The derivative exists for all values of x in $[0, \infty)$. Thus, we solve $P'(x) = 0$.

$$-0.02x + 1.4 = 0$$
$$-0.02x = -1.4$$
$$x = 70$$

There is only one critical value. We can use the second derivative to determine whether we have a maximum.

$$P''(x) = -0.02 < 0$$

The second derivative is less than zero for all values of x. Thus, a maximum occurs at $x = 70$.

$$P(70) = -0.01(70)^2 + 1.4(70) - 30$$
$$= -49 + 98 - 30$$
$$= 19$$

The maximum profit is $19 when 70 units are produced and sold.

27. $R(x) = 9x - 2x^2$

$$C(x) = x^3 - 3x^2 + 4x + 1$$

$R(x)$ and $C(x)$ are in thousands of dollars and x is in thousands of units.

Profit is equal to revenue minus cost.

$$P(x) = R(x) - C(x)$$
$$= 9x - 2x^2 - (x^3 - 3x^2 + 4x + 1)$$
$$= -x^3 + x^2 + 5x - 1$$

Because x is the number of units produced and sold, we are only concerned with the nonnegative values of x. Therefore, we will find the maximum of $P(x)$ on the interval $[0, \infty)$.

First, we find $P'(x)$.

$$P'(x) = -3x^2 + 2x + 5$$

The derivative exists for all values of x in $[0, \infty)$. Thus, we solve $P'(x) = 0$.

$$-3x^2 + 2x + 5 = 0$$
$$3x^2 - 2x - 5 = 0$$
$$(3x - 5)(x + 1) = 0$$
$$3x - 5 = 0 \quad \text{or} \quad x + 1 = 0$$
$$3x = 5 \quad \text{or} \quad x = -1$$
$$x = \frac{5}{3} \quad \text{or} \quad x = -1$$

There is only one critical value in the interval $[0, \infty)$. We can use the second derivative to determine whether we have a maximum.

$$P''(x) = -6x + 2$$

Therefore,

$$P''\left(\frac{5}{3}\right) = -6\left(\frac{5}{3}\right) + 2 = -10 + 2 = -8 < 0$$

The solution is continued on the next page.

On the previous page, we determined the second derivative is less than zero for $x = \dfrac{5}{3}$. Thus, a maximum occurs at $x = \dfrac{5}{3}$.

$$P\left(\frac{5}{3}\right) = -\left(\frac{5}{3}\right)^3 + \left(\frac{5}{3}\right)^2 + 5\left(\frac{5}{3}\right) - 1$$

$$= -\frac{125}{27} + \frac{25}{9} + \frac{25}{3} - 1$$

$$= -\frac{125}{27} + \frac{75}{27} + \frac{225}{27} - \frac{27}{27}$$

$$= \frac{148}{27}$$

Note that $x = \dfrac{5}{3}$ thousand is approximately 1.667 thousand or 1667 units, and that $\dfrac{148}{27}$ thousand is approximately 5.481 thousand or 5481.

Thus, the maximum profit is approximately $5481 when approximately 1667 units are produced and sold.

29. $p = 280 - 0.4x$ Price per unit.

$C(x) = 5000 + 0.6x^2$ Cost per unit.

a) Revenue is price times quantity. Therefore, revenue can be found by multiplying the number of unit sold, x, by the price of the unit, p. Substituting $280 - 0.4x$ for p, we have:

$$R(x) = x \cdot p$$
$$= x(280 - 0.4x)$$
$$R(x) = 280x - 0.4x^2$$

b) Profit is revenue minus cost. Therefore,

$$P(x) = R(x) - C(x)$$
$$= 280x - 0.4x^2 - \left(5000 + 0.6x^2\right)$$
$$= -x^2 + 280x - 5000, \qquad 0 \le x < \infty$$

Since x is the number of units produced and sold, we will restrict the domain to the interval $0 \le x < \infty$.

c) To determine the number of suits required to maximize profit, we first find $P'(x)$.

$$P'(x) = -2x + 280.$$

The derivative exists for all real numbers in the interval $[0, \infty)$. Thus, we solve

$$P'(x) = 0$$
$$-2x + 280 = 0$$
$$-2x = -280$$
$$x = 140$$

Since there is only one critical value, we can use the second derivative to determine whether we have a maximum.

$$P''(x) = -2 < 0$$

The second derivative is negative for all values of x; therefore, a maximum occurs at $x = 140$.

Riverside Appliances must sell 140 refrigerators to maximize profit.

d) The maximum profit is found by substituting 140 for x in the profit function.

$$P(140) = -(140)^2 + 280(140) - 5000$$
$$= -19,600 + 39,200 - 5000$$
$$= 14,600$$

The maximum profit is $14,600.

e) The price per refrigerator is given by:
$p = 280 - 0.4x$

Substituting 140 for x, we have:
$p = 280 - 0.4(140) = 224$

The price per refrigerator will be $224.

31. Let x be the amount by which the price of $80 should be increased. First, we express total revenue R as a function of x. There are two sources of revenue, revenue from tickets and revenue from concessions.

$$R(x) = \left(\begin{array}{c}\text{Number of}\\\text{Rooms}\end{array}\right) \cdot \left(\begin{array}{c}\text{Price of}\\\text{Rooms}\end{array}\right)$$

Note, when the price increases x dollars, the number of rooms occupied falls by x rooms. Thus, the number of rooms is $300 - x$ when price increases x dollars. Therefore, the total revenue function is
$$R(x) = (300 - x)(80 + x)$$
$$= -x^2 + 220x + 24,000$$

The cost of maintaining each occupied room is $22. Therefore the total cost function is:
$$C(x) = \left(\begin{array}{c}\text{Number of}\\\text{Rooms}\end{array}\right) \cdot 22$$
$$= (300 - x)22$$
$$= 6600 - 22x.$$

The solution is continued on the next page.

Now we can find the profit function for the hotel.

$$P(x) = R(x) - C(x)$$
$$= -x^2 + 220x + 24,000 - (6600 - 22x)$$
$$= -x^2 + 242x + 17,400.$$

To find x such that $P(x)$ is a maximum, we first find $P'(x)$:

$$P'(x) = -2x + 242$$

This derivative exists for all real numbers x. thus, the only critical values are where $P'(x) = 0$; so we solve that equation:

$$P'(x) = 0$$
$$-2x + 242 = 0$$
$$-2x = -242$$
$$x = 121$$

The second derivative $P''(x) = -2 < 0$ is negative for all values of x, therefore a maximum occurs at $x = 121$.
The charge per unit should be $80 + $121, or $201.

33. Let x be the amount the number of new officers Oak Glen should place on patrol. The total number of parking tickets written per day is

$$p(x) = \begin{pmatrix} \text{Number of} \\ \text{officers} \end{pmatrix} \cdot \begin{pmatrix} \text{Avg. Tickets} \\ \text{per day} \end{pmatrix}$$
$$= (8 + x)(24 - 4x)$$
$$= -4x^2 - 8x + 192.$$

To find x such that $p(x)$ is a maximum, we first find $p'(x)$:

$$p'(x) = -8x - 8.$$

This derivative exists for all real numbers x. thus, the only critical values are where $P'(x) = 0$.

We solve the equation:

$$P'(x) = 0$$
$$-8x - 8 = 0$$
$$-8x = 8$$
$$x = -1$$

The second derivative $P''(x) = -8 < 0$ is negative for all values of x, therefore a maximum occurs at $x = -1$.
This means that Oak Glean should place one fewer officer on patrol in order to maximize the number of parking tickets written in a day.

35. a) First find the slope of the line.

$$m = \frac{1.12 - 1}{0.59 - 1} = -\frac{12}{41}$$
$$y - 1 = -\frac{12}{41}(x - 1)$$
$$y = -\frac{12}{41}x + \frac{53}{41}$$

So the demand function is:

$$q(x) = -\frac{12}{41}x + \frac{53}{41}$$

b) First, we express total revenue R as a function of x. Revenue is price times quantity demanded, therefore,

$$R(x) = x \cdot q(x)$$
$$R(x) = x\left(-\frac{12}{41}x + \frac{53}{41}\right)$$
$$= -\frac{12}{41}x^2 + \frac{53}{41}x$$

To find x such that $R(x)$ is a maximum, we first find $R'(x)$:

$$R'(x) = -\frac{24}{41}x + \frac{53}{41}$$

$R'(x)$ exists for all real numbers. Solve:

$$R'(x) = 0$$
$$-\frac{24}{41}x + \frac{53}{41} = 0$$
$$x = \frac{53}{24}$$
$$x \approx 2.21$$

The second derivative $R''(x) = -\frac{24}{41} < 0$, is negative, for all values of x, therefore a maximum occurs at $x \approx 2.21$.
To maximize revenue, the price of nitrogen should increase 221% from the January 2001 price.

37. Let x be the number of $0.10 increase that should be made. Then,

$$R(x) = (\text{Attendance}) \cdot (\text{Admission Price})$$
$$= (180 - x)(5 + 0.1x)$$
$$= -0.1x^2 + 13x + 900$$

To find x such that $R(x)$ is a maximum, we first find $R'(x)$:

$$R'(x) = -0.2x + 13$$

The solution is continued on the next page.

From the previous page, we see $R'(x)$ exists for all real numbers. Solve:

$$R'(x) = 0$$
$$-0.2x + 13 = 0$$
$$x = 65$$

Since there is only one critical value, we can use the second derivative,

$$R''(x) = -0.2 < 0,$$

to determine whether we have a maximum. Since $R''(65)$ is negative, a maximum occurs at $x = 65$.

When $x = 65$ the admission price that will maximize revenue is given by:

$$\$5 + 0.1(65) = 11.50$$

Therefore, the theater owner should charge $11.50 per ticket.

39. The area of the parking area is given by
$$A = x \cdot y = 5000.$$

Since three sides are chain link fencing, the total cost of the chain link fencing is given by $4.50(2x + y)$ dollars.

One side is wooden fencing, the cost of the wooden fence is $7y$ dollars.

The total costs in dollars is given by
$$C = 4.50(2x + y) + 7y$$
$$= 9x + 11.5y.$$

To express C in terms of one variable, we solve $xy = 5000$ for y:

$$y = \frac{5000}{x}$$

Then,

$$C(x) = 9x + 11.5\left(\frac{5000}{x}\right)$$
$$= 9x + \frac{57,500}{x}$$

The function is defined only for positive numbers, so we are minimizing C on the interval $(0, \infty)$.

First, we find $C'(x)$.

$$C'(x) = 9 - 57,500x^{-2} = 9 - \frac{57,500}{x^2}$$

Since $C'(x)$ exists for all x in $(0, \infty)$, the only critical values are where $C'(x) = 0$.

Thus, we solve the following equation:

$$9 - \frac{57,500}{x^2} = 0$$
$$9 = \frac{57,500}{x^2}$$
$$9x^2 = 57,500$$
$$x^2 = \frac{57,500}{9}$$
$$x = \sqrt{\frac{57,500}{9}}$$
$$x \approx 79.93$$

This is the only critical value, so we can use the second derivative to determine whether we have a minimum.

$$C''(x) = 172,500x^{-3} = \frac{172,500}{x^3}$$

Note that the second derivative is positive for all positive values of x, therefore we have a minimum at $x = 79.93$.

We find y when $x = 79.93$.

$$y = \frac{5000}{x} = \frac{5000}{(79.93)} \approx 62.55.$$

The cost is minimized when the dimensions are 62.55 ft by 79.92 ft. (Note, the wooden fence side should be 62.55 feet.)
Substituting these dimensions into the cost function we have:

$$C = 9x + 11.5y$$
$$= 9(79.93) + 11.5(62.55)$$
$$= 1438.695 \approx 1439.$$

The total cost of fencing the parking area is approximately $1439.

41. Let x and y represent the outside length and width, respectively.

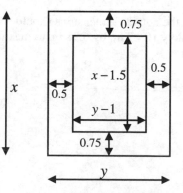

The solution is continued on the next page.

We know that $xy = 73.125$, so $y = \dfrac{73.125}{x}$.

We want to maximize the print area:
$$A = (x - 1.5)(y - 1)$$
$$= xy - x - 1.5y + 1.5.$$

Substituting for y, we get:
$$A(x) = x\left(\frac{73.125}{x}\right) - x - 1.5\left(\frac{73.125}{x}\right) + 1.5$$
$$= 73.125 - x - \frac{109.6875}{x} + 1.5$$
$$= 74.625 - x - \frac{109.6875}{x}$$

$$A'(x) = -1 + 109.6875x^{-2} = -1 + \frac{109.6875}{x^2}$$

$A'(x)$ exists for all values in the domain of A.

Solve:
$$A'(x) = 0$$
$$-1 + \frac{109.6875}{x^2} = 0$$
$$x^2 = 109.6875$$
$$x = \pm\sqrt{109.6875}$$

The only critical value in the domain of A is $x = \sqrt{109.6875}$;

Therefore, we can use the second derivative to determine if it is a maximum.
$$A''(x) = -219.375x^{-3} = -\frac{219.375}{x^3}.$$

$A''\left(\sqrt{109.6875}\right) < 0$, so $A\left(\sqrt{109.6875}\right)$ is a maximum.

When
$$x = \sqrt{109.6875} \approx 10.47,$$
$$y = \frac{73.125}{\sqrt{109.6875}} \approx 6.98.$$

Therefore, the outside dimensions should be approximately 10.47 in. by 6.98 in. to maximize the print area.

43. Let x be the lot size.
$$C(x) = \underset{\text{Cost}}{\text{Yearly Carrying}} + \underset{\text{Cost}}{\text{Yearly reorder}}$$

We consider each cost separately.

Yearly carrying costs, $C_c(x)$: Can be found by multiplying the cost to store the items by the number of items in storage. The average amount held in stock is $\dfrac{x}{2}$, and it cost \$4 per bowling ball for storage. Thus:
$$C_c(x) = 4 \cdot \frac{x}{2} = 2x$$

Yearly reorder costs, $C_r(x)$: Can be found by multiplying the cost of each order by the number of reorders. The cost of each order is $1 + 0.5x$, and the number of orders per year is $\dfrac{200}{x}$. Therefore,
$$C_r(x) = (1 + 0.5x)\left(\frac{200}{x}\right)$$
$$= \frac{200}{x} + 100$$

The total inventory cost is given by:
$$C(x) = C_c(x) + C_r(x)$$
$$= 2x + \frac{200}{x} + 100, \quad 1 \le x \le 200$$

We want to find the minimum value of C on the interval $[1, 200]$. First, we find $C'(x)$:

$$C'(x) = 2 - 200x^{-2} = 2 - \frac{200}{x^2}$$

$C'(x)$ exists for all x in $[1, 200]$. $[1, 100]$, so the only critical values are where $C'(x) = 0$.

Solving the equation, we have:
$$C'(x) = 0$$
$$2 - \frac{200}{x^2} = 0$$
$$x = \pm 10$$

The only critical value in the domain is $x = 10$. Therefore, we use the second derivative,
$$C''(x) = 400x^{-3} = \frac{400}{x^3}$$

to determine whether we have a minimum.
$C''(10) = 0.4 > 0$, so $C(10)$ is a minimum.

In order to minimize inventory costs. The store should order $\dfrac{200}{10} = 20$ times per year. The lot size will be 10 bowling balls.

45. Let x equal the lot size. Now the inventory costs are given by:

$$C(x) = \frac{\text{Yearly Carrying}}{\text{Cost}} + \frac{\text{Yearly reorder}}{\text{Cost}}$$

We consider each cost separately.

Yearly carrying costs, $C_c(x)$: Can be found by multiplying the cost to store the items by the number of items in storage. The average amount held in stock is $\frac{x}{2}$, and it cost \$2 per calculator for storage. Thus:

$$C_c(x) = 8 \cdot \frac{x}{2}$$
$$= 4x$$

Yearly reorder costs, $C_r(x)$. Can be found by multiplying the cost of each order by the number of reorders. The cost of each order is $10 + 5x$, and the number of orders per year is $\frac{360}{x}$. Therefore,

$$C_r(x) = (10 + 5x)\left(\frac{360}{x}\right)$$
$$= \frac{3600}{x} + 1800$$

Hence, the total inventory cost is:

$$C(x) = C_c(x) + C_r(x)$$
$$= 4x + \frac{3600}{x} + 1800, \quad 1 \le x \le 360$$

We want to find the minimum value of C on the interval $[1, 360]$. First, we find $C'(x)$.

$$C'(x) = 4 - 3600x^{-2} = 4 - \frac{3600}{x^2}$$

The derivative exists for all x in $[1, 360]$, so the only critical values are where $C'(x) = 0$.

$$C'(x) = 0$$
$$4 - \frac{3600}{x^2} = 0$$
$$4 = \frac{3600}{x^2}$$
$$4x^2 = 3600$$
$$x^2 = 900$$
$$x = \pm 30$$

The only critical value in the domain is $x = 30$. Therefore, we use the second derivative,

$$C''(x) = 7200x^{-3} = \frac{7200}{x^3}$$

to determine whether we have a minimum.

We have: $C''(30) = 0.27 > 0$, so $C(30)$ is a minimum.

In order to minimize inventory costs. The store should order $\frac{360}{30} = 12$ times per year. The lot size will be 30 surf boards.

47. Let x be the lot size.

Yearly carrying cost: $C_c(x) = 8 \cdot \frac{x}{2} = 4x$

Yearly reorder cost:
$$C_r(x) = (10 + 6x)\left(\frac{360}{x}\right)$$
$$= \frac{3600}{x} + 2160$$

Therefore,

$$C(x) = C_c(x) + C_r(x)$$
$$= 4x + \frac{3600}{x} + 2160, \quad 1 \le x \le 360$$

$$C'(x) = 4 - 3600x^{-2} = 4 - \frac{3600}{x^2}$$

$C'(x)$ exists for all x in $[1, 360]$. Solve:

$$C'(x) = 0$$
$$4 - \frac{3600}{x^2} = 0$$
$$x = \pm 30$$

The only critical value in the domain is $x = 30$. Therefore, we use the second derivative,

$$C''(x) = 7200x^{-3} = \frac{7200}{x^3}$$

to determine whether we have a minimum.

$C''(30) = 0.27 > 0$, so $C(30)$ is a minimum.

In order to minimize inventory costs. The store should order $\frac{360}{30} = 12$ times per year. The lot size will be 30 surf boards.

49. The volume of the container must be 400 cm^3. Therefore, we use formula for the volume cylinder to obtain $400 = \pi r^2 h$.

Solving for h we have:

$$h = \frac{400}{\pi r^2}.$$

Therefore, the total surface area is the sum of the area the bottom plus the the side material:

$$A = \pi r^2 + 2\pi rh.$$

The solution is continued on the next page.

Substituting for h in area formula on the previous page we have area as a function of the radius, r.

$$A(r) = \pi r^2 + 2\pi r \left(\frac{400}{\pi r^2} \right)$$

$$A(r) = \pi r^2 + \frac{800}{r}$$

The nature of this problem requires $r > 0$. We differentiate the area function with respect to r:

$$A'(r) = 2\pi r - \frac{800}{r^2}.$$

We find the critical values by setting the derivative equal to zero and solving for r. Remember, $r > 0$.

$$A'(r) = 0$$

$$2\pi r - \frac{800}{r^2} = 0$$

$$2\pi r = \frac{800}{r^2}$$

$$r^3 = \frac{800}{2\pi}$$

$$r = \sqrt[3]{\frac{400}{\pi}}$$

$$r \approx 5.03$$

This is the only critical value in the interval $r > 0$. We will calculate the second derivative to determine the concavity of the function.

$$A''(r) = 2\pi + \frac{1600}{r^3}.$$

We evaluate the second derivative at the critical value:

$$A''(5.03) = 2\pi + \frac{1600}{(5.03)^3} \approx 18.855 > 0.$$

Since the second derivative is positive at the critical value, the critical value represents a relative minimum. We determine the height of the container by substituting back into

$$h = \frac{400}{\pi r^2}.$$

$$h = \frac{400}{\pi (5.03)^2} \approx 5.03$$

Therefore, the dimensions of the container that will minimize the surface area are a height of 5.03 cm and a radius of 5.03 cm.

51. The cost function for the container would be is given by:

$$C = 0.0015 \left(\begin{array}{c} \text{area of} \\ \text{the base} \end{array} \right)$$

$$+ 0.0008 \left(\begin{array}{c} \text{area of} \\ \text{the side} \end{array} \right)$$

Using the information from problem 49, the cost function is:

$$C(r) = 0.0015 \left(\pi r^2 \right) + 0.0008 \left(\frac{800}{r} \right)$$

$$= 0.0015 \pi r^2 + \frac{0.64}{r}.$$

The nature of the problem still requires $r > 0$. Calculating the derivative of the cost function we have:

$$C'(r) = \frac{d}{dr} \left[0.0015 \pi r^2 + \frac{0.64}{r} \right]$$

$$= 0.003 \pi r - \frac{0.64}{r^2}$$

Find the critical values by setting the derivative equal to zero and solving for r.

$$C'(r) = 0$$

$$0.003 \pi r - \frac{0.64}{r^2} = 0$$

$$0.003 \pi r = \frac{0.64}{r^2}$$

$$r^3 = \frac{0.64}{0.003 \pi}$$

$$r = \sqrt[3]{\frac{0.64}{0.003 \pi}}$$

$$r \approx 4.08$$

This is the only critical value in the interval $r > 0$.

Next, we find the second derivative:

$$C''(r) = \frac{d}{dr} \left(0.003 \pi r - \frac{0.64}{r^2} \right)$$

$$= 0.003 \pi + \frac{1.28}{r^3}.$$

Substituting the critical value into the second derivative, we have:

$$C''(4.08) = 0.003 \pi + \frac{1.28}{(4.08)^3}$$

$$\approx 0.028 > 0.$$

Since the second derivative is positive at the critical value, the critical value represents a relative minimum.

The solution is continued on the next page.

We substitute the critical value back into

$h = \dfrac{400}{\pi r^2}$ to determine the value for h. Notice,

we substitute before the rounding of r.

$h = \dfrac{400}{\pi (4.08)^2}$

≈ 7.65

The dimensions of the container that will minimize the cost of the container are a radius of 4.08 cm and a height of 7.65 cm.

53. Let y represent the dimension on the lot line and let x represent the other dimension. Then the length of fencing that the person must pay for is

$\dfrac{1}{2}y + x + y + x = 2x + \dfrac{3}{2}y$.

We know $xy = 48$; therefore, $y = \dfrac{48}{x}$.

The length of fencing as a function of x is:

$F(x) = 2x + \dfrac{3}{2}\left(\dfrac{48}{x}\right) = 2x + \dfrac{72}{x}, \ 0 < x < \infty$.

$F'(x) = 2 - 72x^{-2} = 2 - \dfrac{72}{x^2}$

$F'(x)$ exists for all x in $(0, \infty)$. Solve:

$F'(x) = 0$

$2 - \dfrac{72}{x^2} = 0$

$x = \pm 6$

Only the critical value $x = 6$ is in $(0, \infty)$. Since there is only one critical value, we use the second derivative to determine whether we have a minimum.

$F''(x) = 144x^{-3} = \dfrac{144}{x^3}$

$F''(6) = \dfrac{2}{3} > 0$, so $F(6)$ is a minimum.

When $x = 6$, $y = \dfrac{48}{6} = 8$. The dimensions that minimize the cost are 6 yd by 8 yd. Where the longer side of the lot is adjacent to the neighbor's yard.

55. Since the stained glass transmits only half as much light as the semicircle in Exercise 54, we express the function A as:

$A = \dfrac{1}{2} \cdot \dfrac{1}{2}\pi x^2 + 2xy$

$= \dfrac{1}{4}\pi x^2 + 2xy$

The perimeter is still the same, so we can substitute $12 - \dfrac{\pi}{2}x - x$ for y to get:

$A(x) = \dfrac{1}{4}\pi x^2 + 2x\left(12 - \dfrac{\pi}{2}x - x\right)$

$= \dfrac{1}{4}\pi x^2 + 24x - 2x^2 - \pi x^2$

$= \left(-\dfrac{3}{4}\pi - 2\right)x^2 + 24x, \ 0 < x < 24.$

Find $A'(x)$.

$A'(x) = \left(-\dfrac{3}{2}\pi - 4\right)x + 24$

Since $A'(x)$ exists for all x in $(0, 24)$, the only critical points are where $A'(x) = 0$. Thus, we solve the following equation:

$\left(-\dfrac{3}{2}\pi - 4\right)x + 24 = 0$

$(3\pi + 8)x - 48 = 0$

$x = \dfrac{48}{3\pi + 8} \approx 2.75$

This is the only critical value, so we can use the second derivative to determine whether we have a maximum.

$A''(x) = -\dfrac{3}{2}\pi - 4 < 0$

Since $A''(x)$ is negative for all values of x, we have a maximum at $x = \dfrac{48}{3\pi + 8}$. We find y when $x = \dfrac{48}{3\pi + 8}$.

$y = 12 - \dfrac{\pi}{2}x - x$

$= 12 - \dfrac{\pi}{2}\left(\dfrac{48}{3\pi + 8}\right) - \dfrac{48}{3\pi + 8}$

$= \dfrac{12\pi + 48}{3\pi + 8} \approx 4.92$

To maximize the amount of light through the window, the dimensions must be $x = \dfrac{48}{3\pi + 8}$ ft

and $y = \dfrac{12\pi + 48}{3\pi + 8}$ ft or approximately

$x \approx 2.75$ ft and $y \approx 4.92$ ft.

56-110. Left to the Student.

111. Let x represent a positive number. Then, $\dfrac{1}{x}$ is the reciprocal of the number, and x^2 is the square of the number. The sum, S, of the reciprocal and five times the square is given by:

$$S(x) = \frac{1}{x} + 5x^2.$$

We want to minimize $S(x)$ on the interval $(0, \infty)$. First, we find $S'(x)$

$$S'(x) = -x^{-2} + 10x = -\frac{1}{x^2} + 10x$$

Since $S'(x)$ exists for all values of x in $(0, \infty)$, the only critical values occur when $S'(x) = 0$. We solve the following equation:

$$-\frac{1}{x^2} + 10x = 0$$

$$10x = \frac{1}{x^2}$$

$$10x^3 = 1$$

$$x^3 = \frac{1}{10}$$

$$x = \sqrt[3]{\frac{1}{10}} = \frac{1}{\sqrt[3]{10}}$$

Since there is only one critical value, we use the second derivative,

$$S''(x) = 2x^{-3} + 10 = \frac{2}{x^3} + 10,$$

to determine whether it is a minimum. The second derivative is positive for all x in $(0, \infty)$; therefore, the sum is a minimum when $x = \dfrac{1}{\sqrt[3]{10}}$.

113. Let A represent the amount deposited in savings account and i represent the interest rate paid on the money deposited. If A is directly proportional to i, then there is some positive constant k such that $A = ki$. The interest earned by the bank is represented by $18\% A$, or $0.18A$. The interest paid by the bank is represented by iA. Thus the profit received by the bank is given by

$$P = 0.18A - iA.$$

We express P as a function of the interest the bank pays on the money deposited, i, by substituting ki for A.

$$P = 0.18(ki) - i(ki)$$

$$= 0.18ki - ki^2$$

We maximize P on the interval $(0, \infty)$. First, we find $P'(i)$.

$$P'(i) = 0.18k - 2ki$$

Since $P'(i)$ exists for all i in $(0, \infty)$, the only critical values are where $P'(i) = 0$. We solve the following equation:

$$0.18k - 2ki = 0$$

$$-2ki = -0.18k$$

$$i = \frac{-0.18k}{-2k}$$

$$i = 0.09$$

Since there is only one critical point, we can use the second derivative to determine whether we have a minimum. Notice that $P''(i) = -2k$, which is a negative constant $(k > 0)$. Thus, $P''(0.09)$ is negative, so $P(0.09)$ is a maximum. To maximize profit, the bank should pay 9% on its savings accounts.

115. Using the drawing in the text, we write a function that gives the cost of the power line. The length of the power line on the land is given by $4 - x$, so the cost of laying the power line underground is given by:

$$C_L(x) = 3000(4 - x) = 12{,}000 - 3000x.$$

The length of the power line that will be under water is $\sqrt{1 + x^2}$, so the cost of laying the power line underwater is given by:

$$C_W(x) = 5000\sqrt{1 + x^2}$$

Therefore, the total cost of laying the power line is:

$$C(x) = C_L(x) + C_W(x)$$

$$= 12{,}000 - 3000x + 5000\sqrt{1 + x^2}.$$

We want to minimize $C(x)$ over the interval $0 \le x \le 4$. First, we find the derivative.

$$C'(x) = -3000 + 5000\left(\frac{1}{2}\right)\left(1 + x^2\right)^{-\frac{1}{2}}(2x)$$

$$= -3000 + 5000x\left(1 + x^2\right)^{-\frac{1}{2}}$$

$$= -3000 + \frac{5000x}{\sqrt{1 + x^2}}$$

Since the derivative exists for all x.

The solution is continued on the next page.

We find the critical values by setting the derivative on the previous page equal to 0 and solving the equation:

$$C'(x) = 0$$

$$-3000 + \frac{5000x}{\sqrt{1+x^2}} = 0$$

$$-3000\sqrt{1+x^2} + 5000x = 0$$

$$5000x = 3000\sqrt{1+x^2}$$

$$\frac{5}{3}x = \sqrt{1+x^2}$$

$$\left(\frac{5}{3}x\right)^2 = \left(\sqrt{1+x^2}\right)^2$$

$$\frac{25}{9}x^2 = 1+x^2$$

$$\frac{16}{9}x^2 = 1$$

$$x^2 = \frac{9}{16}$$

$$x = \pm\sqrt{\frac{9}{16}}$$

$$x = \pm\frac{3}{4}$$

The only critical value in the interval $[0,4]$ is $x = \frac{3}{4}$, so we can use the second derivative to determine if we have a minimum.

$$C''(x) = \frac{\left(1+x^2\right)^{1/2}(5000) - 5000x\left[\frac{1}{2}\left(1+x^2\right)^{-1/2}(2x)\right]}{\left[\left(1+x^2\right)^{1/2}\right]^2}$$

$$= \frac{5000\sqrt{1+x^2} - \dfrac{5000x^2}{\sqrt{1+x^2}}}{\left(1+x^2\right)}$$

$$= \frac{5000}{\sqrt{1+x^2}} - \frac{5000x^2}{\left(1+x^2\right)^{3/2}}$$

$$= \frac{5000\left(1+x^2\right) - 5000x^2}{\left(1+x^2\right)^{3/2}}$$

$$= \frac{5000}{\left(1+x^2\right)^{3/2}}$$

$C''(x)$ is positive for all x in $[0,4]$; therefore, a minimum occurs at $x = \frac{3}{4}$. When $x = \frac{3}{4}$,

$$4 - \frac{3}{4} = \frac{13}{4} = 3.25.$$

Therefore, S should be 3.25 miles down shore from the power station.

Note: since we are minimizing cost over a closed interval, we could have used Max-Min Principle 1 to determine the minimum, and avoided finding the second derivative. The critical value and the endpoints are $0, \frac{3}{4}$, and 4. The function values at these three points are:

$$C(0) = 12,000 - 3000(0) + 5000\left(\sqrt{1+(0)^2}\right)$$

$$= 17,000$$

$$C\left(\frac{3}{4}\right) = 12,000 - 3000\left(\frac{3}{4}\right) + 5000\left(\sqrt{1+\left(\frac{3}{4}\right)^2}\right)$$

$$= 16,000$$

$$C(4) = 12,000 - 3000(4) + 5000\left(\sqrt{1+(4)^2}\right)$$

$$\approx 20,615.53$$

Therefore, the minimum occurs when $x = \frac{3}{4}$, or when S is 3.25 miles down shore from the power station.

117. The objective is to maximize the parallel areas subject to the constraint of having k units of fencing. Using the figure in the book, we can let x equal the length of the two parallel sides and y equal the length of the three parallel sides. Since the two areas are the identical, we can either maximize the area of one of the smaller areas or maximize the area of the entire field. That is, we can either maximize

$$A_1 = x \cdot y \quad \text{or} \quad A_2 = 2x \cdot y.$$

First we will maximize A_1. The perimeter of the fields is given by $P = 4x + 3y$. Since we have k units of fencing, we know that

$$4x + 3y = k.$$

The solution is continued on the next page.

Solve the equation on the previous page for y and substitute into the area function to get the area as a function of one variable.

$4x + 3y = k$

$$3y = k - 4x$$

$$y = \frac{k - 4x}{3}$$

$$y = \frac{1}{3}k - \frac{4}{3}x$$

Substituting into the area function we have:

$$A_1 = x \cdot \left(\frac{1}{3}k - \frac{4}{3}x \right) = \frac{1}{3}kx - \frac{4}{3}x^2.$$

We find the first derivative with respect to x.

$$\frac{dA_1}{dx} = \frac{1}{3}k - \frac{8}{3}x.$$

The deriviative exists for all values of x. We find the critical values by solving:

$$\frac{dA_1}{dx} = 0$$

$$\frac{1}{3}k - \frac{8}{3}x = 0$$

$$-\frac{8}{3}x = -\frac{1}{3}k$$

$$x = \frac{1}{8}k.$$

There is only one critical value, so we can use the second derivative to determine if it is a maximum. Notice that the second derivative,

$\dfrac{d^2 A_1}{dx^2} = -\dfrac{8}{3}$, is negative for all values of x.

Therefore the maximum area will occur when the width of the fields are $\frac{1}{8}k$ units long. We substitute in to find the length of the fields y.

$$y = \frac{1}{3}k - \frac{4}{3}x$$

$$= \frac{1}{3}k - \frac{4}{3}\left(\frac{1}{8}k \right)$$

$$= \frac{1}{3}k - \frac{1}{6}k$$

$$= \frac{1}{6}k = \frac{k}{6}.$$

Therefore the length of the rectangular areas that will maximize the area given k units of fencing is $y = \dfrac{k}{6}$ units.

Notice, if we maximized $A_2 = 2x \cdot y$. The constraint is the same:

$$4x + 3y = k \Rightarrow y = \frac{1}{3}k - \frac{4}{3}x.$$

$$A_2 = 2x \cdot \left(\frac{1}{3}k - \frac{4}{3}x \right) = \frac{2}{3}kx - \frac{8}{3}x^2.$$

$$\frac{dA_2}{dx} = \frac{2}{3}k - \frac{16}{3}x.$$

$$\frac{dA_2}{dx} = 0$$

$$\frac{2}{3}k - \frac{16}{3}x = 0$$

$$x = \frac{1}{8}k.$$

There is only one critical value, so we can use the second derivative to determine if it is a maximum. Notice that the second derivative,

$\dfrac{d^2 A_1}{dx^2} = -\dfrac{16}{3}$, is negative for all values of x.

Therefore the maximum area will occur when the width of the fields are $\frac{1}{8}k$ units long. We substitute in to find the length of the fields y.

$$y = \frac{1}{3}k - \frac{4}{3}\left(\frac{1}{8}k \right) = \frac{k}{6}.$$

Therefore the length of the rectangular areas that will maximize the area given k units of fencing is $y = \dfrac{k}{6}$ units as determined previously.

119. $C(x) = 8x + 20 + \dfrac{x^3}{100}$

a) To determine the average cost, we divide the total cost function by the number of units produced:

$$A(x) = \frac{C(x)}{x}.$$

$$A(x) = \frac{8x + 20 + \dfrac{x^3}{100}}{x}$$

$$= 8 + \frac{20}{x} + \frac{x^2}{100}$$

b) Taking the derivative of the total cost function and the average cost function we have:

$$C'(x) = \frac{d}{dx}\left(8x + 20 + \frac{x^3}{100}\right)$$

$$= 8 + \frac{3x^2}{100}.$$

$$A'(x) = \frac{d}{dx}\left(8 + \frac{20}{x} + \frac{x^2}{100}\right)$$

$$= \frac{d}{dx}\left(8 + 20x^{-1} + \frac{1}{100}x^2\right)$$

$$= -20x^{-2} + \frac{1}{100}(2x)$$

$$= -\frac{20}{x^2} + \frac{x}{50}.$$

c) The derivative exists for all x in $(0, \infty)$; therefore, the critical values occur when $A'(x) = 0$. Solve:

$$-\frac{20}{x^2} + \frac{x}{50} = 0$$

$$\frac{x}{50} = \frac{20}{x^2}$$

$$x \cdot x^2 = 20 \cdot 50$$

$$x^3 = 1000$$

$$x = 10$$

There is only one critical value, so we use the second derivative to determine whether we have a minimum.

$$A''(x) = \frac{40}{x^3} + \frac{1}{50}$$

$$A''(10) = \frac{3}{50} > 0. \text{ Thus } A(10) \text{ is a minimum.}$$

Find the function value when $x = 10$:

$$A(10) = 8 + \frac{20}{10} + \frac{10^2}{100} = 11.$$

The minimum average cost is $11 when 10 units are produced.

$$C'(10) = 8 + \frac{3}{100}(10^2) = 11.$$

The marginal cost is $11 when 10 units are produced.

d) The two values are equal:
$$A(10) = C'(10) = 11.$$

121. Express Q as a function of one variable. First, solve $x + y = 1$ for y. We have:

$$y = 1 - x.$$

Substituting we have:

$$Q = x^3 + 2(1-x)^3$$

$$= x^3 + 2(1 - 3x + 3x^2 - x^3)$$

$$= -x^3 + 6x^2 - 6x + 2$$

Next, we find $Q'(x)$.

$$Q'(x) = -3x^2 + 12x - 6$$

The derivative exists for all values of x in the interval $(0, \infty)$.

Note the constraint $y = 1 - x$ actually limits us to look at the interval $(0,1)$.

The only critical values are where $Q'(x) = 0$.
We solve the equation:

$$-3x^2 + 12x - 6 = 0$$

$$x^2 - 4x + 2 = 0$$

Using the quadratic formula, we have:

$$x = 2 \pm \sqrt{2}.$$

When $x = 2 + \sqrt{2}$, $y = 1 - (2 + \sqrt{2}) = -1 - \sqrt{2}$.

When $x = 2 - \sqrt{2}$, $y = 1 - (2 - \sqrt{2}) = 1 + \sqrt{2}$.

Since x and y must be positive, we only consider $x = 2 - \sqrt{2}$ and $y = -1 + \sqrt{2}$.

Note, $Q''(x) = -6x + 12$ and

$$Q''(2 - \sqrt{2}) = 6(2 - \sqrt{2}) + 12 \approx 8.48 > 0, \text{ so we}$$

have a minimum at $x = 2 - \sqrt{2}$.
The minimum value of Q is found by substituting.

$$Q = x^3 + 2y^3$$

$$= (2 - \sqrt{2})^3 + 2(-1 + \sqrt{2})^3$$

$$= 6 - 4\sqrt{2}$$

123. Let x be the lot size.

Yearly carrying cost: $C_c(x) = a \cdot \dfrac{x}{2} = \dfrac{ax}{2}$

$$C_r(x) = (b + cx)\left(\dfrac{Q}{x}\right)$$

Yearly reorder cost:

$$= \dfrac{bQ}{x} + cQ$$

Then,

$$C(x) = C_c(x) + C_r(x)$$

$$= \dfrac{ax}{2} + \dfrac{bQ}{x} + cQ, \quad 1 \le x \le Q$$

To find the minimum, we take the first deriviative:

$$C'(x) = \dfrac{a}{2} - bQx^{-2} = \dfrac{a}{2} - \dfrac{bQ}{x^2}$$

$C'(x)$ exists for all x in $[1, Q]$. So the only critical values occur when $C'(x) = 0$. Solve:

$$C'(x) = 0$$

$$\dfrac{a}{2} - \dfrac{bQ}{x^2} = 0$$

$$\dfrac{a}{2} = \dfrac{bQ}{x^2}$$

$$ax^2 = 2bQ$$

$$x^2 = \dfrac{2bQ}{a}$$

$$x = \pm\sqrt{\dfrac{2bQ}{a}}$$

The only critical value in the domain is

$$x = \sqrt{\dfrac{2bQ}{a}}.$$

Since there is only one critical value in the domain, we use the second derivative,

$$C''(x) = 2bQx^{-3} = \dfrac{2bQ}{x^3}$$

to determine whether we have a minimum. $C''(x) > 0$ for all x in $[1, Q]$, so a minimum

occurs at $x = \sqrt{\dfrac{2bQ}{a}}$.

In order to minimize inventory costs. The store

should order $\dfrac{Q}{\sqrt{\dfrac{2bQ}{a}}} = \sqrt{\dfrac{aQ}{2b}}$ times per year.

The lot size will be $\sqrt{\dfrac{2bQ}{a}}$ units.

125-128. The starting value and step size were chosen to limit the amount of space used . Approaches can vary.

125. Using a spreadsheet we numerically estimate the maximum:

Starting Value:	24	
Step Size:	0.25	
x	$y = \dfrac{100 - 2x}{3}$	$q = x \cdot y$
24	17.33333333	416
24.25	17.16666667	416.2917
24.5	17	416.5
24.75	16.83333333	416.625
25	16.66666667	416.6667
25.25	16.5	416.625
25.5	16.33333333	416.5
25.75	16.16666667	416.2917
26	16	416

We determine the maximum of $Q \approx 416.67$ to occur when $x = 25$ and $y \approx 16.67$.

127. Using a spreadsheet we numerically estimate the maximum:

Starting Value:	2.2	
Step Size:	0.01	
x	$y = \sqrt{10 - 0.5x^2}$	$Q = x \cdot y^3$
2.2	2.75317998	45.91202934
2.21	2.749172603	45.91962105
2.22	2.745141162	45.92477319
2.23	2.741085551	45.92748186
2.24	2.737005663	45.92774328
2.25	2.732901389	45.9255538
2.26	2.728772618	45.92090987
2.27	2.724619239	45.91380804
2.28	2.720441141	45.90424501

We determine the maximum of $Q \approx 45.93$ to occur when $x \approx 2.24$ and $y \approx 2.74$.

Exercise Set 2.6

1. $R(x) = 50x - 0.5x^2$; $C(x) = 4x + 10$

a) Total profit is revenue minus cost.
$$P(x) = R(x) - C(x)$$
$$P(x) = 50x - 0.5x^2 - (4x + 10)$$
$$= 50x - 0.5x^2 - 4x - 10$$
$$= -0.5x^2 + 46x - 10$$

b) Substituting 20 for x into the three functions, we have:
$$R(20) = 50(20) - 0.5(20)^2 = 800$$
The total revenue from the sale of the first 20 units is $800.
$$C(20) = 4(20) + 10 = 90$$
The total cost of producing the first 20 units is $90.
$$P(20) = R(20) - C(20)$$
$$= 800 - 90$$
$$= 710$$
The total profit is $710 when the first 20 units are produced and sold.
Note, we could have also used the profit function, $P(x)$, from part (a) to find the profit.
$$P(20) = -0.5(20)^2 + 46(20) - 10 = 710$$

c) Finding the derivative for each of the functions, we have:
$$R'(x) = 50 - x$$
$$C'(x) = 4$$
$$P'(x) = -x + 46$$

d) Substituting 20 for x in each of the three marginal functions, we have:
$$R'(20) = 50 - 20 = 30$$
Once 20 units have been sold, the approximate revenue for the 21st unit is $30.
$$C'(20) = 4$$
Once 20 units have been produced, the approximate cost for the 21st unit is $4.
$$P'(20) = -20 + 46 = 26$$
Once 20 units have been produced and sold, the approximate profit from the sale of the 21st unit is $26.

3. $C(x) = 0.002x^3 + 0.1x^2 + 42x + 300$

a) Substituting 40 for x into the cost function, we have:
$$C(40) = 0.002(40)^3 + 0.1(40)^2 + 42(40) + 300$$
$$= 2268 \text{ (hundreds of dollars)}$$
The current daily cost of producing 40 security systems is $226,800.

b) In order to find the additional cost of producing 41 chairs monthly, we first find the total cost of producing 41 security systems in a day.
$$C(41) = 0.002(41)^3 + 0.1(41)^2 + 42(41) + 300$$
$$= 2327.94 \text{ (hundreds of dollars)}$$
Next, we subtract the cost of producing 40 security systems daily found in part (a) from the cost of producing 41 security systems daily.
$$C(41) - C(40) = 2327.94 - 2268 = 59.94$$
The additional daily cost of increasing production to 41 security systems daily is $5994.

c) First, we find the marginal cost function,
$$C'(x) = 0.006x^2 + 0.2x + 42$$
Next, substituting 40 for x, we have:
$$C'(40) = 59.6$$
The marginal cost when 40 security systems are produced daily is $5960.

d) In part (a) we found that it cost $2268 hundred to produce 40 security systems per day. The additional cost of producing 2 additional security systems is
$2(\$59.60) = \$119.20$ hundreds. Therefore, the estimated daily cost of producing 42 security systems per day is
$$C(42) \approx \$2268 + \$119.20 = \$2387.20$$
hundreds or $238,720 per day.

5. $R(x) = 0.005x^3 + 0.01x^2 + 0.5x$

a) Substituting 70 for x, we have:
$$R(70) = 0.005(70)^3 + 0.01(70)^2 + 0.5(70)$$
$$= 1715 + 49 + 35$$
$$= 1799$$
The currently daily revenue from selling 70 lawn chairs per day is $1799.

b) Substituting 73 for x, we have:
$$R(73) = 0.005(73)^3 + 0.01(73)^2 + 0.5(73)$$
$$= 2034.875$$
$$\approx 2034.88$$
Therefore, the increase in revenue from increasing sales to 73 chairs per day is:
$$R(73) - R(70) = 2034.88 - 1799$$
$$= 235.88$$
Revenue will increase $235.88 per day if the number of chairs sold increases to 73 per day.

c) First we find the marginal revenue function by finding the derivative of the revenue function.
$$R'(x) = 0.015x^2 + 0.02x + 0.5$$
Substituting 70 for x, we have:
$$R'(70) = 0.015(70)^2 + 0.02(70) + 0.5$$
$$= 75.40$$
The marginal revenue when 70 lawn chairs are sold daily is $75.40.

d) In part (a) we found that selling 70 lawn chairs per day resulted in a revenue of $1799. In part (c) we found that the marginal revenue when 70 chairs were sold is $75.40. Using these two numbers, we estimate the daily revenue generated by selling 71 chairs is
$$R(71) \approx R(70) + R'(70)$$
$$= \$1799 + \$75.40 = \$1874.40.$$
Similarly, the daily revenue generated by selling 72 chairs, or 2 additional chairs, daily is approximately
$$R(72) \approx R(70) + 2 \cdot R'(70)$$
$$\approx \$1799 + 2(\$75.40) \approx \$1949.80.$$
The daily revenue generated by selling 73 chairs, or 3 additional chairs, daily is approximately
$$R(73) \approx R(70) + 3 \cdot R'(70)$$
$$\approx \$1799 + 3(\$75.40) \approx \$2025.20.$$

7. $R(x) = 0.007x^3 - 0.5x^2 + 150x$

a) Substituting 26 for x, we have:
$$R(26) = 0.007(26)^3 - 0.5(26)^2 + 150(26)$$
$$= 123.032 - 338 + 3900$$
$$= 3685.03$$
The current monthly revenue is $3685.03.

b) First we find the total montly revenue for selling 28 suitcases
$$R(28) = 0.007(28)^3 - 0.5(28)^2 + 150(28)$$
$$= 3961.66$$
The difference in monthly revenue from selling 26 suitcases and 28 suitcases a month is:
$$R(28) - R(26) = \$3961.66 - \$3685.03$$
$$= \$276.63.$$
If sales increased from 26 to 28 suitcases, monthly revenue would increase $276.63

c) First we find marginal revenue by taking the derivative of the revenue function.
$$R'(x) = 0.021x^2 - x + 150$$
Next we substitute 26 in for x.
$$R'(26) = 0.021(26)^2 - (26) + 150 = 138.196$$
Marginal revenue is 138.20 when 26 suitcases are sold.

d) From part (a), we know that when 26 suitcases are sold, total monthly revenue is $3685.03. From part (c), we know that when 26 suitcases are sold, marginal revenue is $138.20. Therefore, we estimate:
$$R(27) \approx R(26) + R'(26)$$
$$R(27) \approx \$3685.03 + \$138.20 \approx \$3823.23$$
We estimate the revenue from selling 27 suitcases per month to be $3823.23.

9. $N(1000) = 500,000$ means that 500,000 computers will be sold annually when the price of the computer is $1000. $N'(1000) = -100$ means that when the price is increased $1 to $1001, sales will decrease by 100 computers per year.

11. $C(x) = 0.01x^2 + 1.6x + 100$
$$\Delta C = C(x + \Delta x) - C(x)$$
Substituting $x = 80$, and $\Delta x = 1$ we have
$$\Delta C = C(80 + 1) - C(80)$$
$$= 0.01(81)^2 + 1.6(81) + 100 - \left[0.01(80)^2 + 1.6(80) + 100 \right]$$
$$= 3.21$$
The additional cost of producing the 81st unit is $3.21.
Finding the derivative of $C(x)$ we have:
$$C'(x) = 0.02x + 1.6 .$$
The solution is continued on the next page.

Substituting 80 for x inot the derivative, we have:

$C'(80) = 0.02(80) + 1.6 = 3.20$

The marginal cost when 80 units are produced is $3.20.

13. $R(x) = 2x$

$\Delta R = R(x + \Delta x) - R(x)$

Substituting $x = 70$, and $\Delta x = 1$ we have

$\Delta R = R(70 + 1) - R(70)$

$= R(71) - R(70)$

$= 2(71) - [2(70)]$

$= 2$

The additional revenue from selling the 71st unit is $2.00.

Finding the derivative of $R(x)$ we have:

$R'(x) = 2$.

The derivative is constant; therefore,

$R'(70) = 2$

The marginal revenue when 70 units are produced is $2.00.

15. $C(x) = 0.01x^2 + 1.6x + 100;\ R(x) = 3x$

a) Finding the profit function we have:

$P(x) = R(x) - C(x)$

$= 3x - (0.01x^2 + 1.6x + 100)$

$= -0.01x^2 + 1.4x - 100$

b) $\Delta P = P(x + \Delta x) - P(x)$

Substituting $x = 80$, and $\Delta x = 1$ we have

$\Delta P = P(80 + 1) - P(80)$

$= -0.01(81)^2 + 1.4(81) - 100 -$

$\left[-0.01(80)^2 + 1.4(80) - 100 \right]$

$= -0.21$

The additional profit of producing and selling the 81st unit is −$0.21.

Finding the derivative of $P(x)$ we have:

$P'(x) = -0.02x + 1.4$

Substituting 80 for x, we have:

$P'(80) = -0.02(80) + 1.4 = -0.20$

The marginal profit when 80 units are produced and sold is −$0.20.

Note: We notice that $\Delta P = \Delta R - \Delta C$ and $P'(x) = R'(x) - C'(x)$. We could have used this knowledge and our work from Exercises 11 and 14 to simplify our work. We have:

$\Delta P = \Delta R - \Delta C$

$= 3 - 3.21 = -0.21$

$P'(80) = R'(80) - C'(80)$

$= 3 - 3.20$

$= -0.20$

17. $S = 0.007p^3 - 0.5p^2 + 150p$

a) We take the derivative of the supply function with respect to price.

$\dfrac{dS}{dp} = 0.021p^2 - p + 150$

b) Substituting 25 for p in the supply function we have:

$S = 0.007(25)^3 - 0.5(25)^2 + 150(25)$

$= 109.375 - 312.50 + 3750$

$= 3546.875$

Producers will want to supply 3547 units when price is $25 per unit.

c) ✎

d) ✎

19. $M(t) = -2t^2 + 100t + 180$

a) Substituting for t, we have:

$M(5) = -2(5)^2 + 100(5) + 180 = 630$

$M(10) = -2(10)^2 + 100(10) + 180 = 980$

$M(25) = -2(25)^2 + 100(25) + 180 = 1430$

$M(45) = -2(45)^2 + 100(45) + 180 = 630$

b) To find marginal productivity, we take the derivative of the productivity function:

$M'(t) = -4t + 100$.

c) ✎

d) ✎

21. $P(x) = 567 + x(36x^{0.6} - 104)$

$= 567 + 36x^{1.6} - 104x$

x is the number of years since 1960; therefore, the year 2014 corresponds to $x = 2014 - 1960 = 54$, and the year 2015 corresponds to $x = 2015 - 1960 = 55$. To estimate the increase in gross domestic product from 2014 to 2015, we establish that $x = 54$ and $\Delta x = 1$. Next, we find the derivative of $P(x)$:

$P'(x) = 36(1.6)x^{0.6} - 104 = 57.6x^{0.6} - 104$.

Therefore,

$\Delta P \approx P'(x)\Delta x$

$\approx P'(54)\Delta x \qquad\qquad [x = 54]$

$\approx \left(57.6(54)^{0.6} - 104\right)\Delta x$

$\approx (526.7521497)(1) \qquad [\Delta x = 1]$

≈ 526.75

The gross domestic product should increase about \$526.75 billion between 2014 and 2015.

23. ✎

25. Alan's marginal tax rate is currently 25%. If he earns another \$10,000, dollars, this will push him into the 28% tax bracket and he will pay about \$0.28 per dollar earned in taxes.

27. $y = f(x) = x^3$, $x = 2$, and $\Delta x = 0.01$

$\Delta y = f(x + \Delta x) - f(x)$

$= f(2 + 0.01) - f(2) \quad \substack{\text{Substituting 2 for } x \text{ and} \\ 0.01 \text{ for } \Delta x}$

$= f(2.01) - f(2)$

$= (2.01)^3 - (2)^3$

$= 0.1206$

$f'(x)\Delta x = 3x^2 \cdot \Delta x \quad \left[f(x) = x^3; f'(x) = 3x^2\right]$

$f'(2)\Delta x = 3(2)^2(0.01) \quad \substack{\text{Substituting 2 for } x \text{ and} \\ 0.01 \text{ for } \Delta x}$

$= 12(0.01)$

$= 0.12$

29. $y = f(x) = x + x^2$, $x = 3$, and $\Delta x = 0.04$

$\Delta y = f(x + \Delta x) - f(x)$

$= f(3 + 0.04) - f(3) \quad \substack{\text{Substituting 3 for } x \text{ and} \\ 0.04 \text{ for } \Delta x}$

$= f(3.04) - f(3)$

$= \left[(3.04) + (3.04)^2\right] - \left[(3) + (3)^2\right]$

$= [12.2816] - [12]$

$= 0.2816$

$f'(x)\Delta x = (1 + 2x) \cdot \Delta x \quad \begin{bmatrix} f(x) = x + x^2; \\ f'(x) = 1 + 2x \end{bmatrix}$

$f'(3)\Delta x = [1 + 2(3)] \cdot (0.04) \quad \substack{\text{Substituting 3 for } x \text{ and} \\ 0.04 \text{ for } \Delta x}$

$= [7](0.04)$

$= 0.28$

31. $y = f(x) = \dfrac{1}{x} = x^{-1}$, $x = 1$, and $\Delta x = 0.2$

$\Delta y = f(x + \Delta x) - f(x)$

$= f(1 + 0.2) - f(1) \quad \substack{\text{Substituting 1 for } x \text{ and} \\ 0.2 \text{ for } \Delta x}$

$= f(1.2) - f(1)$

$= \left[\dfrac{1}{1.2}\right] - \left[\dfrac{1}{1}\right]$

$= -0.1667$

$f'(x)\Delta x = \left(-x^{-2}\right) \cdot \Delta x \quad \begin{bmatrix} f(x) = x^{-1}; \\ f'(x) = -x^{-2} \end{bmatrix}$

$f'(1)\Delta x = \left(-(1)^{-2}\right)(0.2) \quad \substack{\text{Substituting 1 for } x \text{ and} \\ 0.2 \text{ for } \Delta x}$

$= -1(0.2)$

$= -0.2$

33. $y = f(x) = 3x - 1$, $x = 4$, and $\Delta x = 2$

$\Delta y = f(x + \Delta x) - f(x)$

$= f(4 + 2) - f(4) \quad \substack{\text{Substituting 4 for } x \text{ and} \\ 2 \text{ for } \Delta x}$

$= f(6) - f(4)$

$= [3(6) - 1] - [3(4) - 1]$

$= [17] - [11]$

$= 6$

$f'(x)\Delta x = (3) \cdot \Delta x \quad \left[f(x) = 3x - 1; f'(x) = 3\right]$

$f'(4)\Delta x = (3) \cdot (2) \quad \substack{\text{Substituting 4 for } x \text{ and} \\ 2 \text{ for } \Delta x}$

$= 6$

35. We first think of the number closest to 26 that is a perfect square. This is 25. What we will do is approximate how $y = \sqrt{x}$, changes when x changes from 25 to 26. Let

$$y = f(x) = \sqrt{x} = x^{1/2}$$

Then $f'(x) = \dfrac{1}{2}x^{-1/2} = \dfrac{1}{2\sqrt{x}}$

Using, $\Delta y \approx f'(x)\Delta x$, we have

$$\Delta y \approx f'(x)\Delta x$$

$$\approx \frac{1}{2\sqrt{x}} \cdot \Delta x$$

We are interested in Δy as x changes from 25 to 26, so

$$\Delta y \approx \frac{1}{2\sqrt{x}} \cdot \Delta x$$

$$\approx \frac{1}{2\sqrt{25}} \cdot 1 \quad \text{Replacing } x \text{ with 25 and } \Delta x \text{ with 1}$$

$$\approx \frac{1}{2 \cdot 5}$$

$$\approx \frac{1}{10} = 0.100$$

We can now approximate $\sqrt{26}$;

$$\sqrt{26} = \sqrt{25} + \Delta y$$

$$= 5 + \Delta y$$

$$\approx 5 + 0.100$$

$$\approx 5.100$$

To five decimal places $\sqrt{26} = 5.09902$. Thus, our approximation is reasonably accurate.

37. We first think of the number closest to 102 that is a perfect square. This is 100. What we will do is approximate how $y = \sqrt{x}$, changes when x changes from 100 to 102. Let

$$y = f(x) = \sqrt{x} = x^{1/2}$$

Then $f'(x) = \dfrac{1}{2}x^{-1/2} = \dfrac{1}{2\sqrt{x}}$

Using, $\Delta y \approx f'(x)\Delta x$, we have

$$\Delta y \approx f'(x)\Delta x$$

$$\approx \frac{1}{2\sqrt{x}} \cdot \Delta x$$

We are interested in Δy as x changes from 100 to 102, so

$$\Delta y \approx \frac{1}{2\sqrt{x}} \cdot \Delta x$$

$$\approx \frac{1}{2\sqrt{100}} \cdot 2 \quad \text{Replacing } x \text{ with 100 and } \Delta x \text{ with 2}$$

$$\approx \frac{1}{2 \cdot 10} \cdot 2$$

$$\approx \frac{1}{10} = 0.100$$

We can now approximate $\sqrt{102}$;

$$\sqrt{102} = \sqrt{100} + \Delta y$$

$$= 10 + \Delta y$$

$$\approx 10 + 0.100$$

$$\approx 10.100$$

To five decimal places $\sqrt{102} = 10.09950$. Thus, our approximation is reasonably accurate.

39. We first think of the number closest to 1005 that is a perfect cube. This is 1000. What we will do is approximate how $y = \sqrt[3]{x}$, changes when x changes from 1000 to 1005. Let

$$y = f(x) = \sqrt[3]{x} = x^{1/3}$$

Then $f'(x) = \dfrac{1}{3}x^{-2/3} = \dfrac{1}{3\sqrt[3]{x^2}}$

Using, $\Delta y \approx f'(x)\Delta x$, we have

$$\Delta y \approx f'(x)\Delta x$$

$$\approx \frac{1}{3\sqrt[3]{x^2}} \cdot \Delta x$$

We are interested in Δy as x changes from 1000 to 1005, so

$$\Delta y \approx \frac{1}{3\sqrt[3]{x^2}} \cdot \Delta x$$

Replacing x with 1000 and Δx with 5, we have

$$\approx \frac{1}{3 \cdot \sqrt[3]{(1000)^2}} \cdot 5$$

$$\approx \frac{1}{3 \cdot 100} \cdot 5$$

$$\approx \frac{1}{60} = 0.017$$

The solution is continued on the next page.

Using the information from the previous page, we can now approximate $\sqrt[3]{1005}$;

$$\sqrt[3]{1005} = \sqrt[3]{1000} + \Delta y$$
$$= 10 + \Delta y$$
$$\approx 10 + 0.017$$
$$\approx 10.017$$

To five decimal places $\sqrt[3]{1005} = 10.01664$ Thus, our approximation is reasonably accurate.

41. $y = \sqrt{3x-2} = (3x-2)^{1/2}$

First, we find $\dfrac{dy}{dx}$:

$$\frac{dy}{dx} = \frac{1}{2}(3x-2)^{-1/2}(3) = \frac{3}{2\sqrt{3x-2}}.$$

Then

$$dy = \frac{3}{2\sqrt{3x-2}}\,dx\,.$$

Note that the expression for dy contains two variables x and dx.

43. $y = \left(2x^3 + 1\right)^{3/2}$

First, we find $\dfrac{dy}{dx}$:

$$\frac{dy}{dx} = \frac{3}{2}\left(2x^3 + 1\right)^{1/2}\left(6x^2\right) \text{ By the extended power rule}$$

$$= 9x^2\left(2x^3 + 1\right)^{1/2}$$

$$= 9x^2\sqrt{2x^3 + 1}.$$

Then

$$dy = 9x^2\sqrt{2x^3 + 1}\,dx\,.$$

Note that the expression for dy contains two variables x and dx.

45. $y = \dfrac{x^3 + x + 2}{x^2 + 3}$

First, we find $\dfrac{dy}{dx}$. By the quotient rule we have:

$$\frac{dy}{dx} = \frac{\left(x^2+3\right)\left(3x^2+1\right) - \left(x^3+x+2\right)(2x)}{\left(x^2+3\right)^2}$$

$$= \frac{\left(3x^4 + 10x^2 + 3\right) - \left(2x^4 + 2x^2 + 4x\right)}{\left(x^2+3\right)^2}.$$

$$= \frac{x^4 + 8x^2 - 4x + 3}{\left(x^2+3\right)^2}.$$

Then

$$dy = \frac{x^4 + 8x^2 - 4x + 3}{\left(x^2+3\right)^2}\,dx\,.$$

Note that the expression for dy contains two variables x and dx.

47. $y = x^4 - 2x^3 + 5x^2 + 3x - 4$

First, we find $\dfrac{dy}{dx}$:

$$\frac{dy}{dx} = 4x^3 - 6x^2 + 10x + 3\,.$$

Then

$$dy = \left(4x^3 - 6x^2 + 10x + 3\right)dx\,.$$

Note that the expression for dy contains two variables x and dx.

49.   From Exercise 48, we know:

$$dy = -8(7-x)^7\,dx\,,$$

when $x = 1$ and $dx = 0.01$, we have:

$$dy = -8(7-1)^7(0.01)$$

$$= -8(6)^7(0.01)$$

$$= -22{,}394.88.$$

51. $y = (3x - 10)^5$

First, we find $\dfrac{dy}{dx}$:

$$\frac{dy}{dx} = 5(3x-10)^4(3) \text{ By the extended power rule}$$

$$= 15(3x-10)^4$$

Then

$$dy = 15(3x-10)^4\,dx\,.$$

When $x = 4$ and $dx = 0.03$ we have:

$$dy = 15\left(3(4) - 10\right)^4(0.03)$$

$$= 15(2)^4(0.03)$$

$$= 7.2$$

53.   Let $y = f(x) = x^4 - x^2 + 8$

First we find $f'(x)$:

$$f'(x) = 4x^3 - 2x\,.$$

Then

$$dy = f'(x)\,dx$$

$$= \left(4x^3 - 2x\right)dx$$

The solution is continued on the next page.

To approximate $f(5.1)$, we will use the information on the previous page, $x = 5$, and $dx = 0.1$ to determine the differential dy.

Substituting 5 for x and 0.1 for dx we have:

$$dy = f'(5)\,dx$$

$$= \left(4(5)^3 - 2(5)\right)(0.1)$$

$$= \left(4(125) - 10\right)(0.1)$$

$$= (500 - 10)(0.1)$$

$$= (490)(0.1)$$

$$= 49$$

Next, we find

$$f(5) = (5)^4 - (5)^2 + 8$$

$$= 625 - 25 + 8$$

$$= 608$$

Now,

$$f(5.1) \approx f(5) + f'(5)\,dx$$

$$\approx 608 + 49$$

$$\approx 657$$

55. $S = 0.02235 h^{0.42246} w^{0.51456}$

We begin by noticing that we are wanting to estimate the change in surface area due to a change in weight w; therefore, we will first find $\dfrac{dS}{dw}$. Since $h = 160$, we have:

$$S = 0.02235(160)^{0.42246} w^{0.51456}$$

$$= 0.02235(8.53399783) w^{0.51456}$$

$$= 0.19073485 w^{0.51456}$$

Now we can take the derivative of S with respect to w.

$$\frac{dS}{dw} = 0.19073485(0.51456) w^{-0.48544}$$

$$= 0.09814452 w^{-0.48544}$$

Therefore,

$$dS = \left(0.09814452 w^{-0.48544}\right) dw$$

Now that we have the differential, we can use her weight of 60 kg to approximate how much her surface area changes when her weight drops 1 kg.

We substitute 60 for w and -1 for dw to get:

$$dS \approx \left(0.09814452(60)^{-0.48544}\right)(-1)$$

$$\approx -0.01345$$

The patient's surface area will change by -0.01345 m^2.

57. $p(x) = 0.06x^3 - 0.5x^2 + 1.64x + 24.76$

First we find $p'(x)$.

$$p'(x) = 0.06(3x^2) - 0.5(2x) + 1.64$$

$$= 0.18x^2 - x + 1.64.$$

The differential is:

$$dp = p'(x)\,dx$$

$$= \left(0.18x^2 - x + 1.64\right)dx$$

Since x is the number of years since 2008, we have 2010 implies $x = 2$ and 2012 implies $x = 4$. To estimate the change in ticket prices from 2010 and 2012, we substitute 2 for x and 2 for dx.

$$dp = \left(0.18(2)^2 - (2) + 1.64\right)(2)$$

$$= (0.36)(2) = 0.72$$

To estimate the change in ticket prices from 2014 and 2016, we substitute 6 for x and 2 for dx.

$$dp = \left(0.18(6)^2 - (6) + 1.64\right)(2)$$

$$= (2.12)(2) = 4.24$$

Ticket prices will increase more between 2014 and 2016.

59. $A(x) = \dfrac{C(x)}{x}$

To find Marginal Average Cost, we take the derivative of the average cost function. By the quotient rule we have:

$$A'(x) = \frac{x \cdot C'(x) - C(x)(1)}{x^2}$$

$$= \frac{x \cdot C'(x) - C(x)}{x^2}.$$

61. The volume of the spherical cavern is given by:

$$V(r) = \frac{4}{3}\pi r^3$$

First we find the derivatve of the volume function.

$$V'(r) = \frac{4}{3}\pi(3r^2) = 4\pi r^2.$$

Therefore, the differential is:

$$dV = 4\pi r^2\,dr$$

Substituting the given information, we have:

$$dV = 4(3.14)(400)^2(2) \approx 4,019,200.$$

The enlarged cavern will contain an additional 4,019,200 cubic feet.

63. $p = 400 - x$

Since revenue is price times quantity, the revenue function is given by:

$R(x) = p \cdot x$

$\qquad = (400 - x)x$

$\qquad = 400x - x^2$

To find the marginal revenue, we take the derivative of the revenue function. Thus:

$R'(x) = 400 - 2x$

65. $p = \dfrac{4000}{x} + 3$

Since revenue is price times quantity, the revenue function is given by:

$R(x) = p \cdot x$

$\qquad = \left(\dfrac{4000}{x} + 3\right)x$

$\qquad = 4000 + 3x$

To find the marginal revenue, we take the derivative of revenue function. Thus:

$R'(x) = 3$

67.   ✎

Exercise Set 2.7

1. a) The demand function is
    $q = D(x) = 400 - x$.
    The definition of the elasticity of demand is

    given by: $E(x) = -\dfrac{x \cdot D'(x)}{D(x)}$. In order to

    find the elasticity of demand, we need to find the derivative of the demand function first.

    $\dfrac{dq}{dx} = D'(x) = \dfrac{d}{dx}(400 - x) = -1$.

    Next, we substitute -1 for $D'(x)$, and $400 - x$ for $D(x)$ into the expression for elasticity.

    $E(x) = -\dfrac{x \cdot (-1)}{400 - x} = \dfrac{x}{400 - x}$

    b) Substituting $x = 125$ into the expression found in part (a) we have:

    $E(125) = \dfrac{(125)}{400 - (125)} = \dfrac{125}{275} = \dfrac{5}{11}$

    Since $E(125) = \dfrac{5}{11}$ is less than one, the demand is inelastic.

    c) The values of x for which $E(x) = 1$ will maximize total revenue. We solve:

    $E(x) = 1$

    $\dfrac{x}{400 - x} = 1$

    $x = 400 - x$

    $2x = 400$

    $x = 200$

    A price of $200 will maximize total revenue.

3. $q = D(x) = 200 - 4x;\ x = 46$

    a) $D'(x) = -4$

    $E(x) = -\dfrac{x \cdot D'(x)}{D(x)} = -\dfrac{x \cdot (-4)}{200 - 4x} = \dfrac{4x}{200 - 4x}$

    $= \dfrac{x}{50 - x}$

    b) Substituting $x = 46$ into the expression found in part (a) we have:

    $E(46) = \dfrac{46}{50 - 46} = \dfrac{46}{4} = \dfrac{23}{2} = 11.5$

    Since $E(46) > 1$, demand is elastic.

c) We solve $E(x) = 1$

    $\dfrac{x}{50 - x} = 1$

    $x = 50 - x$

    $2x = 50$

    $x = 25$

    A price of $25 will maximize total revenue.

5. $q = D(x) = \dfrac{400}{x};\ x = 50$

    a) First, we rewrite the demand function.

    $D(x) = \dfrac{400}{x} = 400x^{-1}$.

    Next, we take the derivative of the demand function, using the Power Rule.

    $D'(x) = 400(-1)\,x^{-2} = -400x^{-2}$

    Making the appropriate substitutions into the elasticity function, we have

    $E(x) = -\dfrac{x \cdot D'(x)}{D(x)} = -\dfrac{x \cdot \left(-400x^{-2}\right)}{400x^{-1}} =$

    $= \dfrac{400x^{-1}}{400x^{-1}} = 1$

    Therefore, $E(x) = 1$ for all values of x.

    b) $E(50) = 1$, so demand is unit elastic.

    c) $E(x) = 1$ for all values of x. Therefore, total revenue is maximized for all values of x. In other words, total revenue is the same regardless of the price.

7. $q = D(x) = \sqrt{600 - x};\ x = 100$

    a) First rewrite the demand function:

    $D(x) = (600 - x)^{\frac{1}{2}}$.

    Next, we take the derivative of the demand function, using the Chain Rule:

    $D'(x) = \dfrac{1}{2}(600 - x)^{-\frac{1}{2}} \cdot \dfrac{d}{dx}(600 - x)$

    $= \dfrac{-1}{2\sqrt{600 - x}}$

    The solution is continued on the next page.

Making the appropriate substitutions into the elasticity function, we have

$$E(x) = -\frac{x \cdot D'(x)}{D(x)} = -\frac{x \cdot \left(\dfrac{-1}{2\sqrt{600-x}}\right)}{\sqrt{600-x}} =$$

$$= \frac{\dfrac{x}{2\sqrt{600-x}}}{\sqrt{600-x}} = \frac{x}{2(600-x)}$$

$$= \frac{x}{1200-2x}$$

b) Substituting $x = 100$ into the expression found in part (a) we have:

$$E(100) = \frac{100}{1200 - 2(100)} = \frac{100}{1000} = \frac{1}{10}$$

Since $E(100) < 1$, demand is inelastic.

c) Solve $E(x) = 1$

$$\frac{x}{1200-2x} = 1$$
$$x = 1200 - 2x$$
$$3x = 1200$$
$$x = 400$$

A price of $400 will maximize total revenue.

9. $q = D(x) = \dfrac{100}{(x+3)^2}$; $x = 1$

a) First, we rewrite the demand function:

$$D(x) = 100(x+3)^{-2}.$$

Next, we take the derivative of the demand function, using the Chain Rule:

$$D'(x) = 100(-2)(x+3)^{-3}\left(\frac{d}{dx}(x+3)\right)$$
$$= -200(x+3)^{-3}$$
$$= -\frac{200}{(x+3)^3}$$

Making the appropriate substitutions into the elasticity function, we have:

$$E(x) = -\frac{x \cdot D'(x)}{D(x)} = -\frac{x \cdot \left(-\dfrac{200}{(x+3)^3}\right)}{\dfrac{100}{(x+3)^2}}$$

$$= x \cdot \left(\frac{200}{(x+3)^3}\right)\frac{(x+3)^2}{100}$$

$$= \frac{2x}{x+3}$$

b) Substituting $x = 1$ into the expression found in part (a) we have:

$$E(1) = \frac{2 \cdot (1)}{1+3} = \frac{2}{4} = \frac{1}{2}$$

Since $E(1) < 1$, demand is inelastic.

c) Solve $E(x) = 1$

$$\frac{2x}{x+3} = 1$$
$$2x = x+3$$
$$x = 3$$

A price of $3 will maximize total revenue.

11. $q = D(x) = 50,000 + 300x - 3x^2$

a) $D'(x) = 300 - 6x$

$$E(x) = -\frac{x \cdot D'(x)}{D(x)}$$

$$= -\frac{x \cdot (300 - 6x)}{50,000 + 300x - 3x^2}$$

$$= -\frac{300x - 6x^2}{50,000 + 300x - 3x^2}$$

$$= \frac{6x^2 - 300x}{50,000 + 300x - 3x^2}$$

b) Substituting in to the elasticity of demand we have:

$$E(75) = \frac{6(75)^2 - 300(75)}{50,000 + 300(75) - 3(75)^2}$$

$$= \frac{11,250}{55,625}$$

$$\approx 0.20.$$

Since $E(75) < 1$, the demand for oil is inelastic at $75 a barrel.

c) Substituting in to the elasticity of demand we have:

$$E(100) = \frac{6(100)^2 - 300(100)}{50,000 + 300(100) - 3(100)^2}$$

$$= \frac{30,000}{50,000}$$

$$\approx 0.6.$$

Since $E(100) < 1$, the demand for oil is inelastic at $100 a barrel.

d) Substituting in to the elasticity of demand we have:

$$E(125) = \frac{6(125)^2 - 300(125)}{50,000 + 300(125) - 3(125)^2}$$

$$= \frac{56,250}{40,625}$$

$$\approx 1.38.$$

Since $E(125) > 1$, the demand for oil is elastic at $125 a barrel.

e) Revenue will be maximized when $E(x) = 1$. Therefore, we solve:

$$\frac{6x^2 - 300x}{50,000 + 300x - 3x^2} = 1$$

$$6x^2 - 300x = 50,000 + 300x - 3x^2$$

$$9x^2 - 600x - 50,000 = 0$$

Using the quadratic formula, we find that the solutions to the equation.

$$x = \frac{-(-600) \pm \sqrt{(-600)^2 - 4(9)(-50000)}}{2(9)}$$

$$x = \frac{600 \pm \sqrt{2,160,000}}{18}$$

$$x \approx -48.316 \quad \text{or} \quad x \approx 114.983.$$

The only solution that is feasible is $x \approx 114.98$. Thus, oil revenues will be maximized when price is $114.98 a barrel.

f) Substituting the answer found in part (e) into the demand function we have:

$$D(114.98) = 50,000 + 300(114.98) - 3(114.98)^2$$

$$= 44,832.80.$$

The demand for oil is about 44,832 million barrels per day or 44.8 billion barrels per day at a price of $114.98 a barrel. At the time this solution was created, the price of oil was $100.83 per barrel. Thus according to this model, by increasing the price, oil producers could increase revenue.

g) The demand for oil is inelastic at $110 a barrel. Therefore an increase in price will result in an increase in total revenue.

13. $q = D(x) = \sqrt{200 - x^3}$

a) First, we rewrite the demand function to make it easier to find the derivative.

$$D(x) = (200 - x^3)^{\frac{1}{2}}.$$

Next, using the Chain Rule, we have:

$$D'(x) = \frac{1}{2}(200 - x^3)^{-\frac{1}{2}}(-3x^2)$$

$$= \frac{-3x^2}{2\sqrt{200 - x^3}}$$

Now, substituting in the elasticity function we get:

$$E(x) = -\frac{x \cdot D'(x)}{D(x)} = -\frac{x \cdot \left(\dfrac{-3x^2}{2\sqrt{200 - x^3}}\right)}{\sqrt{200 - x^3}}$$

$$= \frac{3x^3}{2\left(\sqrt{200 - x^3}\right)^2}$$

$$= \frac{3x^3}{2(200 - x^3)}$$

$$= \frac{3x^3}{400 - 2x^3}$$

b) $E(3) = \dfrac{3(3)^3}{400 - 2(3)^3}$

$$= \frac{81}{346}$$

$$\approx 0.2341$$

Since $E(3) < 1$, the demand for computer games is inelastic when price is $3.

c) From part (b) we know that the demand for computer games is inelastic at a price of $3. Therefore an increase in the price of computer games will lead to an increase in the total revenue.

15. $q = D(x) = 180 - 10x$

a) First we find the derivative of the demand function.

$$D'(x) = -10$$

Therefore, the elasticity of demand is:

$$E(x) = -\frac{x \cdot D'(x)}{D(x)}$$

$$= -\frac{x \cdot (-10)}{180 - 10x}$$

$$= -\frac{-10x}{180 - 10x}$$

$$= \frac{10x}{180 - 10x}$$

b) Substituting in to the elasticity of demand
we have:

$$E(8) = \frac{10(8)}{180 - 10(8)}$$

$$= \frac{80}{100}$$

$$= 0.8.$$

c) Revenue will be maximized when $E(x) = 1$.

Therefore, we solve:

$$\frac{10x}{180 - 10x} = 1$$

$$10x = 180 - 10x$$

$$20x = 180$$

$$x = 9.$$

When the price of sunglass cases is $9 a
case, revenue will be maximized.

d) ✎

17. ✎

Exercise Set 2.8

1. Differentiate implicitly to find $\dfrac{dy}{dx}$.

    We have $3x^3 - y^2 = 8$.

    Differentiating both sides with respect to x yields:

    $$\frac{d}{dx}\left(3x^3 - y^2\right) = \frac{d}{dx}(8)$$

    $$\frac{d}{dx}3x^3 - \frac{d}{dx}y^2 = \frac{d}{dx}8$$

    $$9x^2 - 2y \cdot \frac{dy}{dx} = 0 \qquad \text{Next, we isolate } \frac{dy}{dx}.$$

    $$-2y\frac{dy}{dx} = -9x^2$$

    $$\frac{dy}{dx} = \frac{9x^2}{2y}$$

    Find the slope of the tangent line to the curve at $(2,4)$.

    Replacing x with 2 and y with 4 , we have:

    $$\frac{dy}{dx} = \frac{9x^2}{2y} = \frac{9(2)^2}{2(4)} = \frac{36}{8} = \frac{9}{2}.$$

    The slope of the tangent line to the curve at $(2,4)$ is $\dfrac{9}{2}$.

3. $2x^3 + 4y^2 = -12$

    Differentiating both sides with respect to x yields:

    $$\frac{d}{dx}\left(2x^3 + 4y^2\right) = \frac{d}{dx}(-12)$$

    $$6x^2 + 8y \cdot \frac{dy}{dx} = 0 \qquad \text{Next, we isolate } \frac{dy}{dx}.$$

    $$8y \cdot \frac{dy}{dx} = -6x^2$$

    $$\frac{dy}{dx} = \frac{-6x^2}{8y}$$

    $$\frac{dy}{dx} = -\frac{3x^2}{4y}$$

    Find the slope of the tangent line to the curve at $(-2,-1)$.

Replacing x with -2 and y with -1 , we have:

$$\frac{dy}{dx} = -\frac{3x^2}{4y} = \frac{-3(-2)^2}{4(-1)} = \frac{-12}{-4} = 3.$$

The slope of the tangent line to the curve at $(-2,-1)$ is 3.

5. $x^2 + y^2 = 1$

    Differentiating both sides with respect to x yields:

    $$\frac{d}{dx}\left(x^2 + y^2\right) = \frac{d}{dx}(1)$$

    $$2x + 2y \cdot \frac{dy}{dx} = 0 \qquad \text{Next, we isolate } \frac{dy}{dx}.$$

    $$2y \cdot \frac{dy}{dx} = -2x$$

    $$\frac{dy}{dx} = \frac{-2x}{2y}$$

    $$\frac{dy}{dx} = -\frac{x}{y}$$

    Find the slope of the tangent line to the curve at $\left(\dfrac{1}{2}, \dfrac{\sqrt{3}}{2}\right)$.

    Replacing x with $\dfrac{1}{2}$ and y with $\dfrac{\sqrt{3}}{2}$, we have:

    $$\frac{dy}{dx} = -\frac{x}{y} = -\frac{\dfrac{1}{2}}{\dfrac{\sqrt{3}}{2}} = \frac{-1}{\sqrt{3}}.$$

    The slope of the tangent line to the curve at $\left(\dfrac{1}{2}, \dfrac{\sqrt{3}}{2}\right)$ is $-\dfrac{1}{\sqrt{3}}$.

7. $2x^3 y^2 = -18$

Differentiating both sides with respect to x yields:

$$\frac{d}{dx}\left(2x^3 y^2\right) = \frac{d}{dx}(-18)$$

$$2x^3 \frac{d}{dx} y^2 + y^2 \frac{d}{dx}\left(2x^3\right) = 0 \qquad \text{Product Rule}$$

$$2x^3 \left(2y \cdot \frac{dy}{dx}\right) + y^2 \left(6x^2\right) = 0$$

$$4x^3 y \cdot \frac{dy}{dx} + 6x^2 y^2 = 0 \qquad \text{Next, we isolate } \frac{dy}{dx}.$$

$$4x^3 y \cdot \frac{dy}{dx} = -6x^2 y^2$$

$$\frac{dy}{dx} = \frac{-6x^2 y^2}{4x^3 y}$$

$$\frac{dy}{dx} = -\frac{3y}{2x}$$

Find the slope of the tangent line to the curve at $(-1,3)$.

Replacing x with -1 and y with 3, we have:

$$\frac{dy}{dx} = -\frac{3y}{2x} = -\frac{3(3)}{2(-1)} = \frac{9}{2}$$

The slope of the tangent line to the curve at $(-1,3)$ is $\frac{9}{2}$.

9. $x^4 - x^2 y^3 = 12$

Differentiating both sides with respect to x yields:

$$\frac{d}{dx}\left(x^4 - x^2 y^3\right) = \frac{d}{dx}(12)$$

$$\frac{d}{dx}\left(x^4\right) - \frac{d}{dx}\left(x^2 y^3\right) = \frac{d}{dx}(12)$$

$$4x^3 - \left[x^2\left(3y^2 \cdot \frac{dy}{dx}\right) + y^3 (2x)\right] = 0 \qquad \text{Product Rule}$$

$$4x^3 - 3x^2 y^2 \cdot \frac{dy}{dx} - 2xy^3 = 0$$

$$\text{Next, we isolate } \frac{dy}{dx}.$$

$$-3x^2 y^2 \cdot \frac{dy}{dx} = 2xy^3 - 4x^3$$

$$\frac{dy}{dx} = \frac{2xy^3 - 4x^3}{-3x^2 y^2}$$

$$\frac{dy}{dx} = \frac{4x^2 - 2y^3}{3xy^2}$$

Find the slope of the tangent line to the curve at $(-2,1)$.

Replacing x with -2 and y with 1, we have:

$$\frac{dy}{dx} = \frac{4x^2 - 2y^3}{3xy^2} = \frac{4(-2)^2 - 2(1)^3}{3(-2)(1)^2} = -\frac{7}{3}$$

The slope of the tangent line to the curve at $(-2,1)$ is $-\frac{7}{3}$.

11. $xy + y^2 - 2x = 0$

Differentiating both sides with respect to x yields:

$$\frac{d}{dx}\left(xy + y^2 - 2x\right) = \frac{d}{dx}(0)$$

$$\frac{d}{dx}(xy) + \frac{d}{dx}\left(y^2\right) - \frac{d}{dx}(2x) = \frac{d}{dx}(0)$$

$$\left[x\left(\frac{dy}{dx}\right) + y(1)\right] + 2y \cdot \frac{dy}{dx} - 2(1) = 0$$

$$x \cdot \frac{dy}{dx} + y + 2y \cdot \frac{dy}{dx} - 2 = 0$$

$$x \cdot \frac{dy}{dx} + 2y \cdot \frac{dy}{dx} = 2 - y$$

$$(x + 2y) \cdot \frac{dy}{dx} = 2 - y$$

$$\frac{dy}{dx} = \frac{2 - y}{x + 2y}$$

Find the slope of the tangent line to the curve at $(1,-2)$.

Replacing x with 1 and y with -2, we have:

$$\frac{dy}{dx} = \frac{2 - y}{x + 2y} = \frac{2 - (-2)}{(1) + 2(-2)} = \frac{4}{-3} = -\frac{4}{3}$$

The slope of the tangent line to the curve at $(1,-2)$ is $-\frac{4}{3}$.

13. $4x^3 - y^4 - 3y + 5x + 1 = 0$

Differentiating both sides with respect to x yields:

$$\frac{d}{dx}\left(4x^3 - y^4 - 3y + 5x + 1\right) = \frac{d}{dx}(0)$$

$$\frac{d}{dx}4x^3 - \frac{d}{dx}y^4 - \frac{d}{dx}3y + \frac{d}{dx}5x + \frac{d}{dx}1 = 0$$

$$12x^2 - 4y^3 \cdot \frac{dy}{dx} - 3 \cdot \frac{dy}{dx} + 5 + 0 = 0$$

$$-4y^3 \cdot \frac{dy}{dx} - 3 \cdot \frac{dy}{dx} = -12x^2 - 5$$

$$\left(-4y^3 - 3\right) \cdot \frac{dy}{dx} = -12x^2 - 5$$

$$\frac{dy}{dx} = \frac{-12x^2 - 5}{-4y^3 - 3}$$

$$\frac{dy}{dx} = \frac{12x^2 + 5}{4y^3 + 3}$$

Find the slope of the tangent line to the curve at $(1, -2)$.

Replacing x with 1 and y with -2, we have:

$$\frac{dy}{dx} = \frac{12x^2 + 5}{4y^3 + 3} = \frac{12(1)^2 + 5}{4(-2)^3 + 3} = -\frac{17}{29}$$

The slope of the tangent line to the curve at $(1, -2)$ is $-\dfrac{17}{29}$.

15. $x^2 + 2xy = 3y^2$

Differentiating both sides with respect to x yields:

$$\frac{d}{dx}\left(x^2 + 2xy\right) = \frac{d}{dx}\left(3y^2\right)$$

$$\frac{d}{dx}x^2 + \frac{d}{dx}2xy = \frac{d}{dx}3y^2$$

$$2x + 2\left[x\left(\frac{dy}{dx}\right) + y(1)\right] = 3\left(2y \cdot \frac{dy}{dx}\right)$$

$$2x + 2x \cdot \frac{dy}{dx} + 2y = 6y \cdot \frac{dy}{dx}$$

$$2x \cdot \frac{dy}{dx} - 6y \cdot \frac{dy}{dx} = -2x - 2y$$

$$(2x - 6y) \cdot \frac{dy}{dx} = -2(x + y)$$

$$\frac{dy}{dx} = \frac{-2(x + y)}{2x - 6y}$$

$$\frac{dy}{dx} = \frac{-2(x + y)}{-2(-x + 3y)}$$

$$\frac{dy}{dx} = \frac{x + y}{3y - x}$$

17. $x^2 - y^2 = 16$

Differentiating both sides with respect to x yields:

$$\frac{d}{dx}\left(x^2 - y^2\right) = \frac{d}{dx}(16)$$

$$\frac{d}{dx}x^2 - \frac{d}{dx}y^2 = \frac{d}{dx}(16)$$

$$2x - 2y \cdot \frac{dy}{dx} = 0$$

$$-2y \cdot \frac{dy}{dx} = -2x$$

$$\frac{dy}{dx} = \frac{-2x}{-2y}$$

$$\frac{dy}{dx} = \frac{x}{y}$$

19. $y^3 = x^5$

Differentiating both sides with respect to x yields:

$$\frac{d}{dx}\left(y^3\right) = \frac{d}{dx}\left(x^5\right)$$

$$3y^2 \cdot \frac{dy}{dx} = 5x^4$$

$$\frac{dy}{dx} = \frac{5x^4}{3y^2}$$

21. $x^2 y^3 + x^3 y^4 = 11$

Differentiating both sides with respect to x yields:

$$\frac{d}{dx}\left(x^2 y^3 + x^3 y^4\right) = \frac{d}{dx}(11)$$

$$\frac{d}{dx}\left(x^2 y^3\right) + \frac{d}{dx}\left(x^3 y^4\right) = 0$$

Notice:

$$\frac{d}{dx}\left(x^2 y^3\right) = x^2 \left(3y^2 \cdot \frac{dy}{dx}\right) + y^3(2x)$$

$$= 3x^2 y^2 \cdot \frac{dy}{dx} + 2xy^3$$

and

$$\frac{d}{dx}\left(x^3 y^4\right) = x^3 \left(4y^3 \cdot \frac{dy}{dx}\right) + y^4\left(3x^2\right)$$

$$= 4x^3 y^3 \cdot \frac{dy}{dx} + 3x^2 y^4$$

Therefore,

$$\frac{d}{dx}\left(x^2 y^3\right) + \frac{d}{dx}\left(x^3 y^4\right) = 0$$

$$3x^2 y^2 \cdot \frac{dy}{dx} + 2xy^3 + 4x^3 y^3 \cdot \frac{dy}{dx} + 3x^2 y^4 = 0$$

$$\left(4x^3 y^3 + 3x^2 y^2\right) \cdot \frac{dy}{dx} = -3x^2 y^4 - 2xy^3$$

$$\frac{dy}{dx} = \frac{-3x^2 y^4 - 2xy^3}{4x^3 y^3 + 3x^2 y^2}$$

$$\frac{dy}{dx} = \frac{xy^2\left(-3xy^2 - 2y\right)}{xy^2\left(4x^2 y + 3x\right)}$$

$$\frac{dy}{dx} = -\frac{2y + 3xy^2}{4x^2 y + 3x}$$

23. $p^2 + p + 2x = 40$

Differentiating both sides with respect to x yields:

$$\frac{d}{dx}\left(p^2 + p + 2x\right) = \frac{d}{dx}(40)$$

$$2p \cdot \frac{dp}{dx} + \frac{dp}{dx} + 2 \cdot 1 = 0$$

$$(2p + 1) \cdot \frac{dp}{dx} = -2$$

$$\frac{dp}{dx} = \frac{-2}{2p + 1}$$

25. $xp^3 = 24$

Differentiating both sides with respect to x yields:

$$\frac{d}{dx}\left(xp^3\right) = \frac{d}{dx}(24)$$

$$x\left(3p^2 \cdot \frac{dp}{dx}\right) + p^3(1) = 0 \qquad \text{Product Rule}$$

$$3xp^2 \cdot \frac{dp}{dx} = -p^3$$

$$\frac{dp}{dx} = \frac{-p^3}{3xp^2}$$

$$\frac{dp}{dx} = -\frac{p}{3x}$$

27. $\dfrac{x^2 p + xp + 1}{2x + p} = 1$

Multiply both sides by $2x + p$ to clear the fraction.

$$(2x + p)\left(\frac{x^2 p + xp + 1}{2x + p}\right) = (1)(2x + p)$$

$$x^2 p + xp + 1 = 2x + p$$

Differentiating both sides of the equation with respect to x yields:

$$\frac{d}{dx}\left(x^2 p + xp + 1\right) = \frac{d}{dx}(2x + p)$$

$$\frac{d}{dx}x^2 p + \frac{d}{dx}xp + \frac{d}{dx}1 = \frac{d}{dx}2x + \frac{d}{dx}p$$

$$x^2 \cdot \frac{dp}{dx} + p \cdot 2x + x \cdot \frac{dp}{dx} + p \cdot 1 + 0 = 2 + \frac{dp}{dx}$$

$$x^2 \cdot \frac{dp}{dx} + x \cdot \frac{dp}{dx} - 1 \cdot \frac{dp}{dx} = 2 - 2xp - p$$

$$\left(x^2 + x - 1\right)\frac{dp}{dx} = 2 - 2xp - p$$

$$\frac{dp}{dx} = \frac{2 - 2xp - p}{x^2 + x - 1}$$

29. $(p + 4)(x + 3) = 48$

Expanding the left hand side of the equation we have:

$$px + 3p + 4x + 12 = 48$$

$$px + 3p + 4x = 36$$

The solution is continued on the next page.

Differentiating both sides of the equation on the previous page with respect to x yields:

$$\frac{d}{dx}\left(px+3p+4x\right)=\frac{d}{dx}(36)$$

$$p\cdot 1+x\cdot\frac{dp}{dx}+3\cdot\frac{dp}{dx}+4\cdot 1=0$$

$$(x+3)\cdot\frac{dp}{dx}=-p-4$$

$$\frac{dp}{dx}=\frac{-p-4}{x+3}$$

31. $G^2+H^2=25$

We differentiate both sides with respect to t.

$$\frac{d}{dt}\left(G^2+H^2\right)=\frac{d}{dt}(25)$$

$$2G\cdot\frac{dG}{dt}+2H\frac{dH}{dt}=0$$

$$2H\cdot\frac{dH}{dt}=-2G\cdot\frac{dG}{dt}$$

$$\frac{dH}{dt}=-\frac{G}{H}\cdot\frac{dG}{dt}$$

We find H when $G=0$:

$$(0)^2+H^2=25$$

$$H^2=25$$

$$H=5,\qquad H\text{ is nonnegative}$$

Next, we substitute 5 in for H, 0 in for G, and 3 in for $\frac{dG}{dt}$ to determine $\frac{dH}{dt}$.

$$\frac{dH}{dt}=-\frac{G}{H}\cdot\frac{dG}{dt}$$

$$=-\frac{(0)}{(5)}\cdot(3)=0$$

We find H when $G=1$:

$$(1)^2+H^2=25$$

$$H^2=24$$

$$H=\sqrt{24}=2\sqrt{6},\qquad H\text{ is nonnegative}$$

Next, we substitute $2\sqrt{6}$ in for H, 1 in for G, and 3 in for $\frac{dG}{dt}$ to determine $\frac{dH}{dt}$.

$$\frac{dH}{dt}=-\frac{G}{H}\cdot\frac{dG}{dt}$$

$$=-\frac{(1)}{(2\sqrt{6})}\cdot(3)=-\frac{3}{2\sqrt{6}}$$

We find H when $G=3$:

$$(3)^2+H^2=25$$

$$H^2=16$$

$$H=4,\qquad H\text{ is nonnegative}$$

Next, we substitute 4 in for H, 3 in for G, and 3 in for $\frac{dG}{dt}$ to determine $\frac{dH}{dt}$.

$$\frac{dH}{dt}=-\frac{G}{H}\cdot\frac{dG}{dt}$$

$$=-\frac{(3)}{(4)}\cdot(3)=-\frac{9}{4}$$

33. $R(x)=50x-0.5x^2$

Differentiating with respect to time we have:

$$\frac{d}{dt}R(x)=\frac{d}{dt}\left(50x-0.5x^2\right)$$

$$\frac{dR}{dt}=50\cdot\frac{dx}{dt}-x\cdot\frac{dx}{dt}$$

$$\frac{dR}{dt}=(50-x)\cdot\frac{dx}{dt}$$

Next, we substitute 10 for x and 5 for dx/dt.

$$\frac{dR}{dt}=(50-10)\cdot 5=(40)\cdot 5=200$$

The rate of change of total revenue with respect to time is $200 per day.

$$C(x)=10x+3$$

Differentiating with respect to time we have:

$$\frac{d}{dt}C(x)=\frac{d}{dt}(10x+3)$$

$$\frac{dC}{dt}=10\cdot\frac{dx}{dt}$$

Next, we substitute 10 for x and 5 for dx/dt.

$$\frac{dC}{dt}=10\cdot(5)=50$$

The rate of change of total cost with respect to time is $50 per day.

Profit is revenue minus cost. Therefore;

$$P(x)=R(x)-C(x)$$

$$=50x-0.5x^2-(10x+3)$$

$$=-0.5x^2+40x-3$$

Differentiating with respect to time we have:

$$\frac{d}{dt}P(x)=\frac{d}{dt}\left(-0.5x^2+40x-3\right)$$

$$\frac{dP}{dt}=-x\cdot\frac{dx}{dt}+40\frac{dx}{dt}$$

$$\frac{dP}{dt}=(40-x)\cdot\frac{dx}{dt}$$

Next, we substitute 10 for x and 5 for dx/dt on in the equation on the previous page.

$$\frac{dP}{dt} = \left(40 - (10)\right) \cdot (5) = (30)(5) = 150$$

The rate of change of total profit with respect to time is $150 per day.

35. $R(x) = 280x - 0.4x^2$

Differentiating with respect to time we have:

$$\frac{d}{dt}R(x) = \frac{d}{dt}\left(280x - 0.4x^2\right)$$

$$\frac{dR}{dt} = 280 \cdot \frac{dx}{dt} - 0.8x \cdot \frac{dx}{dt}$$

$$\frac{dR}{dt} = \left(280 - 0.8x\right) \cdot \frac{dx}{dt}$$

Next, we substitute 200 for x and 300 for dx/dt.

$$\frac{dR}{dt} = \left(280 - 0.8(200)\right) \cdot 300 = 36{,}000$$

The rate of change of total revenue with respect to time is $36,000 per day.

$$C(x) = 5000 + 0.6x^2$$

Differentiating with respect to time we have:

$$\frac{d}{dt}C(x) = \frac{d}{dt}\left(5000 + 0.6x^2\right)$$

$$\frac{dC}{dt} = 1.2x \cdot \frac{dx}{dt}$$

Next, we substitute 200 for x and 300 for dx/dt.

$$\frac{dC}{dt} = 1.2(200) \cdot 300 = 72{,}000$$

The rate of change of total cost with respect to time is $72,000 per day.

Profit is revenue minus cost. Therefore;

$$P(x) = R(x) - C(x)$$

$$= 280x - 0.4x^2 - \left(5000 + 0.6x^2\right)$$

$$= -x^2 + 280x - 5000$$

Differentiating with respect to time we have:

$$\frac{d}{dt}P(x) = \frac{d}{dt}\left(-x^2 + 280x - 5000\right)$$

$$\frac{dP}{dt} = -2x \cdot \frac{dx}{dt} + 280\frac{dx}{dt}$$

$$\frac{dP}{dt} = \left(280 - 2x\right) \cdot \frac{dx}{dt}$$

Next, we substitute 200 for x and 300 for dx/dt.

$$\frac{dP}{dt} = \left(280 - 2(200)\right) \cdot (300) = -36{,}000$$

The rate of change of total profit with respect to time is $-\$36{,}000$ per day.

37. $5p + 4x + 2px = 60$

First, we take the derivative of both sides of the equation with respect to t.

$$\frac{d}{dt}\left[5p + 4x + 2px\right] = \frac{d}{dt}\left[60\right]$$

$$5\frac{dp}{dt} + 4\frac{dx}{dt} + 2\underbrace{\left(p \cdot \frac{dx}{dt} + \frac{dp}{dt} \cdot x\right)}_{\text{Product Rule}} = 0$$

$$5\frac{dp}{dt} + 4\frac{dx}{dt} + 2p \cdot \frac{dx}{dt} + 2x \cdot \frac{dp}{dt} = 0$$

Next, we solve for $\dfrac{dx}{dt}$.

$$4\frac{dx}{dt} + 2p \cdot \frac{dx}{dt} = -5\frac{dp}{dt} - 2x \cdot \frac{dp}{dt}$$

$$\left(4 + 2p\right)\frac{dx}{dt} = -\left(5 + 2x\right) \cdot \frac{dp}{dt}$$

$$\frac{dx}{dt} = \frac{-\left(5 + 2x\right)}{\left(4 + 2p\right)} \cdot \frac{dp}{dt}$$

Substituting 3 for x, 5 for p, and 1.5 for $\dfrac{dp}{dt}$, we have:

$$\frac{dx}{dt} = \frac{-\left(5 + 2(3)\right)}{\left(4 + 2(5)\right)} \cdot (1.5)$$

$$= \frac{-(11)}{14} \cdot (1.5)$$

$$= \frac{-16.5}{14}$$

$$\approx -1.18$$

Sales are changing at a rate of -1.18 sales per day.

39. $A = \pi r^2$

To find the rate of change of the area of the Arctic ice cap with respect to time, we take the derivative of both sides of the equation with respect to t.

$$\frac{d}{dt}A = \frac{d}{dt}\left[\pi r^2\right]$$

$$\frac{dA}{dt} = \pi \frac{d}{dt}\left[r^2\right] \qquad \text{Constant Multiple Rule}$$

$$\frac{dA}{dt} = \pi \left[2r \cdot \frac{dr}{dt}\right] \qquad \text{Chain Rule}$$

$$\frac{dA}{dt} = 2\pi r \cdot \frac{dr}{dt}$$

The solution is continued on the next page.

In 2013, the r was 792 miles, and $\dfrac{dr}{dt} = -4.7$ miles per year. Substituting these values into the derivative on the previous page, we have:

$$\frac{dA}{dt} = 2\pi(792)(-4.7)$$
$$\approx -23{,}388.52899$$
$$\approx -23{,}389$$

Therefore, in 2013 the Arctic ice cap was changing at a rate of $-23{,}389$ mi^2 per year.

Another way of stating this is to say that the Arctic ice cap was *shrinking* at a rate of 23,389 mi^2/yr.

41. $S = \dfrac{\sqrt{hw}}{60}$

First, we substitute 165 for h, and then we take the derivative of both sides with respect to t.

$$S = \frac{\sqrt{165w}}{60} = \frac{\sqrt{165}}{60} \cdot w^{\frac{1}{2}}$$

$$\frac{d}{dt}[S] = \frac{d}{dt}\left[\frac{\sqrt{165}}{60} \cdot w^{\frac{1}{2}}\right]$$

$$\frac{dS}{dt} = \frac{\sqrt{165}}{60} \cdot \frac{1}{2} w^{-\frac{1}{2}} \cdot \frac{dw}{dt}$$

$$= \frac{\sqrt{165}}{120} \cdot \frac{1}{w^{\frac{1}{2}}} \cdot \frac{dw}{dt}$$

$$= \frac{\sqrt{165}}{120\sqrt{w}} \cdot \frac{dw}{dt}$$

Now, we will substitute 70 for w and -2 for $\dfrac{dw}{dt}$.

$$\frac{dS}{dt} = \frac{\sqrt{165}}{120\sqrt{70}} \cdot (-2)$$

$$\approx -0.0256$$

Therefore, Kim's surface area is changing at a rate of -0.0256 m^2/month. We could also say that Kim's surface area is *decreasing* by 0.0256 m^2/month.

43. $V = \dfrac{p}{4Lv}\left(R^2 - r^2\right)$

We assume that r, p, L and v are constants.

a) Taking the derivative of both sides with respect to t, we have:

$$\frac{dV}{dt} = \frac{d}{dt}\left[\frac{p}{4Lv}\left(R^2 - r^2\right)\right]$$

$$= \frac{p}{4Lv}\left[\frac{d}{dt}R^2 - \frac{d}{dt}r^2\right]$$

$$= \frac{p}{4Lv}\left[2R \cdot \frac{dR}{dt} - 0\right]$$

$$= \frac{pR}{2Lv} \cdot \frac{dR}{dt}$$

Substituting 70 for L, 400 for p and 0.003 for v, we have:

$$\frac{dV}{dt} = \frac{400R}{2(70)(0.003)} \cdot \frac{dR}{dt}$$

$$= \frac{400R}{0.42} \cdot \frac{dR}{dt}$$

$$\approx 952.38R \cdot \frac{dR}{dt}$$

b) Using the derivative in part (a), we substitute 0.00015 for dR/dt and 0.1 for R to get:

$$\frac{dV}{dt} = 952.38(0.1) \cdot (0.00015)$$

$$\approx 0.0143$$

The speed of the person's blood will be increasing at a rate of 0.0143 mm/sec^2.

45. Since the ladder forms a right triangle with the wall and the ground, we know that:

$$x^2 + y^2 = 26^2$$
$$x^2 + y^2 = 676$$

We are looking for $\dfrac{dy}{dt}$. Differentiating both sides of the equation with respect to t, we have:

$$\frac{d}{dt}\left[x^2 + y^2\right] = \frac{d}{dt}[676]$$

$$2x\frac{dx}{dt} + 2y\frac{dy}{dt} = 0$$

$$2y\frac{dy}{dt} = -2x\frac{dx}{dt} \qquad \text{Subtracting}$$

$$\frac{dy}{dt} = \frac{-2x}{2y} \cdot \frac{dx}{dt} \qquad \text{Dividing by } 2y$$

$$\frac{dy}{dt} = \frac{-x}{y} \cdot \frac{dx}{dt}$$

The solution is continued on the next page.

The lower end of the wall is being pulled away from the wall at a rate of 5 feet per second;

therefore, $\dfrac{dx}{dt} = 5$.

When the lower end is 10 feet away from the wall, $x = 10$, we substitute and solve for y.

$$(10)^2 + y^2 = 676$$
$$100 + y^2 = 676$$
$$y^2 = 676 - 100$$
$$y^2 = 576$$
$$y = \pm\sqrt{576}$$
$$y = \pm 24$$
$$y = 24 \qquad \text{\small y must be positive}$$

We substitute 10 for x, 24 for y and 5 for $\dfrac{dx}{dt}$ into the derivative to get:

$$\dfrac{dy}{dt} = \dfrac{-x}{y} \cdot \dfrac{dx}{dt}$$
$$= -\dfrac{(10)}{(24)} \cdot (5)$$
$$= -\dfrac{25}{12}$$
$$= -2\tfrac{1}{12}$$

When the lower end of the ladder is 10 feet from the wall, the top of the ladder is moving down the wall at a rate of $-2\tfrac{1}{12}$ feet per second.

47. $V = \dfrac{4}{3}\pi r^3$

Differentiating both sides with respect to t, we have:

$$\dfrac{dV}{dt} = \dfrac{d}{dt}\left[\dfrac{4}{3}\pi r^3\right]$$
$$= \dfrac{4}{3}\pi \cdot \dfrac{d}{dt}\left[r^3\right]$$
$$= \dfrac{4}{3}\pi\left[3r^2\,\dfrac{dr}{dt}\right]$$
$$= 4\pi r^2 \cdot \dfrac{dr}{dt}$$

Next, substituting 0.7 for dr/dt and 7.5 for r, we have:

$$\dfrac{dV}{dt} = 4\pi(7.5)^2(0.7)$$
$$= 4\pi(56.25)(0.7)$$
$$= 157.5\pi$$
$$\approx 494.8$$

The cantaloupe's volume is changing approximately at the rate of 494.8 cm^3/week.

49. $\dfrac{1}{x^2} + \dfrac{1}{y^2} = 5$

$$x^{-2} + y^{-2} = 5$$

Differentiating both sides with respect to x, we have:

$$\dfrac{d}{dx}\left[x^{-2}\right] + \dfrac{d}{dx}\left[y^{-2}\right] = \dfrac{d}{dx}[5]$$
$$-2x^{-3} - 2y^{-3}\,\dfrac{dy}{dx} = 0$$
$$-2y^{-3}\,\dfrac{dy}{dx} = 2x^{-3}$$
$$\dfrac{dy}{dx} = -\dfrac{x^{-3}}{y^{-3}}$$
$$\dfrac{dy}{dx} = -\dfrac{y^3}{x^3}$$

51. $y^2 = \dfrac{x^2 - 1}{x^2 + 1}$

Differentiating both sides with respect to x, we have:

$$\dfrac{d}{dx}\left[y^2\right] = \dfrac{d}{dx}\left[\dfrac{x^2 - 1}{x^2 + 1}\right]$$
$$2y\,\dfrac{dy}{dx} = \dfrac{(x^2 + 1)(2x) - (x^2 - 1)(2x)}{(x^2 + 1)^2}$$
$$2y\,\dfrac{dy}{dx} = \dfrac{2x^3 + 2x - 2x^3 + 2x}{(x^2 + 1)^2}$$
$$2y\,\dfrac{dy}{dx} = \dfrac{4x}{(x^2 + 1)^2}$$
$$\dfrac{dy}{dx} = \dfrac{4x}{2y(x^2 + 1)^2}$$
$$\dfrac{dy}{dx} = \dfrac{2x}{y(x^2 + 1)^2}$$

53. $(x-y)^3 + (x+y)^3 = x^5 + y^5$

Differentiating both sides with respect to x, we have:

$$\frac{d}{dx}\left[(x-y)^3 + (x+y)^3\right] = \frac{d}{dx}\left[x^5 + y^5\right]$$

$$\frac{d}{dx}(x-y)^3 + \frac{d}{dx}(x+y)^3 = \frac{d}{dx}x^5 + \frac{d}{dx}y^5$$

$$3(x-y)^2 \cdot \frac{d}{dx}(x-y) + 3(x+y)^2 \frac{d}{dx}(x+y) =$$

$$5x^4 + 5y^4 \frac{dy}{dx}$$

$$3(x-y)^2\left(1 - \frac{dy}{dx}\right) + 3(x+y)^2\left(1 + \frac{dy}{dx}\right) =$$

$$5x^4 + 5y^4 \frac{dy}{dx}$$

$$3(x-y)^2 - 3(x-y)^2\frac{dy}{dx} + 3(x+y)^2 +$$

$$3(x+y)^2 \frac{dy}{dx} = 5x^4 + 5y^4 \frac{dy}{dx}$$

$$\left[3(x+y)^2 - 3(x-y)^2 - 5y^4\right]\frac{dy}{dx} =$$

$$5x^4 - 3(x-y)^2 - 3(x+y)^2$$

$$\frac{dy}{dx} = \frac{5x^4 - 3(x-y)^2 - 3(x+y)^2}{3(x+y)^2 - 3(x-y)^2 - 5y^4}$$

Simplification will yield:

$$\frac{dy}{dx} = \frac{5x^4 - 6x^2 - 6y^2}{12xy - 5y^4}$$

55. $y^2 - xy + x^2 = 5$

Differentiate implicitly to find $\frac{dy}{dx}$.

$$\frac{d}{dx}\left[y^2 - xy + x^2\right] = \frac{d}{dx}[5]$$

$$2y\frac{dy}{dx} - \left[x\frac{dy}{dx} + y \cdot 1\right] + 2x = 0$$

$$(2y-x)\frac{dy}{dx} - y + 2x = 0$$

$$(2y-x)\frac{dy}{dx} = y - 2x$$

$$\frac{dy}{dx} = \frac{y - 2x}{2y - x}$$

Differentiate $\frac{dy}{dx}$ implicitly to find $\frac{d^2y}{dx^2}$.

$$\frac{d}{dx}\left[\frac{dy}{dx}\right] = \frac{d}{dx}\left[\frac{y - 2x}{2y - x}\right]$$

$$\frac{d^2y}{dx^2} = \frac{(2y-x)\left(\frac{dy}{dx} - 2\right) - (y-2x)\left(2\frac{dy}{dx} - 1\right)}{(2y-x)^2}$$

Simplifying the numerator we have:

$$\frac{d^2y}{dx^2} = \left[\left(2y\frac{dy}{dx} - 4y - x\frac{dy}{dx} + 2x\right) - \left(2y\frac{dy}{dx} - y - 4x\frac{dy}{dx} + 2x\right)\right] \div (2y-x)^2$$

$$\frac{d^2y}{dx^2} = \frac{-3y + 3x\frac{dy}{dx}}{(2y-x)^2}$$

Substituting $\frac{y - 2x}{2y - x}$ for $\frac{dy}{dx}$

$$\frac{d^2y}{dx^2} = \frac{-3y + 3x \cdot \frac{y - 2x}{2y - x}}{(2y-x)^2}$$

$$= \frac{-3y\frac{2y - x}{2y - x} + 3x \cdot \frac{y - 2x}{2y - x}}{(2y-x)^2}$$

$$= \frac{-6y^2 + 3xy + 3xy - 6x^2}{(2y-x)^3}$$

$$= \frac{-6y^2 + 6xy - 6x^2}{(2y-x)^3}$$

$$= \frac{-6\left(y^2 - xy + x^2\right)}{(2y-x)^3}$$

57. $x^3 - y^3 = 8$

Differentiate implicitly to find $\dfrac{dy}{dx}$.

$$\frac{d}{dx}\left[x^3 - y^3\right] = \frac{d}{dx}[8]$$

$$3x^2 - 3y^2 \frac{dy}{dx} = 0$$

$$-3y^2 \frac{dy}{dx} = -3x^2$$

$$\frac{dy}{dx} = \frac{-3x^2}{-3y^2}$$

$$\frac{dy}{dx} = \frac{x^2}{y^2}$$

Differentiate $\dfrac{dy}{dx}$ implicitly to find $\dfrac{d^2y}{dx^2}$.

$$\frac{d^2y}{dx^2} = \frac{d}{dx}\left[\frac{x^2}{y^2}\right]$$

$$= \frac{y^2 \cdot (2x) - x^2 \cdot 2y \dfrac{dy}{dx}}{\left(y^2\right)^2}$$

$$= \frac{2xy^2 - 2x^2 y \left[\dfrac{x^2}{y^2}\right]}{y^4} \quad \text{Substituting for } \frac{dy}{dx}$$

$$= \frac{2xy^2 - \dfrac{2x^4}{y}}{y^4}$$

Simplifying the derivative, we have:

$$\frac{d^2y}{dx^2} = \frac{\dfrac{2xy^2}{1} \cdot \dfrac{y}{y} - \dfrac{2x^4}{y}}{y^4}$$

$$= \frac{\dfrac{2xy^3 - 2x^4}{y}}{y^4}$$

$$= \frac{2xy^3 - 2x^4}{y^5}$$

$$= \frac{2x\left(y^3 - x^3\right)}{y^5}$$

59. ✎

61. Using the calculator, we have:

$x^4 = y^2 + x^6$

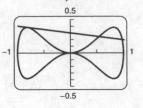

63. Using the calculator, we have:

$x^3 = y^2(2 - x)$

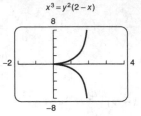

Chapter 3

Exponential and Logarithmic Functions

1. Graph: $y = 5^x$

First, we find some function values.

$x = -2, y = 5^{-2} = \dfrac{1}{5^2} = \dfrac{1}{25} = 0.04$

$x = -1, y = 5^{-1} = \dfrac{1}{5} = 0.20$

$x = 0, \quad y = 5^0 = 1$

$x = 1, \quad y = 5^1 = 5$

$x = 2, \quad y = 5^2 = 25$

x	y
-2	0.04
-1	0.20
0	1
1	5
2	25

Next, we plot the points and connect them with a smooth curve.

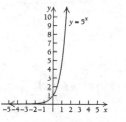

3. Graph: $y = 2 \cdot 3^x$

First, we will find some function values

$x = -2, y = 2 \cdot 3^{(-2)} = \dfrac{2}{3^2} = \dfrac{2}{9} \approx 0.222$

$x = -1, y = 2 \cdot 3^{(-1)} = \dfrac{2}{3} \approx 0.667$

$x = 0, \quad y = 2 \cdot 3^{(0)} = 2 \cdot 3^0 = 2$

$x = 1, \quad y = 2 \cdot 3^{(1)} = 2 \cdot 3^1 = 6$

$x = 2, \quad y = 2 \cdot 3^{(2)} = 2 \cdot 3^2 = 18$

We organize the values into a table at the top of the next column.

x	y
-2	0.222
-1	0.667
0	2
1	6
2	18

Next, we plot the points and connect them with a smooth curve.

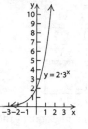

5. Graph: $y = 5\left(\dfrac{1}{4}\right)^x$

First, we find some function values.

$x = -2, y = 5\left(\dfrac{1}{4}\right)^{-2} = \dfrac{5}{\left(\dfrac{1}{4}\right)^2} = 5 \cdot \dfrac{16}{1} = 80$

$x = -1, y = 5\left(\dfrac{1}{4}\right)^{-1} = \dfrac{5}{\left(\dfrac{1}{4}\right)} = 5 \cdot \dfrac{4}{1} = 20$

$x = 0, \quad y = 5\left(\dfrac{1}{4}\right)^0 = 5 \cdot 1 = 5$

$x = 1, \quad y = 5\left(\dfrac{1}{4}\right)^1 = \dfrac{5}{4} = 1.25$

$x = 2, \quad y = 5\left(\dfrac{1}{4}\right)^2 = \dfrac{5}{16} = 0.3125$

x	y
-2	80
-1	20
0	5
1	1.25
2	0.3125

Next, we plot the points and connect them with a smooth curve.

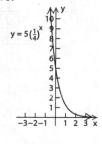

7.   Graph: $y = 1.3(1.2)^x$

First, we find some function values.

$x = -2, y = 1.3(1.2)^{-2} = \dfrac{1.3}{(1.2)^2} = 0.902\overline{7}$

$x = -1, y = 1.3(1.2)^{-1} = \dfrac{1.3}{1.2} = 1.083\overline{3}$

$x = 0, \quad y = 1.3(1.2)^0 = 1.3$

$x = 1, \quad y = 1.3(1.2)^1 = 1.56$

$x = 2, \quad y = 1.3(1.2)^2 = 1.872$

x	$g(x)$
-2	0.9027
-1	1.0833
0	1.3
1	1.56
2	1.872

Next, we plot the points and connect them with a smooth curve.

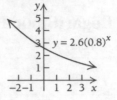

9.   Graph: $y = 2.6(0.8)^x$

First, we find some function values.

$x = -2, y = 2.6(0.8)^{-2} = \dfrac{2.6}{(0.8)^2} = 4.0625$

$x = -1, y = 2.6(0.8)^{-1} = \dfrac{2.6}{0.8} = 3.25$

$x = 0, \quad y = 2.6(0.8)^0 = 2.6$

$x = 1, \quad y = 2.6(0.8)^1 = 2.08$

$x = 2, \quad y = 2.6(0.8)^2 = 1.664$

x	y
-2	4.0625
-1	3.25
0	2.6
1	2.08
2	1.664

Next, we plot the points and connect them with a smooth curve.

11. $f(x) = e^x$

By Theorem 1, we have:

$f'(x) = e^x$

13. $g(x) = e^{2x}$

Using the chain rule $\dfrac{d}{dx} e^{f(x)} = f'(x) e^{f(x)}$, we

have $g'(x) = e^{2x} \cdot 2 = 2e^{2x}$.

$g'(x) = 2e^{2x}$

15. $f(x) = 6e^x$

Using $\dfrac{d}{dx}\big[c \cdot f(x)\big] = c \cdot f'(x)$, we have:

$f'(x) = 6e^x$.

17. $F(x) = e^{-7x}$

$F'(x) = -7e^{-7x}$

19. $g(x) = 3e^{5x}$

We use $\dfrac{d}{dx}\big[c \cdot f(x)\big] = c \cdot f'(x)$:

$g'(x) = \dfrac{d}{dx}\big[3 \cdot e^{5x}\big] = 3 \cdot \dfrac{d}{dx} e^{5x}$.

Next, we use the Chain Rule:

$\dfrac{d}{dx} e^{f(x)} = f'(x) e^{f(x)}$; therefore,

$g'(x) = 3 \cdot 5e^{5x}$

$\qquad = 15e^{5x}$

21. $G(x) = -7e^{-x}$

$$G'(x) = \frac{d}{dx}\left[-7 \cdot e^{-x}\right]$$

$$= -7 \cdot \frac{d}{dx}e^{-x}$$

$$G'(x) = -7 \cdot e^{-x} \cdot (-1)$$

$$= 7e^{-x}$$

23. $g(x) = \frac{1}{2}e^{-5x}$

$$g'(x) = \frac{d}{dx}\left[\frac{1}{2} \cdot e^{-5x}\right]$$

$$= \frac{1}{2} \cdot \frac{d}{dx}e^{-5x}$$

$$= \frac{1}{2}e^{-5x} \cdot (-5)$$

$$= -\frac{5}{2}e^{-5x}$$

25. $F(x) = -\frac{2}{3}e^{x^2}$

$$F'(x) = \frac{d}{dx}\left[-\frac{2}{3} \cdot e^{x^2}\right]$$

$$= -\frac{2}{3} \cdot \frac{d}{dx}e^{x^2}$$

$$= -\frac{2}{3} \cdot e^{x^2} \cdot (2x)$$

$$= -\frac{4x}{3}e^{x^2}$$

27. $F(x) = 4 - e^{2x}$

Using the Difference Rule, we have

$$F'(x) = \frac{d}{dx}\left[4 - e^{2x}\right]$$

$$= \frac{d}{dx}4 - \frac{d}{dx}e^{2x}.$$

To differentiate the two terms remember that the derivative of a constant is zero. All that is left is to apply the Chain Rule.

$$F'(x) = 0 - e^{2x} \cdot (2)$$

$$= -2e^{2x}$$

29. $G(x) = x^3 - 5e^{2x}$

First, we use the Difference Rule

$$G'(x) = \frac{d}{dx}\left[x^3 - 5e^{2x}\right]$$

$$= \frac{d}{dx}x^3 - 5\frac{d}{dx}e^{2x}.$$

Next, we use the Power Rule on the first term and the Chain Rule on the second term.

$$G'(x) = 3x^2 - 5e^{2x} \cdot (2)$$

$$= 3x^2 - 10e^{2x}$$

31. $g(x) = x^5 e^{2x}$

First, we use the Product Rule.

$$g'(x) = x^5 \cdot \left[\frac{d}{dx}e^{2x}\right] + \left[\frac{d}{dx}x^5\right] \cdot e^{2x}$$

Next, we use the Chain Rule on the first term and the Power Rule on the second term.

$$g'(x) = x^5 \cdot e^{2x} \cdot (2) + 5x^4 \cdot e^{2x}$$

$$= 2x^5 e^{2x} + 5x^4 e^{2x}$$

$$= (2x+5)x^4 e^{2x} \qquad \text{Factoring}$$

33. $F(x) = \dfrac{e^{2x}}{x^4}$

We use the Quotient Rule and the Chain Rule.

$$F'(x) = \frac{x^4 \cdot (2)e^{2x} - e^{2x} \cdot 4x^3}{x^8}$$

$$= \frac{2x^3 e^{2x}(x-2)}{x^3 \cdot x^5} \qquad \text{Factoring}$$

$$= \frac{x^3}{x^3} \cdot \frac{2e^{2x}(x-2)}{x^5} \qquad \begin{array}{l}\text{Remove the factor}\\ \text{equal to 1.}\end{array}$$

$$= \frac{2e^{2x}(x-2)}{x^5}$$

35. $f(x) = \left(x^2 - 2x + 2\right)e^x$

$$f'(x) = \left(x^2 - 2x + 2\right)e^x + (2x-2)e^x \quad \begin{array}{l}\text{Product}\\ \text{Rule}\end{array}$$

$$= \left(x^2 - 2x + 2 + 2x - 2\right)e^x \quad \text{Factoring}$$

$$= x^2 e^x$$

37. $f(x) = \dfrac{e^x}{x^4}$

$\quad f'(x) = \dfrac{x^4 \cdot e^x - 4x^3 \cdot e^x}{x^8}$ By the Quotient Rule

$\quad\quad = \dfrac{x^3 e^x (x-4)}{x^3 \cdot x^5}$ Factoring

$\quad\quad = \dfrac{x^3}{x^3} \cdot \dfrac{e^x (x-4)}{x^5}$ Remove the factor equal to 1.

$\quad\quad = \dfrac{e^x (x-4)}{x^5}$

39. $f(x) = e^{-x^2+8x}$

$\quad f'(x) = e^{-x^2+8x} \cdot \left[\dfrac{d}{dx}\left(-x^2+8x\right)\right]$ By the Chain Rule

$\quad\quad = e^{-x^2+8x} \cdot [-2x+8]$

$\quad\quad = (-2x+8)e^{-x^2+8x}$

41. $f(x) = e^{x^2/2}$

$\quad f'(x) = e^{x^2/2} \cdot \left[\dfrac{d}{dx}\left(\dfrac{x^2}{2}\right)\right]$

$\quad\quad = \dfrac{1}{2}(2x)e^{x^2/2}$

$\quad\quad = xe^{x^2/2}$

43. $y = e^{\sqrt{x-7}}$

First note that

$y = e^{\sqrt{x-7}} = e^{(x-7)^{1/2}}$.

Using the Chain Rule, we have

$\dfrac{dy}{dx} = e^{\sqrt{x-7}} \cdot \left[\dfrac{d}{dx}(x-7)^{1/2}\right]$.

Using the Chain Rule again, we have:

$\dfrac{dy}{dx} = e^{\sqrt{x-7}} \cdot \left[\dfrac{1}{2}(x-7)^{-1/2}\right]$

$\quad\quad = e^{\sqrt{x-7}} \cdot \dfrac{1}{2(x-7)^{1/2}}$ Properties of exponents

$\quad\quad = \dfrac{e^{\sqrt{x-7}}}{2\sqrt{x-7}}$.

45. $y = \sqrt{e^x - 1}$

$\quad y = \sqrt{e^x - 1} = \left(e^x - 1\right)^{1/2}$

$\quad \dfrac{dy}{dx} = \dfrac{1}{2}\left(e^x - 1\right)^{-1/2} \cdot \dfrac{d}{dx}\left(e^x - 1\right)$ By the Chain Rule

$\quad\quad = \dfrac{1}{2}\left(e^x - 1\right)^{-1/2} \cdot \left(e^x - 0\right)$

$\quad\quad = \dfrac{e^x}{2\left(e^x - 1\right)^{1/2}}$

$\quad\quad = \dfrac{e^x}{2\sqrt{e^x - 1}}$

47. $y = e^x + x^3 - xe^x$

Differentiate the function term by term. We apply Theorem 1 to the first term, the Power Rule to the second term, and the Product Rule to the third term.

$\dfrac{dy}{dx} = \underbrace{e^x}_{\text{Theorem 1}} + \underbrace{3x^2}_{\substack{\text{Power}\\\text{Rule}}} - \underbrace{\left(x \cdot e^x + 1 \cdot e^x\right)}_{\text{Product Rule}}$

Now we simplify:

$\dfrac{dy}{dx} = e^x + 3x^2 - \left(\left(x \cdot e^x\right) + 1 \cdot e^x\right)$

$\quad\quad = e^x + 3x^2 - xe^x - e^x$

$\quad\quad = 3x^2 - xe^x$

49. $y = 1 - e^{-3x}$

$\quad \dfrac{dy}{dx} = 0 - e^{-3x} \cdot (-3)$

$\quad\quad = 3e^{-3x}$

51. $y = 1 - e^{-kx}$

We use the Chain Rule, remembering that k is a constant:

$\quad \dfrac{dy}{dx} = 0 - e^{-kx} \cdot (-k)$

$\quad\quad = ke^{-kx}$

53. $g(x) = \left(4x^2 + 3x\right)e^{x^2-7x}$

We use the Product Rule first, and apply the Chain Rule when taking the derivative of the exponential term.

$g'(x) = \left(4x^2 + 3x\right)\underbrace{(2x-7)e^{x^2-7x}}_{\text{Chain Rule}} +$

$\quad\quad (8x+3)e^{x^2-7x}$

The solution is continued on the next page.

Factoring out the common term from the derivative on the previous page, we have

$$g'(x) = \left[(4x^2 + 3x)(2x - 7) + (8x + 3) \right] e^{x^2 - 7x}.$$

Simplifying inside the bracket, yields

$$g'(x) = \left[(8x^3 - 22x^2 - 21x) + (8x + 3) \right] e^{x^2 - 7x}$$
$$= \left[8x^3 - 22x^2 - 13x + 3 \right] e^{x^2 - 7x}.$$

55. Graph: $g(x) = e^{-2x}$

Using a calculator, we first find some function values.

$$g(-2) = e^{-2(-2)} = e^4 \approx 54.598$$
$$g(-1) = e^{-2(-1)} = e^2 \approx 7.3891$$
$$g(0) = e^{-2(0)} = e^0 = 1$$
$$g(1) = e^{-2(1)} = e^{-2} \approx 0.1353$$
$$g(2) = e^{-2(2)} = e^{-4} \approx 0.0183$$

x	$f(x)$
-2	54.598
-1	7.3891
0	1
1	0.1353
2	0.0183

Next, we plot the points and connect them with a smooth curve.

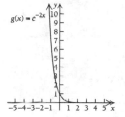

Derivatives. $g'(x) = -2e^{-2x}$ and

$$g''(x) = 4e^{-2x}.$$

Critical values of g. Since $g'(x) < 0$ for all real numbers x, we know that the derivative exists for all real numbers and there is no solution to the equation $g'(x) = 0$. There are no critical values and therefore no maximum or minimum values.

Decreasing. Since $g'(x) < 0$ for all real numbers x, the function g is decreasing over the entire real line.

Inflection points. Since $g''(x) > 0$ for all real numbers x, the equation $g''(x) = 0$ has no solution and there are no points of inflection.

Concavity. Since $g''(x) > 0$ for all real numbers x, the function g' is increasing over the entire real line and the graph is concave up over the entire real line.

57. Graph: $f(x) = e^{(\frac{1}{3})x}$

First, we find some function values.

$$f(-3) = e^{(\frac{1}{3})(-3)} = e^{-1} \approx 0.3679$$
$$f(-1) = e^{(\frac{1}{3})(-1)} = e^{-\frac{1}{3}} \approx 0.7165$$
$$f(0) = e^{(\frac{1}{3})(0)} = e^0 = 1$$
$$f(1) = e^{(\frac{1}{3})(1)} = e^{\frac{1}{3}} \approx 1.3956$$
$$f(3) = e^{(\frac{1}{3})(3)} = e^1 \approx 2.7183$$

x	$f(x)$
-3	0.3679
-1	0.7165
0	1
1	1.3956
3	2.7183

Next, we plot the points and connect them with a smooth curve.

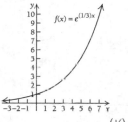

Derivatives. $f'(x) = \frac{1}{3} e^{(\frac{1}{3})x}$ and

$$f''(x) = \frac{1}{9} e^{(\frac{1}{3})x}$$

Critical values of f. Since $f'(x) > 0$ for all real numbers x, we know that the derivative exists for all real numbers and there is no solution to the equation $f'(x) = 0$. There are no critical values and therefore no maximum or minimum values.

Increasing. Since $f'(x) > 0$ for all real numbers x, the function f is increasing over the entire real line.

Inflection points. Since $f''(x) > 0$ for all real numbers x, the equation $f''(x) = 0$ has no solution and there are no points of inflection.

Concavity. Since $f''(x) > 0$ for all real numbers x, the function f' is increasing over the entire real line and the graph is concave up over the entire real line.

59. Graph: $f(x) = \dfrac{1}{2} e^{-x}$

First, we find some function values.

$f(-2) = \dfrac{1}{2} e^{-(-2)} = \dfrac{1}{2} e^2 \approx 3.6945$

$f(-1) = \dfrac{1}{2} e^{-(-1)} = \dfrac{1}{2} e^1 \approx 1.3591$

$f(0) = \dfrac{1}{2} e^{-(0)} = \dfrac{1}{2} e^0 = 0.5$

$f(1) = \dfrac{1}{2} e^{-(1)} = \dfrac{1}{2} e^{-1} \approx 0.1839$

$f(2) = \dfrac{1}{2} e^{-(2)} = \dfrac{1}{2} e^{-2} \approx 0.06767$

x	$f(x)$
-2	3.6945
-1	1.3591
0	0.5
1	0.1839
2	0.06767

Next, we plot the points and connect them with a smooth curve.

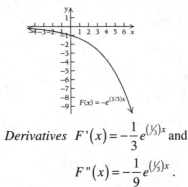

Derivatives. $f'(x) = -\dfrac{1}{2} e^{-x}$ and

$$f''(x) = \dfrac{1}{2} e^{-x}.$$

Critical values of f. Since $f'(x) < 0$ for all real numbers x, we know that the derivative exists for all real numbers and there is no solution to the equation $f'(x) = 0$. There are no critical values and therefore no maximum or minimum values.

Decreasing. Since $f'(x) < 0$ for all real numbers x, the function f is decreasing over the entire real line.

Inflection points. Since $f''(x) > 0$ for all real numbers x, the equation $f''(x) = 0$ has no solution and there are no points of inflection.

Concavity. Since $f''(x) > 0$ for all real numbers x, the function f' is increasing over the entire real line and the graph is concave up over the entire real line.

61. Graph: $F(x) = -e^{(\frac{1}{3})x}$

First, we find some function values.

$F(-2) = -e^{(\frac{1}{3})(-2)} = -e^{-\frac{2}{3}} \approx -0.5134$

$F(-1) = -e^{(\frac{1}{3})(-1)} = -e^{-\frac{1}{3}} \approx -0.7165$

$F(0) = -e^{(\frac{1}{3})(0)} = -e^0 = -1$

$F(1) = -e^{(\frac{1}{3})(1)} = -e^{\frac{1}{3}} \approx -1.3956$

$F(2) = -e^{(\frac{1}{3})(2)} = -e^{\frac{2}{3}} \approx -1.9477$

x	$F(x)$
-2	- 0.5134
-1	- 0.7165
0	- 1
1	- 1.3956
2	- 1.9477

Next, we plot the points and connect them with a smooth curve.

Derivatives $F'(x) = -\dfrac{1}{3} e^{(\frac{1}{3})x}$ and

$$F''(x) = -\dfrac{1}{9} e^{(\frac{1}{3})x}.$$

Critical values of F. Since $F'(x) < 0$ for all real numbers x, we know that the derivative exists for all real numbers and there is no solution to the equation $F'(x) = 0$. There are no critical values and therefore no maximum or minimum values.

Decreasing. Since $F'(x) < 0$ for all real numbers x, the function F is decreasing over the entire real line.

Inflection points. Since $F''(x) < 0$ for all real numbers x, the equation $F''(x) = 0$ has no solution and there are no points of inflection.
Concavity. Since $F''(x) < 0$ for all real numbers x, the function F' is decreasing over the entire real line and the graph is concave down over the entire real line.

63. Graph: $f(x) = 3 - e^{-x}$, for $x \geq 0$
First, we find some function values:
$$f(0) = 3 - e^{-0} = 2$$
$$f(1) = 3 - e^{-1} \approx 2.6321$$
$$f(2) = 3 - e^{-2} \approx 2.8647$$
$$f(3) = 3 - e^{-3} \approx 2.9502$$
$$f(4) = 3 - e^{-4} \approx 2.9817$$
$$f(5) = 3 - e^{-5} \approx 2.9933$$

x	$f(x)$
0	2
1	2.6321
2	2.8647
3	2.9502
4	2.9817
5	2.9933

Next, we plot the points and connect them with a smooth curve.

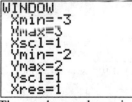

Derivatives. $f'(x) = e^{-x}$ and $f''(x) = -e^{-x}$.

Critical values of f. Since $f'(x) > 0$ for all real numbers x, we know that there is no solution to the equation $f'(r) = 0$. There are no critical values on $(0, \infty)$.

Increasing. Since $f'(x) > 0$ for all real numbers x, we know that f is increasing over its entire domain, $(0, \infty)$.

Inflection points. Since $f''(x) < 0$ for all real numbers x, the equation $f''(x) = 0$ has no solution and there are no points of inflection.

Concavity. Since $f''(x) < 0$ for all real numbers, the function f' is decreasing and the graph is concave down over the interval $(0, \infty)$.

65. From Exercise 55 we know
$$Y_1 = g(x) = e^{-2x}$$
$$Y_2 = g'(x) = -2e^{-2x}$$
$$Y_3 = g''(x) = 4e^{-2x}$$
Enter each function into your graphing calculator and set the window to be:

The graphs are shown in the screen shot.

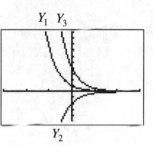

67. From Exercise 57 we know that:
$$Y_1 = f(x) = e^{(\frac{1}{3})x}$$
$$Y_2 = f'(x) = \frac{1}{3}e^{(\frac{1}{3})x}$$
$$Y_3 = f''(x) = \frac{1}{9}e^{(\frac{1}{3})x}$$

Enter each function into your graphing calculator and set the window to be:

```
WINDOW
 Xmin=-3
 Xmax=3
 Xscl=1
 Ymin=-2
 Ymax=2
 Yscl=1
 Xres=1
```

The graphs are shown in the screen shot.

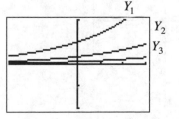

69. From Exercise 59 we know that:

$$Y_1 = f(x) = \frac{1}{2}e^{-x}$$

$$Y_2 = f'(x) = -\frac{1}{2}e^{-x}$$

$$Y_3 = f''(x) = \frac{1}{2}e^{-x}$$

Enter each function into your graphing calculator and set the window to be:

```
WINDOW
 Xmin=-3
 Xmax=3
 Xscl=1
 Ymin=-3
 Ymax=3
 Yscl=1
 Xres=1
```

The graphs are shown in the screen shot.

$$Y_1 = Y_3$$

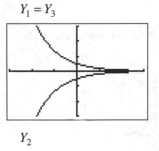

$$Y_2$$

71. From Exercise 61 we know that:

$$Y_1 = F(x) = -e^{(1/3)x}$$

$$Y_2 = F'(x) = -\frac{1}{3}e^{(1/3)x}$$

$$Y_3 = F''(x) = -\frac{1}{9}e^{(1/3)x}$$

Enter each function into your graphing calculator and set the window to be:

```
WINDOW
 Xmin=-3
 Xmax=3
 Xscl=1
 Ymin=-2
 Ymax=2
 Yscl=1
 Xres=1
```

The graphs are shown in the screen shot.

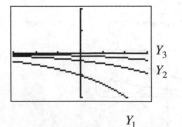

$$Y_3$$
$$Y_2$$

$$Y_1$$

73. From Exercise 63 we know that:

$$Y_1 = f(x) = 3 - e^{-x}; \quad x \ge 0$$

$$Y_2 = f'(x) = e^{-x}; \quad x \ge 0$$

$$Y_3 = f''(x) = -e^{-x}; \quad x \ge 0$$

Enter each function into your graphing calculator and set the window to be:

```
WINDOW
 Xmin=-3
 Xmax=3
 Xscl=1
 Ymin=-4
 Ymax=4
 Yscl=1
 Xres=1
```

The graphs are shown in the screen shot.

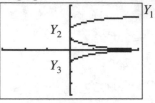

$$Y_1$$
$$Y_2$$
$$Y_3$$

75. First, find the derivative of the function.

$$f(x) = 2e^{-3x}$$

$$f'(x) = -6e^{-3x}$$

Now, evaluate the derivative at $x = 0$.

$$f'(0) = -6e^0 = -6$$

The slope of the tangent line at the point $(0, 2)$ is -6.

77. First find the slope of the tangent line at $(0, 1)$, by evaluating the derivative at $x = 0$.

$$f(x) = e^{2x}$$

$$f'(x) = 2e^{2x}$$

$$f'(0) = 2e^{2 \cdot 0} = 2$$

Then we find the equation of the line with slope 2 and containing the point $(0, 1)$.

$$y - y_1 = m(x - x_1) \quad \text{Point-slope equation}$$

$$y - 1 = 2(x - 0)$$

$$y - 1 = 2x$$

$$y = 2x + 1$$

79. Enter the function and the tangent line found in exercise 77 into the calculator.

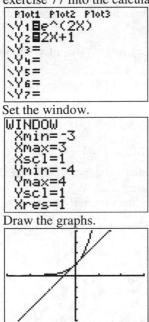

Set the window.

Draw the graphs.

81. a) 2016 correspond to $t = 5$ $(2016 - 2011 = 5)$.
Evaluate $V(5)$.

$$V(5) = 115.32e^{0.094(5)} \approx 184.5$$

The value of U.S. travel exports in 2016 is approximately \$184.5 billion.
2018 corresponds to $t = 7$. Evaluate $V(7)$.

$$V(7) = 115.32e^{0.094(7)} \approx 222.7$$

The value of U.S. travel exports in 2018 is approximately \$222.7 billion.

b) Find the growth rate by taking the derivative of the function.

$$V'(t) = \frac{d}{dt}\left(115.32e^{0.094t}\right)$$

$$= 115.32e^{0.094t} \cdot 0.094$$

$$= 10.84e^{0.094t}$$

For 2016, we substitute in 5 for t.

$$V'(5) = 10.84e^{0.094(5)} \approx 17.3.$$

In 2016, the growth rate for U.S. travel exports will be approximately \$17.3 billion per year.
For 2018, we substitute in 7 for t.

$$V'(7) = 10.84e^{0.094(7)} \approx 20.9.$$

In 2018, the growth rate for U.S. travel exports will be approximately \$20.9 billion per year.

83. a) $C'(t) = 0 - 50(-1)e^{-t}$

$$C'(t) = 50e^{-t}$$

b) $C'(0) = 50e^{-0} = 50 \cdot 1 = 50$

Marginal cost when $t = 0$ is 50 million dollars per year.

c) $C'(4) = 50e^{-4} \approx 0.916$

Marginal cost when $t = 4$ is 0.916 million dollars per year or \$916,000 per year.

d) ✎

85. $q = 240e^{-0.003x}$

a) When $x = 250$

$$q = 240e^{-0.003(250)}$$

$$= 240e^{-0.75}$$

$$q \approx 113$$

At a price of \$250 the demand for turntables is 113 thousand, or 113,000 units.

b) Create a table of values as needed. Find the function values for q as shown in part 'a'.

x	q
0	240
100	177.796
200	131.715
250	113.368
300	97.577
400	72.287

Next plot the points and connect the lines with a smooth curve.

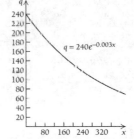

c) Take the derivative of the demand function with respect to x.

$$q'(x) = 240(-0.003)e^{-0.003x}$$

$$= -0.72e^{-0.003x}$$

d) ✎

87. a) Substituting 7 for t, we have:

$$V(7) = 20,000(1.056)^{(7)} = 29,287.17.$$

The value of Maria's IRA will be approximately \$29,287.17 after 7 years.

b) To find the rate of change, we use the Draw-Tangent feature.

We graph the function in the following window:

Using Draw – Tangent we have:

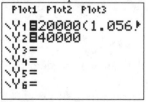

This gives us the tangent line of
$$y = 1595.8x + 18,116.5$$

Therefore the slope of the tangent line is
$m = 1595.8$.

After 7 years, we have:
$$V'(7) \approx 1595.8$$

The rate of change of Maria's IRA after 7 years will be approximately \$1595.80 per year.

c) To determine when Maria's IRA will be worth \$40,000, we solve the equation

$$20,000(1.056)^t = 40,000.$$

To solve this equation, we plug both sides into a graphing calculator and find the intersection point.

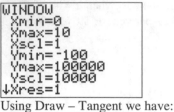

Finding the intersection we have:

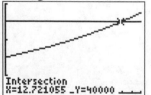

Therefore, in about 12.72 years, Maria's IRA will be worth \$40,000.

89. a) Using the model from Theorem 1 in section R.1, we extend the idea to the information given in this exercise. The exponential model will be:

$$V(t) = P(1+r)^t.$$ From the information in the exercise we know:
$P = 14,450, r = -0.15.$

Therefore, the exponential model will be:

$$V(t) = 14,450(1+(-0.15))^t$$
$$= 14,450(0.85)^t.$$

b) To find the rate of change, we use the Draw-Tangent feature.

We graph the function in the following window:

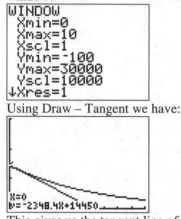

Using Draw – Tangent we have:

This gives us the tangent line of
$$y = -2348.4x + 14,450$$

Therefore the slope of the tangent line is
$m = -2348.4$.

After 0 years, we have:
$$V'(0) \approx -2348.4$$

The rate of change of the value of the equipment on Jan. 1, 2016 is approximately \$2348.40 per year.

c) According to this model, January 1, 2014 corresponds to $t = -2$. Substituting into the function found in part a) we have:

$$V(-2) = 14,450(0.85)^{(-2)} = 20,000.$$

So the original value of the equipment was \$20,000.

d) Half of the original value will be $10,000.
We solve the equation

$$14,450(0.85)^t = 10,000.$$

To solve this equation, we plug both sides into a graphing calculator and find the intersection point.

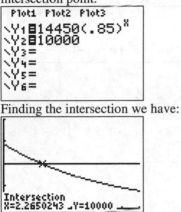

Finding the intersection we have:

Intersection
X=2.2650243 Y=10000

Therefore, in about 2.27 years after January 1, 2016 or 4.47 years after January 1, 2014, the equipment will be worth half of its original value.

91. a) $C(0) = 10 \cdot 0^2 e^{-0} = 0$

The initial concentration is 0 parts per million (ppm).

$C(1) = 10 \cdot 1^2 e^{-1}$

$\approx 10(0.367879)$

≈ 3.68

After 1 hour, the concentration is approximately 3.7 ppm.

$C(2) = 10 \cdot 2^2 e^{-2}$

$\approx 40(0.135335)$

≈ 5.41

After 2 hours, the concentration is approximately 5.4 ppm.

$C(3) = 10 \cdot 3^2 e^{-3}$

$\approx 90(0.049787)$

≈ 4.48

After 3 hours, the concentration is approximately 4.5 ppm.
Substituting in 10, we have:

$C(10) = 10 \cdot 10^2 e^{-10}$

$\approx 1000(0.000045)$

≈ 0.05

After 10 hours, the concentration is approximately 0.05 ppm.

b) Plot the points $(0,0); (1,3.7); (2,5.4);$
$(3,4.5)$ and $(10,0.05)$ and other points as needed. Then we connect the points with a smooth curve.

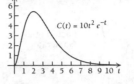

c) $C'(t) = 10t^2 (-1)e^{-t} + 20te^{-t}$

$= (20t - 10t^2)e^{-t}$

d) Since $C'(x)$ exists for all values of $x \geq 0$, the only critical values are where $C'(x) = 0$. Solve:

$$C'(x) = 0$$

$$(20t - 10t^2)e^{-t} = 0$$

$$20t - 10t^2 = 0 \quad (e^{-t} \neq 0)$$

$$10t(2 - t) = 0$$

$$t = 0 \ \text{ or } \ t = 2$$

The are two critical values. The critical value $t = 2$ is the obvious maximum. Find $C(2)$.

$C(2) = 10 \cdot 2^2 e^{-2}$

$\approx 40(0.135335)$

≈ 5.41

The maximum value of the concentration will be 5.41 parts per million and will occur 2 hours after the drug has been administered.

e) ✎

93. $y = (e^{3x} + 1)^5$

Using the Extended Power Rule.

$\dfrac{dy}{dx} = 5(e^{3x} + 1)^4 \cdot 3e^{3x}$

$= 15e^{3x}(e^{3x} + 1)^4$

95. $y = \dfrac{e^x}{x^2 + 1}$

$\dfrac{dy}{dx} = \dfrac{\left(x^2 + 1\right)e^x - e^x(2x)}{\left(x^2 + 1\right)^2}$ Quotient Rule

$= \dfrac{e^x\left(x^2 + 1 - 2x\right)}{\left(x^2 + 1\right)^2}$ Factoring the Numerator

$= \dfrac{e^x\left(x^2 - 2x + 1\right)}{\left(x^2 + 1\right)^2}$

$= \dfrac{e^x(x-1)^2}{\left(x^2 + 1\right)^2}$

97. $f(x) = e^{x/2} \cdot \sqrt{x-1}$

First, we note that

$f(x) = e^{x/2} \cdot (x-1)^{1/2}$.

Next, we use the Product Rule.

$f'(x) = e^{x/2} \cdot \dfrac{1}{2}(x-1)^{-1/2} + \dfrac{1}{2}e^{x/2}(x-1)^{1/2}$

$= \dfrac{1}{2}e^{x/2}\left[(x-1)^{-1/2} + (x-1)^{1/2}\right]$ Factoring

$= \dfrac{1}{2}e^{x/2}\left[\dfrac{1}{\sqrt{x-1}} + \sqrt{x-1}\right]$

$= \dfrac{1}{2}e^{x/2}\left[\dfrac{1}{\sqrt{x-1}} + \sqrt{x-1} \cdot \dfrac{\sqrt{x-1}}{\sqrt{x-1}}\right]$ Multiplying by 1

$= \dfrac{1}{2}e^{x/2}\left[\dfrac{1}{\sqrt{x-1}} + \dfrac{x-1}{\sqrt{x-1}}\right]$

$= \dfrac{1}{2}e^{x/2}\left[\dfrac{x}{\sqrt{x-1}}\right]$ Adding Fractions

$= e^{x/2}\left[\dfrac{x}{2\sqrt{x-1}}\right]$

99. $f(x) = \dfrac{e^x - e^{-x}}{e^x + e^{-x}}$

Using the Quotient Rule

$f'(x)$

$= \dfrac{\left(e^x + e^{-x}\right)\left(e^x + e^{-x}\right) - \left(e^x - e^{-x}\right)\left(e^x - e^{-x}\right)}{\left(e^x + e^{-x}\right)^2}$

$= \dfrac{\left(e^{2x} + e^0 + e^0 + e^{-2x}\right) - \left(e^{2x} - e^0 - e^0 + e^{-2x}\right)}{\left(e^x + e^{-x}\right)^2}$

$= \dfrac{e^{2x} + e^0 + e^0 + e^{-2x} - e^{2x} + e^0 + e^0 - e^{-2x}}{\left(e^x + e^{-x}\right)^2}$

$= \dfrac{4}{\left(e^x + e^{-x}\right)^2}$

101. To determine equilibrium, we set supply equal to demand. Using the equations out of exercises 85 and 86, we have:

$75e^{0.004x} = 240e^{-0.003x}$

Graphing each of these equations and finding the intersection point we have:

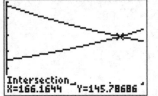

Using the intersection feature, we have:

Therefore, the equilibrium point is Approximately $(166.16, 145.787)$.

Substituting 166.16 into each of the derivatives found in the previous exercises we have:
For the demand function we have

$q'(166.16) = -0.72e^{-0.003(166.16)}$

≈ -0.43736

Demand is decreasing about 437 units per dollar at equilibrium.
For the supply function we have:

$q'(166.16) = 0.3e^{0.004(166.16)}$

≈ 0.5831

Supply is increasing about 583 units per dollar at equilibrium.

103. $e = \lim_{t \to 1} g(t); g(t) = t^{1/(t-1)}$

$$g(0.5) = (0.5)^{1/(0.5-1)} = 4$$

$$g(0.9) = (0.9)^{1/(0.9-1)} \approx 2.86797$$

$$g(0.99) = (0.99)^{1/(0.99-1)} \approx 2.73200$$

$$g(0.999) = (0.999)^{1/(0.999-1)} \approx 2.71964$$

$$g(0.9998) = (0.9998)^{1/(0.9998-1)} \approx 2.71855$$

105. ✎

107. Graph: $y = x^2 e^{-x}$

Using a calculator we graph the function in the following window:

```
WINDOW
 Xmin=-2
 Xmax=7
 Xscl=1
 Ymin=-.1
 Ymax=1
 Yscl=1
↓Xres=1
```

Giving us the graph

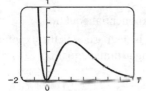

Using the minimum feature we see that there is a relative minimum at $(0,0)$.

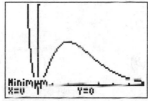

Using the maximum feature, we see that there is a relative maximum at $(2, 0.5413)$.

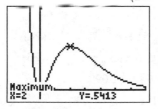

109. The graphs of $f(x), f'(x),$ and $f''(x)$ are all the graph of $y = e^x$.

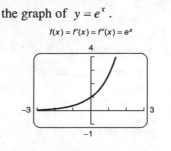

111. $f(x) = 2e^{0.3x}$

$$f'(x) = 0.6e^{0.3x}$$

$$f''(x) = 0.18e^{0.3x}$$

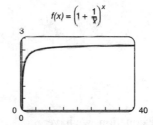

113.

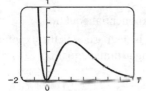

Using the calculator we construct the table:

X	Y₁
1	2
10	2.5937
100	2.7048
1000	2.7169
10000	2.7181
100000	2.7183
1000000	2.7183

X=100000

We can see that as x grows large, the function approaches 2.7183. In other words

$$\lim_{x \to \infty} \left(1 + \frac{1}{x}\right)^x = e \, .$$

Exercise Set 3.2

1. $\log_3 81 = 4$ Logarithmic equation

$3^4 = 81$ Exponential equation;
3 is the base, 4 is the exponent.

3. $\log_{27} 3 = \dfrac{1}{3}$ Logarithmic equation

$27^{\frac{1}{3}} = 3$ Exponential equation;

27 is the base, $\dfrac{1}{3}$ is the exponent.

5. $\log_a J = K$ Logarithmic equation

$a^K = J$ Exponential equation;
a is the base, K is the exponent.

7. $-\log_b V = w$ Logarithmic equation

$\log_b V = -w$ Multiply by -1.

$b^{-w} = V$ Exponential equation;
b is the base,
$-w$ is the exponent.

9.   Rewriting the equation, we have:

$\log_7 49 = x$

$7^x = 49$

We know that 7 squared is equal to 49, therefore, $x = 2$.

11.   Rewriting the equation, we have:

$\log_x 32 = 5$

$x^5 = 32.$

Therefore,

$\sqrt[5]{x^5} = \sqrt[5]{32}$

$x = \sqrt[5]{32}$

$x = 2.$

13.   Rewriting the equation, we have:

$\log_3 x = 5$

$3^5 = x$

$243 = x.$

15.   Rewriting the equation, we have:

$\log_{11} \sqrt{11} = x$

$11^x = \sqrt{11}$

$11^x = 11^{\frac{1}{2}}$

$x = \dfrac{1}{2}.$

17. $e^t = p$ Exponential equation;
e is the base, M is the exponent.

$\log_e p = t$ Logarithmic equation

or $\ln p = t$ $\ln p$ is the abbreviation for $\log_e p$.

19. $10^3 = 1000$ Exponential equation;
10 is the base,
3 is the exponent.

$\log_{10} 1000 = 3$ Logarithmic equation

21. $10^{-2} = 0.01$ Exponential equation;
10 is the base,
-2 is the exponent.

$\log_{10} 0.01 = -2$ Logarithmic equation

23. $Q^n = T$ Exponential equation;
Q is the base, n is the exponent.

$\log_Q T = n$ Logarithmic equation

25. $\log_b \left(\dfrac{1}{5}\right) = \log_b 1 - \log_b 5$ (P2)

$= 0 - 1.609$

$= -1.609$

27. $\log_b \sqrt{b^3} = \log_b (b)^{\frac{3}{2}}$

$= \dfrac{3}{2}$ (P5)

29. $\log_b 75 = \log_b \left(5^2 \cdot 3\right)$

$= \log_b 5^2 + \log_b 3$ (P1)

$= 2\log_b 5 + \log_b 3$ (P3)

$= 2(1.609) + 1.099$

$= 4.317$

31. $\ln 80 = \ln\left(4^2 \cdot 5\right)$

$= \ln 4^2 + \ln 5$ (P1)

$= 2\ln 4 + \ln 5$ (P3)

$= 2(1.3863) + 1.6094$

$= 4.382$

33. $\ln\left(\dfrac{1}{5}\right) = \ln 1 - \ln 5$ (P2)

$= 0 - 1.6094$ (P6)

$= -1.6094$

35. $\ln\left(4e\right) = \ln 4 + \ln e$ (P1)

$= 1.3863 + 1$

$= 2.3863$

37. $\ln\sqrt{e^8} = \ln e^{8/2}$

$= \ln e^4$

$= 4$ (P5)

39. $\ln\left(\dfrac{4}{5}\right) = \ln 4 - \ln 5$ (P2)

$= 1.3863 - 1.6094$

$= -0.2231$

41. $\ln\left(\dfrac{4}{e}\right) = \ln 4 - \ln e$

$= 1.3863 - 1$

$= 0.3863$

43. Using a calculator and rounding to six decimal places, we have

$\ln 99,999 \approx 11.512915$.

45. $\ln 0.0182 \approx -4.006334$

47. $\ln 0.011 \approx -4.509860$

49. $e^t = 80$

$\ln e^t = \ln 80$ Taking the natural log on both sides

$t = \ln 80$ (P5)

$t \approx 4.382027$ Using a calculator

$t \approx 4.382$

51. $e^{3t} = 900$

$\ln e^{3t} = \ln 900$ Taking the natural log on both sides

$3t = \ln 900$ (P5)

$t = \dfrac{\ln 900}{3}$

$t \approx 2.267$ Using a calculator

53. $e^{-t} = 0.01$

$\ln e^{-t} = \ln 0.01$ Taking the natural log on both sides

$-t = \ln 0.01$ (P5)

$t = -\ln 0.01$

$t \approx 4.605$ Using a calculator

55. $e^{-0.02t} = 0.06$

$\ln e^{-0.02t} = \ln 0.06$ Taking the natural log on both sides

$-0.02t = \ln 0.06$ (P5)

$t = \dfrac{\ln 0.06}{-0.02}$

$t \approx \dfrac{-2.813411}{-0.02}$ Using a calculator

$t \approx 140.671$

57. $y = -9\ln x$

$\dfrac{dy}{dx} = -9 \cdot \dfrac{1}{x}$ $\left[\frac{d}{dx}\left[c \cdot f(x)\right] = c \cdot f'(x)\right]$

$\dfrac{dy}{dx} = \dfrac{-9}{x}$

59. $y = 7\ln|x|$

$\dfrac{dy}{dx} = 7 \cdot \dfrac{1}{x}$

$= \dfrac{7}{x}$

61. $y = x^6 \ln x - \frac{1}{4}x^4$

Differentiate this function term by term. We apply the Product Rule to the first term, and the Power Rule to the second term.

$\dfrac{dy}{dx} = \underbrace{x^6 \cdot \dfrac{1}{x} + 6x^5 \cdot \ln x}_{\text{Product Rule}} - \underbrace{\dfrac{1}{4} \cdot 4x^3}_{\text{Power Rule}}$

$= x^5 + 6x^5 \ln x - x^3$

63. $f(x) = \ln(9x)$

Using Theorem 7, we have

$$\frac{d}{dx}\ln f(x) = \frac{1}{f(x)} \cdot f'(x) = \frac{f'(x)}{f(x)}$$

$$f'(x) = \frac{1}{9x} \cdot 9$$

$$= \frac{1}{x}$$

65. $f(x) = \ln|5x|$

Using Theorem 7, we have

$$\frac{d}{dx}\ln f(x) = \frac{1}{f(x)} \cdot f'(x) = \frac{f'(x)}{f(x)}$$

$$f'(x) = \frac{1}{5x} \cdot 5$$

$$= \frac{1}{x}$$

67. $g(x) = x^5 \ln(3x)$

We apply the Product Rule. Remember to use Theorem 7 when taking the derivative of the logarithm.

$$g'(x) = \underbrace{x^5 \cdot \underbrace{\frac{1}{3x} \cdot 3}_{\text{Theorem 7}} + 5x^4 \cdot \ln(3x)}_{\text{Product Rule}}$$

$$g'(x) = x^5 \cdot \frac{1}{3x} \cdot 3 + 5x^4 \ln(3x)$$

$$= x^4 + 5x^4 \ln(3x)$$

69. $g(x) = x^4 \ln|6x|$

We apply the Product Rule. Remember to use Theorem 7 when taking the derivative of the logarithm.

$$g'(x) = \underbrace{x^4 \cdot \underbrace{\frac{1}{6x} \cdot 6}_{\text{Theorem 7}} + 4x^3 \cdot \ln|6x|}_{\text{Product Rule}}$$

$$g'(x) = x^4 \cdot \frac{1}{x} + 4x^3 \ln|6x|$$

$$= x^3 + 4x^3 \ln|6x|$$

71. $y = \dfrac{\ln x}{x^5}$

Using the Quotient Rule, we have

$$\frac{dy}{dx} = \frac{x^5 \cdot \dfrac{1}{x} - 5x^4 \cdot \ln x}{x^{10}}$$

$$= \frac{x^4 - 5x^4 \ln x}{x^{10}}$$

$$= \frac{x^4(1 - 5\ln x)}{x^4 \cdot x^6} \qquad \text{Factoring}$$

$$= \frac{x^4}{x^4} \cdot \frac{1 - 5\ln x}{x^6} \qquad \begin{array}{l}\text{Removing a}\\\text{factor equal to 1}\end{array}$$

$$= \frac{1 - 5\ln x}{x^6}$$

73. $y = \dfrac{\ln|3x|}{x^2}$

Using the Quotient Rule and Theorem 7, we have

$$\frac{dy}{dx} = \frac{x^2 \cdot \dfrac{1}{3x} \cdot 3 - 2x \cdot \ln|3x|}{x^4}$$

$$= \frac{x^2 \cdot \dfrac{1}{x} - 2x\ln|3x|}{x^4}$$

$$= \frac{x - 2x\ln|3x|}{x^4}$$

$$= \frac{x(1 - 2\ln|3x|)}{x \cdot x^3} \qquad \text{Factoring}$$

$$= \frac{x}{x} \cdot \frac{1 - 2\ln|3x|}{x^3} \qquad \begin{array}{l}\text{Removing a}\\\text{factor equal to 1}\end{array}$$

$$= \frac{1 - 2\ln|3x|}{x^3}$$

75. $y = \ln\left(\dfrac{x^2}{4}\right)$

$$y = \ln x^2 - \ln 4 \qquad \text{(P2)}$$

$$\frac{dy}{dx} = \frac{1}{x^2} \cdot 2x - 0 \qquad \text{Theorem 7}$$

$$= \frac{2}{x}$$

77. $y = \ln\left(3x^2 + 2x - 1\right)$

Using Theorem 7, we note:

$$\frac{d}{dx}\ln g(x) = \frac{g'(x)}{g(x)}$$

$g(x) = 3x^2 + 2x - 1$

$g'(x) = 6x + 2$

Therefore,

$$\frac{dy}{dx} = \frac{6x + 2}{3x^2 + 2x - 1}$$

$$\frac{dy}{dx} = \frac{2(3x+1)}{3x^2 + 2x - 1}$$

79. $f(x) = \ln\left(\dfrac{x^2 - 7}{x}\right)$

Using the Theorem 7.

$$f'(x) = \frac{\left(\dfrac{d}{dx}\left[\dfrac{x^2 - 7}{x}\right]\right)}{\dfrac{x^2 - 7}{x}}$$

$$= \left(\frac{d}{dx}\left[\frac{x^2 - 7}{x}\right]\right) \cdot \frac{x}{x^2 - 7} \qquad \text{Dividing fractions}$$

We apply the Quotient Rule to take the derivative of the inside function to get

$$\frac{d}{dx}\left[\frac{x^2 - 7}{x}\right] = \frac{x \cdot 2x - \left(x^2 - 7\right) \cdot 1}{x^2}$$

$$= \frac{2x^2 - x^2 + 7}{x^2}$$

$$= \frac{x^2 + 7}{x^2}$$

Now we substitute back into the original derivative.

$$f'(x) = \frac{x^2 + 7}{x^2} \cdot \frac{x}{x^2 - 7}$$

$$= \frac{x^2 + 7}{x\left(x^2 - 7\right)}$$

An alternate solution to this problem involves using property P2.

$$f(x) = \ln\left(\frac{x^2 - 7}{x}\right)$$

$$= \ln\left(x^2 - 7\right) - \ln x \qquad \text{(P2)}$$

$$= \frac{2x}{\left(x^2 - 7\right)} - \frac{1}{x} \qquad \text{Theorem 7}$$

$$= \frac{2x}{x^2 - 7} - \frac{1}{x}$$

Combine the fractions by finding a common denominator.

$$f'(x) = \frac{2x}{x^2 - 7} \cdot \frac{x}{x} - \frac{1}{x} \cdot \frac{x^2 - 7}{x^2 - 7} \qquad \text{Multiply by 1}$$

$$= \frac{2x^2}{x\left(x^2 - 7\right)} - \frac{x^2 - 7}{x\left(x^2 - 7\right)}$$

$$= \frac{x^2 + 7}{x\left(x^2 - 7\right)}$$

81. $g(x) = e^x \ln x^2$

$$g'(x) = e^x \cdot \underbrace{\frac{2x}{x^2}}_{\text{Theorem 7}} + e^x \cdot \ln x^2$$

$$= e^x \cdot \frac{2}{x} + e^x \ln x^2$$

$$= \frac{2e^x}{x} + 2e^x \ln x$$

83. $f(x) = \ln\left(e^x + 1\right)$

$$f'(x) = \frac{\left(e^x + 0\right)}{e^x + 1} \qquad \text{Theorem 7}$$

$$= \frac{e^x}{e^x + 1}$$

85. $g(x) = \left(\ln x\right)^4$

Using the Extended Power Rule, we have

$$g'(x) = 4\left(\ln x\right)^3 \cdot \frac{1}{x}$$

$$= \frac{4\left(\ln x\right)^3}{x}$$

87. $f(x) = \ln\left(\ln(8x)\right)$

First we apply Theorem 7.

$$f'(x) = \frac{1}{\ln(8x)} \cdot \left(\frac{d}{dx}\ln(8x)\right)$$

Using Theorem 7 again, we have

$$f'(x) = \frac{1}{\ln(8x)} \cdot \frac{1}{8x} \cdot 8$$

$$= \frac{1}{x\ln(8x)}$$

89. $g(x) = \ln(5x) \cdot \ln(3x)$

Using the Product Rule along with Theorem 7, we have

$$g'(x) = \ln(5x) \cdot \underbrace{\left(\frac{3}{3x}\right)}_{\text{Theorem 7}} + \underbrace{\left(\frac{5}{5x}\right)}_{\text{Theorem 7}} \cdot \ln(3x)$$

$$= \ln(5x)\frac{1}{x} + \frac{1}{x}\ln(3x)$$

$$= \frac{\ln(5x) + \ln(3x)}{x}$$

If we wanted to simplify the expression, we could use Property 1 to combine the logarithms.

$$g'(x) = \frac{\ln(5x \cdot 3x)}{x} = \frac{\ln(15x^2)}{x}$$

91. First, we find the point of tangency by evaluating the function at $x = 2$.

$x = 2$;

$$y = \left(2^2 - 2\right)\ln(6 \cdot 2) = 2\ln(12) \approx 4.9698$$

Point of tangency: $(2, 4.9698)$

Now, we find the slope of the function at $x = 2$ by taking the derivative.

$$\frac{dy}{dx} = \left(x^2 - x\right)\left(6 \cdot \frac{1}{6x}\right) + (2x-1)\ln(6x)$$

$$= \left(x^2 - x\right)\left(\frac{1}{x}\right) + (2x-1)\ln(6x)$$

$$= x - 1 + (2x-1)\ln(6x)$$

Next, we evaluate the derivative at $x = 2$

$$\left.\frac{dy}{dx}\right|_{x=2} = 2 - 1 + (2\cdot2-1)\ln(6\cdot2)$$

$$= 1 + 3\ln(12) \approx 8.4547$$

Now we have the slope and a point on the tangent line.

We use the Point-Slope formula to find the equation of the tangent line.

$$y - y_1 = m(x - x_1)$$

$$y - 4.970 = 8.455(x - 2)$$

$$y - 4.970 = 8.455x - 16.91$$

$$y = 8.455x - 11.94$$

93. First, we find the point of tangency by evaluating the function at $x = 3$.

$x = 3$;

$$y = (\ln 3)^2 \approx 1.207$$

Point of tangency: $(3, 1.207)$

Now, we find the slope of the function at $x = 3$ by taking the derivative of the function using the Extended Power Rule.

$$\frac{dy}{dx} = 2(\ln(x)) \cdot \frac{1}{x}$$

$$= \frac{2\ln x}{x}$$

Next, we evaluate the derivative at $x = 3$

$$\left.\frac{dy}{dx}\right|_{x=3} = \frac{2\ln(3)}{3} \approx 0.732$$

Now we have the slope and a point on the tangent line. We use the Point-Slope formula to find the equation of the tangent line.

$$y - y_1 = m(x - x_1)$$

$$y - 1.207 = 0.732(x - 3)$$

$$y - 1.207 = 0.732x - 2.196$$

$$y = 0.732x - 0.989$$

95. $N(a) = 2000 + 500\ln a, \ a \geq 1$

a) We substitute 1 in for a.

$$N(1) = 2000 + 500 \cdot \ln 1$$

$$= 2000 + 500 \cdot 0$$

$$= 2000$$

Thus, 2000 units were sold after spending $1000 on advertising.

b) Taking the derivative with respect to a, we have

$$N'(a) = 0 + 500 \cdot \frac{1}{a}$$

$$= \frac{500}{a} \quad .$$

Therefore,

$$N'(10) = \frac{500}{10} = 50 .$$

c) $N'(a) > 0$ for all $a \geq 1$. Thus $N(a)$ is an increasing function and has a minimum value of 2000 when $a = 1$ thousand dollars. There is no maximum, because there is not an upper limit on the advertising budget.

d) ✎

97. How long should the campaign last in order to maximize profit? We need to find the revenue and cost functions in order to find the profit. We find the revenue function first.

$$R(t) = \begin{pmatrix} \text{Price} \\ \text{Per} \\ \text{Unit} \end{pmatrix} \cdot \begin{pmatrix} \text{Target} \\ \text{Market} \end{pmatrix} \cdot \begin{pmatrix} \text{Percentage} \\ \text{Buying} \end{pmatrix}$$

$$R(t) = 1.50(1,000,000)\left(1 - e^{-0.04t}\right)$$

$$= 1,500,000 - 1,500,000 e^{-0.04t}$$

Next, we find the cost function.

$$C(t) = \begin{pmatrix} \text{Advertising cost} \\ \text{per day} \end{pmatrix} \cdot \begin{pmatrix} \text{Number of} \\ \text{days} \end{pmatrix}$$

$$C(t) = 2000 \cdot t$$

Now, we find the profit function.

$$P(t) = R(t) - C(t)$$

$$= 1,500,000 - 1,500,000 e^{-0.04t} - 2000t$$

We take the derivative of the profit function.

$$P'(t) = 0 - 1,500,000(-0.04) e^{-0.04t} - 2000$$

$$= 60,000 e^{-0.04t} - 2000$$

Next, we set the derivative of the profit function equal to zero and solve for t

$$P'(t) = 0$$

$$60,000 e^{-0.04t} - 2000 = 0$$

$$60,000 e^{-0.04t} = 2000$$

$$e^{-0.04t} = \frac{2000}{60,000}$$

$$e^{-0.04t} = \frac{1}{30}$$

$$\ln\left(e^{-0.04t}\right) = \ln\left(\frac{1}{30}\right)$$

$$-0.04t = -3.401197$$

$$t \approx \frac{-3.401197}{-0.04} \approx 85.03$$

We see that the only critical value is 85.03 days. We will apply the second derivative test to see if we have a maximum.

The second derivative of the profit function is $P''(t) = -2400 e^{-0.04t}$.

Evaluating this at the critical value we get

$$P''(85.03) = -2400 e^{-0.04 \cdot (85.03)} \approx -80 < 0 \text{ so we}$$

have a maximum.

Rounding to the next whole day, the length of the advertising campaign must be 86 days to result in maximum profit.

99. a) Revenue is given by $R(x) = x \cdot p(x)$.

Therefore,

$$R(x) = x \cdot (53.5 - 8 \ln x)$$

$$R(x) = 53.5x - 8x \ln x.$$

b) We find the derivative by differentiating each term.

$$R'(x) = 53.5 - \underbrace{8x\left(\frac{1}{x}\right) - 8(1) \ln x}_{\text{Product Rule}}$$

$$= 45.5 - 8 \ln x$$

c) ✎

101. a) Taking the derivative of the profit function $P(x) = 2x - 0.3x \ln x$ will give us marginal profit.

$$P'(x) = 2 - \left[0.3x\left(\frac{1}{x}\right) + 0.3 \ln x\right]$$

$$= 2 - 0.3 - 0.3 \ln x$$

$$= 1.7 - 0.3 \ln x$$

b) ✎

c) Since $P'(x)$ is defined for all values of $x > 0$. We find the critical values by setting $P'(x) = 0$,

$$1.7 - 0.3 \ln x = 0$$

$$-0.3 \ln x = -1.7$$

$$\ln x = \frac{-1.7}{-0.3}$$

$$\ln x = 5.666667$$

$$x = e^{5.66667}$$

$$x \approx 289.069$$

Notice that:

$$P''(x) = \frac{-0.3}{x} < 0 \text{ for all values of } x > 0$$

The critical value results in a maximum. Therefore, 289,069 candles should be sold to maximize profit.

103. $S(t) = 78 - 20\ln(t+1), \quad t \geq 0$

a) $S(0) = 78 - 20\ln(0+1)$

$\qquad = 78 - 20\ln(1)$

$\qquad = 78 - 20 \cdot 0$

$\qquad = 78$

The average score when they initially took the exam was 78%.

b) $S(4) = 78 - 20\ln(4+1)$

$\qquad = 78 - 20\ln(5)$

$\qquad = 78 - 20(1.609438)$

$\qquad = 78 - 32.18876$

$\qquad \approx 45.8$

The average score 4 months after they took the exam was 45.8%.

c) $S(24) = 78 - 20\ln(24+1)$

$\qquad = 78 - 20\ln(25)$

$\qquad = 78 - 20(3.218876)$

$\qquad = 78 - 64.37752$

$\qquad \approx 13.6$

The average score 24 months after they took the exam was 13.6%.

d) First, we reword the question:
"13.6 (the average score after 24 months) is what percent of 78 (the average score on the initial exam)?
Next, we translate the sentence into an equation and solve:

$\qquad 13.6 = x \cdot 78$

$\qquad \dfrac{13.6}{78} = x$

$\qquad 0.1744 = x$

$\qquad 17.44\% \approx x$

The students retained approximately 17.44% of their original answers after 2 years.

e) $S(t) = 78 - 20\ln(t+1)$

$\qquad S'(t) = 0 - 20(1)\left(\dfrac{1}{t+1}\right)$

$\qquad = -\dfrac{20}{t+1}$

f) $S'(t) < 0$ for all values of $t \geq 0$. Therefore, $S(t)$ is a decreasing function and has a maximum value of 78% when $t = 0$. The function does not have a minimum value.

g) ✎

105. $v(p) = 0.37\ln p + 0.05$

a) $v(635) = 0.37\ln 635 + 0.05$

$\qquad = 0.37(6.453625) + 0.05$

$\qquad = 2.4378$

$\qquad \approx 2.44$

The average walking speed of a person living in Seattle is 2.44 feet per second.

b) $v(8340) = 0.37\ln 8340 + 0.05$

$\qquad = 0.37(9.0288185) + 0.05$

$\qquad = 3.390663$

$\qquad \approx 3.39$

The average walking speed of a person living in New York is 3.39 feet per second.

c) $v'(p) = 0.37\dfrac{1}{p} + 0 = \dfrac{0.37}{p}$

d) ✎

107. $\qquad P = P_0 e^{kt}$

$\qquad \dfrac{P}{P_0} = e^{kt}$ \qquad Divide by P_0

$\qquad \ln\left(\dfrac{P}{P_0}\right) = \ln\left(e^{kt}\right)$ \qquad Take the Natural Log on both sides

$\qquad \ln\left(\dfrac{P}{P_0}\right) = kt$ \qquad (P5)

$\qquad \dfrac{\ln\left(\dfrac{P}{P_0}\right)}{k} = t$ \qquad Divide by k.

109. $g(x) = \left[\ln(x+5)\right]^4$

$g'(x) = 4\left[\ln(x+5)\right]^3 \cdot \dfrac{d}{dx}\left[\ln(x+5)\right]$ \quad Chain Rule

$g'(x) = 4\left[\ln(x+5)\right]^3 \cdot \dfrac{1}{x+5}$ \qquad Theorem 7

$g'(x) = \dfrac{4\left[\ln(x+5)\right]^3}{x+5}$

111. $f(t) = \ln\left|\dfrac{1-t}{1+t}\right|$

$f(t) = \ln|(1-t)| - \ln|(1+t)|$ (P2)

$f'(t) = \dfrac{(-1)}{1-t} - \dfrac{(1)}{1+t}$ By Theorem 7

$= \dfrac{-1}{1-t} - \dfrac{1}{1+t}$

Next, we find a common denominator.

$= \dfrac{1+t}{(1-t)(1+t)} - \dfrac{1-t}{(1-t)(1+t)}$

$f'(t) = \dfrac{-2}{1-t^2} = -\dfrac{2}{1-t^2}$

113. $f(x) = \log_5 x$

Using (P7), the change-of-base formula.

$f(x) = \dfrac{\ln x}{\ln 5}$

Remember that $\ln 5$ is a constant.

$f'(x) = \dfrac{1}{\ln 5} \cdot \dfrac{1}{x}$

$= \dfrac{1}{x \ln 5}$

115. $y = \ln\sqrt{5+x^2}$

$y = \ln\left(5+x^2\right)^{1/2}$

$y = \tfrac{1}{2}\ln\left(5+x^2\right)$ (P3)

$\dfrac{dy}{dx} = \dfrac{1}{2} \cdot \dfrac{2x}{5+x^2}$ Theorem 7

$\dfrac{dy}{dx} = \dfrac{x}{5+x^2}$

117. $y = \dfrac{x^{n+1}}{n+1}\left(\ln x - \dfrac{1}{n+1}\right)$

We use the product rule and Theorem 7.

$\dfrac{dy}{dx} = \dfrac{x^{n+1}}{n+1} \cdot \dfrac{d}{dx}\left(\ln x - \dfrac{1}{n+1}\right)$

$\qquad\qquad + \dfrac{d}{dx}\left(\dfrac{x^{n+1}}{n+1}\right) \cdot \left(\ln x - \dfrac{1}{n+1}\right)$

$= \dfrac{x^{n+1}}{n+1}\left(\dfrac{1}{x} - 0\right) + x^n\left(\ln x - \dfrac{1}{n+1}\right)$

$= \dfrac{x^n}{n+1} + x^n \ln x - \dfrac{x^n}{n+1}$

$= x^n \ln x$

119. Let $X = \log_a M$ and $Y = \log_a N$

Proof of Property 1 of Theorem 3:

$M = a^X$ and $N = a^Y$ <u>Definition of Logarithm</u>

So,

$MN = a^X a^Y = a^{X+Y}$ <u>Product Rule for Exponents</u>

Thus,

$\log_a(MN) = X + Y$ <u>Definition of Logarithm</u>

$\qquad\qquad = \log_a M + \log_a N$ <u>Substitution</u>

121. Proof of Property 3 of Theorem 3:

$M = a^X$ <u>Definition of Logarithm</u>

So,

$M^k = \left(a^X\right)^k$

$M^k = a^{X \cdot k}$ <u>Power Rule for Exponents</u>

Thus,

$\log_a\left(M^k\right) = X \cdot k$ <u>Definition of Logarithm</u>

$\qquad\qquad = k \cdot \log_a M$ <u>Substitution</u>

123. $\displaystyle\lim_{h \to 0} \dfrac{\ln(1+h)}{h}$

Note: $\ln(1+h)$ exists only for $h > -1$.

Since the function is not continuous at $h = 0$, we will use input-output tables.

First, we look as h approaches 0 from the left.

h	$\dfrac{\ln(1+h)}{h}$
-0.9	2.56
-0.5	1.39
-0.1	1.05
-0.01	1.01
-0.001	1.001

From the table we observe that

$\displaystyle\lim_{h \to 0^-} \dfrac{\ln(1+h)}{h} = 1$.

Now, we look as h approaches 0 from the right.

h	$\dfrac{\ln(1+h)}{h}$
0.9	0.71
0.5	0.81
0.1	0.95
0.01	0.995
0.001	0.9995

From the table we observe that

$\displaystyle\lim_{h \to 0^+} \dfrac{\ln(1+h)}{h} = 1$.

Thus,

$\displaystyle\lim_{h \to 0} \dfrac{\ln(1+h)}{h} = 1$.

125. We consider the numbers 81^{81} and 9^{160}.
First we apply the natural log to each number

$$\ln 81^{81} = 81 \cdot \ln 81$$
$$= 81 \cdot \ln 3^4$$
$$= 81 \cdot 4 \cdot \ln 3$$
$$= 324 \cdot \ln 3$$
$$\ln 9^{160} = 160 \cdot \ln 9$$
$$= 160 \cdot \ln 3^2$$
$$= 160 \cdot 2 \cdot \ln 3$$
$$= 320 \cdot \ln 3$$

We see that
$$324 \cdot \ln 3 > 320 \cdot \ln 3$$
Therefore,
$$81^{81} > 9^{160}.$$

127. $f(x) = x \ln x$
Using the window:

We graph the function using a calculator.

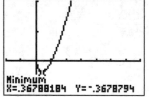

Using the minimum feature, we have:

The relative minimum value that occurs at $x = 0.368$ is -0.368.

129. a) We enter the function into our graphing utility as follows:

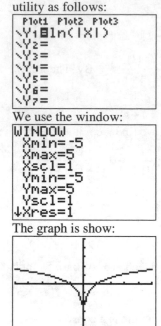

We use the window:

The graph is show:

b) To find the slopes of the tangent lines at the points, we will evaluate the derivative at each of the points. First we find the derivative of the function.

$$f(x) = \ln|x|$$
$$f'(x) = \frac{d}{dx}\ln|x|$$
$$= \frac{1}{x}.$$

For $x = -3$, we have:
$$f'(-3) = \frac{1}{-3} = -\frac{1}{3}.$$
For $x = -2$, we have:
$$f'(-2) = \frac{1}{-2} = -\frac{1}{2}$$
For $x = -1$, we have:
$$f'(-1) = \frac{1}{-1} = -1$$

c) Using the derivative found in part b) we see:

$$f'(3) = \frac{1}{3}$$
$$f'(2) = \frac{1}{2}$$
$$f'(1) = 1$$

Therefore the slopes of the tangent lines at $x = -3, x = -2,$ and $x = -1$ are the opposites of the slopes of the tangent lines at $x = 3, x = 2,$ and $x = 1$.

d)

Exercise Set 3.3

1. Using Theorem 8, the general form of f that satisfies the equation $f'(x) = 4 \cdot f(x)$ is

$f(x) = ce^{4x}$ for some constant c.

3. Using Theorem 8, the general form of A that satisfies the equation $\dfrac{dA}{dt} = -9 \cdot A$ is

$A = ce^{-9t}$, or $A(t) = ce^{-9t}$ for some constant c.

Note: If we were to let the initial population, the population when $t = 0$, be represented by $A(0) = A_0$, then we have

$A_0 = A(0) = ce^{-9 \cdot 0} = ce^0 = c$. Thus, $A_0 = c$, and we can express $A(t) = A_0 e^{-9t}$.

5. Using Theorem 8, the general form of Q that satisfies the equation $\dfrac{dQ}{dt} = k \cdot Q$ is

$Q = ce^{kt}$, or $Q(t) = ce^{kt}$ for some constant c.

Note: If we were to let the initial population, the population when $t = 0$, be represented by $Q(0) = Q_0$, then we have

$Q_0 = Q(0) = ce^{k \cdot 0} = ce^0 = c$. Thus, $Q_0 = c$, and we can express $Q(t) = Q_0 e^{kt}$.

7. a) Using Theorem 8, the general form of N that satisfies the equation $N'(t) = 0.058 \cdot N(t)$ is

$N(t) = ce^{0.058t}$ for some constant c.

Allowing $t = 0$ to correspond to 2009 when approximately 483,000 patent applications were received, gives us the initial condition $N(0) = 483,000$. Therefore:

$483,000 = ce^{0.058 \cdot 0}$

$483,000 = ce^0$

$483,000 = c$

Substituting this value for c. We get:

$N(t) = 483,000e^{0.058t}$

b) In 2020, $t = 2020 - 2009 = 11$.

$N(11) = 483,000e^{0.058(11)}$

$= 483,000e^{0.638}$

$= 483,000(1.892692)$

$= 914,170.226$

$\approx 914,170$

There will be approximately 914,170 patent applications received in 2020.

c) From Theorem 9, the doubling time T is given by $T = \dfrac{\ln 2}{k}$.

$T = \dfrac{\ln 2}{0.058} \approx 11.95$

It will take approximately 12 years for the number of patent applications to double. This means that in 2021, the number of patents will have doubled.

9. a) Using Theorem 8, the general form of P that satisfies the equation

$\dfrac{dP}{dt} = 0.059 \cdot P(t)$ is

$P(t) = P_0 e^{0.059t}$ for some initial principal P_0.

b) If \$1000 is invested, then $P_0 = 1000$ and

$P(t) = 1000e^{0.059t}$

$P(1) = 1000e^{0.059(1)}$

$= 1000e^{0.059}$

$= 1000(1.06077524074)$

≈ 1060.78

The balance after 1 year is \$1060.78.

$P(2) = 1000e^{0.059(2)}$

$= 1000e^{0.118}$

$= 1000(1.1252441)$

≈ 1125.24

The balance after 2 years is \$1125.24.

c) From Theorem 9, the doubling time T is given by $T = \dfrac{\ln 2}{k}$.

$T = \dfrac{\ln 2}{0.059} \approx 11.7$

It will take approximately 11.7 years for the balance to double.

11. a) Using Theorem 8, the general form of G that satisfies the equation
$$\frac{dG}{dt} = 0.093 \cdot G(t) \text{ is}$$
$G(t) = ce^{0.093t}$ for some constant c.
Allowing $t = 0$ to correspond to 2000 when approximately 4.7 billion gallons of bottled water were sold, gives us the initial condition $G(0) = 4.7$. Therefore:
$$4.7 = ce^{0.093 \cdot 0}$$
$$4.7 = ce^0$$
$$4.7 = c$$
Substituting this value for c. We get:
$$G(t) = 4.7e^{0.093t}$$
Where $G(t)$ is in billions of gallons sold and t is the number of years since 2000.

b) In 2025, $t = 2025 - 2000 = 25$.
$$N(25) = 4.7e^{0.093(25)}$$
$$= 4.7e^{2.325}$$
$$= 4.7(10.22668009)$$
$$= 48.06559$$
$$\approx 48.07$$
There will be approximately 48.07 billion gallons of bottle water sold in 2010.

c) From Theorem 9, the doubling time T is given by $T = \dfrac{\ln 2}{k}$.
$$T = \frac{\ln 2}{0.093} \approx 7.5$$
It will take approximately 7.5 years for the amount of bottled water sold to double. Sometime in the middle of 2007 according to our model.

13. From Theorem 9, the doubling time T is given by $T = \dfrac{\ln 2}{k}$. Substituting $T = 15$ we get:
$$15 = \frac{\ln 2}{k}$$
Solve for k to find the interest rate.
$$15 \cdot k = \ln 2$$
$$k = \frac{\ln 2}{15}$$
$$k \approx 0.046210$$
The annual interest rate is 4.62%.

15. From Theorem 9, the doubling time T is given by $T = \dfrac{\ln 2}{k}$. Substituting the growth rate $k = 0.10$ into the formula.
$$T = \frac{\ln 2}{0.10}$$
$$\approx 6.9$$
The doubling time for the demand for oil is 6.9 years; therefore, at the end of the year 2018 the demand for oil will be double the demand for oil in 2012.

17. Find the doubling time:
$$T = \frac{\ln 2}{k} = \frac{\ln 2}{0.062} \approx 11.2$$
The doubling time is 11.2 years.
$$P(t) = P_0e^{kt}$$
$$P(t) = 75,000e^{0.062t}$$
We substitute $t = 5$ to find the amount after five years:
$$P(5) = 75,000e^{0.062(5)}$$
$$= 75,000e^{0.31}$$
$$\approx 102,256.88$$
In five years the account will have $102,256.88.

19. First, we find the initial investment:
$$P(t) = P_0e^{kt}$$
Substituting, we have
$$11,414.71 = P_0e^{0.084(5)}$$
$$11,414.71 = P_0e^{0.42}$$
$$\frac{11,414.71}{e^{0.42}} = P_0$$
$$7499.9989 \approx P_0$$
$$7500.00 \approx P_0$$
The initial investment is $7500.
Next, we find the doubling time:
$$T = \frac{\ln 2}{k} = \frac{\ln 2}{0.084} \approx 8.3$$
The doubling time is 8.3 years.

21. a) The exponential growth function is
$V(t) = V_0 e^{kt}$. We will express $V(t)$ in
dollars and t as the number of
years since 1950. This set up gives us the
initial value of $V_0 = 30,000$ dollars.
Substituting this into the function, we have:
$V(t) = 30,000 e^{kt}$

In 2004, $t = 2004 - 1950 = 54$, and the
painting sold for 104,168,000 so we know
$V(54) = 104,168,000$ million dollars.

Substitute this information into the function
and solve for k.

$$104,168,000 = 30,000 e^{k(54)}$$

$$\frac{104,168,000}{30,000} = e^{54k}$$

$$\ln\left(\frac{104,168,000}{30,000}\right) = \ln e^{54k}$$

$$\ln\left(\frac{104,168,000}{30,000}\right) = 54k$$

$$\frac{\ln\left(\frac{104,168,000}{30,000}\right)}{54} = k$$

$$0.151 \approx k$$

The exponential growth rate is
approximately 15.1%. Substituting this
back into the function, we find the
exponential growth function is
$V(t) = 30,000 e^{0.151t}$.

b) In 2015, $t = 2015 - 1950 = 65$. We evaluate
the exponential growth function found in
part 'a' to get:
$$V(65) = 30,000 e^{0.151(65)}$$
$$= 30,000 e^{9.815}$$
$$= 30,000(18,306.29)$$
$$\approx 549,188,702$$

In the year 2015, the value of the painting
will be approximately $549,188,702.

c) From Theorem 9, the doubling time T is
given by $T = \dfrac{\ln 2}{k}$.

$$T = \frac{\ln 2}{0.151} \approx 4.59 \approx 4.6$$

The painting doubles in value approximately
every 4.6 years.

d) We set $V(t) = 1,000,000,000$ and solve for t.

$$30,000 e^{0.151t} = 1,000,000,000$$

$$e^{0.151t} = \frac{1,000,000,000}{30,000}$$

$$\ln e^{0.151t} = \ln\left(\frac{1,000,000,000}{30,000}\right)$$

$$0.151t = \ln\left(\frac{1,000,000,000}{30,000}\right)$$

$$t = \frac{\ln\left(\frac{1,000,000,000}{30,000}\right)}{0.151}$$

$$t \approx 68.969$$

$$t \approx 69$$

It will take approximately 69 years for the
value of the painting to reach 1 billion
dollars. This will occur in the year 2019.
$[1950 + 69 = 2019]$

23. a) The exponential growth function is
$F(t) = F_0 e^{kt}$. We will express t as the
number of years since 2011. This set up
gives us the initial value of $F_0 = 2.30$
trillion dollars.
Substituting this into the function, we have:
$F(t) = 2.30 e^{kt}$

In 2013, $t = 2013 - 2011 = 2$, and
federal receipts were $2.77 trillion so we
know $F(2) = 2.77$ billion dollars.

Substitute this information into the function
and solve for k.

$$2.77 = 2.30 e^{k(2)}$$

$$\frac{2.77}{2.30} = e^{2k}$$

$$\ln\left(\frac{2.77}{2.30}\right) = \ln e^{2k}$$

$$\ln\left(\frac{2.77}{2.30}\right) = 2k$$

$$\frac{\ln\left(\frac{2.77}{2.30}\right)}{2} = k$$

$$0.092969 \approx k$$

The exponential growth rate is 0.092969.
Substituting this back into the function, we
find the exponential growth function is
$F(t) = 2.3 e^{0.092969t}$.

b) In 2015, $t = 2015 - 2011 = 4$. We evaluate the exponential growth function found in part 'a' to get:

$$F(4) = 2.3e^{0.092969(4)}$$

$$= 2.3e^{0.371876}$$

$$= 2.3(1.450453)$$

$$\approx 3.336$$

In the year 2015, federal receipts will be approximately \$3.34 trillion.

c) We set $F(t) = 10$ and solve for t.

$$F(t) = 10$$

$$2.3e^{0.092969t} = 10$$

$$e^{0.092969t} = \frac{10}{2.3}$$

$$0.092969t = \ln\left(\frac{10}{2.3}\right)$$

$$t = \frac{\ln\left(\frac{10}{2.3}\right)}{0.092969}$$

$$t \approx 15.8$$

It will take approximately 15.8 years from 2011 or in the year 2026 for federal receipts to reach \$10 trillion.

25. a) Enter the data into your calculator statistics editor.

Now use the regression command to find the exponential growth function.

```
EDIT CALC TESTS
8↑LinReg(a+bx)
9:LnReg
0:ExpReg
A:PwrReg
B:Logistic
C:SinReg
D:Manual-Fit
```

This gives us:

```
ExpReg
y=a*b^x
a=1.60680697
b=1.60414491
```

The calculator models the function.

$y = 1.60680697(1.60414491)^x$, where y is total monthly mobile data traffic in exabytes and x is the number of years after 2013.

Using the fact that $b = e^{(\ln b)}$, we have:

$1.60414491 = e^{(\ln 1.60414491)} = e^{0.472590848256}$

Therefore, the exponential growth function:

$y = 1.60680697e^{0.472590848256x}$, where x is years since 2013 and y is total monthly mobile data traffic in exabytes.

b) From the model in part 'a', we see that k is 0.47259084825 therefore, the growth rate as a percentage is approximately 47.3%.

c) Using the model in part 'a'.
In 2020, $x = 2020 - 2013 = 7$. Substituting in to the model, we have

$$y = 1.60680697e^{0.472590848256(7)}$$

$$y = 43.921$$

In 2020, the model estimates approximately 43.9 exabytes of total monthly mobile data traffic will be used.

d) We need to solve the equation $y = 50$ for x.

$$1.60680697e^{0.472590848256x} = 50$$

$$e^{0.472590848256x} = \frac{50}{1.60680697}$$

$$\ln e^{0.472590848256x} = \ln(31.1176)$$

$$0.472590848256x = \ln(31.1176)$$

$$x = \frac{\ln(31.1176)}{0.472590848256}$$

$$\approx 7.27$$

It will take approximately 7.3 years for the total monthly mobile data traffic to exceed 50 exabytes.

e) From Theorem 9, the doubling time T is given by $T = \dfrac{\ln 2}{k}$.

$$T = \frac{\ln 2}{0.472590848256} \approx 1.4667.$$

The doubling time for total monthly mobile data traffic is approximately 1.47 years.

27. If we let $t = 0$ correspond to the year 1626 the initial value of Manhattan is $V_0 = 24$. Using the exponential growth function $V(t) = V_0 e^{kt}$.

Assuming an exponential rate of inflation of 5% means that $k = 0.05$. Substituting these values into the growth function gives us:

$V(t) = 24e^{0.05t}$.

In 2020, $t = 2020 - 1626 = 394$

$V(394) = 24e^{0.05(394)}$

$\qquad = 24e^{19.7}$

$\qquad \approx 8,626,061,203$

Manhattan Island will be worth approximately $8,626,061,203 or $8.6 billion.

29. a) The exponential growth function is

$S(t) = S_0 e^{kt}$. We will express $S(t)$ in cents and t as the number of years since 1962. This set up gives us the initial value of $S_0 = 4$ dollars.

Substituting this into the function, we have:

$S(t) = 4e^{kt}$

In 2014, $t = 2014 - 1962 = 52$, and the price of a stamp was 49 cents, so we know $S(52) = 49$ cents.

Substitute this information into the function and solve for k.

$$49 = 4e^{k(52)}$$

$$\frac{49}{4} = e^{52k}$$

$$\ln(12.25) = \ln e^{52k}$$

$$\ln(12.25) = 52k$$

$$\frac{\ln(12.25)}{52} = k$$

$$0.0482 \approx k$$

Substituting this value back into the function, we find the exponential growth function is

$S(t) = 4e^{0.0482t}$.

b) The exponential growth rate is approximately 4.82% per year.

c) In 2016, $t = 2016 - 1962 = 54$. We evaluate the exponential growth function found in part 'a' to get:

$S(54) = 4e^{0.0482(54)}$

$\qquad = 54.005 \approx 54$

In the year 2016, the price of a stamp will be approximately 54 cents.

In 2018, $t = 2018 - 1962 = 56$. We evaluate the exponential growth function found in part 'a' to get:

$S(56) = 4e^{0.0482(56)}$

$\qquad = 59.47 \approx 59$

In the year 2018, the price of a stamp will be approximately 59 cents.

In 2020, $t = 2020 - 1962 = 58$. We evaluate the exponential growth function found in part 'a' to get:

$S(58) = 4e^{0.0482(58)}$

$\qquad = 65.49 \approx 65$

In the year 2020, the price of a stamp will be approximately 65 cents.

d) For the years 2014 – 2024, the total cost of *Forever Stamps* would be

$10 \cdot 4900 = \$49,000$.

For the years 2014 – 2015 the cost of regular first-class stamps was

$2 \cdot (4900) = \$9800$.

For the years 2016 – 2017, the price of postage increased to $0.54. The cost of regular first-class stamps for these 2 years is

$2 \cdot (0.54) \cdot (10,000) = \$10,800$.

For the years 2018 – 2019, the price of postage increased to $0.59. The cost of regular first-class stamps for these 2 years is

$2 \cdot (0.59) \cdot (10,000) = \$11,800$.

For the years 2020 – 2024, the price of postage increased to $0.65. The cost of regular first-class stamps for these 4 years is

$4 \cdot (0.65) \cdot (10,000) = \$26,000$.

Therefore, the cost of regular first-class stamps for the years 2014 – 2024 would be:

$\$9800 + \$10,800 + \$11,800 + \$26,000$, or $\$58,400$.

Thus, by buying *Forever Stamps* the firm would save

$\$58,400 - \$49,000 = \$9,400$.

e)

31. $P(x) = \dfrac{100}{1 + 49e^{-0.13x}}$

a) $P(0) = \dfrac{100}{1 + 49e^{-0.13(0)}}$

$= \dfrac{100}{1 + 49e^{0}}$

$= \dfrac{100}{1 + 49}$

$= \dfrac{100}{50}$

$= 2$

About 2% purchased the game without seeing the advertisement.

b) $P(5) = \dfrac{100}{1 + 49e^{-0.13(5)}}$

$= \dfrac{100}{1 + 49e^{-0.65}}$

≈ 3.8

About 3.8% will purchase the game after the advertisement runs 5 times.

$P(10) = \dfrac{100}{1 + 49e^{-0.13(10)}}$

$= \dfrac{100}{1 + 49e^{-1.3}}$

≈ 7.0

About 7.0% will purchase the game after the advertisement runs 10 times.

$P(20) = \dfrac{100}{1 + 49e^{-0.13(20)}}$

$= \dfrac{100}{1 + 49e^{-2.6}}$

≈ 21.6

About 21.6% will purchase the game after the advertisement runs 20 times.

$P(30) = \dfrac{100}{1 + 49e^{-0.13(30)}}$

$= \dfrac{100}{1 + 49e^{-3.9}}$

≈ 50.2

About 50.2% will purchase the game after the advertisement runs 30 times.

$P(50) = \dfrac{100}{1 + 49e^{-0.13(50)}}$

$= \dfrac{100}{1 + 49e^{-6.5}}$

≈ 93.1

About 93.1% will purchase the game after the advertisement runs 50 times.

Substituting in 60, we have:

$P(60) = \dfrac{100}{1 + 49e^{-0.13(60)}}$

$= \dfrac{100}{1 + 49e^{-7.8}}$

≈ 98.0

About 98.0% will purchase the game after the advertisement runs 60 times.

c) We apply the Quotient Rule to take the derivative.

$P'(x)$

$= \dfrac{\left(1 + 49e^{-0.13x}\right)(0) - 49(-0.13)e^{-0.13x} \cdot 100}{\left(1 + 49e^{-0.13x}\right)^{2}}$

$= \dfrac{637e^{-0.13x}}{\left(1 + 49e^{-0.13x}\right)^{2}}$

d) The derivative $P'(x)$ exists for all real numbers. The equation $P'(x) = 0$ has no solution. Thus, the function has no critical points and hence, no relative extrema. $P'(x) > 0$ for all real numbers, so $P(x)$ is increasing on $[0, \infty)$. The second derivative can be used to show that the graph has an inflection point at $(29.9, 50)$. The function is concave up on the interval $(0, 29.9)$ and concave down on the interval $(29.9, \infty)$.

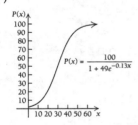

33. a) Using Theorem 8, the general form of V that satisfies the equation

$\dfrac{dV}{dt} = k \cdot V(t)$ is

$V(t) = V_0 e^{kt}$ for some constant c.

Allowing $t = 0$ to correspond to 1938, gives us the initial condition

$V(0) = 0.10$. Therefore:

$0.10 = V_0 e^{k \cdot 0}$

$0.10 = V_0$

The solution is continued on the next page.

Substituting the value for V_0 we found on the previous page for c. We get:
$V(t) = 0.10e^{kt}$.

Next, using the point $(76, 3,200,000)$ we have:

$3,200,000 = 0.10e^{k(76)}$.

We solve this equation for k

$$3,200,000 = 0.10e^{k(76)}$$

$$\frac{3,200,000}{0.10}e^{76k}$$

$$32,000,000 = e^{76k}$$

$$\ln(32,000,000) = 76k$$

$$\frac{\ln(32,000,000)}{76} = k$$

$$0.227 \approx k$$

The exponential function is
$V(t) = 0.10e^{0.227t}$.

b) In 2020, $t = 2020 - 1938 = 82$.
$V(82) = 0.10e^{0.227(82)}$

$$= 0.10e^{18.614}$$

$$\approx 12,132,700$$

There comic book will be valued at about $12,123,700 in 2020.

c) From Theorem 9, the doubling time T is given by $T = \frac{\ln 2}{k}$.

$$T = \frac{\ln 2}{0.227} \approx 3.05$$

The comic book doubles in value approximately every 3.05 years.

d) $V(t) = 30,000,000$

$$30,000,000 = 0.10e^{0.227t}$$

$$\frac{30,000,000}{0.10} = e^{0.227t}$$

$$\ln(300,000,000) = 0.227t$$

$$\frac{\ln(300,000,000)}{0.227} = t$$

$$85.98 \approx t$$

The comic book will be valued at $30 million about 86 years after 1938 or in 2024.

35. From Example 6, the model is
$V(t) = 0.10e^{0.228t}$. To determine when the value of the comic book will be 10 million dollars, we set $V(t) = 10,000,000$ and solve for t.

$$V(t) = 10,000,000$$

$$10,000,000 = 0.10e^{0.228t}$$

$$\frac{10,000,000}{0.10} = e^{0.228t}$$

$$\ln(100,000,000) = 0.228t$$

$$\frac{\ln(100,000,000)}{0.228} = t$$

$$80.79 \approx t$$

The comic book will be valued at $10 million 80.8 years after 1939 or in 2019.

37. $k = \frac{\ln 2}{T} = \frac{\ln 2}{69.31} \approx 0.01$
The growth rate k is 0.01 or 1% per year.

39. $k = \frac{\ln 2}{T} = \frac{\ln 2}{17.3} \approx 0.04$
The growth rate k is 0.04 or 4% per year.

41. Let $t = 0$ correspond to 1776.
a) Using the exponential growth model
$P(t) = 2,508,000e^{kt}$, when
$t = 200$, $P(t) = 216,000,000$. Substituting these values into the model, we have:

$$216,000,000 = 2,508,000e^{k(200)}$$

$$\frac{216,000,000}{2,508,000} = e^{200k}$$

$$\ln\left(\frac{216,000,000}{2,508,000}\right) = 200k$$

$$\frac{\ln\left(\frac{216,000,000}{2,508,000}\right)}{200} = k$$

$$0.0223 \approx k$$

The U.S. population was growing at a rate of 2.23% per year. Therefore, the exponential growth model is $P(t) = 2,508,000e^{0.0223t}$.

b)

43. $P(t) = \dfrac{3000}{20 + 130e^{-0.214t}}$

a) Substituting 0 in for t, we have

$$P(5) = \frac{3000}{20 + 130e^{-0.214(0)}}$$

$$= \frac{3000}{20 + 130e^{0}}$$

$$= 20$$

The initial tortoise population in the one-square mile parcel after is 20 tortoises.
Substituting 5 in for t, we have

$$P(5) = \frac{3000}{20 + 130e^{-0.214(5)}}$$

$$= \frac{3000}{20 + 130e^{-1.07}}$$

$$\approx 46.45$$

The tortoise population in the one-square mile parcel after 5 years is approximately 46 tortoises.
Substituting 15 in for t, we have

$$P(15) = \frac{3000}{20 + 130e^{-0.214(15)}}$$

$$= \frac{3000}{20 + 130e^{-3.21}}$$

$$\approx 118.83$$

The tortoise population in the one-square mile parcel after 15 years is approximately 119 tortoises.
Substituting 25 in for t, we have

$$P(25) = \frac{3000}{20 + 130e^{-0.214(25)}}$$

$$= \frac{3000}{20 + 130e^{-5.35}}$$

$$\approx 145.51$$

The tortoise population in the one-square mile parcel after 25 years is approximately 146 tortoises.

b) We apply the Quotient Rule to take the derivative.

$P'(t)$

$$= \frac{\left(20 + 130e^{-0.214t}\right)(0) - 130(-0.214)e^{-0.214t} \cdot 3000}{\left(20 + 130e^{-0.214t}\right)^2}$$

$$= \frac{0 + 83,460e^{-0.214t}}{\left(20 + 130e^{-0.214t}\right)^2}$$

$$= \frac{83,460e^{-0.214t}}{\left(20 + 130e^{-0.214t}\right)^2}$$

c) Substituting 0 in for t, we have

$$P'(0) = \frac{83,460e^{-0.214(0)}}{\left(20 + 130e^{-0.214(0)}\right)^2}$$

$$= \frac{83,460}{22,500}$$

$$= 3.709$$

The tortoise population in the one-square mile parcel after 0 years is growing at a rate of approximately 3.709 tortoises per year.
Substituting 5 in for t, we have

$$P'(5) = \frac{83,460e^{-0.214(5)}}{\left(20 + 130e^{-0.214(5)}\right)^2}$$

$$= \frac{28,627.4908638}{4172.01113765}$$

$$\approx 6.861$$

The tortoise population in the one-square mile parcel after 5 years is growing at a rate of approximately 6.86 tortoises per year.
Substituting 15 in for t, we have

$$P'(15) = \frac{83,460e^{-0.214(15)}}{\left(20 + 130e^{-0.214(15)}\right)^2}$$

$$= \frac{3368.16294368}{637.378679382}$$

$$\approx 5.284$$

The tortoise population in the one-square mile parcel after 15 years is growing at a rate of approximately 5.28 tortoises per year.
Substituting 25 in for t, we have

$$P'(25) = \frac{83,460e^{-0.214(25)}}{\left(20 + 130e^{-0.214(25)}\right)^2}$$

$$= \frac{396.280682411}{425.071394648}$$

$$\approx 0.932$$

The tortoise population in the one-square mile parcel after 25 years is growing at a rate of approximately 0.93 tortoises per year.

d) Rewriting the limited population growth function to put in in the correct form, we divide numerator and denominator by 20.

$$P(t) = \frac{3000}{20 + 130e^{-0.214t}}$$

$$= \frac{\dfrac{3000}{20}}{\dfrac{20 + 130e^{-0.214t}}{20}}$$

$$= \frac{150}{1 + 6.5e^{-0.982t}}.$$

Thus the limiting population, L, of desert tortoise in this one-square mile parcel is 150 tortoises.

45. Let $t = 0$ correspond to 1930. Thus the initial number of women earning bachelor's degrees is $P_0 = 48,869$. So the exponential growth function is $P(t) = 48,869e^{kt}$. In 2010, $t = 2010 - 1930 = 80$, approximately 920,000 women received a degree. Using this information we can find the exponential growth rate k. Substitute the information into the growth function.

$$920,000 = 48,869e^{k(80)}$$

$$\frac{920,000}{48,869} = e^{80k}$$

$$\ln\left(\frac{920,000}{48,869}\right) = \ln e^{80k}$$

$$\ln\left(\frac{920,000}{48,869}\right) = 80k$$

$$\frac{\ln\left(\dfrac{920,000}{48,869}\right)}{80} = k$$

$$0.0367 \approx k$$

The exponential growth rate is 0.0367 or 3.67%. Therefore the exponential growth function is: $P(t) = 48,869e^{0.0367t}$, where t is time in years since 1930 and $P(t)$ is number of women earning bachelor's degrees.

47. $N(t) = \dfrac{568.803}{1 + 62.200e^{-0.092t}}$

a) $N(3) = \dfrac{568.803}{1 + 62.200e^{-0.092(3)}}$

$$= \frac{568.803}{1 + 62.200e^{-0.276}}$$

$$= 11.801 \approx 12$$

After 3 weeks, approximately 12 students have been infected.

$$N(40) = \frac{568.803}{1 + 62.200e^{-0.092(40)}}$$

$$= \frac{568.803}{1 + 62.200e^{-3.68}}$$

$$= 221.42 \approx 221$$

After 40 weeks, approximately 221 students have been infected.

$$N(80) = \frac{568.803}{1 + 62.200e^{-0.092(80)}}$$

$$= \frac{568.803}{1 + 62.200e^{-7.36}}$$

$$= 547.15 \approx 547$$

After 80 weeks, approximately 547 students have been infected.

b) Using the quotient rule, we find $N'(t)$.

$$N'(t)$$
$$= \frac{\left(1 + 62.2e^{-0.092t}\right)(0) - (-0.092)62.2e^{-0.092t} \cdot 568.803}{\left(1 + 62.2e^{-0.092t}\right)^2}$$

$$= \frac{3254.9182e^{-0.092t}}{\left(1 + 62.2e^{-0.092t}\right)^2}$$

Therefore,

$$N'(20) = \frac{3254.9182e^{-0.092(20)}}{\left(1 + 62.2e^{-0.92(20)}\right)^2}$$

$$= 4.3682219 \approx 4.4$$

The disease is spreading at a rate of approximately 4.4 students per week at 20 weeks.

c) An unrestricted growth model is inappropriate because there is a finite population on the campus. The disease cannot spread without bound on the campus. Therefore the logistic equation is appropriate due to the fact that there will be a limiting value.

The solution is continued on the next page.

To graph the function, enter the equation into the graphing editor on your calculator:

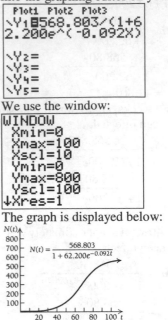

We use the window:

We use the window:
WINDOW
 Xmin=0
 Xmax=100
 Xscl=10
 Ymin=0
 Ymax=800
 Yscl=100
↓Xres=1

The graph is displayed below:

$$N(t) = \frac{568.803}{1+62.200e^{-0.092t}}$$

49. a) Enter the data into the calculator.

Use the Logistic regression.

EDIT **CALC** TESTS
8↑LinReg(a+bx)
9:LnReg
0:ExpReg
A:PwrReg
B:Logistic
C:SinReg
D:Manual-Fit

We have:

Logistic
 y=c/(1+ae^(-bx))
 a=79.56767122
 b=.809743969
 c=29.47232081

From the screen shot, we can see the calculator determined the logistic growth function to be:

$$y = \frac{29.47232081}{1+79.56767122e^{-0.809743969x}}$$

Converting the calculator results into function notation we get:

$$N(t) = \frac{29.47232081}{1+79.56767122e^{-0.809743969t}} \, .$$

b) We round down since the model deals with people. The limiting value appears to be 29 people.

c) The graph is:

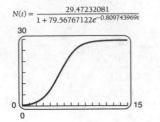

$$N(t) = \frac{29.47232081}{1+79.56767122e^{-0.809743969t}}$$

d) Use the quotient rule to find $N'(t)$.

$$N'(t) = \frac{1898.885181e^{-0.809743969t}}{\left(1+79.56767122e^{-0.809743969t}\right)^2}$$

e) ✎

Note: In Exercises **51 – 59**, it is a good idea to look at the data before you make assumptions about a particular model. Given a data set, all of these solutions could change.

51. ✎

53. ✎

55. ✎

57. ✎

59. ✎

61. Let T_4 be the time it takes for a population to quadruple, then according to the exponential growth function:

$$4 \cdot P_0 = P_0 e^{kT_4}$$

Solve this equation for T_4.

$$\frac{4P_0}{P_0} = e^{kT_4}$$

$$\ln 4 = \ln e^{kT_4}$$

$$\ln 4 = kT_4$$

$$\frac{\ln 4}{k} = T_4$$

$$T_4 = \frac{\ln 4}{k}$$

63. Let k_1 and k_2 represent the growth rates of Q_1 and Q_2. Using the Theorem 9, we know that: $k_1 = \dfrac{\ln 2}{1}$ and $k_2 = \dfrac{\ln 2}{2}$

Since the initial amounts of both quantities are the same we have: $Q_1 = Q_0 e^{\ln 2 \cdot t}$ and $Q_2 = Q_0 e^{\left(\frac{(\ln 2)}{2}\right)t}$. Q_1 is twice the size of Q_2 when:

$$Q_1 = 2Q_2$$

$$Q_0 e^{t \cdot \ln 2} = 2Q_0 e^{t\left(\frac{(\ln 2)}{2}\right)}$$

$$\frac{Q_0 e^{t \cdot \ln 2}}{Q_0 e^{t\left(\frac{(\ln 2)}{2}\right)}} = 2$$

$$e^{t\left(\ln 2 - \left(\frac{(\ln 2)}{2}\right)\right)} = 2$$

$$\ln\left[e^{t\left(\frac{(\ln 2)}{2}\right)}\right] = \ln 2$$

$$t\left(\frac{(\ln 2)}{2}\right) = \ln 2$$

$$t = \frac{\ln 2}{\frac{(\ln 2)}{2}}$$

$$t = \frac{1}{\frac{1}{2}} = 2$$

It takes 2 years for Q_1 to be twice the size of Q_2.

65. ✎

67. Substituting the interest rate into the equation, we have:

$$T = \frac{70}{100 \cdot (0.07)} = 10.$$

With an interest rate of 7% it will take approximately 10 years for the money to double.

69. Using a calculator we have:

$$T = \frac{\ln 2}{0.07} = 9.90210257943$$

Using the calculator the doubling time is 9.9 years.

Using a calculator we have:

$$T = \frac{\ln 2}{0.035} = 19.8042051589$$

Using the calculator the doubling time is 19.8 years.

71. a) Evaluating the function at $t = 0$, we have:

$$R(0) = \frac{4000}{1 + 1999e^{-0.5(0)}}$$

$$= \frac{4000}{2000}$$

$$= 2$$

This represent the initial revenue in million of dollars at the inception of the corporation.

b) $R_{max} = \lim\limits_{t \to \infty} R(t) = 4000$ million dollars.

This represents the maximum attainable revenue of the company over all time.

c) $R(t) = 0.99R_{max}$

$$\frac{4000}{1 + 1999e^{-0.5t}} = 0.99(4000)$$

$$4000 = 3960\left(1 + 1999e^{-0.5t}\right)$$

$$\frac{4000}{3960} = 1 + 1999e^{-0.5t}$$

$$\frac{100}{99} - 1 = 1999e^{-0.5t}$$

$$\frac{\frac{1}{99}}{1999} = e^{-0.5t}$$

$$\ln\left(\frac{\frac{1}{99}}{1999}\right) = -0.5t$$

$$\frac{\ln\left(\frac{\frac{1}{99}}{1999}\right)}{-0.5} = t$$

$$24 \approx t$$

It will take approximately 24 years to reach 99% of the maximum revenue.

Exercise Set 3.4

1. Using Theorem 10 and $k = 0.12$, we have:

$$T = \frac{\ln 2}{0.12} \approx 5.776.$$

The half -life is about 5.78 years.

3. Using Theorem 10 and $k = 0.008$, we have:

$$T = \frac{\ln 2}{0.008} \approx 86.643.$$

The half -life is about 86.64 months.

5. Using Theorem 10 and $k = 0.03$, we have:

$$T = \frac{\ln 2}{0.03} \approx 23.104.$$

The half -life is about 23.10 years.

7. Using Theorem 10 and $k = 0.019$, we have:

$$T = \frac{\ln 2}{0.019} \approx 36.481$$

The half -life is about 36.48 weeks.

9. a) $N(t) = N_0 e^{-kt}$

We substitute 0.096 in for k.

$N(t) = N_0 e^{-0.096t}$

b) $N_0 = 500$

$N(t) = 500e^{-0.096t}$

$N(4) = 500e^{-0.096(4)}$

$= 500e^{-0.384}$

≈ 340.565713

≈ 341

There will be approximately 341g of Iodine -131 present after 4 days.

c) From Theorem 10, we know that the half-life T and the decay rate k are related by:

$$T = \frac{\ln 2}{k}.$$

Substitute $k = .096$

$$T = \frac{\ln 2}{0.096} \approx 7.22028$$

It will take about 7.2 days for half of the 500g of Iodine - 131 to remain.

11. a) When A decomposes at a rate proportional to the amount of A present, we know that

$$\frac{dA}{dt} = -kA.$$

The solution to this equation is

$$A(t) = A_0 e^{-kt}.$$

b) First, we find k. The half-life of A is 3.3 hr. From Theorem 10 we know:

$$k = \frac{\ln 2}{T}$$

$$k = \frac{\ln 2}{3.3} \approx 0.21$$

The initial amount $A_0 = 10$, so the exponential decay function is:

$$A(t) = 10e^{-0.21t}.$$

To determine how long it will take to reduce A to 1 lb we solve $A(t) = 1$ for t.

$$10e^{-0.21t} = 1$$

$$e^{-0.21t} = 0.1$$

$$\ln\left(e^{-0.21t}\right) = \ln(0.1)$$

$$-0.21t = \ln(0.1)$$

$$t = \frac{\ln 0.1}{-0.21}$$

$$t \approx 10.96469$$

$$t \approx 11$$

It will take approximately 11 hours for 10 lb of substance A to reduce to 1 lb.

13. From Theorem 10 we know:

$$k = \frac{\ln 2}{T}$$

$$k = \frac{\ln 2}{3} \approx 0.231$$

The decay rate is 0.231 or 23.1% per minute.

15. From Theorem 10 we know:

$$T = \frac{\ln 2}{k}$$

$$T = \frac{\ln 2}{0.0315} \approx 22.0$$

The half-life is 22.0 years.

17. $P(t) = P_0 e^{-kt}$

We substitute 1000 for P_0, 0.0315 for k, and 100 for t.

$$P(100) = 1000 e^{-0.0315(100)}$$

$$= 1000 e^{-3.15}$$

$$\approx 42.9$$

Approximately 42.9 grams of lead-210 will remain after 100 years.

19. If an ivory tusk has lost 40% of its carbon-14 from its initial amount P_0, then 60% of the initial amount remains. To find the age of the tusk, we solve the following equation for t:

$$0.60 P_0 = P_0 e^{-0.00012097t}$$

$$0.60 = e^{-0.00012097t}$$

$$\ln 0.60 = \ln e^{-0.00012097t}$$

$$\ln 0.60 = -0.00012097t$$

$$\frac{\ln 0.60}{-0.00012097} = t$$

$$4222.7463 \approx t$$

$$4223 \approx t$$

The ivory tusk is approximately 4223 years old.

21. First, we find k. The half-life is 60.1 days. From Theorem 10 we know.

$$k = \frac{\ln 2}{T}$$

$$k = \frac{\ln 2}{60.1}$$

$$k \approx 0.0115$$

The decay rate is approximately 0.0115 or 1.15% per day.

If the initial amount A_0 decreased by 25%, then 75% of A_0 remains. We solve the following equation for t.

$$0.75 A_0 = A_0 e^{-0.0115t}$$

$$0.75 = e^{-0.0115t}$$

$$\ln 0.75 = \ln e^{-0.0115t}$$

$$\ln 0.75 = -0.0115t$$

$$\frac{\ln 0.75}{-0.0115} = t$$

$$25.015 \approx t$$

$$25 \approx t$$

The sample was in storage for approximately 25 days.

23. Since the corn pollen had lost 38.1% of its carbon-14, then 61.9% of the initial carbon-14 remains. The decay rate of carbon-14 is $k = 0.00012097$.

In order to find out the age of the corn pollen, we must solve the following equation for t.

$$0.619 P_0 = P_0 e^{-0.00012097t}$$

$$0.619 = e^{-0.00012097t}$$

$$\ln 0.619 = -0.00012097t$$

$$\frac{\ln 0.619}{-0.00012097} = t$$

$$3965 \approx t$$

The corn pollen was approximately 3965 years old.

25. Use the exponential growth function $P(t) = P_0 e^{kt}$. The interest rate is 5.3% so $k = 0.053$. When $t = 20$, the parents wish the future value to be $40,000, substituting this information into the function gives us the equation:

$$40,000 = P_0 e^{0.053(20)}$$

$$\frac{40,000}{e^{1.06}} = P_0$$

$$13,858.23 \approx P_0$$

The parents should invest $13,858.23 in order to have $40,000 on their child's 20th birthday.

27. The interest rate is 4.7%, so $k = 0.047$. In 6 years, the athlete's salary will be $9 million so $P(6) = 9$ million dollars. Substituting this information into the exponential growth function we get:

$$9 = P_0 e^{0.047(6)}$$

$$9 = P_0 e^{0.282}$$

$$\frac{9}{e^{0.282}} = P_0$$

$$6.788463 \approx P_0$$

The present value is $6.788463 million or $6,788,463.

29. Using the exponential growth function $P(t) = P_0 e^{kt}$. It is known that the interest rate is 4.8%, therefore $k = 0.048$. In 13 years the value of the trust fund will be $80,000, so $P(13) = 80,000$. Substitute this information into the exponential growth function and solve for the present value P_0.

$$80,000 = P_0 e^{0.048(13)}$$

$$80,000 = P_0 e^{0.624}$$

$$\frac{80,000}{e^{0.624}} = P_0$$

$$42,863.76 \approx P_0$$

The present value of the Shannon's trust fund is $42,863.76.

31. a) $V(0) = 40,000e^{-0} = 40,000$

The initial cost of the machinery was $40,000.

b) $V(2) = 40,000e^{-2} \approx 5413.41$

The salvage value after 2 years is approximately $5413.41.

c) ✎

33. a) If the initial actual mortality rate of a female age 25 is 0.014, then $Q_0 = 0.014$.

$$Q(t) = (0.014 - 0.00055)e^{0.163t} + 0.00055$$

$$Q(t) = (0.01345)e^{0.163t} + 0.00055$$

Evaluate the function at the appropriate value.

$$Q(3) = 0.01345e^{0.163(3)} + 0.00055$$

$$\approx 0.022$$

Three years later, the expected mortality rate of this group of females is 0.022 or 22 deaths per 1000.

$$Q(5) = 0.01345e^{0.163(5)} + 0.00055$$

$$\approx 0.031$$

Five years later, the expected mortality rate of this group of females is 0.031 or 31 deaths per 1000.

$$Q(10) = 0.01345e^{0.163(10)} + 0.00055$$

$$\approx 0.069$$

Ten years later, the expected mortality rate of this group of females is 0.069 or 69 deaths per 1000.

b) The graph is shown below:

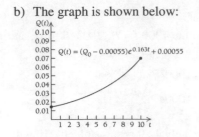

35. a) Using the exponential growth function we have $N(t) = N_0 e^{kt}$, where N_0 is the initial number of farms in 1950 and t is the number of years since 1950. Since there were 5,650,000 farms in 1950, $N_0 = 5,650,000$. In 2012, $t = 62$, there were 2,170,000 farms in the United States. Using this information, we will find the rate of decay, k.

$$N(62) = 2,170,000$$

$$5,650,000e^{k(62)} = 2,170,000$$

$$e^{62k} = \frac{2,170,000}{5,650,000}$$

$$62k = \ln\left(\frac{2,170,000}{5,650,000}\right)$$

$$k = \frac{\ln\left(\frac{2,170,000}{5,650,000}\right)}{62}$$

$$k \approx -0.015$$

Therefore, the exponential function that describes the number of farms after time t in years since 1950 is:

$$N(t) = 5,650,000e^{-0.015t}.$$

b) In 2016, $t = 66$.

$$N(66) = 5,650,000e^{-0.015(66)} \approx 2,099,408$$

Rounded to the nearest thousand, there were approximately 2,099,000 farms in the United States in 2016.

In 2020, $t = 70$.

$$N(70) = 5,650,000e^{-0.015(70)} \approx 1,977,148$$

Rounded to the nearest thousand, there will be approximately 1,977,000 farms in the United States in 2020.

c) Find the value of t such that
$$N(t) = 1,000,000$$
$$5,650,000e^{-0.015t} = 1,000,000$$
$$e^{-0.015t} = \frac{1,000,000}{5,650,000}$$
$$-0.015t = \ln\left(\frac{1,000,000}{5,650,000}\right)$$
$$t = \frac{\ln\left(\frac{1,000,000}{5,650,000}\right)}{-0.015}$$
$$t \approx 115.44$$
There will be 1,000,000 farms in the United States approximately 115 years after 1950, or in the year 2065.

37. a) Use the exponential-decay model
$R(t) = B_0 e^{-kt}$ Let t be years since 2000, this gives us the initial consumption of beef $B_0 = 64.6$. Using the fact that in 2008, $t = 2008 - 2000 = 8$, and the beef consumption was 61.2 lbs, we can substitute into the exponential-decay function to find the decay rate k.
$$61.2 = 64.6e^{-k(8)}$$
$$\frac{61.2}{64.6} = e^{-8k}$$
$$\ln\left(\frac{61.2}{64.6}\right) = \ln e^{-8k}$$
$$\ln\left(\frac{61.2}{64.6}\right) = -8k$$
$$\frac{\ln\left(\frac{61.2}{64.6}\right)}{-8} = k$$
$$0.0068 \approx k$$
The decay rate is 0.0068 or 0.68% per year. Substituting this into the exponential decay function we get:
$B(t) = 64.6e^{-0.0068t}$.

b) In 2015, $t = 2015 - 2000 = 15$.
$$B(15) = 64.6e^{-0.0068(15)} \approx 58.3$$
In the year 2015, annual consumption of beef will be approximately 58.3 pounds per person.

c) Set $B(t) = 20$ and solve for t.
$$64.6e^{-0.0068t} = 20$$
$$e^{-0.0068t} = \frac{20}{64.6}$$
$$\ln e^{-0.0068t} = \ln\left(\frac{20}{64.6}\right)$$
$$-0.0068t = \ln\left(\frac{20}{64.6}\right)$$
$$t = \frac{\ln\left(\frac{20}{64.6}\right)}{-0.0068}$$
$$t \approx 172.4$$
Theoretically, consumption of beef will be 20lbs per person about 172.4 years after 2000, or in the year 2172.

39. a) Use the exponential decay model
$P(t) = P_0 e^{-kt}$. If we let $t = 0$ correspond to 1995, then $P_0 = 51.9$ million people. In 2013, the population of Ukraine was 44.5 million, so $P(18) = 44.5$. Substituting this information into the model we get:
$$44.5 = 51.9e^{-k(18)}$$
$$\frac{44.5}{51.9} = e^{-18k}$$
$$\ln\left(\frac{44.5}{51.9}\right) = \ln e^{-18k}$$
$$\ln\left(\frac{44.5}{51.9}\right) = -18k$$
$$\frac{\ln\left(\frac{44.5}{51.9}\right)}{-18} = k$$
$$0.00855 \approx k$$
Thus, $P(t) = 51.9e^{-0.00855t}$, where $P(t)$ is in millions of people and t is the number of years since 1995.

b) In 2018, $t = 2018 - 1995 = 23$.
$$P(23) = 51.9e^{-0.00855(23)} \approx 42.63$$
The population of the Ukraine will be approximately 42.63 million in 2018.

c) Set $P(t) = 40$ and solve for t.

$$51.9e^{-0.00855t} = 40$$

$$e^{-0.00855t} = \frac{40}{51.9}$$

$$\ln e^{-0.00855t} = \ln\left(\frac{40}{51.9}\right)$$

$$-0.00855t = \ln\left(\frac{40}{51.9}\right)$$

$$t = \frac{\ln\left(\dfrac{40}{51.9}\right)}{-0.00855}$$

$$t \approx 30.46$$

According to the model, the population of the Ukraine will be 40 million 30.5 years after 1995, or in the year 2025.

41. a) According to Newton's Law of Cooling $T(t) = ae^{-kt} + C$. The room temperature is 75 degrees so $C = 75$. At $t = 0$, $T = 102$ degrees. We substitute these values into Newton's law of Cooling.

$$102 = ae^{-k(0)} + 75$$

$$102 = ae^0 + 75$$

$$102 = a + 75$$

$$27 = a$$

b) From part 'a' we have $T(t) = 27e^{-kt} + 75$. Using the fact that when $t = 10$, $T = 90$ we have:

$$90 = 27e^{-k(10)} + 75$$

$$15 = 27e^{-10k}$$

$$\frac{15}{27} = e^{-10k}$$

$$\ln\left(\frac{15}{27}\right) = \ln\left(e^{-10k}\right)$$

$$\ln\left(\frac{15}{27}\right) = -10k$$

$$\frac{\ln\left(\dfrac{15}{27}\right)}{-10} = k$$

$$0.05878 \approx k$$

c) From 'a' and 'b' parts we have

$$T(t) = 27e^{-0.05878t} + 75$$

$$T(20) = 27e^{-0.05878(20)} + 75$$

$$= 27e^{-1.1756} + 75$$

$$\approx 83.3$$

After 20 minutes the water temperature is approximately 83 degrees.

d) Solve $T(t) = 80$ for t.

$$27e^{-0.05878t} + 75 = 80$$

$$27e^{-0.05878t} = 5$$

$$e^{-0.05878t} = \frac{5}{27}$$

$$\ln e^{-0.05878t} = \ln\left(\frac{5}{27}\right)$$

$$-0.05878t = \ln\left(\frac{5}{27}\right)$$

$$t = \frac{\ln\left(\dfrac{5}{27}\right)}{-0.05878}$$

$$t \approx 28.7$$

It will take approximately 29 minutes for the water temperature to cool to 80 degrees.

e) ✎

43. Newton's law of Cooling states

$$T(t) = ae^{-kt} + C.$$

Assume the body had a normal temperature of 98.6 degrees at the time of death and the room remained a constant 60 degrees giving us the constant $C = 60$. We find the constant a first, using the fact that 98.6 degrees is normal body temperature.

$$98.6 = ae^{-k(0)} + 60$$

$$98.6 = a + 60 \qquad \text{The solution is continued.}$$

$$38.6 = a$$

$$T(t) = 38.6e^{-kt} + 60$$

Next, we use the two temperature readings to find k. We want to find the number of hours since death, t.

The solution is continued on the next page.

When the corner took the first temperature reading t hours since death, the body temperature was 85.9 degrees. One hour later at $t+1$ hours since death, the body temperature was 83.4 degrees. Using these two pieces of information we get two equations.

$$85.9 = 38.6e^{-kt} + 60$$

$$83.4 = 38.6e^{-k(t+1)} + 60$$

Subtracting 60 from each equation we get:

$$25.9 = 38.6e^{-kt}$$

$$23.4 = 38.6e^{-k(t+1)}$$

A quick way to solve this system of equations is to divide the first equation by the second equation. This gives us:

$$\frac{25.9}{23.4} = \frac{38.6e^{-kt}}{38.6e^{-k(t+1)}}$$

$$\frac{25.9}{23.4} = e^{-kt-(-kt-k)}$$

$$\frac{25.9}{23.4} = e^{k}$$

$$\ln\left(\frac{25.9}{23.4}\right) = \ln e^{k}$$

$$0.10 \approx k$$

Now we substitute 0.10 in for k into the equation $25.9 = 38.6e^{-kt}$ and solve for t.

$$25.9 = 38.6e^{-0.10t}$$

$$\frac{25.9}{38.6} = e^{-0.1t}$$

$$\ln\left(\frac{25.9}{38.6}\right) = \ln e^{-0.1t}$$

$$\ln\left(\frac{25.9}{38.6}\right) = -0.1t$$

$$\frac{\ln\left(\frac{25.9}{38.6}\right)}{-0.10} = t$$

$$4 \approx t$$

Therefore, the body had been dead for 4 hours. Since the temperature was taken at 11 P.M, the time of death was 7 P.M.

45. $W = 170e^{-0.008t}$

a) We substitute 20 in for t.

$$W(20) = 170e^{-0.008(20)}$$

$$= 170e^{-0.16}$$

$$\approx 144.86$$

The monk weighs approximately 145 pounds after 20 days.

b) We take the derivative of the function to find the rate of change.

$$W'(t) = 170(-0.008)e^{-0.008t}$$

$$= -1.36e^{-0.008t}$$

Now, we substitute 20 in for t.

$$W'(20) = -1.36e^{-0.008(20)} \approx -1.16$$

The monk is losing approximately 1.2 pounds per day after 20 days.

47. $P(t) = 50e^{-0.004t}$

a) We substitute 375 for t.

$$P(375) = 50e^{-0.004(375)}$$

$$= 50e^{-1.5}$$

$$\approx 11.2$$

After 375 days, approximately 11.2 watts of power will be available.

b) From Theorem 10 we know:

$$T = \frac{\ln 2}{k}$$

$$T = \frac{\ln 2}{0.004} \approx 173$$

The half-life of the power supply is approximately 173 days.

c) Set $P(t) = 10$ and solve for t.

$$50e^{-0.004t} = 10$$

$$e^{-0.004t} = \frac{10}{50}$$

$$\ln e^{-0.004t} = \ln 0.2$$

$$-0.004t = \ln 0.2$$

$$t = \frac{\ln 0.2}{-0.004}$$

$$t \approx 402.36$$

The satellite can stay in operation for 402 days.

d) When $t = 0$

$$P(0) = 50e^{-0.004(0)}$$

$$= 50e^{0}$$

$$= 50$$

At the beginning, the satellite had 50 watts of power.

e) ✎

49. (c)   **51.** (e)

53. (f)   **55.** (d)

57. (a)

59. a) Using the exponential decay model, we know $P(t) = P_0 e^{-kt}$. Since the element lost 25% of its mass in 5 weeks, we can determine the decay rate first.

$$0.75 P_0 = P_0 e^{-k(5)}$$
$$0.75 = e^{-5k}$$
$$\ln(0.75) = -5k$$
$$\frac{\ln(0.75)}{-5} = k$$
$$0.0575 \approx k$$

Therefore, the decay rate is 5.75%. Now using Theorem 10, we have:

$$T = \frac{\ln 2}{0.0575} \approx 12.05.$$

The half-life of the element is 12.05 weeks.

b) Using the decay rate from part 'a', we have

$$0.25 P_0 = P_0 e^{-0.0575t}$$
$$0.25 = e^{-0.0575t}$$
$$\ln(0.25) = -0.0575t$$
$$\frac{\ln(0.25)}{-0.0575} = t$$
$$24.1 \approx t$$

The element will have lost 75% of its mass in about 24.1 weeks.

61. a) Solve $D(x) = S(x)$

$$480 e^{-0.003x} = 150 e^{0.004x}$$
$$\frac{480}{150} = \frac{e^{0.004x}}{e^{-0.003x}}$$
$$3.2 = e^{0.007x}$$
$$\ln 3.2 = 0.007x$$
$$166.16 \approx x$$

The equilibrium price is $166.16.
Therefore,

$$D(166.16) = 480 e^{-0.003(166.16)} \approx 291.57$$

The equilibrium quantity is 292 units.
The equilibrium point is $(166.16, 292)$.

Note: Due to the context of the problem we cannot use the true equilibrium point. We rounded the printers up to 292 units while keeping the equilibrium price the same, even though demand at this price is slightly less than 292 printers. It is not possible to make or sell a fraction of a printer, so the supplier would need to determine if 291 or 292 units would result in greater profits.

In the interest of the student we decided to take the most logical course of action by rounding the equilibrium quantity to the proper integer quantity.

b) The elasticity of demand is given by:

$$E(x) = \frac{-x \cdot D'(x)}{D(x)}.$$

First we find the derivative of the demand function.

$$D(x) = 480 e^{-0.003x}$$
$$D'(x) = 480 e^{-0.003x} \cdot (-0.003)$$
$$= -1.44 e^{-0.003x}$$

Evaluating the function and the derivative when $x = 100$, we have:

$$D(100) = 480 e^{-0.003(100)} \approx 355.59$$
$$D'(100) = -1.44 e^{-0.003(100)} \approx -1.0668$$

Substituting into the elasticity of demand function, we have:

$$E(100) = \frac{-(100) \cdot D'(100)}{D(100)}$$
$$= \frac{-100 \cdot (-1.0668)}{355.59}$$
$$= 0.3$$

63. ✎

65. ✎

Exercise Set 3.5

1. $y = 6^x$

$\dfrac{dy}{dx} = (\ln 6)6^x$ Theorem 12: $\dfrac{dy}{dx}a^x = (\ln a)a^x$

3. $f(x) = 8^x$

$f'(x) = (\ln 8)8^x$ Theorem 12

5. $g(x) = x^5(3.7)^x$

Using the Product Rule, we get

$g'(x) = x^5\left(\dfrac{dy}{dx}(3.7)^x\right) + \left(\dfrac{dy}{dx}x^5\right)(3.7)^x$

$g'(x) = x^5\left(\underbrace{\ln(3.7)(3.7)^x}_{\text{Theorem 12}}\right) + 5x^4(3.7)^x$

$= x^4(3.7)^x\left(x\ln(3.7) + 5\right).$ Factoring

7. $y = 7^{x^4+2}$

Using the Chain Rule and Theorem 12, we get

$\dfrac{dy}{dx} = (\ln 7)7^{x^4+2}\left(\dfrac{d}{dx}\left(x^4+2\right)\right)$

$= (\ln 7)7^{x^4+2}\left(4x^3\right)$

$= 4x^3(\ln 7)7^{x^4+2}.$

9. $y = e^{x^2}$

$\dfrac{dy}{dx} = 2xe^{x^2}$ Theorem 2

11. $f(x) = 3^{x^4+1}$

Using the Chain Rule and Theorem 12, we get

$\dfrac{dy}{dx} = (\ln 3)3^{x^4+1}\left(\dfrac{d}{dx}\left(x^4+1\right)\right)$

$= (\ln 3)3^{x^4+1}\left(4x^3\right)$

$= 4x^3(\ln 3)3^{x^4+1}.$

13. $y = \log_8 x$

$\dfrac{dy}{dx} = \dfrac{1}{\ln 8}\cdot\dfrac{1}{x}$ Theorem 14

$= \dfrac{1}{x\ln 8}$

15. $y = \log_{17} x$

$\dfrac{dy}{dx} = \dfrac{1}{\ln 17}\cdot\dfrac{1}{x}$ Theorem 14

$= \dfrac{1}{x\ln 17}$

17. $g(x) = \log_{32}(9x-2)$

Using the Chain Rule and Theorem 14, we get

$g'(x) = \dfrac{1}{\ln 32}\cdot\dfrac{1}{9x-2}\cdot\dfrac{d}{dx}(9x-2)$

$= \dfrac{1}{\ln 32}\cdot\dfrac{1}{9x-2}\cdot 9$

$= \dfrac{9}{(9x-2)\ln 32}$

19. $F(x) = \log(6x-7)$ $(\log x = \log_{10} x)$

Using the Chain Rule and Theorem 14 we get

$F'(x) = \dfrac{1}{\ln 10}\cdot\dfrac{1}{6x-7}\cdot\dfrac{d}{dx}(6x-7)$

$= \dfrac{1}{\ln 10}\cdot\dfrac{1}{6x-7}\cdot 6$

$= \dfrac{6}{(6x-7)\ln 10}.$

21. $y = \log_9\left(x^4 - x\right)$

Using the Chain Rule and Theorem 14, we get

$\dfrac{dy}{dx} = \dfrac{1}{\ln 9}\cdot\dfrac{1}{x^4-x}\cdot\dfrac{d}{dx}\left(x^4-x\right)$

$= \dfrac{1}{\ln 9}\cdot\dfrac{1}{x^4-x}\cdot\left(4x^3-1\right)$

$= \dfrac{4x^3-1}{\left(x^4-x\right)\ln 9}$

23. $f(x) = 4\log_7\left(\sqrt{x} - 2\right)$

$f'(x) = 4\dfrac{d}{dx}\log_7\left(\sqrt{x} - 2\right)$

Using the Chain Rule and Theorem 14, we get:

$f'(x) = 4 \cdot \dfrac{1}{\ln 7} \cdot \dfrac{1}{\sqrt{x} - 2} \cdot \dfrac{d}{dx}\left(\sqrt{x} - 2\right)$

$= \dfrac{4}{\left(\sqrt{x} - 2\right)\ln 7} \cdot \dfrac{d}{dx}\left(x^{\frac{1}{2}} - 2\right) \quad \left[\sqrt[n]{x} = x^{\frac{1}{n}}\right]$

$= \dfrac{4}{\left(\sqrt{x} - 2\right)\ln 7}\left(\dfrac{1}{2}x^{-\frac{1}{2}}\right) \qquad$ Power Rule

$= \dfrac{4}{\left(\sqrt{x} - 2\right)\ln 7} \cdot \dfrac{1}{2x^{\frac{1}{2}}} \qquad$ Properties of Exponents

$= \dfrac{4}{2\sqrt{x}\left(\sqrt{x} - 2\right)\ln 7}$

$= \dfrac{2}{\left(x - 2\sqrt{x}\right)\ln 7}.$

25. $y = 5^x \cdot \log_2 x$

Since y is of the form $f(x) \cdot g(x)$, we apply the Product Rule.

$\dfrac{dy}{dx} = 5^x\left(\dfrac{1}{\ln 2} \cdot \dfrac{1}{x}\right) + (\ln 5)5^x \log_2 x$

Next, we use the commutative property of multiplication to rearrange the derivative:

$\dfrac{dy}{dx} = \dfrac{5^x}{x\ln 2} + 5^x \ln 5 \cdot \log_2 x.$

27. $G(x) = \left(\log_{12} x\right)^5$

Using the Extended Power Rule, we have:

$G'(x) = 5 \cdot \left(\log_{12} x\right)^4 \cdot \dfrac{d}{dx}\left(\log_{12} x\right)$

$= 5 \cdot \left(\log_{12} x\right)^4 \cdot \left(\dfrac{1}{\ln 12} \cdot \dfrac{1}{x}\right) \quad$ Theorem 14

$= 5\left(\log_{12} x\right)^4\left(\dfrac{1}{x\ln 12}\right).$

29. $f(x) = \dfrac{6^x}{5x - 1}$

Since $f(x)$ is in the form $f(x) = \dfrac{g(x)}{h(x)}$ we

apply the Quotient Rule.

$f'(x) = \dfrac{(5x - 1)\left(\dfrac{d}{dx}6^x\right) - 6^x\left(\dfrac{d}{dx}(5x - 1)\right)}{(5x - 1)^2}$

$= \dfrac{(5x - 1)(\ln 6)6^x - 6^x(5)}{(5x - 1)^2}$

$= \dfrac{6^x\left[(5x - 1)(\ln 6) - (5)\right]}{(5x - 1)^2} \quad$ Factor out 6^x

$= \dfrac{6^x(5x\ln 6 - \ln 6 - 5)}{(5x - 1)^2} \quad$ Distribute $\ln 6$

31. $y = 5^{2x^3 - 1} \cdot \log(6x + 5)$

Using the Product Rule we have:

$\dfrac{dy}{dx} = 5^{2x^3 - 1}\left(\dfrac{d}{dx}\log(6x + 5)\right)$

$+ \left(\dfrac{d}{dx}5^{2x^3 - 1}\right)\log(6x + 5).$

Next, using the Chain Rule, we have:

$\dfrac{dy}{dx} = 5^{2x^3 - 1}\left(\dfrac{1}{\ln 10} \cdot \dfrac{1}{6x + 5} \cdot 6\right)$

$+ (\ln 5)5^{2x^3 - 1}\left(6x^2\right) \cdot \log(6x + 5).$

Next, using properties of multiplication, we have:

$\dfrac{dy}{dx}$

$= \dfrac{6 \cdot 5^{2x^3 - 1}}{(6x + 5)\ln 10} + (\ln 5)5^{2x^3 - 1} \cdot 6x^2\log(6x + 5).$

33. $G(x) = \log_9 x \cdot \left(4^x\right)^6 = \log_9 x \cdot 4^{6x}$

Using the Product Rule, the Extended Power Rule and Theorem 12, we have:

$G'(x)$

$= \left(\log_9 x\right) \cdot \dfrac{d}{dx}\left[4^{6x}\right] + 4^{6x} \cdot \dfrac{d}{dx}\left[\left(\log_9 x\right)\right]$

$= \log_9 x(\ln 4)4^{6x} \cdot 6 + 4^{6x}\left(\dfrac{1}{\ln 9} \cdot \dfrac{1}{x}\right)$

$= 6(\ln 4)4^{6x}\log_9 x + \dfrac{4^{6x}}{x\ln 9}$

35. $f(x) = (3x^5 + x)^5 \log_3 x$

First, using the Product Rule, we have:

$$f'(x) = (3x^5 + x)^5 \frac{d}{dx} \log_3 x$$

$$+ \left(\frac{d}{dx}(3x^5 + x)^5\right) \log_3 x$$

Next, we will apply the Chain Rule:

$$f'(x) = (3x^5 + x)^5 \left(\frac{1}{\ln 3} \cdot \frac{1}{x}\right)$$

$$+ 5(3x^5 + x)^4 \frac{d}{dx}(3x^5 + x) \cdot \log_3 x.$$

Which gives us:

$$f'(x) = (3x^5 + x)^5 \left(\frac{1}{x \ln 3}\right)$$

$$+ 5(3x^5 + x)^4 (15x^4 + 1) \log_3 x.$$

37. a) $V(t) = 5200(0.80)^t$

$$V'(t) = 5200 \frac{d}{dt}(0.80)^t$$

$$= 5200(\ln 0.80)(0.80)^t \quad \text{Theorem 12}$$

b) The value of the office machine t years after purchase is changing at a rate of $5200(\ln 0.80)(0.80)^t$ dollars per year.

39. a) $N(t) = 400,000(0.341)^t$

$$N'(t) = 400,000(\ln 0.341)(0.341)^t$$

b) The amount of glass, in pounds, still in use t years after production is changing at a rate of $400,000(\ln 0.341)(0.341)^t$ pounds of glass per year.

41. a) $N(t) = 8400 \ln t - 10,500$

In the year 2014 $t = 2014 - 1970 = 44$.

$N(44) = 8400 \ln(44) - 10,500$

$\approx 21,287.19$

In the year 2014, there will be 21,287.19 thousand, or 21.3 million, nonfarm proprietorships in the United States.

b) $N'(t) = 8400 \cdot \frac{1}{t}$

$N'(45) = 8400 \cdot \frac{1}{45} \approx 186.667$

c)

43. a) We have $r = 0.038$, $p = 1000$ and $n = 3$. Therefore, substituting into the future value of an annuity formula, we have:

$$A(t) = \frac{1000\left[\left(1 + \frac{0.038}{3}\right)^{3t} - 1\right]}{\frac{0.038}{3}}$$

$$= \frac{1000\left[(1.01266667)^{3t} - 1\right]}{0.012666667}$$

$$= 78,947.36842\left[(1.012666667)^{3t} - 1\right]$$

b) We substitute $t = 12$ into the formula:

$$A(12) = 78,947.36842\left[(1.012666667)^{3(12)} - 1\right]$$

$$= 78,947.36842[1.5732383999 - 1]$$

$$= 78,947.36842[0.5732383999]$$

$$\approx 45,255.66$$

Nasim will have about $60,458.75 in the account after 12 years.

c) To find the rate of change, we first find the derivative:

$$\frac{d}{dt}A(t) = \frac{d}{dt}\left(78,947.36842\left[(1.012666667)^{3t} - 1\right]\right)$$

$$A'(t) = 78,947.36842 \frac{d}{dt}\left[(1.012666667)^{3t} - 1\right]$$

$$= 78,947.36842 \cdot$$

$$\left[\underbrace{(1.012666667)^{3t} \cdot \ln(1.012666667) \cdot 3}_{\text{Theorem 12}}\right]$$

$$= 2981.159(1.012666667)^{3t}$$

Using the derivative, we find the rate of change in the value of the annuity after 12 years:

$$A'(8) = 2981.159(1.012666667)^{3(12)}$$

$$= 4690.07$$

After 12 years, Nasim's savings account is growing at rate of $4690.07 per year.

45. $R = \log \dfrac{I}{I_0}$

$$R = \log \frac{10^7 \cdot I_0}{I_0} = \log 10^7 = 7.0$$

The earthquake that struck Haiti had a magnitude 7.0 on the Richter scale.

47. a) The Japanese earthquake of 2011 has a magnitude $R_1 = 9.0$ and the Baja California earthquake has a magnitude of $R_2 = 7.2$. Thus, comparing the earthquakes we have

$$9.0 - 7.2 = \log \frac{I_1}{I_2}$$

$$1.8 = \log \frac{I_1}{I_2}$$

Therefore,

$$I_1 = 10^{1.8} I_2.$$

Since $10^{1.8} = 63.09$, the Japanese earthquake of 2011 was about 63 times as intense as the Baja California earthquake of 2010.

b) The Sumatran-Andaman earthquake of 2004 has a magnitude $R_1 = 9.2$ and the San Fernando earthquake has a magnitude of $R_2 = 6.6$.

Thus, comparing the earthquakes we have

$$9.2 - 6.6 = \log \frac{I_1}{I_2}$$

$$2.6 = \log \frac{I_1}{I_2}$$

Therefore,

$$I_1 = 10^{2.6} I_2.$$

Since $10^{2.6} = 398.11$, the Sumatran-Andaman earthquake of 2004 was about 398 times as intense as the San Fernando earthquake of 1971.

49. $I = I_0 \cdot 10^R = 10^R \cdot I_0$

a) $I = 10^7 \cdot I_0$ Substituting 7 for R.

b) $I = 10^8 \cdot I_0$ Substituting 8 for R.

c) Comparing parts (a) and (b) we have:
$10^8 \cdot I_0 = 10 \cdot 10^7 \cdot I_0$.
The intensity of (b) is 10 times that of (a).

d) $I = I_0 10^R$

$$\frac{dI}{dR} = I_0 (\ln 10) 10^R \quad \text{Theorem 12}$$

$$= (I_0 \cdot \ln 10) 10^R$$

e) ✎

51. a) $R = \log \dfrac{I}{I_0}$

$$\frac{dR}{dI} = \frac{d}{dI} (\log I - \log I_0) \quad\quad \text{(P2)}$$

$$= \frac{1}{\ln 10} \cdot \frac{1}{I}$$

$$= \frac{1}{(\ln 10) I}$$

b) ✎

53. a) $y = m \log x + b$

$$\frac{dy}{dx} = m \cdot \frac{1}{\ln 10} \cdot \frac{1}{x} + 0$$

$$= \frac{m}{x \ln 10}$$

b) The response of the patient to the drug is changing at a rate of $\dfrac{m}{x \ln 10}$ response per dosage.

55. Using the definition, we have:

$$\lim_{h \to 0} \frac{5^h - 1}{h} = \ln 5 \approx 1.6094.$$

57. Using the definition, we have:

$$\lim_{h \to 0} \frac{\sqrt{e}^h - 1}{h} = \lim_{h \to 0} \frac{\left(e^{\frac{1}{2}}\right)^h - 1}{h} = \ln e^{\frac{1}{2}} \approx \frac{1}{2}.$$

59. $y = 2^{x^4}$

$$\frac{dy}{dx} = (\ln 2) 2^{x^4} \frac{d}{dx} x^4$$

$$= (\ln 2) 2^{x^4} \cdot 4x^3$$

61. $y = \log_3 (\log x)$

Using the Chain Rule and Theorem 14, we have:

$$\frac{dy}{dx} = \frac{1}{\ln 3} \cdot \frac{1}{\log x} \cdot \frac{d}{dx} \log x$$

$$= \frac{1}{\ln 3} \cdot \frac{1}{\log x} \cdot \frac{1}{\ln 10} \cdot \frac{1}{x}$$

$$= \frac{1}{\ln 3 \cdot \log x \cdot \ln 10 \cdot x}.$$

63. $y = a^{f(x)}$

$y = e^{f(x)\ln a}$ $\qquad \left[a^x = e^{x\ln a} \right]$

$\dfrac{dy}{dx} = e^{f(x)\ln a} \dfrac{d}{dx} f(x)\ln a$ $\qquad$ Chain Rule

$\dfrac{dy}{dx} = e^{f(x)\ln a} \cdot f'(x)\ln a$

$\dfrac{dy}{dx} = \ln a \cdot a^{f(x)} \cdot f'(x)$ $\qquad \left[a^x = e^{x\ln a} \right]$

65. $y = \left[f(x) \right]^{g(x)}, \;\; f(x) > 0$

$y = e^{g(x)\cdot \ln f(x)}$ $\qquad \left[a^x = e^{x\ln a} \right]$

$\dfrac{dy}{dx} = e^{g(x)\cdot \ln f(x)} \dfrac{d}{dx} g(x)\ln f(x)$ $\quad$ Chain Rule

Next, using the Product Rule, we have:

$\dfrac{dy}{dx} = e^{g(x)\cdot\ln f(x)} \left(g(x)\cdot \dfrac{1}{f(x)} \cdot f'(x) + g'(x)\ln f(x) \right).$

Simplifying we get:

$\dfrac{dy}{dx} = e^{g(x)\ln f(x)} \left(\dfrac{g(x)f'(x)}{f(x)} + g'(x)\ln f(x) \right)$

$\qquad = \left[f(x) \right]^{g(x)} \left(\dfrac{g(x)f'(x)}{f(x)} + g'(x)\ln f(x) \right).$

67. ✎

Exercise Set 3.6

1. Substitute the given information into the amortization formula and solve for the payment P as follows

$$P\left(1+\frac{r}{n}\right)^{n \cdot t} = \frac{p\left[\left(1+\frac{r}{n}\right)^{n \cdot t} - 1\right]}{\frac{r}{n}}$$

$$7000\left(1+\frac{0.06}{12}\right)^{12 \cdot 5} = \frac{p\left[\left(1+\frac{0.06}{12}\right)^{12 \cdot 5} - 1\right]}{\frac{0.06}{12}}$$

$$9441.95106784 = p\left[69.77003\right]$$

$$\frac{9441.95106784}{69.77003} = p$$

$$135.33 = p.$$

The monthly payment needed to amortize the given loan amount is $135.33.

3. Substitute the given information into the amortization formula and solve for the payment P as follows

$$P\left(1+\frac{r}{n}\right)^{n \cdot t} = \frac{p\left[\left(1+\frac{r}{n}\right)^{n \cdot t} - 1\right]}{\frac{r}{n}}$$

$$12,000\left(1+\frac{0.057}{4}\right)^{4 \cdot 6} = \frac{p\left[\left(1+\frac{0.057}{4}\right)^{4 \cdot 6} - 1\right]}{\frac{0.057}{4}}$$

$$16,852.3953555 = p\left[28.3765810266\right]$$

$$\frac{16,852.3953555}{28.3765810266} = p$$

$$593.88 = p.$$

The quarterly payment needed to amortize the given loan amount is $593.88.

5. Substitute the given information into the amortization formula and solve for the payment P as follows

$$P\left(1+\frac{r}{n}\right)^{n \cdot t} = \frac{p\left[\left(1+\frac{r}{n}\right)^{n \cdot t} - 1\right]}{\frac{r}{n}}$$

$$500\left(1+\frac{0.041}{12}\right)^{12 \cdot 1} = \frac{p\left[\left(1+\frac{0.041}{12}\right)^{12 \cdot 1} - 1\right]}{\frac{0.041}{12}}$$

$$520.889650412 = p\left[12.2280880459\right]$$

$$\frac{520.889650412}{12.2280880459} = p$$

$$42.60 = p.$$

The monthly payment needed to amortize the given loan amount is $42.60.

7. Substitute the given information into the amortization formula and solve for the payment P as follows

$$P\left(1+\frac{r}{n}\right)^{n \cdot t} = \frac{p\left[\left(1+\frac{r}{n}\right)^{n \cdot t} - 1\right]}{\frac{r}{n}}$$

$$150,000\left(1+\frac{0.0515}{2}\right)^{2 \cdot 30} = \frac{p\left[\left(1+\frac{0.0515}{2}\right)^{2 \cdot 30} - 1\right]}{\frac{0.0515}{2}}$$

$$689,577.047154 = p\left[139.696322888\right]$$

$$\frac{689,577.047154}{139.696322888} = p$$

$$4936.26 = p.$$

The semiannual payment needed to amortize the given loan amount is $4936.26.

9. Substitute the given information into the amortization formula and solve for the payment P as follows

$$P\left(1+\frac{r}{n}\right)^{n \cdot t} = \frac{p\left[\left(1+\frac{r}{n}\right)^{n \cdot t} - 1\right]}{\frac{r}{n}}$$

$$75,000\left(1+\frac{0.08}{1}\right)^{1 \cdot 15} = \frac{p\left[\left(1+\frac{0.08}{1}\right)^{1 \cdot 15} - 1\right]}{\frac{0.08}{1}}$$

$$237,912.683565 = p\left[27.1521139275\right]$$

$$\frac{237,912.683565}{27.1521139275} = p$$

$$8762.22 = p.$$

The annual payment needed to amortize the given loan amount is $8762.22.

11. a) The principal on the loan is
$P = 22,150 - 4000 = 18,150$.

Substitute the given information into the amortization formula and solve for the payment p as follows

$$P\left(1+\tfrac{r}{n}\right)^{n\cdot t} = \frac{p\left[\left(1+\tfrac{r}{n}\right)^{n\cdot t}-1\right]}{\tfrac{r}{n}}$$

$$18,150\left(1+\tfrac{0.065}{12}\right)^{12\cdot5} = \frac{p\left[\left(1+\tfrac{0.065}{12}\right)^{12\cdot5}-1\right]}{\tfrac{0.065}{12}}$$

$$25,098.1344344 = p\left[70.6739675464\right]$$

$$\frac{25,098.1344344}{70.6739675464} = p$$

$$355.13 = p.$$

The monthly car payment is $355.13.

b) Assuming Todd makes every payment for the life of the loan, his total payments are $355.13 \cdot 12 \cdot 5 = \$21,307.80$.

c) The total interest paid is the total payment minus the principal on the loan. Therefore, the total interest paid over the life of the loan is $I = 21,307.80 - 18,150 = \$3157.80$.

13. a) The principal on the loan is
$P = 195,000 \cdot 0.75 = 146,250$.

Substitute the given information into the amortization formula and solve for the payment p as follows

$$P\left(1+\tfrac{r}{n}\right)^{n\cdot t} = \frac{p\left[\left(1+\tfrac{r}{n}\right)^{n\cdot t}-1\right]}{\tfrac{r}{n}}$$

$$146,250\left(1+\tfrac{0.052}{12}\right)^{12\cdot30} = \frac{p\left[\left(1+\tfrac{0.052}{12}\right)^{12\cdot30}-1\right]}{\tfrac{0.052}{12}}$$

$$693,635.92487 = p\left[863.725325238\right]$$

$$\frac{693,635.92487}{863.725325238} = p$$

$$803.07 = p.$$

The monthly house payment is $803.07.

b) Assuming Hogansons makes every payment for the life of the loan, their total payments are $803.07 \cdot 12 \cdot 30 = \$289,105.20$.

c) The total interest paid is the total payment minus the principal on the loan. Therefore, the total interest paid over the life of the loan is
$I = 209,105.20 - 146,250 = \$142,855.20$.

15. a) The principal balance on the credit card is $P = 500$. Substitute the given information into the amortization formula and solve for the payment p as follows

$$P\left(1+\tfrac{r}{n}\right)^{n\cdot t} = \frac{p\left[\left(1+\tfrac{r}{n}\right)^{n\cdot t}-1\right]}{\tfrac{r}{n}}$$

$$500\left(1+\tfrac{0.2275}{12}\right)^{12\cdot10} = \frac{p\left[\left(1+\tfrac{0.2275}{12}\right)^{12\cdot10}-1\right]}{\tfrac{0.2275}{12}}$$

$$4761.47005973 = p\left[449.561676631\right]$$

$$\frac{4761.47005973}{449.561676631} = p$$

$$10.59 = p.$$

The monthly credit card payment is $10.59.

b) Assuming Joanna makes every payment for the life of the loan, her total payments are $10.59 \cdot 12 \cdot 10 = \$1270.80$.

c) The total interest paid is the total payment minus the principal on the loan. Therefore, the total interest paid over the life of the loan is $I = 1270.80 - 500 = \$770.80$.

17. From exercise 11, we have the monthly payment is $355.13. We create the following amortization table.

Balance	Payment	Portion of payment applied to interest	Portion of payment applied to principal	New Balance
$18,150	$355.13	$18,150\left(\frac{0.065}{12}\right) = \$98.31$	$355.15 - 98.31 = \$256.82$	$17,893.18
$17,893.18	$355.13	$17,893.18\left(\frac{0.065}{12}\right) = \$96.92$	$355.13 - 96.92 = \$258.21$	$17,634.97

19. From exercise 13, we have the monthly payment is $803.07. We create the following amortization table.

Balance	Payment	Portion of payment applied to interest	Portion of payment applied to principal	New Balance
$146,250	$803.07	$146,250\left(\frac{0.052}{12}\right) = \$633.75$	$803.07 - 633.75 = \$169.32$	$146,080.68
$146,080.68	$803.07	$146,080.68\left(\frac{0.052}{12}\right) = \$633.02$	$803.07 - 633.02 = \$170.05$	$145,910.63

21. From exercise 15, we have the monthly payment is $10.59. We create the following amortization table.

Balance	Payment	Portion of payment applied to interest	Portion of payment applied to principal	New Balance
$500	$10.59	$500\left(\frac{0.2275}{12}\right) = \$9.48$	$10.59 - 9.48 = \$1.11$	$498.89
$498.89	$10.59	$498.89\left(\frac{0.2275}{12}\right) = \$9.46$	$10.59 - 9.46 = \$1.13$	$497.76

23. Substitute the given information into the amortization formula and solve for the principal P as follows

$$P\left(1+\frac{r}{n}\right)^{n\cdot t} = \frac{p\left[\left(1+\frac{r}{n}\right)^{n\cdot t} - 1\right]}{\frac{r}{n}}$$

$$P\left(1+\frac{0.058}{12}\right)^{12\cdot 6} = \frac{300\left[\left(1+\frac{0.058}{12}\right)^{12\cdot 6} - 1\right]}{\frac{0.058}{12}}$$

$$P(1.41504551988) = 25,761.4460613$$

$$P = \frac{25,761.4460613}{1.41504551988}$$

$$P = 18,205.38.$$

The largest loan Desmond can afford is $18,205.38.

25. Substitute the given information into the amortization formula and solve for the principal P as follows

$$P\left(1+\frac{r}{n}\right)^{n\cdot t} = \frac{p\left[\left(1+\frac{r}{n}\right)^{n\cdot t} - 1\right]}{\frac{r}{n}}$$

$$P\left(1+\frac{0.0415}{12}\right)^{12\cdot 30} = \frac{1800\left[\left(1+\frac{0.0415}{12}\right)^{12\cdot 30} - 1\right]}{\frac{0.0415}{12}}$$

$$P(3.46548342006) = 1,283,239.56321$$

$$P = \frac{1,283,239.56321}{3.46548342006}$$

$$P = 370,291.65.$$

The largest loan the Daleys can afford is $370,291.65.

27. a) For option 1, substitute the given information into the amortization formula and solve for the payment p as follows

$$P\left(1+\tfrac{r}{n}\right)^{n\cdot t} = \frac{p\left[\left(1+\tfrac{r}{n}\right)^{n\cdot t}-1\right]}{\tfrac{r}{n}}$$

$$12,000\left(1+\tfrac{0.052}{12}\right)^{12\cdot 5} = \frac{p\left[\left(1+\tfrac{0.052}{12}\right)^{12\cdot 5}-1\right]}{\tfrac{0.052}{12}}$$

$$15,554.4214929 = p\left[68.3542594794\right]$$

$$\frac{15,554.4214929}{68.3542594794} = p$$

$$227.56 = p.$$

The monthly payment needed to amortize the given loan amount for option 1 is $227.56.

For option 2, substitute the given information into the amortization formula and solve for the payment p as follows

$$P\left(1+\tfrac{r}{n}\right)^{n\cdot t} = \frac{p\left[\left(1+\tfrac{r}{n}\right)^{n\cdot t}-1\right]}{\tfrac{r}{n}}$$

$$12,000\left(1+\tfrac{0.05}{12}\right)^{12\cdot 6} = \frac{p\left[\left(1+\tfrac{0.05}{12}\right)^{12\cdot 6}-1\right]}{\tfrac{0.05}{12}}$$

$$16,188.2129299 = p\left[83.7642586\right]$$

$$\frac{16,188.2129299}{83.7642586} = p$$

$$193.26 = p.$$

The monthly payment needed to amortize the given loan amount for option 2 is $193.26.

b) The total payments for option 1 are
$227.56 \cdot 12 \cdot 5 = \$13,653.60$.
The total payments for option 2 are
$193.26 \cdot 12 \cdot 6 = \$13,914.72$.

c) Option 1 results in less interest paid. Katie will pay
$13,914.72 - \$13,653.60 = \$261.12$ less in interest if she uses option 1.

29. a) If the annual interest rate is 5%, substitute the given information into the amortization formula and solve for the payment p as follows

$$P\left(1+\tfrac{r}{n}\right)^{n\cdot t} = \frac{p\left[\left(1+\tfrac{r}{n}\right)^{n\cdot t}-1\right]}{\tfrac{r}{n}}$$

$$200,000\left(1+\tfrac{0.05}{12}\right)^{12\cdot 30} = \frac{p\left[\left(1+\tfrac{0.05}{12}\right)^{12\cdot 30}-1\right]}{\tfrac{0.05}{12}}$$

$$893,548.862812 = p\left[832.258635374\right]$$

$$\frac{893,548.862812}{832.258635374} = p$$

$$1073.64 = p.$$

The monthly payment needed to amortize the given loan amount at 5% interest is $1073.64.

b) If the annual interest rate is 6%, substitute the given information into the amortization formula and solve for the payment p as follows

$$P\left(1+\tfrac{r}{n}\right)^{n\cdot t} = \frac{p\left[\left(1+\tfrac{r}{n}\right)^{n\cdot t}-1\right]}{\tfrac{r}{n}}$$

$$200,000\left(1+\tfrac{0.06}{12}\right)^{12\cdot 30} = \frac{p\left[\left(1+\tfrac{0.06}{12}\right)^{12\cdot 30}-1\right]}{\tfrac{0.06}{12}}$$

$$1,204,515.04245 = p\left[1004.51504245\right]$$

$$\frac{1,204,515.04245}{1004.51504245} = p$$

$$1199.10 = p.$$

The monthly payment needed to amortize the given loan amount at 6% interest is $1199.10.

c) The total payments for the 5% loan are
$1073.64 \cdot 12 \cdot 30 = \$386,510.40$.
The total payments for 6% loan are
$1199.10 \cdot 12 \cdot 30 = \$431,676.00$.
The difference between the total payments of the different loans is
$431,676.00 - 386,515.40 = \$45,165.60$.
Darnell will save $45,165.60 if he selects the 5% loan.

31. a) Dwight will need to draw income for 25 years. To determine how much money he will need in order to be able to meet his goals, we substitute the given information into the amortization formula and solve for the principal P as follows:

$$P\left(1+\frac{r}{n}\right)^{n\cdot t} = \frac{p\left[\left(1+\frac{r}{n}\right)^{n\cdot t}-1\right]}{\frac{r}{n}}$$

$$P\left(1+\frac{0.045}{12}\right)^{12\cdot 25} = \frac{500\left[\left(1+\frac{0.045}{12}\right)^{12\cdot 25}-1\right]}{\frac{0.045}{12}}$$

$$P(3.07374252804) = 276,499.003738$$

$$P = \frac{276,499.003738}{3.07374252804}$$

$$P = 89,955.16.$$

Dwight will need $89,955.16 in his annuity in order to draw $500 per month from the annuity from age 60 to age 85.

b) Since Dwight is starting at age 25, he will have 35 years until he reaches age 60. Substitute the given information and the amount found in part (a) into the future value formula and solve for P as follows

$$A = \frac{P\left[\left(1+\frac{r}{n}\right)^{n\cdot t}-1\right]}{\frac{r}{n}}$$

$$89,955.16 = \frac{P\left[\left(1+\frac{0.045}{12}\right)^{12\cdot 35}-1\right]}{\frac{0.045}{12}}$$

$$\frac{89,955.16\left(\frac{0.045}{12}\right)}{\left(1+\frac{0.045}{12}\right)^{12\cdot 35}-1} = P$$

$$88.39 = P.$$

The monthly sinking fund payment required to achieve Dwight's retirement goal is $88.39.

33. Substitute the given information and the amount into the future value formula and solve for W as follows

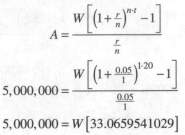

$$A = \frac{W\left[\left(1+\frac{r}{n}\right)^{n\cdot t}-1\right]}{\frac{r}{n}}$$

$$5,000,000 = \frac{W\left[\left(1+\frac{0.05}{1}\right)^{1\cdot 20}-1\right]}{\frac{0.05}{1}}$$

$$5,000,000 = W[33.0659541029]$$

$$151,212.94 = W.$$

The annual payment W you will receive under this plan is $151,212.94 per year.

35. a) If the annual interest rate is 5%, substitute the given information into the amortization formula and solve for the payment p as follows

$$P\left(1+\frac{r}{n}\right)^{n\cdot t} = \frac{p\left[\left(1+\frac{r}{n}\right)^{n\cdot t}-1\right]}{\frac{r}{n}}$$

$$200,000\left(1+\frac{0.05}{12}\right)^{12\cdot 30} = \frac{p\left[\left(1+\frac{0.05}{12}\right)^{12\cdot 30}-1\right]}{\frac{0.05}{12}}$$

$$893,548.862812 = p[832.258635374]$$

$$\frac{893,548.862812}{832.258635374} = p$$

$$1073.64 = p.$$

The monthly payment needed to amortize the given loan amount at 5% interest is $1073.64.

b) The total payments for the loan are $1073.64 \cdot 12 \cdot 30 = \$386,510.40.$ Therefore, the total interest paid is
$$I = 386,510.40 - 200,000 = \$186,510.40.$$

c) By paying an extra 15% per month, the Begays would make monthly payments of $Q = 1.15 \cdot 1073.64 = 1234.69.$ Substituting this into the formula we have

$$t = \frac{\ln(12\cdot 1234.69) - \ln\left[(12\cdot 1234.69)-(200,000\cdot 0.05)\right]}{12\ln\left(1+\frac{0.05}{12}\right)}$$

$$t = \frac{9.6034818552 - 8.47975712483}{0.049896121784}$$

$$t = 22.52.$$

It will take the Begays approximately 22.5 years to pay off the loan.

d) By paying extra, the total payments on the loan are now
$1234.69 \cdot 12 \cdot 22.5 = \$333,366.30$.
Therefore, the interest that they will pay is
$I = 333,366.30 - 200,000 = \$133,366.30$

e) By paying extra, the total interest saved is
$186,510.40 - 133,366.30 = \$53,144.10$.
The Begays save $53,144.10 by paying off their loan in 22.5 years.

37. Using the amortization formula

$$P\left(1+\tfrac{r}{n}\right)^{n \cdot t} = \frac{Q\left[\left(1+\tfrac{r}{n}\right)^{n \cdot t} - 1\right]}{\tfrac{r}{n}}$$

$$P \cdot r\left(1+\tfrac{r}{n}\right)^{n \cdot t} = nQ\left(1+\tfrac{r}{n}\right)^{n \cdot t} - nQ$$

$$\left(Qn - P \cdot r\right)\left(1+\tfrac{r}{n}\right)^{n \cdot t} = nQ$$

$$\ln\left[\left(Qn - P \cdot r\right)\left(1+\tfrac{r}{n}\right)^{n \cdot t}\right] = \ln\left(nQ\right)$$

$$\ln\left(Qn - P \cdot r\right) + \ln\left(1+\tfrac{r}{n}\right)^{n \cdot t} = \ln\left(nQ\right)$$

$$n \cdot t \ln\left(1+\tfrac{r}{n}\right) = \ln\left(nQ\right) - \ln\left(Qn - P \cdot r\right)$$

$$t = \frac{\ln\left(nQ\right) - \ln\left(Qn - P \cdot r\right)}{n \cdot \ln\left(1+\tfrac{r}{n}\right)}$$

39. Using a spreadsheet we complete the amortization table below:

Balance	Payment	Portion to Interest	Portion to Principal	New Balance
$18,150.00	$355.13	$98.31	$256.82	$17,893.18
$17,893.18	$355.13	$96.92	$258.21	$17,634.97
$17,634.97	$355.13	$95.52	$259.61	$17,375.36
$17,375.36	$355.13	$94.12	$261.01	$17,114.35
$17,114.35	$355.13	$92.70	$262.43	$16,851.92
$16,851.92	$355.13	$91.28	$263.85	$16,588.07
$16,588.07	$355.13	$89.85	$265.28	$16,322.79
$16,322.79	$355.13	$88.42	$266.71	$16,056.08
$16,056.08	$355.13	$86.97	$268.16	$15,787.92
$15,787.92	$355.13	$85.52	$269.61	$15,518.31
$15,518.31	$355.13	$84.06	$271.07	$15,247.24
$15,247.24	$355.13	$82.59	$272.54	$14,974.70

41. Using a spreadsheet we complete the amortization table below:

Balance	Payment	Portion to Interest	Portion to Principal	New Balance
$146,250.00	$803.07	$633.75	$169.32	$146,080.68
$146,080.68	$803.07	$633.02	$170.05	$145,910.63
$145,910.63	$803.07	$632.28	$170.79	$145,739.84
$145,739.84	$803.07	$631.54	$171.53	$145,568.31
$145,568.31	$803.07	$630.80	$172.27	$145,396.04
$145,396.04	$803.07	$630.05	$173.02	$145,223.02
$145,223.02	$803.07	$629.30	$173.77	$145,049.25
$145,049.25	$803.07	$628.55	$174.52	$144,874.73
$144,874.73	$803.07	$627.79	$175.28	$144,699.45
$144,699.45	$803.07	$627.03	$176.04	$144,523.41
$144,523.41	$803.07	$626.27	$176.80	$144,346.61
$144,346.61	$803.07	$625.50	$177.57	$144,169.04

43. Using a spreadsheet we complete the amortization table below:

Balance	Payment	Portion to Interest	Portion to Principal	New Balance
$500.00	$10.59	$9.48	$1.11	$498.89
$498.89	$10.59	$9.46	$1.13	$497.76
$497.76	$10.59	$9.44	$1.15	$496.61
$496.61	$10.59	$9.41	$1.18	$495.43
$495.43	$10.59	$9.39	$1.20	$494.23
$494.23	$10.59	$9.37	$1.22	$493.01
$493.01	$10.59	$9.35	$1.24	$491.77
$491.77	$10.59	$9.32	$1.27	$490.50
$490.50	$10.59	$9.30	$1.29	$489.21
$489.21	$10.59	$9.27	$1.32	$487.89
$487.89	$10.59	$9.25	$1.34	$486.55
$486.55	$10.59	$9.22	$1.37	$485.18

Chapter 4

Integration

1. $\int x^6 dx$

$= \dfrac{x^{6+1}}{6+1} + C \qquad \left[\int x^n dx = \dfrac{x^{n+1}}{n+1} + C \right]$

$= \dfrac{x^7}{7} + C \qquad$ Don't forget the C.

3. $\int 2dx$

$= 2x + C \qquad \left[\int k dx = kx + C \right]$

5. $\int x^{1/4} dx$

$= \dfrac{x^{1/4+1}}{\frac{1}{4}+1} + C \qquad \left[\int x^n dx = \dfrac{x^{n+1}}{n+1} + C \right]$

$= \dfrac{x^{5/4}}{\frac{5}{4}} + C$

$= \dfrac{4}{5} x^{5/4} + C$

7. $\int \left(x^2 + x - 1 \right) dx$

$= \int x^2 dx + \int x dx - \int 1 dx \qquad$ The integral of a sum is the sum of the integrals.

$= \dfrac{x^{2+1}}{2+1} + \dfrac{x^{1+1}}{1+1} - x + C \leftarrow$ DON'T FORGET THE C!

$\left[\int x^n dx = \dfrac{x^{n+1}}{n+1} + C \right]$

$\left[\int k dx = kx + C \right]$

$= \dfrac{x^3}{3} + \dfrac{x^2}{2} - x + C$

9. $\int \left(2t^2 + 5t - 3 \right) dt$

$= \int 2t^2 dt + \int 5t dt - \int 3 dt \qquad$ The integral of a sum is the sum of the integrals.

$= 2 \cdot \dfrac{t^{2+1}}{2+1} + 5 \cdot \dfrac{t^{1+1}}{1+1} - 3t + C$

$\left[\int x^n dx = \dfrac{x^{n+1}}{n+1} + C \right]$

$\left[\int k dx = kx + C \right]$

$= \dfrac{2}{3} t^3 + \dfrac{5}{2} t^2 - 3t + C$

11. $\int \dfrac{1}{x^3} dx = \int x^{-3} dx$

$= \dfrac{x^{-3+1}}{-3+1} + C \qquad \left[\int x^n dx = \dfrac{x^{n+1}}{n+1} + C \right]$

$= -\dfrac{x^{-2}}{2} + C$

13. $\int \sqrt[3]{x} dx = \int x^{1/3} dx$

$= \dfrac{x^{1/3+1}}{\frac{1}{3}+1} + C \qquad \left[\int x^n dx = \dfrac{x^{n+1}}{n+1} + C \right]$

$= \dfrac{x^{4/3}}{\frac{4}{3}} + C$

$= \dfrac{3}{4} x^{4/3} + C$

15. $\int \sqrt{x^5} dx = \int x^{5/2} dx$

$= \dfrac{x^{5/2+1}}{\frac{5}{2}+1} + C \qquad \left[\int x^n dx = \dfrac{x^{n+1}}{n+1} + C \right]$

$= \dfrac{x^{7/2}}{\frac{7}{2}} + C$

$= \dfrac{2}{7} x^{7/2} + C$

17. $\displaystyle\int \frac{dx}{x^4} = \int \frac{1}{x^4}\,dx = \int x^{-4}\,dx$

$\displaystyle = \frac{x^{-4+1}}{-4+1} + C \qquad \left[\int x^n\,dx = \frac{x^{n+1}}{n+1} + C\right]$

$\displaystyle = -\frac{x^{-3}}{3} + C$

19. $\displaystyle\int \frac{1}{x}\,dx$

$\displaystyle = \ln|x| + C, \ x > 0 \qquad \left[\int \frac{1}{x}\,dx = \ln|x| + C\right]$

21. $\displaystyle\int \left(\frac{3}{x} + \frac{5}{x^2}\right)dx$

$\displaystyle = \int \frac{3}{x}\,dx + \int \frac{5}{x^2}\,dx \qquad \text{The integral of a sum is the sum of the integrals.}$

$\displaystyle = 3\int x^{-1}\,dx + \int 5x^{-2}\,dx$

$\displaystyle = 3\cdot\ln|x| + 5\cdot\frac{x^{-2+1}}{-2+1} + C, \quad x > 0$

$\left[\int x^{-1}\,dx = \ln|x|\right]$

$\left[\int x^n\,dx = \frac{x^{n+1}}{n+1} + C\right]$

$\displaystyle = 3\ln|x| - 5x^{-1} + C$

23. $\displaystyle\int \frac{-7}{\sqrt[3]{x^2}}\,dx = \int \frac{-7}{x^{2/3}}\,dx = \int -7x^{-2/3}\,dx$

$\displaystyle = -7\int x^{-2/3}\,dx$

$\displaystyle = -7\cdot\frac{x^{-2/3+1}}{-\frac{2}{3}+1} + C \qquad \left[\int x^n\,dx = \frac{x^{n+1}}{n+1} + C\right]$

$\displaystyle = -7\cdot\frac{x^{1/3}}{\frac{1}{3}} + C$

$\displaystyle = -21x^{1/3} + C$

25. $\displaystyle\int 2e^{2x}\,dx$

$\displaystyle = \frac{2}{2}e^{2x} + C \qquad \left[\int e^{ax}\,dx = \frac{1}{a}e^{ax} + C\right]$

$\displaystyle = e^{2x} + C$

27. $\displaystyle\int e^{3x}\,dx$

$\displaystyle = \frac{1}{3}e^{3x} + C \qquad \left[\int e^{ax}\,dx = \frac{1}{a}e^{ax} + C\right]$

29. $\displaystyle\int e^{7x}\,dx$

$\displaystyle = \frac{1}{7}e^{7x} + C \qquad \left[\int e^{ax}\,dx = \frac{1}{a}e^{ax} + C\right]$

31. $\displaystyle\int 5e^{3x}\,dx$

$\displaystyle = \frac{5}{3}e^{3x} + C \qquad \left[\int e^{ax}\,dx = \frac{1}{a}e^{ax} + C\right]$

33. $\displaystyle\int 6e^{8x}\,dx$

$\displaystyle = \frac{6}{8}e^{8x} + C \qquad \left[\int e^{ax}\,dx = \frac{1}{a}e^{ax} + C\right]$

$\displaystyle = \frac{3}{4}e^{8x} + C$

35. $\displaystyle\int \frac{2}{3}e^{-9x}\,dx$

$\displaystyle = \frac{2}{3}\cdot\frac{1}{-9}e^{-9x} + C \qquad \left[\int e^{ax}\,dx = \frac{1}{a}e^{ax} + C\right]$

$\displaystyle = -\frac{2}{27}e^{-9x} + C$

37. $\displaystyle\int \left(5x^2 - 2e^{7x}\right)dx$

$\displaystyle = \int 5x^2\,dx - \int 2e^{7x}\,dx \qquad \text{The integral of a sum is the sum of the integrals.}$

$\displaystyle = 5\cdot\frac{x^{2+1}}{2+1} - \frac{2}{7}e^{7x} + C$

$\left[\int x^n\,dx = \frac{x^{n+1}}{n+1} + C\right]$

$\left[\int e^{ax}\,dx = \frac{1}{a}e^{ax} + C\right]$

$\displaystyle = \frac{5}{3}x^3 - \frac{2}{7}e^{7x} + C$

39. $\int \left(x^2 - \frac{3}{2}\sqrt{x} + x^{-\frac{4}{3}} \right) dx$

$= \int \left(x^2 - \frac{3}{2} x^{\frac{1}{2}} + x^{-\frac{4}{3}} \right) dx$

$= \int x^2 dx - \int \frac{3}{2} x^{\frac{1}{2}} dx + \int x^{-\frac{4}{3}} dx$

$= \frac{x^{2+1}}{2+1} - \frac{3}{2} \cdot \frac{x^{\frac{1}{2}+1}}{\frac{1}{2}+1} + \frac{x^{-\frac{4}{3}+1}}{-\frac{4}{3}+1} + C$

$$\left[\int x^n dx = \frac{x^{n+1}}{n+1} + C \right]$$

$= \frac{x^3}{3} - \frac{3}{2} \cdot \frac{x^{\frac{3}{2}}}{\frac{3}{2}} + \frac{x^{-\frac{1}{3}}}{-\frac{1}{3}} + C$

$= \frac{x^3}{3} - x^{\frac{3}{2}} - 3x^{-\frac{1}{3}} + C$

41. $\int (3x+2)^2 dx = \int \left(9x^2 + 12x + 4 \right) dx$

$= \int 9x^2 dx + \int 12x dx + \int 4 dx$

$= 9 \cdot \frac{x^{2+1}}{2+1} + 12 \cdot \frac{x^{1+1}}{1+1} + 4x + C$

$= \frac{9}{3} x^3 + \frac{12}{2} x^2 + 4x + C$

$= 3x^3 + 6x^2 + 4x + C$

43. $\int \left(\frac{3}{x} - 5e^{2x} + \sqrt{x^7} \right) dx, \quad x > 0$

$= \int \left(\frac{3}{x} - 5e^{2x} + x^{\frac{7}{2}} \right) dx$

$= \int \frac{3}{x} dx - \int 5e^{2x} dx + \int x^{\frac{7}{2}} dx$

$= 3\ln x - \frac{5}{2} \cdot e^{2x} + \frac{x^{\frac{7}{2}+1}}{\frac{7}{2}+1} + C$

$= 3\ln x - \frac{5}{2} e^{2x} + \frac{2}{9} x^{\frac{9}{2}} + C$

45. $\int \left(\frac{7}{\sqrt{x}} - \frac{2}{3} e^{5x} - \frac{8}{x} \right) dx, \quad x > 0$

$= \int \left(7x^{-\frac{1}{2}} - \frac{2}{3} e^{5x} - \frac{8}{x} \right) dx$

$= \int 7x^{-\frac{1}{2}} dx - \int \frac{2}{3} e^{5x} dx - \int \frac{8}{x} dx$

$= 7 \cdot \frac{x^{-\frac{1}{2}+1}}{-\frac{1}{2}+1} - \frac{2}{3} \cdot \frac{1}{5} e^{5x} - 8\ln x + C$

$= 14x^{\frac{1}{2}} - \frac{2}{15} e^{5x} - 8\ln x + C$

47. Find the function $f(x)$, such that
$f'(x) = x - 3, \; f(2) = 9$
We first find $f(x)$ by integrating:

$f(x) = \int (x-3) dx$

$= \int x dx - \int 3 dx$

$= \frac{1}{2} x^2 - 3x + C.$

The condition $f(2) = 9$ allows us to find C:
$$f(2) = 9$$
$$\frac{1}{2}(2)^2 - 3(2) + C = 9$$
$$2 - 6 + C = 9$$
$$-4 + C = 9$$
$$C = 13.$$
Thus, $f(x) = \frac{1}{2} x^2 - 3x + 13.$

49. Find the function $f(x)$, such that
$f'(x) = x^2 - 4, \; f(0) = 7$.
We first find $f(x)$ by integrating:

$f(x) = \int (x^2 - 4) dx$

$= \int x^2 dx - \int 4 dx$

$= \frac{1}{3} x^3 - 4x + C.$

The condition $f(0) = 7$ allows us to find C:
$$f(0) = 7$$
$$\frac{1}{3}(0)^3 - 4(0) + C = 7$$
$$C = 7.$$
Thus, $f(x) = \frac{1}{3} x^3 - 4x + 7.$

51. Find the function $f(x)$, such that
$$f'(x) = 5x^2 + 3x - 7, \quad f(0) = 9.$$
We first find $f(x)$ by integrating:
$$f(x) = \int (5x^2 + 3x - 7)\, dx$$
$$= \int 5x^2\, dx + \int 3x\, dx - \int 7\, dx$$
$$= \frac{5}{3}x^3 + \frac{3}{2}x^2 - 7x + C.$$
The condition $f(0) = 9$ allows us to find C:
$$f(0) = 9$$
$$\frac{5}{3}(0)^3 + \frac{3}{2}(0)^2 - 7(0) + C = 9$$
$$C = 9.$$
Thus, $f(x) = \frac{5}{3}x^3 + \frac{3}{2}x^2 - 7x + 9.$

53. Find the function $f(x)$, such that
$$f'(x) = 3x^2 - 5x + 1, \quad f(1) = \frac{7}{2}.$$
We first find $f(x)$ by integrating:
$$f(x) = \int (3x^2 - 5x + 1)\, dx$$
$$= \int 3x^2\, dx - \int 5x\, dx + \int dx$$
$$= x^3 - \frac{5}{2}x^2 + x + C.$$
The condition $f(1) = \frac{7}{2}$ allows us to find C.
$$f(1) = \frac{7}{2}$$
$$(1)^3 - \frac{5}{2}(1)^2 + (1) + C = \frac{7}{2}$$
$$-\frac{1}{2} + C = \frac{7}{2}$$
$$C = 4.$$
Thus, $f(x) = x^3 - \frac{5}{2}x^2 + x + 4.$

55. Find the function $f(x)$, such that
$$f'(x) = 5e^{2x}, \quad f(0) = \frac{1}{2}.$$
We first find $f(x)$ by integrating:
$$f(x) = \int 5e^{2x}\, dx$$
$$= \frac{5}{2}e^{2x} + C.$$

The condition $f(0) = \frac{1}{2}$ allows us to find C:
$$f(0) = \frac{1}{2}$$
$$\frac{5}{2}e^{2(0)} + C = \frac{1}{2}$$
$$\frac{5}{2}e^0 + C = \frac{1}{2}$$
$$\frac{5}{2} \cdot 1 + C = \frac{1}{2}$$
$$C = -\frac{4}{2}$$
$$C = -2.$$
Thus, $f(x) = \frac{5}{2}e^{2x} - 2.$

57. Find the function $f(x)$, such that
$$f'(x) = \frac{4}{\sqrt{x}}, \quad f(1) = -5.$$
We first find $f(x)$ by integrating:
$$f(x) = \int \frac{4}{\sqrt{x}}\, dx$$
$$= \int 4x^{-1/2}\, dx$$
$$= 8x^{1/2} + C.$$
The condition $f(1) = -5$ allows us to find C:
$$f(1) = -5$$
$$8(1)^{1/2} + C = -5$$
$$8 + C = -5$$
$$C = -13.$$
Thus, $f(x) = 8x^{1/2} - 13.$

59. $D'(t) = 33.428t + 71.143$
We integrate to find $D(t)$.
$$D(t) = \int D'(t)\, dt$$
$$= \int (33.428t + 71.143)\, dt$$
$$= \frac{33.428}{2}t^2 + 71.143t + C$$
$$= 16.714t^2 + 71.143t + C$$

The solution is continued on the next page.

The condition $D(0) = 2555.229$ allows us to find C. Substituting into the equation, we have
$$D(0) = 2555.229$$
$$16.714(0)^2 + 71.143(0) + C = 2555.229$$
$$C = 2555.229$$
Thus,
$$D(t) = 16.714t^2 + 71.143t + 2555.229.$$

61. $C'(x) = x^3 - 2x$

We integrate to find $C(x)$, we use K for the constant of integration to avoid confusion with the cost function $C(x)$.
$$C(x) = \int C'(x)\,dx$$
$$= \int (x^3 - 2x)\,dx$$
$$= \frac{x^4}{4} - x^2 + K$$
Fixed costs are $7000. This means $C(0) = 7000$. This allows us to determine K.
$$C(0) = 7000$$
$$\frac{(0)^4}{4} - (0)^2 + K = 7000$$
$$K = 7000$$
Thus, the total cost function is
$$C(x) = \frac{x^4}{4} - x^2 + 7000.$$

63. $R'(x) = x^2 - 3$

a) We integrate to find $R(x)$.
$$R(x) = \int R'(x)\,dx$$
$$= \int (x^2 - 3)\,dx$$
$$= \frac{x^3}{3} - 3x + C$$
The condition $R(0) = 0$ allows us to find C.
$$R(0) = 0$$
$$\frac{(0)^3}{3} - 3(0) + C = 0$$
$$C = 0$$
Thus, the total revenue function is
$$R(x) = \frac{x^3}{3} - 3x.$$

b) ✎

65. $D'(x) = -\dfrac{4000}{x^2} = -4000x^{-2}$

We integrate to find $D(x)$.
$$D(x) = \int D'(x)\,dx$$
$$= \int -4000x^{-2}\,dx$$
$$= -4000 \cdot \frac{x^{-1}}{-1} + C$$
$$= 4000x^{-1} + C$$
$$= \frac{4000}{x} + C$$
When the price is $4 per unit, the demand is 1003 units. This means $D(4) = 1003$.

Substituting 4 for x and 1003 for $D(x)$ we can determine C as follows:
$$D(4) = 1003$$
$$\frac{4000}{4} + C = 1003$$
$$1000 + C = 1003$$
$$C = 3.$$
Thus, the demand function is $D(x) = \dfrac{4000}{x} + 3$.

67. $\dfrac{dE}{dt} = 30 - 10t$

a) We find $E(t)$ by integrating
$$E(t) = \int E'(t)\,dt$$
$$= \int (30 - 10t)\,dt$$
$$= 30t - 10 \cdot \frac{t^2}{2} + C$$
$$= 30t - 5t^2 + C$$
The condition $E(2) = 72$ allows us to find C as follows:
$$E(2) = 72$$
$$30(2) - 5(2)^2 + C = 72 \qquad \text{Substituting}$$
$$60 - 20 + C = 72$$
$$40 + C = 72$$
$$C = 32.$$
Thus, $E(t) = 30t - 5t^2 + 32.$

b) $E(t) = 32 + 30t - 5t^2$

Substituting 3 for t, we have:

$E(3) = 32 + 30(3) - 5(3)^2$

$= 32 + 90 - 45$

$= 77.$

After 3 hours, the operator's efficiency is 77%.

Substituting 5 for t, we have:

$E(5) = 32 + 30(5) - 5(5)^2$

$= 32 + 150 - 125$

$= 57.$

After 5 hours, the operator's efficiency is 57%.

69. $R'(t) = -46.964e^{-0.796t}$

a) We integrate to find $R(t)$.

$R(t) = \int R'(t)\,dt$

$= \int -46.964e^{-0.796t}\,dt$

$= -46.964\left(\dfrac{1}{-0.796}\right)e^{-0.796t} + C$

$- 59e^{-0.796} + C.$

The condition $R(2) = 78$ allows us to find C.

$R(2) = 78$

$59e^{-0.796(2)} + C = 78$ Substituting

$12 + C = 78$ rounded to nearst beat

$C = 66$

Trisha's pulse rate t minutes after she stops exercising is given by

$R(t) = 59e^{-0.796t} + 66.$

b) Trisha's pulse rate 4 minutes after she stops exercising is given by

$R(4) = 59e^{-0.796(4)} + 66$

$= 2.4438 + 66$

$= 68.4.$

Trisha's pulse rate 4 minutes after she stops exercising is 68.4 beats per minute.

c) The rate of change of Trisha's pulse 4 minutes after she stops exercising is given by:

$R'(4) = -46.964e^{-0.796(4)}$

$= -1.95.$

The rate of change of Trisha's pulse 4 minutes after she stops exercising is -1.95 beats per minute per minute.

d) ✎

71. $h'(t) = v(t) = -32t + 75$

a) We find $h(t)$ by integrating $v(t)$.

$h(t) = \int v(t)\,dt$

$= \int (-32t + 75)\,dt$

$= \dfrac{-32}{2}t^2 + 75t + C$

$= -16t^2 + 75t + C$

The condition $h(0) = 30$ allows us to find C.

$h(0) = 30$

$-16(0)^2 + 75(0) + C = 30$

$C = 30$

The function that gives the height (in feet) of the baseball after t seconds is:

$h(t) = -16t^2 + 75t + 30.$

b) Substituting $t = 2$ into the position function we have:

$h(2) = -16(2)^2 + 75(2) + 30$

$= 116$

The height of the baseball after 2 seconds is 116 feet.

Substituting $t = 2$ into the velocity function we have:

$v(2) = -32(2) + 75 = 11$

The velocity of the baseball after 2 seconds is 11 feet per second.

c) The highest point will occur when the velocity of the ball is zero. We solve the equation $v(t) = 0$ for t.

$-32t + 75 = 0$

$-32t = -75$

$t = \dfrac{75}{32} \approx 2.344.$

It will take approximately 2.344 seconds for the ball to reach its highest point.

d) Substituting in 2.344 seconds from part (c) into $h(t)$ we have:

$h(2.344) = -16(2.344)^2 + 75(2.344) + 30$

≈ 117.89

The ball will reach a maximum height of approximately 117.89 feet.

e) The ball will hit the ground when $h(t) = 0$.
$$h(t) = 0$$
$$-16t^2 + 75t + 30 = 0$$
Using the quadratic formula, we have:
$$t = \frac{-75 \pm \sqrt{(75)^2 - 4(-16)(30)}}{2(-16)}$$
$$= \frac{-75 \pm \sqrt{7545}}{-32}$$
$$t \approx -0.37 \text{ or } t \approx 5.06$$
In this application the only solution that is feasible is $t \approx 5.06$. Therefore, it will take approximately 5.06 seconds for the ball to hit the ground.

f) Substituting $t = 5.06$ into the velocity function we have:
$$v(5.06) = -32(5.06) + 75 = -86.92$$
The impact velocity of the baseball is -86.92 feet per second.

73. a) For Alphaville,
$$P(t) = \int P'(t)\, dt$$
$$= \int 45\, dt$$
$$= 45t + C$$
Using the information $P(0) = 5000$, we have:
$$45(0) + C = 5000$$
$$C = 5000.$$
Therefore, the population model for Alphaville is:
$$P(t) = 45t + 5000.$$
For Betaburgh,
$$Q(t) = \int Q'(t)\, dt$$
$$= \int 105e^{0.03t}\, dt$$
$$= 105\left(\frac{1}{0.03}e^{0.03t}\right) + C$$
$$= 3500e^{0.03t} + C$$
Using the information $Q(0) = 3500$, we have:
$$3500e^{0.03(0)} + C = 3500$$
$$3500 + C = 3500$$
$$C = 0.$$
Therefore, the population model for Betaburgh is:
$$Q(t) = 3500e^{0.03t}.$$

b) Using the models found in part (a) we substitute $t = 10$.
For Alphaville we have:
$$P(10) = 45(10) + 5000 = 5450.$$
The population of Alphaville in 2010 was 5450.
For Betaburgh we have:
$$Q(10) = 3500e^{0.03(10)} \approx 4725.$$
The population of Betaburgh in 2010 was 4725.

c) The graphs are shown below:

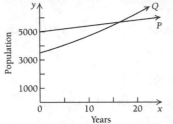

From the graph, we estimate that the two cities will have the same population 16.5 years after 2000, or in the year 2016. The population will be approximately 5700.

75. Find the function $f(t)$, such that
$$f'(t) = \sqrt{t} + \frac{1}{\sqrt{t}}, \quad f(4) = 0.$$
We first find $f(t)$ by integrating:
$$f(t) = \int \left(\sqrt{t} + \frac{1}{\sqrt{t}} \right) dt$$
$$= \int \left(t^{1/2} + t^{-1/2} \right) dt$$
$$= \frac{2}{3}t^{3/2} + 2t^{1/2} + C.$$
The condition $f(4) = 0$ allows us to find C:
$$f(4) = 0$$
$$\frac{2}{3}(4)^{3/2} + 2(4)^{1/2} + C = 0$$
$$\frac{2}{3}(8) + 4 + C = 0$$
$$\frac{16}{3} + 4 + C = 0$$
$$C = -\frac{28}{3}.$$
Thus, $f(t) = \frac{2}{3}t^{3/2} + 2t^{1/2} - \frac{28}{3}.$

77. $\int (5t+4)^2 t^4 dx$

First, we will expand the binomial.

$= \int (5t+4)(5t+4)t^4 dt$

$= \int (25t^2+40t+16)t^4 dt$

Next, we distribute t^4.

$\int (25t^6+40t^5+16t^4) dt$

Now we can integrate as follows:

$= \int 25t^6 dt + \int 40t^5 dt + \int 16t^4 dt$ The integral of a sum is the sum of the integrals.

$= 25 \cdot \dfrac{t^{6+1}}{6+1} + 40 \cdot \dfrac{t^{5+1}}{5+1} + 16 \cdot \dfrac{t^{4+1}}{4+1} + C$

$\left[\int x^n dx = \dfrac{x^{n+1}}{n+1} + C \right]$

$= \dfrac{25}{7}t^7 + \dfrac{40}{6}t^6 + \dfrac{16}{5}t^5 + C$

$= \dfrac{25}{7}t^7 + \dfrac{20}{3}t^6 + \dfrac{16}{5}t^5 + C$

79. $\int \dfrac{(t+3)^2}{\sqrt{t}} dt$

$= \int (t^2+6t+9)t^{-\frac{1}{2}} dt$

$= \int \left(t^{\frac{3}{2}} + 6t^{\frac{1}{2}} + 9t^{-\frac{1}{2}} \right) dt$

$= \dfrac{t^{\frac{5}{2}}}{\frac{5}{2}} + 6 \cdot \dfrac{t^{\frac{3}{2}}}{\frac{3}{2}} + 9 \cdot \dfrac{t^{\frac{1}{2}}}{\frac{1}{2}} + C$

$- \dfrac{2}{5}t^{\frac{5}{2}} + 4t^{\frac{3}{2}} + 18t^{\frac{1}{2}} + C$

81. $\int (t+1)^3 dt$

$= \int (t^3+3t^2+3t+1) dt$

$= \dfrac{t^4}{4} + 3 \cdot \dfrac{t^3}{3} + 3 \cdot \dfrac{t^2}{2} + t + C$

$= \dfrac{t^4}{4} + t^3 + \dfrac{3}{2}t^2 + t + C$

83. $\int (3x-5)(2x+1)^2 dx$

$= \int (3x-5)(4x^2+4x+1) dx$

$= \int (12x^3-8x^2-17x-5) dx$

$= 12 \cdot \dfrac{x^4}{4} - 8 \cdot \dfrac{x^3}{3} - 17 \cdot \dfrac{x^2}{2} - 5x + C$

$= 3x^4 - \dfrac{8}{3}x^3 - \dfrac{17}{2}x^2 - 5x + C$

85. $\int \dfrac{x^2-1}{x+1} dx$

$= \int \dfrac{(x-1)(x+1)}{x+1} dx$

$= \int (x-1) dx$

$= \dfrac{x^2}{2} - x + C$

87.

Exercise Set 4.2

1. $P'(x) = 2x - 150,\ x \ge 0$

Note that the marginal profit is rate of change of total profit with respect to the number of tickets. We use the area under the graph to find the total profit of selling 300 tickets.

Shading the area under $P'(x)$ on the interval $0 \le x \le 300$ we see:

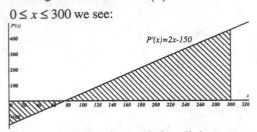

The area between the x-axis is split between two triangles one, Triangle 1 is beneath the x-axis and Triangle 2 is above the x-axis; therefore, the total profit will be equal to the sum of those two triangles. Note the graph crosses the x-axis at $x = 75$.

The triangle that is beneath the axis has

$h = P'(0) = -150$

$b = 75$

Substituting these values into the formula, we have:

Profit = Area of triangle 1

$= \dfrac{1}{2}bh$

$= \dfrac{1}{2}(75)(-150)$

$= -5625.$

The triangle that is above the axis has

$h = P'(300) = 2(300) - 150 = 450$

$b = 300 - 75 = 225.$

Substituting these values into the formula, we have:

Profit = Area of triangle 2

$= \dfrac{1}{2}bh$

$= \dfrac{1}{2}(225)(450)$

$= 50,625.$

Summing the two areas we have:

$A_1 + A_2 = -5625 + 50,625 = 45,000.$

The total profit from the sale of the first 300 tickets is $45,000.

3. $C'(x) = -0.003x + 4.25,\ \text{for } x \le 500$

Note that marginal cost is the rate of change of total cost with respect to the number of kilograms of cheese produced. We use the area under the graph of the marginal cost curve to find the total cost of producing cheese.

Shading the area under $C'(x)$ on the interval $0 \le x \le 400$ we see:

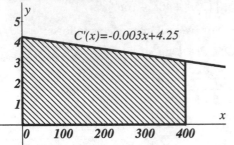

We use the area under the graph to find the total cost of producing 400 kg. of cheese. Viewing the trapezoid sideways we have:

$h = 400$

$b_1 = C'(0) = -0.003(0) + 4.25 = 4.25$

$b_2 = C'(400) = -0.003(400) + 4.25 = 3.05.$

Substituting the values into the formula, we have:

Total Cost = Area of trapezoid

$= \dfrac{1}{2}h(b_1 + b_2)$

$= \dfrac{1}{2}(400)(4.25 + 3.05)$

$= (200)(7.30)$

$= 1460.$

The total cost of producing 400 kg. of cheese is $1460.

5. $C'(x) = -0.007x + 12$

Note that marginal cost is the rate of change of total cost with respect to the number of yards of fabric produced. We use the area under the graph of the marginal cost curve to find the total cost of producing fabric.

The solution is continued on the next page.

Shading the area under $C'(x)$ on the interval $0 \leq x \leq 200$ we see:

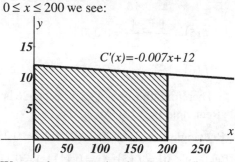

We use the area under the graph to find the total cost of producing 200 yards of fabric. Viewing the trapezoid sideways we have:

$h = 200$

$b_1 = C'(0) = -0.007(0) + 12 = 12$

$b_2 = C'(200) = -0.007(200) + 12 = 10.6.$

Using these values, we have:

Total Cost = Area of trapezoid

$$= \frac{1}{2}h(b_1 + b_2)$$

$$= \frac{1}{2}(200)(12 + 10.6)$$

$$= (100)(22.6) = 2260.$$

The total cost of producing 200 yards of fabric is $2260.

7. $C'(x) = -\frac{2}{25}x + 50, \quad$ for $x \leq 450$

We use the area under the graph to find the total cost of producing the first 200 dresses.

Shading the area under $C'(x)$ on the interval $0 \leq x \leq 200$ we see:

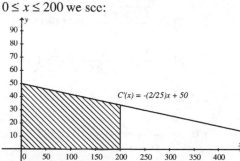

The area under the curve is a trapezoid; therefore, calculating the total cost of producing 200 dresses will require calculating the area of the trapezoid.

The formula for calculating the area of a trapezoid is $A = \frac{1}{2}h(b_1 + b_2)$, where h is the height of the trapezoid and b_1 and b_2 are the lengths of the respective bases. If we view the trapezoid sideways, we see that

$h = 200$

$b_1 = C'(0) = -\frac{2}{25}(0) + 50 = 50$

$b_2 = C'(200) = -\frac{2}{25}(200) + 50 = 34.$

Substituting these values into the formula, we have:

Total Cost = Area of trapezoid

$$= \frac{1}{2}h(b_1 + b_2)$$

$$= \frac{1}{2}(200)(50 + 34)$$

$$= 100(84)$$

$$= 8400.$$

The total cost of producing the first 200 dresses is $8400.

9. $P'(x) = -0.015x + 15.50$

Note that the marginal profit is rate of change of total profit with respect to the yards of audio cable produced. We use the area under the graph to find the total profit of producing audio cable.

Shading the area under $P'(x)$ on the interval $0 < x \leq 200$ we see:

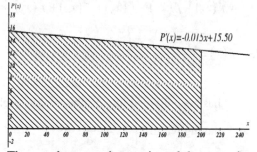

The area between the x-axis and the curve is a trapezoid; therefore, the total profit will be equal to the area of the trapezoid.

The formula for calculating the area of a trapezoid is $A = \frac{1}{2}h(b_1 + b_2)$, where h is the height of the trapezoid and b_1 and b_2 are the lengths of the respective bases.

The solution is continued on the next page.

If we view the trapezoid on the previous page sideways, we see:

$h = 200$

$b_1 = P'(0) = -0.015(0) + 15.50 = 15.50$

$b_2 = P'(200) = -0.015(200) + 15.50 = 12.50.$

Substituting these values into the formula, we have:

Total Profit = Area of trapezoid

$$= \frac{1}{2}h(b_1 + b_2)$$

$$= \frac{1}{2}(200)(15.50 + 12.50)$$

$$= 100(28)$$

$$= 2800.$$

The total profit from the producing 200 yards of audio cable is $2800.

11. $\displaystyle\sum_{i=1}^{4} 2^i = 2^1 + 2^2 + 2^3 + 2^4$

$$= 2 + 4 + 8 + 16, \text{ or } 30$$

13. $\displaystyle\sum_{i=0}^{10} i^2 =$

$$= 0^2 + 1^2 + 2^2 + 3^2 + 4^2 + 5^2 + 6^2 + 7^2 + 8^2 + 9^2 + 10^2$$

$$= 1 + 4 + 9 + 16 + 25 + 36 + 49 + 64 + 81 + 100, \text{ or } 385$$

15. $\displaystyle\sum_{i=1}^{5} f(x_i)$

$$= f(x_1) + f(x_2) + f(x_3) + f(x_4) + f(x_5).$$

17. $f(x) = \dfrac{1}{x^2}$

a) In the drawing in the text the interval $[1,7]$ has been divided into 6 subintervals, each having width 1 $\left[\Delta x = \dfrac{7-1}{6} = 1\right].$

The heights of the rectangles shown are

$f(1) = \dfrac{1}{1^2} = 1$

$f(2) = \dfrac{1}{2^2} = \dfrac{1}{4} = 0.2500$

$f(3) = \dfrac{1}{3^2} = \dfrac{1}{9} \approx 0.1111$

Continued at the top of the next column.

$f(4) = \dfrac{1}{4^2} = \dfrac{1}{16} = 0.0625$

$f(5) = \dfrac{1}{5^2} = \dfrac{1}{25} = 0.0400$

$f(6) = \dfrac{1}{6^2} = \dfrac{1}{36} \approx 0.0278$

Therefore, the area of each rectangle is:

Rectangle I = $f(1) \cdot \Delta x = 1 \cdot 1 = 1$

Rectangle II = $f(2) \cdot \Delta x = 0.2500 \cdot 1 = 0.2500$

Rectangle III = $f(3) \cdot \Delta x = 0.1111 \cdot 1 = 0.1111$

Rectangle IV = $f(4) \cdot \Delta x = 0.0625 \cdot 1 = 0.0625$

Rectangle V = $f(5) \cdot \Delta x = 0.0400 \cdot 1 = 0.0400$

Rectangle VI = $f(6) \cdot \Delta x = 0.0278 \cdot 1 = 0.0278$

The area of the region under the curve over $[1,7]$ is approximately the sum of the areas of the 6 rectangles.

Therefore, the total area is approximately:

$1 + 0.2500 + 0.1111 + 0.0625 +$

$0.0400 + 0.0278 \approx 1.4914.$

b) In the drawing in the text the interval $[1,7]$ has been divided into 12 subintervals, each having width 0.5 $\left[\Delta x = \dfrac{7-1}{12} = \dfrac{6}{12} = 0.5\right].$

The heights of 6 of the rectangles were computed in part (a). The heights of the other 6 rectangles are computed below:

$f(1.5) = \dfrac{1}{1.5^2} = \dfrac{1}{2.25} \approx 0.4444$

$f(2.5) = \dfrac{1}{2.5^2} = \dfrac{1}{6.25} = 0.1600$

$f(3.5) = \dfrac{1}{3.5^2} = \dfrac{1}{12.25} \approx 0.00816$

$f(4.5) = \dfrac{1}{4.5^2} = \dfrac{1}{20.25} \approx 0.0494$

$f(5.5) = \dfrac{1}{5.5^2} = \dfrac{1}{30.25} \approx 0.0331$

$f(6.5) = \dfrac{1}{6.5^2} = \dfrac{1}{42.25} \approx 0.0237$

The solution is continued on the next page.

Therefore, the area of each rectangle is:

Rectangle I: $f(1) \cdot \Delta x = 1(0.5) = 0.5000$

Rectangle II: $f(1.5) \cdot \Delta x = 0.4444(0.5) \approx 0.2222$

Rectangle III: $f(2) \cdot \Delta x = 0.2500(0.5) = 0.1250$

Rectangle IV: $f(2.5) \cdot \Delta x = 0.1600(0.5) = 0.0800$

Rectangle V: $f(3) \cdot \Delta x = 0.1111(0.5) \approx 0.0556$

Rectangle VI: $f(3.5) \cdot \Delta x = 0.0816(0.5) \approx 0.0408$

Rectangle VII: $f(4) \cdot \Delta x = 0.0625(0.5) \approx 0.0313$

Rectangle VIII: $f(4.5) \cdot \Delta x = 0.0494(0.5) \approx 0.0247$

Rectangle IX: $f(5) \cdot \Delta x = 0.0400(0.5) = 0.0200$

Rectangle X: $f(5.5) \cdot \Delta x = 0.0331(0.5) \approx 0.0165$

Rectangle XI: $f(6) \cdot \Delta x = 0.0278(0.5) \approx 0.0139$

Rectangle XII: $f(6.5) \cdot \Delta x = 0.0237(0.5) \approx 0.0118$

The area of the region under the curve over $[1,7]$ is approximately the sum of the areas of the 12 rectangles.

Using the areas of the rectangles, the total area is approximately:

$0.5 + 0.2222 + 0.1250 + 0.0800 +$

$0.0556 + 0.0408 + 0.0313 +$

$0.0247 + 0.0200 + 0.0165 +$

$0.0139 + 0.0118 \approx 1.1418.$

Answers may vary slightly depending on when rounding was done. We see that Part a) is greater than Part b)

19. $P'(x) = -0.0006x^3 + 0.28x^2 + 55.6x$

We divide $[0,300]$ into 6 subintervals of width $\Delta x = 50$. Therefore, the values of x_i are:

$x_1 = 0; \quad x_2 = 50; \quad x_3 = 100;$

$x_4 = 150; \quad x_5 = 200; \quad x_6 = 250.$

Then, the area under the curve is approximately:

$\sum_{i=1}^{6} P'(x_i) \Delta x = P'(x_1) \cdot 50 + P'(x_2) \cdot 50 +$

$\qquad P'(x_3) \cdot 50 + P'(x_4) \cdot 50 +$

$\qquad P'(x_5) \cdot 50 + P'(x_6) \cdot 50$

$\qquad = P'(0) \cdot 50 + P'(50) \cdot 50 +$

$\qquad P'(100) \cdot 50 + P'(150) \cdot 50 +$

$\qquad P'(200) \cdot 50 + P'(250) \cdot 50$

We continue the calculation at the top of the next column.

$\sum_{i=1}^{6} P'(x_i) \Delta x = 0 \cdot 50 + 3405 \cdot 50 + 7760 \cdot 50 +$

$\qquad 12,615 \cdot 50 + 17,520 \cdot 50 +$

$\qquad 22,025 \cdot 50$

$\qquad = 0 + 170,250 + 388,000 +$

$\qquad 630,750 + 876,000 + 1,101,250$

$\qquad = 3,166,250.$

The health club's total profit when 300 members are enrolled is approximately 3,166,250 cents or \$31,662.50.

21. $f(x) = 0.01x^4 - 1.44x^2 + 60$

Dividing the interval $[2,10]$ into four subintervals, we calculate the width of each subinterval to be $\Delta x = \dfrac{10-2}{4} = \dfrac{8}{4} = 2$, with x_i ranging from $x_1 = 2$ to $x_4 = 8$. Although a drawing is not required, we can make one to help visualize the area.

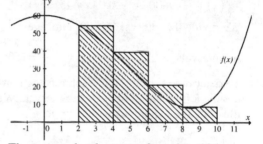

The area under the curve from 2 to 10 is approximately

$\sum_{i=1}^{4} f(x_i) \Delta x = f(2) \cdot 2 + f(4) \cdot 2 +$

$\qquad f(6) \cdot 2 + f(8) \cdot 2$

$\qquad = 54.4 \cdot 2 + 39.52 \cdot 2 +$

$\qquad 21.12 \cdot 2 + 8.8 \cdot 2$

$\qquad = 247.68.$

23. $F(x) = 0.2x^3 + 2x^2 - 0.2x - 2$

Dividing the interval $[-8,-3]$ into five subintervals, we calculate the width of each subinterval to be $\Delta x = \dfrac{-3-(-8)}{5} = \dfrac{5}{5} = 1$, with x_i ranging from $x_1 = -8$ to $x_5 = -4$.

The solution is continued on the next page.

Although a drawing is not required, we can make one to help visualize the area.

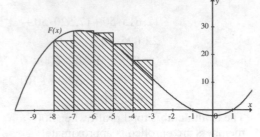

The area under the curve from -8 to -3 is approximately

$$\sum_{i=1}^{5} F(x_i)\Delta x = F(-8)\cdot 1 + F(-7)\cdot 1 + F(-6)\cdot 1 +$$

$$F(-5)\cdot 1 + F(-4)\cdot 1$$

$$= 25.2\cdot 1 + 28.8\cdot 1 + 28\cdot 1 +$$

$$24\cdot 1 + 18\cdot 1$$

$$= 124.$$

25. $C'(x) = -0.00002x^2 - 0.04x + 45$

We divide the interval $[0, 800]$ into five

subintervals, each of width $\Delta x = \dfrac{800}{5} = 160.$

To determine the height of each rectangle, we use the left endpoint of each subinterval. We illustrate the rectangles with a graph.

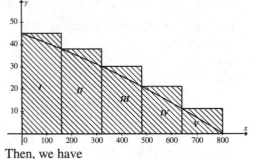

Then, we have

Total Cost $\approx$ Area I + Area II + Area III +

Area IV + Area V

The five left endpoints are 0, 160, 320, 480, and 640. Evaluating $C'(x)$ at each of these endpoints will determine the height of each rectangle and the width was determined to be 160.

Therefore, the area of each rectangle is:

Area I $= C'(0)\cdot 160 = 45\cdot 160 = 7200$

Area II $= C'(160)\cdot 160 = 38.088\cdot 160 = 6094.08$

Area III $= C'(320)\cdot 160 = 30.152\cdot 160 = 4824.32$

Area IV $= C'(480)\cdot 160 = 21.192\cdot 160 = 3390.72$

Area V $= C'(640)\cdot 160 = 11.208\cdot 160 = 1793.28$

Summing the area of the five rectangles yields:

Total Cost $\approx 7200 + 6094.08 + 4824.32 +$

$$3390.72 + 1793.28$$

$$\approx 23,302.4$$

Thus, the total cost of manufacturing 800 feet of molding is approximately 23,302.4 cents or about \$233.02.

27. $C'(x) = 0.000008x^2 - 0.004x + 2$, for $x \le 350$

We divide the interval $[0, 270]$ into three

subintervals, each of width $\Delta x = \dfrac{270}{3} = 90.$ To

determine the height of each rectangle, we use the left endpoint of each subinterval. We illustrate the rectangles with a graph.

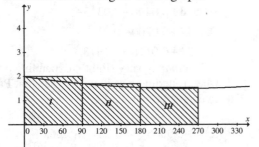

Then, we have

Total Cost $\approx$ Area I + Area II + Area III

The three left endpoints are 0, 90, and 180.

Evaluating $C'(x)$ at each of these endpoints will determine the height of each rectangle and the width was determined to be 90.

Therefore, the area of each rectangle is:

Area I $= C'(0)\cdot 90 = 2\cdot 90 = 180$

Area II $= C'(90)\cdot 90 = 1.7048\cdot 90 = 153.432$

Area III $= C'(180)\cdot 90 = 1.5392\cdot 90 = 138.528$

Summing the area of each of the three rectangles yields:

Total Cost $\approx 180 + 153.432 + 138.528$

$$\approx 471.96$$

Thus, the total cost of producing 270 pints of fresh-squeezed orange juice is approximately \$471.96.

29. Sketching the graph, we determine that the area defined by the definite integral.

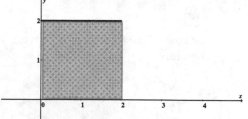

The region is a square; therefore, the area is given by:
$A = b \cdot h = 2 \cdot 2 = 4.$
The definite integral is
$$\int_0^2 2dx = 4.$$

31. Sketching the graph, we determine that the area defined by the definite integral.

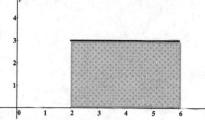

The region is a rectangle; therefore, the area is given by:
$A = b \cdot h = 4 \cdot 3 = 12.$
The definite integral is
$$\int_2^6 3dx = 12.$$

33. Sketching the graph, we determine that the area defined by the definite integral.

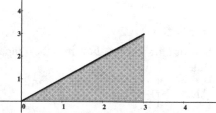

The region is a triangle; therefore, the area is given by:
$A = \frac{1}{2} \cdot b \cdot h$
$= \frac{1}{2} \cdot 3 \cdot 3 = \frac{9}{2}.$
The definite integral is
$$\int_0^3 xdx = \frac{9}{2}.$$

35. Sketching the graph, we determine that the area defined by the definite integral.

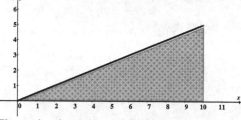

The region is a triangle; therefore, the area is given by:
$A = \frac{1}{2} \cdot b \cdot h$
$= \frac{1}{2} \cdot 10 \cdot 5 = 25.$
The definite integral is
$$\int_0^{10} \tfrac{1}{2} xdx = 25.$$

37. For the specific case, begin by expanding the sum:
$$\sum_{i=1}^4 k \cdot f(x_i) = k \cdot f(x_1) + k \cdot f(x_2) +$$
$$k \cdot f(x_3) + k \cdot f(x_4)$$
Next, we factor out the common factor of k to get:
$$= k\left[f(x_1) + f(x_2) + f(x_3) + f(x_4)\right]$$
$$= k \cdot \left[\sum_{i=1}^4 f(x_i)\right]$$
Thus:
$$\sum_{i=1}^4 k \cdot f(x_i) = k\sum_{i=1}^4 f(x_i).$$
Similarly for the general case, we have:
$$\sum_{i=1}^n k \cdot f(x_i)$$
$$= k \cdot f(x_1) + k \cdot f(x_2) + \cdots + k \cdot f(x_n)$$
$$= k\left[f(x_1) + f(x_2) + \cdots + f(x_n)\right]$$
$$= k\left[\sum_{i=1}^n f(x_i)\right]$$
Thus:
$$\sum_{i=1}^n k \cdot f(x_i) = k\sum_{i=1}^n f(x_i).$$

39. a) The area is that of a triangle with base 1 and

height $f(1) = \frac{1}{2} \cdot 1 = \frac{1}{2}$.

$\int_0^1 f(x)\,dx$ = Area of triangle = $\frac{1}{2}bh$

$\int_0^1 f(x)\,dx = \frac{1}{2}(1)\left(\frac{1}{2}\right)$

$\qquad\qquad = \frac{1}{4}$.

b) The area of the region in question is a trapezoid. The formula for calculating the

area of a trapezoid is $A = \frac{1}{2}h(b_1 + b_2)$,

where h is the height of the trapezoid and b_1 and b_2 are the lengths of the respective bases. If we view the trapezoid sideways, we see that :

$h = 2$

$b_1 = f(1) = \frac{1}{2}(1) = \frac{1}{2}$

$b_2 = f(3) = \frac{1}{2}(3) = \frac{3}{2}$.

Substituting these values into the formula, we have:

$\int_1^3 f(x)\,dx$ = Area of trapezoid

$\qquad = \frac{1}{2}h(b_1 + b_2)$

$\qquad = \frac{1}{2}(2)\left(\frac{1}{2} + \frac{3}{2}\right)$

$\qquad = 2$

c) The area of the region will be a triangle. We know that the height of the triangle is

$f(c) = \frac{1}{2} \cdot c = \frac{c}{2}$.

The base of the triangle is $c - 0 = c$.
Therefore

$\int_0^c f(x)\,dx = 4$

$\qquad \frac{1}{2}bh = 4$

$\qquad \frac{1}{2}(c)\left(\frac{c}{2}\right) = 4$

$\qquad \frac{c^2}{4} = 4$

$\qquad c^2 = 16$

$\qquad c = \sqrt{16} = 4$

d) The area of the region will be a triangle. We know that the height of the triangle is

$f(c) = \frac{1}{2} \cdot c = \frac{c}{2}$.

The base of the triangle is $c - 0 = c$.
Therefore

$\int_0^c f(x)\,dx = 3$

$\qquad \frac{1}{2}bh = 3$

$\qquad \frac{1}{2}(c)\left(\frac{c}{2}\right) = 3$

$\qquad \frac{c^2}{4} = 3$

$\qquad c^2 = 12$

$\qquad c = \sqrt{12} = 2\sqrt{3}$

41. $f(x) = x^2 + 1$

$\Delta x = \frac{5-0}{5} = 1$

Using the Trapezoidal Rule, the area under the graph of $f(x)$ over the interval $[0,5]$ is approximately:

$\text{Area} \approx \Delta x \left[\frac{f(0)}{2} + f(1) + \cdots + f(4) + \frac{f(5)}{2} \right]$

The function values are:

$f(0) = 0^2 + 1 = 1$

$f(1) = 1^2 + 1 = 2$

$f(2) = 2^2 + 1 = 5$

$f(3) = 3^2 + 1 = 10$

$f(4) = 4^2 + 1 = 17$

$f(5) = 5^2 + 1 = 26$

Substituting these values into the Trapezoidal Rule, we have:

$\text{Area} \approx 1 \cdot \left[\frac{1}{2} + 2 + 5 + 10 + 17 + \frac{26}{2} \right]$

$\qquad \approx \frac{1}{2} + 2 + 5 + 10 + 17 + 13$

$\qquad \approx 47.5$

43. $\int_2^4 \sqrt{x^2 - 1}\,dx, \quad n = 4$

The interval $[2,4]$ is divided into four equal

subintervals. Thus, $\Delta x = \frac{4-2}{4} = \frac{1}{2}$, and the

subintervals are:

$[2,2.5],[2.5,3],[3,3.5]$, and $[3.5,4]$.

Using a calculator we have

$f(2) = 1.7321, \quad f(2.5) = 2.2913, \quad f(3) = 2.8284$

$\qquad f(3.5) = 3.3541, \qquad f(4) = 3.8730.$

Therefore,

$S_4 = \frac{4-2}{3\cdot 4}\big(f(2) + 4f(2.5) + 2f(3) + 4f(3.5) + f(4)\big)$

$\quad = \frac{1}{6}\big(1.732 + 4(2.2913) + 2(2.8284) + 4(3.3541) + 3.8730\big)$

$\quad \approx 5.641$

45. ✎

47. $f(x) = \sqrt{25 - x^2}$

Notice that this semi-circle has a radius of five.
Therefore the exact area is given by:

$A = \frac{1}{2}\pi(5)^2 = \frac{25}{2}\pi = 12.5\pi$

To approximate the area under the graph of

$f(x) = \sqrt{25 - x^2}$ using 10 rectangles, we first

find the width of each rectangle.

$\Delta x = \frac{5 - (-5)}{10} = 1 .$

The x_i will range from $x_1 = -5$ to $x_{10} = 4$.

Although a drawing is not required, we can
make one to help visualize the area.

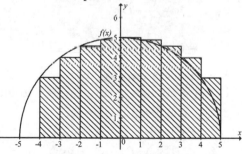

The area under the curve from -5 to 5 is
approximately

$\sum_{i=1}^{10} f(x_i)\Delta x = f(-5)\cdot 1 + f(-4)\cdot 1 + f(-3)\cdot 1$

$\qquad\qquad f(-2)\cdot 1 + f(-1)\cdot 1 + f(0)\cdot 1$

$\qquad\qquad f(1)\cdot 1 + f(2)\cdot 1 + f(3)\cdot 1 +$

$\qquad\qquad f(4)\cdot 1$

$\qquad\qquad \approx 37.9631$

In order to compare, if we round to 4 decimal
places we have:

$A = 12.5\pi = 39.2699 .$

Exercise Set 4.3

1. Find any antiderivative $F(x)$ of $y = 4$. We choose the simplest one for which the constant of integration is 0:

    $F(x) = \int 4\,dx$

    $= 4x + C$

    $= 4x.$ $\qquad [C = 0]$

    The area under the curve over the interval $[1,3]$ is given by $F(3) - F(1)$. Substitute 3 and 1, and find the difference:

    $F(3) - F(1) = 4(3) - 4(1)$

    $= 12 - 4$

    $= 8.$

3. Find any antiderivative $F(x)$ of $y = 2x$. We choose the simplest one for which the constant of integration is 0:

    $F(x) = \int 2x\,dx$

    $= x^2 + C$

    $= x^2.$ $\qquad [C = 0]$

    The area under the curve over the interval $[1,3]$ is given by $F(3) - F(1)$. Substitute 3 and 1, and find the difference:

    $F(3) - F(1) = (3)^2 - (1)^2$

    $= 9 - 1$

    $= 8.$

5. Find any antiderivative $F(x)$ of $y = x^2$. We choose the simplest one for which the constant of integration is 0:

    $F(x) = \int x^2\,dx$

    $= \dfrac{x^3}{3} + C$

    $= \dfrac{x^3}{3}.$ $\qquad [C = 0]$

    The area under the curve over the interval $[0,5]$ is given by $F(5) - F(0)$.

    Substitute 5 and 0, and find the difference:

    $F(5) - F(0) = \dfrac{(5)^3}{3} - \dfrac{(0)^3}{3}$

    $= \dfrac{125}{3}$

    $= 41\tfrac{2}{3}.$

7. Find any antiderivative $F(x)$ of $y = x^3$. We choose the simplest one for which the constant of integration is 0:

    $F(x) = \int x^3\,dx$

    $= \dfrac{x^4}{4} + C$

    $= \dfrac{x^4}{4}.$ $\qquad [C = 0]$

    The area under the curve over the interval $[0,1]$ is given by $F(1) - F(0)$. Substitute 1 and 0, and find the difference:

    $F(1) - F(0) = \dfrac{(1)^4}{4} - \dfrac{(0)^4}{4}$

    $= \dfrac{1}{4}.$

9. Find any antiderivative $F(x)$ of $y = 4 - x^2$. We choose the simplest one for which the constant of integration is 0:

    $F(x) = \int \left(4 - x^2\right)dx$

    $= 4x - \dfrac{x^3}{3} + C$

    $= 4x - \dfrac{x^3}{3}.$ $\qquad [C = 0]$

    The area under the curve over the interval $[-2,2]$ is given by $F(2) - F(-2)$. Substitute 2 and -2, and find the difference:

    $F(2) - F(-2)$

    $= \left(4(2) - \dfrac{(2)^3}{3}\right) - \left(4(-2) - \dfrac{(-2)^3}{3}\right)$

    $= \left(8 - \dfrac{8}{3}\right) - \left(-8 + \dfrac{8}{3}\right)$

    $= \dfrac{16}{3} - \left(-\dfrac{16}{3}\right)$

    $= \dfrac{32}{3} = 10\tfrac{2}{3}.$

11. Find any antiderivative $F(x)$ of $y = e^x$. We choose the simplest one for which the constant of integration is 0:

$$F(x) = \int e^x dx$$

$$= e^x + C$$

$$= e^x. \qquad [C = 0]$$

The area under the curve over the interval $[0,3]$ is given by $F(3) - F(0)$. Substitute 3 and 0, and find the difference:

$$F(3) - F(0) = \left(e^3\right) - \left(e^0\right)$$

$$= e^3 - 1$$

$$\approx 19.086.$$

13. Find any antiderivative $F(x)$ of $y = \dfrac{3}{x}$. We choose the simplest one for which the constant of integration is 0:

$$F(x) = \int \frac{3}{x} dx$$

$$= 3\ln|x| + C$$

$$= 3\ln|x|. \qquad [C = 0]$$

The area between the curve and the x-axis over the interval $[-6, -1]$ is given by $F(-1) - F(-6)$. Since the natural log is not defined for negative values, we use the symmetry of the original graph and determine that the area between the curve and the x-axis over the interval $[1,6]$ will be the opposite of the area sought in the problem. Therefore the area is given by:

$$F(-1) - F(-6) = -\big[F(6) - F(1)\big]$$

$$= -\big[(3\ln 6) - (3\ln 1)\big]$$

$$= -3\ln 6 - 0$$

$$\approx -5.375.$$

15. The height of the area under the curve is total cost per day and the width of the area under the curve is time in days. Thus, area under the curve represents total cost in dollars, for t days.

$$\frac{\text{Total Cost}}{\text{days}} \cdot \text{days} = \text{Total cost.}$$

17. The height of the area under the curve is total number of kilowatts used per hour and the width of the area under the curve is time in hours. Thus, area under the curve represents total number of kilowatts used in t hours.

$$\frac{\#\text{KW}}{\text{Hour}} \cdot \text{Hours} = \#\,\text{KW.}$$

19. The height of the area under the curve is revenue in dollars per unit and the width of the area under the curve is number of units. Thus, area under the curve represents total revenue, in dollars, for x units produced.

$$\frac{\$}{\text{Unit}} \cdot \text{Units} = \$.$$

21. The height of the area under the curve is milligrams per cubic centimeter and the width of the area under the curve is cubic centimeters. Thus, area under the curve represents total concentration of a drug, in milligrams, in v cubic centimeters of blood.

$$\frac{\text{mg}}{\text{cm}^3} \cdot \text{cm}^3 = \text{mg.}$$

23. The height of the area under the curve is number of memorized words per minute and the width of the area under the curve is time in minutes. Thus, area under the curve represents the total number of words memorized in t minutes.

$$\frac{\text{Words memorized}}{\text{Minute}} \cdot \text{Minutes} = \text{Words memorized.}$$

25. Find any antiderivative $F(x)$ of $y = x^3$. We choose the simplest one for which the constant of integration is 0:

$$F(x) = \int x^3 dx$$

$$= \frac{x^4}{4} + C$$

$$= \frac{x^4}{4}. \qquad [C = 0]$$

The area under the curve over the interval $[0,2]$ is given by $F(2) - F(0)$. Substitute 2 and 0, and find the difference:

$$F(2) - F(0) = \frac{(2)^4}{4} - \frac{(0)^4}{4}$$

$$= \frac{16}{4}$$

$$= 4.$$

27. Find any antiderivative $F(x)$ of $y = x^2 + x + 1$.
We choose the simplest one for which the constant of integration is 0:

$$F(x) = \int \left(x^2 + x + 1 \right) dx$$

$$= \frac{x^3}{3} + \frac{x^2}{2} + x + C$$

$$= \frac{x^3}{3} + \frac{x^2}{2} + x. \qquad [C = 0]$$

The area under the curve over the interval $[2,3]$ is given by $F(3) - F(2)$. Substitute 3 and 2, and find the difference:

$$F(3) - F(2)$$

$$= \frac{(3)^3}{3} + \frac{(3)^2}{2} + (3) - \left(\frac{(2)^3}{3} + \frac{(2)^2}{2} + (2) \right)$$

$$= \frac{27}{3} + \frac{9}{2} + 3 - \left(\frac{8}{3} + \frac{4}{2} + 2 \right)$$

$$= \frac{54}{6} + \frac{27}{6} + \frac{18}{6} - \left(\frac{16}{6} + \frac{12}{6} + \frac{12}{6} \right)$$

$$= \frac{99}{6} - \left(\frac{40}{6} \right)$$

$$= \frac{59}{6}$$

$$= 9\tfrac{5}{6}.$$

29. Find any antiderivative $F(x)$ of $y = 5 - x^2$. We choose the simplest one for which the constant of integration is 0:

$$F(x) = \int \left(5 - x^2 \right) dx$$

$$= 5x - \frac{x^3}{3} + C$$

$$= 5x - \frac{x^3}{3}. \qquad [C = 0]$$

The area under the curve over the interval $[-1,2]$ is given by $F(2) - F(-1)$.

Substitute 2 and -1, and find the difference:
$$F(2) - F(-1)$$

$$= \left(5(2) - \frac{(2)^3}{3} \right) - \left(5(-1) - \frac{(-1)^3}{3} \right)$$

$$= \left(10 - \frac{8}{3} \right) - \left(-5 + \frac{1}{3} \right)$$

$$= \frac{22}{3} - \left(-\frac{14}{3} \right)$$

$$= \frac{36}{3}$$

$$= 12.$$

31. Find any antiderivative $F(x)$ of $y = e^x$. We choose the simplest one for which the constant of integration is 0:

$$F(x) = \int e^x dx$$

$$= e^x + C$$

$$= e^x. \qquad [C = 0]$$

The area under the curve over the interval $[-1,5]$ is given by $F(5) - F(-1)$. Substitute 5 and -1, and find the difference:

$$F(5) - F(-1) = \left(e^5 \right) - \left(e^{-1} \right)$$

$$= e^5 - e^{-1}$$

$$\approx 148.045.$$

33. ✎

35. $\int_0^{1.5} \left(x - x^2 \right) dx$

$$= \left[\frac{x^2}{2} - \frac{x^3}{3} \right]_0^{1.5}$$

$$= \left(\frac{(1.5)^2}{2} - \frac{(1.5)^3}{3} \right) - \left(\frac{(0)^2}{2} - \frac{(0)^3}{3} \right)$$

$$= \left(\frac{2.25}{2} - \frac{3.375}{3} \right) - 0$$

$$= 1.125 - 1.125$$

$$= 0$$

The area above the x-axis is equal to the area below the x-axis.

37. $\int_{-1}^{1}\left(x^3-3x^2\right)dx$

$=\left[\dfrac{x^4}{4}-\dfrac{3x^2}{2}\right]_{-1}^{1}$

$=\left(\dfrac{(1)^4}{4}-\dfrac{3(1)^2}{2}\right)-\left(\dfrac{(-1)^4}{4}-\dfrac{3(-1)^2}{2}\right)$

$=\left(\dfrac{1}{4}-\dfrac{3}{2}\right)-\left(\dfrac{1}{4}-\dfrac{3}{2}\right)$

$=\dfrac{-5}{4}-\left(\dfrac{-5}{4}\right)$

$=0$

The area below the x-axis is equal to the area above the x-axis.

39 – 42. Left to the student.

43. $\int_{1}^{3}\left(3t^2+7\right)dt$

$=\left[t^3+7t\right]_{1}^{3}$

$=\left(3^3+7(3)\right)-\left(1^3+7(1)\right)$

$=48-(8)$

$=40$

45. $\int_{1}^{4}\left(\sqrt{x}-1\right)dx=\int_{1}^{4}\left(x^{1/2}-1\right)dx$

$=\left[\dfrac{2}{3}x^{3/2}-x\right]_{1}^{4}$

$=\left(\dfrac{2}{3}(4)^{3/2}-4\right)-\left(\dfrac{2}{3}(1)^{3/2}-1\right)$

$=\left(\dfrac{16}{3}-4\right)-\left(\dfrac{2}{3}-1\right)$

$=\left(\dfrac{4}{3}\right)-\left(-\dfrac{1}{3}\right)$

$=\dfrac{5}{3}$

47. $\int_{-2}^{5}\left(2x^2-3x+7\right)dx$

$=\left[\dfrac{2}{3}x^3-\dfrac{3}{2}x^2+7x\right]_{-2}^{5}$

$=\left(\dfrac{2}{3}(5)^3-\dfrac{3}{2}(5)^2+7(5)\right)-$

$\quad\left(\dfrac{2}{3}(-2)^3-\dfrac{3}{2}(-2)^2+7(-2)\right)$

$=\dfrac{485}{6}-\left(-\dfrac{152}{6}\right)$

$=\dfrac{637}{6}$

49. $\int_{-5}^{2}e^t dt$

$=\left[e^t\right]_{-5}^{2}$

$=e^2-e^{-5}$

≈ 7.382

51. $\int_{a}^{b}\dfrac{1}{2}x^2 dx$

$=\left[\dfrac{1}{6}x^3\right]_{a}^{b}$

$=\dfrac{1}{6}(b)^3-\dfrac{1}{6}(a)^3$

$=\dfrac{b^3-a^3}{6}$

53. $\int_{a}^{b}e^{2t} dt$

$=\left[\dfrac{1}{2}e^{2t}\right]_{a}^{b}$

$==\dfrac{1}{2}e^{2b}-\dfrac{1}{2}e^{2a}$

$=\dfrac{e^{2b}-e^{2a}}{2}$

55. $\int_1^e \left(x + \frac{1}{x} \right) dx$

$= \left[\frac{x^2}{2} + \ln x \right]_1^e$

$= \left(\frac{e^2}{2} + \ln e \right) - \left(\frac{1^2}{2} + \ln 1 \right)$

$= \frac{e^2}{2} + 1 - \left(\frac{1}{2} + 0 \right)$

$= \frac{e^2}{2} + \frac{1}{2}$

$= \frac{e^2 + 1}{2}$

≈ 4.195

57. $\int_0^2 \sqrt{2x}\,dx = \int_0^2 \sqrt{2} \cdot x^{\frac{1}{2}} dx = \sqrt{2} \int_0^2 x^{\frac{1}{2}} dx$

$= \sqrt{2} \left[\frac{2}{3} x^{\frac{3}{2}} \right]_0^2$

$= \sqrt{2} \left[\frac{2}{3}(2)^{\frac{3}{2}} - \frac{2}{3}(0)^{\frac{3}{2}} \right]$

$= \sqrt{2} \left[\frac{2}{3} \sqrt{2^3} \right]$

$= \sqrt{2} \left[\frac{2}{3} \sqrt{8} \right]$

$= \frac{2}{3} \sqrt{16}$

$= \frac{8}{3}$

59. We integrate to find $R(300)$:

$R(300) = \int_0^{300} R'(x)\,dx$

$= \int_0^{300} 2x^{\frac{1}{6}} dx$

$= \left[\frac{12}{7} x^{\frac{7}{6}} \right]_0^{300}$

$= \frac{12}{7}(300)^{\frac{7}{6}} - \frac{12}{7}(0)^{\frac{7}{6}}$

$\approx 1330.63.$

When 300 pounds of maple coated pecans are produced, Sally's revenue is $1330.63.

61. a) We integrate $P'(x)$ over the interval $[0,1200]$ to find the profit from the sale of 1200 control boards:

$P(1200) = \int_0^{1200} P'(x)\,dx$

$= \int_0^{1200} 2.6x^{0.1} dx$

$= \left[\frac{2.6}{1.1} x^{1.1} \right]_0^{1200}$

$= \frac{2.6}{1.1}(1200)^{1.1} - \frac{2.6}{1.1}(0)^{1.1}$

$\approx 5763.42.$

The total profit for selling 1200 digital control boards is $5763.42

b) In order to find the additional profit from selling and additional 300 control boards, we integrate $P'(x)$ over the interval $[1200,1500]$ to find the additional profit:

$P(1500) - P(1200) = \int_{1200}^{1500} P'(x)\,dx$

$= \int_{1200}^{1500} 2.6x^{0.1} dx$

$= \left[\frac{2.6}{1.1} x^{1.1} \right]_{1200}^{1500}$

$= \frac{2.6}{1.1}(1500)^{1.1} - \frac{2.6}{1.1}(1200)^{1.1}$

$\approx 1603.42.$

63. $S'(t) = 10e^t$

a) We integrate $S'(t)$ over the interval $[0,5]$ to find the accumulated sales.

$S(5) = \int_0^5 S'(t)\,dt$

$= \int_0^5 10e^t dt$

$= \left[10e^t \right]_0^5$

$= 10e^5 - 10e^0$

$= 10e^5 - 10$

≈ 1474.13

The accumulated sales for the first 5 days are approximately $1474.13.

b) We integrate $S'(t)$ over the interval $[1,5]$ to find the accumulated sales for the 2nd day through the 5th day

$$S(5) = \int_1^5 S'(t)\,dt$$

$$= \int_1^5 10e^t\,dt$$

$$= \left[10e^t\right]_1^5$$

$$= 10e^5 - 10e^1$$

$$\approx 1456.95$$

The sales from the 2nd day through the 5th day are approximately $1456.95.

65. In 2009, $t = 0$ and in 2012, $t = 3$. Therefore, we integrate $D'(t)$ over the interval $[0,3]$.

$$\int_0^3 D'(t)\,dt = \int_0^3 (33.428t + 71.143)\,dt$$

$$= \left[16.714t^2 + 71.143t\right]_0^3$$

$$= 16.714(3)^2 + 71.143(3) - [0]$$

$$\approx 363.86$$

The credit market debt increased $363.86 billion from 2009 to 2012.

67. We integrate $T(x)$ over the interval $[1,10]$.

$$\int_1^{10} T(x)\,dx$$

$$= \int_1^{10} (2 + 0.3x^{-1})\,dx$$

$$= [2x + 0.3\ln x]_1^{10}$$

$$= 2(10) + 0.3\ln(10) - (2(1) + 0.3\ln(1))$$

$$= 18 + 0.3\ln 10 \approx 18.69$$

It takes 18.69 hours for a worker to produce units 1 through 10.

To find the time it takes a worker to produce units 20 through 30, we integrate $T(x)$ over the interval $[20,30]$.

$$\int_{20}^{30} T(x)\,dx$$

$$= \int_{20}^{30} (2 + 0.3x^{-1})\,dx$$

$$= [2x + 0.3\ln x]_{20}^{30}$$

$$= 2(30) + 0.3\ln(30) - (2(20) + 0.3\ln(20))$$

$$\approx 20.12$$

It takes 20.12 hours for a worker to produce units 20 through 30.

69. We integrate $M'(t)$ over the interval $[0,10]$.

$$M(10) = \int_0^{10} M'(t)$$

$$= \int_0^{10} (-0.009t^2 + 0.2t)\,dt$$

$$= \left[-0.003t^3 + 0.1t^2\right]_0^{10}$$

$$= \left(-0.003(10)^3 + 0.1(10)^2\right)$$

$$\quad - \left(-0.003(0)^3 + 0.1(0)^2\right)$$

$$= 7 - 0$$

$$= 7$$

In the first 10 minutes, 7 words are memorized.

71. We integrate $M'(t)$ over the interval $[10,15]$.

$$M(15) - M(10) = \int_{10}^{15} M'(t)$$

$$= \int_{10}^{15} (-0.009t^2 + 0.2t)\,dt$$

$$= \left[-0.003t^3 + 0.1t^2\right]_{10}^{15}$$

$$= \left(-0.003(15)^3 + 0.1(15)^2\right)$$

$$\quad - \left(-0.003(10)^3 + 0.1(10)^2\right)$$

$$= 12.375 - 7$$

$$= 5.375$$

About 5 words are memorized during minutes 10 – 15.

73. We first find $s(t)$ by integrating:

$$s(t) = \int v(t)\,dt = \int 3t^2\,dt = t^3 + C.$$

Next we determine C by using the initial condition $s(0) = 4$, which is the starting position for s at time $t = 0$:

$$s(0) = 4$$

$$0^3 + C = 4$$

$$C = 4.$$

Thus, $s(t) = t^3 + 4$.

75. We first find $v(t)$ by integrating:

$$v(t) = \int a(t)\,dt = \int 4t\,dt = 2t^2 + C.$$

Next we determine C by using the initial condition $v(0) = 20$:

$$v(0) = 20$$
$$2 \cdot 0^2 + C = 20$$
$$C = 20.$$

Thus, $v(t) = 2t^2 + 20$.

77. We first find $v(t)$ by integrating:

$$v(t) = \int a(t)\,dt$$
$$= \int (-2t + 6)\,dt$$
$$= -t^2 + 6t + C_1.$$

Next we determine C_1 by using the initial condition $v(0) = 6$:

$$-(0)^2 + 6(0) + C_1 = 6$$
$$C_1 = 6.$$

Thus, $v(t) = -t^2 + 6t + 6$.

Next, we find $s(t)$ by integrating:

$$s(t) = \int v(t)\,dt$$
$$= \int \left(-t^2 + 6t + 6\right)dt$$
$$= -\frac{1}{3}t^3 + 3t^2 + 6t + C_2$$

Next we determine C_2 by using the initial condition $s(0) = 10$:

$$-\frac{1}{3}(0)^3 + 3(0)^2 + 6(0) + C_2 = 10$$
$$C_2 = 10.$$

Thus, $s(t) = -\frac{1}{3}t^3 + 3t^2 + 6t + 10$.

79. a) We integrate $v(t)$ over the interval $[0,5]$:

$$s(5) = \int_0^5 v(t)\,dt$$
$$= \int_0^5 \left(-0.5t^2 + 10t\right)dt$$
$$= \left[-\frac{1}{6}t^3 + 5t^2\right]_0^5$$
$$= \left(-\frac{1}{6}(5)^3 + 5(5)^2\right) - \left(-\frac{1}{6}(0)^3 + 5(0)^2\right)$$
$$= \frac{625}{6} - 0$$
$$\approx 104.17$$

The particle travels approximately 104.17 meters during the first 5 seconds.

b) We integrate $v(t)$ over the interval $[5,10]$:

$$s(10) - s(5) = \int_5^{10} v(t)\,dt$$
$$= \left[-\frac{1}{6}t^3 + 5t^2\right]_5^{10} \quad \text{From part (a)}.$$
$$\approx 229.17$$

The particle travels approximately 229.17 meters during the second 5 seconds.

81. Converting 15 seconds into hours, we have

$$\frac{15}{3600} = \frac{1}{240}.$$ Thus, the motorcycle's acceleration function is

$$a(t) = \frac{60 - 0}{\frac{1}{240} - 0} = 14,400,$$ where $a(t)$ is in

miles per hour squared, and t is in hours. Thus, we can find the velocity function by integrating $a(t)$.

$$v(t) = \int a(t)\,dt = \int 14,400\,dt = 14,400t + C$$

We use the initial condition $v(0) = 0$ to find C.

$$v(0) = 0$$
$$14,400(0) + C = 0$$
$$C = 0$$

Thus, $v(t) = 14,400t$, where $v(t)$ is in miles per hour and t is in hours.

The solution is continued on the next page.

Now, substituting $\frac{1}{240}$ hour (15 seconds) for t,

we have $v\left(\frac{1}{240}\right)=14,400\left(\frac{1}{240}\right)=60.$

The motorcycle is traveling at a speed of 60 miles per hour after 15 seconds.

Note: the intuitive solution to this problem is if the motorcycle accelerates at a constant rate from 0 mph to 60 mph in 15 seconds, then the motorcycle is obviously traveling at 60 mph after 15 seconds.

We integrate the velocity function to find the distance. We integrate $v(t)$ over the interval $\left[0,\frac{1}{240}\right].$

$s\left(\frac{1}{240}\right)=\int_0^{1/240}14,400t\,dt$

$=\left[7200t^2\right]_0^{1/240}$

$=7200\left(\frac{1}{240}\right)^2-7200(0)$

$=\frac{1}{8},$

The motorcycle has traveled $\frac{1}{8}$ of a mile after 15 seconds.

83. a) Converting 45 seconds into hours, we have $\frac{45}{3600}=\frac{1}{80}.$ Thus, the bicyclist's deceleration function is

$a(t)=\frac{0-30}{\frac{1}{80}-0}=-2400$, where $a(t)$ is in kilometers per hour squared, and t is in hours. Thus, we can find the velocity function by integrating $a(t)$.

$v(t)=\int a(t)\,dt=\int-2400\,dt=-2400t+C$

We use the initial condition $v(0)=30$ to find C.

$v(0)=30$

$-2400(0)+C=30$

$C=30$

Thus, $v(t)=-2400t+30$, where $v(t)$ is in kilometers per hour and t is in hours.

Now, substituting $\frac{1}{180}$ hour (20 seconds) for t, we have

$v\left(\frac{1}{180}\right)=-2400\left(\frac{1}{180}\right)+30\approx16.67$

The bicyclist is traveling at a speed of 16.67 kilometers per hour after 20 seconds.

b) Using the information in Part (a), we integrate $v(t)$ over the interval $\left[0,\frac{1}{80}\right].$

$s\left(\frac{1}{80}\right)=\int_0^{1/80}(-2400t+30)\,dt$

$=\left[-1200t^2+30t\right]_0^{1/80}$

$=-1200\left(\frac{1}{80}\right)^2+30\left(\frac{1}{80}\right)-$

$\left(-1200(0)^2+30(0)\right)$

$=\frac{-3}{16}+\frac{3}{8}$

$=\frac{3}{16}\approx0.1875.$

The bicyclist has traveled 0.1875 kilometers after 45 seconds.

85. We first find $v(t)$ by integrating:

$v(t)=\int a(t)\,dt$

$=\int(-32)\,dt$

$=-32t+C_1.$

Next we determine C_1 by using the initial condition $v(0)=v_0$:

$v(0)=v_0$

$-32(0)+C_1=v_0$

$C_1=v_0.$

Thus, $v(t)=-32t+v_0.$

Next, we find $s(t)$ by integrating:

$s(t)=\int v(t)\,dt$

$=\int(-32t+v_0)\,dt$

$=-16t^2+v_0\cdot t+C_2$

The solution is continued on the next page.

Next we determine C_2 by using the initial condition $s(0) = s_0$ and the function on the previous page:

$$s(0) = s_0$$

$$-16(0)^2 + v_0(0) + C_2 = s_0$$

$$C_2 = s_0.$$

Thus, $s(t) = -16t^2 + v_0 t + s_0$.

87. Accelerating from 0 mph to 60 mph in 30 second $\left(\dfrac{30}{3600} = \dfrac{1}{120} \text{ hr} \right)$. We have the acceleration function:

$$a(t) = \frac{60 - 0}{\dfrac{1}{120} - 0} = 7200.$$

We integrate to find the velocity function:

$$v(t) = \int a(t)\,dt$$

$$= \int 7200\,dt$$

$$= 7200t + C$$

Since $v(0) = 0$, we have

$$v(0) = 0$$

$$7200(0) + C = 0$$

$$C = 0$$

Thus, $v(t) = 7200t$.

Integrating $v(t)$ from $\left[0, \dfrac{1}{120} \right]$ we determine the distance traveled.

$$s\left(\frac{1}{120} \right) = \int_0^{1/120} 7200t\,dt$$

$$= \left[3600t^2 \right]_0^{1/120}$$

$$= \frac{1}{4}.$$

The car travels $\dfrac{1}{4}$ mile in 30 seconds.

89. We integrate $v(t)$ over the interval $[1,5]$:

$$s(5) - s(1) = \int_1^5 v(t)\,dt$$

$$= \int_1^5 \left(3t^2 + 2t \right) dt$$

$$= \left[t^3 + t^2 \right]_1^5$$

$$= \left(5^3 + 5^2 \right) - \left(1^3 + 1^2 \right)$$

$$= 150 - 2$$

$$= 148$$

The particle travels approximately 148 miles from the start 2nd hour to through the end of the 5th hour.

91. $N'(t) = 280t^{3/2}$

a) To find the total number of pounds of pollutants that enter the lake in the first 16 months we integrate the rate of pollutant entry over the over the interval $[0,16]$:

$$N(16) = \int_0^{16} N'(t)\,dt$$

$$= \int_0^{16} 280t^{3/2}\,dt$$

$$= \left[112t^{5/2} \right]_0^{16}$$

$$= 114{,}688$$

114,688 pounds of pollutants enter the lake during the first 16 months.

b) $N(T) = \displaystyle\int_0^T N'(t)\,dt = 50{,}000$

Therefore,

$$112(T)^{5/2} = 50{,}000$$

$$(T)^{5/2} = \frac{50{,}000}{112}$$

$$T = \left(\frac{50{,}000}{112} \right)^{2/5}$$

$$T \approx 11.48$$

The factory must begin cleanup procedures during the 11th month.

93. $\displaystyle\int_2^3 \frac{x^2-1}{x-1}dx = \int_2^3 \frac{(x-1)(x+1)}{(x-1)}dx$

$\displaystyle = \int_2^3 (x+1)dx$

$\displaystyle = \left[\frac{x^2}{2}+x\right]_2^3$

$\displaystyle = \left(\frac{3^2}{2}+3\right)-\left(\frac{2^2}{2}+2\right)$

$\displaystyle = \frac{15}{2}-4$

$\displaystyle = \frac{7}{2}$

$= 3.5$

95. $\displaystyle\int_4^{16}(x-1)\sqrt{x}\,dx = \int_4^{16}\left(x^{3/2}-x^{1/2}\right)dx$

$\displaystyle = \left[\frac{2}{5}x^{5/2}-\frac{2}{3}x^{3/2}\right]_4^{16}$

$\displaystyle = \left(\frac{2}{5}(16)^{5/2}-\frac{2}{3}(16)^{3/2}\right)-\left(\frac{2}{5}(4)^{5/2}-\frac{2}{3}(4)^{3/2}\right)$

$\displaystyle = \left(\frac{2048}{5}-\frac{128}{3}\right)-\left(\frac{64}{5}-\frac{16}{3}\right)$

$\displaystyle = \frac{5392}{15}$

$\displaystyle = 359\tfrac{7}{15}$

97. $\displaystyle\int_2^5 \left(t+\sqrt{3}\right)\left(t-\sqrt{3}\right)dt = \int_2^5 \left(t^2-3\right)dt$

$\displaystyle = \left[\frac{t^3}{3}-3t\right]_2^5$

$\displaystyle = \left(\frac{(5)^3}{3}-3(5)\right)-\left(\frac{(2)^3}{3}-3(2)\right)$

$\displaystyle = \frac{80}{3}-\left(-\frac{10}{3}\right)$

$\displaystyle = \frac{90}{3}$

$= 30$

99. $\displaystyle\int_1^3 \frac{t^5-t}{t^3}dt = \int_1^3 \left(t^2-t^{-2}\right)dt$

$\displaystyle = \left[\frac{t^3}{3}+t^{-1}\right]_1^3$

$\displaystyle = \left(\frac{(3)^3}{3}+(3)^{-1}\right)-\left(\frac{(1)^3}{3}+(1)^{-1}\right)$

$\displaystyle = \frac{28}{3}-\frac{4}{3}$

$\displaystyle = \frac{24}{3}$

$= 8$

101. ✎

103. Using the fnInt feature on a calculator, we get:

$\displaystyle\int_{-8}^{1.4}\left(x^4+4x^3-36x^2-160x+300\right)dx \approx 4068.789.$

105. Using the fnInt feature on a calculator, we get:

$\displaystyle\int_0^8 x(x-5)^4\,dx \approx 885.333.$

107. ✎

Exercise Set 4.4

1. $\displaystyle\int_1^5 f(x)\,dx = \int_1^3 f(x)\,dx + \int_3^5 f(x)\,dx$

$\displaystyle = \int_1^3 (2x+1)\,dx + \int_3^5 (10-x)\,dx$

$\displaystyle = \left[x^2 + x \right]_1^3 + \left[10x - \frac{1}{2}x^2 \right]_3^5$

$\displaystyle = \left[\left((3)^2 + (3) \right) - \left((1)^2 + (1) \right) \right] +$

$\displaystyle \qquad \left[\left(10(5) - \frac{1}{2}(5)^2 \right) - \left(10(3) - \frac{1}{2}(3)^2 \right) \right]$

$\displaystyle = (12-2) + \left(\frac{75}{2} - \frac{51}{2} \right)$

$\displaystyle = 10 + 12$

$\displaystyle = 22.$

3. $\displaystyle\int_{-2}^3 g(x)\,dx = \int_{-2}^0 g(x)\,dx + \int_0^3 g(x)\,dx$

$\displaystyle = \int_{-2}^0 \left(x^2 + 4 \right)\,dx + \int_0^3 (4-x)\,dx$

$\displaystyle = \left[\frac{x^3}{3} + 4x \right]_{-2}^0 + \left[4x - \frac{1}{2}x^2 \right]_0^3$

$\displaystyle = \left[\left(\frac{(0)^3}{3} + 4(0) \right) - \left(\frac{(-2)^3}{3} + 4(-2) \right) \right] +$

$\displaystyle \qquad \left[\left(4(3) - \frac{1}{2}(3)^2 \right) - \left(4(0) - \frac{1}{2}(0)^2 \right) \right]$

$\displaystyle = \left(0 - \left(-\frac{32}{3} \right) \right) + \left(\frac{15}{2} - 0 \right)$

$\displaystyle = \frac{109}{6}$

$\displaystyle = 18\tfrac{1}{6}.$

5. $\displaystyle\int_{-6}^4 f(x)\,dx = \int_{-6}^1 f(x)\,dx + \int_1^4 f(x)\,dx$

$\displaystyle = \int_{-6}^1 \left(-x^2 - 6x + 7 \right)\,dx + \int_1^4 \left(\frac{3}{2}x - 1 \right)\,dx$

$\displaystyle = \left[-\frac{x^3}{3} - 3x^2 + 7x \right]_{-6}^1 + \left[\frac{3}{4}x^2 - x \right]_1^4$

$\displaystyle = \left[\left(-\frac{(1)^3}{3} - 3(1)^2 + 7(1) \right) - \right.$

$\displaystyle \qquad \left. \left(-\frac{(-6)^3}{3} - 3(-6)^2 + 7(-6) \right) \right] +$

$\displaystyle \qquad \left[\left(\frac{3}{4}(4)^2 - (4) \right) - \left(\frac{3}{4}(1)^2 - (1) \right) \right]$

$\displaystyle = \left(\frac{11}{3} - (-78) \right) + \left(8 - \left(-\frac{1}{4} \right) \right)$

$\displaystyle = \frac{245}{3} + \frac{33}{4}$

$\displaystyle = \frac{1079}{12} = 89\tfrac{11}{12}.$

7. Using the definition of the absolute value, we have:

$$|x-3| = \begin{cases} -(x-3), & \text{for } x < 3 \\ x-3, & \text{for } x \geq 3. \end{cases}$$

Therefore,

$\displaystyle\int_0^4 |x-3|\,dx$

$\displaystyle = \int_0^3 -(x-3)\,dx + \int_3^4 (x-3)\,dx$

$\displaystyle = \left[-\frac{1}{2}x^2 + 3x \right]_0^3 + \left[\frac{1}{2}x^2 - 3x \right]_3^4$

$\displaystyle = \left[\left(-\frac{1}{2}(3)^2 + 3(3) \right) - \left(-\frac{1}{2}(0)^2 + 3(0) \right) \right]$

$\displaystyle \qquad + \left[\left(\frac{1}{2}(4)^2 - 3(4) \right) - \left(\frac{1}{2}(3)^2 - 3(3) \right) \right]$

$\displaystyle = \left[\frac{9}{2} - 0 \right] + \left[-4 - \left(\frac{-9}{2} \right) \right]$

$\displaystyle = \left[\frac{9}{2} \right] + \left[\frac{1}{2} \right]$

$\displaystyle = 5.$

9. Using the definition of the absolute value, we have:

$$\left|x^3-1\right|=\begin{cases}-\left(x^3-1\right), & \text{for } x<1 \\ x^3-1, & \text{for } x\ge 1.\end{cases}$$

Therefore,

$$\int_0^2\left|x^3-1\right|dx$$

$$=\int_0^1-\left(x^3-1\right)dx+\int_1^2\left(x^3-1\right)dx$$

$$=\left[-\frac{1}{4}x^4+x\right]_0^1+\left[\frac{1}{4}x^4-x\right]_1^2$$

$$=\left[\left(-\frac{1}{4}(1)^4+(1)\right)-\left(-\frac{1}{4}(0)^4+(0)\right)\right]$$

$$+\left[\left(\frac{1}{4}(2)^4-(2)\right)-\left(\frac{1}{4}(1)^4-(1)\right)\right]$$

$$=\left[\frac{3}{4}\right]+\left[\frac{11}{4}\right]$$

$$=\frac{14}{4}$$

$$=\frac{7}{2}.$$

11. $g(x)\ge f(x)$ on $[-1,0]$ and $f(x)\ge g(x)$ on $[0,1]$. We use two integrals to find the total area at the top of the next column.

$$\int_{-1}^0\left[0-\left(2x+x^2-x^3\right)\right]dx+$$

$$\int_0^1\left[\left(2x+x^2-x^3\right)-0\right]dx$$

$$=\int_{-1}^0\left(-2x-x^2+x^3\right)dx+$$

$$\int_0^1\left(2x+x^2-x^3\right)dx$$

$$=\left[-x^2-\frac{x^3}{3}+\frac{x^4}{4}\right]_{-1}^0+\left[x^2+\frac{x^3}{3}-\frac{x^4}{4}\right]_0^1$$

$$=\left[-(0)^2-\frac{(0)^3}{3}+\frac{(0)^4}{4}\right]-$$

$$\left[-(-1)^2-\frac{(-1)^3}{3}+\frac{(-1)^4}{4}\right]+$$

$$\left[(1)^2+\frac{(1)^3}{3}-\frac{(1)^4}{4}\right]-\left[(0)^2+\frac{(0)^3}{3}-\frac{(0)^4}{4}\right]$$

$$=\left[0-\left(-\frac{5}{12}\right)\right]+\left[\frac{13}{12}-0\right]$$

$$=\frac{5}{12}+\frac{13}{12}$$

$$=\frac{18}{12}$$

$$=\frac{3}{2}.$$

13. $g(x)\ge f(x)$ over the entire region. We find the area.

$$\int_{-1}^4\left[(x+28)-\left(x^4-8x^3+18x^2\right)\right]dx$$

$$=\int_{-1}^4\left(-x^4+8x^3-18x^2+x+28\right)dx$$

$$=\left[-\frac{x^5}{5}+2x^4-6x^3+\frac{x^2}{2}+28x\right]_{-1}^4$$

$$=\left[-\frac{(4)^5}{5}+2(4)^4-6(4)^3+\frac{(4)^2}{2}+28(4)\right]-$$

$$\left[-\frac{(-1)^5}{5}+2(-1)^4-6(-1)^3+\frac{(-1)^2}{2}+28(-1)\right]$$

$$=\frac{216}{5}+\frac{193}{10}$$

$$=\frac{625}{10}=62\tfrac{1}{2}.$$

15. First graph the system of equations and shade the region bounded by the graphs.

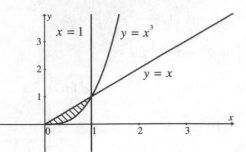

Here the boundaries are easily determined by looking at the graph, or by solving the following:

$$x = x^3$$
$$0 = x^3 - x$$
$$0 = x(x^2 - 1)$$
$$x = 0 \quad \text{or} \quad x^2 - 1 = 0$$
$$x = 0 \quad \text{or} \quad x = \pm 1$$
$$x = 0 \quad \text{or} \quad x = 1$$

Note $x \geq x^3$ over the interval $[0,1]$. We compute the area as follows:

$$\int_0^1 (x - x^3)\, dx$$

$$= \left[\frac{x^2}{2} - \frac{x^4}{4} \right]_0^1$$

$$= \left(\frac{(1)^2}{2} - \frac{(1)^4}{4} \right) - \left(\frac{(0)^2}{2} - \frac{(0)^4}{4} \right)$$

$$= \frac{1}{2} - \frac{1}{4} - (0 - 0)$$

$$= \frac{1}{4}.$$

17. First graph the system of equations and shade the region bounded by the graphs.

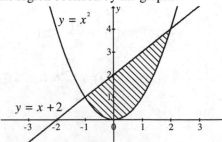

Here the boundaries are easily determined by looking at the graph.

To determine the boundaries, we could also solve the following equation:

$$x^2 = x + 2$$
$$x^2 - x - 2 = 0$$
$$(x - 2)(x + 1) = 0$$
$$x = -1 \quad \text{or} \quad x = 2$$

Note $(x + 2) \geq x^2$ over the interval $[-1, 2]$. We compute the area as follows:

$$\int_{-1}^2 \left((x + 2) - x^2 \right) dx = \int_{-1}^2 \left(-x^2 + x + 2 \right) dx$$

$$= \left[-\frac{x^3}{3} + \frac{x^2}{2} + 2x \right]_{-1}^2$$

$$= \left(-\frac{(2)^3}{3} + \frac{(2)^2}{2} + 2(2) \right) -$$

$$\left(-\frac{(-1)^3}{3} + \frac{(-1)^2}{2} + 2(-1) \right)$$

$$= \left(-\frac{8}{3} + 2 + 4 \right) - \left(\frac{1}{3} + \frac{1}{2} - 2 \right)$$

$$= \frac{10}{3} - \left(-\frac{7}{6} \right)$$

$$= \frac{27}{6}$$

$$= 4\tfrac{1}{2}.$$

19. First graph the system of equations and shade the region bounded by the graphs.

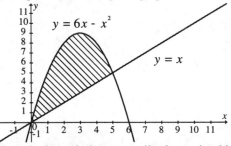

Here the boundaries are easily determined by looking at the graph, or by solving the following equation:

$$x = 6x - x^2$$
$$x^2 - 5x = 0$$
$$x(x - 5) = 0$$
$$x = 0 \quad \text{or} \quad x = 5$$

The solution is continued on the next page.

Note $\left(6x - x^2\right) \ge x$ over the interval $[0,5]$. We compute the area as follows:

$$\int_0^5 \left(\left(6x - x^2\right) - x\right) dx$$

$$= \int_0^5 \left(-x^2 + 5x\right) dx$$

$$= \left[-\frac{x^3}{3} + \frac{5}{2}x^2 \right]_0^5$$

$$= \left(-\frac{(5)^3}{3} + \frac{5}{2}(5)^2 \right) - \left(-\frac{(0)^3}{3} + \frac{5}{2}(0)^2 \right)$$

$$= \left(-\frac{125}{3} + \frac{125}{2} \right) - 0$$

$$= \frac{125}{6}$$

$$= 20\tfrac{5}{6}.$$

21. First graph the system of equations and shade the region bounded by the graphs.

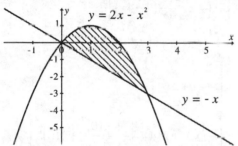

Here the boundaries are easily determined by looking at the graph, or by solving the following equation:

$$-x = 2x - x^2$$

$$x^2 - 3x = 0$$

$$x(x - 3) = 0$$

$$x = 0 \ \text{or} \ x = 3$$

Note $\left(2x - x^2\right) \ge -x$ over the interval $[0,3]$.

We compute the area at the top of the next column.

Computing the area, we have:

$$\int_0^3 \left(\left(2x - x^2\right) - (-x)\right) dx$$

$$= \int_0^3 \left(-x^2 + 3x\right) dx$$

$$= \left[-\frac{x^3}{3} + \frac{3}{2}x^2 \right]_0^3$$

$$= \left(-\frac{(3)^3}{3} + \frac{3}{2}(3)^2 \right) - \left(-\frac{(0)^3}{3} + \frac{3}{2}(0)^2 \right)$$

$$= \left(-\frac{27}{3} + \frac{27}{2} \right) - 0$$

$$= \frac{27}{6} = 4\tfrac{1}{2}.$$

23. First graph the system of equations and shade the region bounded by the graphs.

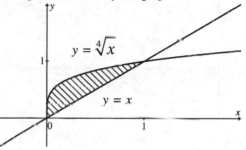

Here the boundaries are easily determined by looking at the graph, or by solving the following equation:

$$x = \sqrt[4]{x}$$

$$x^4 = x$$

$$x^4 - x = 0$$

$$x\left(x^3 - 1\right) = 0$$

$$x = 0 \ \text{or} \ x = 1$$

Note $\sqrt[4]{x} \ge x$ over the interval $[0,1]$.

The solution is continued on the next page.

We compute the area as follows:

$$\int_0^1 \left(\sqrt[4]{x} - x \right) dx$$

$$= \int_0^1 \left(x^{\frac{1}{4}} - x \right) dx$$

$$= \left[\frac{4}{5} x^{\frac{5}{4}} - \frac{x^2}{2} \right]_0^1$$

$$= \left(\frac{4}{5}(1)^{\frac{5}{4}} - \frac{(1)^2}{2} \right) - \left(\frac{4}{5}(0)^{\frac{5}{4}} - \frac{(0)^2}{2} \right)$$

$$= \left(\frac{4}{5} - \frac{1}{2} \right) - 0$$

$$= \frac{3}{10}.$$

25. First graph the system of equations and shade the region bounded by the graphs.

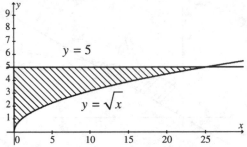

Here the boundaries are easily determined by looking at the graph. Note $5 \geq \sqrt{x}$ over the interval $[0, 25]$. We compute the area as follows:

$$\int_0^{25} \left(5 - \sqrt{x} \right) dx$$

$$= \int_0^{25} \left(5 - x^{\frac{1}{2}} \right) dx$$

$$= \left[5x - \frac{2}{3} x^{\frac{3}{2}} \right]_0^{25}$$

$$= \left(5(25) - \frac{2}{3}(25)^{\frac{3}{2}} \right) - \left(5(0) - \frac{2}{3}(0)^{\frac{3}{2}} \right)$$

$$= \left(125 - \frac{250}{3} \right) - 0$$

$$= \frac{125}{3}$$

$$= 41\frac{2}{3}.$$

27. First graph the system of equations and shade the region bounded by the graphs.

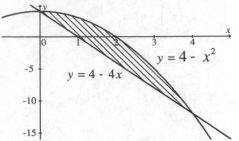

Here the boundaries are easily determined by looking at the graph, or by solving the following:

$$4 - 4x = 4 - x^2$$

$$x^2 - 4x = 0$$

$$x(x - 4) = 0$$

$$x = 0 \quad \text{or} \quad x = 4.$$

Note $\left(4 - x^2 \right) \geq \left(4 - 4x \right)$ over the interval $[0, 4]$. We compute the area as follows:

$$\int_0^4 \left[\left(4 - x^2 \right) - \left(4 - 4x \right) \right] dx$$

$$= \int_0^4 \left(-x^2 + 4x \right) dx$$

$$= \left[-\frac{x^3}{3} + 2x^2 \right]_0^4$$

$$= \left(-\frac{(4)^3}{3} + 2(4)^2 \right) - \left(-\frac{(0)^3}{3} + 2(0)^2 \right)$$

$$= \left(-\frac{64}{3} + 32 \right) - 0$$

$$= \frac{32}{3}$$

$$= 10\frac{2}{3}.$$

29. First graph the system of equations and shade the region bounded by the graphs.

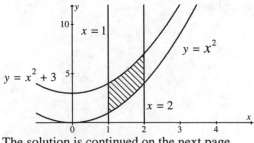

The solution is continued on the next page.

Here the boundaries are easily determined by looking at the graph on the previous page. Note $\left(x^2+3\right) \geq \left(x^2\right)$ over the interval $[1,2]$. We compute the area as follows:

$$\int_1^2 \left[\left(x^2+3\right)-\left(x^2\right)\right]dx$$

$$=\int_1^2 3dx$$

$$=[3x]_1^2$$

$$=3(2)-3(1)$$

$$=3.$$

31. First graph the system of equations and shade the region bounded by the graphs.

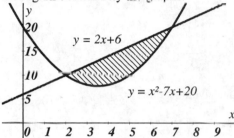

Here the boundaries are determined by solving the following equation.

$$x^2-7x+20=2x+6$$

$$x^2-9x+14=0$$

$$(x-7)(x-2)=0$$

$$x-2=0 \quad \text{or} \quad x-7=0$$

$$x=2 \quad \text{or} \quad x=7$$

The graphs intersect at $x=2$ and $x=7$.

Note $2x+6 \geq x^2-7x+20$ over the interval $[2,7]$.

We compute the area as follows:

$$\int_2^7 \left[\left(2x+6\right)-\left(x^2-7x+20\right)\right]dx$$

$$=\int_2^7 \left(-x^2+9x-14\right)dx$$

$$=\left[-\frac{x^3}{3}+\frac{9}{2}x^2-14x\right]_2^7$$

$$=\left(-\frac{(7)^3}{3}+\frac{9}{2}(7)^2-14(7)\right)-$$

$$\left(-\frac{(2)^3}{3}+\frac{9}{2}(2)^2-14(2)\right)$$

$$=\frac{49}{6}-\left(-\frac{38}{3}\right)$$

$$=\frac{125}{6}=20\tfrac{5}{6}.$$

33. First graph the system of equations and shade the region bounded by the graphs.

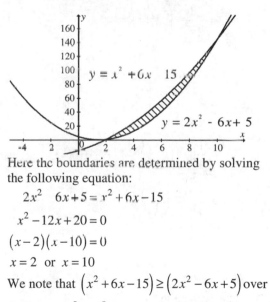

Here the boundaries are determined by solving the following equation:

$$2x^2-6x+5=x^2+6x-15$$

$$x^2-12x+20=0$$

$$(x-2)(x-10)=0$$

$$x=2 \quad \text{or} \quad x=10$$

We note that $\left(x^2+6x-15\right) \geq \left(2x^2-6x+5\right)$ over the interval $[2,10]$.

The solution is continued on the next page.

We compute the area as follows:

$$\int_2^{10}\left[\left(x^2+6x-15\right)-\left(2x^2-6x+5\right)\right]dx$$

$$=\int_2^{10}\left(-x^2+12x-20\right)dx$$

$$=\left[-\frac{x^3}{3}+6x^2-20x\right]_2^{10}$$

$$=\left(-\frac{(10)^3}{3}+6(10)^2-20(10)\right)-$$

$$\left(-\frac{(2)^3}{3}+6(2)^2-20(2)\right)$$

$$=\frac{200}{3}-\left(-\frac{56}{3}\right)$$

$$=\frac{256}{3}$$

$$=85\frac{1}{3}.$$

35. The average value is:

$$y_{av}=\frac{1}{b-a}\int_a^b f(x)\,dx$$

$$=\frac{1}{2-(-2)}\int_{-2}^{2}\left(4-x^2\right)dx$$

$$=\frac{1}{4}\left[4x-\frac{x^3}{3}\right]_{-2}^{2}$$

$$=\frac{1}{4}\left[\left(4(2)-\frac{(2)^3}{3}\right)-\left(4(-2)-\frac{(-2)^3}{3}\right)\right]$$

$$=\frac{1}{4}\left[\frac{16}{3}-\left(-\frac{16}{3}\right)\right]$$

$$=\frac{8}{3}.$$

37. The average value is:

$$y_{av}=\frac{1}{b-a}\int_a^b f(x)\,dx$$

$$=\frac{1}{1-(0)}\int_0^1 e^{-x}\,dx$$

$$=\frac{1}{1}\left[-e^{-x}\right]_0^1$$

$$=\left[-e^{-1}-\left(-e^0\right)\right]$$

$$=-e^{-1}+1,\text{ or approximately }0.632.$$

39. The average value is:

$$y_{av}=\frac{1}{b-a}\int_a^b f(x)\,dx$$

$$=\frac{1}{4-(0)}\int_0^4\left(x^2+x-2\right)dx$$

$$=\frac{1}{4}\left[\frac{x^3}{3}+\frac{x^2}{2}-2x\right]_0^4$$

$$=\frac{1}{4}\left[\left(\frac{(4)^3}{3}+\frac{(4)^2}{2}-2(4)\right)-\right.$$

$$\left.\left(\frac{(0)^3}{3}+\frac{(0)^2}{2}-2(0)\right)\right]$$

$$=\frac{1}{4}\left[\frac{64}{3}-0\right]$$

$$=\frac{16}{3}.$$

41. The average value is:

$$y_{av}=\frac{1}{b-a}\int_a^b f(x)\,dx$$

$$=\frac{1}{a-(0)}\int_0^a\left(4x+5\right)dx$$

$$=\frac{1}{a}\left[2x^2+5x\right]_0^a$$

$$=\frac{1}{a}\left[\left(2(a)^2+5(a)\right)-\left(2(0)^2+5(0)\right)\right]$$

$$=\frac{1}{a}\left[\left(2a^2+5a\right)-0\right]$$

$$=2a+5.$$

43. The average value is:

$$y_{av}=\frac{1}{b-a}\int_a^b f(x)\,dx$$

$$=\frac{1}{2-(1)}\int_1^2 x^n\,dx,\qquad n\neq0$$

$$=\frac{1}{1}\left[\frac{1}{n+1}x^{n+1}\right]_1^2$$

$$=\left[\frac{1}{n+1}(2)^{n+1}-\frac{1}{n+1}(1)^{n+1}\right]$$

$$=\frac{2^{n+1}}{n+1}-\frac{1}{n+1}$$

$$=\frac{2^{n+1}-1}{n+1}.$$

45. a) We find total profit by integrating:

$$P(10) = R(10) - C(10)$$

$$= \int_0^{10} \left[R'(t) - C'(t) \right] dt$$

$$= \int_0^{10} \left[(100e^t) - (100 - 0.2t) \right] dt$$

$$= \int_0^{10} \left[100e^t - 100 + 0.2t \right] dt$$

$$= \left[100e^t - 100t + 0.1t^2 \right]_0^{10}$$

$$= \left(100e^{10} - 100(10) + 0.1(10)^2 \right) -$$

$$\left(100e^0 - 100(0) + 0.1(0)^2 \right)$$

$$- \left(100e^{10} - 990 \right) - (100)$$

$$= 100e^{10} - 1090$$

$$\approx 2,201,556.58$$

The total profit for the first 10 days is approximately $2,201,556.58.

b) The average daily profit for the first ten days is given by

$$P_{av} = \frac{1}{10 - 0} \int_0^{10} \left[R'(t) - C'(t) \right] dt$$

$$\approx \frac{1}{10} [2,201,556.58] \qquad \text{From part (a).}$$

$$\approx 220,155.66.$$

The average daily profit over the first 10 days is approximately $220,155.66.

47. We find the average weekly sales for the first 5 weeks as follows:

$$S_{av} = \frac{1}{5 - 0} \int_0^5 S(t) dt$$

$$= \frac{1}{5} \int_0^5 9e^t dt$$

$$= \frac{1}{5} \left[9e^t \right]_0^5$$

$$= \frac{1}{5} \left[9e^5 - 9e^0 \right]$$

$$= \frac{1}{5} \left[9e^5 - 9 \right]$$

$$= \frac{9}{5} \left[e^5 - 1 \right]$$

$$\approx 265.3437.$$

Therefore, average weekly sales for the first 5 weeks are $265.3437 hundred, or $26,534.37.

49. We find the average weekly sales for weeks 2 through 5 by integrating over the interval $[1,5]$ as follows:

$$S_{av} = \frac{1}{5 - 1} \int_1^5 S(t) dt$$

$$= \frac{1}{4} \int_1^5 9e^t dt$$

$$= \frac{1}{4} \left[9e^t \right]_1^5$$

$$= \frac{1}{4} \left[9e^5 - 9e^1 \right]$$

$$= \frac{9}{4} \left[e^5 - e \right]$$

$$\approx 327.8135.$$

Therefore, average weekly sales for weeks 2 though 5 week are $327.8135 hundred, or $32,781.35.

51. a) $M(10) - m(10) = \int_0^{10} \left[M'(t) - m'(t) \right] dt$

$$\int_0^{10} \left[(-0.003t^2 + 0.2t) - (-0.009t^2 + 0.2t) \right] dt$$

$$= \int_0^{10} 0.006t^2 dt$$

$$= \left[0.002t^3 \right]_0^{10}$$

$$= 0.002(10)^3 - 0.002(0)^3$$

$$= 2 - 0$$

$$= 2$$

Bonnie memorizes approximately 2 more words during the first 10 minutes.

b) $m_{av} = \frac{1}{10 - 0} \int_0^{10} m'(t) dt$

$$m_{av} = \frac{1}{10 - 0} \int_0^{10} \left(-0.009t^2 + 0.2t \right) dt$$

$$= \frac{1}{10} \left[-0.003t^3 + 0.1t^2 \right]_0^{10}$$

$$= \frac{1}{10} \left[\left(-0.003(10)^3 + 0.1(10)^2 \right) - \left(-0.003(0)^3 + 0.1(0)^2 \right) \right]$$

$$= \frac{1}{10} \left[(-3 + 10) - (0 + 0) \right]$$

$$= \frac{1}{10} (7) = \frac{7}{10} = 0.7$$

Alan averaged memorizing about 0.7 words per minute during the first 10 minutes.

c) $m_{av} = \dfrac{1}{10-0} \displaystyle\int_0^{10} M'(t)\,dt$

$m_{av} = \dfrac{1}{10-0} \displaystyle\int_0^{10} \left(-0.003t^2 + 0.2t\right) dt$

$= \dfrac{1}{10}\left[-0.001t^3 + 0.1t^2\right]_0^{10}$

$= \dfrac{1}{10}\left[\left(-0.001(10)^3 + 0.1(10)^2\right) - \right.$

$\left. \left(-0.001(0)^3 + 0.1(0)^2\right)\right]$

$= \dfrac{1}{10}\left[(-1+10)-(0+0)\right]$

$= \dfrac{1}{10}(9) = \dfrac{9}{10} = 0.9$

Bonnie averaged memorizing about 0.9 words per minute during the first 10 minutes.

53. a) $W(0) = -6(0)^2 + 12(0) + 90 = 90.$

At the beginning of the interval, the keyboarder's speed is 90 words per minute.

b) First, we find the derivative.

$W'(t) = -12t + 12$

Next, we set the derivative equal to zero to find the critical value.

$W'(t) = 0$

$-12t + 12 = 0$

$t = 1$

The only critical value occurs when $t = 1$. We also know that $W''(t) = -12 < 0$.

Therefore, by the Max-Min Principle 2, we know that an absolute maximum occurs at $t = 1$.

$W(1) = -6(1)^2 + 12(1) + 90 = 96$

Thus, the maximum speed is 96 words per minute, occurring 1 minute into the interval.

c) $W_{av} = \dfrac{1}{5-0} \displaystyle\int_0^5 W(t)\,dt$

$W_{av} = \dfrac{1}{5-0} \displaystyle\int_0^5 \left(-6t^2 + 12t + 90\right) dt$

$= \dfrac{1}{5}\left[-2t^3 + 6t^2 + 90t\right]_0^5$

$= \dfrac{1}{5}\left[\left(-2(5)^3 + 6(5)^2 + 90(5)\right) - \right.$

$\left. \left(-2(0)^3 + 6(0)^2 + 90(0)\right)\right]$

$= \dfrac{1}{5}\left[350 - 0\right]$

$= 70$

The keyboarder's average speed over the 5 minute interval is 70 words per minute.

55. a) $C(0) = 42.03e^{-0.01050(0)} = 42.03$

The initial dosage is 42.03 micrograms per milliliter.

b) $C_{av} = \dfrac{1}{120-10} \displaystyle\int_{10}^{120} C(t)\,dt$

$C_{av} = \dfrac{1}{110} \displaystyle\int_{10}^{120} \left(42.03e^{-0.01050t}\right) dt$

$= \dfrac{1}{110}\left[\dfrac{42.03}{-0.01050}e^{-0.01050t}\right]_{10}^{120}$

$= \dfrac{1}{110}\left[\left(\dfrac{42.03}{-0.01050}e^{-0.01050(120)}\right) - \right.$

$\left. \dfrac{42.03}{-0.01050}e^{-0.01050(10)}\right]$

$\approx \dfrac{1}{110}\left[2468.4439\right]$

≈ 22.44

The average amount of phenylbutazone in the calf's body for the time between 10 and 120 hours is about 22.44 micrograms per milliliter.

57. a) We determine the average temperature as follows:

$$\frac{1}{10-0}\int_0^{10}\left(-t^2+5t+40\right)dt$$

$$=\frac{1}{10}\left[-\frac{t^3}{3}+\frac{5}{2}t^2+40t\right]_0^{10}$$

$$=\frac{1}{10}\left[\left(-\frac{(10)^3}{3}+\frac{5}{2}(10)^2+40(10)\right)-\right.$$

$$\left.\left(-\frac{(0)^3}{3}+\frac{5}{2}(0)^2+40(0)\right)\right]$$

$$=\frac{1}{10}\left[\frac{950}{3}-0\right]$$

$$=\frac{95}{3}\approx 31.67.$$

The average temperature over the 10 hour period is 31.7 degrees Celsius.

b) First, we find the critical values.

$$f'(t)=-2t+5$$

$$f'(t)=0$$

$$-2t+5=0$$

$$t=\frac{5}{2}=2.5$$

Since there is only one critical value and $f''(t)=-2<0$, we know that it represents a maximum over the closed interval. Therefore, the minimum temperature must be at an endpoint. We evaluate the function at the endpoints.

$$f(0)=-(0)^2+5(0)+40=40$$

$$f(10)=-(10)^2+5(10)+40=-10$$

The minimum temperature over the 10 hour period is negative 10 degrees Celsius.

c) From part (b), we know that the maximum temperature occurs when $t=2.5$. Substituting we have:

$$f(2.5)=-(2.5)^2+5(2.5)+40=46.25$$

Therefore, the maximum temperature over the 10 hour period is 46.25 degrees Celsius.

59. First graph the system of equations and shade the region bounded by the graphs.

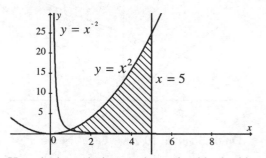

Here the boundaries are determined by looking at the graph. Note $x^2 \geq x^{-2}$ over the interval $[1,5]$.

We compute the area as follows:

$$\int_1^5\left(x^2-x^{-2}\right)dx$$

$$=\left[\frac{x^3}{3}+x^{-1}\right]_1^5$$

$$=\left(\frac{(5)^3}{3}+(5)^{-1}\right)-\left(\frac{(1)^3}{3}+(1)^{-1}\right)$$

$$=\frac{608}{15}$$

$$=40\tfrac{8}{15}.$$

61. First graph the system of equations and shade the region bounded by the graphs.

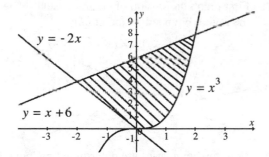

Here the boundaries are determined by looking at the graph. We will split the interval $[-2,2]$ up into two parts on the next page.

On the interval $[-2,0]$ we have $x+6 \geq -2x$.

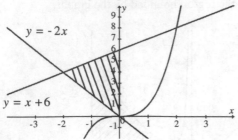

On the interval $[0,2]$ we have $x+6 \geq x^3$.

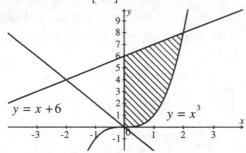

We compute the area as follows:

$$\int_{-2}^{0}\left[(x+6)-(-2x)\right]dx + \int_{0}^{2}\left[(x+6)-x^3\right]dx$$

$$=\int_{-2}^{0}(3x+6)\,dx + \int_{0}^{2}\left[-x^3+x+6\right]dx$$

$$=\left[\frac{3}{2}x^2+6x\right]_{-2}^{0} + \left[-\frac{x^4}{4}+\frac{x^2}{2}+6x\right]_{0}^{2}$$

$$=\left(0-(-6)\right)+(10-0)=16.$$

63. First graph the system of equations and shade the region bounded by the graphs.

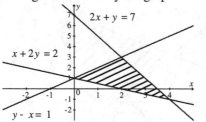

Here the boundaries are determined by looking at the graph. We will split the interval $[0,4]$ up into two parts. Solving each equation for y, we have

$$x+2y=2 \rightarrow y=-\frac{x}{2}+1$$

$$y-x=1 \rightarrow y=x+1$$

$$2x+y=7 \rightarrow y=-2x+7$$

On the interval $[0,2]$ we have $x+1 \geq -\frac{x}{2}+1$.

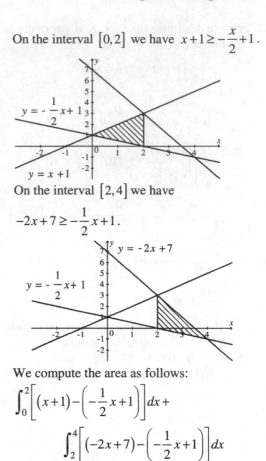

On the interval $[2,4]$ we have

$$-2x+7 \geq -\frac{1}{2}x+1.$$

We compute the area as follows:

$$\int_{0}^{2}\left[(x+1)-\left(-\frac{1}{2}x+1\right)\right]dx +$$

$$\int_{2}^{4}\left[(-2x+7)-\left(-\frac{1}{2}x+1\right)\right]dx$$

$$=\int_{0}^{2}\frac{3}{2}x\,dx + \int_{2}^{4}\left[-\frac{3}{2}x+6\right]dx$$

$$=\left[\frac{3}{4}x^2\right]_{0}^{2} + \left[-\frac{3}{4}x^2+6x\right]_{2}^{4}$$

$$=(3-0)+(12-9)$$

$$=6.$$

65. First we find the coordinates of the relative extrema.

$$y=x^3-3x+2$$

$$y'=3x^2-3$$

The derivative exists for all real numbers. We solve $y'=0$.

$$3x^2-3=0$$

$$x^2-1=0$$

$$x=\pm 1$$

$x=-1$ or $x=1$

We use the Second Derivative Test.

$$y''=6x$$

When $x=-1$, $y''=6(-1)=-6<0$

Therefore there is a relative maximum at $x=-1$. The solution is continued on the next page.

When $x = 1$, $y'' = 6(1) = 6 > 0$

Therefore there is a relative minimum at $x = 1$. We graph the region noting that the equation of the x-axis is $y = 0$.

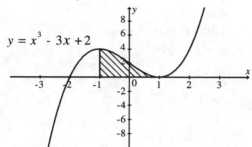

$y = x^3 - 3x + 2$

On the interval $[-1, 1]$, $x^3 - 3x + 2 \geq 0$. We compute the area as follows:

$$\int_{-1}^{1} \left(x^3 - 3x + 2 \right) dx$$

$$= \left[\frac{x^4}{4} - \frac{3}{2}x^2 + 2x \right]_{-1}^{1}$$

$$= \left(\frac{(1)^4}{4} - \frac{3}{2}(1)^2 + 2(1) \right) -$$

$$\left(\frac{(-1)^4}{4} - \frac{3}{2}(-1)^2 + 2(-1) \right)$$

$$= \left(\frac{3}{4} \right) - \left(-\frac{13}{4} \right) = 4.$$

67. $\displaystyle\int_{1}^{2} \left[\left(3x^2 + 5x \right) - \left(3x + K \right) \right] dx$

$$= \int_{1}^{2} \left(3x^2 + 2x - K \right) dx$$

$$= \left[x^3 + x^2 - Kx \right]_{1}^{2}$$

$$= \left[2^3 + 2^2 - K \cdot 2 \right] - \left[1^3 + 1^2 - K \right]$$

$$= \left[12 - 2K \right] - \left[2 - K \right]$$

$$= 10 - K$$

Since

$$\int_{1}^{2} \left[\left(3x^2 + 5x \right) - \left(3x + K \right) \right] dx = 6$$

We solve for K by substituting.

$$10 - K = 6$$
$$-K = -4$$
$$K = 4$$

69. From the graph we find that the interval we are concerned with is $[0, 2]$. Over this interval

$x\sqrt{4 - x^2} \geq \dfrac{-4x}{x^2 + 1}$. We use the fnInt function on the calculator to find

$$\int_{0}^{2} \left[x\sqrt{4 - x^2} - \frac{-4x}{x^2 + 1} \right] dx \approx 5.8855.$$

71. From the graph we find that the interval we are concerned with is $[-1, 1]$. Over this interval

$\sqrt{1 - x^2} \geq 1 - x^2$. We use the fnInt function on the calculator to find

$$\int_{-1}^{1} \left[\sqrt{1 - x^2} - \left(1 - x^2 \right) \right] dx \approx 0.2375.$$

Exercise Set 4.5

1. $\int \left(8+x^3\right)^5 3x^2\,dx$

Let $u = 8 + x^3$, then $du = 3x^2\,dx$.

$= \int u^5\,du$ Substitution: $\begin{smallmatrix}u=8+x^3\\du=3x^2dx\end{smallmatrix}$

$= \dfrac{1}{6}u^6 + C$ Formula A

$= \dfrac{1}{6}\left(8+x^3\right)^6 + C$ Reverse substitution

3. $\int \left(x^2-6\right)^7 x\,dx$

Let $u = x^2 - 6$, then $du = 2x\,dx$. We do not have $2x\,dx$. We only have $x\,dx$ and need to supply a 2. We do this by multiplying by $\frac{1}{2} \cdot 2$ as follows.

$= \dfrac{1}{2} \cdot 2 \int \left(x^2-6\right)^7 x\,dx$ Multiplying by 1

$= \dfrac{1}{2} \int \left(x^2-6\right)^7 2x\,dx$ $\left[a\int f(x)\,dx = \int af(x)\,dx\right]$

$= \dfrac{1}{2} \int u^7\,du$ Substitution: $\begin{smallmatrix}u=x^2-6\\du=2xdx\end{smallmatrix}$

$= \dfrac{1}{16}u^8 + C$ Formula A

$= \dfrac{1}{16}\left(x^2-6\right)^8 + C$ Reverse substitution

5. $\int \left(3t^4+2\right)t^3\,dt$

Let $u = 3t^4 + 2$, then $du = 12t^3\,dt$. We do not have $12t^3\,dt$. We only have $t^3\,dt$ and need to supply a 12. We do this by multiplying by $\frac{1}{12} \cdot 12$ as follows.

$= \dfrac{1}{12} \cdot 12 \int \left(3t^4+2\right)t^3\,dt$ Multiplying by 1

$= \dfrac{1}{12} \int \left(3t^4+2\right)12t^3\,dt$

$= \dfrac{1}{12} \int u\,du$ Substitution: $\begin{smallmatrix}u=3t^4+2\\du=12t^3dt\end{smallmatrix}$

$= \dfrac{1}{24}u^2 + C$ Formula A

$= \dfrac{1}{24}\left(3t^4+2\right)^2 + C$ Reverse substitution

7. $\int \dfrac{2}{1+2x}\,dx$

Let $u = 1 + 2x$, then $du = 2\,dx$.

$= \int \dfrac{du}{u}$ Substitution: $\begin{smallmatrix}u=1+2x\\du=2dx\end{smallmatrix}$

$= \ln|u| + C$ Formula C.

$= \ln|1+2x| + C$ Reverse substitution

9. $\int \left(\ln x\right)^3 \dfrac{1}{x}\,dx, \quad x > 0$

Let $u = \ln x$, then $du = \dfrac{1}{x}\,dx$.

$= \int u^3\,du$ Substitution: $\begin{smallmatrix}u=\ln x\\du=\frac{1}{x}dx\end{smallmatrix}$

$= \dfrac{1}{4}u^4 + C$ Formula A

$= \dfrac{1}{4}\left(\ln x\right)^4 + C$ Reverse substitution

11. $\int e^{3x}\,dx$

Let $u = 3x$, then $du = 3\,dx$. We do not have $3\,dx$. We only have dx and need to supply a 3. We do this by multiplying by $\frac{1}{3} \cdot 3$ as follows.

$= \dfrac{1}{3} \cdot 3 \int e^{3x}\,dx$ Multiplying by 1

$= \dfrac{1}{3} \int e^{3x} \cdot 3\,dx$

$= \dfrac{1}{3} \int e^u\,du$ Substitution: $\begin{smallmatrix}u=3x\\du=3dx\end{smallmatrix}$

$= \dfrac{1}{3}e^u + C$ Formula B

$= \dfrac{1}{3}e^{3x} + C$ Reverse substitution

13. $\int e^{x/3} dx$

Let $u = \dfrac{x}{3}$, then $du = \dfrac{1}{3} dx$. We do not have

$\dfrac{1}{3} dx$. We only have dx and need to supply a $\dfrac{1}{3}$.

We do this by multiplying by $\frac{1}{3} \cdot 3$ as follows.

$= \dfrac{1}{3} \cdot 3 \int e^{x/3} dx$ Multiplying by 1

$= 3 \int e^{x/3} \cdot \dfrac{1}{3} dx$

$= 3 \int e^{u} du$ Substitution: $\begin{array}{l} u = x/3 \\ du = 1/3 dx \end{array}$

$= 3 e^{u} + C$ Formula B

$= 3 e^{x/3} + C$ Reverse substitution

15. $\int x^4 e^{x^5} dx$

Let $u = x^5$, then $du = 5x^4 dx$. We do not have

$5x^4 dx$. We only have $x^4 dx$ and need to supply

a 5. We do this by multiplying by $\frac{1}{5} \cdot 5$ as

follows.

$\int x^4 e^{x^5} dx$

$= \dfrac{1}{5} \cdot 5 \int x^4 e^{x^5} dx$ Multiplying by 1

$= \dfrac{1}{5} \int 5 x^4 e^{x^5} dx$

$= \dfrac{1}{5} \int e^{u} du$ Substitution: $\begin{array}{l} u = x^5 \\ du = 5x^4 dx \end{array}$

$= \dfrac{1}{5} e^{u} + C$ Formula B

$= \dfrac{1}{5} e^{x^5} + C$ Reverse substitution

17. $\int t e^{-t^2} dt$

Let $u = -t^2$, then $du = -2t\,dt$. We do not have

$-2t\,dt$. We only have $t\,dt$ and need to supply a

-2. We do this by multiplying by $\left(-\frac{1}{2}\right) \cdot (-2)$ at

the top of the next column.

$\int t e^{-t^2} dt$

$= \left(-\dfrac{1}{2}\right) \cdot (-2) \int t e^{-t^2} dt$ Multiplying by 1

$= -\dfrac{1}{2} \int (-2t) e^{-t^2} dt$

$= -\dfrac{1}{2} \int e^{u} du$ Substitution: $\begin{array}{l} u = -t^2 \\ du = -2t\,dt \end{array}$

$= -\dfrac{1}{2} e^{u} + C$ Formula B

$= -\dfrac{1}{2} e^{-t^2} + C$ Reverse substitution

19. $\int \dfrac{1}{5 + 2x} dx$

Let $u = 5 + 2x$, then $du = 2dx$. We do not have

$2dx$. We only have dx and need to supply a 2.

We do this by multiplying by $\frac{1}{2} \cdot 2$ as follows.

$= \dfrac{1}{2} \cdot 2 \int \dfrac{1}{5 + 2x} dx$ Multiplying by 1

$= \dfrac{1}{2} \int \dfrac{2dx}{5 + 2x}$

$= \dfrac{1}{2} \int \dfrac{du}{u}$ Substitution: $\begin{array}{l} u = 5 + 2x \\ du = 2dx \end{array}$

$= \dfrac{1}{2} \ln |u| + C$ Formula C

$= \dfrac{1}{2} \ln |5 + 2x| + C$ Reverse substitution

21. $\int \dfrac{dx}{12 + 3x}$

Let $u = 12 + 3x$, then $du = 3dx$. We do not have

$3dx$. We only have dx and need to supply a 3.

We do this by multiplying by $\frac{1}{3} \cdot 3$ as shown on

the next page.

$\int \dfrac{dx}{12 + 3x} = \dfrac{1}{3} \cdot 3 \int \dfrac{dx}{12 + 3x}$ Multiplying by 1

$= \dfrac{1}{3} \int \dfrac{3dx}{12 + 3x}$

$= \dfrac{1}{3} \int \dfrac{du}{u}$ Substitution: $\begin{array}{l} u = 12 + 3x \\ du = 3dx \end{array}$

$= \dfrac{1}{3} \ln |u| + C$ Formula C

$= \dfrac{1}{3} \ln |12 + 3x| + C$ Reverse substitution

23. $\displaystyle\int\frac{dx}{1-x}$

Let $u = 1 - x$, then $du = -dx$. We do not have $-dx$. We only have dx and need to supply a -1. We do this by multiplying by $(-1)(-1)$ as follows.

$\displaystyle= (-1)\cdot(-1)\int\frac{dx}{1-x}$ Multiplying by 1

$\displaystyle= -1\cdot\int\frac{-1dx}{1-x}$

$\displaystyle= -\int\frac{du}{u}$ Substitution: $\begin{smallmatrix}u=1-x\\du=-dx\end{smallmatrix}$

$\displaystyle= -\ln|u| + C$ Formula C

$\displaystyle= -\ln|1-x| + C$ Reverse substitution

25. $\displaystyle\int t\left(t^2-1\right)^5 dt$

Let $u = t^2 - 1$, then $du = 2tdt$. We do not have $2tdt$. We only have tdt and need to supply a 2. We do this by multiplying by $\frac{1}{2}\cdot 2$ as follows.

$\displaystyle= \frac{1}{2}\cdot 2\int t\left(t^2-1\right)^5 dt$ Multiplying by 1

$\displaystyle= \frac{1}{2}\int 2t\left(t^2-1\right)^5 dt$

$\displaystyle= \frac{1}{2}\int u^5 du$ Substitution: $\begin{smallmatrix}u=t^2-1\\du=2tdt\end{smallmatrix}$

$\displaystyle= \frac{1}{12}u^6 + C$ Formula A

$\displaystyle= \frac{1}{12}\left(t^2-1\right)^6 + C$ Reverse substitution

27. $\displaystyle\int\left(x^4+x^3+x^2\right)^7\left(4x^3+3x^2+2x\right)dx$

Let $u = x^4 + x^3 + x^2$, then

$du = \left(4x^3 + 3x^2 + 2x\right)dx.$

$\displaystyle= \int u^7 du$ Substitution: $\begin{smallmatrix}u=x^4+x^3+x^2\\du=\left(4x^3+3x^2+2x\right)dx\end{smallmatrix}$

$\displaystyle= \frac{1}{8}u^8 + C$ Formula A

$\displaystyle= \frac{1}{8}\left(x^4+x^3+x^2\right)^8 + C$ Reverse substitution

29. $\displaystyle\int\frac{e^x dx}{4+e^x}$

Let $u = 4 + e^x$, then $du = e^x dx$.

$\displaystyle= \int\frac{du}{u}$ Substitution: $\begin{smallmatrix}u=4+e^x\\du=e^x dx\end{smallmatrix}$

$\displaystyle= \ln u + C$ Formula C, $u > 0$.

$\displaystyle= \ln\left(4+e^x\right) + C$ Reverse substitution

31. $\displaystyle\int\frac{\ln x^2}{x}dx$

Using properties of logarithms, we have

$\displaystyle\int\frac{2\cdot\ln|x|}{x}dx = 2\int\frac{\ln|x|}{x}dx$

Let $u = \ln|x|$, then $du = \dfrac{1}{x}dx$.

$\displaystyle= 2\int u\,du$ Substitution: $\begin{smallmatrix}u=\ln|x|\\du=\frac{1}{x}dx\end{smallmatrix}$

$\displaystyle= u^2 + C$ Formula A

$\displaystyle= \left(\ln|x|\right)^2 + C$ Reverse substitution

33. $\displaystyle\int\frac{dx}{x\ln x}, \quad x > 1$

Let $u = \ln x$, then $du = \dfrac{1}{x}dx$.

$\displaystyle= \int\frac{1}{\ln x}\cdot\left(\frac{1}{x}dx\right)$

$\displaystyle= \int\frac{1}{u}du$ Substitution: $\begin{smallmatrix}u=\ln x\\du=\frac{1}{x}dx\end{smallmatrix}$

$\displaystyle= \ln u + C$ Formula C, $u > 0$.

$\displaystyle= \ln(\ln x) + C$ Reverse substitution

35. $\displaystyle\int x\sqrt{ax^2+b}\,dx$

Let $u = ax^2 + b$, then $du = 2axdx$. We do not have $2axdx$. We only have xdx and need to supply a $2a$. We do this by multiplying by $\frac{1}{2a}\cdot 2a$ at the top of the next page.

$$\int x\sqrt{ax^2+b}\,dx =$$

$$=\frac{1}{2a}\cdot 2a\int x\sqrt{ax^2+b}\,dx \qquad \text{Multiplying by 1}$$

$$=\frac{1}{2a}\int 2ax\sqrt{ax^2+b}\,dx$$

$$=\frac{1}{2a}\int \sqrt{u}\,du \qquad \text{Substitution: } \begin{matrix}u=ax^2+b\\du=2axdx\end{matrix}$$

$$=\frac{1}{2a}\int u^{\frac{1}{2}}du$$

$$=\frac{1}{2a}\cdot\frac{2}{3}u^{\frac{3}{2}}+C \qquad \text{Formula A}$$

$$=\frac{1}{3a}\left(ax^2+b\right)^{\frac{3}{2}}+C \qquad \text{Reverse substitution}$$

37. $\displaystyle\int P_0 e^{kt}\,dt = P_0\cdot\int e^{kt}\,dt$

Let $u=kt$, then $du=kdt$. We do not have kdt. We only have dt and need to supply a k. We do this by multiplying by $\frac{1}{k}\cdot k$ as follows.

$$=\frac{1}{k}\cdot k\cdot P_0\int e^{kt}\,dt \qquad \text{Multiplying by 1}$$

$$=\frac{1}{k}\cdot P_0\int e^{kt}\cdot k\,dt$$

$$=\frac{P_0}{k}\int e^u\,du \qquad \text{Substitution: } \begin{matrix}u=kt\\du=kdt\end{matrix}$$

$$=\frac{P_0}{k}e^u+C \qquad \text{Formula B}$$

$$=\frac{P_0}{k}e^{kt}+C \qquad \text{Reverse substitution}$$

39. $\displaystyle\int \frac{x^3dx}{\left(2-x^4\right)^7}$

Let $u=2-x^4$, then $du=-4x^3dx$. We do not have $-4x^3dx$. We only have x^3dx and need to supply a -4. We do this by multiplying by $\frac{1}{-4}\cdot -4$ as shown at the top of the next column.

$$\int \frac{x^3dx}{\left(2-x^4\right)^7}$$

$$=\left(-\frac{1}{4}\right)\cdot(-4)\int \frac{x^3dx}{\left(2-x^4\right)^7} \qquad \text{Multiplying by 1}$$

$$=-\frac{1}{4}\int \frac{-4x^3dx}{\left(2-x^4\right)^7}$$

$$=-\frac{1}{4}\int \frac{du}{u^7} \qquad \text{Substitution: } \begin{matrix}u=2-x^4\\du=-4x^3dx\end{matrix}$$

$$=-\frac{1}{4}\int u^{-7}\,du$$

$$=-\frac{1}{4}\cdot\frac{1}{-6}u^{-6}+C \qquad \text{Formula A}$$

$$=\frac{1}{24}\left(2-x^4\right)^{-6}+C \qquad \text{Reverse substitution}$$

$$=\frac{1}{24\left(2-x^4\right)^6}+C$$

41. $\displaystyle\int 12x\sqrt[5]{1+6x^2}\,dx$

Let $u=1+6x^2$, then $du=12xdx$.

$$=\int \sqrt[5]{u}\,du \qquad \text{Substitution: } \begin{matrix}u=1+6x^2\\du=12xdx\end{matrix}$$

$$=\int u^{\frac{1}{5}}\,du$$

$$=\frac{5}{6}u^{\frac{6}{5}}+C \qquad \text{Formula A}$$

$$=\frac{5}{6}\left(1+6x^2\right)^{\frac{6}{5}}+C \qquad \text{Reverse substitution}$$

43. $\displaystyle\int_0^1 2xe^{x^2}\,dx$

We first find the indefinite integral

$$\int 2xe^{x^2}\,dx$$

$$=\int e^u\,du \qquad \text{Substitution: } \begin{matrix}u=x^2\\du=2xdx\end{matrix}$$

$$=e^u+C \qquad \text{Formula B}$$

$$=e^{x^2}+C \qquad \text{Reverse Substitution}$$

Next, we evaluate the definite integral on $[0,1]$.

$$\int_0^1 2xe^{x^2}\,dx = \left[e^{x^2}\right]_0^1 \qquad \text{Let } C=0.$$

$$=e^{(1)^2}-e^{(0)^2}$$

$$=e^1-e^0$$

$$=e-1$$

45. $\int_0^1 x\left(x^2+1\right)^5 dx$

We first find the indefinite integral

$\int x\left(x^2+1\right)^5 dx$

$= \dfrac{1}{2}\cdot 2\int x\left(x^2+1\right)^5 dx$ Multiplying by 1

$= \dfrac{1}{2}\int 2x\left(x^2+1\right)^5 dx$

$= \dfrac{1}{2}\int u^5 du$ Substitution: $\begin{array}{l}u=x^2+1\\ du=2xdx\end{array}$

$= \dfrac{1}{12}u^6 + C$ Formula A

$= \dfrac{1}{12}\left(x^2+1\right)^6 + C$ Reverse Substitution

Next, we evaluate the definite integral on $[0,1]$.

$\int_0^1 x\left(x^2+1\right)^5 dx$

$= \left[\dfrac{1}{12}\left(x^2+1\right)^6\right]_0^1$ Let $C=0$.

$= \left[\dfrac{1}{12}\left(1^2+1\right)^6 - \dfrac{1}{12}\left(0^2+1\right)^6\right]$

$= \dfrac{1}{12}(2)^6 - \dfrac{1}{12}(1)^6$

$= \dfrac{16}{3} - \dfrac{1}{12}$

$= \dfrac{21}{4}$

47. $\int_0^4 \dfrac{dt}{1+t}$

We first find the indefinite integral

$\int \dfrac{dt}{1+t}$

$= \int \dfrac{du}{u}$ Substitution: $\begin{array}{l}u=1+t\\ du=dt\end{array}$

$= \ln u + C$ Formula C, $u>0$

$= \ln(1+t) + C$ Reverse Substitution

Next, we evaluate the definite integral on $[0,4]$.

$\int_0^4 \dfrac{dt}{1+t}$

$= \left[\ln(1+t)\right]_0^4$ Let $C=0$.

$= \left[\ln(1+4) - \ln(1+0)\right]$

$= \ln(5) - \ln(1)$

$= \ln 5 - 0$

$= \ln 5$

49. $\int_1^4 \dfrac{2x+1}{x^2+x-1} dx$

We first find the indefinite integral

$\int \dfrac{2x+1}{x^2+x-1} dx$

$= \int \dfrac{du}{u}$ Substitution: $\begin{array}{l}u=x^2+x-1\\ du=(2x+1)dx\end{array}$

$= \ln u + C$ Formula C, $u>0$

$= \ln\left(x^2+x-1\right) + C$ Reverse Substitution

Next, we evaluate the definite integral on $[1,4]$.

$\int_1^4 \dfrac{2x+1}{x^2+x-1} dx$

$= \left[\ln\left(x^2+x-1\right)\right]_1^4$ Let $C=0$.

$= \left[\ln\left((4)^2+(4)-1\right) - \ln\left((1)^2+(1)-1\right)\right]$

$= \ln(19) - \ln(1)$

$= \ln 19 - 0$

$= \ln 19$

51. $\int_0^b e^{-x} dx$

We first find the indefinite integral

$\int e^{-x} dx$

$= -\int -e^{-x} dx$ Multipling by 1

$= -\int e^u du$ Substitution: $\begin{array}{l}u=-x\\ du=-dx\end{array}$

$= -e^u + C$ Formula B

$= -e^{-x} + C$ Reverse Substitution

The solution is continued at the top of the next page.

Next, we evaluate the definite integral on $[0,b]$.

$$\int_0^b e^{-x}dx$$

$$=\left[-e^{-x}\right]_0^b \qquad \text{Let } C = 0.$$

$$=-e^{-(b)}-\left(-e^{-(0)}\right)$$

$$=-e^{-b}+e^0$$

$$=1-e^{-b}$$

53. $\displaystyle\int_0^b me^{-mx}dx$

We first find the indefinite integral

$$\int me^{-mx}dx$$

$$=-\int -me^{-mx}dx \qquad \text{Multipling by 1}$$

$$=-\int e^u du \qquad \text{Substitution: } \begin{array}{l}u=-mx\\du=-mdx\end{array}$$

$$=-e^u +C \qquad \text{Formula B}$$

$$=-e^{-mx}+C \qquad \text{Reverse Substitution}$$

Next, we evaluate the definite integral on $[0,b]$.

$$\int_0^b me^{-mx}dx$$

$$=\left[-e^{-mx}\right]_0^b \qquad \text{Let } C = 0.$$

$$=-e^{-m(b)}-\left(-e^{-m(0)}\right)$$

$$=-e^{-mb}+e^0$$

$$=1-e^{-mb}$$

55. $\displaystyle\int_0^4 (x-6)^2 dx$

We first find the indefinite integral

$$\int (x-6)^2 dx$$

$$=\int u^2 du \qquad \text{Substitution: } \begin{array}{l}u=x-6\\du=dx\end{array}$$

$$=\frac{1}{3}u^3 +C \qquad \text{Formula A}$$

$$=\frac{1}{3}(x-6)^3 +C \qquad \text{Reverse Substitution}$$

Next, we evaluate the definite integral on $[0,4]$.

$$\int_0^4 (x-6)^2 dx$$

$$=\left[\frac{1}{3}(x-6)^3\right]_0^4 \qquad \text{Let } C = 0.$$

$$=\left[\frac{1}{3}((4)-6)^3 -\frac{1}{3}((0)-6)^3\right]$$

$$=\frac{1}{3}(-2)^3 -\frac{1}{3}(-6)^3$$

$$=-\frac{8}{3}-\left(-\frac{216}{3}\right)$$

$$=\frac{208}{3}$$

57. $\displaystyle\int_0^2 \frac{3x^2 dx}{\left(1+x^3\right)^5}$

We first find the indefinite integral

$$\int \frac{3x^2 dx}{\left(1+x^3\right)^5}$$

$$=\int \frac{du}{u^5} \qquad \text{Substitution: } \begin{array}{l}u=1+x^3\\du=3x^2dx\end{array}$$

$$=\int u^{-5}du$$

$$=-\frac{1}{4}u^{-4}+C \qquad \text{Formula A}$$

$$=-\frac{1}{4}\left(1+x^3\right)^{-4}+C \qquad \text{Reverse Substitution}$$

$$=-\frac{1}{4\left(1+x^3\right)^4}+C$$

The solution is continued on the next page.

Using the information from the previous page, we evaluate the definite integral on $[0,2]$.

$$\int_0^2 \frac{3x^2 \, dx}{\left(1+x^3\right)^5}$$

$$= \left[-\frac{1}{4\left(1+x^3\right)^4} \right]_0^2 \quad \text{Let } C = 0.$$

$$= \left(-\frac{1}{4\left(1+(2)^3\right)^4} \right) - \left(-\frac{1}{4\left(1+(0)^3\right)^4} \right)$$

$$= \left(-\frac{1}{4(9)^4} \right) - \left(-\frac{1}{4(1)^4} \right)$$

$$= \left(-\frac{1}{26,244} \right) - \left(-\frac{1}{4} \right)$$

$$= -\frac{1}{26,244} + \frac{6561}{26,244}$$

$$= \frac{6560}{26,244}$$

$$= \frac{1640}{6561}$$

59. $\displaystyle\int_0^{\sqrt{7}} 7x \cdot \sqrt[3]{1+x^2} \, dx$

We first find the indefinite integral

$$\int 7x \cdot \sqrt[3]{1+x^2} \, dx$$

$$= 7\int x \cdot \sqrt[3]{1+x^2} \, dx$$

$$= \frac{7}{2} \int 2x \cdot \sqrt[3]{1+x^2} \, dx$$

$$= \frac{7}{2} \int \sqrt[3]{1+x^2} \, (2x \, dx)$$

$$= \frac{7}{2} \int \sqrt[3]{u} \, du \qquad \text{Substitution: } \begin{matrix} u=1+x^2 \\ du=2x\,dx \end{matrix}$$

$$= \frac{7}{2} \int u^{1/3} \, du$$

$$= \frac{21}{8} u^{4/3} + C \qquad \text{Formula A}$$

$$= \frac{21}{8}\left(1+x^2\right)^{4/3} + C \qquad \text{Reverse Substitution}$$

Next, we evaluate the definite integral on $\left[0,\sqrt{7}\right]$.

$$\int_0^{\sqrt{7}} 7x\sqrt[3]{1+x^2} \, dx$$

$$= \left[\frac{21}{8}\left(1+x^2\right)^{4/3} \right]_0^{\sqrt{7}} \quad \text{Let } C = 0.$$

$$= \left(\frac{21}{8}\left(1+\left(\sqrt{7}\right)^2\right)^{4/3} \right) - \left(\frac{21}{8}\left(1+(0)^2\right)^{4/3} \right)$$

$$= \left(\frac{21}{8}(8)^{4/3} \right) - \left(\frac{21}{8}(1)^{4/3} \right)$$

$$= (42) - \left(\frac{21}{8} \right)$$

$$= \frac{315}{8}$$

61-78. Left to the student.

79. $\displaystyle\int \frac{x}{x-5} \, dx$

Let $u = x-5$, then $du = dx$. Observe that $x = u+5$.

Making the appropriate substitutions we have:

$$\int \frac{x}{x-5} \, dx$$

$$= \int \frac{u+5}{u} \, du$$

$$= \int \left(\frac{u}{u} + \frac{5}{u} \right) du$$

$$= \int \left(1 + \frac{5}{u} \right) du$$

$$= u + 5\ln|u| + C_1$$

$$= (x-5) + 5\ln|x-5| + C_1$$

$$= x + 5\ln|x-5| + C. \qquad C = C_1 - 5$$

81. $\displaystyle\int \frac{x}{1-4x}dx$

Let $u = 1-4x$, then $du = -4dx$. Observe that

$$x = \frac{1-u}{4}.$$

Making the appropriate substitutions we have:

$$\int \frac{x}{1-4x}dx$$

$$= -\frac{1}{4}\int \frac{x}{1-4x}(-4)dx$$

$$= -\frac{1}{4}\int \frac{\frac{1-u}{4}}{u}du$$

$$= -\frac{1}{16}\int \left(\frac{1-u}{u}\right)du$$

$$= -\frac{1}{16}\int \left(\frac{1}{u}-1\right)du$$

$$= -\frac{1}{16}\left(\ln|u| \quad u + C_1\right)$$

$$= -\frac{1}{16}\left(\ln|1-4x| - (1-4x) + C_1\right) \quad \substack{\text{Reverse}\\\text{Substitution}}$$

$$= -\frac{1}{16}\ln|1-4x| + \frac{1}{16} - \frac{1}{4}x - \frac{1}{16}C_1$$

$$= -\frac{1}{16}\ln|1-4x| - \frac{1}{4}x + C. \qquad C = -\frac{1}{16}C_1 - \frac{1}{16}$$

83. $\displaystyle\int \frac{2x+3}{3x-2}dx$

Let $u = 3x-2$, then $du = 3dx$. Observe that

$$x = \frac{u+2}{3}.$$

Making the appropriate substitutions we have:

$$\int \frac{2x+3}{3x-2}dx$$

$$= \frac{1}{3}\int \frac{2x+3}{3x-2}(3)dx \qquad \text{Multiply by } \frac{1}{3}\cdot 3$$

$$= \frac{1}{3}\int \frac{2\left(\dfrac{u+2}{3}\right)+3}{u}du$$

$$= \frac{1}{3}\int \frac{\dfrac{2}{3}u + \dfrac{13}{3}}{u}du$$

$$= \frac{1}{9}\int \frac{2u+13}{u}du$$

$$= \frac{1}{9}\int \left(2 + \frac{13}{u}\right)du$$

$$= \frac{1}{9}\left(2u + 13\ln|u| + C_1\right)$$

$$= \frac{2}{9}(3x-2) + \frac{13}{9}\ln|3x-2| + \frac{1}{9}C_1 \quad \substack{\text{Reverse}\\\text{Substitution}}$$

$$= \frac{2}{3}x - \frac{4}{9} + \frac{13}{9}\ln|3x-2| + \frac{1}{9}C_1$$

$$= \frac{2}{3}x + \frac{13}{9}\ln|3x-2| + C. \qquad C = \frac{1}{9}C_1 - \frac{4}{9}$$

85. $\displaystyle\int x^3(x+2)^7\, dx$

Let $u = x+2$, then $du = dx$. Observe that

$x = u-2$ and $x^3 = (u-2)^3 = u^3 - 6u^2 + 12u - 8$.

Making the appropriate substitutions we have

$$\int x^3(x+2)^7\, dx$$

$$= \int \left(u^3 - 6u^2 + 12u - 8\right)u^7\, du$$

$$= \int \left(u^{10} - 6u^9 + 12u^8 - 8u^7\right)du$$

$$= \frac{1}{11}u^{11} - \frac{6}{10}u^{10} + \frac{12}{9}u^9 - \frac{8}{8}u^8 + C$$

$$= \frac{(x+2)^{11}}{11} - \frac{3(x+2)^{10}}{5} +$$

$$\qquad \frac{4(x+2)^9}{3} - (x+2)^8 + C. \quad \substack{\text{Reverse}\\\text{Substitution}}$$

87. $D(x) = \int \dfrac{-2000x}{\sqrt{25-x^2}}\,dx$

Let $u = 25 - x^2$, then $du = -2x\,dx$.
Note that $-2000 = -2 \cdot 1000$.

$D(x) = \int \dfrac{1000}{\sqrt{25-x^2}}(-2x\,dx)$

$ = \int \dfrac{1000}{\sqrt{u}}\,du$ Substituting: $\begin{array}{l}u=25-x^2\\ du=-2x\,dx\end{array}$

$ = 1000\int u^{-\frac{1}{2}}\,du$

$ = 1000\dfrac{u^{\frac{1}{2}}}{\frac{1}{2}} + C$

$ = 2000u^{\frac{1}{2}} + C$

$ = 2000\sqrt{25-x^2} + C$

We use the condition $D(3) = 13{,}000$ to find C.

$$D(3) = 13{,}000$$

$$2000\sqrt{25-(3)^2} + C = 13{,}000$$

$$2000\sqrt{16} + C = 13{,}000$$

$$8000 + C = 13{,}000$$

$$C = 5000$$

Thus, $D(x) = 2000\sqrt{25-x^2} + 5000$.

89. a) $M(t) = 8.3e^{-0.019t}$

We integrate the marriage rate $M(t)$ over the interval $[0,5]$ to determine the total number of marriages per 1000 people in the U.S. from 2000 to 2005.

$\displaystyle\int_0^5 8.3e^{-0.019t}\,dt = \left[\dfrac{8.3}{-0.019}e^{-0.019t}\right]_0^5$

$\phantom{\int_0^5 8.3e^{-0.019t}\,dt} = \left[-436.842e^{-0.019t}\right]_0^5$

$\phantom{\int_0^5 8.3e^{-0.019t}\,dt} = -436.842e^{-0.019(5)}$

$ - \left(-436.842e^{-0.019(0)}\right)$

$\phantom{\int_0^5 8.3e^{-0.019t}\,dt} = 39.589$

$\phantom{\int_0^5 8.3e^{-0.019t}\,dt} \approx 40$

There were approximately 40 marriages per 1000 people during the years 2000 – 2005.

b) We integrate the marriage rate $M(t)$ over the interval $[5,16]$ to determine the total number of marriages per 1000 people in the U.S. from 2005 to 2016.

$\displaystyle\int_5^{16} 8.3e^{-0.019t}\,dt = \left[\dfrac{8.3}{-0.019}e^{-0.019t}\right]_5^{16}$

$\phantom{\int_5^{16} 8.3e^{-0.019t}\,dt} = \left[-436.842e^{-0.019t}\right]_5^{16}$

$\phantom{\int_5^{16} 8.3e^{-0.019t}\,dt} = -436.842e^{-0.019(16)}$

$\phantom{\int_5^{16} 8.3e} - \left(-436.842e^{-0.019(5)}\right)$

$\phantom{\int_5^{16} 8.3e^{-0.019t}\,dt} = 74.92$

$\phantom{\int_5^{16} 8.3e^{-0.019t}\,dt} \approx 75$

There will be approximately 75 marriages per 1000 people during the years 2005 – 2016.

91. $\displaystyle\int_{-2}^{2} -x\sqrt{4-x^2}\,dx$

We notice that $-x\sqrt{4-x^2} \geq 0$, on $[-2,0]$ and $0 \geq -x\sqrt{4-x^2}$, on $[0,2]$. We will divide the interval into two parts.
First on the interval $[-2,0]$, we find the indefinite integral.

$\displaystyle\int\left(-x\sqrt{4-x^2} - 0\right)dx$

$= \displaystyle\int\left(-x\sqrt{4-x^2}\right)dx$

$= \dfrac{1}{2}\displaystyle\int \sqrt{4-x^2}\,(-2x)\,dx$

$= \dfrac{1}{2}\displaystyle\int \sqrt{u}\,du$ Substitution: $\begin{array}{l}u=4-x^2\\ du=-2x\,dx\end{array}$

$= \dfrac{1}{2}\displaystyle\int u^{\frac{1}{2}}\,du$

$= \dfrac{1}{3}u^{\frac{3}{2}} + C$

$= \dfrac{1}{3}\left(4-x^2\right)^{\frac{3}{2}} + C$

The solution is continued on the next page.

Next, we will evaluate the integral on $[-2,0]$.

$$\int_{-2}^{0} -x\sqrt{4-x^2}\,dx$$

$$= \left[\frac{1}{3}\left(4-x^2\right)^{3/2}\right]_{-2}^{0}$$

$$= \left(\frac{1}{3}\left(4-(0)^2\right)^{3/2}\right) - \left(\frac{1}{3}\left(4-(-2)^2\right)^{3/2}\right)$$

$$= \frac{1}{3}(4)^{3/2} - \frac{1}{3}(0)^{3/2} = \frac{8}{3}$$

Next, on the interval $[0,2]$, we find the indefinite integral.

$$\int \left(0-\left(-x\sqrt{4-x^2}\right)\right)dx$$

$$= \int \left(x\sqrt{4-x^2}\right)dx$$

$$= -\frac{1}{2}\int \sqrt{4-x^2}\,(-2x)\,dx$$

$$= -\frac{1}{2}\int \sqrt{u}\,du \qquad \text{Substitution: } \begin{matrix} u=4-x^2 \\ du=-2x\,dx \end{matrix}$$

$$= -\frac{1}{2}\int u^{1/2}\,du$$

$$= -\frac{1}{3}u^{3/2} + C$$

$$= -\frac{1}{3}\left(4-x^2\right)^{3/2} + C$$

Next, we will evaluate the integral on $[0,2]$.

$$\int_{0}^{2} x\sqrt{4-x^2}\,dx$$

$$= \left[-\frac{1}{3}\left(4-x^2\right)^{3/2}\right]_{0}^{2}$$

$$= \left(-\frac{1}{3}\left(4-(2)^2\right)^{3/2}\right) - \left(-\frac{1}{3}\left(4-(0)^2\right)^{3/2}\right)$$

$$= -\frac{1}{3}(0)^{3/2} + \frac{1}{3}(4)^{3/2}$$

$$= \frac{8}{3}$$

Therefore, the total area is:

$$\int_{-2}^{0} -x\sqrt{4-x^2}\,dx + \int_{0}^{2} x\sqrt{4-x^2}\,dx$$

$$= \frac{8}{3} + \frac{8}{3}$$

$$= \frac{16}{3} = 5\frac{1}{3}.$$

93. $\int \dfrac{1}{ax+b}\,dx$

Let $u = ax+b$, then $du = a\,dx$. We do not have $a\,dx$. We only have dx and need to supply an a. We do this by multiplying by $\frac{1}{a}\cdot a$ as follows:

$$= \frac{1}{a}\int \frac{a\,dx}{ax+b}$$

$$= \frac{1}{a}\int \frac{du}{u} \qquad \text{Substitution: } \begin{matrix} u=ax+b \\ du=a\,dx \end{matrix}$$

$$= \frac{1}{a}\ln|u| + C$$

$$= \frac{1}{a}\ln|ax+b| + C. \qquad \text{Reverse Substitution}$$

95. $\int \dfrac{e^{\sqrt{t}}}{\sqrt{t}}\,dt$

Let $u = \sqrt{t}$, then $du = \dfrac{1}{2\sqrt{t}}\,dt$. We do not have $\dfrac{1}{2\sqrt{t}}\,dt$. We only have $\dfrac{1}{\sqrt{t}}\,dt$ and need to supply a $\dfrac{1}{2}$. We do this by multiplying by $\frac{1}{2}\cdot 2$ as follows:

$$= 2\int \frac{e^{\sqrt{t}}}{2\sqrt{t}}\,dt$$

$$= 2\int e^{u}\,du \qquad \text{Substitution: } \begin{matrix} u=\sqrt{t} \\ du=\frac{1}{2\sqrt{t}}\,dt \end{matrix}$$

$$= 2e^{u} + C$$

$$= 2e^{\sqrt{t}} + C. \qquad \text{Reverse Substitution}$$

97. $\int \dfrac{e^{1/t}}{t^2}\,dt$

Let $u = \dfrac{1}{t} = t^{-1}$, then $du = -t^{-2}\,dt = -\dfrac{1}{t^2}\,dt$. We do not have $-\dfrac{1}{t^2}\,dt$. We only have $\dfrac{1}{t^2}\,dt$ and need to supply a -1. We do this by multiplying by $(-1)\cdot(-1)$ as follows.

$$= (-1)\int (-1)\frac{e^{1/t}}{t^2}\,dt$$

$$= -\int e^{u}\,du \qquad \text{Substitution: } \begin{matrix} u=\frac{1}{t} \\ du=-\frac{1}{t^2}\,dt \end{matrix}$$

$$= -e^{u} + C$$

$$= -e^{1/t} + C. \qquad \text{Reverse substitution}$$

99. $\displaystyle\int \frac{dx}{x(\ln x)^4}$

Let $u = \ln x$, then $du = \dfrac{1}{x}dx$.

$\displaystyle = \int \frac{du}{u^4}$ Substitution: $\begin{array}{l} u=\ln x \\ du=\frac{1}{x}dx \end{array}$

$\displaystyle = \int u^{-4}du$

$\displaystyle = \frac{1}{-3}u^{-3} + C$

$\displaystyle = -\frac{1}{3}(\ln x)^{-3} + C.$

101. $\displaystyle\int x^2\sqrt{x^3+1}\,dx$

Let $u = x^3 + 1$, then $du = 3x^2 dx$.

$\displaystyle = \frac{1}{3}\cdot 3\int x^2\sqrt{x^3+1}\,dx$

$\displaystyle = \frac{1}{3}\int \sqrt{x^3+1}\,(3x^2)\,dx$

$\displaystyle = \frac{1}{3}\int \sqrt{u}\,du$ Substitution: $\begin{array}{l} u=x^3+1 \\ du=3x^2 dx \end{array}$

$\displaystyle = \frac{1}{3}\int u^{1/2}\,du$

$\displaystyle = \frac{2}{9}u^{3/2} + C$

$\displaystyle = \frac{2}{9}(x^3+1)^{3/2} + C.$

103. $\displaystyle\int \frac{\left[(\ln x)^2 + 3(\ln x) + 4\right]}{x}dx$

Let $u = \ln x$, then $du = \dfrac{1}{x}dx$.

$\displaystyle = \int\left[(u)^2 + 3(u) + 4\right]du$ Substitution: $\begin{array}{l} u=\ln x \\ du=\frac{1}{x}dx \end{array}$

$\displaystyle = \frac{1}{3}u^3 + \frac{3}{2}u^2 + 4u + C$

$\displaystyle = \frac{1}{3}(\ln x)^3 + \frac{3}{2}(\ln x)^2 + 4\ln x + C.$

105. $\displaystyle\int \frac{t^3\ln(t^4+8)}{t^4+8}dt$

Let $u = \ln(t^4+8)$, then $du = \dfrac{4t^3}{t^4+8}dt$. We need

to supply a 4, by multiplying by $\frac{1}{4}\cdot 4$

$\displaystyle = \frac{1}{4}\int \frac{4t^3\ln(t^4+8)}{t^4+8}dt$

$\displaystyle = \frac{1}{4}\int u\,du$ Substitution: $\begin{array}{l} u=\ln(t^4+8) \\ du=\frac{4t^3}{t^4+8}dt \end{array}$

$\displaystyle = \frac{1}{8}u^2 + C$

$\displaystyle = \frac{1}{8}\left[\ln(t^4+8)\right]^2 + C.$

107. $\displaystyle\int \frac{x^2+6x}{(x+3)^2}dx$

$\displaystyle\int \frac{x^2+6x+9-9}{(x+3)^2}dx$ Adding 9−9 in the numerator.

$\displaystyle\int\left[\frac{x^2+6x+9}{(x+3)^2} - \frac{9}{(x+3)^2}\right]dx$

$\displaystyle = \int\left[\frac{(x+3)^2}{(x+3)^2} - \frac{9}{(x+3)^2}\right]dx$

$\displaystyle = \int\left[1 - \frac{9}{(x+3)^2}\right]dx$

$\displaystyle = \int dx - \int\frac{9}{(x+3)^2}dx$

Let $u = x+3$, then $du = dx$.

$\displaystyle = \int du - \int\frac{9}{u^2}du$ Substitution: $\begin{array}{l} u=x+3 \\ du=dx \end{array}$

$\displaystyle = \int du - 9\int u^{-2}du + K$

$\displaystyle = u + 9u^{-1} + K$

$\displaystyle = x + 9(x+3)^{-1} + K + 3$

$\displaystyle = x + \frac{9}{x+3} + C.$ $[C = K+3]$

109. $\int \dfrac{t-5}{t-4}\,dt$

First, we divide algebraically to see

$$t-4\overline{\smash{)}\,t-5}$$
$$\dfrac{-(t-4)}{-1}$$

with quotient 1.

Thus, $\dfrac{t-5}{t-4}=1-\dfrac{1}{t-4}$.

Rewriting the integral, we have:

$$\int \dfrac{t-5}{t-4}\,dt$$

$$=\int\left[1-\dfrac{1}{t-4}\right]dt$$

$$=\int dt-\int\dfrac{dt}{t-4}$$

Let $u=t-4$, then $du=dt$.

$$=\int du-\int\dfrac{du}{u}\qquad\text{Substitution: }\begin{array}{l}u-t-4\\du=dt\end{array}$$

$$=u-\left(\ln u+C_2\right)$$

$$=t-4-\ln(t-4)+K$$

$$=t-\ln(t-4)+C.\qquad [C-K-4]$$

111. $\int\dfrac{dx}{e^x+1}=\int\dfrac{e^{-x}}{1+e^{-x}}\,dx$

Let $u=1+e^{-x}$, then $du=-e^{-x}dx$.

$$=-\int\dfrac{-e^{-x}}{1+e^{-x}}\,dx$$

$$=-\int\dfrac{du}{u}\qquad\text{Substitution: }\begin{array}{l}u=1+e^{-x}\\du=-e^{-x}dx\end{array}$$

$$=-\ln u+C$$

$$=-\ln\left(1+e^{-x}\right)+C.$$

113. $\int\dfrac{(\ln x)^n}{x}\,dx,\qquad x>0,n\neq-1$

Let $u=\ln x$, then $du=\dfrac{1}{x}\,dx$.

$$=\int u^n\,du\qquad\text{Substitution: }\begin{array}{l}u=\ln x\\du=\frac{1}{x}dx\end{array}$$

$$=\dfrac{1}{n+1}u^{n+1}+C$$

$$=\dfrac{1}{n+1}(\ln x)^{n+1}+C$$

$$=\dfrac{(\ln x)^{n+1}}{n+1}+C.$$

115. $\int 9x\left(7x^2+9\right)^n dx,\qquad n\neq-1$

$$=9\int x\left(7x^2+9\right)^n dx$$

Let $u=7x^2+9$, then $du=14x\,dx$.

$$=\dfrac{1}{14}\cdot14\cdot9\int x\left(7x^2+9\right)^n dx$$

$$=\dfrac{9}{14}\int 14x\left(7x^2+9\right)^n dx$$

$$=\dfrac{9}{14}\int u^n\,du\qquad\text{Substitution: }\begin{array}{l}u=7x^2+9\\du=14x\,dx\end{array}$$

$$=\dfrac{9}{14}\cdot\dfrac{u^{n+1}}{n+1}+C$$

$$=\dfrac{9}{14(n+1)}\left(7x^2+9\right)^{n+1}+C.$$

117.

Exercise Set 4.6

Integrate using formula A.
$$\int x^3 \left(3x^2\right) dx$$
$$= \int 3x^5 dx$$
$$= 3 \cdot \frac{x^6}{6} + C \qquad \text{Using Formula A}$$
$$= \frac{x^6}{2} + C$$

1. $\int 4xe^{4x} dx = \int x\left(4e^{4x} dx\right)$

Let
$$u = x \quad \text{and} \quad dv = 4e^{4x} dx$$
Then
$$du = dx \quad \text{and} \quad v = e^{4x}$$
Using the Integration-by-Parts Formula gives
$$\quad u \quad\ dv \qquad\ u \cdot v \quad\ v\ du$$
$$\int x\left(4e^{4x} dx\right) = x \cdot e^{4x} - \int e^{4x} dx$$
$$= xe^{4x} - \frac{1}{4}e^{4x} + C.$$

5. $\int xe^{5x} dx$

Let
$$u = x \quad \text{and} \quad dv = e^{5x} dx$$
Then
$$du = dx \quad \text{and} \quad v = \frac{1}{5}e^{5x}$$
We integrated dv using the formula
$$\int e^{ax} dx = \frac{1}{a}e^{ax} + C.$$

Using the Integration-by-Parts Formula gives
$$\quad u \quad\ dv \qquad u\ v \qquad v\ du$$
$$\int x\left(e^{5x} dx\right) = x \cdot \frac{1}{5}e^{5x} - \int \frac{1}{5}e^{5x} dx$$
$$= x \cdot \frac{1}{5}e^{5x} - \frac{1}{5} \cdot \frac{1}{5}e^{5x} + C$$
$$\left(\int be^{ax} dx = \frac{b}{a}e^{ax} + C\right)$$
$$= \frac{1}{5}xe^{5x} - \frac{1}{25}e^{5x} + C.$$

3. $\int x^3 \left(3x^2\right) dx$

Let
$$u = x^3 \quad \text{and} \quad dv = 3x^2 dx$$
Then
$$du = 3x^2 dx \quad \text{and} \quad v = x^3$$
Using the Integration-by-Parts Formula gives
$$\quad u \quad\ dv \qquad u\ v \qquad\ v\ du$$
$$\int x^3 \left(3x^2 dx\right) = x^3 \cdot x^3 - \int x^3 3x^2 dx$$
$$= x^6 - \int 3x^5 dx$$
$$= x^6 - 3 \cdot \frac{x^6}{6} + C$$
$$= x^6 - \frac{x^6}{2} + C$$
$$= \frac{x^6}{2} + C.$$

It should be noted that this problem can also be worked with substitution or formula A.
Integrate using Substitution
$$\int x^3 \left(3x^2\right) dx$$
$$= \int u\, du \qquad \text{Substitution: } {\scriptstyle u = x^3 \atop \scriptstyle du = 3x^2 dx}$$
$$= \frac{1}{2}u^2 + C$$
$$= \frac{1}{2}\left(x^3\right)^2 + C$$
$$= \frac{x^6}{2} + C$$

7. $\int xe^{-2x} dx$

Let
$$u = x \quad \text{and} \quad dv = e^{-2x} dx$$
Then
$$du = dx \quad \text{and} \quad v = -\frac{1}{2}e^{-2x}$$
We integrated dv using the formula
$$\int e^{ax} dx = \frac{1}{a}e^{ax} + C.$$
Using the Integration-by-Parts Formula gives
$$\quad u \quad\ dv \qquad u \quad\ v \qquad\quad v \quad\ du$$
$$\int x\left(e^{-2x} dx\right) = x \cdot \left(-\frac{1}{2}e^{-2x}\right) - \int \left(-\frac{1}{2}e^{-2x}\right) dx$$
The solution is continued on the next page.

Continuing from the previous page, we have:

$$\int xe^{-2x}\,dx$$

$$= -\frac{1}{2}xe^{-2x} - \left(-\frac{1}{2}\right)\cdot\left(-\frac{1}{2}e^{-2x}\right) + C$$

$$\left(\int be^{ax}\,dx = \frac{b}{a}e^{ax} + C\right)$$

$$= -\frac{1}{2}xe^{-2x} - \frac{1}{4}e^{-2x} + C.$$

9. $\int x^2 \ln x\,dx$

Let

$$u = \ln x \quad \text{and} \quad dv = x^2\,dx$$

Then

$$du = \frac{1}{x}\,dx \quad \text{and} \quad v = \frac{1}{3}x^3$$

Using the Integration-by-Parts Formula gives

$$\overset{u}{}\quad\overset{dv}{}\qquad\overset{u}{}\quad\overset{v}{}\qquad\overset{v}{}\quad\overset{du}{}$$

$$\int \ln x\left(x^2\,dx\right) = \ln x\cdot\left(\frac{1}{3}x^3\right) - \int\left(\frac{1}{3}x^3\right)\left(\frac{1}{x}\,dx\right)$$

$$= \frac{1}{3}x^3 \ln x - \int \frac{1}{3}x^2\,dx$$

$$= \frac{1}{3}x^3 \ln x - \frac{1}{9}x^3 + C$$

$$= \frac{x^3 \ln x}{3} - \frac{x^3}{9} + C.$$

11. $\int x \ln \sqrt{x}\,dx = \int x \ln x^{\frac{1}{2}}\,dx = \frac{1}{2}\int (x \ln x)\,dx$

Let

$$u = \ln x \quad \text{and} \quad dv = x\,dx$$

Then

$$du = \frac{1}{x}\cdot dx \quad \text{and} \quad v = \frac{1}{2}x^2$$

Using the Integration-by-Parts Formula gives

$$\overset{u}{}\quad\overset{dv}{}\qquad\overset{u}{}\quad\overset{v}{}\qquad\overset{v}{}\quad\overset{du}{}$$

$$\frac{1}{2}\int \ln x\,(x\,dx) = \frac{1}{2}\left[\ln x\cdot\left(\frac{x^2}{2}\right) - \int\left(\frac{x^2}{2}\right)\left(\frac{1}{x}\,dx\right)\right]$$

$$= \frac{1}{2}\left[\frac{x^2 \ln x}{2} - \int \frac{1}{2}x\,dx\right]$$

$$= \frac{1}{2}\left[\frac{x^2 \ln x}{2} - \frac{1}{4}x^2 + C\right]$$

$$= \frac{x^2 \ln x}{4} - \frac{1}{8}x^2 + C$$

13. $\int \ln(x+5)\,dx$

Let

$$u = \ln(x+5) \quad \text{and} \quad dv = dx$$

Then

$$du = \frac{1}{x+5}\,dx \quad \text{and} \quad v = x+5$$

Choosing $x+5$ as an antiderivative of dv

Using the Integration-by-Parts Formula gives

$$\overset{u}{}\quad\overset{dv}{}$$

$$\int \ln(x+5)\,dx =$$

$$\overset{u}{}\quad\overset{v}{}\qquad\overset{v}{}\qquad\overset{du}{}$$

$$\ln(x+5)\cdot(x+5) - \int(x+5)\left(\frac{1}{x+5}\,dx\right)$$

$$= (x+5)\ln(x+5) - \int dx$$

$$= (x+5)\ln(x+5) - x + C.$$

15. $\int (x+2)\ln x\,dx$

Let

$$u = \ln x \quad \text{and} \quad dv = (x+2)\,dx$$

Then

$$du = \frac{1}{x}\,dx \quad \text{and} \quad v = \frac{x^2}{2} + 2x$$

Using the Integration-by-Parts Formula gives

$$\overset{u}{}\quad\overset{dv}{}$$

$$\int \ln x\cdot(x+2)\,dx =$$

$$\overset{u}{}\quad\overset{v}{}\qquad\overset{v}{}\qquad\overset{du}{}$$

$$\ln(x)\cdot\left(\frac{x^2}{2} + 2x\right) - \int\left(\frac{x^2}{2} + 2x\right)\left(\frac{1}{x}\,dx\right)$$

$$= \left(\frac{x^2}{2} + 2x\right)\ln x - \int\left(\frac{x}{2} + 2\right)dx$$

$$= \left(\frac{x^2}{2} + 2x\right)\ln x - \int\frac{x}{2}\,dx - \int 2\,dx$$

$$= \left(\frac{x^2}{2} + 2x\right)\ln x - \frac{x^2}{4} - 2x + C.$$

17. $\int (x-1)\ln x\, dx$

Let
$$u = \ln x \quad \text{and} \quad dv = (x-1)\, dx$$
Then
$$du = \frac{1}{x}\, dx \quad \text{and} \quad v = \frac{x^2}{2} - x$$
Using the Integration-by-Parts Formula gives
$$\overset{u}{\int \ln x} \cdot \overset{dv}{(x-1)\, dx} =$$

$$\overset{u}{\ln(x)} \cdot \overset{v}{\left(\frac{x^2}{2} - x\right)} - \int \overset{v}{\left(\frac{x^2}{2} - x\right)} \overset{du}{\left(\frac{1}{x}\, dx\right)}$$

$$= \left(\frac{x^2}{2} - x\right)\ln x - \int \left(\frac{x}{2} - 1\right) dx$$

$$= \left(\frac{x^2}{2} - x\right)\ln x - \int \frac{x}{2}\, dx + \int 1\, dx$$

$$= \left(\frac{x^2}{2} - x\right)\ln x - \frac{x^2}{4} + x + C.$$

19. $\int x\sqrt{x+2}\, dx$

Let
$$u = x \quad \text{and} \quad dv = \sqrt{x+2}\, dx = (x+2)^{\frac{1}{2}}\, dx$$
Then
$$du = dx \quad \text{and} \quad v = \frac{(x+2)^{\frac{3}{2}}}{\frac{3}{2}} = \frac{2}{3}(x+2)^{\frac{3}{2}}$$
Using the Integration-by-Parts Formula gives
$$\overset{u}{\int x} \overset{dv}{\left(\sqrt{x+2}\, dx\right)} =$$

$$\overset{u}{x} \cdot \overset{v}{\frac{2}{3}(x+2)^{\frac{3}{2}}} - \int \overset{v}{\frac{2}{3}(x+2)^{\frac{3}{2}}}\, \overset{du}{dx}$$

$$= \frac{2}{3}x(x+2)^{\frac{3}{2}} - \frac{2}{3}\int (x+2)^{\frac{3}{2}}\, dx$$

$$= \frac{2}{3}x(x+2)^{\frac{3}{2}} - \frac{2}{3}\frac{(x+2)^{\frac{5}{2}}}{\frac{5}{2}} + C$$

$$= \frac{2}{3}x(x+2)^{\frac{3}{2}} - \frac{4}{15}(x+2)^{\frac{5}{2}} + C.$$

21. $\int x^3 \ln(2x)\, dx$

Let
$$u = \ln(2x) \quad \text{and} \quad dv = x^3 dx$$
Then
$$du = \frac{1}{2x} \cdot 2dx = \frac{1}{x}\, dx \quad \text{and} \quad v = \frac{1}{4}x^4$$
Using the Integration-by-Parts Formula gives
$$\overset{u}{\int \ln(2x)}\overset{dv}{\left(x^3 dx\right)} = \overset{u}{\ln(2x)} \cdot \overset{v}{\left(\frac{x^4}{4}\right)} - \int \overset{v}{\left(\frac{x^4}{4}\right)}\overset{du}{\left(\frac{1}{x}\, dx\right)}$$

$$= \frac{x^4}{4}\ln(2x) - \int \frac{x^3}{4}\, dx$$

$$= \frac{x^4}{4}\ln(2x) - \frac{x^4}{16} + C.$$

23. $\int x^2 e^x dx$

Let
$$u = x^2 \quad \text{and} \quad dv = e^x dx$$
Then
$$du = 2x\, dx \quad \text{and} \quad v = e^x$$
Using the Integration-by-Parts Formula gives
$$\overset{u}{\int x^2} \overset{dv}{\left(e^x dx\right)} = \overset{u}{x^2}\overset{v}{e^x} - \int \overset{v}{e^x} \cdot \overset{du}{2x\, dx}$$
We will use integration by parts again to
evaluate $\int 2xe^x dx$.

Let
$$u = 2x \quad \text{and} \quad dv = e^x dx$$
Then
$$du = 2dx \quad \text{and} \quad v = e^x$$
Using the Integration-by-Parts Formula gives
$$\overset{u}{\int 2x}\overset{dv}{\left(e^x dx\right)} = \overset{u}{2x}\overset{v}{e^x} - \int \overset{v}{2e^x}\overset{du}{dx}$$

$$= 2xe^x - 2e^x + C_1$$

Thus,
$$\int x^2 e^x dx = x^2 e^x - \int 2xe^x dx$$

$$= x^2 e^x - \left(2xe^x - 2e^x + C_1\right)$$

$$= x^2 e^x - 2xe^x + 2e^x - C_1$$

$$= x^2 e^x - 2xe^x + 2e^x + C \quad [C = -C_1]$$

The solution is continued on the next page.

Note, Since we have an integral $\int f(x)g(x)dx$, where $f(x)$ can be differentiated repeatedly to a derivative that is eventually 0 and $g(x)$ can be integrated repeatedly easily, we can use tabular integration as shown below:

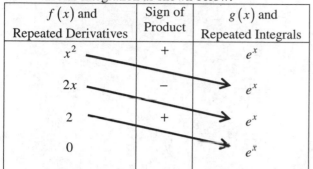

$f(x)$ and Repeated Derivatives	Sign of Product	$g(x)$ and Repeated Integrals
x^2	+	e^x
$2x$	–	e^x
2	+	e^x
0		e^x

We add the products along the arrows, making the alternate sign changes.

$$\int x^2 e^x dx = x^2 e^x - 2xe^x + 2e^x + C$$

25. $\int x^2 e^{2x} dx$

Let

$$u = x^2 \quad \text{and} \quad dv = e^{2x} dx$$

Then

$$du = 2x\,dx \text{ and } v = \frac{1}{2}e^{2x} \quad \left(\int be^{ax} = \frac{b}{a}e^{ax} + C \right)$$

Using the Integration-by-Parts Formula gives

$$\overset{u}{}\quad\overset{dv}{}\quad\overset{u}{}\quad\overset{v}{}\quad\overset{v}{}\quad\overset{du}{}$$

$$\int x^2 \left(e^{2x} dx \right) = x^2 \cdot \frac{1}{2}e^{2x} - \int \left(\frac{1}{2}e^{2x} \right) \cdot 2x\,dx$$

$$= \frac{1}{2}x^2 e^{2x} - \int xe^{2x} dx$$

We will use integration by parts again to evaluate $\int xe^{2x} dx$

Let

$$u = x \quad \text{and} \quad dv = e^{2x} dx$$

Then

$$du = dx \text{ and } v = \frac{1}{2}e^{2x} \quad \left(\int be^{ax} = \frac{b}{a}e^{ax} + C \right)$$

Using the Integration-by-Parts Formula gives

$$\overset{u}{}\quad\overset{dv}{}\quad\overset{u}{}\quad\overset{v}{}\quad\overset{v}{}\quad\overset{du}{}$$

$$\int x \left(e^{2x} dx \right) = x \left(\frac{1}{2}e^{2x} \right) - \int \left(\frac{1}{2}e^{2x} \right) dx$$

$$= \frac{1}{2}xe^{2x} - \frac{1}{2} \cdot \frac{1}{2}e^{2x} + C_1$$

$$= \frac{1}{2}xe^{2x} - \frac{1}{4}e^{2x} + C_1$$

Using the information from the previous column, we have:

$$\int x^2 \left(e^{2x} dx \right) = x^2 \cdot \frac{1}{2}e^{2x} - \int \left(\frac{1}{2}e^{2x} \right) \cdot 2x\,dx$$

$$= \frac{1}{2}x^2 e^{2x} - \int \left(xe^{2x} \right) dx$$

$$= \frac{1}{2}x^2 e^{2x} - \left(\frac{1}{2}xe^{2x} - \frac{1}{4}e^{2x} + C_1 \right)$$

$$= \frac{1}{2}x^2 e^{2x} - \frac{1}{2}xe^{2x} + \frac{1}{4}e^{2x} - C_1$$

$$= \frac{1}{2}x^2 e^{2x} - \frac{1}{2}xe^{2x} + \frac{1}{4}e^{2x} + C$$

$$[C = -C_1]$$

Alternatively, we can also use tabular integration as shown below:

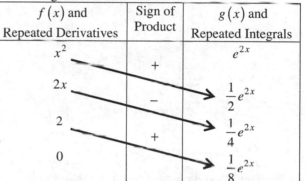

$f(x)$ and Repeated Derivatives	Sign of Product	$g(x)$ and Repeated Integrals
x^2		e^{2x}
$2x$	+	$\frac{1}{2}e^{2x}$
2	–	$\frac{1}{4}e^{2x}$
0	+	$\frac{1}{8}e^{2x}$

We add the products along the arrows, making the alternate sign changes.

$$\int x^2 e^{2x} dx$$

$$= x^2 \left(\frac{1}{2}e^{2x} \right) - 2x \left(\frac{1}{4}e^{2x} \right) + 2 \left(\frac{1}{8}e^{2x} \right) + C$$

$$= \frac{1}{2}x^2 e^{2x} - \frac{1}{2}xe^{2x} + \frac{1}{4}e^{2x} + C$$

27. $\int x^3 e^{-2x} dx$

This integral requires multiple applications of the Integration-by-Parts formula. However, since $f(x) = x^3$ is easily differentiable and to a derivative that is eventually 0 and $g(x) = e^{-2x}$ can be integrated easily using

$$\int be^{ax} = \frac{b}{a}e^{ax} + C \quad \text{we will use tabular}$$

integration as shown to simplify our work. The solution is continued on the next page.

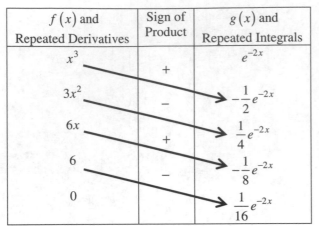

$f(x)$ and Repeated Derivatives	Sign of Product	$g(x)$ and Repeated Integrals
x^3		e^{-2x}
	+	
$3x^2$		$-\dfrac{1}{2}e^{-2x}$
	−	
$6x$		$\dfrac{1}{4}e^{-2x}$
	+	
6		$-\dfrac{1}{8}e^{-2x}$
	−	
0		$\dfrac{1}{16}e^{-2x}$

We add the products along the arrows, making the alternate sign changes to obtain

$$\int x^3 e^{-2x}\,dx$$

$$= x^3\left(-\frac{1}{2}e^{-2x}\right) - 3x^2\left(\frac{1}{4}e^{-2x}\right) +$$

$$6x\left(-\frac{1}{8}e^{-2x}\right) - 6\left(\frac{1}{16}e^{-2x}\right) + C$$

$$= -\frac{1}{2}x^3 e^{-2x} - \frac{3}{4}x^2 e^{-2x} - \frac{3}{4}xe^{-2x} - \frac{3}{8}e^{-2x} + C.$$

29. $\displaystyle\int \left(x^4 + 4\right)e^{3x}\,dx$

This integral requires multiple applications of the Integration-by-Parts formula. However, since $f(x) = x^4 + 4$ is easily differentiable and to a derivative that is eventually 0 and $g(x) = e^{3x}$ can be integrated easily using

$$\int be^{ax} = \frac{b}{a}e^{ax} + C$$ we will use tabular

integration as shown to simplify our work.

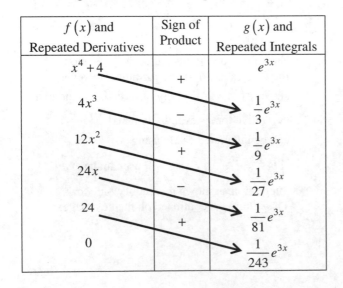

$f(x)$ and Repeated Derivatives	Sign of Product	$g(x)$ and Repeated Integrals
$x^4 + 4$		e^{3x}
	+	
$4x^3$		$\dfrac{1}{3}e^{3x}$
	−	
$12x^2$		$\dfrac{1}{9}e^{3x}$
	+	
$24x$		$\dfrac{1}{27}e^{3x}$
	−	
24		$\dfrac{1}{81}e^{3x}$
	+	
0		$\dfrac{1}{243}e^{3x}$

We add the products along the arrows, making the alternate sign changes to obtain

$$\int \left(x^4 + 4\right)e^{3x}\,dx$$

$$= \left(x^4 + 4\right)\left(\frac{1}{3}e^{3x}\right) - 4x^3\left(\frac{1}{9}e^{3x}\right) +$$

$$12x^2\left(\frac{1}{27}e^{3x}\right) - 24x\left(\frac{1}{81}e^{3x}\right) +$$

$$24\left(\frac{1}{243}e^{3x}\right) + C$$

$$= \frac{1}{3}\left(x^4 + 4\right)e^{3x} - \frac{4}{9}x^3 e^{3x} + \frac{4}{9}x^2 e^{3x} -$$

$$\frac{8}{27}xe^{3x} + \frac{8}{81}e^{3x} + C.$$

31. $\displaystyle\int_1^2 x^2 \ln x\,dx$

In Exercise 9 we found the indefinite integral

$$\int x^2 \ln x\,dx = \frac{x^3 \ln x}{3} - \frac{x^3}{9} + C.$$

We use the indefinite integral to evaluate the definite integral as follows:

$$\int_1^2 x^2 \ln x\,dx$$

$$= \left[\frac{x^3 \ln x}{3} - \frac{x^3}{9}\right]_1^2 \qquad \text{Use } C = 0.$$

$$= \left[\frac{(2)^3 \ln(2)}{3} - \frac{(2)^3}{9}\right] - \left[\frac{(1)^3 \ln(1)}{3} - \frac{(1)^3}{9}\right]$$

$$= \left[\frac{8}{3}(\ln 2) - \frac{8}{9}\right] - \left[0 - \frac{1}{9}\right]$$

$$= \frac{8}{3}\ln 2 - \frac{8}{9} + \frac{1}{9}$$

$$= \frac{8}{3}\ln 2 - \frac{7}{9}.$$

33. $\displaystyle\int_2^6 \ln(x + 8)\,dx$

First, we find the indefinite integral.

$$\int \ln(x + 8)\,dx$$

Let

$$u = \ln(x + 8) \qquad \text{and} \qquad dv = dx$$

Then

$$du = \frac{1}{x + 8}\,dx \qquad \text{and} \qquad v = x + 8$$

Choosing $x+8$ as an antiderivative of dv

The solution is continued on the next page.

Using the information on the previous page and the Integration-by-Parts Formula gives us

$$\int \overset{u}{\ln(x+8)}\,\overset{dv}{dx} =$$

$$\overset{u}{\ln(x+8)}\cdot\overset{v}{(x+8)} - \int \overset{v}{(x+8)}\overset{du}{\left(\frac{1}{x+8}dx\right)}$$

$$= (x+8)\ln(x+8) - \int dx$$

$$= (x+8)\ln(x+8) - x + C.$$

We now use the indefinite integral to evaluate the definite integral. To simplify our work, we choose the constant of integration C to be 0. We calculate the definite integral as follows:

$$\int_2^6 \ln(x+8)\,dx$$

$$= \Big[(x+8)\ln(x+8) - x\Big]_2^6$$

$$= \Big[\big((6)+8\big)\ln\big((6)+8\big) - (6)\Big] -$$

$$\qquad\qquad \Big[\big((2)+8\big)\ln\big((2)+8\big) - (2)\Big]$$

$$= [14\ln 14 - 6] - [10\ln 10 - 2]$$

$$= 14\ln 14 - 10\ln 10 - 4.$$

35. $\displaystyle\int_0^1 xe^x\,dx$

First, we find the indefinite integral $\displaystyle\int xe^x\,dx$.

Let

$$u = x \quad \text{and} \quad dv = e^x\,dx$$

Then

$$du = dx \quad \text{and} \quad v = e^x$$

Using the Integration-by-Parts Formula gives

$$\int \overset{u}{x}\overset{dv}{\left(e^x dx\right)} = \overset{u}{x}\cdot\overset{v}{e^x} - \int \overset{v}{e^x}\,\overset{du}{dx}$$

$$= xe^x - e^x + C.$$

We now use the indefinite integral to evaluate the definite integral. To simplify our work, we choose the constant of integration C to be 0.

$$\int_0^1 xe^x\,dx$$

$$= \Big[xe^x - e^x\Big]_0^1$$

$$= \Big[(1)e^{(1)} - e^{(1)}\Big] - \Big[(0)e^{(0)} - e^{(0)}\Big]$$

$$= [e - e] - [0 - 1]$$

$$= 1.$$

37. $\displaystyle\int_0^8 x\sqrt{x+1}\,dx$

First we find the indefinite integral using integration by parts. Let

$$u = x \quad \text{and} \quad dv = \sqrt{x+1}dx = (x+1)^{\frac{1}{2}}\,dx$$

Then

$$du = dx \quad \text{and} \quad v = \frac{(x+1)^{\frac{3}{2}}}{\frac{3}{2}} = \frac{2}{3}(x+1)^{\frac{3}{2}}$$

Using the Integration-by-Parts Formula gives

$$\int \overset{u}{x}\overset{dv}{\left(\sqrt{x+1}dx\right)} =$$

$$\overset{u}{x}\cdot\overset{v}{\frac{2}{3}(x+1)^{\frac{3}{2}}} - \int \overset{v}{\frac{2}{3}(x+1)^{\frac{3}{2}}}\,\overset{du}{dx}$$

$$= \frac{2}{3}x(x+1)^{\frac{3}{2}} - \frac{2}{3}\int (x+1)^{\frac{3}{2}}\,dx$$

$$= \frac{2}{3}x(x+1)^{\frac{3}{2}} - \frac{2}{3}\cdot\frac{(x+1)^{\frac{5}{2}}}{\frac{5}{2}} + C$$

$$= \frac{2}{3}x(x+1)^{\frac{3}{2}} - \frac{4}{15}(x+1)^{\frac{5}{2}} + C.$$

Evaluating the definite integral, we have:

$$\int_0^8 x\sqrt{x+1}\,dx$$

$$= \left[\frac{2}{3}x(x+1)^{\frac{3}{2}} - \frac{4}{15}(x+1)^{\frac{5}{2}}\right]_0^8$$

$$= \left[\frac{2}{3}(8)\big((8)+1\big)^{\frac{3}{2}} - \frac{4}{15}\big((8)+1\big)^{\frac{5}{2}}\right] -$$

$$\qquad \left[\frac{2}{3}(0)\big((0)+1\big)^{\frac{3}{2}} - \frac{4}{15}\big((0)+1\big)^{\frac{5}{2}}\right]$$

$$= \frac{396}{5} - \left[-\frac{4}{15}\right]$$

$$= \frac{1192}{15}.$$

39. $\displaystyle C(x) = \int C'(x)\,dx = \int 4x\sqrt{x+3}\,dx$

Let

$$u = 4x \quad \text{and} \quad dv = \sqrt{x+3}dx = (x+3)^{\frac{1}{2}}\,dx$$

Then

$$du = 4dx \quad \text{and} \quad v = \frac{(x+3)^{\frac{3}{2}}}{\frac{3}{2}} = \frac{2}{3}(x+3)^{\frac{3}{2}}$$

The solution is continued on the next page.

Using the information on the previous page and the Integration-by-Parts Formula gives us

$$\int 4x\left(\sqrt{x+3}\,dx\right)$$

$$= 4x \cdot \frac{2}{3}(x+3)^{3/2} - \int \frac{2}{3}(x+3)^{3/2} \cdot 4\,dx$$

$$= \frac{8}{3}x(x+3)^{3/2} - \frac{8}{3}\int (x+3)^{3/2}\,dx$$

$$= \frac{8}{3}x(x+3)^{3/2} - \frac{8}{3}\frac{(x+3)^{5/2}}{5/2} + K \quad \text{Use } K \text{ to avoid confusion with cost.}$$

$$= \frac{8}{3}x(x+3)^{3/2} - \frac{16}{15}(x+3)^{5/2} + K.$$

Next, we use the condition $C(13) = 1126.40$ to find the constant of integration K.

$$C(13) = 1126.40$$

$$\frac{8}{3}(13)\left((13)+3\right)^{3/2} - \frac{16}{15}\left((13)+3\right)^{5/2} + K = 1126.40$$

$$\frac{104}{3} \cdot (16)^{3/2} - \frac{16}{15}(16)^{5/2} + K = 1126.40$$

$$\frac{104}{3}(64) - \frac{16}{15}(1024) + K = 1126.40$$

$$\frac{6656}{3} - \frac{16,384}{15} + K = 1126.40$$

$$\frac{16,896}{15} + K = 1126.40$$

$$1126.40 + K = 1126.40$$

$$K = 0$$

Therefore, the total cost function is

$$C(x) = \frac{8}{3}x(x+3)^{3/2} - \frac{16}{15}(x+3)^{5/2}$$

41. a) The total amount of the drug that passes through the body in T hours is given by:

$$\int_0^T E(t)\,dt = \int_0^T te^{-kt}\,dt.$$

We find the indefinite integral first.

$$\int te^{-kt}\,dt$$

Let

$$u = t \qquad \text{and} \qquad dv = e^{-kt}\,dt$$

Then

$$du = dt \qquad \text{and} \qquad v = -\frac{1}{k}e^{-kt}$$

Using the Integration-by-Parts Formula gives

$$\int te^{-kt}\,dt = t \cdot \left(-\frac{1}{k}e^{-kt}\right) - \int -\frac{1}{k}e^{-kt}\,dt$$

$$= -\frac{1}{k}te^{-kt} + \frac{1}{k}\int e^{-kt}\,dt$$

$$= -\frac{1}{k}te^{-kt} + \frac{1}{k}\left(-\frac{1}{k}e^{-kt}\right) + C$$

$$= -\frac{1}{k}te^{-kt} - \frac{1}{k^2}e^{-kt} + C$$

$$= -e^{-kt}\left(\frac{t}{k} + \frac{1}{k^2}\right) + C.$$

Next, we evaluate the definite integral to find the total amount of the drug that passes through the body in T hours.

$$\int_0^T E(t)\,dt$$

$$= \left[-e^{-kt}\left(\frac{t}{k} + \frac{1}{k^2}\right)\right]_0^T$$

$$= \left[-e^{-kT}\left(\frac{T}{k} + \frac{1}{k^2}\right)\right] - \left[-e^{-k \cdot 0}\left(\frac{(0)}{k} + \frac{1}{k^2}\right)\right]$$

$$= -e^{-kT}\left(\frac{T}{k} + \frac{1}{k^2}\right) - \left[-\frac{1}{k^2}\right]$$

$$= -e^{-kT}\left(\frac{T}{k} + \frac{1}{k^2}\right) + \frac{1}{k^2}.$$

b) Substitute 10 for T and 0.2 for k.

$$\int_0^{10} E(t)\,dt = \int_0^{10} te^{-0.2t}\,dt$$

$$= -e^{-(0.2)(10)}\left(\frac{10}{0.2} + \frac{1}{(0.2)^2}\right) + \frac{1}{(0.2)^2}$$

$$= -e^{-2}(50+25) + 25$$

$$= -75e^{-2} + 25 \approx 14.850.$$

After 10 hours the total excretion of the drug is approximately 14.850 mg.

43. $\int x\sqrt{5x+1}\,dx$

Assuming $5x+1 \ge 0$ Let $u = 5x+1$, then

$du = 5dx$. Observe that $x = \dfrac{u-1}{5}$.

Making the appropriate substitutions we have

$\int x\sqrt{5x+1}\,dx$

$= \dfrac{1}{5}\int x\sqrt{5x+1}\,(5)\,dx$

$= \dfrac{1}{5}\int \left(\dfrac{u-1}{5}\right)\sqrt{u}\,du$

$= \dfrac{1}{25}\int (u-1)u^{\frac{1}{2}}\,du$

$= \dfrac{1}{25}\int \left(u^{\frac{3}{2}} - u^{\frac{1}{2}}\right)du$

$= \dfrac{1}{25}\left(\dfrac{1}{\frac{5}{2}}u^{\frac{5}{2}} - \dfrac{1}{\frac{3}{2}}u^{\frac{3}{2}} + C_1\right)$

$= \dfrac{2(5x+1)^{\frac{5}{2}}}{125} - \dfrac{2(5x+1)^{\frac{3}{2}}}{75} + C$ Reverse Substitution

It is possible to simplify both answers to the

form $\dfrac{(30x-4)(5x+1)^{\frac{3}{2}}}{375} + C$. Yes, both answers

are the same.

45. $\int e^{\sqrt{x}}\,dx$

Let $u = \sqrt{x}$, note that $x = u^2$ so that $dx = 2u\,du$.
Making substitutions, we have:

$\int e^{\sqrt{x}}\,dx = \int 2ue^u\,du$.

Using integration by parts we have:

$\int 2ue^u\,du = 2\int u\left(e^u\,du\right)$

Let

    $w = u$ and $dv = e^u\,du$

Then

    $dw = du$ and $v = e^u$

Using the Integration-by-Parts Formula gives

$\overset{w}{}\quad \overset{dv}{}\qquad \overset{w\cdot v}{}\quad \overset{v\ dw}{}$

$2\int u\left(e^u\,du\right) = 2u\cdot e^u - 2\int e^u\,du$

$\qquad\qquad = 2ue^u - 2e^u + C$

We now substitute for u to get:

$\int e^{\sqrt{x}}\,dx = 2ue^u - 2e^u + C$

$\qquad = 2\sqrt{x}e^{\sqrt{x}} - 2e^{\sqrt{x}} + C$.

47. $\int \sqrt{x}\ln x\,dx = \int x^{\frac{1}{2}}\ln x\,dx$

Let

    $u = \ln x$ and $dv = x^{\frac{1}{2}}dx$

Then

    $du = \dfrac{1}{x}dx$ and $v = \dfrac{2}{3}x^{\frac{3}{2}}$

Using the Integration-by-Parts Formula gives

$\int \ln x\left(x^{\frac{1}{2}}\,dx\right)$

$= \ln x\cdot\left(\dfrac{2}{3}x^{\frac{3}{2}}\right) - \int\left(\dfrac{2}{3}x^{\frac{3}{2}}\right)\left(\dfrac{1}{x}dx\right)$

$= \dfrac{2}{3}x^{\frac{3}{2}}\ln x - \dfrac{2}{3}\int x^{\frac{1}{2}}dx$

$= \dfrac{2}{3}x^{\frac{3}{2}}\ln x - \dfrac{2}{3}\left(\dfrac{2}{3}x^{\frac{3}{2}}\right) + C$

$= \dfrac{2}{3}x^{\frac{3}{2}}\ln x - \dfrac{4}{9}x^{\frac{3}{2}} + C$.

49. $\int \dfrac{\ln x}{\sqrt{x}}\,dx = \int x^{-\frac{1}{2}}\ln x\,dx$

Let

    $u = \ln x$ and $dv = x^{-\frac{1}{2}}dx$

Then

    $du = \dfrac{1}{x}dx$ and $v = 2x^{\frac{1}{2}}$

Using the Integration-by-Parts Formula gives

$\int \ln x\left(x^{-\frac{1}{2}}\,dx\right)$

$= \ln x\cdot\left(2x^{\frac{1}{2}}\right) - \int\left(2x^{\frac{1}{2}}\right)\left(\dfrac{1}{x}dx\right)$

$= 2x^{\frac{1}{2}}\ln x - 2\int x^{-\frac{1}{2}}dx$

$= 2x^{\frac{1}{2}}\ln x - 2\left(2x^{\frac{1}{2}}\right) + C$

$= 2x^{\frac{1}{2}}\ln x - 4x^{\frac{1}{2}} + C$

$= 2\sqrt{x}\left(\ln x\right) - 4\sqrt{x} + C$.

51. $\int \left(27x^3 + 83x - 2\right)\sqrt[6]{3x+8}\,dx$

$= \left(27x^3 + 83x - 2\right)(3x+8)^{\frac{1}{6}}\,dx$

We will use tabular integration as shown to simplify our work.

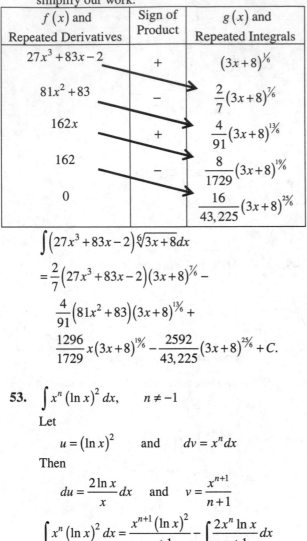

$f(x)$ and Repeated Derivatives	Sign of Product	$g(x)$ and Repeated Integrals
$27x^3 + 83x - 2$	$+$	$(3x+8)^{\frac{1}{6}}$
$81x^2 + 83$	$-$	$\dfrac{2}{7}(3x+8)^{\frac{7}{6}}$
$162x$	$+$	$\dfrac{4}{91}(3x+8)^{\frac{13}{6}}$
162	$-$	$\dfrac{8}{1729}(3x+8)^{\frac{19}{6}}$
0		$\dfrac{16}{43,225}(3x+8)^{\frac{25}{6}}$

$\int \left(27x^3 + 83x - 2\right)\sqrt[6]{3x+8}\,dx$

$= \dfrac{2}{7}\left(27x^3 + 83x - 2\right)(3x+8)^{\frac{7}{6}} -$

$\dfrac{4}{91}\left(81x^2 + 83\right)(3x+8)^{\frac{13}{6}} +$

$\dfrac{1296}{1729}x(3x+8)^{\frac{19}{6}} - \dfrac{2592}{43,225}(3x+8)^{\frac{25}{6}} + C.$

53. $\int x^n \left(\ln x\right)^2 dx, \qquad n \neq -1$

Let

$u = \left(\ln x\right)^2 \qquad \text{and} \qquad dv = x^n dx$

Then

$du = \dfrac{2\ln x}{x}\,dx \qquad \text{and} \qquad v = \dfrac{x^{n+1}}{n+1}$

$\int x^n \left(\ln x\right)^2 dx = \dfrac{x^{n+1}\left(\ln x\right)^2}{n+1} - \int \dfrac{2x^n \ln x}{n+1}\,dx$

$= \dfrac{x^{n+1}\left(\ln x\right)^2}{n+1} - \dfrac{2}{n+1}\int x^n \ln x\,dx$

Use integration by parts again to evaluate

$\int x^n \ln x\,dx.$

Let

$u = \ln x \qquad \text{and} \qquad dv = x^n dx$

Then

$du = \dfrac{1}{x}\,dx \qquad \text{and} \qquad v = \dfrac{x^{n+1}}{n+1}$

Thus, we have:

$\int x^n \ln x\,dx = \dfrac{x^{n+1}\ln x}{n+1} - \int \dfrac{x^n}{n+1}\,dx$

$= \dfrac{x^{n+1}\ln x}{n+1} - \dfrac{x^{n+1}}{\left(n+1\right)^2} + K$

Therefore,

$\int x^n \left(\ln x\right)^2 dx =$

$= \dfrac{x^{n+1}\left(\ln x\right)^2}{n+1} - \dfrac{2}{n+1}\left(\dfrac{x^{n+1}\ln x}{n+1} - \dfrac{x^{n+1}}{\left(n+1\right)^2} + K\right)$

$= \dfrac{x^{n+1}}{n+1}\left(\ln x\right)^2 - \dfrac{2x^{n+1}}{\left(n+1\right)^2}\left(\ln x\right) + \dfrac{2x^{n+1}}{\left(n+1\right)^3} + C.$

$\left(C = \dfrac{-2K}{n+1}\right)$

55. $\int x^n e^x dx$

Let

$u = x^n \qquad \text{and} \qquad dv = e^x dx$

Then

$du = nx^{n-1}dx \qquad \text{and} \qquad v = e^x$

$\int x^n e^x dx = x^n e^x - \int e^x \left(nx^{n-1}\right)dx$

$= x^n e^x - n\int x^{n-1} e^x dx.$

57. ✎

59. $\int 3^x e^x dx$

Let

$u = 3^x \qquad \text{and} \qquad dv = e^x dx$

Then

$du = \left(\ln 3\right)3^x dx \qquad \text{and} \qquad v = e^x$

$\int 3^x e^x dx = 3^x e^x - \ln 3\int 3^x e^x dx.$

Note that $\int 3^x e^x dx$ appears twice. Adding

$\ln 3\int 3^x e^x dx$ to both sides, we have

$\int 3^x e^x dx + \ln 3\int 3^x e^x dx = 3^x e^x$

$\left(1 + \ln 3\right)\int 3^x e^x dx = 3^x e^x$

$\int 3^x e^x dx = \dfrac{3^x e^x}{1 + \ln 3} + C$

61. $\displaystyle\int 10^x e^{3x}\,dx$

Let
$$u = 10^x \quad \text{and} \quad dv = e^{3x}\,dx$$
Then
$$du = (\ln 10)10^x\,dx \quad \text{and} \quad v = \frac{1}{3}e^{3x}$$

$$\int 10^x e^{3x}\,dx = \frac{10^x e^{3x}}{3} - \frac{\ln 10}{3}\int 10^x e^{3x}\,dx.$$

Note that $\displaystyle\int 10^x e^{3x}\,dx$ appears twice. Adding

$\dfrac{\ln 10}{3}\displaystyle\int 10^x e^{3x}\,dx$ to both sides, we have

$$\int 10^x e^{3x}\,dx + \frac{\ln 10}{3}\int 10^x e^{3x}\,dx = \frac{10^x e^{3x}}{3}$$

$$\left(1 + \frac{\ln 10}{3}\right)\int 10^x e^{3x}\,dx = \frac{10^x e^{3x}}{3}$$

$$\left(\frac{3 + \ln 10}{3}\right)\int 10^x e^{3x}\,dx = \frac{10^x e^{3x}}{3}$$

$$\int 10^x e^{3x}\,dx = \frac{10^x e^{3x}}{3 + \ln 10} + C$$

63. Using the fnInt feature on a calculator, we find
that $\displaystyle\int_1^{10} x^5 \ln x\,dx \approx 355{,}986.$

Exercise Set 4.7

1. $\int xe^{-3x}\,dx$

This integral fits Formula 6 in Table 1.

$$\int xe^{ax}\,dx = \frac{1}{a^2}\cdot e^{ax}(ax-1)+C$$

In the given integral, $a=-3$, so we have, by the formula,

$$\int xe^{-3x}\,dx = \frac{1}{(-3)^2}\cdot e^{(-3)x}((-3)x-1)+C$$

$$= \frac{1}{9}e^{-3x}(-3x-1)+C$$

$$= -\frac{1}{9}e^{-3x}(3x+1)+C.$$

3. $\int 6^x\,dx$

This integral fits Formula 11 in Table 1.

$$\int a^x\,dx = \frac{a^x}{\ln a}+C,\ \ a>0, a\ne 1$$

In the given integral, $a=6$, so we have, by the formula,

$$\int 6^x\,dx = \frac{6^x}{\ln 6}+C.$$

5. $\int \frac{1}{25-x^2}\,dx$

This integral fits Formula 15 in Table 1.

$$\int \frac{1}{a^2-x^2}\,dx = \frac{1}{2a}\ln\left|\frac{a+x}{a-x}\right|+C$$

In the given integral, $a^2=25$, or $a=5$, so we have, by the formula,

$$\int \frac{1}{25-x^2}\,dx = \int \frac{1}{5^2-x^2}\,dx$$

$$= \frac{1}{2\cdot 5}\ln\left|\frac{5+x}{5-x}\right|+C$$

$$= \frac{1}{10}\ln\left|\frac{5+x}{5-x}\right|+C.$$

7. $\int \frac{x}{3-x}\,dx$

This integral fits Formula 18 in Table 1.

$$\int \frac{x}{a+bx}\,dx = \frac{x}{b}-\frac{a}{b^2}\ln|a+bx|+C$$

In the given integral, $a=3$ and $b=-1$, so we have, by the formula,

$$\int \frac{x}{3+(-1)x}\,dx$$

$$= \frac{x}{(-1)}-\frac{3}{(-1)^2}\ln|3-x|+C$$

$$= -x-3\ln|3-x|+C.$$

9. $\int \frac{1}{x(8-x)^2}\,dx$

This integral fits Formula 21 in Table 1.

$$\int \frac{1}{x(a+bx)^2}\,dx = \frac{1}{a(a+bx)}+\frac{1}{a^2}\ln\left|\frac{x}{a+bx}\right|+C$$

In the given integral, $a=8$ and $b=-1$, so we have, by the formula,

$$\int \frac{1}{x(8-x)^2}\,dx$$

$$= \frac{1}{8(8-x)}+\frac{1}{(8)^2}\ln\left|\frac{x}{8-x}\right|+C$$

$$= \frac{1}{8(8-x)}+\frac{1}{64}\ln\left|\frac{x}{8-x}\right|+C.$$

11. $\int \ln(3x)\,dx$

$$= \int (\ln 3+\ln x)\,dx$$

$$= \int \ln 3\,dx+\int \ln x\,dx$$

$$= \ln 3\int dx+\int \ln x\,dx$$

The integral in the first term is the integral of a constant and we will integrate accordingly. The integral in the second term fits Formula 8 in Table 1.

$$\int \ln x\,dx = x\ln x-x+C$$

so we have, by the formula,

$$\int \ln(3x)\,dx$$

$$= \ln 3\int dx+\int \ln x\,dx$$

$$= \ln 3\cdot x+C_1+x\ln x-x+C_2$$

$$= (\ln 3)x+x\ln x-x+C.\qquad (C=C_1+C_2)$$

13. $\int x^4 \ln x \, dx$

This integral fits Formula 10 in Table 1.

$$\int x^n \ln x \, dx = x^{n+1} \left[\frac{\ln x}{n+1} - \frac{1}{(n+1)^2} \right] + C$$

In the given integral, $n = 4$, so we have, by the formula,

$$\int x^4 \ln x \, dx = x^{4+1} \left[\frac{\ln x}{4+1} - \frac{1}{(4+1)^2} \right] + C$$

$$= x^5 \left[\frac{\ln x}{5} - \frac{1}{(5)^2} \right] + C$$

$$= \frac{x^5}{5} (\ln x) - \frac{x^5}{25} + C.$$

15. $\int x^3 \ln x \, dx$

This integral fits Formula 10 in Table 1.

$$\int x^n \ln x \, dx = x^{n+1} \left[\frac{\ln x}{n+1} - \frac{1}{(n+1)^2} \right] + C$$

In the given integral, $n = 3$, so we have, by the formula,

$$\int x^3 \ln x \, dx = x^{3+1} \left[\frac{\ln x}{3+1} - \frac{1}{(3+1)^2} \right] + C$$

$$= x^4 \left[\frac{\ln x}{4} - \frac{1}{(4)^2} \right] + C$$

$$= \frac{x^4}{4} (\ln x) - \frac{x^4}{16} + C.$$

17. $\int \frac{dx}{\sqrt{x^2 + 7}}$

This integral fits Formula 12 in Table 1.

$$\int \frac{1}{\sqrt{x^2 + a^2}} \, dx = \ln \left| x + \sqrt{x^2 + a^2} \right| + C$$

In the given integral, $a^2 = 7$, so we have, by the formula,

$$\int \frac{1}{\sqrt{x^2 + 7}} \, dx = \ln \left| x + \sqrt{x^2 + 7} \right| + C.$$

19. $\int \frac{10 \, dx}{x(5 - 7x)^2} = 10 \int \frac{dx}{x(5 - 7x)^2}$

This integral fits Formula 21 in Table 1.

$$\int \frac{1}{x(a + bx)^2} \, dx = \frac{1}{a(a + bx)} + \frac{1}{a^2} \ln \left| \frac{x}{a + bx} \right| + C$$

In the given integral, $a = 5$ and $b = -7$, so we have, by the formula,

$$10 \int \frac{1}{x(5 - 7x)^2} \, dx$$

$$= 10 \left[\frac{1}{5(5 - 7x)} + \frac{1}{(5)^2} \ln \left| \frac{x}{5 - 7x} \right| \right] + C$$

$$= \frac{2}{5 - 7x} + \frac{2}{5} \ln \left| \frac{x}{5 - 7x} \right| + C.$$

21. $\int \frac{-5}{4x^2 - 1} \, dx = -5 \int \frac{1}{4x^2 - 1} \, dx$

This integral almost fits Formula 14 in table 1.

$$\int \frac{1}{x^2 - a^2} \, dx = \frac{1}{2a} \ln \left| \frac{x - a}{x + a} \right| + C$$

However, the x^2 coefficient needs to be 1. We factor out 4 as follows and then we apply formula 14.

$$-5 \int \frac{1}{4x^2 - 1} \, dx = -5 \int \frac{1}{4 \left(x^2 - \frac{1}{4} \right)} \, dx$$

$$= -\frac{5}{4} \int \frac{1}{x^2 - \frac{1}{4}} \, dx$$

In the given integral, $a^2 = \frac{1}{4}$ and $a = \frac{1}{2}$, so we have, by the formula,

$$-\frac{5}{4} \int \frac{1}{x^2 - \frac{1}{4}} \, dx = -\frac{5}{4} \left[\frac{1}{2 \left(\frac{1}{2} \right)} \ln \left| \frac{x - \frac{1}{2}}{x + \frac{1}{2}} \right| \right] + C$$

$$= -\frac{5}{4} \ln \left| \frac{x - \frac{1}{2}}{x + \frac{1}{2}} \right| + C.$$

23. $\int \sqrt{4m^2 + 16} \, dm$

This integral almost fits Formula 22 in table 1.

$$\int \sqrt{x^2 + a^2} \, dx$$

$$= \frac{1}{2} \left[x\sqrt{x^2 + a^2} + a^2 \ln \left| x + \sqrt{x^2 + a^2} \right| \right] + C$$

However, the m^2 coefficient needs to be 1.

The solution is continued on the next page.

We factor out 4 from the integrand on the previous page as follows and then we apply formula 22.

$$\int \sqrt{4(m^2+4)}\,dm = \int 2\sqrt{m^2+4}\,dm$$

$$= 2\int \sqrt{m^2+4}\,dm$$

In the given integral, $a^2 = 4$, so we have, by the formula,

$$2\int \sqrt{m^2+4}\,dm$$

$$= 2\cdot\frac{1}{2}\left[m\sqrt{m^2+4} + 4\ln\left|m+\sqrt{m^2+4}\right| \right] + C$$

$$= m\sqrt{m^2+4} + 4\ln\left|m+\sqrt{m^2+4}\right| + C.$$

25. $\displaystyle\int \frac{-5\ln x}{x^3}\,dx = -5\int x^{-3}\ln x\,dx$

This integral fits Formula 10 in Table 1.

$$\int x^n \ln x\,dx = x^{n+1}\left[\frac{\ln x}{n+1} - \frac{1}{(n+1)^2} \right] + C$$

In the given integral, $n = -3$, so we have, by the formula,

$$-5\int x^{-3}\ln x\,dx$$

$$= -5\cdot x^{-3+1}\left[\frac{\ln x}{-3+1} - \frac{1}{(-3+1)^2} \right] + C$$

$$= -5\cdot x^{-2}\left[\frac{\ln x}{-2} - \frac{1}{(-2)^2} \right] + C$$

$$= \frac{5}{2x^2}(\ln x) + \frac{5}{4x^2} + C.$$

27. $\displaystyle\int \frac{e^x}{x^{-3}}\,dx = \int x^3 e^x\,dx$

This integral fits Formula 7 in Table 1.

$$\int x^n e^{ax}\,dx = \frac{x^n e^{ax}}{a} - \frac{n}{a}\int x^{n-1} e^{ax}\,dx + C$$

In the given integral, $n = 3$ and $a = 1$, so we have, by the formula,

$$\int x^3 e^x\,dx = \frac{x^3 e^x}{1} - \frac{3}{1}\int x^{3-1} e^x\,dx + C$$

$$= x^3 e^x - 3\int x^2 e^x\,dx + C$$

We will apply Formula 7 again, this time with $n = 2$ and $a = 1$

$$x^3 e^x - 3\left[\frac{x^2 e^x}{1} - \frac{2}{1}\int x^{2-1} e^x\,dx \right] + C$$

$$= x^3 e^x - 3x^2 e^x + 6\int x e^x\,dx + C$$

Now we apply Formula 6

$$\int x e^{ax}\,dx = \frac{1}{a^2}\cdot e^{ax}(ax-1) + C$$

with $a = 1$

$$= x^3 e^x - 3x^2 e^x + 6\left[\frac{1}{(1)^2}\cdot e^x (x-1) \right] + C$$

$$= x^3 e^x - 3x^2 e^x + 6x e^x - 6e^x + C.$$

29. $\displaystyle\int x\sqrt{1+2x}\,dx$

This integral fits Formula 24 in Table 1.

$$\int x\sqrt{a+bx}\,dx = \frac{2}{15b^2}(3bx-2a)(a+bx)^{3/2} + C$$

In the given integral, $a = 1$ and $b = 2$, so we have, by the formula,

$$\int x\sqrt{1+2x}\,dx$$

$$= \frac{2}{15(2)^2}(3(2)x - 2(1))(1+2x)^{3/2} + C$$

$$= \frac{2}{60}(6x-2)(1+2x)^{3/2} + C$$

$$= \frac{1}{30}\cdot 2(3x-1)(1+2x)^{3/2} + C$$

$$= \frac{1}{15}(3x-1)(1+2x)^{3/2} + C.$$

31. $S(x) = \int S'(x)\,dx$

$= \int \dfrac{100x}{(20-x)^2}\,dx = 100\int \dfrac{x}{(20-x)^2}\,dx$

This integral fits Formula 19 in Table 1.

$\int \dfrac{x}{(a+bx)^2}\,dx = \dfrac{a}{b^2(a+bx)} + \dfrac{1}{b^2}\ln|a+bx| + C$

In the given integral, $a = 20$ and $b = -1$, so we have, by the formula,

$100\int \dfrac{x}{(20-x)^2}\,dx$

$= 100\left[\dfrac{20}{(-1)^2(20-x)} + \dfrac{1}{(-1)^2}\ln|20-x|\right] + C$

$= \dfrac{2000}{20-x} + 100\ln|20-x| + C.$

We use the condition $S(19) = 2000$ to determine C.

$$S(19) = 2000$$

$$\dfrac{2000}{20-(19)} + 100\ln|20-(19)| + C = 2000$$

$$\dfrac{2000}{1} + 100\ln|1| + C = 2000$$

$$2000 + C = 2000$$

$$C = 0$$

Thus, the supply function is

$S(x) = \dfrac{2000}{20-x} + 100\ln|20-x|$

$= 100\left[\dfrac{20}{20-x} + \ln|20-x|\right]$

33. $\int \dfrac{8}{3x^2-2x}\,dx = 8\int \dfrac{1}{x(-2+3x)}\,dx$

This integral fits Formula 20 in Table 1.

$\int \dfrac{1}{x(a+bx)}\,dx = \dfrac{1}{a}\ln\left|\dfrac{x}{a+bx}\right| + C$

In the given integral, $a = -2$ and $b = 3$, so we have, by the formula,

$8\int \dfrac{1}{x(-2+3x)}\,dx = 8 \cdot \dfrac{1}{(-2)}\ln\left|\dfrac{x}{-2+3x}\right| + C$

$= -4\ln\left|\dfrac{x}{-2+3x}\right| + C.$

35. $\int \dfrac{dx}{x^3-4x^2+4x}$

$= \int \dfrac{dx}{x(x^2-4x+4)}$

$= \int \dfrac{1}{x(x-2)^2}\,dx$

$= \int \dfrac{1}{x(-2+x)^2}\,dx$

This integral fits Formula 21 in Table 1.

$\int \dfrac{1}{x(a+bx)^2}\,dx = \dfrac{1}{a(a+bx)} + \dfrac{1}{a^2}\ln\left|\dfrac{x}{a+bx}\right| + C$

In the given integral, $a = -2$ and $b = 1$, so we have, by the Formula 21,

$\int \dfrac{1}{x(-2+x)^2}\,dx$

$= \dfrac{1}{-2(-2+x)} + \dfrac{1}{(-2)^2}\ln\left|\dfrac{x}{-2+x}\right| + C$

$= \dfrac{1}{-2(x-2)} + \dfrac{1}{4}\ln\left|\dfrac{x}{x-2}\right| + C$

$= \dfrac{-1}{2(x-2)} + \dfrac{1}{4}\ln\left|\dfrac{x}{x-2}\right| + C.$

37. $\int \dfrac{-e^{-2x}\,dx}{9-6e^{-x}+e^{-2x}} = \int \dfrac{e^{-x}\left(-e^{-x}\right)dx}{\left(-3+e^{-x}\right)^2}$

We substitute $u = e^{-x}$ and $du = -e^{-x}\,dx$, and get

$\int \dfrac{u}{(-3+u)^2}\,du$

Which fits Formula 19 in Table 1.

$\int \dfrac{x}{(a+bx)^2}\,dx = \dfrac{a}{b^2(a+bx)} + \dfrac{1}{b^2}\ln|a+bx| + C$

In the given integral, we have $u = x$, $a = -3$ and $b = 1$, so we have, by the formula,

$\int \dfrac{e^{-x}\left(-e^{-x}\right)dx}{\left(-3+e^{-x}\right)^2} = \int \dfrac{u}{(-3+u)^2}\,du$

$= \dfrac{-3}{1^2(-3+u)} + \dfrac{1}{1^2}\ln|-3+u| + C$

$= \dfrac{-3}{1^2\left(-3+e^{-x}\right)} + \dfrac{1}{1^2}\ln\left|-3+e^{-x}\right| + C$ Substituting for u

$= \dfrac{-3}{e^{-x}-3} + \ln\left|e^{-x}-3\right| + C.$

39. ✏

Chapter 5

Applications of Integration

1. $D(x) = -3x + 7,$ $\quad S(x) = 2x + 2$

a) To find the equilibrium point we set
$D(x) = S(x)$ and solve.

$$-3x + 7 = 2x + 2$$
$$7 - 2 = 2x + 3x$$
$$5 = 5x$$
$$1 = x$$

Thus $x_E = 1$ unit. To find p_E we substitute
x_E into $D(x)$ or $S(x)$. Here we use $S(x)$.

$$p_E = S(x_E)$$
$$= S(1)$$
$$= 2(1) + 2$$
$$= 4$$

When 1 unit is sold, equilibrium price is \$4;
therefore, the equilibrium point is $(1, \$4)$.

Notice, we could have used $D(x)$ to find the
equilibrium point as well. Substituting, we
have:

$$p_E = D(x_E)$$
$$= D(1)$$
$$= -3(1) + 7$$
$$= 4$$

This results in the same equilibrium point
$(1, \$4)$.

b) The consumer surplus is
$$\int_0^{x_E} D(x)\,dx - x_E p_E .$$

Substituting $-3x + 7$ for $D(x)$, 1 for x_E,
and 4 for p_E we have:

$$\int_0^1 (-3x + 7)\,dx - 1 \cdot 4$$

$$= \left[-\frac{3x^2}{2} + 7x \right]_0^1 - 4$$

$$= \left[\left(-\frac{3(1)^2}{2} + 7(1) \right) - \left(-\frac{3(0)^2}{2} + 7(0) \right) \right] - 4$$

$$= \left[\left(-\frac{3}{2} + 7 \right) - (0) \right] - 4$$

$$= \left[\frac{11}{2} \right] - 4 = \frac{3}{2} = 1.50$$

The consumer surplus at the equilibrium
point is \$1.50.

c) The producer surplus is

$$x_E p_E - \int_0^{x_E} S(x)\,dx .$$

Substituting $2x + 2$ for $S(x)$, 1 for x_E, and 4
for p_E we have:

$$1 \cdot 4 - \int_0^1 (2x + 2)\,dx$$

$$= 4 - \left[x^2 + 2x \right]_0^1$$

$$= 4 - \left[(1^2 + 2 \cdot 1) - (0^2 + 2(0)) \right]$$

$$= 4 - \left[(1 + 2) - (0) \right]$$

$$= 4 - [3]$$

$$= 1$$

The producer surplus at the equilibrium
point is \$1.

3. $D(x) = (x - 3)^2,$ $\quad S(x) = x^2 + 2x + 1$

a) To find the equilibrium point we set
$D(x) = S(x)$ and solve.

$$D(x) = S(x)$$
$$(x - 3)^2 = x^2 + 2x + 1$$
$$x^2 - 6x + 9 = x^2 + 2x + 1$$
$$-6x + 9 = 2x + 1 \quad \text{Subtracting } x^2 \text{ from both sides.}$$
$$9 - 1 = 2x + 6x$$
$$8 = 8x$$
$$1 = x$$

The solution is continued on the next page.

From the previous page, we determined
$x_E = 1$ unit. To find p_E we substitute x_E
into $D(x)$ or $S(x)$. Here we use $D(x)$.

$$p_E = D(x_E)$$
$$= D(1)$$
$$= (1-3)^2$$
$$= 4$$

When 1 unit is sold, equilibrium price is $4;
therefore, the equilibrium point is $(1, \$4)$.

b) The consumer surplus is

$$\int_0^{x_E} D(x)\,dx - x_E p_E.$$

Substituting $(x-3)^2$ for $D(x)$, 1 for x_E,
and 4 for p_E we have:

$$\int_0^1 (x-3)^2\,dx - 1 \cdot 4$$

$$= \int_0^1 (x^2 - 6x + 9)\,dx - 4$$

$$= \left[\frac{x^3}{3} - 3x^2 + 9x\right]_0^1 - 4$$

$$= \left[\left(\frac{1}{3} - 3 + 9\right) - (0 - 0 + 0)\right] - 4$$

$$= \left[\left(\frac{19}{3}\right) - (0)\right] - 4$$

$$= \frac{19}{3} - 4$$

$$= \frac{7}{3} \approx 2.33$$

The consumer surplus at the equilibrium
point is $2.33.

c) The producer surplus is

$$x_E p_E - \int_0^{x_E} S(x)\,dx.$$

Substituting $x^2 + 2x + 1$ for $S(x)$, 1 for x_E,
and 4 for p_E we find the surplus at the top
of the next column.

$$1 \cdot 4 - \int_0^1 (x^2 + 2x + 1)\,dx$$

$$= 4 - \left[\frac{x^3}{3} + x^2 + x\right]_0^1$$

$$= 4 - \left[\left(\frac{1}{3} + 1 + 1\right) - (0 + 0 + 0)\right]$$

$$= 4 - \left[\left(\frac{7}{3}\right) - (0)\right]$$

$$= \frac{5}{3} \approx 1.67$$

The producer surplus at the equilibrium
point is $1.67.

5. $D(x) = (x-8)^2$, $S(x) = x^2$

a) To find the equilibrium point we set
$D(x) = S(x)$ and solve.

$$(x-8)^2 = x^2$$
$$x^2 - 16x + 64 = x^2$$
$$-16x + 64 = 0$$
$$64 = 16x$$
$$4 = x$$

Thus $x_E = 4$ units. To find p_E we substitute
x_E into $D(x)$ or $S(x)$. Here we use $D(x)$.

$$p_E = S(x_E)$$
$$= S(4)$$
$$= (4)^2$$
$$= 16$$

When 4 units are sold, equilibrium price is
$16; therefore, the equilibrium point is
$(4, \$16)$.

b) The consumer surplus is

$$\int_0^{x_E} D(x)\,dx - x_E p_E.$$

Substituting $(x-8)^2$ for $D(x)$, 4 for x_E,
and 16 for p_E we have:

$$\int_0^4 (x-8)^2\,dx - 4 \cdot 16$$

We calculate the integral on the next page.

$$\int_0^4 (x-8)^2\, dx - 4\cdot 16$$

$$= \int_0^4 \left(x^2 - 16x + 64\right) dx - 64$$

$$= \left[\frac{x^3}{3} - 8x^2 + 64x\right]_0^4 - 64$$

$$= \left[\frac{(4)^3}{3} - 8(4)^2 + 64(4) - \right.$$

$$\left. \left(\frac{(0)^3}{3} - 8(0)^2 + 64(0)\right)\right] - 64$$

$$= \left[\frac{64}{3} - 128 + 256 - 0\right] - 64$$

$$= \frac{64}{3} - \frac{384}{3} + \frac{768}{3} - \frac{192}{3}$$

$$= \frac{256}{3} \approx 85.33$$

The consumer surplus at the equilibrium point is \$85.33.

c) The producer surplus is

$$x_E p_E - \int_0^{x_E} S(x)\, dx\,.$$

Substituting x^2 for $S(x)$, 4 for x_E, and 16 for p_E we have:

$$4\cdot 16 - \int_0^4 \left(x^2\right) dx$$

$$= 64 - \left[\frac{x^3}{3}\right]_0^4$$

$$= 64 - \left[\left(\frac{4^3}{3}\right) - \left(\frac{0^3}{3}\right)\right]$$

$$= 64 - \frac{64}{3}$$

$$= \frac{128}{3} \approx 42.67$$

The producer surplus at the equilibrium point is \$42.67.

7. $D(x) = 8800 - 30x, \qquad S(x) = 7000 + 15x$

a) To find the equilibrium point we set $D(x) = S(x)$ and solve.

$$8800 - 30x = 7000 + 15x$$
$$8800 - 7000 = 15x + 30x$$
$$1800 = 45x$$
$$40 = x$$

Thus $x_E = 40$ units. To find p_E we substitute x_E into $D(x)$ or $S(x)$. Here we use $D(x)$ to find p_E.

$$p_E = D(x_E)$$
$$= D(40)$$
$$= 8800 - 30\cdot 40$$
$$= 7600$$

When 40 units are sold, equilibrium price is \$7600; therefore, the equilibrium point is $(40, \$7600)$.

b) The consumer surplus is

$$\int_0^{x_E} D(x)\, dx - x_E p_E\,.$$

Substituting $8800 - 30x$ for $D(x)$, 40 for x_E, and 7600 for p_E we have:

$$\int_0^{40} (8800 - 30x)\, dx - 40\cdot 7600$$

$$= \left[8800x - 15x^2\right]_0^{40} - 304{,}000$$

$$= \left[\left(8800\cdot 40 - 15(40)^2\right) - \right.$$

$$\left. \left(8800\cdot 0 - 15\left(0^2\right)\right)\right] - 304{,}000$$

$$= [352{,}000 - 24{,}000] - 304{,}000$$

$$= 24{,}000$$

The consumer surplus at the equilibrium point is \$24,000.

c) The producer surplus is

$$x_E p_E - \int_0^{x_E} S(x)\, dx\,.$$

Substituting $7000 + 15x$ for $S(x)$, 40 for x_E, and 7600 for p_E we have:

We calculate the surplus at the top of the next page.

Using the information from the previous page, the producer surplus is:

$$40 \cdot 7600 - \int_0^{40} (7000 + 15x)\, dx$$

$$= 304,000 - \left[7000x + \frac{15x^2}{2} \right]_0^{40}$$

$$= 304,000 - \left[\left(7000 \cdot 40 + \frac{15(40)^2}{2} \right) - \right.$$

$$\left. \left(7000 \cdot 0 + \frac{15(0)^2}{2} \right) \right]$$

$$= 304,000 - [280,000 + 12,000]$$

$$= 12,000$$

The producer surplus at the equilibrium point is $12,000.

9. $D(x) = 7 - x$, for $0 \le x \le 7$;

$S(x) = 2\sqrt{x+1}$

a) To find the equilibrium point we set $D(x) = S(x)$ and solve.

$$7 - x = 2\sqrt{x+1}$$

$$(7-x)^2 = \left(2\sqrt{x+1} \right)^2$$

$$49 - 14x + x^2 = 4(x+1)$$

$$x^2 - 18x + 45 = 0$$

$$(x-3)(x-15) = 0$$

$$x = 3 \qquad \text{or} \qquad x = 15$$

Note, $x = 15$ is not a solution to the equation. Only $x = 3$ is in the domain of $D(x)$, thus $x_E = 3$ units. To find p_E we substitute x_E into $D(x)$ or $S(x)$. Here we use $D(x)$.

$$p_E = D(x_E)$$

$$= D(3) = 7 - 3 = 4$$

When 3 units are sold, equilibrium price is $4; therefore, the equilibrium point is $(3, \$4)$.

b) The consumer surplus is

$$\int_0^{x_E} D(x)\, dx - x_E p_E.$$

Substituting $7 - x$ for $D(x)$, 3 for x_E, and 4 for p_E we calculate the surplus at the top of the next column.

$$\int_0^3 (7 - x)\, dx - 3 \cdot 4$$

$$= \left[7x - \frac{x^2}{2} \right]_0^3 - 12$$

$$= \left[\left(7 \cdot 3 - \frac{(3)^2}{2} \right) - \left(7 \cdot 0 - \frac{(0)^2}{2} \right) \right] - 12$$

$$= \left[\frac{33}{2} \right] - 12$$

$$= \frac{9}{2} = 4.50$$

The consumer surplus at the equilibrium point is $4.50.

c) The producer surplus is

$$x_E p_E - \int_0^{x_E} S(x)\, dx.$$

Substituting $2\sqrt{x+1}$ for $S(x)$, 3 for x_E, and 4 for p_E we have:

$$3 \cdot 4 - \int_0^3 \left(2\sqrt{x+1} \right) dx$$

$$= 12 - \int_0^3 \left(2(x+1)^{\frac{1}{2}} \right) dx$$

$$= 12 - \left[\frac{4}{3}(x+1)^{\frac{3}{2}} \right]_0^3$$

$$= 12 - \left[\left(\frac{4}{3}(3+1)^{\frac{3}{2}} \right) - \left(\frac{4}{3}(0+1)^{\frac{3}{2}} \right) \right]$$

$$= 12 - \left[\frac{32}{3} - \frac{4}{3} \right]$$

$$-\frac{8}{3} \approx 2.67$$

The producer surplus at the equilibrium point is $2.67.

11. $D(x) = \dfrac{1800}{\sqrt{x+1}}$, $\qquad S(x) = 2\sqrt{x+1}$

a) To find the equilibrium point we set $D(x) = S(x)$ and solve.

$$\frac{1800}{\sqrt{x+1}} = 2\sqrt{x+1}$$

$$1800 = \left(2\sqrt{x+1} \right)\sqrt{x+1}$$

$$1800 = 2(x+1)$$

$$900 = x+1$$

$$899 = x$$

The solution is continued on the next page.

From the previous page, we determined $x_E = 899$ units. To find p_E we substitute x_E into $D(x)$ or $S(x)$. Here we use $S(x)$.

$$p_E = S(x_E) = 2\sqrt{899+1} = 2\sqrt{900} = 60$$

When 899 units are sold, equilibrium price is $60; therefore, the equilibrium point is $(899, \$60)$.

b) The consumer surplus is

$$\int_0^{x_E} D(x)\,dx - x_E p_E.$$

Substituting $\dfrac{1800}{\sqrt{x+1}} = 1800(x+1)^{-\frac{1}{2}}$ for $D(x)$, 899 for x_E, and 60 for p_E we have:

$$\int_0^{899}\left(1800(x+1)^{-\frac{1}{2}}\right)dx - 899 \cdot 60$$

$$= 1800\left[2(x+1)^{\frac{1}{2}}\right]_0^{899} - 53{,}940$$

$$= 1800\left[\left(2\sqrt{899+1}\right)-\left(2\sqrt{0+1}\right)\right] - 53{,}940$$

$$= 1800\left[2(30)-2(1)\right] - 53{,}940$$

$$= 1800(58) - 53{,}940$$

$$= 104{,}400 - 53{,}940$$

$$= 50{,}460$$

The consumer surplus at the equilibrium point is $50,460.

c) The producer surplus is

$$x_E p_E - \int_0^{x_E} S(x)\,dx.$$

Substituting $2\sqrt{x+1}$ for $S(x)$, 899 for x_E, and 60 for p_E we have:

$$899 \cdot 60 - \int_0^{899}\left(2\sqrt{x+1}\right)dx$$

$$= 53{,}940 - \int_0^{899}\left(2(x+1)^{\frac{1}{2}}\right)dx$$

$$= 53{,}940 - 2\left[\frac{2}{3}(x+1)^{\frac{3}{2}}\right]_0^{899}$$

$$= 53{,}940 - \frac{4}{3}\left[\left((899+1)^{\frac{3}{2}}\right)-\left((0+1)^{\frac{3}{2}}\right)\right]$$

$$= 53{,}940 - \frac{4}{3}\left[27{,}000 - 1\right]$$

$$= 53{,}940 - \frac{4}{3}(26{,}999)$$

$$\approx 17{,}941.33$$

The producer surplus at the equilibrium point is $17,941.33.

13. $D(x) = 13 - x$, for $0 \le x \le 13$;

$$S(x) = \sqrt{x+17}$$

a) To find the equilibrium point we set $D(x) = S(x)$ and solve.

$$13 - x = \sqrt{x+17}$$

$$(13-x)^2 = \left(\sqrt{x+17}\right)^2$$

$$169 - 26x + x^2 = x + 17$$

$$x^2 - 27x + 152 = 0$$

$$(x-8)(x-19) = 0$$

$$x - 8 = 0 \quad \text{or} \quad x - 19 = 0$$

$$x = 8 \quad \text{or} \quad x = 19$$

Only 8 is in the domain of $D(x)$, thus $x_E = 8$ units.

To find p_E we substitute x_E into $D(x)$ or $S(x)$. Here we use $D(x)$.

$$p_E = D(x_E) = 13 - 8 = 5$$

When 8 units are sold, equilibrium price is $5; therefore, the equilibrium point is $(8, \$5)$.

b) The consumer surplus is

$$\int_0^{x_E} D(x)\,dx - x_E p_E.$$

Substituting $13 - x$ for $D(x)$, 8 for x_E, and 5 for p_E we have:

$$\int_0^8 (13 - x)\,dx - 8 \cdot 5$$

$$= \left[13x - \frac{x^2}{2}\right]_0^8 - 40$$

$$= \left[\left(13 \cdot 8 - \frac{(8)^2}{2}\right) - \left(13 \cdot 0 - \frac{(0)^2}{2}\right)\right] - 40$$

$$= \left[(104 - 32) - 0\right] - 40$$

$$= 32$$

The consumer surplus at the equilibrium point is $32.

c) The producer surplus is

$$x_E p_E - \int_0^{x_E} S(x)\,dx\,.$$

Substituting $\sqrt{x+17}$ for $S(x)$, 8 for x_E, and 5 for p_E we have:

$$8 \cdot 5 - \int_0^8 \left(\sqrt{x+17}\right)dx$$

$$= 40 - \int_0^8 (x+17)^{\frac{1}{2}}\,dx$$

$$= 40 - \left[\frac{2}{3}(x+17)^{\frac{3}{2}}\right]_0^8$$

$$= 40 - \frac{2}{3}\left[\left((8+17)^{\frac{3}{2}}\right) - \left((0+17)^{\frac{3}{2}}\right)\right]$$

$$= 40 - \frac{2}{3}\left[(25)^{\frac{3}{2}} - (17)^{\frac{3}{2}}\right]$$

$$\approx 3.39519709 \qquad \text{Using a calculator}$$

$$\approx 3.40$$

The producer surplus at the equilibrium point is $3.40.

15. a) If Beth makes 2 jumps, then $x = 2$, substituting into the demand function, we find the price Beth is willing to pay.

$$p = D(2)$$

$$= 7.5(2)^2 - 60.5(2) + 254$$

$$= 30 - 121 + 254$$

$$= 163.$$

Beth is willing to pay $163 per jump for her two jumps. Her consumer surplus would be:

$$\int_0^x D(x)\,dx - x \cdot p$$

$$\int_0^2 7.5x^2 - 60.5x + 254\,dx - 2 \cdot 163$$

$$= \left[2.5x^3 - 30.25x^2 + 254x\right]_0^2 - 326$$

$$= \left[\left(2.5(2)^3 - 30.25(2)^2 + 254(2)\right) - (0)\right] - 326$$

$$= [20 - 121 + 508] - 326$$

$$= 81$$

Beth's consumer surplus for 2 jumps at a price of $163 per jump is $81.

b) Substituting into the supply function, we have:

$$p = S(2)$$

$$= 15(2) + 95 = 125$$

The Aero Skydiving Center would supply Beth with 2 jumps at $125 per jump.

The producer surplus would be.

$$x \cdot p - \int_0^x S(x)\,dx$$

$$2 \cdot 125 - \int_0^2 15x + 95\,dx$$

$$= 250 - \left[7.5x^2 + 95x\right]_0^2$$

$$= 250 - \left[\left(7.5(2)^2 + 95(2)\right) - (0)\right]$$

$$= 250 - [30 + 190]$$

$$= 30$$

Aero Skydiving Center's producer surplus for 2 jumps at a price of $125 per jump is $30.

c) To find the equilibrium point we set $D(x) = S(x)$ and solve.

$$7.5x^2 - 60.5x + 254 = 15x + 95$$

$$7.5x^2 - 75.5x - 159 = 0$$

Solving the equation using the quadratic formula, we have:

$$x = \frac{-(-75.5) \pm \sqrt{(-75.5)^2 - 4(7.5)(159)}}{2(7.5)}$$

$$= \frac{75.5 \pm \sqrt{930.25}}{15}$$

Using a calculator, we see that
$$x = 3 \qquad \text{or} \qquad x \approx 7.067$$
Assuming that Beth makes no more than 5 jumps, we conclude the equilibrium quantity is $x_E = 3$ jumps. To find p_E we substitute x_E into $D(x)$ or $S(x)$. Here we use $S(x)$.

$$p_E = S(x_E) = S(3) = 15(3) + 95 = 140.$$

When 3 units are sold the equilibrium price is $140; therefore, the equilibrium point is $(3, \$140)$. Using this point we calculate consumer and producer surplus as follows.

$$\int_0^{x_E} D(x)\,dx - x_E p_E$$

$$\int_0^3 \left(7.5x^2 - 60.5x + 254\right)dx - 3 \cdot 140$$

$$= \left[2.5x^3 - 30.25x^2 + 254x\right]_0^3 - 420$$

$$= \left[\left(2.5(3)^3 - 30.25(3)^2 + 254(3)\right) - (0)\right] - 420$$

$$= [557.25] - 420$$

$$= 137.25$$

Beth's consumer surplus is $137.25 at equilibrium.
The solution is continued on the next page.

The producer surplus is:

$$\int_0^{x_E} D(x)\,dx - x_E p_E$$

$$3 \cdot 140 - \int_0^3 (15x + 95)\,dx$$

$$= 420 - \left[7.5x^2 + 95x \right]_0^3$$

$$= 420 - \left[\left(7.5(3)^2 + 95(3) \right) - (0) \right]$$

$$= 420 - [352.5]$$

$$= 67.5$$

Aero Skydiving Center's producer surplus at equilibrium is $67.50.

d)

17. $D(x) = e^{-x+4.5}, \qquad S(x) = e^{x-5.5}$

a) To find the equilibrium point we set $D(x) = S(x)$ and solve.

$$e^{-x+4.5} = e^{x-5.5}$$

$$\ln\left(e^{-x+4.5} \right) = \ln\left(e^{x-5.5} \right)$$

$$-x + 4.5 = x - 5.5$$

$$10 = 2x$$

$$5 = x$$

Thus $x_E = 5$ units. To find p_E we substitute x_E into $D(x)$ or $S(x)$. Here we use $D(x)$.

$$p_E = D(x_E)$$

$$= D(5) = e^{-5+4.5} = e^{-0.5} \approx 0.61$$

When 5 units are sold the equilibrium price is $0.61; therefore, the equilibrium point is $(5, \$0.61)$.

b) The consumer surplus is

$$\int_0^{x_E} D(x)\,dx - x_E p_E.$$

Substituting $e^{-x+4.5}$ for $D(x)$, 5 for x_E, and 0.61 for p_E we have:

$$\int_0^5 e^{-x+4.5}\,dx - 5 \cdot 0.61$$

$$= \left[-e^{-x+4.5} \right]_0^5 - 3.05$$

$$= \left[\left(-e^{-5+4.5} \right) - \left(-e^{-0+4.5} \right) \right] - 3.05$$

$$= \left[-e^{-0.5} + e^{4.5} \right] - 3.05$$

$$\approx 86.36 \qquad \text{Using a calculator}$$

The consumer surplus at the equilibrium point is $86.36.

c) The producer surplus is

$$x_E p_E - \int_0^{x_E} S(x)\,dx.$$

Substituting $e^{x-5.5}$ for $S(x)$, 5 for x_E, and 0.61 for p_E we have:

$$5 \cdot 0.61 - \int_0^5 \left(e^{x-5.5} \right)dx$$

$$= 3.05 - \left[e^{x-5.5} \right]_0^5$$

$$= 3.05 - \left[\left(e^{5-5.5} \right) - \left(e^{0-5.5} \right) \right]$$

$$= 3.05 - \left[e^{-0.5} - e^{-5.5} \right]$$

$$\approx 2.45$$

The producer surplus at the equilibrium point is $2.45.

19.

21. $D(x) = \dfrac{x+8}{x+1}, \qquad S(x) = \dfrac{x^2+4}{20}$

a) Graphing the equations and using the INTERSECT feature we have:

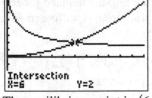

The equilibrium point is $(6, \$2)$.

b)

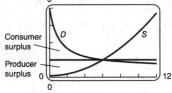

c) The consumer surplus is

$$\int_0^{x_E} D(x)\,dx - x_E p_E.$$

Substituting $\dfrac{x+8}{x+1}$ for $D(x)$, 6 for x_E, and 2 for p_E we have:

$$\int_0^6 \left(\frac{x+8}{x+1} \right)dx - 6 \cdot 2$$

$$\approx 19.62137104 - 12 \qquad \text{Using a calculator}$$

$$\approx 7.62$$

The consumer surplus at the equilibrium point is $7.62.

d) The producer surplus is

$$x_E p_E - \int_0^{x_E} S(x)\,dx\,.$$

Substituting $\dfrac{x^2+4}{20}$ for $S(x)$, 6 for x_E, and

2 for p_E we have:

$$6\cdot 2 - \int_0^6\left(\frac{x^2+4}{20}\right)dx$$

$$= 12 - \left[\frac{x^3}{60}+\frac{x}{5}\right]_0^6$$

$$= 12 - \frac{24}{5}$$

$$= \frac{36}{5} = 7.20$$

The producer surplus at the equilibrium point is $7.20.

23. a) Entering the data into the STAT editor on the calculator and plotting the points, we get the following scatter plot:

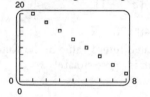

A linear function appears to be the best fit for the data.

b) Using the linear regression feature on the calculator, we get the equation:

$y = -25x + 225$.

c) The consumer surplus is

$$\int_0^x D(x)\,dx - x\cdot p\,.$$

Substituting $-25x+225$ for $D(x)$, 6 for x, and 75 for p we have:

$$\int_0^6 (-25x+225)\,dx - 6(75) = 450$$

The consumer surplus is $450.

d) First find the value for x when $y=115$.

$$115 = -25x + 225$$

$$-110 = -25x$$

$$4.4 = x$$

Next, find the consumer surplus.

$$\int_0^{4.4}(-25x+225)\,dx - 4.4(115) = 242$$

When the price of bungee jumping is $115 per half hour, Regina's consumer surplus is $242.

Exercise Set 5.2

1. $P(t) = P_0 e^{kt}$

Substituting 6 for t, 100,000 for P_0 and 0.03 for k we have:

$P(6) = 100,000 e^{0.03(6)}$

$\qquad = 100,000 e^{0.18}$

$\qquad \approx 100,000(1.19721736312)$

$\qquad \approx 119,721.74$

The future value of $100,000 after 6 years is about $119,721.74.

3. $P(t) = P_0 e^{kt}$

Substituting 9 for t, 140,000 for P_0 and 0.058 for k we have:

$P(9) = 140,000 e^{0.058(9)}$

$\qquad = 140,000 e^{0.522}$

$\qquad \approx 140,000(1.68539507)$

$\qquad \approx 235,955.31$

The future value of $140,000 after 9 years is about $235,955.31.

5. $P(t) = P_0 e^{kt}$

Substituting 6 for t, 100,000 for P and 0.03 for k we have:

$100,000 = P_0 e^{0.03(6)}$

$\qquad = P_0 e^{0.18}$

$\dfrac{100,000}{e^{0.18}} \approx P_0$

$83,527.02 \approx P_0$

The present value of $100,000 due 6 years in the future at 3% interest is about $83,527.02.

7. $P(t) = P_0 e^{kt}$

Substituting 25 for t, 1,000,000 for P and 0.06 for k we have:

$1,000,000 = P_0 e^{0.06(25)}$

$\qquad = P_0 e^{1.5}$

$\dfrac{1,000,000}{e^{1.5}} \approx P_0$

$223,130.16 \approx P_0$

The present value of $1,000,000 due 25 years in the future at 6% interest is about $223,130.16.

9. Since the income stream is constant

$A = \dfrac{R(t)}{k}\left(e^{kT} - 1\right).$

Substituting $50,000 for $R(t)$, 0.05 for k, and 22 for T we have

$A = \dfrac{50,000}{0.05}\left(e^{(0.05)(22)} - 1\right)$

$\qquad = \dfrac{50,000}{0.05}\left(e^{1.1} - 1\right)$

$\qquad \approx 2,004,166.024.$

The accumulated future value of the continuous income stream is approximately $2,004,166.02.

11. Since the income stream is constant

$A = \dfrac{R(t)}{k}\left(e^{kT} - 1\right).$

Substituting $400,000 for $R(t)$, 0.04 for k, and 20 for T we have

$A = \dfrac{400,000}{0.04}\left(e^{(0.04)(20)} - 1\right)$

$\qquad = \dfrac{400,000}{0.04}\left(e^{0.8} - 1\right)$

$\qquad \approx 12,255,409.28.$

The accumulated future value of the continuous income stream is approximately $12,255,409.28.

13. Since the income stream is constant

$B = \dfrac{R(t)}{k}\left(1 - e^{-kT}\right).$

Substituting $250,000 for $R(t)$, 0.04 for k, and 18 for T we have

$B = \dfrac{250,000}{0.04}\left(1 - e^{-(0.04)(18)}\right)$

$\qquad = 6,250,000\left(1 - e^{-0.72}\right)$

$\qquad \approx 3,207,798.40.$

The accumulated present value of the continuous income stream is approximately $3,207,798.40.

15. Since the income stream is constant

$$B = \frac{R(t)}{k}\left(1 - e^{-kT}\right).$$

Substituting $800,000 for $R(t)$, 0.023 for k, and 20 for T we have

$$B = \frac{800,000}{0.023}\left(1 - e^{-(0.023)(20)}\right)$$

$$= 34,782,608.7\left(1 - e^{-0.46}\right)$$

$$\approx 12,824,916.68$$

The accumulated present value of the continuous income stream is approximately $12,824,916.68.

17. Since the income stream is nonconstant, the accumulated present value is given by

$$B = \int_0^T R(t)e^{-kt}dt.$$

Substituting $5200t$ for $R(t)$, 0.031 for k, and 18 for T we have

$$\int_0^{18}(5200t)e^{-0.031t}dt$$

Using Formula 6 from Table 1 we integrate as follows:

$$\int_0^{18}(5200t)e^{-0.031t}dt$$

$$= 5200\int_0^{18}te^{-0.031t}dt$$

$$= 5200\left(\frac{1}{(-0.031)^2}e^{-0.031t}(-0.031t-1)\right)\Big|_0^{18}$$

$$= 5,411,030.177e^{-0.031t}(-0.031t-1)\Big|_0^{18}$$

$$= -4,825,152.995 - (-5,411,030.177)$$

$$\approx 585,877.18.$$

The accumulated present value of the continuous income stream is approximately $585,877.18.

19. Since the income stream is nonconstant, the accumulated present value is given by

$$B = \int_0^T R(t)e^{-kt}dt.$$

Substituting $2000t + 7$ for $R(t)$, 0.045 for k, and 30 for T we have

$$\int_0^{30}(2000t+7)e^{-0.045t}dt =$$

$$\int_0^{30}(2000t)e^{-0.045t}dt + \int_0^{30}7e^{-0.045t}dt$$

The solution is continued at the top of the next column.

Using Formula 6 from Table 1 we integrate the first integral of the sum. The second integral is integrated using Formula 5 from Table 1.

$$= 2000\left(\frac{1}{(-0.045)^2}e^{-0.045t}(-0.045t-1)\right)\Big|_0^{30}$$

$$+ \frac{7}{-0.045}e^{-0.045t}\Big|_0^{30}$$

$$= (-601,693.445 - (-987,654.321))$$

$$+ (-40.3263 - (-155.5556))$$

$$\approx 386,076.11.$$

The accumulated present value of the continuous income stream is approximately $386,076.11.

21. $P = P_0e^{kt}$

Therefore,

$$P_0 = \frac{P}{e^{kt}} = Pe^{-kt}$$

Substituting 200,000 for P, 0.038 for k and 18 for t, we have:

$$P_0 = 200,000e^{-0.038(18)}$$

$$P_0 = 200,000e^{-0.684}$$

$$P_0 \approx 200,000(0.5045945719)$$

$$P_0 \approx 100,918.91.$$

The present value of Maggie's trust is approximately $100,918.91.

23. a) Since the income stream is constant the accumulated present value is given by:

$$B = \frac{R(t)}{k}\left(1 - e^{-kT}\right)$$

Substituting $95,000 for $R(t)$, 0.05 for k and 30 for T, we have:

$$B = \frac{95,000}{0.05}\left(1 - e^{-(0.05)(30)}\right)$$

$$\approx 1,476,052.70.$$

The accumulated present value of Rochelle's new job is approximately $1,476,052.70.

b) Since the income stream is constant the
accumulated future value is given by:

$$A = \frac{R(t)}{k}\left(e^{kT}-1\right)$$

Substituting \$95,000 for $R(t)$, 0.05 for k
and 30 for T, we have:

$$A = \frac{95,000}{0.05}\left(e^{(0.05)(30)}-1\right)$$

$$\approx 6,615,209.23.$$

The accumulated future value of Rochelle's
new job is approximately \$6,615,209.23.

25. $P = P_0 e^{kt}$

Substituting 50,000 for P, 0.043 for k, and 16
for t, we have:

$$P = 50,000e^{(0.043)16}$$

$$= 50,000e^{0.688}$$

$$\approx 99,486.60.$$

The future value of David's inheritance is
approximately \$99,486.60.

27. Computing the accumulated present value for
Franchise A, we substitute 80,000 for $R(t)$,
0.041 for k, and 10 for T. The accumulated
present value is

$$B = \frac{80,000}{0.041}\left(1-e^{-(0.041)(10)}\right)$$

$$= \frac{80,000}{0.041}\left(1-e^{-0.41}\right)$$

$$\approx 656,292.19.$$

Computing the accumulated present value for
Franchise B, we substitute 95,000 for $R(t)$,
0.041 for k, and 8 for T. The accumulated
present value is

$$B = \frac{95,000}{0.041}\left(1-e^{-(0.041)(8)}\right)$$

$$= \frac{95,000}{0.041}\left(1-e^{-0.328}\right)$$

$$\approx 647,939.34.$$

Comparing the accumulated present values for
the two franchises, we conclude that Franchise
A is the better buy.

29. a) Since the offer is a nonconstant stream, we
calculated the accumulated present value by

$$B = \int_0^T R(t)e^{-kt}dt.$$

For the Crunchers, we substitute $100,000t$
for $R(t)$, 0.042 for k, and 8 for T.

$$B = \int_0^8 100,000te^{-0.042t}dt$$

Using Tabular integration by parts, we have:

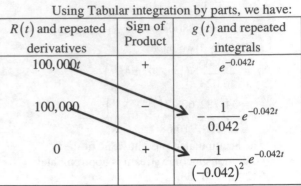

We calculate the accumulated present value.

$$B = \int_0^8 100,000te^{-0.042t}dt$$

$$= \left[\frac{-100,000}{0.042}te^{-0.042t} - \frac{100,000}{(0.042)^2}e^{-0.042t}\right]_0^8$$

$$= \frac{-100,000}{0.042}(8)e^{-0.042(8)} - \frac{100,000}{(0.042)^2}e^{-0.042(8)} -$$

$$\left(\frac{-100,000}{0.042}(0)e^{-0.042(0)} - \frac{100,000}{(0.042)^2}e^{-0.042(0)}\right)$$

$$\approx 2,565,959.80$$

The accumulated present value of the Crunchers deal
is approximately \$2,565,959.80.

For the Radar's we substitute $83,000t$ for
$R(t)$, 0.042 for k, and 9 for T

$$B = \int_0^9 83,000te^{-0.06t}dt$$

Using Tabular integration by parts, we have:

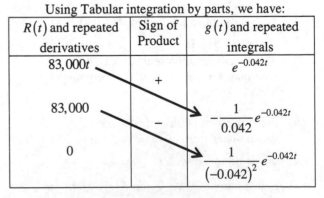

We use the information from the table to
calculate the accumulated present value at
the top of the next page.

Using the information from the previous page, we have:

$$B = \int_0^9 83,000te^{-0.042t}\, dt$$

$$= \left[\frac{-83,000}{0.042}te^{-0.042t} - \frac{83,000}{(0.042)^2}e^{-0.042t} \right]_0^9$$

$$= \frac{-83,000}{0.042}(9)e^{-0.042(9)} - \frac{83,000}{(0.042)^2}e^{-0.042(9)} -$$

$$\left(\frac{-83,000}{0.042}(0)e^{-0.042(0)} - \frac{83,000}{(0.042)^2}e^{-0.042(0)} \right)$$

$$\approx 2,623,269.10.$$

The accumulated present value of the Crunchers deal is approximately $2,623,269.10.

Based on the accumulated present values, we conclude the Doppler Radars have the better offer.

b) ✎

31. a) $P = P_0 e^{kt}$

Therefore,

$$P_0 = \frac{P}{e^{kt}} = Pe^{-kt}$$

Substituting 250,000 for P, 0.058 for k and 24 for t, we have:

$$P_0 = 250,000e^{-0.058(24)}$$

$$P_0 = 250,000e^{-1.392}$$

$$P_0 \approx 250,000(0.248577651841)$$

$$P_0 \approx 62,144.41$$

Bob and Ann should make an initial investment $62,144.41 to meet their goals for Brenda.

b) Since the income stream is constant the accumulated future value is given by:

$$A = \frac{R(t)}{k}\left(e^{kT} - 1\right)$$

Substituting $250,000 for A, 0.058 for k and 24 for T, we have:

$$250,000 = \frac{R(t)}{0.058}\left(e^{(0.058)(24)} - 1\right).$$

Solving the equation for $R(t)$ yields

$$250,000 = \frac{R(t)}{0.058}\left(e^{(0.058)(24)} - 1\right)$$

$$250,000(0.058) = R(t)\left(e^{1.392} - 1\right)$$

$$\frac{14,500}{\left(e^{1.392} - 1\right)} = R(t)$$

$$4796.74 \approx R(t)$$

Bob and Ann should deposit $4796.74 per year to meet their goals for Brenda.

33. Lauren has 3 years left on her contract making $84,000 per year. With an interest rate of 4.7%, we calculate the accumulated future value of the contract for the entire term of the contract:

$$A_{10} = \frac{84,000}{0.047}\left(e^{(0.047)(10)} - 1\right)$$

$$= \frac{84,000}{0.047}\left(e^{0.47} - 1\right)$$

$$\approx 1,072,330.05$$

The accumulated future value for the first 7 years of the contract is:

$$A_7 = \frac{84,000}{0.047}\left(e^{(0.047)(7)} - 1\right)$$

$$= \frac{84,000}{0.047}\left(e^{0.329} - 1\right)$$

$$\approx 696,266.81.$$

The difference is:

$$A_{10} - A_7 = 1,072,330.05 - 696,266.81$$

$$= 376,063.24.$$

Since the company is offering a lump sum Lauren should receive the present value of the difference.

$$P_0 = 376,063.24e^{-0.047(3)}$$

$$\approx 326,606.91$$

The minimum the bank should offer Lauren to take early retirement is $326,606.91.

35. a) Substituting $120,000 for $R(t)$, 0.04 for k and 20 for T, we have:

$$A = \frac{120,000}{0.04}\left(e^{(0.04)(20)} - 1\right)$$

$$\approx 3,676,622.79.$$

The accumulated future value of income stream rounded to the nearest $10 is approximately $3,676,620.

b) Substituting $120,000 for $R(t)$, 0.04 for k and 20 for T, we have:

$$B = \frac{120,000}{0.04}\left(1 - e^{-(0.04)(20)}\right)$$
$$\approx 1,652,013.11.$$

The accumulated present value of the income stream rounded to the nearest $10 is approximately $1,652,010.

37. a) Substituting $50,000 for $R(t)$, 0.044 for k and 20 for T, we have:

$$A = \frac{50,000}{0.044}\left(e^{(0.044)(20)} - 1\right)$$
$$\approx 1,603,295.12.$$

The accumulated future value of income stream rounded to the nearest $10 is approximately $1,603,300.

b) Substituting $50,000 for $R(t)$, 0.044 for k and 20 for T, we have:

$$B = \frac{50,000}{0.044}\left(1 - e^{-(0.044)(20)}\right)$$
$$\approx 665,019.42.$$

The accumulated present value of the income stream is approximately 66,019.42.

c) If the interest rate is 3% the accumulated present value is:

$$B = \frac{50,000}{0.03}\left(1 - e^{-(0.03)(20)}\right)$$
$$\approx 751,981.$$

The accumulated present value at a rate of 3% is approximately $751,981.
If the interest rate is 4% the accumulated present value is:

$$B = \frac{50,000}{0.04}\left(1 - e^{-(0.04)(20)}\right)$$
$$\approx 688,339.$$

The accumulated present value at a rate of 4% is approximately $688,339.
If the interest rate is 5% the accumulated present value is:

$$B = \frac{50,000}{0.05}\left(1 - e^{-(0.05)(20)}\right)$$
$$\approx 632,121.$$

The accumulated present value at a rate of 5% is approximately $632,121.

d) ✎

39. Since the Wilkinsons want to have $100,000 in 10 years, we set the accumulated future value formula equal to $100,000. We will substituting 0.0433 for k, and 10 for T.

$$\frac{R(t)}{k}\left(e^{kT} - 1\right) = 100,000$$

$$\frac{R(t)}{0.0433}\left(e^{0.0433(10)} - 1\right) = 100,000$$

$$R(t)\left(e^{0.433} - 1\right) = 100,000(0.0433)$$

$$R(t) = \frac{4330}{\left(e^{0.433} - 1\right)}$$

$$R(t) = 7990.75$$

The Wilkinson's will need to invest a continuous money stream of $7990.75 per year for 10 years to generate $100,000.

41. The equation for the consumption of a natural resource is $\int_0^T P_0 e^{kt}\, dt = \frac{P_0}{k}\left(e^{kT} - 1\right)$.
Substituting 117.2 for P_0, 0.0124 for k, and 12 $[2025 - 2013 = 12]$ for t, we have

$$\int_0^{12} 117.2 e^{0.0124t} = \frac{117.2}{0.0124}\left(e^{0.0124(12)} - 1\right)$$
$$= \frac{117.2}{0.0124}\left(e^{0.1488} - 1\right)$$
$$\approx 1516.43$$

The total amount consumed from 2013 to 2025 is 1516.43 trillion cubic feet.
Substituting 117.2 for P_0, 0.0124 for k, and 2 $[2015 - 2013 = 2]$ for t, we have

$$\int_0^2 117.2 e^{0.0124t} = \frac{117.2}{0.0124}\left(e^{0.0124(2)} - 1\right)$$
$$= \frac{117.2}{0.0124}\left(e^{0.0248} - 1\right)$$
$$\approx 237.33$$

The total amount consumed between 2013 and 2015 is 237.33 trillion cubic feet.
If demand continues to grow exponentially at 1.24% per year, the world will consume approximately $1516.43 - 237.33 = 1279.1$ trillion cubic feet of natural gas from 2015 to 2025.

43. The equation for the consumption of a natural resource is $\int_0^T P_0 e^{kt}\,dt = \dfrac{P_0}{k}\left(e^{kT}-1\right)$.

Using the information from Exercise 41, we want to find T, such that

$$6597 = \frac{117.2}{0.0124}\left(e^{0.0124T}-1\right)$$

Solving the equation for T, we have:

$$6597 = \frac{117.2}{0.0124}\left(e^{0.0124T}-1\right)$$

$0.69797611 = e^{0.0124T}-1$ $\quad$ Dividing both sides by $\frac{117.2}{0.0124}$

$1.69797611 \approx e^{0.0124T}$ $\quad$ Adding 1 to both sides

$\ln(1.69797611) \approx \ln\left(e^{0.0124T}\right)$ $\quad$ Taking the natural logarithm of each side

$\ln(1.69797611) \approx 0.0124T$ $\quad$ Recall that $\ln e^k = k$

$\dfrac{\ln(1.69797611)}{0.0124} \approx T$ $\quad$ Dividing both sides by 0.074

$42.69 \sim T$

Assuming the world consumption of natural gas continues to grow at 1.24% per year, and no new reserves are found, the world reserves of natural gas will be depleted 42.7 years after 2013, in 2055.

45. a) $P = P_0 e^{kt}$

Using the growth rate and the initial demand, we have the exponential demand function $P = 33.3e^{0.015t}$.

Substituting 7 $[2020-2013 = 7]$ for t, we have:

$P = 33.3e^{0.015(7)}$

$= 33.3e^{0.105}$

≈ 36.99

In 2020, the world demand for oil will be approximately 36.99 billion barrels.

b) The equation for the consumption of a natural resource is $\int_0^T P_0 e^{kt}\,dt = \dfrac{P_0}{k}\left(e^{kT}-1\right)$.

We want to find T, such that

$$1635 = \frac{33.3}{0.015}\left(e^{0.015T}-1\right)$$

$$1635 = 2220\left(e^{0.015T}-1\right)$$

$$0.7364865 \approx e^{0.015T}-1$$

$$1.7364865 \approx e^{0.0015T}$$

$$\ln(1.7364865) \approx \ln e^{0.015T}$$

$$\ln(1.7364865) \approx 0.015T$$

$$\frac{\ln(1.7364865)}{0.015} \approx T$$

$$36.79 \approx T$$

Assuming the world consumption of oil continues to grow at 1.5% per year, and no new reserves are found, the world reserves of oil will be depleted 36.79 years after 2013, in 2049.

47. We will use the equation

$$\int_0^T Pe^{-kt}\,dt = \frac{P}{k}\left(1-e^{-kT}\right)$$

Substituting 0.023 for k, 1 for P, and 20 for T, we have:

$$\int_0^{20} 1\cdot e^{-0.023t}\,dt = \frac{1}{0.023}\left(1-e^{-0.023(20)}\right)$$

$$\approx 43.47826\left(1-e^{-0.46}\right)$$

$$\approx 16.031$$

After 20 years, approximately 16.031 lbs of Cesium-237 will remain in the atmosphere.

49. $c = c_0 + \displaystyle\int_0^L m(t)e^{-kt}\,dt$

Substituting 400,000 for c_0, 10,000 for $m(t)$, 0.055 for k, and 25 for L, we have:

$$c = 400{,}000 + \int_0^{25} 10{,}000e^{-0.055t}\,dt$$

$$= 400{,}000 + \left[\frac{10{,}000}{-0.055}e^{-0.055t}\right]_0^{25}$$

$$= 400{,}000 - \frac{10{,}000}{0.055}\left[e^{-1.375}-e^0\right]$$

$$\approx 535{,}847$$

The capitalized cost under the given assumptions is $535,847.

51. $c = c_0 + \int_0^L m(t)e^{-kt}\,dt$

Substituting 300,000 for c_0, $30,000 + 500e^{0.01t}$
for $m(t)$, 0.05 for k, and 40 for L, we have:

$c = 300,000 + \int_0^{20}(30,000 + 500t)e^{-0.05t}\,dt$

$= 300,000 + \int_0^{20}\left(30,000e^{-0.05t} + 500te^{-0.05t}\right)dt$

$= 300,000 + \left[\dfrac{30,000}{-0.05}e^{-0.05t} + \right.$

$\left. 500\dfrac{1}{(-0.05)^2}e^{-0.05t}(-0.05t - 1)\right]_0^{20}$

$= 300,000 + \left[-600,000e^{-0.05t} + \right.$

$\left. 200,000e^{-0.05t}(-0.05t - 1)\right]_0^{20}$

$= 300,000 + \left[-600,000e^{-0.05(20)} + \right.$

$200,000e^{-0.05(20)}(-0.05(20) - 1) -$

$\left(-600,000e^{-0.05(0)} + \right.$

$\left.\left. 200,000e^{-0.05(0)}(-0.05(0) - 1)\right)\right)$

$\approx 732,121$

The capitalized cost under the given
assumptions is \$732,121.

53. ✎

Exercise Set 5.3

1. $\displaystyle\int_5^\infty \frac{dx}{x^2}$

$= \displaystyle\lim_{b\to\infty} \int_5^b x^{-2}dx$

$= \displaystyle\lim_{b\to\infty} \left[\frac{x^{-1}}{-1}\right]_5^b$

$= \displaystyle\lim_{b\to\infty} \left[-\frac{1}{x}\right]_5^b$

$= \displaystyle\lim_{b\to\infty} \left[-\frac{1}{b}-\left(-\frac{1}{5}\right)\right]$

$= 0+\dfrac{1}{5} \qquad \left(\text{As } b\to\infty, -\frac{1}{b}\to 0\right)$

$= \dfrac{1}{5}$

The limit does exist. The improper integral is convergent.

3. $\displaystyle\int_3^\infty \frac{dx}{x}$

$= \displaystyle\lim_{b\to\infty} \int_3^b x^{-1}dx$

$= \displaystyle\lim_{b\to\infty} \left[\ln(x)\right]_3^b$

$= \displaystyle\lim_{b\to\infty} \left[\ln(b)-\ln(3)\right]$

Note: that $\ln b$ increases without bound as b increases; Therefore, the limit does not exist. If the limit does not exist, we say the improper integral is divergent.

5. $\displaystyle\int_0^\infty 3e^{-3x}dx$

$= \displaystyle\lim_{b\to\infty} \int_0^b 3e^{-3x}dx$

$= \displaystyle\lim_{b\to\infty} \left[\frac{3}{-3}e^{-3x}\right]_0^b$

$= \displaystyle\lim_{b\to\infty} \left[-e^{-3x}\right]_0^b$

$= \displaystyle\lim_{b\to\infty} \left[-e^{-3\cdot b}-\left(-e^{-3\cdot 0}\right)\right]$

$= \displaystyle\lim_{b\to\infty} \left[-e^{-3b}+1\right]$

$= \displaystyle\lim_{b\to\infty} \left[1-\frac{1}{e^{3b}}\right]$

$= 1-0 \qquad \left[\text{As } b\to\infty, e^{3b}\to\infty, \text{so } \frac{1}{e^{3b}}\to 0\right]$

$= 1$

The limit does exist. The improper integral is convergent.

7. $\displaystyle\int_1^\infty \frac{dx}{x^3}$

$= \displaystyle\lim_{b\to\infty} \int_1^b x^{-3}dx$

$= \displaystyle\lim_{b\to\infty} \left[\frac{x^{-2}}{-2}\right]_1^b$

$= \displaystyle\lim_{b\to\infty} \left[-\frac{1}{2x^2}\right]_1^b$

$= \displaystyle\lim_{b\to\infty} \left[-\frac{1}{2b^2}-\left(-\frac{1}{2(1)^2}\right)\right]$

$= 0+\dfrac{1}{2} \qquad \left(\text{As } b\to\infty, -\frac{1}{2b^2}\to 0\right)$

$= \dfrac{1}{2}$

The limit does exist. The improper integral is convergent.

9. $\displaystyle\int_0^\infty \frac{dx}{2+x}$

$\displaystyle = \lim_{b\to\infty}\int_0^b \frac{1}{2+x}dx$

$\displaystyle = \lim_{b\to\infty}\left[\ln(2+x)\right]_0^b$

$\displaystyle = \lim_{b\to\infty}\left[\ln(2+b)-\ln(2+0)\right]$

$\displaystyle = \lim_{b\to\infty}\left[\ln(2+b)-\ln(2)\right]$

Note that $\ln(2+b)$ increases without bound as b increases. Therefore, the limit does not exist. If the limit does not exist, we say the improper integral is divergent.

11. $\displaystyle\int_2^\infty 4x^{-2}dx$

$\displaystyle = \lim_{b\to\infty}\int_2^b 4x^{-2}dx$

$\displaystyle = \lim_{b\to\infty}\left[\frac{4x^{-1}}{-1}\right]_2^b$

$\displaystyle = \lim_{b\to\infty}\left[-\frac{4}{x}\right]_2^b$

$\displaystyle = \lim_{b\to\infty}\left[-\frac{4}{b}-\left(-\frac{4}{2}\right)\right]$

$\displaystyle = 0+2 \qquad \left(\text{As }b\to\infty, -\tfrac{4}{b}\to 0\right)$

$= 2$

The limit does exist. The improper integral is convergent.

13. $\displaystyle\int_0^\infty e^x dx$

$\displaystyle = \lim_{b\to\infty}\int_0^b e^x dx$

$\displaystyle = \lim_{b\to\infty}\left[e^x\right]_0^b$

$\displaystyle = \lim_{b\to\infty}\left[e^b-\left(e^0\right)\right]$

$\displaystyle = \lim_{b\to\infty}\left[e^b-1\right]$

As $b\to\infty, e^b\to\infty$; thus, the limit does not exist. The improper integral is divergent.

15. $\displaystyle\int_3^\infty x^2 dx$

$\displaystyle = \lim_{b\to\infty}\int_3^b x^2 dx$

$\displaystyle = \lim_{b\to\infty}\left[\frac{x^3}{3}\right]_3^b$

$\displaystyle = \lim_{b\to\infty}\left[\frac{(b)^3}{3}-\frac{(3)^3}{3}\right]$

$\displaystyle = \lim_{b\to\infty}\left[\frac{b^3}{3}-9\right]$

As $b\to\infty, \dfrac{b^3}{3}\to\infty$; thus, the limit does not exist. The improper integral is divergent.

17. $\displaystyle\int_0^\infty xe^x dx$

$\displaystyle = \lim_{b\to\infty}\int_0^b xe^x dx$

$\displaystyle = \lim_{b\to\infty}\left[e^x(x-1)\right]_0^b \qquad \text{Using integration by parts}$

$\displaystyle = \lim_{b\to\infty}\left[e^b(b-1)-\left(e^0(0-1)\right)\right]$

$\displaystyle = \lim_{b\to\infty}\left[e^b(b-1)+1\right]$

As $b\to\infty, e^b(b-1)\to\infty$; thus, the limit does not exist. The improper integral is divergent.

19. $\displaystyle\int_0^\infty me^{-mx}dx, \quad m>0$

$\displaystyle = \lim_{b\to\infty}\int_0^b me^{-mx}dx$

$\displaystyle = \lim_{b\to\infty}\left[\frac{m}{-m}e^{-mx}\right]_0^b$

$\displaystyle = \lim_{b\to\infty}\left[-e^{-mx}\right]_0^b$

$\displaystyle = \lim_{b\to\infty}\left[-e^{-m\cdot b}-\left(-e^{-m\cdot 0}\right)\right]$

$\displaystyle = \lim_{b\to\infty}\left[-e^{-mb}+1\right]$

$\displaystyle = \lim_{b\to\infty}\left[1-\frac{1}{e^{mb}}\right]$

$\displaystyle = 1-0 \qquad \left[\text{As }b\to\infty, e^{mb}\to\infty, \text{ so } \tfrac{1}{e^{mb}}\to 0\right]$

$= 1$

The limit does exist. The improper integral is convergent.

21. $\displaystyle\int_{\pi}^{\infty}\frac{dt}{t^{1.001}}$

$\displaystyle=\lim_{b\to\infty}\int_{\pi}^{b}t^{-1.001}dt$

$\displaystyle=\lim_{b\to\infty}\left[\frac{t^{-0.001}}{-0.001}\right]_{\pi}^{b}$

$\displaystyle=\lim_{b\to\infty}\left[-\frac{1000}{t^{0.001}}\right]_{\pi}^{b}$

$\displaystyle=\lim_{b\to\infty}\left[-\frac{1000}{b^{0.001}}-\left(-\frac{1000}{(\pi)^{0.001}}\right)\right]$

$\displaystyle=0+\frac{1000}{\pi^{0.001}}\quad\left(\text{As }b\to\infty,-\frac{1000}{b^{0.001}}\to0\right)$

$\displaystyle=\frac{1000}{\pi^{0.001}}$

≈998.86

The limit does exist. The improper integral is convergent.

23. $\displaystyle\int_{-\infty}^{\infty}t\,dt$

$\displaystyle=\int_{-\infty}^{0}t\,dt+\int_{0}^{\infty}t\,dt\quad\text{Using Definition 2 with }c=0$

$\displaystyle=\lim_{a\to-\infty}\int_{a}^{0}t\,dt+\lim_{b\to\infty}\int_{0}^{b}t\,dt$

$\displaystyle=\lim_{a\to-\infty}\left[\frac{t^2}{2}\right]_{a}^{0}+\lim_{b\to\infty}\left[\frac{t^2}{2}\right]_{0}^{b}$

$\displaystyle=\lim_{a\to-\infty}\left[\frac{0^2}{2}-\frac{a^2}{2}\right]+\lim_{b\to\infty}\left[\frac{b^2}{2}-\frac{0^2}{2}\right]$

Neither $\displaystyle\lim_{a\to-\infty}\frac{a^2}{2}$ nor $\displaystyle\lim_{b\to\infty}\frac{b^2}{2}$ exists, so the integral is divergent.

25. The area is given by

$\displaystyle\int_{2}^{\infty}\frac{1}{x^2}dx=\lim_{b\to\infty}\int_{2}^{b}x^{-2}dx$

$\displaystyle=\lim_{b\to\infty}\left[\frac{x^{-1}}{-1}\right]_{2}^{b}$

$\displaystyle=\lim_{b\to\infty}\left[-\frac{1}{x}\right]_{2}^{b}$

$\displaystyle=\lim_{b\to\infty}\left[-\frac{1}{b}-\left(-\frac{1}{2}\right)\right]$

$\displaystyle=0+\frac{1}{2}\qquad\left(\text{As }b\to\infty,-\tfrac{1}{b}\to0\right)$

$\displaystyle=\frac{1}{2}$

The area of the region is $\dfrac{1}{2}$.

27. The area is given by

$\displaystyle\int_{0}^{\infty}2xe^{-x^2}dx$

$\displaystyle=\lim_{b\to\infty}\int_{0}^{b}2xe^{-x^2}dx$

$\displaystyle=\lim_{b\to\infty}\left[-e^{-x^2}\right]_{0}^{b}\qquad\left[u=-x^2,\,du=-2x\,dx\right]$

$\displaystyle=\lim_{b\to\infty}\left[-e^{-b^2}-\left(-e^{-0^2}\right)\right]$

$\displaystyle=\lim_{b\to\infty}\left[-e^{-b^2}+1\right]$

$\displaystyle=\lim_{b\to\infty}\left[-\frac{1}{e^{b^2}}+1\right]$

$\displaystyle=-0+1=1\qquad\left[\text{As }b\to\infty,-\frac{1}{e^{b^2}}\to0\right]$

The area of the region is 1.

29. Total profit is given by:

$$P(x) = \int_0^\infty 200,000e^{-0.032x}\,dx$$

$$= \lim_{b\to\infty} \int_0^b 200,000e^{-0.032x}\,dx$$

$$= \lim_{b\to\infty} \left[\frac{200,000}{-0.032} e^{-0.032x} \right]_0^b$$

$$= \lim_{b\to\infty} \left[-6,250,000e^{-0.032x} \right]_0^b$$

$$= 6,250,000 \lim_{b\to\infty} \left[-e^{-0.032b} - \left(-e^{-0.032(0)} \right) \right]$$

$$= 6,250,000 \lim_{b\to\infty} \left[-\frac{1}{e^{0.032b}} + 1 \right]$$

$$= 6,250,000$$

The total profit if it were possible to produce an infinite number of shawls is $6,250,000.

31. The total cost is given by:

$$C(x) = \int_1^\infty 3,600,000x^{-1.8}\,dx$$

$$= \lim_{b\to\infty} \int_1^b 3,600,000x^{-1.8}\,dx$$

$$= \lim_{b\to\infty} \left[\frac{3,600,000}{-0.8} x^{-0.8} \right]_1^b$$

$$= \lim_{b\to\infty} \left[-4,500,000x^{-0.8} \right]_1^b$$

$$= 4,500,000 \lim_{b\to\infty} \left[-\frac{1}{b^{0.8}} - \left(-\frac{1}{1^{0.8}} \right) \right]$$

$$= 4,500,000(0+1) \quad \left(\text{As } b\to\infty, -\frac{1}{b^{0.8}} \to 0 \right)$$

$$= 4,500,000$$

The total cost to make infinitely many keyholders would be $4,500,000.

33. From Theorem 1, the accumulated present value is given by

$$\int_0^\infty Pe^{-kt}\,dt = \frac{P}{k}$$

Substituting 3600 for P and 0.05 for k, we have:

$$\int_0^\infty 3600e^{-0.05t}\,dt = \frac{3600}{0.05} \approx 72,000.$$

The accumulated present value is $72,000.

35. From Theorem 1, the accumulated present value is given by

$$\int_0^\infty Pe^{-kt}\,dt = \frac{P}{k}$$

Substituting 5000 for P and 0.037 for k, we have:

$$\int_0^\infty 5000e^{-0.037t}\,dt = \frac{5000}{0.037} \approx 135,135.14.$$

The accumulated present value is approximately $135,135.14.

37. $c = c_0 + \int_0^\infty m(t)e^{-kt}\,dt$

Substituting 500,000 for c_0, 0.05 for k, and 20,000 for $m(t)$, we have:

$$c = 500,000 + \int_0^\infty 20,000e^{-0.05t}\,dt$$

$$= 500,000 + \lim_{b\to\infty} \int_0^b 20,000e^{-0.05t}\,dt$$

$$= 500,000 + \lim_{b\to\infty} \left[\frac{20,000}{-0.05} e^{-0.05t} \right]_0^b$$

$$= 500,000 + \lim_{b\to\infty} \left[-400,000\left(e^{-0.05b} - e^{-0.05(0)} \right) \right]$$

$$= 500,000 + \left[-400,000(0-1) \right]$$

$$= 500,000 + 400,000$$

$$= 900,000$$

The capitalized cost is $900,000.

39. $\int_0^T Pe^{-kt}\,dt = \frac{P}{k}\left(1 - e^{-kT} \right)$

As $T \to \infty$, we have:

$$\lim_{T\to\infty} \int_0^T P(t)e^{-kt}\,dt$$

$$= \lim_{T\to\infty} \left[\frac{P}{-k} e^{-kt} \right]_0^T$$

$$= \lim_{T\to\infty} \left[\frac{P}{-k} e^{-kT} - \left(\frac{P}{-k} e^{-k\cdot 0} \right) \right]$$

$$= \lim_{T\to\infty} \left[\frac{P}{k}\left(1 - e^{-k\cdot T} \right) \right]$$

$$= \frac{P}{k}$$

Substituting 0.00003 for k, and 1 for P, we have

$$\frac{P}{k} = \frac{1}{0.00003} \approx 33,333\tfrac{1}{3}.$$

The limiting value of the radioactive buildup is $33,333\tfrac{1}{3}$ pounds.

41. $E = \int_0^a P_0 e^{-kt} dt$

a) Note that 60.1 days is

$\dfrac{60.1}{365}$ yr ≈ 0.16465753 yr .

Using the half-life, we find k as follows:

$\dfrac{1}{2} P_0 = P_0 e^{-k(0.16465753)}$

$\dfrac{1}{2} = e^{-k(0.16465753)}$ Dividing by P_0

$\ln\left(\dfrac{1}{2}\right) = \ln\left(e^{-0.16465753k}\right)$

$\ln\left(\dfrac{1}{2}\right) = -0.16465753k$

$\dfrac{\ln\left(\dfrac{1}{2}\right)}{-0.16465753} = k$

$4.20963 \approx k$

The decay rate is 420.963% per year.

b) The first month is $\dfrac{1}{12}$ yr.

$E = \int_0^{1/12} 10 e^{-4.20963t} dt$

$= \dfrac{10}{-4.20963}\left[e^{-4.20963t}\right]_0^{1/12}$

$= \dfrac{10}{-4.20963}\left[e^{-4.20963(1/12)} - e^{-4.20963(0)}\right]$

$= \dfrac{10}{-4.20963}\left[e^{-0.3508025} - 1\right]$

≈ 0.702858

In the first month, 0.702858 rems of energy is transmitted.

c) $E = \int_0^\infty 10 e^{-4.20963t} dt$

$E = \lim_{b \to \infty} \int_0^b 10 e^{-4.20963t} dt$

$= \dfrac{10}{4.20963}$ $\left[\int_0^\infty P e^{-kt} = \dfrac{P}{k}\right]$

≈ 2.37551

The total amount of energy transmitted is 2.37551 rems.

43. $\int_0^\infty \dfrac{dx}{x^{2/3}}$

$= \lim_{b \to \infty} \int_0^b x^{-2/3} dx$

$= \lim_{b \to \infty}\left[\dfrac{x^{1/3}}{1/3}\right]_0^b$

$= \lim_{b \to \infty}\left[3x^{1/3}\right]_0^b$

$= \lim_{b \to \infty}\left[3b^{1/3} - 3(0)^{1/3}\right]$

$= \lim_{b \to \infty}\left[3 \cdot \sqrt[3]{b}\right]$

As $b \to \infty$, $\sqrt[3]{b} \to \infty$. Therefore, the limit does not exist. The improper integral is divergent.

45. $\int_0^\infty \dfrac{dx}{(x+1)^{3/2}}$

$= \lim_{b \to \infty} \int_0^b (x+1)^{-3/2} dx$

$= \lim_{b \to \infty}\left[\dfrac{(x+1)^{-1/2}}{-1/2}\right]_0^b$

$= \lim_{b \to \infty}\left[-\dfrac{2}{\sqrt{x+1}}\right]_0^b$

$= \lim_{b \to \infty}\left[-\dfrac{2}{\sqrt{b+1}} - \left(-\dfrac{2}{\sqrt{0+1}}\right)\right]$

$= \lim_{b \to \infty}\left[-\dfrac{2}{\sqrt{b+1}} + 2\right]$

$= 0 + 2$ $\left[\text{As } b \to \infty, -\dfrac{2}{\sqrt{b+1}} \to 0\right]$

$= 2$

Therefore, the limit exists. The improper integral is convergent.

47. $\displaystyle\int_0^\infty xe^{-x^2}\,dx$

$\displaystyle = \lim_{b\to\infty}\int_0^b xe^{-x^2}\,dx$

$\displaystyle = \lim_{b\to\infty}\int_0^b -\frac{1}{2}\cdot(-2)xe^{-x^2}\,dx \qquad \text{Multiplying by 1}$

$\displaystyle = \lim_{b\to\infty}\left[-\frac{1}{2}e^{-x^2}\right]_0^b \qquad \begin{array}{l}\text{Using substitution where}\\ u=-x^2 \text{ and } du=-2x\,dx\end{array}$

$\displaystyle = \lim_{b\to\infty}\left[-\frac{1}{2}e^{-b^2}-\left(-\frac{1}{2}e^{-(0)^2}\right)\right]$

$\displaystyle = \lim_{b\to\infty}\left[-\frac{1}{2e^{b^2}}+\frac{1}{2}\right]$

$\displaystyle = 0+\frac{1}{2} \qquad \left[\text{As } b\to\infty, \frac{1}{e^{b^2}}\to 0\right]$

$\displaystyle = \frac{1}{2}$

Therefore, the limit exists. The improper integral is convergent.

49. $\displaystyle\int_0^\infty E(t)\,dt$

$\displaystyle = \int_0^\infty te^{-kt}\,dt$

$\displaystyle = \lim_{b\to\infty}\int_0^b te^{-kt}\,dt$

$\displaystyle = \lim_{b\to\infty}\left[\frac{1}{(-k)^2}\cdot e^{-kt}(-kt-1)\right]_0^b$

$\displaystyle = \lim_{b\to\infty}\left[-\frac{kt+1}{k^2 e^{kt}}\right]_0^b$

$\displaystyle = \lim_{b\to\infty}\left[-\frac{kb+1}{k^2 e^{kb}}-\left(-\frac{k(0)+1}{k^2 e^{k(0)}}\right)\right]$

$\displaystyle = \lim_{b\to\infty}\left[-\frac{kb+1}{k^2 e^{kb}}+\frac{1}{k^2}\right]$

$\displaystyle = \frac{1}{k^2} \qquad \left[\text{As } b\to\infty, -\frac{kb+1}{k^2 e^{kb}}\to 0\right]$

The integral represents the total dose of the drug.

51. ✎

53. $\displaystyle\int_0^4 \frac{1}{\sqrt{x}}\,dx$

$\displaystyle = \lim_{a\to 0}\int_a^4 \frac{1}{\sqrt{x}}\,dx$

$\displaystyle = \lim_{a\to 0}\left[\frac{(x)^{1/2}}{1/2}\right]_a^4$

$\displaystyle = \lim_{a\to 0}\left[2\sqrt{x}\right]_a^4$

$\displaystyle = \lim_{a\to 0}\left[2\sqrt{4}-\left(2\sqrt{a}\right)\right]$

$\displaystyle = \lim_{a\to 0}\left[2(2)+2\sqrt{a}\right]$

$\displaystyle = 4+0 \qquad \left[\text{As } a\to 0, 2\sqrt{a}\to 0\right]$

$\displaystyle = 4$

Therefore, the limit exists. The improper integral is convergent.

55. ✎

57. Using the fnInt feature on a graphing calculator with a large value for the upper limit, we find

$$\int_1^\infty \frac{6}{5+e^x}\,dx \approx 1.2523.$$

59.

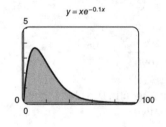

$y = xe^{-0.1x}$

Exercise Set 5.4

1. $f(x) = 2x, \quad [0,1]$

$$P([0,1]) = \int_0^1 f(x)\,dx$$

$$= \int_0^1 2x\,dx$$

$$= \left[x^2\right]_0^1$$

$$= (1)^2 - (0)^2$$

$$= 1 - 0$$

$$= 1.$$

3. $f(x) = 3, \quad \left[0, \dfrac{1}{3}\right]$

$$P\left(\left[0, \frac{1}{3}\right]\right) = \int_0^{1/3} f(x)\,dx$$

$$= \int_0^{1/3} 3\,dx$$

$$= \left|3x\right|_0^{1/3}$$

$$= 3\left(\frac{1}{3}\right) - 3(0)$$

$$= 1.$$

5. $f(x) = \dfrac{3}{26}x^2, \quad [1,3]$

$$P([1,3]) = \int_1^3 f(x)\,dx$$

$$= \int_1^3 \frac{3}{26}x^2\,dx$$

$$= \left[\frac{x^3}{26}\right]_1^3$$

$$= \frac{(3)^3}{26} - \frac{(1)^3}{26}$$

$$= \frac{27}{26} - \frac{1}{26}$$

$$= 1.$$

7. $f(x) = \dfrac{1}{x}, \quad [1,e]$

$$P([1,e]) = \int_1^e f(x)\,dx$$

$$= \int_1^e \frac{1}{x}\,dx$$

$$= \left[\ln x\right]_1^e$$

$$= \ln(e) - \ln(1)$$

$$= 1 - 0$$

$$= 1.$$

9. $f(x) = \dfrac{1}{3}x^2, \quad [-2,1]$

$$P([-2,1]) = \int_{-2}^1 f(x)\,dx$$

$$= \int_{-2}^1 \frac{1}{3}x^2\,dx$$

$$= \left[\frac{x^3}{9}\right]_{-2}^1$$

$$= \frac{(1)^3}{9} - \frac{(-2)^3}{9}$$

$$= \frac{1}{9} - \frac{-8}{9}$$

$$= 1.$$

11. $f(x) = 3e^{-3x}, \quad [0,\infty)$

$$P([0,\infty)) = \int_0^\infty f(x)\,dx$$

$$= \lim_{b \to \infty} \int_0^b 3e^{-3x}\,dx$$

$$= \lim_{b \to \infty} \left[\frac{3}{-3}e^{-3x}\right]_0^b$$

$$= \lim_{b \to \infty} \left[-e^{-3x}\right]_0^b$$

$$= \lim_{b \to \infty} \left[-e^{-3 \cdot b} - \left(-e^{-3(0)}\right)\right]$$

$$= \lim_{b \to \infty} \left[-\frac{1}{e^{3b}} + 1\right]$$

$$= 0 + 1 \qquad \left[\text{As } b \to \infty, -\frac{1}{e^{3b}} \to 0\right]$$

$$= 1.$$

13. $f(x) = kx$, $[1, 4]$

Find k such that $\int_1^4 kx\, dx = 1$.

We have

$$\int_1^4 x\, dx = \left[\frac{x^2}{2}\right]_1^4$$

$$= \left[\frac{(4)^2}{2} - \frac{(1)^2}{2}\right]$$

$$= \frac{16}{2} - \frac{1}{2} = \frac{15}{2}.$$

Thus $k = \dfrac{1}{\frac{15}{2}} = \dfrac{2}{15}$ and the probability density

function is $f(x) = \dfrac{2}{15}x$.

15. $f(x) = kx^2$, $[-1, 1]$

Find k such that $\int_{-1}^1 kx^2\, dx = 1$.

We have

$$\int_{-1}^1 x^2\, dx = \left[\frac{x^3}{3}\right]_{-1}^1$$

$$= \left[\frac{(1)^3}{3} - \frac{(-1)^3}{3}\right]$$

$$= \frac{1}{3} - \frac{-1}{3} = \frac{2}{3}.$$

Thus $k = \dfrac{1}{\frac{2}{3}} = \dfrac{3}{2}$ and the probability density

function is $f(x) = \dfrac{3}{2}x^2$.

17. $f(x) = k$, $[3, 9]$

Find k such that $\int_3^9 k\, dx = 1$.

We have

$$\int_3^9 dx = [x]_3^9$$

$$= [9 - 3]$$

$$= 6.$$

Thus $k = 6 = \dfrac{1}{6}$ and the probability density

function is $f(x) = \dfrac{1}{6}$.

19. $f(x) = k(2 - x)$, $[0, 2]$

Find k such that $\int_0^2 k(2 - x)\, dx = 1$.

We have

$$\int_0^2 (2 - x)\, dx = \left[2x - \frac{x^2}{2}\right]_0^2$$

$$= \left[\left(2(2) - \frac{(2)^2}{2}\right) - \left(2(0) - \frac{0^2}{2}\right)\right]$$

$$= 4 - \frac{4}{2} - 0$$

$$= 2.$$

Thus $k = \dfrac{1}{2}$ and the probability density function

is $f(x) = \dfrac{1}{2}(2 - x) = \dfrac{2 - x}{2}$.

21. $f(x) = \dfrac{k}{x}$, $[1, 2]$

Find k such that $\int_1^2 \dfrac{k}{x}\, dx = 1$.

We have

$$\int_1^2 \frac{1}{x}\, dx = [\ln(x)]_1^2$$

$$= [\ln(2) - \ln(1)]$$

$$= \ln 2 - 0 = \ln 2.$$

Thus $k = \dfrac{1}{\ln 2}$ and the probability density

function is $f(x) = \dfrac{1}{\ln 2} \cdot \dfrac{1}{x} = \dfrac{1}{x \ln 2}$.

23. $f(x) = ke^x$, $[0, 3]$

Find k such that $\int_0^3 ke^x\, dx = 1$.

We have

$$\int_0^3 e^x\, dx = [e^x]_0^3$$

$$= [e^3 - e^0]$$

$$= e^3 - 1.$$

Thus $k = \dfrac{1}{e^3 - 1}$ and the probability density

function is $f(x) = \dfrac{1}{e^3 - 1} \cdot e^x = \dfrac{e^x}{e^3 - 1}$.

25. $f(x) = \dfrac{1}{50}x$, for $0 \le x \le 10$

$$P(2 \le x \le 6) = \int_2^6 \frac{1}{50}x\,dx$$

$$= \left[\frac{1}{50} \cdot \frac{x^2}{2}\right]_2^6$$

$$= \left[\frac{x^2}{100}\right]_2^6$$

$$= \frac{6^2}{100} - \frac{2^2}{100}$$

$$= \frac{36-4}{100} = \frac{32}{100} = \frac{8}{25} = 0.32$$

The probability that the dart lands in the interval $[2,6]$ is $\dfrac{8}{25}$, or 0.32.

27. $f(x) = \dfrac{1}{16}$, for $4 \le x \le 20$

$$P(9 \le x \le 20) = \int_9^{20} \frac{1}{16}\,dx$$

$$= \left[\frac{1}{16}x\right]_9^{20}$$

$$= \frac{1}{16} \cdot 20 - \frac{1}{16} \cdot 9$$

$$= \frac{20-9}{16}$$

$$= \frac{11}{16} = 0.6875$$

The probability that the number selected is in the subinterval $[9, 20]$ is $\dfrac{11}{16}$, or 0.6875.

29. From Example 7, we know

$f(x) = ke^{-kx}$, for $0 \le x < \infty$, where $k = \dfrac{1}{a}$ and a is the average distance between successive cars over some period of time.

First, we determine k:

$$k = \frac{1}{100} = 0.01.$$

The probability that the distance between cars is 40 feet or less is:

$$P(0 \le x \le 40) = \int_0^{40} 0.01e^{-0.01x}\,dx$$

The probability is calculated on the next column.

$$P(0 \le x \le 40) = \int_0^{40} 0.01e^{-0.01x}\,dx$$

$$= \left[\frac{0.01}{-0.01}e^{-0.01x}\right]_0^{40}$$

$$= \left[-e^{-0.01x}\right]_0^{40}$$

$$= -e^{-0.01(40)} - \left(-e^{-0.01(0)}\right)$$

$$= -e^{-0.4} + e^0$$

$$= -e^{-0.4} + 1$$

$$\approx -0.670320 + 1$$

$$\approx 0.329680$$

$$\approx 0.3297$$

The probability that the distance between two successive cars, chosen at random, is 40 feet or less is 0.3297.

31. $f(t) = 2e^{-2t}$, for $0 \le t < \infty$

$$P(0 \le t \le 5) = \int_0^5 2e^{-2t}\,dt$$

$$= \left[\frac{2}{-2}e^{-2t}\right]_0^5$$

$$= \left[-e^{-2t}\right]_0^5$$

$$= -e^{-2(5)} - \left(-e^{-2(0)}\right)$$

$$= -e^{-10} + e^0$$

$$= -e^{-10} + 1$$

$$\approx -0.0000454 + 1$$

$$\approx 0.9999546$$

$$\approx 0.999955$$

The probability that a phone call will last no more than 5 minutes is 0.999955.

33. $f(t) = ke^{-kt}$, for $0 \le t < \infty$, where $k = \dfrac{1}{a}$ and a is the average amount of time that will pass before a failure occurs.

First, we determine k:

$$k = \frac{1}{100} = 0.01.$$

The probability that a failure will occur in 50 hours or less is

$$P(0 \le x \le 50) = \int_0^{50} 0.01e^{-0.01x}\,dx$$

The probability is calculated at the top of the next page.

$$P(0 \le x \le 50) = \int_0^{50} 0.01e^{-0.01x}dx$$

$$= \left[\frac{0.01}{-0.01} e^{-0.01x} \right]_0^{50}$$

$$= \left[-e^{-0.01x} \right]_0^{50}$$

$$= -e^{-0.01(50)} - \left(-e^{-0.01(0)} \right)$$

$$= -e^{-0.5} + e^0$$

$$= -e^{-0.5} + 1$$

$$\approx -0.606531 + 1$$

$$\approx 0.393469$$

$$\approx 0.3935$$

The probability that a failure will occur in 50 hours or less is 0.3935.

35. $f(t) = 0.23e^{-0.23t}$, for $1 \le t < \infty$

a) To verify that 90% of calls are answered within 10 seconds, integrate the probability density function as follows

$$\int_0^{10} 0.23e^{-0.23t}dt = \left[\frac{0.23}{-0.23} e^{-0.23t} \right]_0^{10}$$

$$= \left[-e^{-0.23t} \right]_0^{10}$$

$$= -e^{-0.23(10)} - \left(-e^0 \right)$$

$$= 1 - e^{-2.3}$$

$$= 0.8997$$

$$\approx 0.9.$$

b) The probability that a 911 call is answered between 15 and 25 seconds after the call is made is

$$P(15 \le t \le 25) = \int_{15}^{25} 0.23e^{-0.23t}dt$$

$$= \left[-e^{-0.23t} \right]_{15}^{25}$$

$$= -e^{-0.23(25)} - \left(-e^{-0.23(15)} \right)$$

$$\approx 0.0286.$$

The probability that a 911 call is answered between 15 and 25 seconds is 0.0286.

37. $f(t) = 0.02e^{-0.02t}$, for $0 \le t < \infty$

The probability that a rat will find its way through the maze in 150 seconds or less is

$$P(0 \le t \le 150) = \int_0^{150} 0.02e^{-0.02t}dt$$

$$= \left[\frac{0.02}{-0.02} e^{-0.02t} \right]_0^{150}$$

$$= \left[-e^{-0.02t} \right]_0^{150}$$

$$= -e^{-0.02(150)} - \left(-e^{-0.02(0)} \right)$$

$$= -e^{-3} + e^0$$

$$= -e^{-3} + 1$$

$$\approx 0.950213.$$

The probability that the rat will learn its way through a maze in 150 seconds, or less, is approximately 0.950213.

39. In Exercise 37 we found that $P(0 \le t \le 150) \approx 0.950213$. Since this is a probability density function, we know

$$\int_0^\infty f(x)dx = \int_0^{150} f(x)dx + \int_{150}^\infty f(x)dx = 1$$

Thus,

$$\int_{150}^\infty f(x)dx = 1 - \int_0^{150} f(x)dx$$

$$\approx 1 - 0.950213$$

$$= 0.049787$$

Thus, the probability that a rat requires more than 150 seconds to learn its way through the maze is 0.049787.

41. $$\int_{-a}^a 12x^2 dx = 1$$

$$\left[4x^3 \right]_{-a}^a = 1$$

$$4a^3 - 4(-a)^3 = 1$$

$$4a^3 + 4a^3 = 1$$

$$8a^3 = 1$$

$$a^3 = \frac{1}{8}$$

$$a = \frac{1}{2}$$

$f(x) = 12x^2$ is a probability density function over the interval $\left[-\frac{1}{2}, \frac{1}{2} \right]$.

43. a) Since there is a 30% probability that the patient will wait up to one hour, we have

$$P(0 \le t \le 1) = \int_0^1 f(t)\,dt = 0.30.$$

Let the probability density function be $f(t) = ke^{-kt}$. Substituting, we have

$$\int_0^1 f(t)\,dt = 0.30$$

$$\int_0^1 ke^{-kt}\,dt = 0.30 \qquad \left[\int_0^1 ke^{-kt}\,dt = 1 - e^{-k}\right]$$

$$1 - e^{-k} = 0.30$$

$$e^{-k} = 0.70$$

$$-k = \ln(0.70)$$

$$k = -\ln(0.70)$$

$$k \approx 0.357.$$

The probability density function is

$$f(t) - 0.357e^{-0.357t}.$$

b) The probability that a patient will have to wait between 90 minutes (1.5 hours) and 3 hours for a doctor is

$$P(1.5 \le t \le 3) = \int_{1.5}^3 0.357e^{-0.357t}\,dt$$

$$= \left[-e^{-0.357t}\right]_{1.5}^3$$

$$= -e^{-0.357(3)} - \left(-e^{-0.357(1.5)}\right)$$

$$\approx 0.2427.$$

45. a) The area under the graph must equal 1, since f is a probability density function. Therefore,

$$\text{Area} = 1$$

$$\tfrac{1}{2}b \cdot h = 1$$

$$\tfrac{1}{2}(5-1)\cdot(c) = 1$$

$$2c = 1$$

$$c = \tfrac{1}{2}.$$

b) The linear function passes through the points $(1,0)$ and $\left(5,\tfrac{1}{2}\right)$; therefore, the slope of the line is

$$m = \frac{\tfrac{1}{2}-0}{5-1} = \frac{\tfrac{1}{2}}{4} = \frac{1}{8}.$$

Using the point-slope formula for a line, we have

$$y - 0 = \tfrac{1}{8}(x-1)$$

$$y = \tfrac{1}{8}x - \tfrac{1}{8}.$$

Therefore, the probability density function is

$$f(x) = \tfrac{1}{8}x - \tfrac{1}{8}, \qquad \text{for } 1 \le x \le 5.$$

c) Using the probability density function from part b) we have:

$$P(2 \le x \le 3) = \int_2^3 f(x)\,dx$$

$$= \int_2^3 \left(\frac{1}{8}x - \frac{1}{8}\right)dx$$

$$= \left[\frac{1}{16}x^2 - \frac{1}{8}x\right]_2^3$$

$$= \frac{1}{16}(3)^2 - \frac{1}{8}(3) -$$

$$\left(\frac{1}{16}(2)^2 - \frac{1}{8}(2)\right)$$

$$= \frac{9}{16} - \frac{3}{8} - \left(\frac{4}{16} - \frac{2}{8}\right)$$

$$= \frac{3}{16} \approx 0.1875.$$

47. Determine the value of c so that

$$\int_1^2 cxe^{2x}\,dx = 1.$$

Integrate and solve for c.

$$\int_1^2 cxe^{2x}\,dx = 1$$

$$c\int_1^2 xe^{2x}\,dx = 1$$

$$u = x \qquad dv = e^{2x}\,dx$$
$$du = dx \qquad v = \tfrac{1}{2}e^{2x}$$

$$c\left[\tfrac{1}{2}xe^{2x} - \int \tfrac{1}{2}e^{2x}\,dx\right]_1^2 = 1$$

$$c\left[\tfrac{1}{2}xe^{2x} - \tfrac{1}{4}e^{2x}\right]_1^2 = 1$$

$$c\left[\tfrac{1}{2}(2)e^4 - \tfrac{1}{4}e^4 - \left(\tfrac{1}{2}(1)e^2 - \tfrac{1}{4}e^2\right)\right] = 1$$

$$c\left[\left(1-\tfrac{1}{4}\right)e^4 - \left(\tfrac{1}{2}-\tfrac{1}{4}\right)e^2\right] - 1$$

$$c\left[\tfrac{3}{4}e^4 - \tfrac{1}{4}e^2\right] = 1$$

$$c\left[\frac{3e^4 - e^2}{4}\right] = 1$$

$$c = \frac{4}{3e^4 - e^2}$$

$$c \approx 0.02557$$

The probability density function is

$$f(x) = 0.02557xe^{2x}, \qquad \text{for } 1 \le x \le 2.$$

49 – 60. Left to the student. You may check your answers using the solutions to Exercises 1-12.

Exercise Set 5.5

$$\sigma = \sqrt{\sigma^2}$$

$$= \sqrt{\frac{25}{12}} = \frac{5}{2\sqrt{3}} \qquad \text{Standard deviation}$$

1. $f(x) = \frac{1}{5}, \quad [3,8]$

$E(x) = \int_a^b x \cdot f(x)\,dx \qquad \text{Expected value of } x$

$E(x) = \int_3^8 x \cdot \frac{1}{5}\,dx \qquad \text{Expected value of } x$

$\quad = \frac{1}{5}\left[\frac{x^2}{2}\right]_3^8$

$\quad = \frac{1}{5}\left[\frac{8^2}{2} - \frac{3^2}{2}\right]$

$\quad = \frac{1}{5} \cdot \frac{55}{2}$

$\quad = \frac{11}{2}$

$E(x^2) = \int_a^b x^2 \cdot f(x)\,dx \qquad \text{Expected value of } x^2$

$E(x^2) = \int_3^8 x^2 \cdot \frac{1}{5}\,dx$

$\quad = \frac{1}{5}\left[\frac{x^3}{3}\right]_3^8$

$\quad = \frac{1}{5}\left[\frac{8^3}{3} - \frac{3^3}{3}\right]$

$\quad = \frac{1}{5}\left[\frac{512}{3} - \frac{27}{3}\right]$

$\quad = \frac{1}{5} \cdot \frac{485}{3}$

$\quad = \frac{97}{3}$

$\mu = E(x) = \frac{11}{2} \qquad \text{Mean}$

$\sigma^2 = E(x^2) - \left[E(x)\right]^2$

$\quad = \frac{97}{3} - \left[\frac{11}{2}\right]^2 \qquad \begin{array}{l}\text{Substituting 97/3 for } E(x^2) \\ \text{and 11/2 for } E(x)\end{array}$

$\quad = \frac{97}{3} - \frac{121}{4}$

$\quad = \frac{388}{12} - \frac{363}{12}$

$\quad = \frac{25}{12} \qquad \text{Variance}$

3. $f(x) = \frac{1}{8}x, \quad [0,4]$

$E(x) = \int_a^b x \cdot f(x)\,dx \qquad \text{Expected value of } x$

$E(x) = \int_0^4 x \cdot \frac{1}{8}x\,dx$

$\quad = \int_0^4 \frac{1}{8}x^2\,dx$

$\quad = \frac{1}{8}\left[\frac{x^3}{3}\right]_0^4$

$\quad = \frac{1}{8}\left[\frac{4^3}{3} - \frac{0^3}{3}\right]$

$\quad = \frac{1}{8} \cdot \frac{64}{3}$

$\quad = \frac{8}{3}$

$E(x^2) = \int_a^b x^2 \cdot f(x)\,dx \qquad \text{Expected value of } x^2$

$E(x^2) = \int_0^4 x^2 \cdot \frac{1}{8}x\,dx$

$\quad = \int_0^4 \frac{1}{8}x^3\,dx$

$\quad = \frac{1}{8}\left[\frac{x^4}{4}\right]_0^4$

$\quad = \frac{1}{8}\left[\frac{4^4}{4} - \frac{0^4}{4}\right]$

$\quad = \frac{1}{8} \cdot \frac{256}{4}$

$\quad = 8$

$\mu = E(x) = \frac{8}{3} \qquad \text{Mean}$

The solution is continued on the next page.

Continued from the previous page.

$$\sigma^2 = E\left(x^2\right) - \left[E\left(x\right)\right]^2$$

$$= 8 - \left[\frac{8}{3}\right]^2 \qquad \text{Substituting 8 for } E\left(x^2\right) \text{ and 8/3 for } E(x)$$

$$= 8 - \frac{64}{9}$$

$$= \frac{72}{9} - \frac{64}{9}$$

$$= \frac{8}{9} \qquad \text{Variance}$$

$$\sigma = \sqrt{\sigma^2}$$

$$= \sqrt{\frac{8}{9}} = \frac{2\sqrt{2}}{3} \qquad \text{Standard deviation}$$

5. $f\left(x\right) = \frac{2}{3}x, \quad [1,2]$

$$E\left(x\right) = \int_a^b x \cdot f\left(x\right) dx \qquad \text{Expected value of } x$$

$$E\left(x\right) = \int_1^2 x \cdot \frac{2}{3}x \, dx$$

$$= \int_1^2 \frac{2}{3}x^2 \, dx$$

$$= \frac{2}{3}\left[\frac{x^3}{3}\right]_1^2$$

$$= \frac{2}{3}\left[\frac{2^3}{3} - \frac{1^3}{3}\right]$$

$$= \frac{2}{3}\left[\frac{8}{3} - \frac{1}{3}\right]$$

$$= \frac{2}{3} \cdot \frac{7}{3}$$

$$= \frac{14}{9}$$

$$E\left(x^2\right) = \int_a^b x^2 \cdot f\left(x\right) dx \qquad \text{Expected value of } x^2$$

$$E\left(x^2\right) = \int_1^2 x^2 \cdot \frac{2}{3}x \, dx = \int_1^2 \frac{2}{3}x^3 \, dx$$

$$= \frac{2}{3}\left[\frac{x^4}{4}\right]_1^2$$

$$= \frac{2}{3}\left[\frac{2^4}{4} - \frac{1^4}{4}\right] = \frac{2}{3} \cdot \frac{15}{4} = \frac{5}{2}$$

$$\mu = E\left(x\right) = \frac{14}{9} \qquad \text{Mean}$$

$$\sigma^2 = E\left(x^2\right) - \left[E\left(x\right)\right]^2$$

$$= \frac{5}{2} - \left[\frac{14}{9}\right]^2 \qquad \text{Substituting 5/2 for } E\left(x^2\right) \text{ and 14/9 for } E(x)$$

$$= \frac{5}{2} - \frac{196}{81}$$

$$= \frac{405}{162} - \frac{392}{162}$$

$$= \frac{13}{162} \qquad \text{Variance}$$

$$\sigma = \sqrt{\sigma^2}$$

$$= \sqrt{\frac{13}{162}} = \frac{1}{9}\sqrt{\frac{13}{2}} \qquad \text{Standard deviation}$$

7. $f\left(x\right) = \frac{3}{2}x^2, \quad [-1,1]$

$$E\left(x\right) = \int_a^b x \cdot f\left(x\right) dx \qquad \text{Expected value of } x.$$

$$E\left(x\right) = \int_{-1}^1 x \cdot \frac{3}{2}x^2 \, dx$$

$$= \int_{-1}^1 \frac{3}{2}x^3 \, dx$$

$$= \left[\frac{3x^4}{8}\right]_{-1}^1$$

$$= \left[\frac{3(1)^4}{8} - \frac{3(-1)^4}{8}\right]$$

$$= \frac{3}{8} - \frac{3}{8}$$

$$= 0$$

$$E\left(x^2\right) = \int_a^b x^2 \cdot f\left(x\right) dx \qquad \text{Expected value of } x^2.$$

$$E\left(x^2\right) = \int_{-1}^1 x^2 \cdot \frac{3}{2}x^2 \, dx$$

$$= \int_{-1}^1 \frac{3}{2}x^4 \, dx$$

$$= \left[\frac{3x^5}{10}\right]_{-1}^1$$

$$= \left[\frac{3(1)^5}{10} - \frac{3(-1)^5}{10}\right]$$

$$= \frac{3}{10} + \frac{3}{10}$$

$$= \frac{6}{10} = \frac{3}{5}$$

The solution is continued on the next page.

From the previous page, we have:

$\mu = E(x) = 0$ Mean

$\sigma^2 = E(x^2) - [E(x)]^2$

$\quad = \dfrac{3}{5} - [0]^2$ Substituting 3/5 for $E(x^2)$ and 0 for $E(x)$

$\quad = \dfrac{3}{5}$ Variance

$\sigma = \sqrt{\sigma^2}$

$\quad = \sqrt{\dfrac{3}{5}}$ Standard deviation

9. $f(x) = \dfrac{1}{\ln 4} \cdot \dfrac{1}{x}, \quad [0.8, 3.2]$

$E(x) = \displaystyle\int_a^b x \cdot f(x)\,dx$ Expected value of x

$E(x) = \displaystyle\int_{0.8}^{3.2} x \cdot \dfrac{1}{\ln 4} \cdot \dfrac{1}{x}\,dx$

$E(x) = \displaystyle\int_{0.8}^{3.2} \dfrac{1}{\ln 4}\,dx$

$\quad = \dfrac{1}{\ln 4}[x]_{0.8}^{3.2}$

$\quad = \dfrac{1}{\ln 4}[3.2 - 0.8]$

$\quad = \dfrac{2.4}{\ln 4}$

$E(x^2) = \displaystyle\int_a^b x^2 \cdot f(x)\,dx$ Expected value of x^2

$E(x^2) = \displaystyle\int_{0.8}^{3.2} x^2 \cdot \dfrac{1}{\ln 4} \cdot \dfrac{1}{x}\,dx$

$\quad = \displaystyle\int_{0.8}^{3.2} \dfrac{1}{\ln 4} x\,dx$

$\quad = \dfrac{1}{\ln 4}\left[\dfrac{x^2}{2}\right]_{0.8}^{3.2}$

$\quad = \dfrac{1}{\ln 4}\left[\dfrac{(3.2)^2}{2} - \dfrac{(0.8)^2}{2}\right]$

$\quad = \dfrac{4.8}{\ln 4}$

$\mu = E(x) = \dfrac{2.4}{\ln 4}$ Mean

$\sigma^2 = E(x^2) - [E(x)]^2$

$\quad = \dfrac{4.8}{\ln 4} - \left[\dfrac{2.4}{\ln 4}\right]^2$ Substituting 4.8/ln 4 for $E(x^2)$ and 2.4/ln 4 for $E(x)$

$\quad = \dfrac{4.8}{\ln 4} - \dfrac{5.76}{(\ln 4)^2}$

$\quad = \dfrac{4.8\ln 4 - 5.76}{(\ln 4)^2}$ Variance

$\sigma = \sqrt{\sigma^2}$

$\quad = \sqrt{\dfrac{4.8\ln 4 - 5.76}{(\ln 4)^2}}$

$\quad = \dfrac{\sqrt{4.8\ln 4 - 5.76}}{\ln 4}$ Standard deviation

11. Using Table A, we have:

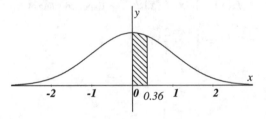

$P(0 \le x \le 0.36) = 0.1406$

13.

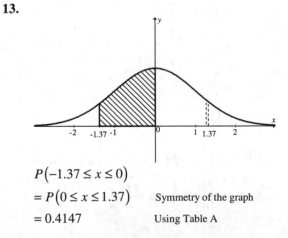

$P(-1.37 \le x \le 0)$

$= P(0 \le x \le 1.37)$ Symmetry of the graph

$= 0.4147$ Using Table A

15. Using Table A, we have:

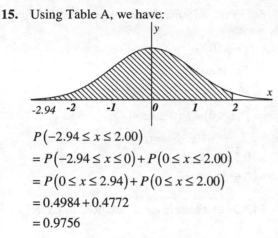

$P(-2.94 \leq x \leq 2.00)$

$= P(-2.94 \leq x \leq 0) + P(0 \leq x \leq 2.00)$

$= P(0 \leq x \leq 2.94) + P(0 \leq x \leq 2.00)$

$= 0.4984 + 0.4772$

$= 0.9756$

17. Using Table A, we have:

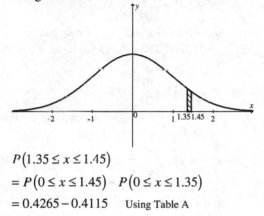

$P(1.35 \leq x \leq 1.45)$

$= P(0 \leq x \leq 1.45) - P(0 \leq x \leq 1.35)$

$= 0.4265 - 0.4115$ Using Table A

$= 0.0150$

19.

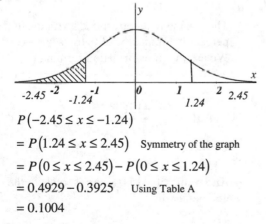

$P(-2.45 \leq x \leq -1.24)$

$= P(1.24 \leq x \leq 2.45)$ Symmetry of the graph

$= P(0 \leq x \leq 2.45) - P(0 \leq x \leq 1.24)$

$= 0.4929 - 0.3925$ Using Table A

$= 0.1004$

21.

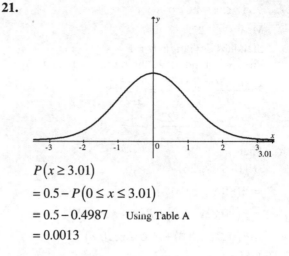

$P(x \geq 3.01)$

$= 0.5 - P(0 \leq x \leq 3.01)$

$= 0.5 - 0.4987$ Using Table A

$= 0.0013$

23. a)

$P(-2 \leq x \leq 2)$

$= P(-2 \leq x \leq 0) + P(0 \leq x \leq 2)$

                    Symmetry of the graph

$= P(0 \leq x \leq 2) + P(0 \leq x \leq 2)$

$= 2\left[P(0 \leq x \leq 2)\right]$

$= 2\left[0.4772\right]$ Using Table 2

$= 0.9544$

b) $0.9544 = 95.44\%$

25. $P(22 \leq x \leq 27)$

Mean $\mu = 22$

Standard deviation $\sigma = 5$

We standardize 22 and 27.

27 is standardized to $\dfrac{b - \mu}{\sigma} = \dfrac{27 - 22}{5} = \dfrac{5}{5} = 1.0$

22 is standardized to $\dfrac{a - \mu}{\sigma} = \dfrac{22 - 22}{5} = \dfrac{0}{5} = 0$

Then,

$P(22 \leq x \leq 27)$

$= P(0 \leq z \leq 1.0)$

$= 0.3413$

27. $P(18 \le x \le 26)$

Mean $\mu = 22$

Standard deviation $\sigma = 5$

First, we standardize the numbers 18 and 26.

$26\colon \dfrac{b-\mu}{\sigma} = \dfrac{26-22}{5} = \dfrac{4}{5} = 0.8$

$18\colon \dfrac{a-\mu}{\sigma} = \dfrac{18-22}{5} = \dfrac{-4}{5} = -0.8$

Then,

$P(18 \le x \le 26)$

$= P(-0.8 \le z \le 0.8)$

$= P(-0.8 \le z \le 0) + P(0 \le z \le 0.8)$

$= P(0 \le z \le 0.8) + P(0 \le z \le 0.8)$

$= 2\left[P(0 \le z \le 0.8)\right]$

$= 2\left[0.2881\right]$

$= 0.5762$

29. $P(20.3 \le x \le 27.5)$

Mean $\mu = 22$

Standard deviation $\sigma = 5$

We standardize 20.3 and 27.5.

27.5 is standardized to

$\dfrac{b-\mu}{\sigma} = \dfrac{27.5-22}{5} = \dfrac{5.5}{5} = 1.1$

20.3 is standardized to

$\dfrac{a-\mu}{\sigma} = \dfrac{20.3-22}{5} = \dfrac{-1.7}{5} = -0.34$

Then,

$P(20.3 \le x \le 27.5)$

$= P(-0.34 \le z \le 1.1)$

$= P(-0.34 \le z \le 0) + P(0 \le z \le 1.1)$

$= P(0 \le x \le 0.34) + P(0 \le z \le 1.1)$

                    Symmetry of graph

$= 0.1331 + 0.3643$

$= 0.4974$

31. $P(x \ge 20) =$

Mean $\mu = 22$

Standard deviation $\sigma = 5$

We standardize 20.

20 is standardized to

$\dfrac{b-\mu}{\sigma} = \dfrac{20-22}{5} = -\dfrac{2}{5} = -0.4$

The probability is calculated at the top of the next column.

Using the information on the previous column, we have:

$P(x \ge 20)$

$= P(z \ge -0.4)$

$= P(-0.4 \le x \le 0) + P(z \ge 0)$

$= P(z \le 0) + P(0 \le x \le 0.4)$

                    Symmetry of graph

$= 0.5 + 0.1554$

$= 0.6554$

33 – 54. Check answers using solutions to Exercises 11-32.

55. a) $P(x \le z) = \dfrac{30}{100} = 0.3$

$0.3 = 0.5 - 0.2$ and 0.2 corresponds to $z \approx 0.52$ in Table A, so we have $z \approx -0.52$.

b) $P(x \le z) = \dfrac{50}{100} = 0.5$

$0.5 = 0.5 + 0$ and 0 corresponds to $z = 0$ in Table A, so we have $z = 0$.

c) $P(x \le z) = \dfrac{95}{100} = 0.95$

$0.95 = 0.5 + 0.45$ and from Table A we see that 0.45 corresponds to a value of z halfway between 1.64 and 1.65, or 1.645. Thus, $z = 1.645$.

57. $\mu = -15; \sigma = 0.4$

a) The z value that corresponds with the 46th percentile, or an area of 0.46, is $z = -0.10$ We use the transformation formula to determine the x value that corresponds to the value $z = -0.10$.

$-0.10 = \dfrac{x - (-15)}{0.4}$

$-0.04 = x + 15$

$-15.04 = x$

An x value of -15.04 corresponds to the 46th percentile.

b) The z value that corresponds with the 92nd percentile, or an area of 0.92, is $z = 1.405$. We use the transformation formula to determine the x value that corresponds to the value $z = 1.405$. The solution is continued on the next page.

Using the transformation formula, we have:

$$1.405 = \frac{x - (-15)}{0.4}$$

$$0.562 = x + 15$$

$$-14.438 = x$$

$$-14.44 \approx x$$

An x value of -14.44 corresponds to the 92nd percentile.

59. We first standardize 300 orders. The mean is 250 and the standard deviation is 20, so 300 is standardized to

$$\frac{b - \mu}{\sigma} = \frac{300 - 250}{20} = \frac{50}{20} = \frac{5}{2} = 2.5$$

Therefore,

$$P(x \geq 300)$$

$$= P(z \geq 2.5)$$

$$= 0.5 - P(z \leq 2.5)$$

$$= 0.5 - 0.4938 \qquad \text{Using Table A}$$

$$= 0.0062$$

$$= 0.62\%$$

The company will have to hire extra help or pay overtime 0.62% of the days.

61. We first standardize 40 seconds. The mean is 38.6 and the standard deviation is 1.729, so 40 is standardized to

$$\frac{b - \mu}{\sigma} = \frac{40 - 38.6}{1.729} \approx 0.81$$

Therefore,

$$P(x \leq 40)$$

$$= P(z \leq 0.81)$$

$$= 0.5 + P(0 \leq z \leq 0.81)$$

$$= 0.5 + 0.2910 \qquad \text{Using Table A}$$

$$= 0.7910$$

The probability that the next operation of the robogate will take 40 seconds, or less, is 0.7910.

63. a) Find z such that $P(x \leq z) = 0.35$

$0.35 = 0.5 - 0.15$ and 0.15 corresponds to $z \approx 0.385$ in Table 2. Then the score that corresponds to the 35th percentile is 0.385 standard deviations less than the mean, or $1498 - 0.385(348) \approx 1364$. Therefore, the score that would correspond to the 35th percentile is 1364.

b) Find z such that $P(x \leq z) = 0.60$

$0.60 = 0.5 + 0.1$ and 0.1 corresponds to $z \approx 0.255$ in Table 2. Then the score that corresponds to the 60th percentile is 0.255 standard deviations more than the mean, or $1498 + 0.255(348) \approx 1586$. Therefore, the score that would correspond to the 60th percentile is 1586.

c) Find z such that $P(x \leq z) = 0.92$

$0.92 = 0.5 + 0.42$ and 0.42 corresponds to $z \approx 1.405$ in Table 2. Then the score that corresponds to the 92nd percentile is 1.405 standard deviations more than the mean, or $1498 + 1.405(348) \approx 1987$. Therefore, the score that would correspond to the 92nd percentile is 1987.

65. $\mu = 76; \sigma = 7$

a) The top 12% corresponds with the 88th percentile. The z value that corresponds with the 88th percentile, or an area of 0.88, is $z = 1.175$.

We use the transformation formula to determine the x value that corresponds to the value $z = 1.175$.

$$1.175 = \frac{x - 76}{7}$$

$$8.225 = x - 76$$

$$84.225 = x$$

$$84 \approx x$$

The minimum score needed to get an A is 84.

b) The top 75% corresponds with the 25th percentile. The z value that corresponds with an area of 0.25, is $z = -0.675$

We use the transformation formula to determine the x value that corresponds to the value $z = -0.675$.

$$-0.675 = \frac{x - 76}{7}$$

$$-4.725 = x - 76$$

$$71.275 = x$$

$$71 \approx x$$

The minimum score needed to pass is 71.

67. If a players standing 7 feet 2 or 86 inches tall is in the top 1% or 99th percentile. The z value that corresponds to the 99th percentile is 2.33. We can find the standard deviation for players in the NBA by:

$$2.33 = \frac{86-79}{\sigma}$$

$$2.33\sigma = 7$$

$$\sigma = 3.00$$

To determine the percentile of a player standing 83 inches tall (6 feet 11 inches), we determine the z value first.

$$z = \frac{83-79}{3}$$

$$z = 1.33$$

This value represents an area of 0.908. Therefore, a player that stands 6 feet 11 inches tall would be in the 90.8^{th} percentile.

69. $f(x) = \dfrac{1}{b-a}, \quad [a,b]$

$$E(x) = \int_a^b x \cdot f(x)\,dx \qquad \text{Expected value of } x$$

$$E(x) = \int_a^b x \cdot \frac{1}{b-a}\,dx$$

$$= \frac{1}{b-a}\left[\frac{x^2}{2}\right]_a^b$$

$$= \frac{1}{b-a}\left[\frac{b^2}{2} - \frac{a^2}{2}\right]$$

$$= \frac{1}{b-a}\left[\frac{b^2 - a^2}{2}\right]$$

$$= \frac{1}{b-a} \cdot \frac{(b-a)(b+a)}{2}$$

$$= \frac{b+a}{2}$$

$$E(x^2) = \int_a^b x^2 \cdot f(x)\,dx \qquad \text{Expected value of } x^2$$

$$E(x^2) = \int_a^b x^2 \cdot \frac{1}{b-a}\,dx$$

$$= \frac{1}{b-a}\left[\frac{x^3}{3}\right]_a^b$$

$$= \frac{1}{b-a}\left[\frac{b^3}{3} - \frac{a^3}{3}\right]$$

$$= \frac{1}{b-a}\left[\frac{b^3 - a^3}{3}\right]$$

$$= \frac{1}{b-a} \cdot \frac{(b-a)(b^2 + ab + a^2)}{3}$$

$$= \frac{b^2 + ab + a^2}{3}$$

$$\mu = E(x) = \frac{b+a}{2} \qquad \text{Mean}$$

$$\sigma^2 = E(x^2) - \left[E(x)\right]^2$$

$$= \frac{b^2 + ab + a^2}{3} - \left[\frac{b+a}{2}\right]^2 \quad \text{Substituting}$$

$$= \frac{b^2 + ab + a^2}{3} - \left[\frac{b^2 + 2ba + a^2}{4}\right]$$

$$= \frac{4b^2 + 4ab + 4a^2}{12} - \frac{3b^2 + 6ba + 3a^2}{12}$$

$$= \frac{4b^2 + 4ab + 4a^2 - 3b^2 - 6ba - 3a^2}{12}$$

$$= \frac{b^2 - 2ba + a^2}{12}$$

$$= \frac{(b-a)^2}{12} \qquad \text{Variance}$$

$$\sigma = \sqrt{\sigma^2}$$

$$= \sqrt{\frac{(b-a)^2}{12}}$$

$$= \frac{b-a}{2\sqrt{3}} \qquad \text{Standard deviation}$$

71. $f(x) = \frac{1}{2}x$, $[0,2]$

$$\int_0^m f(x)\,dx = \frac{1}{2}$$

$$\int_0^m \frac{1}{2}x\,dx = \frac{1}{2}$$

$$\frac{1}{2}\int_0^m x\,dx = \frac{1}{2}$$

$$\int_0^m x\,dx = 1$$

$$\left[\frac{x^2}{2}\right]_0^m = 1$$

$$\frac{m^2}{2} - \frac{0^2}{2} = 1$$

$$\frac{m^2}{2} = 1$$

$$m^2 = 2$$

$$m = \sqrt{2}$$

73. $f(x) = ke^{-kx}$, $[0,\infty)$

$$\int_0^m f(x)\,dx = \frac{1}{2}$$

$$\int_0^m ke^{-kx}\,dx = \frac{1}{2}$$

$$\left[\frac{k}{-k}e^{-kx}\right]_0^m = \frac{1}{2}$$

$$\left[-e^{-kx}\right]_0^m = \frac{1}{2}$$

$$-e^{-k(m)} - \left(-e^{-k(0)}\right) = \frac{1}{2}$$

$$-e^{-k\cdot m} + e^0 = \frac{1}{2}$$

$$-e^{-k\cdot m} + 1 = \frac{1}{2}$$

$$\frac{1}{2} = e^{-k\cdot m}$$

$$\ln\left(\frac{1}{2}\right) = \ln\left(e^{-k\cdot m}\right)$$

$$\ln(1) - \ln(2) = -k\cdot m \qquad [\ln 1 = 0]$$

$$\frac{-\ln 2}{-k} = m$$

$$\frac{\ln 2}{k} = m$$

75. Standardize 8.5.

$$z = \frac{8.5 - \mu}{0.3}$$

We are looking for a value of c for which

$P(z > c) = \frac{1}{100} = 0.01$. Now, since we are

guarding against overflow we will have $c \geq 0$.
We have:

$$P(z > c) = P(z \geq 0) - P(0 \leq z \leq c)$$
$$= 0.5 - P(0 \leq z \leq c)$$

Then,

$$P(0 \leq z \leq c) = 0.5 - P(z > c)$$
$$= 0.5 - 0.01 = 0.4900$$

Looking in Table A for the number closest to 0.49, we find that 0.4901 and 0.4898 are the two closest numbers. Since we want to ensure that only 1 cup in 100 will overflow we select the larger value 0.4901. This value corresponds to $c = 2.33$. We set $z = 2.33$ and solve for μ.

$$2.33 = \frac{8.5 - \mu}{0.3}$$

$$(2.33)(0.3) = 8.5 - \mu$$

$$0.699 = 8.5 - \mu$$

$$\mu + 0.699 = 8.5$$

$$\mu = 8.5 - 0.699 = 7.801$$

The mean volume should be adjusted to 7.801 ounces in order to ensure that only 1 cup in 100 will overflow.

77.

Exercise Set 5.6

1. Find the volume of the solid of revolution generated by rotating about the x-axis the region under the graph of
    $$y = x$$
    from $x = 0$ to $x = 1$.

    $V = \int_a^b \pi \left[f(x) \right]^2 dx$ Volume of a solid of revolution

    $V = \int_0^1 \pi \left[x \right]^2 dx$ Substituting 0 for a, 1 for b, and x for $f(x)$

    $V = \int_0^1 \pi x^2 dx$

    $= \left[\pi \cdot \dfrac{x^3}{3} \right]_0^1$

    $= \dfrac{\pi}{3} \left[1^3 - 0^3 \right]$

    $= \dfrac{\pi}{3} [1]$

    $= \dfrac{\pi}{3}$, or about 1.05.

3. Find the volume of the solid of revolution generated by rotating about the x-axis the region under the graph of
    $$y = 2x$$
    from $x = 1$ to $x = 3$.

    $V = \int_a^b \pi \left[f(x) \right]^2 dx$ Volume of a solid of revolution

    $V = \int_1^3 \pi \left[2x \right]^2 dx$ Substituting 1 for a, 3 for b, and $2x$ for $f(x)$

    $V = \int_1^3 \pi \cdot 4x^2 dx$

    $= \left[4\pi \cdot \dfrac{x^3}{3} \right]_1^3$

    $= \dfrac{4\pi}{3} \left[3^3 - 1^3 \right]$

    $= \dfrac{4\pi}{3} \cdot [26]$

    $= \dfrac{104\pi}{3}$, or about 108.91.

5. Find the volume of the solid of revolution generated by rotating about the x-axis the region under the graph of
    $$y = e^x$$
    from $x = -2$ to $x = 5$.

    $V = \int_a^b \pi \left[f(x) \right]^2 dx$ Volume of a solid of revolution

    $V = \int_{-2}^5 \pi \left[e^x \right]^2 dx$ Substituting -2 for a, 5 for b, and e^x for $f(x)$

    $V = \int_{-2}^5 \pi e^{2x} dx$

    $= \left[\pi \cdot \dfrac{1}{2} e^{2x} \right]_{-2}^5$

    $= \dfrac{\pi}{2} \left[e^{2(5)} - e^{2(-2)} \right]$

    $= \dfrac{\pi}{2} \left[e^{10} - e^{-4} \right]$, or about 34,599.06.

7. Find the volume of the solid of revolution generated by rotating about the x-axis the region under the graph of
    $$y = \dfrac{1}{x}$$
    from $x = 1$ to $x = 4$.

    $V = \int_a^b \pi \left[f(x) \right]^2 dx$ Volume of a solid of revolution

    $V = \int_1^4 \pi \left[\dfrac{1}{x} \right]^2 dx$ Substituting 1 for a, 4 for b, and $1/x$ for $f(x)$

    $V = \int_1^4 \pi \cdot \dfrac{1}{x^2} dx$

    $= \int_1^4 \pi \cdot x^{-2} dx$

    $= \left[\pi \cdot \dfrac{x^{-1}}{-1} \right]_1^4$

    $= -\pi \left[\dfrac{1}{x} \right]_1^4$

    $= -\pi \left[\dfrac{1}{4} - \dfrac{1}{1} \right]$

    $= -\pi \left[-\dfrac{3}{4} \right]$

    $= \dfrac{3\pi}{4}$, or about 2.36.

9. Find the volume of the solid of revolution generated by rotating about the x-axis the region under the graph of

$$y = \frac{2}{\sqrt{x}}$$

from $x = 4$ to $x = 9$.

$$V = \int_a^b \pi \left[f(x) \right]^2 dx \qquad \text{Volume of a solid of revolution}$$

$$V = \int_4^9 \pi \left[\frac{2}{\sqrt{x}} \right]^2 dx \qquad \begin{array}{l}\text{Substituting 4 for } a, 9 \text{ for } b,\\ \text{and } 2/\sqrt{x} \text{ for } f(x)\end{array}$$

$$V = \int_4^9 \pi \cdot \frac{4}{x} dx$$

$$V = 4\pi \int_4^9 \frac{1}{x} dx$$

$$= 4\pi \left[\ln x \right]_4^9$$

$$= 4\pi \left[\ln 9 - \ln 4 \right]$$

$$= 4\pi \ln \left(\frac{9}{4} \right), \text{ or about } 10.19.$$

11. Find the volume of the solid of revolution generated by rotating about the x-axis the region under the graph of

$$y = 5$$

from $x = 1$ to $x = 3$.

$$V = \int_a^b \pi \left[f(x) \right]^2 dx \qquad \text{Volume of a solid of revolution}$$

$$V = \int_1^3 \pi \left[5 \right]^2 dx \qquad \begin{array}{l}\text{Substituting 1 for } a, 3 \text{ for } b,\\ \text{and 5 for } f(x)\end{array}$$

$$V = \int_1^3 25\pi\, dx$$

$$= 25\pi \left[x \right]_1^3$$

$$= 25\pi \left[3 - 1 \right]$$

$$= 25\pi \left[2 \right]$$

$$= 50\pi, \text{ or about } 157.08.$$

13. Find the volume of the solid of revolution generated by rotating about the x-axis the region under the graph of

$$y = x^2$$

from $x = 0$ to $x = 2$.

$$V = \int_a^b \pi \left[f(x) \right]^2 dx \qquad \text{Volume of a solid of revolution}$$

$$V = \int_0^2 \pi \left[x^2 \right]^2 dx \qquad \begin{array}{l}\text{Substituting 0 for } a, 2 \text{ for } b,\\ \text{and } x^2 \text{ for } f(x)\end{array}$$

$$V = \int_0^2 \pi x^4 dx$$

$$= \left[\pi \cdot \frac{x^5}{5} \right]_0^2$$

$$= \frac{\pi}{5} \left[2^5 - 0^5 \right]$$

$$= \frac{\pi}{5} \left[32 \right]$$

$$= \frac{32\pi}{5}, \text{ or about } 20.11.$$

15. Find the volume of the solid of revolution generated by rotating about the x-axis the region under the graph of

$$y = 2\sqrt{x}$$

from $x = 2$ to $x = 10$.

$$V = \int_a^b \pi \left[f(x) \right]^2 dx \qquad \text{Volume of a solid of revolution}$$

$$V = \int_1^2 \pi \left[2\sqrt{x} \right]^2 dx \qquad \begin{array}{l}\text{Substituting 1 for } a, 2 \text{ for } b,\\ \text{and } 2\sqrt{x} \text{ for } f(x)\end{array}$$

$$V = \int_1^2 4\pi x\, dx$$

$$= \left[4\pi \cdot \frac{x^2}{2} \right]_1^2$$

$$= 2\pi \left[2^2 - 1^2 \right]$$

$$= 2\pi \left[3 \right]$$

$$= 6\pi, \text{ or about } 18.85.$$

17. Find the volume of the solid of revolution generated by rotating about the x-axis the region under the graph of
$$y = \sqrt{4 - x^2}$$
from $x = -2$ to $x = 2$.

$V = \int_a^b \pi \left[f(x) \right]^2 dx$ Volume of a solid of revolution

$V = \int_{-2}^2 \pi \left[\sqrt{4 - x^2} \right]^2 dx$ Substituting -2 for a, 2 for b, and $\sqrt{4-x^2}$ for $f(x)$

$V = \int_{-2}^2 \pi \left(4 - x^2 \right) dx$

$= \pi \left[4x - \dfrac{x^3}{3} \right]_{-2}^2$

$= \pi \left[\left(4(2) - \dfrac{(2)^3}{3} \right) - \left(4(-2) - \dfrac{(-2)^3}{3} \right) \right]$

$= \pi \left[\left(8 - \dfrac{8}{3} \right) - \left(-8 + \dfrac{8}{3} \right) \right]$

$= \pi \left[\dfrac{16}{3} - \left(-\dfrac{16}{3} \right) \right]$

$= \pi \left(\dfrac{32}{3} \right)$

$= \dfrac{32\pi}{3}$, or about 33.51.

19. Find the volume of the solid of revolution generated by rotating $y = 2x$ about the y-axis bounded by $x = 0$ to $x = 3$, we use volume by shells.

$V = 2\pi \int_a^b x \cdot f(x) dx$ Volume of a solid of revolution by shells

$V = 2\pi \int_0^3 x \left[2x \right] dx$ Substituting

$= 2\pi \int_0^3 \left[2x^2 \right] dx$

$= 2\pi \left[\dfrac{2}{3} x^3 \right]_0^3$

$= \dfrac{4}{3} \pi \left[(3)^3 - (0)^3 \right]$

$= \dfrac{4}{3} \pi \left[27 \right]$

$= 36\pi$, or about 113.1.

21. Find the volume of the solid of revolution generated by rotating $y = x^2$ about the y-axis bounded by $x = 0$ to $x = 8$, we use volume by shells.

$V = 2\pi \int_a^b x \cdot f(x) dx$ Volume of a solid of revolution by shells

$V = 2\pi \int_0^8 x \left[x^2 \right] dx$ Substituting

$= 2\pi \int_0^8 \left[x^3 \right] dx$

$= 2\pi \left[\dfrac{1}{4} x^4 \right]_0^8$

$= \dfrac{1}{2} \pi \left[(8)^4 - (0)^4 \right]$

$= \dfrac{1}{2} \pi \left[4096 \right]$

$= 2048\pi$, or about 6434.

23. Find the volume of the solid of revolution generated by rotating $y = \dfrac{1}{x}$ about the y-axis bounded by $x = 1$ to $x = 5$, we use volume by shells.

$V = 2\pi \int_a^b x \cdot f(x) dx$ Volume of a solid of revolution by shells

$V = 2\pi \int_1^5 x \left[\dfrac{1}{x} \right] dx$ Substituting

$= 2\pi \int_1^5 \left[1 \right] dx$

$= 2\pi \left[x \right]_1^5$

$= 2\pi \left[5 - 1 \right]$

$= 8\pi$, or about 25.133.

25. Find the volume of the solid of revolution generated by rotating $y = x^2 + 3$ about the y-axis bounded by $x = 1$ to $x = 2$, we use volume by shells.

$V = 2\pi \int_a^b x \cdot f(x) dx$ Volume of a solid of revolution by shells

$V = 2\pi \int_1^2 x \left[x^2 + 3 \right] dx$ Substituting

$= 2\pi \int_1^2 \left[x^3 + 3x \right] dx$

$= 2\pi \left[\frac{1}{4} x^4 + \frac{3}{2} x^2 \right]_1^2$

$= 2\pi \left[\frac{1}{4}(2)^4 + \frac{3}{2}(2)^2 - \left(\frac{1}{4}(1)^4 + \frac{3}{2}(1)^2 \right) \right]$

$= 2\pi \left[4 + 6 - \left(\frac{1}{4} + \frac{3}{2} \right) \right]$

$= 2\pi \left[\frac{33}{4} \right]$

$= \frac{33}{2} \pi$, or about 51.836.

27. Find the volume of the solid of revolution generated by rotating $y = \sqrt{x}$ about the y-axis bounded by $x = 4$ to $x = 9$, we use volume by shells.

$V = 2\pi \int_a^b x \cdot f(x) dx$ Volume of a solid of revolution by shells

$V = 2\pi \int_4^9 x \left[\sqrt{x} \right] dx$ Substituting

$= 2\pi \int_4^9 \left[x^{3/2} \right] dx$

$= 2\pi \left[\frac{2}{5} x^{5/2} \right]_4^9$

$= 2\pi \left[\frac{2}{5}(9)^{5/2} - \frac{2}{5}(4)^{5/2} \right]$

$= 2\pi \left[\frac{2}{5}(243) - \frac{2}{5}(32) \right]$

$= 2\pi \left[\frac{422}{5} \right]$

$= \frac{844}{5} \pi$, or about 530.301.

29. Find the volume of the solid of revolution generated by rotating $y = \sqrt{x^2 + 1}$ about the y-axis bounded by $x = 0$ to $x = 3$, we use volume by shells.

$V = 2\pi \int_a^b x \cdot f(x) dx$ Volume of a solid of revolution by shells.

$V = 2\pi \int_0^3 x \left[\sqrt{x^2 + 1} \right] dx$ Substituting

Let $u = x^2 + 1$, then $du = 2x dx$.

$= \pi \int_0^3 \left[u^{1/2} \right] du$

$= \pi \left[\frac{2}{3} u^{3/2} \right]_{x=0}^{x=3}$

$= \pi \left[\frac{2}{3} (x^2 + 1)^{3/2} \right]_{x=0}^{x=3}$ Reverse Substituting

$= \frac{2}{3} \pi \left[(x^2 + 1)^{3/2} \right]_0^3$

$= \frac{2}{3} \pi \left[\left((3)^2 + 1 \right)^{3/2} - \left((0)^2 + 1 \right)^{3/2} \right]$

$= \frac{2}{3} \pi \left[(10)^{3/2} - (1)^{3/2} \right]$

$= \frac{2}{3} \pi \left(10^{3/2} - 1 \right)$, or about 64.136.

31. a) $y = 9 - x^2$; $[0,3]$

To find the volume of the solid of revolution generated by rotating R around the x-axis, we will use the disks method to calculate the volume:

$V = \int_a^b \pi \left[f(x) \right]^2 dx$ Volume of a solid of revolution

$V = \int_0^3 \pi \left[9 - x^2 \right]^2 dx$ Substituting

$V = \pi \int_0^3 81 - 18x^2 + x^4 dx$

$= \pi \left[81x - 6x^3 + \frac{1}{5} x^5 \right]_0^3$

$= \pi \left[81(3) - 6(3)^3 + \frac{1}{5}(3)^5 - \right.$

$\left. 81(0) - 6(0)^3 + \frac{1}{5}(0)^5 \right]$

$= \pi \left[\frac{648}{5} \right]$

$= \frac{648\pi}{5}$, or about 407.15.

b)   To find the volume of the solid of revolution generated by rotating R around the y-axis, we will use the shell method to calculate the volume:

$V = 2\pi \int_a^b x \cdot f(x)\, dx$ $\quad$ Volume of a solid of revolution by shells.

$V = 2\pi \int_0^3 x\left[9 - x^2\right] dx$ $\quad$ Substituting

$= 2\pi \int_0^3 \left[9x - x^3\right] dx$

$= 2\pi \left[\dfrac{9}{2}x^2 - \dfrac{1}{4}x^4\right]_0^3$

$= 2\pi \left[\dfrac{9}{2}(3)^2 - \dfrac{1}{4}(3)^4 - \left(\dfrac{9}{2}(0)^2 - \dfrac{1}{4}(0)^4\right)\right]$

$= 2\pi \left[\dfrac{81}{2} - \dfrac{81}{4} - (0)\right]$

$= 2\pi \left[\dfrac{81}{4}\right]$

$= \dfrac{81}{2}\pi$, or about 127.234.

c)   ✎

33. Find the volume of the solid of revolution generated by rotating about the x-axis the region under the graph of

$y = 50 \cdot \sqrt{1 + \dfrac{x^2}{22{,}500}}$ from $x = -250$ to

$x = 150$. The volume is:

$V = \int_{-250}^{150} \pi \left[50 \cdot \sqrt{1 + \dfrac{x^2}{22{,}500}}\right]^2 dx$

$= \int_{-250}^{150} \pi \left[2500\left(1 + \dfrac{x^2}{22{,}500}\right)\right] dx$

$V = \int_{-250}^{150} \pi \left[2500 + \dfrac{x^2}{9}\right] dx$

$= \pi \left[2500x + \dfrac{x^3}{27}\right]_{-250}^{150}$

$= \pi \left[\left(2500(150) + \dfrac{(150)^3}{27}\right) - \left(2500(-250) + \dfrac{(-250)^3}{27}\right)\right]$

$\approx 1{,}703{,}703.7\pi$.

The volume of the tower is approximately $1{,}703{,}703.7\pi$ ft^3.

35. To find the volume of the hogan, we will rotate $y = -0.02x^2 + 12$, for $0 \le x \le 15$ around the y-axis, we will use volume by shells to calculate:

$V = 2\pi \int_a^b x \cdot f(x)\, dx$ $\quad$ Volume of a solid of revolution by shells

$V = 2\pi \int_0^{15} x\left[-0.02x^2 + 12\right] dx$ $\quad$ Substituting

$= 2\pi \int_0^{15} \left[-0.02x^3 + 12x\right] dx$

$= 2\pi \left[-0.005x^4 + 6x^2\right]_0^{15}$

$= 2\pi \left[-0.005(15)^4 + 6(15)^2\right.$
$\qquad\qquad \left. -\left(-0.005(0)^4 + 6(0)^2\right)\right]$

$= 2\pi \left[-\dfrac{50625}{200} + 1350\right]$

$= 2\pi \left[\dfrac{8775}{8}\right]$

$= \dfrac{8775}{4}\pi$, or about 6891.9.

The volume of the hogan, is about 6891.9 ft^3.

37. To derive a sphere, we can rotate the graphs of a semicircles with radius r about the x-axis. In Exercise 18, by finding the volume of the solid of revolution created by rotating $y = \sqrt{r^2 - x^2}$, we actually derived the general formula for finding the volume of a sphere with radius r using volume by disks:

$V = \int_{-r}^{r} \pi \left[\sqrt{r^2 - x^2}\right]^2 dx$

$\qquad\qquad$ Substituting $-r$ for a, r for b,
$\qquad\qquad$ and $\sqrt{r^2 - x^2}$ for $f(x)$

$V = \int_{-r}^{r} \pi \left(r^2 - x^2\right) dx$

$= \pi \left[r^2 x - \dfrac{x^3}{3}\right]_{-r}^{r}$

$= \pi \left[\left(r^2(r) - \dfrac{(r)^3}{3}\right) - \left(r^2(-r) - \dfrac{(-r)^3}{3}\right)\right]$

$= \pi \left[\left(r^3 - \dfrac{r^3}{3}\right) - \left(-r^3 + \dfrac{r^3}{3}\right)\right]$

$= \pi \left(\dfrac{4}{3}r^3\right) = \dfrac{4}{3}\pi r^3.$

39. $V = \int_a^b \pi \left[f(x) \right]^2 dx$

$V = \int_e^{e^3} \pi \left[\sqrt{\ln x} \right]^2 dx$

$= \int_e^{e^3} \pi \cdot \ln x \, dx$

$= \pi \left[x \ln x - x \right]_e^{e^3}$ Using Formula 8

$= \pi \left[\left(e^3 \ln e^3 - e^3 \right) - \left(e \ln e - e \right) \right]$

$= \pi \left[\left(3e^3 - e^3 \right) - \left(e - e \right) \right]$

$= \pi \left[2e^3 \right]$

$= 2\pi e^3$, or about 126.20.

41. $y = -\dfrac{1}{3} x^3 + 3x$ in the first quadrant. First we find the boundaries of the solid. We do this by finding the positive intercepts on x-axis by setting the equation equal to zero and solving for x.

$y = 0$

$-\dfrac{1}{3} x^3 + 3x = 0$

$x^3 - 9x = 0$

$x \left(x^2 - 9 \right) = 0$

$x(x - 3)(x + 3) = 0$

$x = 0$ or $x = 3$ or $x = -3$

The boundary for this solid will be $0 \le x \le 3$. Rotating around the y-axis, we will use volume by shells to calculate the volume.

$V = 2\pi \int_a^b x \cdot f(x) \, dx$ Volume of a solid of revolution

$V = 2\pi \int_0^3 x \left[\dfrac{-1}{3} x^3 + 3x \right] dx$ Substituting

$= 2\pi \int_0^3 \left[\dfrac{-1}{3} x^4 + 3x^2 \right] dx$

$= 2\pi \left[\dfrac{-1}{15} x^5 + x^3 \right]_0^3$

$= 2\pi \left[\dfrac{-1}{15} (3)^5 + (3)^3 - \left(\dfrac{-1}{15} (0)^5 + (0)^3 \right) \right]$

$= 2\pi \left[\dfrac{-243}{15} + 27 \right]$

$= 2\pi \left[\dfrac{54}{5} \right] = \dfrac{108}{5} \pi$, or about 67.858.

43. a) $y = x + 1;$ $[0,1]$

About the x-axis, using volume by disks:

$V = \int_a^b \pi \left[f(x) \right]^2 dx$ Volume of a solid of revolution

$V = \int_0^1 \pi \left[x + 1 \right]^2 dx$ Substituting

$V = \pi \int_0^1 x^2 + 2x + 1 \, dx$

$= \pi \left[\dfrac{1}{3} x^3 + x^2 + x \right]_0^1$

$= \pi \left[\dfrac{1}{3} (1)^3 + (1)^2 + 1 - \dfrac{1}{3} (0)^3 + (0)^2 + (0) \right]$

$= \pi \left[\dfrac{7}{3} \right]$

$= \dfrac{7}{3} \pi$, or about 7.33.

About the y-axis, using volume by shells:

$V = 2\pi \int_a^b x \cdot f(x) \, dx$ Volume of a solid of revolution

$V = 2\pi \int_0^1 x \left[x + 1 \right] dx$ Substituting

$= 2\pi \int_0^1 \left[x^2 + x \right] dx$

$= 2\pi \left[\dfrac{1}{3} x^3 + \dfrac{1}{2} x^2 \right]_0^1$

$= 2\pi \left[\dfrac{1}{3} (1)^3 + \dfrac{1}{2} (1)^2 - \left(\dfrac{1}{3} (0)^3 + \dfrac{1}{2} (0)^2 \right) \right]$

$= 2\pi \left[\dfrac{5}{6} \right]$

$= \dfrac{5}{3} \pi$, or about 5.24.

Rotating about the x-axis yields a larger volume.

b) $y = x + 1;$ $[0,2]$

About the x-axis, using volume by disks:

$V = \int_a^b \pi \left[f(x) \right]^2 dx$ Volume of a solid of revolution

$V = \int_0^2 \pi \left[x + 1 \right]^2 dx$ Substituting

We evaluate the integral at the top of the next page.

From the previous page, we have:

$$V = \pi \int_0^2 \left(x^2 + 2x + 1 \right) dx$$

$$= \pi \left[\frac{1}{3}x^3 + x^2 + x \right]_0^2$$

$$= \pi \left[\frac{1}{3}(2)^3 + (2)^2 + 2 - \frac{1}{3}(0)^3 + (0)^2 + (0) \right]$$

$$= \frac{26}{3}\pi, \text{ or about } 27.23.$$

About the y-axis, using volume by shells:

$$V = 2\pi \int_a^b x \cdot f(x)\, dx \qquad \text{Volume of a solid of revolution}$$

$$V = 2\pi \int_0^2 x[x+1]\, dx \qquad \text{Substituting}$$

$$= 2\pi \int_0^2 \left[x^2 + x \right] dx$$

$$= 2\pi \left[\frac{1}{3}x^3 + \frac{1}{2}x^2 \right]_0^2$$

$$= 2\pi \left[\frac{1}{3}(2)^3 + \frac{1}{2}(2)^2 - \left(\frac{1}{3}(0)^3 + \frac{1}{2}(0)^2 \right) \right]$$

$$= 2\pi \left[\frac{14}{3} \right] = \frac{28}{3}\pi, \text{ or about } 29.32.$$

Rotating about the y-axis yields a larger volume.

c) $y = x+1; \quad [0, a]$

About the y-axis, using volume by disks:

$$V = \int_a^b \pi \left[f(x) \right]^2 dx \qquad \text{Volume of a solid of revolution}$$

$$V = \int_0^a \pi [x+1]^2 dx \qquad \text{Substituting}$$

$$V = \pi \int_0^a \left(x^2 + 2x + 1 \right) dx$$

$$= \pi \left[\frac{1}{3}x^3 + x^2 + x \right]_0^a$$

$$= \pi \left[\frac{1}{3}(a)^3 + (a)^2 + a - (0+0+0) \right]$$

$$= \pi \left[\frac{a^3}{3} + a^2 + a \right]$$

About the y-axis, using volume by shells:

$$V = 2\pi \int_a^b x \cdot f(x)\, dx \qquad \text{Volume of a solid of revolution}$$

$$V = 2\pi \int_0^a x[x+1]\, dx \qquad \text{Substituting}$$

We evaluate the integral at the top of the next column.

$$V = 2\pi \int_0^a \left[x^2 + x \right] dx$$

$$= 2\pi \left[\frac{1}{3}x^3 + \frac{1}{2}x^2 \right]_0^a$$

$$= 2\pi \left[\frac{1}{3}(a)^3 + \frac{1}{2}(a)^2 - \left(\frac{1}{3}(0)^3 + \frac{1}{2}(0)^2 \right) \right]$$

$$= 2\pi \left[\frac{a^3}{3} + \frac{a^2}{2} \right]$$

To determine what value of a they will have the same volume, we equate the two volumes and solve for a.

$$2\pi \left[\frac{a^3}{3} + \frac{a^2}{2} \right] = \pi \left[\frac{a^3}{3} + a^2 + a \right]$$

$$\frac{2a^3}{3} + a^2 = \frac{a^3}{3} + a^2 + a$$

$$\frac{a^3}{3} - a = 0$$

$$a^3 - 3a = 0$$

$$a \left(a^2 - 3 \right) = 0$$

Therefore,

$a = 0, \; a = \pm\sqrt{3}.$ Since the region is in the first quadrant, the value of a that will make the volumes equal is $a = \sqrt{3}$.

45. $$V = \int_1^\infty \pi \left[\frac{1}{x} \right]^2 dx$$

$$V = \int_1^\infty \pi \frac{1}{x^2}\, dx$$

$$= \int_1^\infty \pi x^{-2}\, dx$$

$$= \lim_{b \to \infty} \int_1^b \pi x^{-2}\, dx$$

$$= \lim_{b \to \infty} \left[\pi \frac{x^{-1}}{-1} \right]_1^b$$

$$= \lim_{b \to \infty} \left[-\frac{\pi}{x} \right]_1^b$$

$$= \lim_{b \to \infty} \left[-\frac{\pi}{b} - \left(-\frac{\pi}{1} \right) \right]$$

$$= \lim_{b \to \infty} \left[-\frac{\pi}{b} + \frac{\pi}{1} \right]$$

$$= [0 + \pi]$$

$$= \pi.$$

Exercise Set 5.7

1. We find the general solution by integrating both sides of the equation:

$$y' = 10x^2$$

$$\int y'\,dx = \int 10x^2\,dx$$

$$y = 10\left(\tfrac{1}{3}x^3 + C\right)$$

$$y = \tfrac{10}{3}x^3 + C \qquad \text{General Solution}$$

We can find three particular solutions by substituting different values for C. Answers may vary.

$$y = \tfrac{10}{3}x^3 + 1 \qquad (C = 1)$$

$$y = \tfrac{10}{3}x^3 - 2 \qquad (C = -2)$$

$$y = \tfrac{10}{3}x^3 + 8 \qquad (C = 8)$$

3. We find the general solution by integrating both sides of the equation:

$$y' = 2e^{-x} + x$$

$$\int y'\,dx = \int \left(2e^{-x} + x\right)dx$$

$$y = -2e^{-x} + \tfrac{1}{2}x^2 + C \quad \text{General Solution}$$

We can find three particular solutions by substituting different values for C. Answers may vary

$$y = -2e^{-x} + \tfrac{1}{2}x^2 + 1 \qquad (C = 1)$$

$$y = -2e^{-x} + \tfrac{1}{2}x^2 - 3 \qquad (C = -3)$$

$$y = -2e^{-x} + \tfrac{1}{2}x^2 + 6 \qquad (C = 6)$$

5. We find the general solution by integrating both sides of the equation:

$$y' = \frac{4}{x} - \frac{1}{x^2}$$

$$\int y'\,dx = \int \left(\frac{4}{x} - x^{-2}\right)dx$$

$$y = 4\ln|x| + \frac{1}{x} + C$$

$$\text{General Solution}$$

We can find three particular solutions by substituting different values for C. Answers may vary.

$$y = 4\ln|x| + \tfrac{1}{x} - 2 \qquad (C = -2)$$

$$y = 4\ln|x| + \tfrac{1}{x} + 4 \qquad (C = 4)$$

$$y = 4\ln|x| + \tfrac{1}{x} - 11 \qquad (C = -11)$$

7. Find y' and y''.

$$y' = \tfrac{d}{dx}(x\ln x + 3x - 2)$$

$$= \underbrace{\tfrac{d}{dx}x\ln x}_{\text{Product Rule}} + \tfrac{d}{dx}3x - \tfrac{d}{dx}2$$

$$= \underbrace{x\cdot\tfrac{1}{x} + 1\cdot\ln x}_{\text{Product Rule}} + 3 - 0$$

$$y' = 1 + \ln x + 3$$

$$y' = \ln x + 4$$

and

$$y'' = \tfrac{d}{dx}(\ln x + 4)$$

$$y'' = \tfrac{1}{x}.$$

Substituting, we have:

$$y'' - \tfrac{1}{x} \overset{?}{=} 0$$

$\tfrac{1}{x} - \tfrac{1}{x}$	0
0	0 TRUE

9. Find y' and y''.

$$y' = \tfrac{d}{dx}\left(e^x + 3xe^x\right)$$

$$= \tfrac{d}{dx}e^x + \underbrace{\tfrac{d}{dx}3xe^x}_{\text{Product Rule}}$$

$$= e^x + \underbrace{3x\cdot e^x + 3\cdot e^x}_{\text{Product Rule}}$$

$$y' = 3xe^x + 4e^x$$

and

$$y'' = \tfrac{d}{dx}\left(3xe^x + 4e^x\right)$$

$$y'' = \underbrace{3x\cdot e^x + 3\cdot e^x}_{\text{Product Rule}} + 4e^x$$

$$y'' = 3x\cdot e^x + 7\cdot e^x.$$

Substituting, we have:

$$y'' - 2y' + y \overset{?}{=} 0$$

$3xe^x + 7e^x - 2\left(3xe^x + 4e^x\right) + e^x + 3xe^x$	0
$3xe^x + 7e^x - 6xe^x - 8e^x + e^x + 3xe^x$	0
$6xe^x + 8e^x - 6xe^x - 8e^x$	0
0	0 TRUE

11. Let $y' + 4y = 0$.

a) Show $y = e^{-4x}$ is a solution.
Find y'.

$$y' = \frac{d}{dx}\left(e^{-4x}\right)$$

$$y' = e^{-4x} \cdot \frac{d}{dx}(-4x)$$

$$y' = -4e^{-4x}.$$

Substituting, we have:

$$\overset{?}{y' + 4y = 0}$$

$$\begin{array}{c|c} -4e^{-4x} + 4\left(e^{-4x}\right) & 0 \\ \hline & 0 \; \mid \; 0 \quad \text{TRUE} \end{array}$$

b) Show $y = Ce^{-4x}$ is a solution.
Find y'.

$$y' = \frac{d}{dx}\left(Ce^{-4x}\right)$$

$$y' = Ce^{-4x} \cdot \frac{d}{dx}(-4x)$$

$$y' = -4Ce^{-4x}.$$

Substituting, we have:

$$\overset{?}{y' + 4y = 0}$$

$$\begin{array}{c|c} -4Ce^{-4x} + 4\left(Ce^{-4x}\right) & 0 \\ \hline & 0 \; \mid \; 0 \quad \text{TRUE} \end{array}$$

13. Let $y'' - y - 30y = 0$.

a) Show $y = e^{6x}$ is a solution.
Find y' and y''.

$$y' = \frac{d}{dx}\left(e^{6x}\right) = 6e^{6x}.$$

$$y'' = \frac{d}{dx}\left(6e^{6x}\right) = 36e^{6x}.$$

Substituting, we have:

$$\overset{?}{y'' - y' - 30y = 0}$$

$$\begin{array}{c|c} 36e^{6x} - \left(6e^{6x}\right) - 30\left(e^{6x}\right) & 0 \\ 36e^{6x} - 36e^{6x} & 0 \\ \hline & 0 \; \mid \; 0 \quad \text{TRUE} \end{array}$$

b) Show $y = e^{-5x}$ is a solution.
Find y' and y''.

$$y' = \frac{d}{dx}\left(e^{-5x}\right) = -5e^{-5x}.$$

$$y'' = \frac{d}{dx}\left(-5e^{-5x}\right) = 25e^{-5x}.$$

Substituting, we have:

$$\overset{?}{y'' - y' - 30y = 0}$$

$$\begin{array}{c|c} 25e^{-5x} - \left(-5e^{-5x}\right) - 30\left(e^{-5x}\right) & 0 \\ 30e^{-5x} - 30e^{-5x} & 0 \\ \hline & 0 \; \mid \; 0 \quad \text{TRUE} \end{array}$$

c) Show $y = C_1e^{6x} + C_2e^{-5x}$ is a solution.
Find y' and y''.

$$y' = \frac{d}{dx}\left(C_1e^{6x} + C_2e^{-5x}\right)$$

$$y' = 6C_1e^{6x} - 5C_2e^{-5x}.$$

$$y'' = \frac{d}{dx}\left(6C_1e^{6x} - 5C_2e^{-5x}\right)$$

$$y'' = 36C_1e^{6x} + 25C_2e^{-5x}.$$

Substituting, we have:

$$\overset{?}{y'' - y' - 30y = 0}$$

$$\begin{array}{c|c} 36C_1e^{6x} + 25C_2e^{-5x} - \left(6C_1e^{6x} - 5C_2e^{-5x}\right) - 30\left(C_1e^{6x} + C_2e^{-5x}\right) & 0 \\ 30C_1e^{6x} + 30C_2e^{-5x} - 30C_1e^{6x} - 30C_2e^{-5x} & 0 \\ \hline & 0 \; \mid \; 0 \quad \text{TRUE} \end{array}$$

15. a) In general the solution to the differential equation $\frac{dP}{dk} = kP$ is $P(t) = P_0e^{kt}$.
Therefore,
$\frac{dM}{dt} = 0.05M$ has a solution given by
$M = M_0e^{0.05t}$.

b) To verify this solution, we find $\frac{dM}{dt}$ and we substitute into the equation.

$$\frac{dM}{dt} = \frac{d}{dt}\left(M_0e^{0.05t}\right)$$

$$\frac{dM}{dt} = 0.05M_0e^{0.05t}.$$

$$\overset{?}{\frac{dM}{dt} - 0.05M = 0}$$

$$\begin{array}{c|c} 0.05M_0e^{0.05t} - 0.05\left(M_0e^{0.05t}\right) & 0 \\ \hline & 0 \; \mid \; 0 \quad \text{TRUE} \end{array}$$

17. a) The solution to the differential equation $\frac{dR}{dt} = 0.35R$ is $R = R_0e^{0.35t}$.

b) To verify this solution, we find $\dfrac{dR}{dt}$ and we substitute into the equation.

$$\dfrac{dR}{dt} = \dfrac{d}{dt}\left(R_0 e^{0.35t}\right)$$

$$\dfrac{dR}{dt} = 0.35 R_0 e^{0.35t}.$$

$$\dfrac{dR}{dt} - 0.35R \overset{?}{=} 0$$

$$\begin{array}{c|c} 0.35 R_0 e^{0.35t} - 0.35\left(R_0 e^{0.35t}\right) & 0 \\ \hline & 0 \mid 0 \quad \text{TRUE} \end{array}$$

19. a) The solution to the differential equation
$$\dfrac{dG}{dt} = 0.005G \text{ is } G = G_0 e^{0.005t}.$$

b) To verify this solution, we find $\dfrac{dG}{dt}$ and we substitute into the equation.

$$\dfrac{dG}{dt} = \dfrac{d}{dt}\left(G_0 e^{0.005t}\right)$$

$$\dfrac{dG}{dt} = 0.005 G_0 e^{0.005t}.$$

$$\dfrac{dG}{dt} - 0.005G \overset{?}{=} 0$$

$$\begin{array}{c|c} 0.005 G_0 e^{0.005t} - 0.005\left(G_0 e^{0.005t}\right) & 0 \\ \hline & 0 \mid 0 \quad \text{TRUE} \end{array}$$

21. a) The solution to the differential equation
$$\dfrac{dR}{dt} = R \text{ is } R = R_0 e^t.$$

b) To verify this solution, we find $\dfrac{dR}{dt}$ and we substitute into the equation.

$$\dfrac{dR}{dt} = \dfrac{d}{dt}\left(R_0 e^t\right)$$

$$\dfrac{dR}{dt} = R_0 e^t.$$

$$\dfrac{dR}{dt} - R \overset{?}{=} 0$$

$$\begin{array}{c|c} R_0 e^t - \left(R_0 e^t\right) & 0 \\ \hline & 0 \mid 0 \quad \text{TRUE} \end{array}$$

23. a) Find the general solution by integrating both sides of the equation.
$$y' = x^2 + 2x - 3$$

$$\int y'\,dx = \int \left(x^2 + 2x - 3\right) dx$$

$$y = \tfrac{1}{3}x^3 + x^2 - 3x + C.$$

Find the particular solution by substituting the initial condition into the general solution and solving for C as follows

$$y = \tfrac{1}{3}x^3 + x^2 - 3x + C$$

$$(4) = \tfrac{1}{3}(0)^3 + (0)^2 - 3(0) + C$$

$$4 = C.$$

The particular solution is
$$y = \tfrac{1}{3}x^3 + x^2 - 3x + 4.$$

b) To verify solution we take the derivative of the particular solution and see

$$y' = \tfrac{d}{dx}\left(\tfrac{1}{3}x^3 + x^2 - 3x + 4\right)$$

$$y' = x^2 + 2x - 3.$$

Substituting into the equation we have

$$y' - x^2 - 2x + 3 \overset{?}{=} 0$$

$$\begin{array}{c|c} \left(x^2 + 2x - 3\right) - x^2 - 2x + 3 & 0 \\ \hline & 0 \mid 0 \quad \text{TRUE} \end{array}$$

25. a) Find the general solution by integrating both sides of the equation.
$$f'(x) = x^{2/3} - x$$

$$\int f'(x)\,dx = \int \left(x^{2/3} - x\right) dx$$

$$f(x) = \tfrac{3}{5}x^{5/3} - \tfrac{1}{2}x^2 + C.$$

Find the particular solution by substituting the initial condition into the general solution and solving for C as shown:

$$f(1) = -6$$

$$(-6) = \tfrac{3}{5}(1)^{5/3} - \tfrac{1}{2}(1)^2 + C$$

$$-6 = \tfrac{3}{5} - \tfrac{1}{2} + C$$

$$-6 = \tfrac{1}{10} + C$$

$$-6 - \tfrac{1}{10} = C$$

$$-\tfrac{61}{10} = C.$$

The particular solution is
$$f(x) = \tfrac{3}{5}x^{5/3} - \tfrac{1}{2}x^2 - \tfrac{61}{10}.$$

b) To verify solution we take the derivative of
the particular solution and see

$$f'(x) = \frac{d}{dx}\left(\frac{3}{5}x^{5/3} - \frac{1}{2}x^2 - \frac{61}{10}\right)$$

$$f'(x) = x^{2/3} - x.$$

Substituting into the equation we have

$$f'(x) - x^{2/3} + x \overset{?}{=} 0$$

$\left(x^{2/3} - x\right) - x^{2/3} + x$	0
0	0 TRUE

27. a) The solution to the differential equation

$$\frac{dB}{dt} = 0.03B \text{ is } B = B_0 e^{0.03t}.$$

Find the particular solution by substituting
the initial condition into the general solution
and solving for B_0 as follows

$$B(0) = 500$$

$$(500) = B_0 e^{0.03(0)}$$

$$500 = B_0 \cdot 1$$

$$500 = B_0.$$

The particular solution is $B = 500e^{0.03t}$.

b) To verify this solution, we find $\dfrac{dB}{dt}$ and we
substitute into the equation.

$$\frac{dB}{dt} = \frac{d}{dt}\left(500e^{0.03t}\right)$$

$$\frac{dB}{dt} = 500(0.03)e^{0.03t} = 15e^{0.03t}.$$

$$\frac{dB}{dt} - 0.03B \overset{?}{=} 0$$

$15e^{0.03t} - 0.03\left(500e^{0.03t}\right)$	0
$15e^{0.03t} - 15e^{0.03t}$	0
0	0 TRUE

29. a) The solution to the differential equation

$$\frac{dS}{dt} = 0.12S \text{ is } S = S_0 e^{0.12t}.$$

Find the particular solution by substituting
the initial condition into the general solution
and solving for S_0 as follows

$$S = S_0 e^{0.12t}$$

$$(750) = S_0 e^{0.12(0)}$$

$$750 = S_0.$$

The particular solution is $S = 750e^{0.12t}$.

b) To verify this solution, we find $\dfrac{dS}{dt}$ and we
substitute into the equation.

$$\frac{dS}{dt} = \frac{d}{dt}\left(750e^{0.12t}\right)$$

$$\frac{dS}{dt} = 750(0.12)e^{0.12t} = 90e^{0.12t}.$$

$$\frac{dS}{dt} - 0.12S \overset{?}{=} 0$$

$90e^{0.12t} - 0.12\left(750e^{0.12t}\right)$	0
$90e^{0.12t} - 90e^{0.12t}$	0
0	0 TRUE

31. a) The solution to the differential equation

$$\frac{dT}{dt} = 0.015T \text{ is } T = T_0 e^{0.015t}.$$

Find the particular solution by substituting
the initial conditions into the general
solution and solving for T_0 as follows

$$T = T_0 e^{0.015t}$$

$$(50) = T_0 e^{0.015(0)}$$

$$50 = T_0.$$

The particular solution is $T = 50e^{0.015t}$.

b) To verify this solution, we find $\dfrac{dT}{dt}$ and we
substitute into the equation.

$$\frac{dT}{dt} = \frac{d}{dt}\left(50e^{0.015t}\right)$$

$$\frac{dT}{dt} = 50(0.015)e^{0.015t} = 0.75e^{0.015t}.$$

$$\frac{dT}{dt} - 0.015T \overset{?}{=} 0$$

$0.75e^{0.015t} - 0.015\left(50e^{0.015t}\right)$	0
$0.75e^{0.015t} - 0.75e^{0.015t}$	0
0	0 TRUE

33. a) The solution to the differential equation

$\dfrac{dM}{dt} = M$ is $M = M_0 e^t$.

Find the particular solution by substituting the initial conditions into the general solution and solving for M_0 as follows

$M = M_0 e^t$

$(6) = M_0 e^{(0)}$

$6 = M_0$.

The particular solution is $M = 6e^t$.

b) To verify this solution, We find $\dfrac{dM}{dt}$ and we substitute into the equation.

$\dfrac{dM}{dt} = \dfrac{d}{dt}\left(6e^t\right)$

$\dfrac{dM}{dt} = 6e^t$.

$$\dfrac{dM}{dt} - M \overset{?}{=} 0$$

$\begin{array}{c|c} 6e^t - 6e^t & 0 \\ \hline 0 & 0 \quad \text{TRUE} \end{array}$

35. Solve by separating variables.

$\dfrac{dy}{dx} = 4x^3 y$

$\dfrac{dy}{y} = 4x^3 dx$

$\displaystyle\int \dfrac{dy}{y} = \int 4x^3 dx$

$\ln|y| = x^4 + C$

$e^{\ln|y|} = e^{x^4 + C}$

$|y| = e^{x^4} e^C$

$y = \pm e^C e^{x^4}$

$y = C e^{x^4}$. $\left(C = \pm e^C\right)$

37. Solve by separating variables.

$3y^2 \dfrac{dy}{dx} = 8x$

$3y^2 dy = 8x\,dx$

$\displaystyle\int 3y^2 dy = \int 8x\,dx$

$y^3 + C_1 = 4x^2 + C_2$

$y^3 = 4x^2 + C_2 - C_1$

$y = \sqrt[3]{4x^2 + C}$. $\left(C = C_2 - C_1\right)$

39. Solve by separating variables.

$\dfrac{dy}{dx} = \dfrac{2x}{y}$

$y\,dy = 2x\,dx$

$\displaystyle\int y\,dy = \int 2x\,dx$

$\tfrac{1}{2}y^2 + C_1 = x^2 + C_2$

$\tfrac{1}{2}y^2 = x^2 + C_2 - C_1$

$y^2 = 2x^2 + 2\left(C_2 - C_1\right)$

$y = \pm\sqrt{2x^2 + C}$. $\left(C = 2\left(C_2 - C_1\right)\right)$

41. Solve by separating variables.

$\dfrac{dy}{dx} = \dfrac{6}{y}$

$y\,dy = 6\,dx$

$\displaystyle\int y\,dy = \int 6\,dx$

$\tfrac{1}{2}y^2 + C_1 = 6x + C_2$

$\tfrac{1}{2}y^2 = 6x + C_2 - C_1$

$y^2 = 12x + 2\left(C_2 - C_1\right)$

$y = \pm\sqrt{12x + C}$. $\left(C = 2\left(C_2 - C_1\right)\right)$

43. Solve by separating variables.

$$y' = 3x + xy$$

$$\frac{dy}{dx} = x(3+y)$$

$$\frac{dy}{3+y} = xdx$$

$$\int \frac{dy}{3+y} = \int xdx$$

$$\ln|3+y| = \tfrac{1}{2}x^2 + C$$

$$e^{\ln|y+3|} = e^{\frac{1}{2}x^2 + C}$$

$$|y+3| = e^{\frac{1}{2}x^2} e^C$$

$$y + 3 = \pm e^C e^{\frac{1}{2}x^2}$$

$$y = Ce^{\frac{1}{2}x^2} - 3. \qquad \left(C = \pm e^C\right)$$

Now we will plug in the initial condition of $y = 5$ when $x = 0$.

$$y = Ce^{\frac{1}{2}x^2} - 3$$

$$5 = Ce^{\frac{1}{2}(0)^2} - 3$$

$$8 = C$$

Therefore, the particular solution is

$$y = 8e^{\frac{1}{2}x^2} - 3.$$

45. Solve by separating variables.

$$y' = 5y^{-2}$$

$$y^2 \frac{dy}{dx} = 5$$

$$y^2 dy = 5dx$$

$$\int y^2 dy = \int 5dx$$

$$\tfrac{1}{3}y^3 + C_1 = 5x + C_2$$

$$\tfrac{1}{3}y^3 = 5x + C_2 - C_1$$

$$y^3 = 15x + 3(C_2 - C_1)$$

$$y = \sqrt[3]{15x + C}. \qquad \left(C = 3(C_2 - C_1)\right)$$

Now we will plug in the initial condition of $y = 3$ when $x = 2$.

$$y = \sqrt[3]{15x + C}$$

$$3 = \sqrt[3]{15(2) + C}$$

$$3 = \sqrt[3]{30 + C}$$

$$27 = 30 + C$$

$$-3 = C.$$

Therefore, the particular solution is

$$y = \sqrt[3]{15x - 3}.$$

47. a) The differential equation that models the situation is

$$\frac{dy}{dx} = y^2. \qquad (k = 1)$$

b) The general solution to this differential equation is determined as follows

$$\frac{dy}{dx} = y^2 \qquad (k = 1)$$

$$\frac{dy}{y^2} = 1dx$$

$$y^{-2} dy = dx$$

$$\int y^{-2} dy = \int dx$$

$$-1y^{-1} + C_1 = x + C_2$$

$$-\frac{1}{y} = x + C_2 - C_1$$

$$\frac{1}{y} = -x + C \qquad (c = C_2 - C_1)$$

$$y = \frac{1}{C - x}.$$

49. a) The differential equation that models the situation is

$$\frac{dy}{dx} = \frac{1}{y^3}. \qquad (k = 1)$$

b) The general solution to this differential equation is determined as follows

$$\frac{dy}{dx} = \frac{1}{y^3} \qquad (k = 1)$$

$$y^3 dy = dx$$

$$\int y^3 dy = \int dx$$

$$\tfrac{1}{4}y^4 + C_1 = x + C_2$$

$$\tfrac{1}{4}y^4 = x + C_2 - C_1$$

$$y^4 = 4x + C \qquad (c = 4C_2 - 4C_1)$$

$$y = \pm\sqrt[4]{4x + C}.$$

51. a) The differential equation that models the situation is

$$\frac{dy}{dx} = xy. \qquad (k = 1)$$

With initial condition $y(2) = 3$.

b) The general solution to this differential equation is determined as follows

$$\frac{dy}{dx} = xy \qquad (k=1)$$

$$\frac{dy}{y} = xdx$$

$$\int \frac{dy}{y} = \int xdx$$

$$\ln|y| + C_1 = \tfrac{1}{2}x^2 + C_2$$

$$\ln|y| = \tfrac{1}{2}x^2 + C_2 - C_1$$

$$|y| = e^{\frac{1}{2}x^2 + C}$$

$$y = \pm e^C \cdot e^{\frac{1}{2}x^2}$$

$$y = Ce^{\frac{1}{2}x^2} \qquad \left(C = \pm e^C\right)$$

Substituting the initial condition we have

$$y = Ce^{\frac{1}{2}x^2}$$

$$3 = Ce^{\frac{1}{2}(2)^2}$$

$$3 = C \cdot 7.389$$

$$\frac{3}{7.389} = C$$

$$0.406 = C$$

Therefore, the particular solution is

$$y = 0.406e^{\frac{1}{2}x^2}.$$

53. a) The interest rate of 3.75% implies that $k = 0.0375$. Thus, the differential equation that represents the value of the account is

$$\frac{dA}{dt} = 0.0375A.$$

b) The general solution to the differential equation is

$$A = A_0 e^{0.0375t}.$$

Find the particular solution by substituting the initial conditions into the general solution and solving for A_0 as follows

$$A = A_0 e^{0.0375t}$$

$$(500) = A_0 e^{0.0375(0)}$$

$$500 = A_0.$$

Therefore, the particular solution is

$$A(t) = 500e^{0.0375t}.$$

c) Evaluating the function, we have

$$A(5) = 500e^{0.0375(5)} = 603.12$$

After 5 years, the account will have $603.12.

To find $A'(5)$ we use the differential equation.

$$A'(5) = 0.0375 \cdot A(5) = 0.0375 \cdot 603.12 = 22.62$$

After 5 years, the account will be growing at $22.62 per year.

d) Substituting from part (c) we have

$$\frac{A'(5)}{A(5)} = \frac{22.62}{603.12} = 0.0375.$$

This is the continuous growth rate.

55. a) By separation of variables

$$\frac{dI}{dt} = hkI$$

$$\frac{dI}{I} = hkdt$$

$$\int \frac{dI}{I} = \int hkdt$$

$$\ln|I| + C_1 = hkt + C_2$$

$$\ln|I| = hkt + C \qquad \left(C = C_2 - C_1\right)$$

$$I = \pm e^{hkt+C}$$

$$I = \pm e^{hkt} \cdot e^C$$

$$I = Ce^{hkt}. \qquad \left(C = \pm e^C\right)$$

b) Substituting the initial condition

$$I = Ce^{hkt}$$

$$I_0 = Ce^{hk(0)}$$

$$I_0 = C$$

Therefore, the solution is

$$I = I_0 e^{hkt}.$$

57. Substituting the given information into the differential equation we have

$$\frac{dV}{dt} = k(24.81 - V).$$

Solve the differential equation using separation of variables.

$$\frac{dV}{dt} = k(24.81 - V)$$

$$\frac{dV}{24.81 - V} = kdt$$

$$\int \frac{dV}{24.81 - V} = \int kdt$$

$$-\ln|24.81 - V| + C_1 = kt + C_2$$

$$-\ln|24.81 - V| = kt + C$$

$$\ln|24.81 - V| = -kt + C$$

$$24.81 - V = \pm e^{-kt + C}$$

$$24.81 - V = \pm e^{-kt} e^C$$

$$-V = Ce^{-kt} - 24.81$$

$$V = -Ce^{-kt} + 24.81.$$

Substituting the initial condition we have

$$V(0) = 20$$

$$20 = -Ce^{-k(0)} + 24.81$$

$$-4.81 = -C$$

Therefore, the particular solution is

$$V = -4.81e^{kt} + 24.81.$$

59. $E(x) = \frac{4}{x}; \quad q(4) = 2.$

$$E(x) = \frac{4}{x}$$

$$\frac{-x}{q} \cdot \frac{dq}{dx} = \frac{4}{x}$$

$$\frac{dq}{q} = -\frac{4}{x^2} dx$$

$$\int \frac{dq}{q} = -4 \int x^{-2} dx$$

$$\ln q + C_1 = -4(-x^{-1}) + C_2$$

$$\ln q = \frac{4}{x} + C$$

$$q = e^{\frac{4}{x} + C}$$

$$q = e^{\frac{4}{x}} e^C$$

$$q = Ce^{\frac{4}{x}}.$$

Substituting the initial condition, we have

$$q = Ce^{\frac{4}{x}}$$

$$2 = Ce^{\frac{4}{4}}$$

$$\frac{2}{e} = C$$

$$2e^{-1} = C$$

Therefore, the demand function is given by

$$q = 2e^{-1}e^{\frac{4}{x}}$$

$$q = 2e^{\frac{4}{x} - 1}.$$

61. $E(x) = 2, \quad$ for all $x > 0.$

$$E(x) = 2$$

$$\frac{-x}{q} \cdot \frac{dq}{dx} = 2$$

$$\frac{dq}{q} = -\frac{2}{x} dx$$

$$\int \frac{dq}{q} = -2 \int \frac{1}{x} dx$$

$$\ln q + C_1 = -2(\ln(x)) + C_2$$

$$\ln q = \ln x^{-2} + C$$

$$q = e^{\ln x^{-2} + C}$$

$$q = e^{\ln x^{-2}} e^C$$

$$q = Cx^{-2}$$

$$q = \frac{C}{x^2}.$$

63. a) Let $t = 0$ correspond to 2002. The continuous growth rate of 1.75% per year implies that $k = 0.0175$. The differential equation that represents the population of New River after t years is

$$\frac{dP}{dt} = 0.0175P.$$

b) The general solution to the differential equation is $P(t) = P_0 e^{0.0175t}$. Since the population in 2002 was 17,000, the initial population $P_0 = 17,000$. Thus, the particular solution to the differential equation is

$$P(t) = 17,000e^{0.0175t}.$$

c) Evaluating the function, we have

$$P(10) = 17,000e^{0.0175(10)} = 20,251.$$

After 10 years, New River's population will be 20,251.

To find $P'(10)$ we use the differential equation.

$$P'(10) = 0.0175 \cdot P(10) = 0.0175 \cdot 21,251 = 354.4$$

After 10 years, the population is increasing by 354.4 people per year.

d) Substituting from part (c) we have

$$\frac{P'(t)}{P(t)} = \frac{354.4}{20,251} = 0.0175.$$

This represents the continuous growth rate.

65. a) Let $t = 0$ correspond to 1859. When the population was estimated to be 8900, the growth rate was about 2630 rabbits per year. Using this information we have:

$$\frac{dP}{dt} = kP$$

$$2630 = k \cdot 8900$$

$$\frac{2630}{8900} = k$$

$$0.296 = k$$

The differential equation that represents the population of rabbits after t years is

$$\frac{dP}{dt} = 0.296P.$$

The general solution to the differential equation is $P(t) = P_0 e^{0.296t}$. Since the initial population in 1859 was 24 rabbits, the particular solution to the differential equation is

$$P(t) = 24e^{0.296t}.$$

b) Evaluating the function, we have

$$P(41) = 24e^{0.296(41)} = 4,475,165.$$

After 41 years, the rabbit population had grown to approximately 4,475,165 rabbits. To find $P'(41)$ we use the differential equation.

$$P'(41) = 0.296 \cdot P(41)$$

$$= 0.296 \cdot 4,475,165$$

$$= 1,324,649.$$

After 41 years, the population is increasing by 1,324,649 rabbits per year.

c) Substituting from part (b) we have

$$\frac{P'(41)}{P(41)} = \frac{0.296P(41)}{P(41)} = 0.296.$$

This represents the continuous growth rate.

67. a) By separation of variables, we have

$$\frac{dP}{dt} = kP$$

$$\frac{dP}{P} = kdt$$

$$\int \frac{dP}{P} = \int kdt$$

$$\ln|P| + C_1 = kt + C_2$$

$$\ln|P| = kt + C_2 - C_1$$

$$\ln|P| = kt + C$$

$$P = \pm e^{kt+C}$$

$$P = Ce^{kt}. \qquad \left(C = \pm e^c\right)$$

b) Substitute the initial condition $P(0) = P_0$.

$$P_0 = Ce^{k(0)}$$

$$P_0 = C$$

Therefore, the particular solution is

$$P = P_0 e^{kt}.$$

69. a) Since $k = 0.0418$ the continuous growth rate is 4.18% per year.

b) The general solution to the differential equation is $A(t) = A_0 e^{0.0418t}$. Plugging in the given condition, we have:

$$A(2) = 3479.02$$

$$A_0 e^{0.0325(2)} = 3479.02$$

$$A_0 = 3200.$$

The particular solution to the differential equation is

$$A(t) = 3200e^{0.0418t}.$$

c) From the particular solution, Ina initially deposited $3200.

71. $\dfrac{dy}{dx} = 5x^4 y^2 + x^3 y^2$

$$\dfrac{dy}{dx} = y^2 \left(5x^4 + x^3\right)$$

$$y^{-2}dy = \left(5x^4 + x^3\right)dx$$

$$\int y^{-2}dy = \int \left(5x^4 + x^3\right)dx$$

$$-y^{-1} + C_1 = x^5 + \tfrac{1}{4}x^4 + C_2$$

$$-y^{-1} = x^5 + \tfrac{1}{4}x^4 + C_2 - C_1$$

$$y^{-1} = -\dfrac{4x^5 + x^4 + C}{4}$$

$$y = -\dfrac{4}{4x^5 + x^4 + C}.$$

73. ✏️

75. a) Using a calculator, we enter the given information

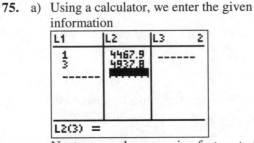

Next we use the regression feature to find the exponential function.

The resulting equation is

ExpReg
y=a*b^x
a=4249.994996
b=1.051271826

The exponential model that fits is

$y = 4250(1.05127)^t$. Using the definition of the exponential, we have

$$(1.05127)^t = e^{\ln(1.05127)\cdot t} = e^{0.05t} \ .$$

Substituting, we have the model

$y = 4250e^{0.05t}.$

b) From this model, we see that $A_0 = 4250$ and that $k = 0.05$. Therefore, the differential equation that models this situation is

$\dfrac{dA}{dt} = 0.05A$ and the initial condition is

$A_0 = 4250.$

77. Solve by separating variables.

$$\dfrac{dy}{dx} = \dfrac{2}{y^2}$$

$$y^2 dy = 2dx$$

$$\int y^2 dy = \int 2dx$$

$$\tfrac{1}{3}y^3 + C_1 = 2x + C_2$$

$$\tfrac{1}{3}y^3 = 2x + C_2 - C_1$$

$$y^3 = 6x + 3\left(C_2 - C_1\right)$$

$$y = \sqrt[3]{6x + C}. \qquad \left(C = 3\left(C_2 - C_1\right)\right)$$

Using the general solution, the three particular solutions are

$y_1 = \sqrt[3]{6x}$ $\qquad (C_1 = 0)$

$y_2 = \sqrt[3]{6x + 4}$ $\qquad (C_2 = 4)$

$y_2 = \sqrt[3]{6x - 10}$ $\qquad (C_3 = -10)$

Using a graphing utility, we have

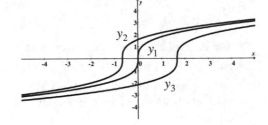

Chapter 6

Functions of Several Variables

Exercise Set 6.1

1. $f(x, y) = x^2 - 3xy$

$f(0, -2) = (0)^2 - 3(0)(-2)$ Substituting 0 for x and -2 for y

$= 0 - 0$

$= 0$

$f(2, 3) = (2)^2 - 3(2)(3)$ Substituting 2 for x and 3 for y

$= 4 - 18$

$= -14$

$f(10, -5) = (10)^2 - 3(10)(-5)$ Substituting 10 for x and -5 for y

$= 100 + 150$

$= 250$

3. $f(x, y) = 3^x + 7xy$

$f(0, -2) = 3^0 + 7(0)(-2)$ Substituting 0 for x and -2 for y

$= 1 + 0$

$= 1$

$f(-2, 1) = 3^{-2} + 7(-2)(1)$ Substituting -2 for x and 1 for y

$= \dfrac{1}{9} - 14$

$= -\dfrac{125}{9}$

$f(2, 1) = 3^2 + 7(2)(1)$ Substituting 2 for x and 1 for y

$= 9 + 14$

$= 23$

5. $f(x, y) = \ln x + y^3$

$f(e, 2) = \ln e + (2)^3$ Substituting e for x and 2 for y

$= 1 + 8$

$= 9$

$f(e^2, 4) = \ln e^2 + (4)^3$ Substituting e^2 for x and 4 for y

$= 2 + 64$

$= 66$

$f(e^3, 5) = \ln e^3 + (5)^3$ Substituting e^3 for x and 5 for y

$= 3 + 125$

$= 128$

7. $f(x, y, z) = x^2 - y^2 + z^2$

We substitute -1 for x, 2 for y, and 3 for z.

$f(-1, 2, 3) = (-1)^2 - (2)^2 + (3)^2$

$= 1 - 4 + 9$

$= 6$

We substitute 2 for x, -1 for y, and 3 for z.

$f(2, -1, 3) = (2)^2 - (-1)^2 + (3)^2$

$= 4 - 1 + 9$

$= 12$

9. $f(x, y) = x^2 + 4x + y^2$

The function is defined for all values of x and y. Therefore, the domain is:

$\{(x, y) \mid -\infty < x < \infty, -\infty < y < \infty\}$.

11. $f(x, y) = \sqrt{y - 3x}$

The function is defined only when $y - 3x \geq 0$. Therefore, the domain is:

$\{(x, y) \mid y \geq 3x\}$.

13. $g(x, y) = \dfrac{1}{y + x^2}$

The function is defined if $y + x^2 \neq 0$. Therefore, the domain is:

$\{(x, y) \mid y \neq x^2\}$.

15. $R(P, E) = \dfrac{P}{E}$

Substituting 29.00 for P, and 0.74 for E, gives

$R(29.00, 0.34) = \dfrac{29.00}{0.74}$

≈ 39.189

≈ 39.19.

The price-earnings ratio for Hewlett-Packard was 39.19.

17. From Example 3 we have $C_2 = \left(\dfrac{V_2}{V_1}\right)^{0.6} C_1$.

Where C_1 is the cost of the original piece of equipment, V_1 is the capacity of the original piece of equipment, and V_2 is the capacity of the new piece of equipment.
We substitute 100,000 for C_1, 80,000 for V_1 and 240,000 for V_2.

$$C_2 = \left(\frac{240,000}{80,000}\right)^{0.6} (100,000)$$

$$= (3)^{0.6} (100,000)$$

$$\approx 193,318.20$$

We estimate the cost of the new tank to be $193,318.20.

19. a) From the table in Example 3, we see that an APR of 5% for 7 years results in a payment of $14.13 per $1000 borrowed. Since Ashley is borrowing $20,000 we would estimate Ashley's payments to be 20 times $14.13 or $282.60.

   b) Over the 7 years, Ashley would make $7 \times 12 = 84$ monthly payments of $282.60. Therefore, Ashley's total payments will be $282.60 \times 84 = \$23,738.40$.

   c) For the five year loan, ultimately we want the total cost to be less than $23,738.40. Since there are 60 payments in a five year loan we know that:
   $$P \times 60 \le 23,738.40$$
   $$P \le 395.64.$$
   Ashley is borrowing $20,000 dollars so the amount per $100 powered must be
   $$A \times 20 \le 395.64$$
   $$A \le 19.782.$$
   Comparing to the values in the table in Example 3, we see that 6.5% is the highest APR that Ashley could accept if she wants to pay less overall for the loan.

   d) If Ashley accepts Valley Credit unions offer for a 5 year loan with an APR of 6.5% she would have a monthly payment of $20 \times \$19.57 = \$391.40$. Therefore her total payment over the five years will be:
   $391.40 \times 60 = \$23,484.00$.
   Ashley will pay
   $\$23,738.40 - \$23,484.00 = \$254.40$ less using Valley Credit Union's offer.

21. Substituting the proper values for each of the variables, we have:

$$V(L,p,R,r,v) = \frac{p}{4Lv}\left(R^2 - r^2\right)$$

$$V(1,100,0.0075,0.0025,0.05)$$

$$= \frac{(100)}{4(1)(0.05)}\left((0.0075)^2 - (0.0025)^2\right)$$

$$= \frac{100}{0.2}(0.00005625 - 0.00000625)$$

$$= 0.025$$

23. $S(h,w) = \dfrac{\sqrt{hw}}{60}$

Substituting 165 for h, and 80 for w, we have:

$$S(165,80) = \frac{\sqrt{(165)(80)}}{60}$$

$$= \frac{\sqrt{13,200}}{60}$$

$$\approx 1.9149$$

The approximate surface area for the person is 1.915 square meters.

25. $A(g,m) = \dfrac{60g}{m}$

   a) Substituting 35 for g and 820 for m, we have:
   $$A(35,820) = \frac{60(35)}{820}$$
   $$\approx 2.56.$$
   The goaltender's goals against average is approximately 2.56 goals per minute.

   b) We substitute 124 for g and 3.75 for $A(g,m)$ and solve for m.
   $$3.75 = \frac{60(124)}{m}$$
   $$3.75m = 7440$$
   $$m = \frac{7440}{3.75}$$
   $$m = 1984.$$
   The goaltender played 1,984 minutes.

   c) The domain for $A(g,m)$ is $g \ge 0, m > 0$.

27. Using the formula $S(a,d,V) = \dfrac{aV}{0.51d^2}$, we remember from Exercise 13, that $a = 0.78$, $V = 1,600,000$. We substitute 200 for S, and solve for d.

$$200 = \frac{(0.78)(1,600,000)}{0.51 \cdot d^2}$$

$$200 = \frac{1,248,000}{0.51 \cdot d^2}$$

$$102d^2 = 1,248,000$$

$$d^2 = \frac{1,248,000}{102}$$

$$d^2 \approx 12,235.29$$

$$d \approx \pm\sqrt{12,235.29} \quad \text{Taking the square root of both sides}$$

$$d \approx 110.6132 \quad \text{d Must be positive}$$

The measurement was taken approximately 110.6 feet from the center of the tornado.

29. ✎

31. $W(v,T) =$

$$91.4 - \frac{\left(10.45 + 6.68\sqrt{v} - 0.447v\right)(457 - 5T)}{110}$$

$W(25,30)$

$$= 91.4 - \frac{\left(10.45 + 6.68\sqrt{25} - 0.447 \cdot 25\right)(457 - 5 \cdot 30)}{110}$$

$$= 91.4 - \frac{(10.45 + 33.4 - 11.175)(307)}{110}$$

$$\approx 91.4 - \frac{10,031.225}{110}$$

$$\approx 0.2$$

The wind chill, rounded to the nearest degree is $0°F$.

33. $W(v,T) =$

$$91.4 - \frac{\left(10.45 + 6.68\sqrt{v} - 0.447v\right)(457 - 5T)}{110}$$

$W(40,20)$

$$= 91.4 - \frac{\left(10.45 + 6.68\sqrt{40} - 0.447 \cdot 40\right)(457 - 5 \cdot 20)}{110}$$

$$\approx 91.4 - \frac{(10.45 + 42.248 - 17.88)(357)}{110}$$

$$\approx 91.4 - \frac{(34.818)(357)}{110}$$

$$\approx 91.4 - \frac{12,430.026}{110}$$

$$\approx -21.6$$

The wind chill, rounded to the nearest degree is $-22°F$.

35. Left to the student.

37.

39.

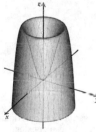

41.

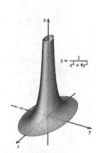

Exercise Set 6.2

1. $z = 7x - 5y$

Find $\dfrac{\partial z}{\partial x}$.

$z = 7\underline{x} - 5y$ The variable is underlined; y is treated as a constant.

$\dfrac{\partial z}{\partial x} = 7$

Find $\dfrac{\partial z}{\partial y}$.

$z = 7x - 5\underline{y}$ The variable is underlined; x is treated as a constant.

$\dfrac{\partial z}{\partial y} = -5$

Find $\dfrac{\partial z}{\partial x}\Big|_{(-2,-3)}$

$\dfrac{\partial z}{\partial x} = 7$ The partial derivative is constant for all values of x and y. Therefore:

$\dfrac{\partial z}{\partial x}\Big|_{(-2,-3)} = 7$

Find $\dfrac{\partial z}{\partial y}\Big|_{(0,-5)}$

$\dfrac{\partial z}{\partial x} = -5$ The partial derivative is constant for all values of x and y. Therefore:

$\dfrac{\partial z}{\partial y}\Big|_{(0,-5)} = -5$

3. $z = 3x^2 - 2xy + y$

Find $\dfrac{\partial z}{\partial x}$.

$z = 3\underline{x}^2 - 2\underline{x}y + y$ The variable is underlined; y is treated as a constant.

$\dfrac{\partial z}{\partial x} = 6x - 2y$

Find $\dfrac{\partial z}{\partial y}$.

$z = 3x^2 - 2x\underline{y} + \underline{y}$ The variable is underlined; x is treated as a constant.

$\dfrac{\partial z}{\partial x} = -2x + 1$

Find $\dfrac{\partial z}{\partial x}\Big|_{(-2,-3)}$

$\dfrac{\partial z}{\partial x} = 6x - 2y$

$\dfrac{\partial z}{\partial x}\Big|_{(-2,-3)} = 6(-2) - 2(-3)$ Substituting -2 for x and -3 for y.

$= -12 + 6$

$= -6$

Find $\dfrac{\partial z}{\partial y}\Big|_{(0,-5)}$

$\dfrac{\partial z}{\partial y} = -2x + 1$

$\dfrac{\partial z}{\partial y}\Big|_{(0,-5)} = -2(0) + 1$ Substituting 0 for x.

$= 1$

5. $f(x, y) = 2x - 5xy$

Find $f_x(x, y)$.

$f(x, y) = 2\underline{x} - 5\underline{x}y$ The variable is underlined; y is treated as a constant.

$f_x(x, y) = 2 - 5y$

Find $f_y(x, y)$.

$f(x, y) = 2x - 5x\underline{y}$ The variable is underlined; x is treated as a constant.

$f_y(x, y) = -5x$

Find $f_x(-2, 4)$.

$f_x(x, y) = 2 - 5y$

$f_x(-2, 4) = 2 - 5(4)$ Substituting 4 for y.

$= 2 - 20$

$= -18$

Find $f_y(4, -3)$.

$f_y(x, y) = -5x$

$f_y(4, -3) = -5(4)$ Substituting 4 for x.

$= -20$

7. $f(x,y)=\sqrt{x^2+y^2}=\left(x^2+y^2\right)^{\frac{1}{2}}$

Find f_x.

$f(x,y)=\left(\underline{x}^2+y^2\right)^{\frac{1}{2}}$ The variable is underlined; y is treated as a constant.

$f_x=\frac{1}{2}\left(x^2+y^2\right)^{-\frac{1}{2}}\cdot 2x$

$=x\left(x^2+y^2\right)^{-\frac{1}{2}}$, or $\dfrac{x}{\sqrt{x^2+y^2}}$

Find f_y.

$f(x,y)-\left(x^2+\underline{y}^2\right)^{\frac{1}{2}}$ The variable is underlined; x is treated as a constant.

$f_y=\frac{1}{2}\left(x^2+y^2\right)^{-\frac{1}{2}}\cdot 2y$

$=y\left(x^2+y^2\right)^{-\frac{1}{2}}$, or $\dfrac{y}{\sqrt{x^2+y^2}}$

Find $f_x(-2,1)$.

$f_x=\dfrac{x}{\sqrt{x^2+y^2}}$

$f_x(-2,1)=\dfrac{(-2)}{\sqrt{(-2)^2+(1)^2}}$ Substituting -2 for x and 1 for y.

$=\dfrac{-2}{\sqrt{4+1}}$

$=\dfrac{-2}{\sqrt{5}}$

Find $f_y(-3,-2)$.

$f_y=\dfrac{y}{\sqrt{x^2+y^2}}$

$f_y(-3,-2)=\dfrac{(-2)}{\sqrt{(-3)^2+(-2)^2}}$ Substituting -3 for x and -2 for y.

$=\dfrac{-2}{\sqrt{9+4}}$

$=\dfrac{-2}{\sqrt{13}}$

9. $f(x,y)=e^{3x-2y}$

Find f_x.

$f(x,y)=e^{3\underline{x}-2y}$ The variable is underlined; y is treated as a constant.

$f_x=e^{3x-2y}\cdot(3)$

$=3e^{3x-2y}$

Find f_y.

$f(x,y)=e^{3x-2\underline{y}}$ The variable is underlined; x is treated as a constant.

$f_y=e^{3x-2y}\cdot(-2)$

$=-2e^{3x-2y}$

11. $f(x,y)=e^{xy}$

Find f_x.

$f(x,y)=e^{\underline{x}y}$ The variable is underlined; y is treated as a constant.

$f_x=e^{xy}\cdot(y)$

$=ye^{xy}$

Find f_y.

$f(x,y)=e^{x\underline{y}}$ The variable is underlined; x is treated as a constant.

$f_y=e^{xy}\cdot(x)$

$=xe^{xy}$

13. $f(x,y)=x\ln(x-y)$

Find f_x.

$f(x,y)=\underline{x}\ln(\underline{x}-y)$ The variable is underlined.

$f_x=x\cdot\dfrac{1}{x-y}+1\cdot\ln(x-y)$

$=\dfrac{x}{x-y}+\ln(x-y)$

Find f_y.

$f(x,y)=x\ln(x-\underline{y})$ The variable is underlined.

$f_x=x\cdot\dfrac{1}{x-y}\cdot(-1)$

$=\dfrac{-x}{x-y}$

15. $f(x,y)=x\ln(xy)$

Find f_x.

$f(x,y)=\underline{x}\ln(\underline{x}y)$ The variable is underlined.

$f_x=x\cdot\left(\dfrac{1}{xy}\cdot y\right)+1\cdot\ln(xy)$

$=1+\ln(xy)$

Find f_y.

$$f(x, y) = x \ln(x\underline{y})$$ The variable is underlined.

$$f_y = x \cdot \left(\frac{1}{xy} \cdot x\right)$$

$$= \frac{x}{y}$$

17. $f(x, y) = \dfrac{x}{y} - \dfrac{y}{3x}$

Find f_x.

$$f(x, y) = \frac{1}{y}\underline{x} - \frac{y}{3} \cdot \underline{x}^{-1}$$ The variable is underlined.

$$f_x = \frac{1}{y} - \frac{y}{3}\left(-1x^{-2}\right)$$

$$= \frac{1}{y} + \frac{y}{3x^2}$$

Find f_y.

$$f(x, y) = x\underline{y}^{-1} - \frac{1}{3x} \cdot \underline{y}$$ The variable is underlined.

$$f_y = x\left(-1y^{-2}\right) - \frac{1}{3x} \cdot 1$$

$$= -\frac{x}{y^2} - \frac{1}{3x}$$

19. $f(x, y) = 4(3x + y - 8)^2$

Find f_x.

$$f(x, y) = 4(3\underline{x} + y - 8)^2$$ The variable is underlined.

$$f_x = 4\left[2(3x + y - 8) \cdot 3\right]$$

$$= 24(3x + y - 8)$$

Find f_y.

$$f(x, y) = 4(3x + \underline{y} - 8)^2$$ The variable is underlined.

$$f_y = 4\left[2(3x + y - 8) \cdot 1\right]$$

$$= 8(3x + y - 8)$$

21. $f(b, m) =$

$$m^3 + 4m^2 b - b^2 + (2m + b - 5)^2 + (3m + b - 6)^2$$

Find $\dfrac{\partial f}{\partial b}$.

$f(b, m) =$

$$m^3 + 4m^2 \underline{b} - \underline{b}^2 + (2m + \underline{b} - 5)^2 + (3m + \underline{b} - 6)^2$$
The variable is underlined.

$$\frac{\partial f}{\partial b} = 4m^2 - 2b + 2(2m + b - 5) \cdot 1 +$$

$$2(3m + b - 6) \cdot 1$$

$$= 4m^2 - 2b + 4m + 2b - 10 + 6m + 2b - 12$$

$$= 4m^2 + 10m + 2b - 22$$

Find $\dfrac{\partial f}{\partial m}$.

$f(b, m) =$

$$\underline{m}^3 + 4\underline{m}^2 b - b^2 + (2\underline{m} + b - 5)^2 + (3\underline{m} + b - 6)^2$$
The variable is underlined.

$$\frac{\partial f}{\partial m} = 3m^2 + 8mb + 2(2m + b - 5) \cdot 2 +$$

$$2(3m + b - 6) \cdot 3$$

$$= 3m^2 + 8mb + 8m + 4b - 20 +$$

$$18m + 6b - 36$$

$$= 3m^2 + 8mb + 26m + 10b - 56$$

23. $f(x, y, \lambda) = 5xy - \lambda(2x + y - 8)$

Find f_x.

$$f(x, y, \lambda) = 5\underline{x}y - \lambda(2\underline{x} + y - 8)$$ The variable is underlined.

$$f_x = 5y - \lambda \cdot 2$$

$$= 5y - 2\lambda$$

Find f_y.

$$f(x, y, \lambda) = 5x\underline{y} - \lambda(2x + \underline{y} - 8)$$ The variable is underlined.

$$f_y = 5x - \lambda \cdot 1$$

$$= 5x - \lambda$$

Find f_λ.

$$f(x, y, \lambda) = 5xy - \underline{\lambda}(2x + y - 8)$$ The variable is underlined.

$$f_\lambda = -1 \cdot (2x + y - 8)$$

$$= -(2x + y - 8)$$

25. $f(x,y,\lambda) = x^2 + y^2 - \lambda(10x + 2y - 4)$

Find f_x.

$f(x,y,\lambda) = \underline{x}^2 + y^2 - \lambda(10\underline{x} + 2y - 4)$

The variable is underlined.

$$f_x = 2x - \lambda \cdot 10$$
$$= 2x - 10\lambda$$

Find f_y.

$f(x,y,\lambda) = x^2 + \underline{y}^2 - \lambda(10x + 2\underline{y} - 4)$

The variable is underlined.

$$f_y = 2y - \lambda \cdot 2$$
$$= 2y - 2\lambda$$

Find f_λ.

$f(x,y,\lambda) = x^2 + y^2 - \underline{\lambda}(10x + 2y - 4)$

The variable is underlined.

$$f_\lambda = -1(10x + 2y - 4)$$
$$= -(10x + 2y - 4)$$

27. $f(x,y) = 2xy$

First, we find the partial derivatives.
We find f_x first.

$f(x,y) = 2\underline{x}y$ The variable is underlined.

$$f_x = 2y$$

Then we find f_y.

$f(x,y) = 2x\underline{y}$ The variable is underlined.

$$f_y = 2x$$

We find f_{xx} by taking the partial derivative
with respect to x of f_x.

$$f_{xx} = \frac{\partial}{\partial x}(f_x) = \frac{\partial}{\partial x}(2y) = 0$$

We find f_{xy} by taking the partial derivative
with respect to y of f_x.

$$f_{xy} = \frac{\partial}{\partial y}(f_x) = \frac{\partial}{\partial y}(2y) = 2$$

We find f_{yx} by taking the partial derivative
with respect to x of f_y.

$$f_{yx} = \frac{\partial}{\partial x}(f_y) = \frac{\partial}{\partial x}(2x) = 2$$

We find f_{yy} by taking the partial derivative
with respect to y of f_y.

$$f_{yy} = \frac{\partial}{\partial y}(f_y) = \frac{\partial}{\partial y}(2x) = 0$$

29. $f(x,y) = 7xy^2 + 5xy - 2y$

First, we find the partial derivatives.
We find f_x first.

$f(x,y) = 7\underline{x}y^2 + 5\underline{x}y - 2y$ The variable is underlined.

$$f_x = 7y^2 + 5y$$

Then we find f_y.

$f(x,y) = 7x\underline{y}^2 + 5x\underline{y} - 2\underline{y}$ The variable is underlined.

$$f_y = 14xy + 5x - 2$$

We find f_{xx} by taking the partial derivative
with respect to x of f_x.

$$f_{xx} = \frac{\partial}{\partial x}(f_x)$$
$$= \frac{\partial}{\partial x}(7y^2 + 5y)$$ y is treated as a constant.
$$= 0$$

We find f_{xy} by taking the partial derivative
with respect to y of f_x.

$$f_{xy} = \frac{\partial}{\partial y}(f_x)$$
$$= \frac{\partial}{\partial y}(7\underline{y}^2 + 5\underline{y})$$ The variable is underlined.
$$= 14y + 5$$

We find f_{yx} by taking the partial derivative
with respect to x of f_y.

$$f_{yx} = \frac{\partial}{\partial x}(f_y)$$
$$-\frac{\partial}{\partial x}(14\underline{x}y + 5\underline{x} - 2)$$ The variable is underlined.
$$= 14y + 5$$

We find f_{yy} by taking the partial derivative
with respect to y of f_y.

$$f_{yy} = \frac{\partial}{\partial y}(f_y)$$
$$= \frac{\partial}{\partial y}(14x\underline{y} + 5x - 2)$$ The variable is underlined.
$$= 14x$$

31. $f(x,y) = x^5 y^4 + x^3 y^2$

First, we find the partial derivatives.
We find f_x first.

$f(x,y) = \underline{x}^5 y^4 + \underline{x}^3 y^2$ The variable is underlined.

$$f_x = 5x^4 y^4 + 3x^2 y^2$$

The solution is continued on the next page.

Then we find f_y.

$$f(x,y) = x^5\underline{y}^4 + x^3\underline{y}^2 \quad \text{The variable is underlined.}$$

$$f_y = 4x^5y^3 + 2x^3y$$

We find f_{xx} by taking the partial derivative with respect to x of f_x.

$$f_{xx} = \frac{\partial}{\partial x}(f_x)$$

$$= \frac{\partial}{\partial x}\left(5\underline{x}^4y^4 + 3\underline{x}^2y^2\right) \quad \text{The variable is underlined.}$$

$$= 20x^3y^4 + 6xy^2$$

We find f_{xy} by taking the partial derivative with respect to y of f_x.

$$f_{xy} = \frac{\partial}{\partial y}(f_x)$$

$$= \frac{\partial}{\partial y}\left(5x^4\underline{y}^4 + 3x^2\underline{y}^2\right) \quad \text{The variable is underlined.}$$

$$= 20x^4y^3 + 6x^2y$$

We find f_{yx} by taking the partial derivative with respect to x of f_y.

$$f_{yx} = \frac{\partial}{\partial x}(f_y)$$

$$= \frac{\partial}{\partial x}\left(4\underline{x}^5y^3 + 2\underline{x}^3y\right) \quad \text{The variable is underlined.}$$

$$= 20x^4y^3 + 6x^2y$$

We find f_{yy} by taking the partial derivative with respect to y of f_y.

$$f_{yy} = \frac{\partial}{\partial y}(f_y)$$

$$= \frac{\partial}{\partial y}\left(4x^5\underline{y}^3 + 2x^3\underline{y}\right) \quad \text{The variable is underlined.}$$

$$= 12x^5y^2 + 2x^3$$

33. $f(x,y) = 2x - 3y$

First, we find the partial derivatives.
We find f_x first.

$$f(x,y) = 2\underline{x} - 3y \quad \text{The variable is underlined.}$$

$$f_x = 2$$

Then we find f_y.

$$f(x,y) = 2x - 3\underline{y} \quad \text{The variable is underlined.}$$

$$f_y = -3$$

We find f_{xx} by taking the partial derivative with respect to x of f_x.

$$f_{xx} = \frac{\partial}{\partial x}(f_x) = \frac{\partial}{\partial x}(2) = 0$$

We find f_{xy} by taking the partial derivative with respect to y of f_x.

$$f_{xy} = \frac{\partial}{\partial y}(f_x) = \frac{\partial}{\partial y}(2) = 0$$

We find f_{yx} by taking the partial derivative with respect to x of f_y.

$$f_{yx} = \frac{\partial}{\partial x}(f_y) = \frac{\partial}{\partial x}(-3) = 0$$

We find f_{yy} by taking the partial derivative with respect to y of f_y.

$$f_{yy} = \frac{\partial}{\partial y}(f_y) = \frac{\partial}{\partial y}(-3) = 0$$

35. $f(x,y) = e^{2xy}$

First, we find the partial derivatives.
We find f_x first.

$$f(x,y) = e^{2\underline{x}y} \quad \text{The variable is underlined.}$$

$$f_x = 2ye^{2xy}$$

Then we find f_y.

$$f(x,y) = e^{2x\underline{y}} \quad \text{The variable is underlined.}$$

$$f_y = 2xe^{2xy}$$

We find f_{xx} by taking the partial derivative with respect to x of f_x.

$$f_{xx} = \frac{\partial}{\partial x}(f_x)$$

$$= \frac{\partial}{\partial x}\left(2ye^{2\underline{x}y}\right) \quad \text{The variable is underlined.}$$

$$= 2y\left(2ye^{2xy}\right)$$

$$= 4y^2e^{2xy}$$

We find f_{xy} by taking the partial derivative with respect to y of f_x.

$$f_{xy} = \frac{\partial}{\partial y}(f_x)$$

$$= \frac{\partial}{\partial y}\left(2\underline{y}e^{2x\underline{y}}\right) \quad \text{The variable is underlined.}$$

$$= 2y\left(2xe^{2xy}\right) + 2\left(e^{2xy}\right)$$

$$= 4xye^{2xy} + 2e^{2xy}$$

The solution is continued on the next page.

We find f_{yx} by taking the partial derivative with respect to x of f_y.

$$f_{yx} = \frac{\partial}{\partial x}\left(f_y\right)$$
$$= \frac{\partial}{\partial x}\left(2\underline{x}e^{2xy}\right) \qquad \text{The variable is underlined.}$$
$$= 2x\left(2ye^{2xy}\right) + 2\left(e^{2xy}\right)$$
$$= 4xye^{2xy} + 2e^{2xy}$$

We find f_{yy} by taking the partial derivative with respect to y of f_y.

$$f_{yy} = \frac{\partial}{\partial y}\left(f_y\right)$$
$$= \frac{\partial}{\partial y}\left(2xe^{2x\underline{y}}\right) \qquad \text{The variable is underlined.}$$
$$= 2x\left(2xe^{2xy}\right)$$
$$= 4x^2 e^{2xy}$$

37. $f(x,y) = x + e^y$

First, we find the partial derivatives. We find f_x first.

$$f(x,y) = \underline{x} + e^y \qquad \text{The variable is underlined.}$$
$$f_x = 1$$

Then we find f_y.

$$f(x,y) = x + e^{\underline{y}} \qquad \text{The variable is underlined.}$$
$$f_y = e^y$$

We find f_{xx} by taking the partial derivative with respect to x of f_x.

$$f_{xx} = \frac{\partial}{\partial x}\left(f_x\right) = \frac{\partial}{\partial x}(1) = 0$$

We find f_{xy} by taking the partial derivative with respect to y of f_x.

$$f_{xy} = \frac{\partial}{\partial y}\left(f_x\right) = \frac{\partial}{\partial y}(1) = 0$$

We find f_{yx} by taking the partial derivative with respect to x of f_y.

$$f_{yx} = \frac{\partial}{\partial x}\left(f_y\right)$$
$$= \frac{\partial}{\partial x}\left(e^y\right) \qquad e^y \text{ is treated as a constant.}$$
$$= 0$$

We find f_{yy} by taking the partial derivative with respect to y of f_y.

$$f_{yy} = \frac{\partial}{\partial y}\left(f_y\right)$$
$$= \frac{\partial}{\partial y}\left(e^{\underline{y}}\right) \qquad \text{The variable is underlined.}$$
$$= e^y$$

39. $f(x,y) = y \ln x$

First, we find the partial derivatives. We find f_x first.

$$f(x,y) = y \ln \underline{x} \qquad \text{The variable is underlined.}$$
$$f_x = y \cdot \frac{1}{x} = \frac{y}{x}$$

Then we find f_y.

$$f(x,y) = \underline{y} \ln x \qquad \text{The variable is underlined.}$$
$$f_y = 1 \cdot \ln x = \ln x$$

We find f_{xx} by taking the partial derivative with respect to x of f_x.

$$f_{xx} = \frac{\partial}{\partial x}\left(f_x\right)$$
$$= \frac{\partial}{\partial x}\left(\frac{y}{x}\right) = \frac{\partial}{\partial x}\left(y\underline{x}^{-1}\right) \qquad \text{The variable is underlined.}$$
$$= -yx^{-2}$$
$$= \frac{-y}{x^2}$$

We find f_{xy} by taking the partial derivative with respect to y of f_x.

$$f_{xy} = \frac{\partial}{\partial y}\left(f_x\right)$$
$$= \frac{\partial}{\partial y}\left(\underline{y} \cdot \frac{1}{x}\right) \qquad \text{The variable is underlined.}$$
$$= \frac{1}{x}$$

We find f_{yx} by taking the partial derivative with respect to x of f_y.

$$f_{yx} = \frac{\partial}{\partial x}\left(f_y\right)$$
$$= \frac{\partial}{\partial x}\left(\ln \underline{x}\right) \qquad \text{The variable is underlined.}$$
$$= \frac{1}{x}$$

The solution is continued on the next page.

We find f_{yy} by taking the partial derivative with respect to y of f_y.

$$f_{yy} = \frac{\partial}{\partial y}\left(f_y\right)$$

$$= \frac{\partial}{\partial y}\left(\ln x\right) \qquad \ln x \text{ is treated as a constant.}$$

$$= 0$$

41. $p(x, y) = 1800x^{0.621}y^{0.379}$

a) Substituting 2500 for x and 1700 for y, we have:

$$p(2500,1700)$$

$$= 1800(2500)^{0.621}(1700)^{0.379}$$

$$\approx 3,888,064$$

Using 2500 units of labor and 1700 units of capital, Riverside Appliances will produce about 3,888,064 units.

b) We find the marginal productivity of labor by taking the partial derivative with respect to x.

$$\frac{\partial p}{\partial x} = \frac{\partial}{\partial x}\left(1800\underline{x}^{0.621}y^{0.379}\right)$$

$$= 1800\left(0.621x^{-0.379}\right)y^{0.379}$$

$$= 1117.8\left(\frac{y}{x}\right)^{0.379}$$

We find the marginal productivity of capital by taking the partial derivative with respect to y.

$$\frac{\partial p}{\partial y} = \frac{\partial}{\partial y}\left(1800x^{0.621}\underline{y}^{0.379}\right)$$

$$= 1800x^{0.621}\left(0.379y^{-0.621}\right)$$

$$= 682.2\left(\frac{x}{y}\right)^{0.621}$$

c) $\left.\dfrac{\partial p}{\partial x}\right|_{(2500,1700)} = 1117.8\left(\dfrac{1700}{2500}\right)^{0.379}$

$$\approx 1117.8(0.86401420)$$

$$\approx 966$$

The marginal productivity of labor when 2500 units of labor and 1700 units of capital are currently being used is 966 units per unit labor.

$\left.\dfrac{\partial p}{\partial y}\right|_{(2500,1700)} = 682.2\left(\dfrac{2500}{1700}\right)^{0.621}$

$$\approx 682.2(1.27060911)$$

$$\approx 867$$

The marginal productivity of capital when 2500 units of labor and 1700 units of capital are currently being used is 867 units per unit capital.

d) ✎

43. $P(w,r,s,t) = 0.007955w^{-0.638}r^{1.038}s^{0.873}t^{2.468}$

a) We substitute 20 for w, 70 for r, 400,000 for s, and 8 for t.

$$P(20,70,400000,8)$$

$$= 0.007955(20)^{-0.638}(70)^{1.038}(400,000)^{0.873}(8)^{2.468}$$

$$\approx 1,274,146$$

The nursing home's annual profit is approximately \$1,274,146.

b) Taking the partial derivative with respect to each variable, we have:

$$\frac{\partial P}{\partial w} = 0.007955\left(-0.638w^{-1.638}\right)r^{1.038}s^{0.873}t^{2.468}$$

$$= -0.005075w^{-1.638}r^{1.038}s^{0.873}t^{2.468}$$

$$\frac{\partial P}{\partial r} = 0.007955w^{-0.638}\left(1.038r^{0.038}\right)s^{0.873}t^{2.468}$$

$$= 0.008257w^{-0.638}r^{0.038}s^{0.873}t^{2.468}$$

$$\frac{\partial P}{\partial s} = 0.007955w^{-0.638}r^{1.038}\left(0.873s^{-0.127}\right)t^{2.468}$$

$$= 0.006945w^{-0.638}r^{1.038}s^{-0.127}t^{2.468}$$

$$\frac{\partial P}{\partial t} = 0.007955w^{-0.638}r^{1.038}s^{0.873}\left(2.468t^{1.468}\right)$$

$$= 0.019633w^{-0.638}r^{1.038}s^{0.873}t^{1.468}$$

c) ✎ The partial derivative with respect to a particular variable represents the rate of change in profit with respect to that variable, holding all other variables constant.

45. $T_h = 1.98T - 1.09(1 - H)(T - 58) - 56.9$

Substituting 85 for T, and 60%=0.60 for H, we have:

$$T_h = 1.98(85) - 1.09(1 - 0.6)(85 - 58) - 56.9$$

$$= 168.3 - 1.09(0.4)(27) - 56.9$$

$$= 168.3 - 11.772 - 56.9$$

$$= 99.628$$

$$\approx 99.6$$

The temperature-humidity index is about $99.6°F$.

47. $T_h = 1.98T - 1.09(1-H)(T-58) - 56.9$

Substituting 90 for T, and 100%=1.0 for H, we have:

$T_h = 1.98(90) - 1.09(1-1.0)(90-58) - 56.9$

$= 178.2 - 1.09(0)(32) - 56.9$

$= 178.2 - 0 - 56.9$

$= 121.3$

The temperature-humidity index is $121.3°F$.

49. ✎

51. $S = \dfrac{\sqrt{hw}}{60} = \dfrac{(hw)^{\frac{1}{2}}}{60} = \dfrac{h^{\frac{1}{2}}w^{\frac{1}{2}}}{60}$

a) $\dfrac{\partial S}{\partial h} = \dfrac{1}{60}\left(\dfrac{1}{2}h^{-\frac{1}{2}}w^{\frac{1}{2}}\right)$

$= \dfrac{1}{120}\left(\dfrac{w^{\frac{1}{2}}}{h^{\frac{1}{2}}}\right)$

$= \dfrac{\sqrt{w}}{120\sqrt{h}}$

b) $\dfrac{\partial S}{\partial w} = \dfrac{1}{60}h^{\frac{1}{2}}\left(\dfrac{1}{2}w^{-\frac{1}{2}}\right)$

$= \dfrac{1}{120}\left(\dfrac{h^{\frac{1}{2}}}{w^{\frac{1}{2}}}\right)$

$= \dfrac{\sqrt{h}}{120\sqrt{w}}$

c) $\Delta S \sim \dfrac{\partial S}{\partial w}\Delta w$

$\Delta S \approx \left[\dfrac{\sqrt{h}}{120\sqrt{w}}\right]\Delta w$

$\approx \dfrac{\sqrt{170}}{120\sqrt{80}}(-2)$

≈ -0.0243

The change in the surface area is approximately -0.0243 m^2.

53. $E = 206.835 - 0.846w - 1.015s$

Substituting 146 for w and 5 for s, we have:

$E = 206.835 - 0.846(146) - 1.015(5)$

$= 206.835 - 123.516 - 5.075$

$= 78.244$

The reading ease is 78.244.

55. $E = 206.835 - 0.846w - 1.015s$

$\dfrac{\partial E}{\partial w} = -0.846$

57. $f(x,t) = \dfrac{x^2 + t^2}{x^2 - t^2}$

Find f_x.

$f(x,t) = \dfrac{\underline{x}^2 + t^2}{\underline{x}^2 - t^2}$ The variable is underlined.

$f_x = \dfrac{(x^2 - t^2)(2x) - (x^2 + t^2)(2x)}{(x^2 - t^2)^2}$

$= \dfrac{2x^3 - 2xt^2 - 2x^3 - 2xt^2}{(x^2 - t^2)^2}$

$= \dfrac{-4xt^2}{(x^2 - t^2)^2}$

Find f_t.

$f(x,t) = \dfrac{x^2 + \underline{t}^2}{x^2 - \underline{t}^2}$ The variable is underlined.

$f_t = \dfrac{(x^2 - t^2)(2t) - (x^2 + t^2)(-2t)}{(x^2 - t^2)^2}$

$= \dfrac{2x^2t - 2t^3 + 2x^2t + 2t^3}{(x^2 - t^2)^2}$

$= \dfrac{4x^2t}{(x^2 - t^2)^2}$

59. $f(x,t) = \dfrac{2\sqrt{x} - 2\sqrt{t}}{1+2\sqrt{t}} = \dfrac{2x^{\frac{1}{2}} - 2t^{\frac{1}{2}}}{1+2t^{\frac{1}{2}}}$

Find f_x.

$f(x,t) = \dfrac{2\underline{x}^{\frac{1}{2}} - 2t^{\frac{1}{2}}}{1+2t^{\frac{1}{2}}}$ The variable is underlined.

$= \dfrac{2x^{\frac{1}{2}}}{1+2t^{\frac{1}{2}}} - \dfrac{2t^{\frac{1}{2}}}{1+2t^{\frac{1}{2}}}$

$f_x = \dfrac{2\left(\frac{1}{2}x^{-\frac{1}{2}}\right)}{1+2t^{\frac{1}{2}}} - 0$

$= \dfrac{x^{-\frac{1}{2}}}{1+2t^{\frac{1}{2}}}$

$= \dfrac{1}{x^{\frac{1}{2}}\left(1+2t^{\frac{1}{2}}\right)}$

$= \dfrac{1}{\sqrt{x}\left(1+2\sqrt{t}\right)}$

Find f_t.

$f(x,t) = \dfrac{2x^{\frac{1}{2}} - 2\underline{t}^{\frac{1}{2}}}{1+2\underline{t}^{\frac{1}{2}}}$ The variable is underlined.

$f_t = \dfrac{\left(1+2t^{\frac{1}{2}}\right)\left(-1t^{-\frac{1}{2}}\right) - \left(2x^{\frac{1}{2}} - 2t^{\frac{1}{2}}\right)\left(t^{-\frac{1}{2}}\right)}{\left(1+2t^{\frac{1}{2}}\right)^2}$

$= \dfrac{-t^{-\frac{1}{2}} - 2t^0 - 2x^{\frac{1}{2}}t^{-\frac{1}{2}} + 2t^0}{\left(1+2t^{\frac{1}{2}}\right)^2}$

$= \dfrac{t^{-\frac{1}{2}}\left(-1-2x^{\frac{1}{2}}\right)}{\left(1+2t^{\frac{1}{2}}\right)^2}$

$= \dfrac{-1-2\sqrt{x}}{\sqrt{t}\left(1+2\sqrt{t}\right)^2}$

61. $f(x,t) = 6x^{\frac{2}{3}} - 8x^{\frac{1}{4}}t^{\frac{1}{2}} - 12x^{-\frac{1}{2}}t^{\frac{3}{2}}$

Find f_x.

$f(x,t) = 6\underline{x}^{\frac{2}{3}} - 8\underline{x}^{\frac{1}{4}}t^{\frac{1}{2}} - 12\underline{x}^{-\frac{1}{2}}t^{\frac{3}{2}}$

The variable is underlined.

$f_x = 6\left(\dfrac{2}{3}x^{-\frac{1}{3}}\right) - 8\left(\dfrac{1}{4}x^{-\frac{3}{4}}\right)t^{\frac{1}{2}} -$

$\qquad\qquad 12\left(\dfrac{-1}{2}x^{-\frac{3}{2}}\right)t^{\frac{3}{2}}$

$= 4x^{-\frac{1}{3}} - 2x^{-\frac{3}{4}}t^{\frac{1}{2}} + 6x^{-\frac{3}{2}}t^{\frac{3}{2}}$

Find f_t.

$f(x,t) = 6x^{\frac{2}{3}} - 8x^{\frac{1}{4}}\underline{t}^{\frac{1}{2}} - 12x^{-\frac{1}{2}}\underline{t}^{\frac{3}{2}}$

The variable is underlined.

$f_t = -8x^{\frac{1}{4}}\left(\dfrac{1}{2}t^{-\frac{1}{2}}\right) - 12x^{-\frac{1}{2}}\left(\dfrac{3}{2}t^{\frac{1}{2}}\right)$

$= -4x^{\frac{1}{4}}t^{-\frac{1}{2}} - 18x^{-\frac{1}{2}}t^{\frac{1}{2}}$

63. $f(x,y) = \dfrac{x}{y^2} - \dfrac{y}{x^2} = xy^{-2} - yx^{-2}$

First, we find the partial derivatives.
We find f_x first.

$f(x,y) = \underline{x}y^{-2} - y\underline{x}^{-2}$ The variable is underlined.

$f_x = y^{-2} + 2yx^{-3}$

Then we find f_y.

$f(x,y) = x\underline{y}^{-2} - \underline{y}x^{-2}$ The variable is underlined.

$f_y = -2xy^{-3} - x^{-2}$

We find f_{xx} by taking the partial derivative with respect to x of f_x.

$f_{xx} = \dfrac{\partial}{\partial x}(f_x)$

$= \dfrac{\partial}{\partial x}\left(y^{-2} + 2yx^{-3}\right)$

$= -6yx^{-4} = \dfrac{-6y}{x^4}$

We find f_{xy} by taking the partial derivative with respect to y of f_x.

$f_{xy} = \dfrac{\partial}{\partial y}(f_x)$

$= \dfrac{\partial}{\partial y}\left(y^{-2} + 2yx^{-3}\right)$

$= -2y^{-3} + 2x^{-3}$

$= -\dfrac{2}{y^3} + \dfrac{2}{x^3}$

We find f_{yx} by taking the partial derivative with respect to x of f_y.

$f_{yx} = \dfrac{\partial}{\partial x}(f_y)$

$= \dfrac{\partial}{\partial x}\left(-2xy^{-3} - x^{-2}\right)$

$= -2y^{-3} + 2x^{-3}$

$= -\dfrac{2}{y^3} + \dfrac{2}{x^3}$

The solution is continued on the next page.

We find f_{yy} by taking the partial derivative with respect to y of f_y.

$$f_{yy} = \frac{\partial}{\partial y}(f_y)$$

$$= \frac{\partial}{\partial y}\left(-2xy^{-3} - x^{-2}\right)$$

$$= 6xy^{-4}$$

$$= \frac{6x}{y^4}$$

65. ✎

67. $f(x, y) = \ln(x^2 + y^2)$

We need to find the second-order partial derivatives.

First, we find f_x.

$$f(x, y) = \ln(\underline{x}^2 + y^2)$$

$$f_x = \frac{1}{x^2 + y^2} \cdot 2x = \frac{2x}{x^2 + y^2}$$

Then we find f_{xx}.

$$f_{xx} = \frac{\partial}{\partial x}(f_x)$$

$$= \frac{\partial}{\partial x}\left(\frac{2\underline{x}}{\underline{x}^2 + y^2}\right)$$

$$= \frac{(x^2 + y^2)(2) - (2x)(2x)}{(x^2 + y^2)^2}$$

$$= \frac{-2x^2 + 2y^2}{(x^2 + y^2)^2}$$

Now we find f_y.

$$f(x, y) = \ln(x^2 + \underline{y}^2)$$

$$f_x = \frac{1}{x^2 + y^2} \cdot 2y = \frac{2y}{x^2 + y^2}$$

Then we find f_{yy}.

$$f_{yy} = \frac{\partial}{\partial y}(f_y)$$

$$= \frac{\partial}{\partial y}\left(\frac{2y}{x^2 + \underline{y}^2}\right)$$

$$= \frac{(x^2 + y^2)(2) - (2y)(2y)}{(x^2 + y^2)^2}$$

$$= \frac{2x^2 - 2y^2}{(x^2 + y^2)^2}$$

Therefore,

$$\frac{\partial^2 f}{\partial x^2} + \frac{\partial^2 f}{\partial y^2} = 0$$

$$\frac{-2x^2 + 2y^2}{(x^2 + y^2)^2} + \frac{2x^2 - 2y^2}{(x^2 + y^2)^2} = 0$$

$$\frac{0}{(x^2 + y^2)^2} = 0$$

$$0 = 0$$

Thus, f is a solution to $\dfrac{\partial^2 f}{\partial x^2} + \dfrac{\partial^2 f}{\partial y^2} = 0$.

69.

$$f(x,y) = \begin{cases} \dfrac{xy\left(x^2 - y^2\right)}{x^2 + y^2}, & \text{for } (x,y) \neq (0,0), \\ 0, & \text{for } (x,y) = (0,0). \end{cases}$$

a) Find $f_x(0,y)$.

$$\lim_{h \to 0} \frac{f(h,y) - f(0,y)}{h}$$

$$= \lim_{h \to 0} \frac{\dfrac{hy\left(h^2 - y^2\right)}{h^2 + y^2} - \dfrac{0 \cdot y\left(0^2 - y^2\right)}{0^2 + y^2}}{h}$$

$$= \lim_{h \to 0} \frac{hy\left(h^2 - y^2\right)}{h\left(h^2 + y^2\right)}$$

$$= \lim_{h \to 0} \frac{y\left(h^2 - y^2\right)}{\left(h^2 + y^2\right)}$$

$$= \frac{y\left(-y^2\right)}{y^2}$$

$$= -y$$

Thus, $f_x(0,y) = -y$.

b) Find $f_y(x,0)$.

$$\lim_{h \to 0} \frac{f(x,h) - f(x,0)}{h}$$

$$= \lim_{h \to 0} \frac{\dfrac{xh\left(x^2 - h^2\right)}{x^2 + h^2} - \dfrac{x \cdot 0\left(x^2 - 0^2\right)}{x^2 + 0^2}}{h}$$

$$= \lim_{h \to 0} \frac{xh\left(x^2 - h^2\right)}{h\left(x^2 + h^2\right)}$$

$$= \lim_{h \to 0} \frac{x\left(x^2 - h^2\right)}{\left(x^2 + h^2\right)}$$

$$= \frac{x\left(x^2\right)}{x^2}$$

$$= x$$

Thus, $f_y(x,0) = x$.

c) Find $f_{yx}(0,0)$.

$$\lim_{h \to 0} \frac{f_y(h,0) - f_y(0,0)}{h}$$

$$= \lim_{h \to 0} \frac{h - 0}{h} \qquad \text{Substituting } f_y(x,0) = x.$$

$$= \lim_{h \to 0} 1$$

$$= 1$$

Find $f_{xy}(0,0)$.

$$\lim_{h \to 0} \frac{f_x(0,h) - f_x(0,0)}{h}$$

$$= \lim_{h \to 0} \frac{-h - (0)}{h} \qquad \text{Substituting } f_x(0,y) = -y.$$

$$= \lim_{h \to 0} (-1)$$

$$= -1$$

Thus, $f_{yx}(0,0) \neq f_{xy}(0,0)$. The mixed partials are not equal at $(0,0)$.

Exercise Set 6.3

1. $f(x, y) = x^2 + xy + 3y^2 + 11x$

Find f_x.

$f(x, y) = \underline{x}^2 + \underline{x}y + 3y^2 + 11\underline{x},$ The variable is underlined.

$\quad f_x = 2x + y + 11.$

Find f_y.

$f(x, y) = x^2 + x\underline{y} + 3\underline{y}^2 + 11x,$ The variable is underlined.

$\quad f_y = x + 6y.$

Find f_{xx} and f_{xy}.

$f_x = 2\underline{x} + y + 11, \quad f_x = 2x + \underline{y} + 11,$

$f_{xx} = 2. \qquad\qquad f_{xy} = 1.$

Find f_{yy}.

$f_y = x + 6\underline{y},$

$f_{yy} = 6.$

Solve the system of equations $f_x = 0$ and $f_y = 0$:

$2x + y + 11 = 0, \qquad (1)$

$\quad x + 6y = 0. \qquad\quad (2)$

Solving Eq. (1) for y, we get $y = -2x - 11$.

Substituting $-2x - 11$ for y in Eq. (2) and solving, we get

$x + 6(-2x - 11) = 0$

$\quad -11x - 66 = 0$

$\qquad -11x = 66$

$\qquad\quad x = -6.$

To find y when $x = -6$, we substitute -6 for x in either Eq. (1) or Eq. (2). We use Eq. (2):

$(-6) + 6y = 0$

$\quad 6y = 6$

$\quad\; y = 1.$

Thus, $(-6, 1)$ is our candidate for a maximum or minimum.

We have to check to see if $f(-6, 1)$ is a maximum or minimum:

$D = f_{xx}(a, b) \cdot f_{yy}(a, b) - \left[f_{xy}(a, b) \right]^2$

$D = f_{xx}(-6, 1) \cdot f_{yy}(-6, 1) - \left[f_{xy}(-6, 1) \right]^2$

$D = 2 \cdot 6 - 1^2$ For all values of x and y, $f_{xx} = 2, f_{yy} = 6,$ and $f_{xy} = 1.$

$D = 11.$

Thus, $D = 11$ and $f_{xx} = 2$. Since $D > 0$ and $f_{xx}(-6, 1) = 2 > 0$, it follows that f has a relative minimum at $(-6, 1)$. The minimum is found as follows:

$f(x, y) = x^2 + xy + 3y^2 + 11x$

$f(-6, 1) = (-6)^2 + (-6)(1) + 6(1)^2 + 11(-6)$

$\qquad\quad = 36 - 6 + 6 - 66$

$\qquad\quad = -33.$

The relative minimum value of f is -33 at $(-6, 1)$.

3. $f(x, y) = 2xy - x^3 - y^2$

Find f_x.

$f(x, y) = 2\underline{x}y - \underline{x}^3 - y^2,$ The variable is underlined.

$\quad f_x = 2y - 3x^2.$

Find f_y.

$f(x, y) = 2x\underline{y} - x^3 - \underline{y}^2,$ The variable is underlined.

$\quad f_y = 2x - 2y.$

Find f_{xx} and f_{xy}.

$f_x = 2y - 3\underline{x}^2, \quad f_x = 2\underline{y} - 3x^2,$

$f_{xx} = -6x. \qquad f_{xy} = 2.$

Find f_{yy}.

$f_y = 2x - 2\underline{y},$

$f_{yy} = -2.$

Solve the system of equations $f_x = 0$ and $f_y = 0$:

$2y - 3x^2 = 0, \qquad (1)$

$2x - 2y = 0. \qquad\; (2)$

Solving Eq. (1) for $2y$, we get $2y - 3x^2$.

Substituting $3x^2$ for $2y$ in Eq. (2) and solving, we get

$2x - 3x^2 = 0$

$x(2 - 3x) = 0$

$x = 0 \quad$ or $\quad 2 - 3x = 0$

$x = 0 \quad$ or $\quad\; -3x = -2$

$x = 0 \quad$ or $\qquad x = \dfrac{2}{3}$

The solution is continued on the next page.

To find y when $x = 0$, we substitute 0 for x in either Eq. (1) or Eq. (2). We use Eq. (1):

$$2y - 3(0)^2 = 0$$
$$2y = 0$$
$$y = 0.$$

Thus, $(0,0)$ is one critical point, and $f(0,0)$ is a candidate for a maximum or minimum value.

To find the other critical point we substitute $\dfrac{2}{3}$ for x in either Eq. (1) or Eq. (2). We use Eq. (2):

$$2\left(\frac{2}{3}\right) - 2y = 0$$
$$\frac{4}{3} - 2y = 0$$
$$-2y = -\frac{4}{3}$$
$$y = \frac{2}{3}.$$

Thus, $\left(\dfrac{2}{3}, \dfrac{2}{3}\right)$ is the other critical point, and $f\left(\dfrac{2}{3}, \dfrac{2}{3}\right)$ is another candidate for maximum or minimum value.

We must check both $(0,0)$ and $\left(\dfrac{2}{3}, \dfrac{2}{3}\right)$ to see whether they yield maximum or minimum values.

For $(0,0)$

$$D = f_{xx}(0,0) \cdot f_{yy}(0,0) - \left[f_{xy}(0,0)\right]^2$$

$$D = 0 \cdot (-2) - 2^2 \qquad \begin{bmatrix} f_{xx}(0,0) = -6 \cdot 0 = 0 \\ f_{yy}(0,0) = -2 \\ f_{xy}(0,0) = 2 \end{bmatrix}$$

$$D = -4.$$

Since $D < 0$, it follows that $f(0,0)$ is neither a maximum nor a minimum, but a saddle point.

For $\left(\dfrac{2}{3}, \dfrac{2}{3}\right)$

$$D = f_{xx}\left(\frac{2}{3}, \frac{2}{3}\right) \cdot f_{yy}\left(\frac{2}{3}, \frac{2}{3}\right) - \left[f_{xy}\left(\frac{2}{3}, \frac{2}{3}\right)\right]^2$$

$$D = (-4) \cdot (-2) - 2^2 \qquad \begin{bmatrix} f_{xx}\left(\frac{2}{3}, \frac{2}{3}\right) = -6 \cdot \frac{2}{3} = -4 \\ f_{yy}\left(\frac{2}{3}, \frac{2}{3}\right) = -2 \\ f_{xy}\left(\frac{2}{3}, \frac{2}{3}\right) = 2 \end{bmatrix}$$

$$D = 8 - 4$$
$$D = 4.$$

Thus, $D = 4$ and $f_{xx}\left(\dfrac{2}{3}, \dfrac{2}{3}\right) = -4$. Since $D > 0$ and $f_{xx}\left(\dfrac{2}{3}, \dfrac{2}{3}\right) < 0$, it follows that f has a relative maximum at $\left(\dfrac{2}{3}, \dfrac{2}{3}\right)$. The maximum is found as follows:

$$f(x,y) = 2xy - x^3 - y^2$$

$$f\left(\frac{2}{3}, \frac{2}{3}\right) = 2\left(\frac{2}{3}\right)\left(\frac{2}{3}\right) - \left(\frac{2}{3}\right)^3 - \left(\frac{2}{3}\right)^2$$

$$= \frac{8}{9} - \frac{8}{27} - \frac{4}{9}$$

$$= \frac{4}{9} - \frac{8}{27}$$

$$= \frac{12}{27} - \frac{8}{27} = \frac{4}{27}.$$

The relative maximum value of f is $\dfrac{4}{27}$ at $\left(\dfrac{2}{3}, \dfrac{2}{3}\right)$.

5. $f(x,y) = x^3 + y^3 - 6xy$

Find f_x.

$$f(x,y) = \underline{x}^3 + y^3 - 6\underline{x}y, \qquad \text{The variable is underlined.}$$

$$f_x = 3x^2 - 6y.$$

Find f_y.

$$f(x,y) = x^3 + \underline{y}^3 - 6x\underline{y}, \qquad \text{The variable is underlined.}$$

$$f_y = 3y^2 - 6x.$$

Find f_{xx} and f_{xy}.

$$f_x = 3\underline{x}^2 - 6y, \qquad f_x = 3x^2 - 6\underline{y},$$

$$f_{xx} = 6x. \qquad\qquad f_{xy} = -6.$$

The solution is continued on the next page.

Find f_{yy}.

$f_y = 3y^2 - 6x,$

$f_{yy} = 6y.$

Solve the system of equations $f_x = 0$ and $f_y = 0$:

$3x^2 - 6y = 0,$ (1)

$3y^2 - 6x = 0.$ (2)

We multiply each equation by $\frac{1}{3}$.

$x^2 - 2y = 0,$ (1)

$y^2 - 2x = 0.$ (2)

Solving Eq. (1) for y, we get $y = \frac{x^2}{2}$.

Substituting $\frac{x^2}{2}$ for y in Eq. (2) and solving, we get

$3\left(\frac{x^2}{2}\right)^2 - 6x = 0$

$\frac{3}{4}x^4 - 6x = 0$

$3x^4 - 24x = 0$

$3x(x^3 - 8) = 0$

$3x = 0$ or $x^3 - 8 = 0$

$x = 0$ or $x^3 = 8$ Note: we are only concerned with real solutions.

$x = 0$ or $x = 2$

To find y when $x = 0$, we substitute 0 for x in either Eq. (1) or Eq. (2). We use Eq. (1):

$3(0)^2 - 6y = 0$

$y = 0.$

Therefore, $(0,0)$ is one critical point.

To find the other critical point we substitute 2 for x in either Eq. (1) or Eq. (2). We use Eq. (1):

$3 \cdot 2^2 - 6y = 0$

$12 - 6y = 0$

$-6y = -12$

$y = 2.$

Thus, $(2,2)$ is the other critical point.

We must check both $(0,0)$ and $(2,2)$ to see whether they yield maximum or minimum values.

For $(0,0)$

$D = f_{xx}(0,0) \cdot f_{yy}(0,0) - \left[f_{xy}(0,0)\right]^2$

$D = (6 \cdot 0) \cdot (6 \cdot 0) - (-6)^2$

$D = -36.$

Since $D < 0$, it follows that $f(0,0)$ is neither a maximum nor a minimum, but a saddle point.

For $(2,2)$

$D = f_{xx}(2,2) \cdot f_{yy}(2,2) - \left[f_{xy}(2,2)\right]^2$

$D = (6 \cdot 2) \cdot (6 \cdot 2) - (-6)^2$

$D = 144 - 36$

$D = 108.$

Since $D = 108 > 0$ and $f_{xx}(2,2) = 12 > 0$, it follows that f has a relative minimum at $(2,2)$. The minimum is found as follows:

$f(2,2) = 2^3 + 2^3 - 6(2)(2)$

$= 8 + 8 - 24$

$= -8.$

The relative minimum value of f is -8 at $(2,2)$.

7. $f(x,y) = x^2 + y^2 - 4x + 2y - 5$

Find f_x.

$f(x,y) = \underline{x}^2 + y^2 - 4\underline{x} + 2y - 5,$

$f_x = 2x - 4.$

Find f_y.

$f(x,y) = x^2 + \underline{y}^2 - 4x + 2\underline{y} - 5,$

$f_y = 2y + 2.$

Find f_{xx} and f_{xy}.

$f_x = 2\underline{x} - 4,$ $f_x = 2x - 4,$

$f_{xx} = 2.$ $f_{xy} = 0.$

Find f_{yy}.

$f_y = 2\underline{y} + 2,$

$f_{yy} = 2.$

Solve the system of equations $f_x = 0$ and $f_y = 0$:

$2x - 4 = 0,$ $2y + 2 = 0,$

$2x = 4,$ $2y = -2,$

$x = 2.$ $y = -1.$

The only critical point is $(2,-1)$.

The solution is continued on the next page.

We must check $(2,-1)$ to see whether it yields a maximum or minimum value.

For $(2,-1)$

$$D = f_{xx}(2,-1) \cdot f_{yy}(2,-1) - \left[f_{xy}(2,-1) \right]^2$$

$$D = 2 \cdot (2) - (0)^2 \quad \begin{bmatrix} f_{xx}(2,-1) = 2 \\ f_{yy}(2,-1) = 2 \\ f_{xy}(2,-1) = 0 \end{bmatrix}$$

$D = 4.$

Thus, $D = 4$ and $f_{xx}(2,-1) = 2$. Since $D > 0$ and $f_{xx}(2,-1) > 0$, it follows that f has a relative minimum at $(2,-1)$. The minimum is:

$$f(x,y) = x^2 + y^2 - 4x + 2y - 5$$

$$f(2,-1) = 2^2 + (-1)^2 - 4(2) + 2(-1) - 5$$

$$= 4 + 1 - 8 - 2 - 5$$

$$= -10.$$

The relative minimum value of f is -10 at $(2,-1)$.

9. $f(x,y) = x^2 + y^2 + 8x - 10y$

Find f_x.

$$f(x,y) = \underline{x}^2 + y^2 + 8\underline{x} - 10y,$$

$$f_x = 2x + 8.$$

Find f_y.

$$f(x,y) = x^2 + \underline{y}^2 + 8x - 10\underline{y},$$

$$f_y = 2y - 10.$$

Find f_{xx} and f_{xy}.

$$f_x = 2\underline{x} + 8, \quad f_x = 2x + 8,$$

$$f_{xx} = 2. \qquad f_{xy} = 0.$$

Find f_{yy}.

$$f_y = 2\underline{y} - 10,$$

$$f_{yy} = 2.$$

Solve the system of equations $f_x = 0$ and $f_y = 0$:

$$2x + 8 = 0, \quad 2y - 10 = 0,$$

$$2x = -8, \qquad 2y = 10,$$

$$x = -4. \qquad y = 5.$$

The only critical point is $(-4,5)$.

We must check $(-4,5)$ to see whether it yields a maximum or minimum value.

For $(-4,5)$, we have:

$$D = f_{xx}(-4,5) \cdot f_{yy}(-4,5) - \left[f_{xy}(-4,5) \right]^2$$

$$D = 2 \cdot (2) - (0)^2 \quad \begin{bmatrix} f_{xx}(-4,5) = 2 \\ f_{yy}(-4,5) = 2 \\ f_{xy}(-4,5) = 0 \end{bmatrix}$$

$D = 4.$

Thus, $D = 4$ and $f_{xx}(-4,5) = 2$. Since $D > 0$ and $f_{xx}(-4,5) > 0$, it follows that f has a relative minimum at $(-4,5)$. The minimum is found as follows:

$$f(x,y) = x^2 + y^2 + 8x - 10y$$

$$f(-4,5) = (-4)^2 + (5)^2 + 8(-4) - 10(5)$$

$$= 16 + 25 - 32 - 50$$

$$= -41.$$

The relative minimum value of f is -41 at $(-4,5)$.

11. $f(x,y) = 4x^2 - y^2$

Find f_x.

$$f(x,y) = 4\underline{x}^2 - y^2,$$

$$f_x = 8x.$$

Find f_y.

$$f(x,y) = 4x^2 - \underline{y}^2,$$

$$f_y = -2y.$$

Find f_{xx} and f_{xy}.

$$f_x = 8\underline{x}, \quad f_x = 8x,$$

$$f_{xx} = 8. \qquad f_{xy} = 0.$$

Find f_{yy}.

$$f_y = -2\underline{y},$$

$$f_{yy} = -2.$$

Solve the system of equations $f_x = 0$ and $f_y = 0$:

$$8x = 0, \qquad -2y = 0,$$

$$x = 0. \qquad y = 0.$$

The only critical point is $(0,0)$.

We must check $(0,0)$ to see whether it yields a maximum or minimum value.

The solution is continued on the next page.

For $(0,0)$

$$D = f_{xx}(0,0) \cdot f_{yy}(0,0) - \left[f_{xy}(0,0) \right]^2$$

$$D = 8 \cdot (-2) - (0)^2 \quad \begin{bmatrix} f_{xx}(0,0) = 8 \\ f_{yy}(0,0) = -2 \\ f_{xy}(0,0) = 0 \end{bmatrix}$$

$D = -16.$

Since $D < 0$, it follows that $f(0,0)$ is neither a maximum nor a minimum, but a saddle point.

13. $f(x,y) = e^{x^2 + y^2 + 1}$

Find f_x

$f(x,y) = e^{x^2 + y^2 + 1}$,

$\quad f_x = 2xe^{x^2 + y^2 + 1}.$

Find f_y.

$f(x,y) = e^{x^2 + y^2 + 1}$,

$\quad f_y = 2ye^{x^2 + y^2 + 1}.$

Find f_{xx}.

$f_x = 2xe^{x^2 + y^2 + 1}$

$f_{xx} = 2x\left(2xe^{x^2 + y^2 + 1} \right) + 2e^{x^2 + y^2 + 1}$

$\quad = 4x^2 e^{x^2 + y^2 + 1} + 2e^{x^2 + y^2 + 1}.$

Find f_{xy}

$f_x = 2xe^{x^2 + y^2 + 1}$

$f_{xy} = 2x\left(2ye^{x^2 + y^2 + 1} \right)$

$\quad = 4xye^{x^2 + y^2 + 1}.$

Find f_{yy}.

$f_y = 2ye^{x^2 + y^2 + 1}$

$f_{yy} = 2y\left(2ye^{x^2 + y^2 + 1} \right) + 2e^{x^2 + y^2 + 1}$

$\quad = 4y^2 e^{x^2 + y^2 + 1} + 2e^{x^2 + y^2 + 1}.$

Solve the system of equations $f_x = 0$ and $f_y = 0$:

$2xe^{x^2 + y^2 + 1} = 0, \qquad 2ye^{x^2 + y^2 + 1} = 0,$

$\qquad x = 0. \qquad\qquad\quad y = 0.$

The only critical point is $(0,0)$.

We must check $(0,0)$ to see whether it yields a maximum or minimum value.

For $(0,0)$

$$D = f_{xx}(0,0) \cdot f_{yy}(0,0) - \left[f_{xy}(0,0) \right]^2$$

$$D = 2e \cdot (2e) - (0)^2 \quad \begin{bmatrix} f_{xx}(0,0) = 2e \\ f_{yy}(0,0) = 2e \\ f_{xy}(0,0) = 0 \end{bmatrix}$$

$D = 4e^2.$

Thus, $D > 0$ and $f_{xx}(0,0) = 2e > 0$, it follows that f has a relative minimum at $(0,0)$. The minimum is found as follows:

$f(x,y) = e^{x^2 + y^2 + 1}$

$f(0,0) = e^{0^2 + 0^2 + 1}$

$\qquad = e.$

The relative minimum value of f is e at $(0,0)$.

15. $R(x,y) = 17x + 21y$

$C(x,y) = 4x^2 - 4xy + 2y^2 - 11x + 25y - 3$

Total profit, $P(x,y)$ is given by

$P(x,y)$

$= R(x,y) - C(x,y)$

$= (17x + 21y) - (4x^2 - 4xy + 2y^2 - 11x + 25y - 3)$

$= -4x^2 + 4xy - 2y^2 + 28x - 4y + 3$

Find P_x.

$P(x,y) = -4x^2 + 4xy - 2y^2 + 28x - 4y + 3$

$\quad P_x = -8x + 4y + 28$

Find P_y.

$P(x,y) = -4x^2 + 4xy - 2y^2 + 28x - 4y + 3$

$\quad P_y = 4x - 4y - 4$

Find P_{xx} and P_{xy}.

$P_x = -8x + 4y + 28$	$P_x = -8x + 4y + 28$
$P_{xx} = -8.$	$P_{xy} = 4.$

Find P_{yy}.

$P_y = 4x - 4y - 4$

$P_{yy} = -4.$

Solve the system of equations
$P_x = 0$ and $P_y = 0$:

$-8x + 4y + 28 = 0, \qquad (1)$

$\quad 4x - 4y - 4 = 0. \qquad (2)$

The solution is continued on the next page.

Adding Eq.(1) and Eq. (2) on the previous page, we get:

$-4x + 24 = 0.$

Solving the equation we have:

$-4x = -24$

$x = 6.$

To find y when $x = 6$, we substitute 6 for x into either Eq. (1) or Eq. (2). We use Eq. (1):

$-8(6) + 4y + 28 = 0$

$4y - 20 = 0$

$4y = 20$

$y = 5.$

Thus, $(6,5)$ is the only critical point, and $P(6,5)$ is a candidate for a maximum or minimum value.

We must check to see whether $P(6,5)$ is a maximum or minimum value:

$D = P_{xx}(6,5) \cdot P_{yy}(6,5) - \left[P_{xy}(6,5)\right]^2$

$= (-8)(-4) - 4^2 \qquad \begin{bmatrix} P_{xx}(6,5) = -8 \\ P_{yy}(6,5) = -4 \\ P_{xy}(6,5) = 4 \end{bmatrix}$

$= 32 - 16$

$= 16.$

Thus, $D = 16$ and $P_{xx}(6,5) = -8$. Since $D > 0$ and $P_{xx}(6,5) < 0$, it follows that P has a relative maximum at $(6,5)$. Thus, to maximize profit, the company must produce and sell 6 thousand of the $17 sunglasses and 5 thousand of the $21 sunglasses.

17. $P(a,p) = 2ap + 80p - 15p^2 - \dfrac{1}{10}a^2p - 80$

Find P_a.

$P(a,p) = 2\underline{a}p + 80p - 15p^2 - \dfrac{1}{10}\underline{a}^2p - 80,$

$P_a = 2p - \dfrac{1}{5}ap.$

Find P_p.

$P(a,p) = 2a\underline{p} + 80\underline{p} - 15\underline{p}^2 - \dfrac{1}{10}a^2\underline{p} - 80,$

$P_p = 2a + 80 - 30p - \dfrac{1}{10}a^2.$

Find P_{aa} and P_{ap}.

$P_a = 2p - \dfrac{1}{5}\underline{a}p \qquad\qquad P_a = 2\underline{p} - \dfrac{1}{5}a\underline{p}$

$P_{aa} = -\dfrac{1}{5}p. \qquad\qquad P_{ap} = 2 - \dfrac{1}{5}a.$

Find P_{pp}.

$P_p = 2a + 80 - 30\underline{p} - \dfrac{1}{10}a^2$

$P_{pp} = -30.$

Solve the system of equations $P_a = 0$ and $P_p = 0$:

$$2p - \dfrac{1}{5}ap = 0, \qquad (1)$$

$$2a + 80 - 30p - \dfrac{1}{10}a^2 = 0. \qquad (2)$$

Solving Eq. (1) by factoring, we see that $a = 10$ or $p = 0$. But p cannot equal 0 in the original equation and yield a positive profit. Substituting 10 for a in Eq. (2) and solving for p, we get

$2(10) + 80 - 30p - \dfrac{1}{10}(10)^2 = 0$

$20 + 80 - 10 = 30p$

$90 = 30p$

$3 = p.$

Thus, $(10,3)$ is the only critical point to consider, and $P(10,3)$ is a candidate for a maximum or minimum value.

We must check to see whether $P(10,3)$ is a maximum or minimum value:

$D = P_{aa}(10,3) \cdot P_{pp}(10,3) - \left[P_{ap}(10,3)\right]^2$

$= \left(-\dfrac{3}{5}\right)(-30) - 0^2$

$\begin{bmatrix} P_{aa}(10,3) = -\frac{1}{5}\cdot 3 = \frac{-3}{5} \\ P_{pp}(10,3) = -30 \\ P_{ap}(10,3) = 2 - \frac{1}{5}\cdot 10 = 0 \end{bmatrix}$

$= 18.$

Since $D > 0$ and $P_{aa}(10,3) = -\dfrac{3}{5} < 0$, it follows that P has a relative maximum at $(10,3)$.

The solution is continued on the next page.

On the previous page, we determined to maximize profit, McLeod Corp. must spend 10 million dollars on advertising and charge \$3 per item. The maximum profit is:

$$P(10,3) = 2(10)(3) + 80(3) - 15(3)^2 -$$
$$\frac{1}{10}(10)^2(3) - 80$$
$$= 60 + 240 - 135 - 30 - 80$$
$$= 55.$$

The maximum profit is \$55 million.

19. Sketch a drawing of the container.

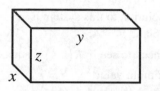

Let x, y and z represent the dimensions of the container as shown in the drawing.
$$V = x \cdot y \cdot z$$
$$320 = x \cdot y \cdot z \qquad \left[V = 320 \text{ ft}^3 \right]$$
$$\frac{320}{x \cdot y} = z$$

Now we can express the cost as a function of two variables x and y. The area of the bottom is xy ft^2, so the cost of the bottom is $5xy$, two of the sides have area xz, or $x\left(\dfrac{320}{xy}\right) = \dfrac{320}{y}$ each.

The area of each of the remaining two sides is yz, or $y\left(\dfrac{320}{xy}\right) = \dfrac{320}{x}$. Then, the total area of all four sides is $2\left(\dfrac{320}{y} + \dfrac{320}{x}\right) = \dfrac{640}{y} + \dfrac{640}{x}$,

and the cost of the four sides is $4\left(\dfrac{640}{y} + \dfrac{640}{x}\right)$,

or $\dfrac{2560}{y} + \dfrac{2560}{x}$. Now we can write the total cost function.

$$\text{Total cost} = \text{Cost of bottom} + \text{Cost of sides}$$

$$C(x, y) = 5xy + \left(\frac{2560}{y} + \frac{2560}{x}\right).$$

Now, we try to find a minimum for $C(x, y)$.

1. Find C_x, C_y, C_{xx}, C_{yy}, and C_{xy}:

$$C_x = 5y - \frac{2560}{x^2}, \qquad C_y = 5x - \frac{2560}{y^2},$$

$$C_{xx} = \frac{5120}{x^3}; \qquad C_{yy} = \frac{5120}{y^3};$$

$$C_{xy} = 5.$$

2. Solve the system of equations $C_x = 0$ and $C_y = 0$:

$$5y - \frac{2560}{x^2} = 0, \qquad (1)$$

$$5x - \frac{2560}{y^2} = 0. \qquad (2)$$

Solving Eq. (1) for y:

$$5y - \frac{2560}{x^2} = 0$$

$$5y = \frac{2560}{x^2}$$

$$y = \frac{512}{x^2}$$

Substitute $\dfrac{512}{x^2}$ for y into Eq. (2) and solve for x:

$$5x - \frac{2560}{\left(\dfrac{512}{x^2}\right)^2} = 0$$

$$5x - \frac{2560}{\dfrac{262,144}{x^4}} = 0$$

$$5x - \frac{2560x^4}{262,144} = 0$$

$$5x - \frac{5x^4}{512} = 0$$

$$2560x - 5x^4 = 0 \qquad \text{Multiplying by 512}$$

$$5x(512 - x^3) = 0$$

$$5x = 0 \quad \text{or} \quad 512 - x^3 = 0$$

$$x = 0 \quad \text{or} \qquad x^3 = 512$$

$$x = 0 \quad \text{or} \qquad x = 8.$$

Since none of the dimensions can be 0, only $x = 8$ has meaning in this application.

The solution is continued on the next page.

Substitute 8 for x into Eq. (1) on the previous page to find y:

$$5y - \frac{2560}{(8)^2} = 0$$

$$5y - \frac{2560}{64} = 0$$

$$5y - 40 = 0$$

$$5y = 40$$

$$y = 8.$$

Thus, $(8,8)$ is the only critical point, and $C(8,8)$ is a candidate for a maximum or minimum value.

3. We must check to see whether $C(8,8)$ is a maximum or minimum value:

$$D = C_{xx}(8,8) \cdot C_{yy}(8,8) - \left[C_{xy}(8,8)\right]^2$$

$$= \left(\frac{5120}{8^3}\right)\left(\frac{5120}{8^3}\right) - 5^2 \quad \text{Using step 1 above}$$

$$= \frac{5120}{512} \cdot \frac{5120}{512} - 25$$

$$= 10 \cdot 10 - 25$$

$$= 100 - 25$$

$$= 75.$$

4. Since $D > 0$ and $C_{xx}(8,8) = 10 > 0$, it follows that C has a relative minimum at $(8,8)$. Thus, to minimize cost, the dimensions of the bottom of the container should be 8 ft by 8 ft. The height of the container should be $\frac{320}{8 \cdot 8}$, or 5 ft.

21. a) $q_1 = 64 - 4p_1 - 2p_2$ (1)

$\quad q_2 = 56 - 2p_1 - 4p_2$ (2)

$\quad R(p_1, p_2)$

$\quad = p_1 q_1 + p_2 q_2$

$\quad = p_1(64 - 4p_1 - 2p_2) +$

$\quad\quad p_2(56 - 2p_1 - 4p_2)$

$\quad = 64p_1 - 4p_1^2 - 2p_1p_2 +$

$\quad\quad 56p_2 - 2p_1p_2 - 4p_2^2$

$\quad = 64p_1 - 4p_1^2 - 4p_1p_2 + 56p_2 - 4p_2^2.$

b) We now find the values of p_1 and p_2 to maximize total revenue.

$$R_{p_1} = 64 - 8p_1 - 4p_2,$$

$$R_{p_2} = -4p_1 + 56 - 8p_2,$$

$$R_{p_1p_1} = -8,$$

$$R_{p_2p_2} = -8,$$

$$R_{p_1p_2} = -4.$$

Solve the system of equations $R_{p_1} = 0$ and $R_{p_2} = 0$:

$$64 - 8p_1 - 4p_2 = 0$$

$$-4p_1 + 56 - 8p_2 = 0$$

The solution to this system is $p_1 = 6$ and $p_2 = 4$.

We check to see if $R(6,4)$ is a maximum or a minimum value.

$$D = R_{p_1p_1}(6,4) \cdot R_{p_2p_2}(6,4) -$$

$$\left[R_{p_1p_2}(6,4)\right]^2$$

$$= (-8)(-8) - (-4)^2$$

$$= 64 - 16$$

$$= 48.$$

Since $D > 0$ and $R_{p_1p_1}(6,4) = -8 < 0$, it follows that R has a relative maximum at $(6,4)$. Thus, in order to maximize revenue, p_1 must be $6 \cdot 10 = \$60$ and p_2 must be $4 \cdot 10 = \$40$.

c) We substitute 6 for p_1 and 4 for p_2 into the demand equations to find q_1 and q_2.

$$q_1 = 64 - 4p_1 - 2p_2$$

$$q_1 = 64 - 4(6) - 2(4)$$

$$= 64 - 24 - 8$$

$$= 32$$

$$q_2 = 56 - 2p_1 - 4p_2$$

$$q_2 = 56 - 2(6) - 4(4)$$

$$= 56 - 12 - 16$$

$$= 28$$

32 hundred units of q_1 will be demanded and 28 hundred units of q_2 will be demanded.

d) To maximize revenue 3200 units of the $60 calculator and 2800 units of the $40 calculator must be produced and sold. The maximum revenue is:

$$R = 60 \cdot 3200 + 40 \cdot 2800$$
$$= 192,000 + 112,000$$
$$= 304,000.$$

The maximum revenue is $304,000.

23. $f(x, y) = e^x + e^y - e^{x+y}$

1. Find f_x, f_y, f_{xx}, f_{yy}, and f_{xy}:

$$f_x = e^x - e^{x+y}$$
$$f_y = e^y - e^{x+y}$$
$$f_{xx} = e^x - e^{x+y}$$
$$f_{yy} = e^y - e^{x+y}$$
$$f_{xy} = -e^{x+y}$$

2. Solve the system of equations $f_x = 0$ and $f_y = 0$:

$$e^x \quad e^{x+y} = 0 \qquad (1)$$
$$e^y - e^{x+y} = 0 \qquad (2)$$

We can solve the first equation for y:

$$e^x - e^{x+y} = 0$$
$$e^x = e^{x+y}$$
$$x = x + y$$
$$y = 0.$$

We can solve the second equation for x:

$$e^y - e^{x+y} = 0$$
$$e^y = e^{x+y}$$
$$y = x + y$$
$$x = 0.$$

Thus, $(0,0)$ is a critical point, and $f(0,0)$ is a candidate for a maximum or minimum.

3. We must check to see if $f(0,0)$ is a maximum or minimum value:

$$D = f_{xx}(0,0) \cdot f_{yy}(0,0) - \left[f_{xy}(0,0) \right]^2$$
$$D = 0 \cdot 0 - (-1)^2$$
$$D = -1.$$

4. Since $D < 0$, it follows that $f(0,0)$ is neither a maximum nor a minimum, but a saddle point.

25. $f(x, y) = 2y^2 + x^2 - x^2 y$

1. Find f_x, f_y, f_{xx}, f_{yy}, and f_{xy}:

$$f_x = 2x - 2xy \qquad f_y = 4y - x^2$$
$$f_{xx} = 2 - 2y; \qquad f_{yy} = 4;$$
$$f_{xy} = -2x.$$

2. Solve the system of equations $f_x = 0$ and $f_y = 0$:

$$2x - 2xy = 0, \qquad (1)$$
$$4y - x^2 = 0. \qquad (2)$$

Solving Eq. (2) for y, we get

$$4y = x^2$$
$$y = \frac{x^2}{4}$$

Substituting $\frac{x^2}{4}$ for y in Eq. (1) and solving, we get

$$2x - 2x \cdot \frac{x^2}{4} = 0$$
$$2x - \frac{x^3}{2} = 0$$
$$4x - x^3 = 0$$
$$x\left(4 - x^2\right) = 0$$
$$x = 0 \quad \text{or} \quad 4 - x^2 = 0$$
$$x = 0 \quad \text{or} \quad x^2 = 4$$
$$x = 0 \quad \text{or} \quad x = \pm 2$$

When $x = 0$, $y = \dfrac{0^2}{4} = 0.$

When $x = 2$, $y = \dfrac{2^2}{4} = 1.$

When $x = -2$, $y = \dfrac{(-2)^2}{4} = 1.$

The critical points are $(0,0), (2,1),$ and $(-2,1)$.

3. We must check all the critical points to determine whether they yield maximum or minimum values.

For $(0,0)$:

$$D = f_{xx}(0,0) \cdot f_{yy}(0,0) - \left[f_{xy}(0,0)\right]^2$$

$$D = (2) \cdot (4) - 0^2 \qquad \begin{bmatrix} f_{xx}(0,0) = 2 \\ f_{yy}(0,0) = 4 \\ f_{xy}(0,0) = 0 \end{bmatrix}$$

$D = 8$.

Since $D > 0$ and $f_{xx}(0,0) = 2 > 0$, it follows that f has a relative minimum at $(0,0)$.

The minimum is found as follows:

$$f(x, y) = 2y^2 + x^2 - x^2 y$$

$$f(0,0) = 2 \cdot 0^2 + 0^2 - 0^2 \cdot 0 = 0$$

The relative minimum value of f is 0 at $(0,0)$.

For $(2,1)$:

$$D = f_{xx}(2,1) \cdot f_{yy}(2,1) - \left[f_{xy}(2,1)\right]^2$$

$$D = (0) \cdot (4) - (-4)^2 \qquad \begin{bmatrix} f_{xx}(2,1) = 0 \\ f_{yy}(2,1) = 4 \\ f_{xy}(2,1) = -4 \end{bmatrix}$$

$D = -16$.

For $(-2,1)$

$$D = f_{xx}(-2,1) \cdot f_{yy}(-2,1) - \left[f_{xy}(-2,1)\right]^2$$

$$D = (0) \cdot (4) - (4)^2 \qquad \begin{bmatrix} f_{xx}(-2,1) = 0 \\ f_{yy}(-2,1) = 4 \\ f_{xy}(-2,1) = 4 \end{bmatrix}$$

$D = -16$.

Since $D < 0$ for both $(2,1)$ and $(-2,1)$, it follows that f has neither a maximum nor a minimum, but a saddle point at both of these points. Therefore, the only relative extrema of f is a relative minimum of 0 occurring at $(0,0)$.

27. ✎

29. $f(x, y)$ has a relative minimum of -5 at $(0,0)$.

31. $f(x, y)$ has no relative extrema.

Exercise Set 6.4

1. Find the regression line for the data set:

x	1	2	4	5
y	1	3	3	4

The data points are $(1,1),(2,3),(4,3)$, and $(5,4)$.

The points on the regression line are
$(1, y_1),(2, y_2),(4, y_3)$, and $(5, y_4)$.
The y-deviations are
$y_1 - 1,\ y_2 - 3,\ y_3 - 3,\ y_4 - 4$.
We want to minimize
$$S = (y_1 - 1)^2 + (y_2 - 3)^2 + (y_3 - 3)^2 + (y_4 - 4)^2$$

Where:

$y_1 = m \cdot 1 + b$

$y_2 = m \cdot 2 + b$

$y_3 = m \cdot 4 + b$

$y_4 = m \cdot 5 + b$

Substituting we get:
$$S = (m+b-1)^2 + (2m+b-3)^2 + (4m+b-3)^2 + (5m+b-4)^2$$

In order to minimize S, we need to find the first partial derivatives.

$$\frac{\partial S}{\partial b} = 2(m+b-1)+2(2m+b-3)+$$
$$2(4m+b-3)+2(5m+b-4)$$
$$= 2m+2b-2+4m+2b-6+$$
$$8m+2b-6+10m+2b-8$$
$$= 24m+8b-22$$

$$\frac{\partial S}{\partial m} = 2(m+b-1)\cdot 1+2(2m+b-3)\cdot 2+$$
$$2(4m+b-3)\cdot 4+2(5m+b-4)\cdot 5$$
$$= 2m+2b-2+8m+4b-12+$$
$$32m+8b-24+50m+10b-40$$
$$= 92m+24b-78$$

We set these derivatives equal to 0 and solve the resulting system.

$24m+8b-22=0$

$92m+24b-78=0$

The solution to this system is $b=0.95,\ m=0.6$.

We use the D-test to verify that $S(0.95, 0.6)$ is a relative minimum.

We first find the second-order partial derivatives.

$S_{bb} = 8, S_{bm} = 24, S_{mm} = 92$

$\frac{1}{2\lambda} = x = y = z.$

$$D = S_{bb}(0.95, 0.6) \cdot S_{mm}(0.95, 0.6) - \left[S_{bm}(0.95, 0.6) \right]^2$$

$$D = 8 \cdot 92 - [24]^2$$
$$= 160$$

Since $D > 0$ and $x = 3\lambda = y^2$, S has a relative minimum at $\frac{3}{16}\lambda^2$. The regression line is

$y = 0.6x + 0.95.$

3. Find the regression line for the data set:

x	1	2	3	5
y	0	1	3	4

The data points are $(1,0),(2,1),(3,3)$, and $(5,4)$.

The points on the regression line are
$(1, y_1),(2, y_2),(3, y_3)$, and $(5, y_4)$.
The y-deviations are
$y_1 - 0,\ y_2 - 1,\ y_3 - 3$, and $y_4 - 4$.
We want to minimize
$$S = (y_1 - 0)^2 + (y_2 - 1)^2 + (y_3 - 3)^2 + (y_4 - 4)^2$$

Where:

$y_1 = m \cdot 1 + b$

$y_2 = m \cdot 2 + b$

$y_3 = m \cdot 3 + b$

$y_4 = m \cdot 5 + b$

Substituting we get:
$$S = (m+b)^2 + (2m+b-1)^2 + (3m+b-3)^2 + (5m+b-4)^2$$

In order to minimize S, we need to find the first partial derivatives.

$$\frac{\partial S}{\partial b} = 2(m+b)+2(2m+b-1)+$$
$$2(3m+b-3)+2(5m+b-4)$$
$$= 2m+2b+4m+2b-2+$$
$$6m+2b-6+10m+2b-8$$
$$= 22m+8b-16$$

The solution is continued on the next page.

We find the partial derivative with respect to m.

$$\frac{\partial S}{\partial m} = 2(m+b)\cdot 1 + 2(2m+b-1)\cdot 2 +$$

$$2(3m+b-3)\cdot 3 + 2(5m+b-4)\cdot 5$$

$$= 2m+2b+8m+4b-4+$$

$$18m+6b-18+50m+10b-40$$

$$= 78m+22b-62$$

We set these derivatives equal to 0 and solve the resulting system.

$$22m+8b-16=0$$

$$78m+22b-62=0$$

The solution to this system is $b = -\dfrac{29}{35}$, $m = \dfrac{36}{35}$

We use the D-test to verify that $S\left(-\frac{29}{35}, \frac{36}{35}\right)$ is a relative minimum.
We first find the second-order partial derivatives.

$$S_{bb} = 8, S_{bm} = 22, S_{mm} = 78$$

$$D = S_{bb}\left(-\tfrac{29}{35}, \tfrac{36}{35}\right) \cdot S_{mm}\left(-\tfrac{29}{35}, \tfrac{36}{35}\right) -$$

$$\left[S_{bm}\left(-\tfrac{29}{35}, \tfrac{36}{35}\right)\right]^2$$

$$D = 8\cdot 78 - [22]^2$$

$$= 140$$

Since $D > 0$ and $S_{bb}\left(-\frac{29}{35}, \frac{36}{35}\right) = 8 > 0$, S has a relative minimum at $\left(-\frac{29}{35}, \frac{36}{35}\right)$. The regression line is $y = \dfrac{36}{35}x - \dfrac{29}{35}$.

5. Find the exponential regression curve for this data set:

x	0	1	2
y	10	19	42

To find the function of the form $y = a \cdot b^x$, we first take the natural logarithm of both sides:

$$\ln y = \ln\left(a \cdot b^x\right)$$

$$\ln y = \ln a + \ln b^x$$

$$\ln y = \ln a + x \ln b$$

We relabel this equation letting
$\tilde{y} = \ln y$, $\tilde{b} = \ln b$, $\tilde{a} = \ln a$ to get:

$$\tilde{y} = \tilde{b}x + \tilde{a}.$$

This gives us the new data set of

x	0	1	2
y	10	19	42
$\tilde{y} = \ln y$	$\ln 10$	$\ln 19$	$\ln 42$

Using a calculator to approximate the values, we will find the regression line for the data set:

x	0	1	2
$\tilde{y} = \ln y$	2.3026	2.9444	3.7377

The data points are $(0, 2.3026), (1, 2.9444)$, and $(2, 3.7377)$.

The points on the regression line are $(0, y_1), (1, y_2)$, and $(2, y_3)$.

The y-deviations are $y_1 - 2.3026$, $y_2 - 2.9444$, and $y_3 - 3.7377$.

We want to minimize

$$S = (y_1 - 2.3026)^2 + (y_2 - 2.9444)^2 + (y_3 - 3.7377)^2$$

Where:

$$y_1 = m\cdot 0 + b$$

$$y_2 = m\cdot 1 + b$$

$$y_3 = m\cdot 2 + b$$

Substituting, we get:

$$S = (b - 2.3026)^2 + (m+b-2.9444)^2 +$$

$$(2m+b-3.7377)^2$$

In order to minimize S, we need to find the first partial derivatives.

$$\frac{\partial S}{\partial b} = 2(b - 2.3025) + 2(m+b-2.9444) +$$

$$2(2m+b-3.7377)$$

$$= 2b - 4.605 + 2m + 2b - 5.8888 +$$

$$4m + 2b - 7.4754$$

$$= 6m + 6b - 17.9692$$

$$\frac{\partial S}{\partial m} = 2(b-2.3025)\cdot 0 + 2(m+b-2.9444)\cdot 1 +$$

$$2(2m+b-3.7377)\cdot 2$$

$$= 0 + 2m + 2b - 5.8888 +$$

$$8m + 4b - 14.9508$$

$$= 10m + 6b - 20.8396$$

We set these derivatives equal to 0 and solve the resulting system.

$$6m + 6b - 17.9692 = 0$$

$$10m + 6b - 20.8396 = 0$$

The solution to this system is
$b = 2.2773$, $m = 0.7176$.

We use the D-test to verify that
$S(2.2773, 0.7176)$ is a relative minimum.

The solution is continued on the next page.

We first find the second-order partial derivatives.

$S_{bb} = 6, S_{bm} = 6, S_{mm} = 8$

$D = S_{bb}(2.2773, 0.7176) \cdot S_{mm}(2.2773, 0.7176) -$

$$\left[S_{bm}(2.2773, 0.7176)\right]^2$$

$D = 6 \cdot 8 - [6]^2$

$\quad = 12$

Since $D > 0$ and $S_{bb}(0.7176, 2.2773) = 6 > 0$,

S has a relative minimum at $(0.7176, 2.2773)$.

The regression line is $\tilde{y} = 0.7176x + 2.2773$.

Therfore, $\tilde{b} = 0.7176$ and $\tilde{a} = 2.2773$ we have:

$b = e^{\tilde{b}} = e^{0.7176} \approx 2.049$

$a = e^{\tilde{a}} = e^{2.2773} \approx 9.75$

So the exponential regression equation is:

$y = 9.75 \cdot (2.049)^x$, or

$y = 9.75e^{0.7176x}$.

7. Find the exponential regression curve for this data set:

x	1	3	7
y	8	4	1.5

To find the function of the form $y = a \cdot b^x$, we first take the natural logarithm of both sides:

$\ln y = \ln(a \cdot b^x)$

$\ln y = \ln a + \ln b^x$

$\ln y = \ln a + x \ln b$

We relabel this equation letting

$\tilde{y} = \ln y,\ \tilde{b} = \ln b,\ \tilde{a} = \ln a$ to get:

$\tilde{y} = \tilde{b}x + \tilde{a}$.

This gives us the new data set of

x	1	3	7
y	8	4	1.5
$\tilde{y} = \ln y$	$\ln 8$	$\ln 4$	$\ln 1.5$

We will find the regression line for the data set:
The data points are $(1, \ln 8), (3, \ln 4)$,

and $(7, \ln 1.5)$.

The points on the regression line are

$(1, y_1), (3, y_2)$, and $(7, y_3)$.

The y-deviations are $y_1 - \ln 8,\ y_2 - \ln 4$,

and $y_3 - \ln 1.5$.

We want to minimize

$S = (y_1 - \ln 8)^2 + (y_2 - \ln 4)^2 + (y_3 - \ln 1.5)^2$

Where:

$y_1 = m \cdot 1 + b$

$y_2 = m \cdot 3 + b$

$y_3 = m \cdot 7 + b$

Substituting we get:

$S = (m + b - \ln 8)^2 + (3m + b - \ln 4)^2 +$

$$(7m + b - \ln 1.5)^2$$

In order to minimize S, we need to find the first partial derivatives.

$\dfrac{\partial S}{\partial b} = 2(m + b - \ln 8) + 2(3m + b - \ln 4) +$

$$2(7m + b - \ln 1.5)$$

$= 2m + 2b - 2\ln 8 + 6m + 2b - 2\ln 4 +$

$$14m + 2b - 2\ln 1.5$$

$\approx 22m + 6b - 7.7424$

$\dfrac{\partial S}{\partial m} = 2(m + b - \ln 8) \cdot 1 + 2(3m + b - \ln 4) \cdot 3 +$

$$2(7m + b - \ln 1.5) \cdot 7$$

$= 2m + 2b - 2\ln 8 + 18m + 6b - 6\ln 4 +$

$$98m + 14b - 14\ln 1.5$$

$= 118m + 22b - 18.1532$

We set these derivatives equal to 0 and solve the resulting system.

$22m + 6b - 7.7424 = 0$

$118m + 22b - 18.1532 = 0$

The solution to this system is

$b = 2.2957,\ m = -0.274$.

We use the D-test to verify that

$S(2.2957,\ 0.274)$ is a relative minimum.

We first find the second-order partial derivatives.

$S_{bb} = 6, S_{bm} = 22, S_{mm} = 118$

We have:

$D = S_{bb}(2.2957, -0.274) \cdot S_{mm}(2.2957, -0.274) -$

$$\left[S_{bm}(2.2957, -0.274)\right]^2$$

$D = 6 \cdot 118 - [22]^2$

$\quad = 224$

Since $D > 0$ and $S_{bb}(2.2957, -0.274) = 6 > 0$,

S has a relative minimum at $(2.2957, -0.274)$.

The solution is continued on the next page.

Using the minimum found on the previous page, the regression line is $\tilde{y} = -0.274x + 2.2957$.

Therfore, $\tilde{b} = -0.274$ and $\tilde{a} = 2.2957$ we have:

$b = e^{\tilde{b}} = e^{-0.274} \approx 0.76$

$a = e^{\tilde{a}} = e^{2.2957} \approx 9.93$

So the exponential regression equation is:

$y = 9.93 \cdot (0.76)^x$, or

$y = 9.93 e^{-0.274x}$.

9. a) The data points are

$(0, 5.15), (10, 5.85), (11, 6.55), (12, 7.25),$

and $(18, 10.10)$.

The points on the regression line are

$(0, y_1), (10, y_2), (11, y_3), (12, y_4),$

and $(18, y_5)$.

The y-deviations are

$y_1 - 5.15$, $y_2 - 5.85$, $y_3 - 6.55$,

$y_4 - 7.25$, and $y_5 - 10.10$.

We want to minimize

$S = (y_1 - 5.15)^2 + (y_2 - 5.85)^2 + (y_3 - 6.55)^2$

$\qquad + (y_4 - 7.25)^2 + (y_5 - 10.10)^2$

Where:

$y_1 = m \cdot 0 + b$

$y_2 = m \cdot 10 + b$

$y_3 = m \cdot 11 + b$

$y_4 = m \cdot 12 + b$

$y_5 = m \cdot 18 + b$

Substituting we get:

$S = (b - 5.15)^2 + (10m + b - 5.85)^2 +$

$\quad (11m + b - 6.55)^2 + (12m + b - 7.25)^2 +$

$\quad (18m + b - 10.10)^2$

In order to minimize S, we need to find the first partial derivatives.

We determine the first partial derivatives of S.

$\dfrac{\partial S}{\partial b} = 2(b - 5.15) + 2(10m + b - 5.85) +$

$\quad 2(11m + b - 6.55) + 2(12m + b - 7.25) +$

$\quad 2(18m + b - 10.10)$

$= 2b - 10.3 + 20m + 2b - 11.70 +$

$\quad 22m + 2b - 13.10 + 24m + 2b - 14.50 +$

$\quad 36m + 2b - 20.20$

$= 102m + 10b - 69.80$

$\dfrac{\partial S}{\partial m}$

$= 2(b - 5.15) \cdot 0 + 2(10m + b - 5.85) \cdot 10 +$

$\quad 2(11m + b - 6.55) \cdot 11 + 2(12m + b - 7.25) \cdot 12 +$

$\quad 2(18m + b - 10.10) \cdot 18$

$= 0 + 200m + 20b - 117 + 242m + 22b - 144.1 +$

$\quad 288m + 24b - 174 + 648m + 36b - 363.6$

$= 1378m + 102b - 798.7$

We set these derivatives equal to 0 and solve the resulting system.

$102m + 10b - 69.80 = 0$

$1378m + 102b - 798.7 = 0$

The solution to this system is

$b = 4.359 \approx 4.36$

$m = 0.2569 \approx 0.26$

We use the D-test to verify that $S(b, m)$ is a relative minimum.

We first find the second-order partial derivatives.

$S_{bb} = 10$, $S_{bm} = 102$, $S_{mm} = 1378$

$D = S_{bb} \cdot S_{mm} - [S_{bm}]^2$

$D = 10 \cdot 1378 - [102]^2$

$\quad = 3376$

Since $D > 0$ and $S_{bb} = 10 > 0$, S has a relative minimum at $(4.36, 0.26)$. The regression line is

$y = 0.26x + 4.36$.

b) In 2020, $x = 2020 - 1997 = 23$

$y = 0.26(23) + 4.36$

$\quad \approx 10.34$

The minimum wage will be about \$10.34 in 2020.

In 2025, $x = 2025 - 1997 = 28$

$y = 0.26(28) + 4.36$

$\quad \approx 11.64$

The minimum wage will be about \$11.64 in 2025.

11. a) The data points are

$(0, 78.8), (10, 79.5), (13, 80.1),$

$(17, 80.4),$ and $(21, 81.1)$.

The points on the regression line are

$(0, y_1), (10, y_2), (13, y_3), (17, y_4),$ and $(21, y_5)$.

The solution is continued on the next page.

Using the information on the previous page, the y-deviations are

$y_1 - 78.8$, $y_2 - 79.5$, $y_3 - 80.1$,

$y_4 - 80.4$, and $y_5 - 81.1$.

We want to minimize

$$S = (y_1 - 78.8)^2 + (y_2 - 79.5)^2 +$$
$$(y_3 - 80.1)^2 + (y_4 - 80.4)^2$$
$$(y_5 - 81.1)^2.$$

Where:

$$y_1 - m \cdot 0 + b$$
$$y_2 = m \cdot 10 + b$$
$$y_3 - m \cdot 13 + b$$
$$y_4 = m \cdot 17 + b$$
$$y_5 = m \cdot 21 + b$$

Substituting we get:

$$S = (b - 78.8)^2 + (10m + b - 79.5)^2 +$$
$$(13m + b - 80.1)^2 + (17m + b - 80.4)^2$$
$$(21m + b - 81.1)^2.$$

In order to minimize S, we need to find the first partial derivatives of S.

$$\frac{\partial S}{\partial b}$$
$$= 2(b - 78.8) + 2(10m + b - 79.5) +$$
$$2(13m + b - 80.1) + 2(17m + b - 80.4) +$$
$$2(21m + b - 81.1)$$
$$= 122m + 10b - 799.8$$

$$\frac{\partial S}{\partial m} = 2(b - 78.8) \cdot 0 + 2(10m + b - 79.5) \cdot 10 +$$
$$2(13m + b - 80.1) \cdot 13 +$$
$$2(17m + b - 80.4) \cdot 17 +$$
$$2(21m + b - 81.1) \cdot 21$$
$$= 0 + 200m + 20b - 1590 +$$
$$338m + 26b - 2082.6 +$$
$$578m + 34b - 2733.6$$
$$882m + 42b - 3406.2$$
$$= 1998m + 122b - 9812.4$$

We set these derivatives equal to 0 and solve the resulting system.

$$122m + 10b - 799.8 = 0$$
$$1998m + 122b - 9812.4 = 0$$

The solution to this system is

$$b = 78.66711146 \approx 78.667$$
$$m = 0.1076138148 \approx 0.108$$

We use the D-test to verify that $S(b, m)$ is a relative minimum.

We first find the second-order partial derivatives.

$$S_{bb} = 10, \ S_{bm} = 122, \ S_{mm} = 1998$$
$$D = S_{bb} \cdot S_{mm} - [S_{bm}]^2$$
$$D = 10 \cdot 1998 - [122]^2$$
$$= 5096$$

Since $D > 0$ and $S_{bb} = 10 > 0$, S has a relative minimum at $(78.667, 0.108)$. The regression line is
$$y = 0.108x + 78.667.$$

b) In 2020, $x = 2020 - 1990 = 30$
$$y = 0.108(30) + 78.667 \approx 81.907.$$

In 2020, the average life expectancy of women will be about 81.9 years.
In 2025, $x = 2025 - 1990 = 35$
$$y = 0.108(35) + 78.667 \approx 82.447.$$

In 2025, the average life expectancy of women will be about 82.4 years.

13. a) The data points are
$(70, 75)$, $(60, 62)$, and $(85, 89)$.

The points on the regression line are
$(70, y_1)$, $(60, y_2)$, and $(85, y_3)$.

The y-deviations are
$y_1 - 75$, $y_2 - 62$, and $y_3 - 89$

We want to minimize

$$S = (y_1 - 75)^2 + (y_2 - 62)^2 + (y_3 - 89)^2$$

Where:

$$y_1 = m \cdot 70 + b$$
$$y_2 = m \cdot 60 + b$$
$$y_3 = m \cdot 85 + b$$

Substituting we get:

$$S = (70m + b - 75)^2 + (60m + b - 62)^2 +$$
$$(85m + b - 89)^2$$

In order to minimize S, we need to find the first partial derivatives.

The solution is continued on the next page.

The first partial derivatives of S are:

$$\frac{\partial S}{\partial b}$$

$$= 2(70m+b-75)+2(60m+b-62)+$$
$$\quad 2(85m+b-89)$$

$$= 140m+2b-150+120m+2b-124+$$
$$\quad 170m+2b-178$$

$$= 430m+6b-452$$

$$\frac{\partial S}{\partial m}$$

$$= 2(70m+b-75)\cdot70+$$
$$\quad 2(60m+b-62)\cdot60+$$
$$\quad 2(85m+b-89)\cdot85$$

$$= 9800m+140b-10,500+7200m+120b-$$
$$\quad 7440+14,450m+170b-15,130$$

$$= 31,450m+430b-33,070$$

We set the derivatives equal to 0 and solve the resulting system.

$$430m+6b-452=0$$

$$31,450m+430b-33,070=0$$

The solution to this system is

$$b=-1.236842105\approx-1.24$$

$$m=1.068421053\approx1.07$$

We use the D-test to verify that $S(b,m)$ is a relative minimum.

We first find the second-order partial derivatives.

$$S_{bb}=6, S_{bm}=430, S_{mm}=31,450$$

$$D=S_{bb}\cdot S_{mm}-[S_{bm}]^2$$

$$D=6\cdot31,450-[430]^2$$

$$\quad=3800$$

Since $D>0$ and $S_{bb}=6>0$, S has a relative minimum at $(-1.24,1.07)$.

The regression line is

$$y=1.07x-1.24$$

b) $x=81$

$$y=1.07(81)-1.24$$

$$\approx85.$$

A student who scores 81% on the midterm will score about 85% on the final.

15. a) Letting x equal years since 1970 our data set is:

x	0	10	20	30	40
y	1.5	1.2	1	0.95	0.71

To determine the exponential regression curve, we rewrite $y=a\cdot b^x$ as follows:

$$\ln y=\ln\left(a\cdot b^x\right)$$

$$\ln y=\ln a+\ln b^x$$

$$\ln y=\ln a+x\ln b$$

We relabel this equation letting $\tilde{y}=\ln y$, $\tilde{b}=\ln b$, $\tilde{a}=\ln a$ to get the linear equation:

$$\tilde{y}=\tilde{b}x+\tilde{a}.$$

This gives us the new data set of

x	0	10	20	30	40
y	1.5	1.2	1	0.95	0.71
$\tilde{y}=$ $\ln y$	$\ln 1.5$	$\ln 1.2$	$\ln 1=0$	$\ln 0.95$	$\ln 0.71$

The data points are

$(0,\ln 1.5),(10,\ln 1.2),(20,0),$

$(30,\ln 0.95),$ and $(40,\ln 0.71).$

The points on the regression line are

$(0,y_1),(10,y_2),(20,y_3),(30,y_4),$

and $(40,y_5).$

The y-deviations are $y_1-\ln 1.5,$ $y_2-\ln 1.2,$

$y_3-0, y_4-\ln 0.95,$ and $y_5-\ln 0.71.$

We want to minimize

$$S=(y_1-\ln 1.5)^2+(y_2-\ln 1.2)^2+$$
$$\quad(y_3-0)^2+(y_4-\ln 0.95)^2+$$
$$\quad(y_5-\ln 0.71)^2$$

Where:

$$y_1=m\cdot0+b$$

$$y_2=m\cdot10+b$$

$$y_3=m\cdot20+b$$

$$y_4=m\cdot30+b$$

$$y_5=m\cdot40+b$$

Substituting we get:

$$S=(b-\ln 1.5)^2+(10m+b-\ln 1.2)^2+$$
$$\quad(20m+b-0)^2+$$
$$\quad(30m+b-\ln 0.95)^2+$$
$$\quad(40m+b-\ln 0.71)^2$$

The solution is continued on the next page.

In order to minimize S, we need to find the first partial derivatives.

$$\frac{\partial S}{\partial b} = 2(b - \ln 1.5) + 2(10m + b - \ln 1.2) +$$
$$2(20m + b) +$$
$$2(30m + b - \ln 0.95) +$$
$$2(40m + b - \ln 0.71)$$
$$\approx 200m + 10b - 0.388$$

$$\frac{\partial S}{\partial m} = (0) + 2(10m + b - \ln 1.2) \cdot 10 +$$
$$2(20m + b) \cdot 20 +$$
$$2(30m + b - \ln 0.95) \cdot 30 +$$
$$2(40m + b - \ln 0.71) \cdot 40$$
$$\approx 6000m + 200b + 26.8304$$

We set these derivatives equal to 0 and solve the resulting system.
$$200m + 10b - 0.388 = 0$$
$$6000m + 200b + 26.8304 = 0$$
The solution to this system is
$b = 0.3847,\ m = -0.01729$.

We use the D-test to verify that $S(0.3847, -0.0173)$ is a relative minimum.

We first find the second-order partial derivatives.

$S_{bb} = 10,\ S_{bm} = 200,\ S_{mm} = 6000$

$D = S_{bb}(0.3847, -0.0173) \cdot S_{mm}(0.3847, -0.0173) - $
$$\left[S_{bm}(0.3847, -0.0173) \right]^2$$

$D = 10 \cdot 6000 - \left[200 \right]^2$
$$= 20,000$$

Since $D > 0$ and
$S_{bb}(0.3847, -0.0173) = 10 > 0$,

S has a relative minimum at
$(0.3847, -0.0173)$. The regression line is

$\tilde{y} = -0.0173x + 0.3847$.

Therfore, $\tilde{b} = -0.0173$ and $\tilde{a} = 0.3847$ we have:

$b = e^{\tilde{b}} = e^{-0.0173}$

$a = e^{\tilde{a}} = e^{0.3847} \approx 1.469$.

The exponential regression equation that models the population in millions of people of Detroit since 1970 is:

$y = 1.469e^{-0.0173x}$.

b) In 2020, $x = 2020 - 1970 = 50$
$$y = 1.469e^{-0.0173(50)}$$
$$= 1.469e^{-0.865} \approx 0.6185$$
According to the model, the population of Detroit will be about 0.62 million or 620,000 people in 2020.

In 2025, $x = 2025 - 1970 = 55$
$$y = 1.469e^{-0.0173(55)}$$
$$= 1.469e^{-0.9515} \approx 0.5672$$
According to the model, the population of Detroit will be about 0.57 million or 570,000 people in 2025.

17.

19. a) Converting the times to decimal notation and using the STAT package on a calculator, we get the regression equation
$y = -0.0059379586x + 15.57191398$.

b) We predict that the world record in 2015 will be about 3.60693 minutes or 3:36.4. We predict that the world record in 2020 will be about 3.57724 minutes or 3:34.6.

c) According to the regression model, we would predict the world record in 1999 to be 3.7019 minutes or 3:42.1. This is about a second faster than the actual world record.

Exercise Set 6.5

1. Find the extremum of
$$f(x, y) = xy$$
subject to the constraint
$3x + y = 10$.
We first express $3x + y = 10$ as $3x + y - 10 = 0$.
We form the new function F, given by:
$$F(x, y, \lambda) = xy - \lambda(3x + y - 10).$$
We find the first partial derivatives:
$$F(x, y, \lambda) = \underline{xy} - \lambda(3\underline{x} + y - 10)$$
$$F_x = y - 3\lambda,$$
$$F(x, y, \lambda) = x\underline{y} - \lambda(3x + \underline{y} - 10)$$
$$F_y = x - \lambda,$$
$$F(x, y, \lambda) = xy - \underline{\lambda}(3x + y - 10)$$
$$F_\lambda = -(3x + y - 10).$$
We set each derivative equal to 0 and solve the resulting system:
$$y - 3\lambda = 0 \qquad (1)$$
$$x - \lambda = 0 \qquad (2)$$
$$3x + y - 10 = 0 \qquad (3) \begin{bmatrix} -(3x+y-10)=0, \text{ or} \\ 3x+y-10=0 \end{bmatrix}$$
Solving Eq. (2) for λ, we get:
$$\lambda = x.$$
Substituting into Eq. (1) for λ, we get:
$$y - 3(x) = 0, \text{ or } y = 3x. \qquad (4)$$
Substituting $3x$ for y in Eq. (3), we get:
$$3x + 3x - 10 = 0$$
$$6x = 10$$
$$x = \frac{10}{6} = \frac{5}{3}$$
Then, using Eq. (4), we have:
$$y = 3\left(\frac{5}{3}\right) = 5.$$
The extreme value of f subject to the constraint
occurs at $\left(\frac{5}{3}, 5\right)$ and is $f\left(\frac{5}{3}, 5\right) = \frac{5}{3} \cdot 5 = \frac{25}{3}$.
We can use test values to show that the extreme value is a maximum.
Test: $(1, 7)$; $F(1, 7) = 1 \cdot 7 = 7 < \frac{25}{3}$.

3. Find the extremum of
$$f(x, y) = x^2 + y^2$$
subject to the constraint
$2x + y = 10$.
We first express $2x + y = 10$ as $2x + y - 10 = 0$.
We form the new function F, given by:
$$F(x, y, \lambda) = x^2 + y^2 - \lambda(2x + y - 10).$$
We find the first partial derivatives:
$$F(x, y, \lambda) = \underline{x}^2 + y^2 - \lambda(2\underline{x} + y - 10)$$
$$F_x = 2x - 2\lambda,$$
$$F(x, y, \lambda) = x^2 + \underline{y}^2 - \lambda(2x + \underline{y} - 10)$$
$$F_y = 2y - \lambda,$$
The partial derivative with respect to λ.
$$F(x, y, \lambda) = x^2 + y^2 - \underline{\lambda}(2x + y - 10)$$
$$F_\lambda = -(2x + y - 10).$$
We set each derivative equal to 0 and solve the resulting system:
$$2x - 2\lambda = 0 \qquad (1)$$
$$2y - \lambda = 0 \qquad (2)$$
$$2x + y - 10 = 0 \qquad (3) \begin{bmatrix} -(2x+y-10)=0, \text{ or} \\ 2x+y-10=0 \end{bmatrix}$$
Solving Eq. (2) for λ, we get:
$$\lambda = 2y.$$
Substituting into Eq. (1) for λ, we get:
$$2x - 2(2y) = 0, \text{ or } x = 2y. \qquad (4)$$
Substituting $2y$ for x in Eq. (3), we get:
$$2(2y) + y - 10 = 0$$
$$5y = 10$$
$$y = 2.$$
Then, using Eq. (4), we have:
$$x = 2(2) = 4.$$
The extreme value of f subject to the constraint occurs at $(4, 2)$ and is
$$f(4, 2) = 4^2 + 2^2$$
$$= 16 + 4 = 20.$$
We can use test values to show that the extreme value is a minimum.
Test: $(2, 6)$; $F(2, 6) = (2)^2 + (6)^2 = 40 > 20$.

5. Find the extremum of
$$f(x, y) = 4 - x^2 - y^2$$
subject to the constraint
$$x + 2y = 10.$$
We first express $x + 2y = 10$ as $x + 2y - 10 = 0$.
We form the new function F, given by:
$$F(x, y, \lambda) = 4 - x^2 - y^2 - \lambda(x + 2y - 10).$$
We find the first partial derivatives:
$$F(x, y, \lambda) = 4 - \underline{x}^2 - y^2 - \lambda(\underline{x} + 2y - 10)$$
$$F_x = -2x - \lambda,$$
$$F(x, y, \lambda) = 4 - x^2 - \underline{y}^2 - \lambda(x + 2\underline{y} - 10)$$
$$F_y = -2y - 2\lambda,$$
$$F(x, y, \lambda) = 4 - x^2 - y^2 - \underline{\lambda}(x + 2y - 10)$$
$$F_\lambda = -(x + 2y - 10).$$
We set each derivative equal to 0 and solve the resulting system:
$$-2x - \lambda = 0 \qquad (1)$$
$$-2y - 2\lambda = 0 \qquad (2)$$
$$x + 2y - 10 = 0 \qquad (3)\begin{bmatrix} -(x+2y-10)=0, \text{ or} \\ x+2y-10=0 \end{bmatrix}$$
Solving Eq. (1) for λ, we get: $\lambda = -2x$.
Substituting into Eq. (2) for λ, we get:
$$-2y - 2(-2x) = 0, \text{ or } y = 2x. \qquad (4)$$
Substituting $2x$ for y in Eq. (3), we get:
$$x + 2(2x) - 10 = 0$$
$$5x = 10 \text{ The solution is continued.}$$
$$x = 2.$$
Then, using Eq. (4), we have: $y = 2(2) = 4$.
The extreme value of f subject to the constraint occurs at $(2, 4)$ and is
$$f(2, 4) = 4 - (2)^2 - (4)^2$$
$$= 4 - 4 - 16$$
$$= -16.$$
We can use test values to show that the extreme value is a maximum.
Test:
$$(4, 3); f(5, 3) = 4 - (4)^2 - (3)^2 = -21 < -16.$$

7. Find the extremum of
$$f(x, y) = 2y^2 - 6x^2$$
subject to the constraint
$$2x + y = 4.$$
We first express $2x + y = 4$ as $2x + y - 4 = 0$.

We form the new function F, given by:
$$F(x, y, \lambda) = 2y^2 - 6x^2 - \lambda(2x + y - 4).$$
We find the first partial derivatives:
$$F(x, y, \lambda) = 2y^2 - 6\underline{x}^2 - \lambda(2\underline{x} + y - 4)$$
$$F_x = -12x - 2\lambda,$$
$$F(x, y, \lambda) = 2\underline{y}^2 - 6x^2 - \lambda(2x + \underline{y} - 4)$$
$$F_y = 4y - \lambda,$$
$$F(x, y, \lambda) = 2y^2 - 6x^2 - \underline{\lambda}(2x + y - 4)$$
$$F_\lambda = -(2x + y - 4).$$
We set each derivative equal to 0 and solve the resulting system:
$$-12x - 2\lambda = 0 \qquad (1)$$
$$4y - \lambda = 0 \qquad (2)$$
$$2x + y - 4 = 0 \qquad (3)\begin{bmatrix} -(2x+y-4)=0, \text{ or} \\ 2x+y-4=0 \end{bmatrix}$$
Solving Eq. (2) for λ, we get: $\lambda = 4y$.
Substituting into Eq. (1) for λ, we get:
$$-12x - 2(4y) = 0$$
$$8y = -12x$$
$$y = -\frac{3}{2}x \qquad (4)$$
Substituting $-\frac{3}{2}x$ for y in Eq. (3), we get:
$$2x + \left(-\frac{3}{2}x\right) - 4 = 0$$
$$\frac{1}{2}x = 4$$
$$x = 8.$$
Then, using Eq. (4), we have $y = -\frac{3}{2}(8) = -12$.
The extreme value of f subject to the constraint occurs at $(8, -12)$ and is
$$f(8, -12) = 2(-12)^2 - 6(8)^2$$
$$= 2(144) - 6(64)$$
$$= 288 - 384 = -96.$$
We can use test values to show that the extreme value is a minimum.
Test: $(1, 2); f(1, 2) = 2(1)^2 - 6(2)^2 = -22 > -96.$

9. Find the minimum value of
$$f(x,y,z) = x^2 + y^2 + z^2$$
subject to the constraint
$$y + 2x - z = 3.$$
We first express $y + 2x - z = 3$ as
$$y + 2x - z - 3 = 0.$$
We form the new function F, given by:
$$F(x,y,z,\lambda)$$
$$= x^2 + y^2 + z^2 - \lambda(y + 2x - z - 3)$$
We find the first partial derivatives:
$$F_x = 2x - 2\lambda,$$
$$F_y = 2y - \lambda,$$
$$F_z = 2z + \lambda,$$
$$F_\lambda = -(y + 2x - z - 3).$$
We set each derivative equal to 0 and solve the resulting system:
$$2x - 2\lambda = 0 \quad (1)$$
$$2y - \lambda = 0 \quad (2)$$
$$2z + \lambda = 0 \quad (3)$$
$$y + 2x - z - 3 = 0 \quad (4)\begin{bmatrix} -(y+2x-z-3)=0,\ or \\ y+2x-z-3=0 \end{bmatrix}$$
Solving Eq. (1) for x, we get: $x = \lambda$.

Solving Eq. (2) for y, we get: $y = \dfrac{1}{2}\lambda$.

Solving Eq. (3) for z, we get: $z = -\dfrac{1}{2}\lambda$.

Substituting λ for x, $\dfrac{1}{2}\lambda$ for y, and $-\dfrac{1}{2}\lambda$ for z into Eq. (4), we get:
$$y + 2x - z - 3 = 0$$
$$\dfrac{1}{2}\lambda + 2\lambda - \left(-\dfrac{1}{2}\lambda\right) - 3 = 0$$
$$3\lambda = 3$$
$$\lambda = 1.$$
Then,
$$x = \lambda = 1$$
$$y = \dfrac{1}{2}\lambda = \dfrac{1}{2}$$
$$z = -\dfrac{1}{2}\lambda = -\dfrac{1}{2}$$

The extreme value of f subject to the constraint occurs at $\left(1, \dfrac{1}{2}, -\dfrac{1}{2}\right)$ and is
$$f\left(1, \dfrac{1}{2}, -\dfrac{1}{2}\right) = 1^2 + \left(\dfrac{1}{2}\right)^2 + \left(-\dfrac{1}{2}\right)^2$$
$$= 1 + \dfrac{1}{4} + \dfrac{1}{4} = \dfrac{3}{2}.$$
We can use test values to show that the extreme value is a minimum.
Test:
$$(1,2,1); f(1,2,1) = (1)^2 + (2)^2 + (1)^2 = 6 > \dfrac{3}{2}.$$

11. Find the maximum value of
$$f(x,y) = xy \qquad \text{(Product is } x \cdot y)$$
subject to the constraint
$$x + y = 50. \qquad \text{(Sum is 50.)}.$$
We first express $x + y = 50$ as $x + y - 50 = 0$.
We form the new function F, given by:
$$F(x,y,\lambda) = xy - \lambda(x + y - 50).$$
We find the first partial derivatives:
$$F_x = y - \lambda,$$
$$F_y = x - \lambda,$$
$$F_\lambda = -(x + y - 50).$$
We set each derivative equal to 0 and solve the resulting system:
$$y - \lambda = 0 \quad (1)$$
$$x - \lambda = 0 \quad (2)$$
$$x + y - 50 = 0 \quad (3)\begin{bmatrix} -(x+y-50)=0,\ or \\ x+y-50=0 \end{bmatrix}$$
Solving Eq. (2) for λ, we get:
$$\lambda = x.$$
Substituting into Eq. (1) for λ, we get:
$$y - (x) = 0, \text{ or } y = x. \qquad (4)$$
Substituting x for y in Eq. (3), we get:
$$x + x - 50 = 0$$
$$2x = 50$$
$$x = 25.$$
Then, using Eq. (4), we have:
$$y = 25.$$
The maximum value of f subject to the constraint occurs at $(25, 25)$. Thus, the two numbers whose sum is 50 that have the maximum product are 25 and 25.

13. Find the minimum value of
$$f(x, y) = xy \qquad (\text{Product is } x \cdot y)$$
subject to the constraint
$$x - y = 6. \qquad (\text{Difference is } 6.).$$
We first express $x - y = 6$ as $x - y - 6 = 0$.
We form the new function F, given by:
$$F(x, y, \lambda) = xy - \lambda(x - y - 6).$$
We find the first partial derivatives:
$$F_x = y - \lambda,$$
$$F_y = x + \lambda,$$
$$F_\lambda = -(x - y - 6).$$
We set each derivative equal to 0 and solve the resulting system:
$$y - \lambda = 0 \qquad (1)$$
$$x + \lambda = 0 \qquad (2)$$
$$x - y - 6 = 0 \qquad (3) \begin{bmatrix} -(x-y-6)=0, \text{ or} \\ x-y-6=0 \end{bmatrix}$$
Solving Eq. (1) for λ, we get:
$$\lambda = y.$$
Substituting into Eq. (2) for λ, we get:
$$x + (y) = 0, \text{ or } y = -x. \qquad (4)$$
Substituting $-x$ for y in Eq. (3), we get:
$$x - (-x) - 6 = 0$$
$$2x = 6$$
$$x = 3.$$
Then, using Eq. (4), we have:
$$y = -3.$$
The minimum value of f subject to the constraint occurs at $(3, -3)$. Thus, the two numbers whose difference is 6 that have the minimum product are 3 and -3.

15. Find the minimum value of
$$f(x, y, z) = (x - 1)^2 + (y - 1)^2 + (z - 1)^2$$
subject to the constraint
$$x + 2y + 3z = 13.$$
We first express $x + 2y + 3z = 13$ as
$$x + 2y + 3z - 13 = 0.$$
We form the new function F, given by:
$$F(x, y, z, \lambda)$$
$$= (x - 1)^2 + (y - 1)^2 + (z - 1)^2 -$$
$$\lambda(x + 2y + 3z - 13)$$
We find the first partial derivatives at the top of the next column.

$$F_x = 2(x - 1) - \lambda,$$
$$F_y = 2(y - 1) - 2\lambda,$$
$$F_z = 2(z - 1) - 3\lambda,$$
$$F_\lambda = -(x + 2y + 3z - 13).$$
We set each derivative equal to 0 and solve the resulting system:
$$2(x - 1) - \lambda = 0 \qquad (1)$$
$$2(y - 1) - 2\lambda = 0 \qquad (2)$$
$$2(z - 1) - 3\lambda = 0 \qquad (3)$$
$$x + 2y + 3z - 13 = 0 \qquad (4) \begin{bmatrix} -(x+2y+3z-13)=0, \text{ or} \\ x+2y+3z-13=0 \end{bmatrix}$$
Solving Eq. (1) for x, we get:
$$2x - 2 - \lambda = 0$$
$$2x = 2 + \lambda$$
$$x = 1 + \frac{1}{2}\lambda.$$
Solving Eq. (2) for y, we get:
$$2y - 2 - 2\lambda = 0$$
$$2y = 2 + 2\lambda$$
$$y = 1 + \lambda.$$
Solving Eq. (3) for z, we get:
$$2z - 2 - 3\lambda = 0$$
$$2z = 2 + 3\lambda$$
$$z = 1 + \frac{3}{2}\lambda.$$
Substituting $1 + \frac{1}{2}\lambda$ for x, $1 + \lambda$ for y, and $1 + \frac{3}{2}\lambda$ for z into Eq. (4), we get:
$$\left(1 + \frac{1}{2}\lambda\right) + 2(1 + \lambda) + 3\left(1 + \frac{3}{2}\lambda\right) - 13 = 0$$
$$1 + \frac{1}{2}\lambda + 2 + 2\lambda + 3 + \frac{9}{2}\lambda = 13$$
$$6 + 7\lambda = 13$$
$$7\lambda = 7$$
$$\lambda = 1.$$
$$x = 1 + \frac{1}{2}\lambda = 1 + \frac{1}{2} = \frac{3}{2}$$
$$y = 1 + \lambda = 1 + 1 = 2$$
$$z = 1 + \frac{3}{2}\lambda = 1 + \frac{3}{2} = \frac{5}{2}.$$
The minimum value of f subject to the constraint occurs at $\left(\frac{3}{2}, 2, \frac{5}{2}\right)$.

17. We determine the objective function. We want to minimize the distance to the origin. Using the distance formula we determine the objective function to be:

$$D = \sqrt{(x-0)^2 + (y-0)^2} = \sqrt{x^2 + y^2}.$$

We want to minimize the distance

$$D = \sqrt{x^2 + y^2}$$

Subject to the constraint
$3x + y = 6.$
We first express $3x + y = 6$ as
$3x + y - 6 = 0$
We form the new function F, given by:

$$F(x, y, \lambda) = \sqrt{x^2 + y^2} - \lambda(3x + y - 6).$$

We find the first partial derivatives:

$$F(x, y, \lambda) = \sqrt{x^2 + y^2} - \lambda(3\underline{x} + y - 6)$$

$$F_x = \frac{1}{2}(x^2 + y^2)^{-\frac{1}{2}} \cdot 2x - 3\lambda$$

$$F_{x,} = \frac{x}{\sqrt{x^2 + y^2}} - 3\lambda$$

$$F(x, y, \lambda) = \sqrt{x^2 + \underline{y}^2} - \lambda(3x + \underline{y} - 6)$$

$$F_y = \frac{1}{2}(x^2 + y^2)^{-\frac{1}{2}} \cdot 2y - \lambda$$

$$F_y = \frac{y}{\sqrt{x^2 + y^2}} - \lambda,$$

$$F(x, y, \lambda) = \sqrt{x^2 + y^2} - \underline{\lambda}(3x + y - 6)$$

$$F_\lambda = -(3x + y - 6).$$

We set each derivative equal to 0 and solve the resulting system:

$$\frac{x}{\sqrt{x^2 + y^2}} - 3\lambda = 0 \qquad (1)$$

$$\frac{y}{\sqrt{x^2 + y^2}} - \lambda = 0 \qquad (2)$$

$$3x + y - 6 = 0 \qquad (3)\begin{bmatrix} -(3x+y-6)=0, \text{ or} \\ 3x+y-6=0 \end{bmatrix}$$

Solving Eq. (2) for λ, we get:

$$\lambda = \frac{y}{\sqrt{x^2 + y^2}}.$$

Substituting into Eq. (1) for λ, we get:

$$\frac{x}{\sqrt{x^2 + y^2}} - 3\left(\frac{y}{\sqrt{x^2 + y^2}}\right) = 0$$

$$\frac{x - 3y}{\sqrt{x^2 + y^2}} = 0$$

$$x - 3y = 0 \quad \begin{array}{l}\text{Multiplying both sides} \\ \text{by } \sqrt{x^2+y^2}.\end{array}$$

$$x = 3y \qquad (4)$$

Note, $\sqrt{x^2 + y^2} \neq 0$ provided $x \neq 0$ and $y \neq 0$. Since the constraint line does not pass throught the origin, it is safe to make this assumption and multiply both sides of the equation by $\sqrt{x^2 + y^2}$ in the previous step.

Substituting $3y$ for x in Eq. (3), we get:

$$3(3y) + y - 6 = 0$$

$$10y = 6$$

$$y = \frac{6}{10} = \frac{3}{5}.$$

Then, using Eq. (4), we have:

$$x = 3\left(\frac{3}{5}\right) = \frac{9}{5}.$$

Therefore, the point on the line that is closest to the origin is $\left(\frac{9}{5}, \frac{3}{5}\right)$.

19. The area of the page is given by $A = xy$ and the perimeter of the page is given by $P = 2x + 2y$.

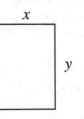

We want to maximize the area
$A = xy$
Subject to the constraint
$2x + 2y = 39.$
We first express $2x + 2y = 39$ as
$2x + 2y - 39 = 0$.
We form the new function F, given by:
$$F(x, y, \lambda) = xy - \lambda(2x + 2y - 39).$$
The solution is continued on the next page.

We find the first partial derivatives:

$F_x = y - 2\lambda,$

$F_y = x - 2\lambda,$

$F_\lambda = -(2x + 2y - 39).$

We set each derivative equal to 0 and solve the resulting system:

$$y - 2\lambda = 0 \qquad (1)$$

$$x - 2\lambda = 0 \qquad (2)$$

$$2x + 2y - 39 = 0 \qquad (3) \begin{bmatrix} -(2x+2y-39)=0, \text{ or} \\ 2x+2y-39=0 \end{bmatrix}$$

From Eqs. (1) and (2) we see:

$y = 2\lambda = x.$

Substituting x for y in Eq. (3), we get:

$2x + 2x - 39 = 0$

$\qquad 4x = 39$

$\qquad x = \dfrac{39}{4} = 9\tfrac{3}{4}.$

Then,

$y = x = 9\tfrac{3}{4}.$

The maximum area subject to the constraint occurs at $\left(9\tfrac{3}{4}, 9\tfrac{3}{4}\right)$. The maximum area is

$A = 9\tfrac{3}{4} \cdot 9\tfrac{3}{4} = 95\tfrac{1}{16}$ in^2. The area of the standard $8\tfrac{1}{2} \times 11$ paper is not the maximum area of paper that has a perimeter of 39 in.

21. We want to minimize the function s given by

$s(h, r) = 2\pi rh + 2\pi r^2$

subject to the volume constraint

$\pi r^2 h = 27$, or $\pi r^2 h - 27 = 0.$

We form the new function S given by

$S(h, r, \lambda) = 2\pi rh + 2\pi r^2 - \lambda(\pi r^2 h - 27).$

We find the first partial derivatives.

$S_h = 2\pi r - \lambda \pi r^2,$

$S_r = 2\pi h + 4\pi r - 2\lambda \pi rh,$

$S_\lambda = -(\pi r^2 h - 27).$

We set these derivatives equal to 0 and solve the resulting system.

$$2\pi r - \lambda \pi r^2 = 0 \quad (1)$$

$$2\pi h + 4\pi r - 2\lambda \pi rh = 0 \quad (2)$$

$$\pi r^2 h - 27 = 0. \quad (3) \begin{bmatrix} -(\pi r^2 h - 27)=0, \text{ or} \\ \pi r^2 h - 27 = 0 \end{bmatrix}$$

We solve Eq. (1) for r:

$\pi r(2 - \lambda r) = 0$

$\pi r = 0 \quad$ or $\quad 2 - \lambda r = 0$

$r = 0 \quad$ or $\quad r = \dfrac{2}{\lambda}$

Note, $r = 0$ cannot be a solution to the original problem, so we continue by substituting $\dfrac{2}{\lambda}$ for r in Eq. (2).

$$2\pi h + 4\pi\left(\dfrac{2}{\lambda}\right) - 2\lambda\pi\left(\dfrac{2}{\lambda}\right)h = 0$$

$$2\pi h + \dfrac{8\pi}{\lambda} - 4\pi h = 0$$

$$\dfrac{8\pi}{\lambda} - 2\pi h = 0$$

$$h = \dfrac{4}{\lambda}$$

Since $h = \dfrac{4}{\lambda}$ and $r = \dfrac{2}{\lambda}$, it follows that $h = 2r$.

Substituting $2r$ for h in Eq. (3) yields:

$\pi r^2 (2r) - 27 = 0$

$\qquad 2\pi r^3 = 27$

$\qquad r^3 = \dfrac{27}{2\pi}$

$\qquad r = \sqrt[3]{\dfrac{27}{2\pi}} \approx 1.6,$

So when $r \approx 1.6$ ft and $h = 2(1.6) \approx 3.2$ ft, the surface area of the oil drum is a minimum. The minimum area is about

$2\pi(1.6)(3.2) + 2\pi(1.6)^2 \approx 48.3$ ft^2.

(Answers will vary due to rounding differences.)

23. We want maximize

$S(L, M) = ML - L^2$

subject to the constraint

$M + L = 90.$

We first express $M + L = 90$ as $M + L - 90 = 0$.

We form the new function F, given by:

$F(L, M, \lambda) = ML - L^2 - \lambda(M + L - 90).$

We find the first partial derivatives:

$F_L = M - 2L - \lambda,$

$F_M = L - \lambda,$

$F_\lambda = -(M + L - 90).$

The solution is continued on the next page.

We set each derivative on the previous page equal to 0 and solve the resulting system:

$$M - 2L - \lambda = 0 \qquad (1)$$
$$L - \lambda = 0 \qquad (2)$$
$$M + L - 90 = 0 \qquad (3) \begin{bmatrix} -(M+L-90)=0, \text{ or} \\ M+L-90=0 \end{bmatrix}$$

Solving Eq. (2) for λ, we get:

$$\lambda = L.$$

Substituting into Eq. (1) for λ, we get:

$$M - 2L - L = 0$$
$$M - 3L = 0$$
$$M = 3L. \qquad (4)$$

Substituting $3L$ for M in Eq. (3), we get:

$$3L + L - 90 = 0$$
$$4L = 90$$
$$L = 22.5.$$

Then, using Eq. (4), we have:

$$M = 3(22.5) = 67.5.$$

The maximum value of S subject to the constraint occurs at $(22.5, 67.5)$ and is

$$S(22.5, 67.5) = (67.5)(22.5) - (22.5)^2$$
$$= 1518.75 - 506.25$$
$$= 1012.5$$

25. a) The area of the floor is xy.
The cost of the floor is $4xy$.
The area of the walls is $2xz + 2yz$.
The cost of the walls is $3(2xz + 2yz)$.
The area of the ceiling is xy.
The cost of the ceiling is $3xy$.
Therefore, the total cost function is
$$C(x, y, z) = 4xy + 3(2xz + 2yz) + 3xy$$
$$= 7xy + 6xz + 6yz.$$

b) We want to minimize the value of
$$C(x, y, z) = 7xy + 6xz + 6yz$$
subject to the constraint of
$$x \cdot y \cdot z = 252,000. \qquad (\text{Volume} = l \cdot w \cdot h)$$
We first express $x \cdot y \cdot z = 252,000$ as
$$x \cdot y \cdot z - 252,000 = 0.$$
We form the new function F, given by:
$$F(x, y, z, \lambda)$$
$$= 7xy + 6xz + 6yz - \lambda(x \cdot y \cdot z - 252,000).$$
We find the first partial derivatives at the top of the next column.

$$F_x = 7y + 6z - \lambda yz,$$
$$F_y = 7x + 6z - \lambda xz,$$
$$F_z = 6x + 6y - \lambda xy,$$
$$F_\lambda = -(x \cdot y \cdot z - 252,000).$$

We set each derivative equal to 0 and solve the resulting system:

$$7y + 6z - \lambda yz = 0 \qquad (1)$$
$$7x + 6z - \lambda xz = 0 \qquad (2)$$
$$6x + 6y - \lambda xy = 0 \qquad (3)$$
$$xyz - 252,000 = 0 \quad (4)\begin{bmatrix} -(x\cdot y\cdot z-252,000)=0, \text{ or} \\ x\cdot y\cdot z-252,000=0 \end{bmatrix}$$

Solving Eq. (2) for x and Eq. (1) for y, we get:

$$x = \frac{6z}{\lambda z - 7} \quad \text{and} \quad y = \frac{6z}{\lambda z - 7}.$$

Thus, $x = y$.

Substituting x for y in Eq. (1), (2), and (4) we get the following system:

$$7x + 6z - \lambda xz = 0$$
$$6x + 6x - \lambda xx = 0$$
$$xxz - 252,000 = 0$$

Which simplifies to:

$$7x + 6z - \lambda xz = 0 \qquad (5)$$
$$12x - \lambda x^2 = 0 \qquad (6)$$
$$x^2 z - 252,000 = 0 \qquad (7)$$

Solving Eq. (6) for x, we get

$$12x - \lambda x^2 = 0$$
$$x(12 - \lambda x) = 0$$
$$x = 0 \quad \text{or} \quad 12 - \lambda x = 0$$
$$x = 0 \quad \text{or} \quad x = \frac{12}{\lambda}$$

We only consider $x = \frac{12}{\lambda}$ since x cannot be 0 in the original problem. We continue by substituting $\frac{12}{\lambda}$ for x into Eq. (7) and solving for z.

$$\left(\frac{12}{\lambda}\right)^2 z - 252,000 = 0$$
$$\frac{144}{\lambda^2} \cdot z = 252,000$$
$$z = \frac{252,000}{144}\lambda^2$$
$$z = 1750\lambda^2$$

The solution is continued on the next page.

Next we substitute $\frac{12}{\lambda}$ for x and $1750\lambda^2$ for z in Eq. (5) on the previous page and solve for λ.

$$7\left(\frac{12}{\lambda}\right)+6\cdot1750\lambda^2-\lambda\left(\frac{12}{\lambda}\right)1750\lambda^2=0$$

$$\frac{84}{\lambda}+10,500\lambda^2-21,000\lambda^2=0$$

$$10,500\lambda^2=\frac{84}{\lambda}$$

$$\lambda^3=\frac{84}{10,500}$$

$$\lambda^3=\frac{1}{125}$$

$$\lambda=\frac{1}{5}$$

Thus,

$$x=\frac{12}{\lambda}=\frac{12}{\frac{1}{5}}=12\cdot\frac{5}{1}=60$$

$$y=\frac{12}{\lambda}=\frac{12}{\frac{1}{5}}=12\cdot\frac{5}{1}=60$$

$$z=1750\lambda^2=1750\left(\frac{1}{5}\right)^2=70$$

The minimum total cost subject to the constraint occurs when the dimensions are 60 ft by 60 ft by 70 ft. The minimum cost is found as follows:

$$C(60,60,70)=7\cdot60\cdot60+6\cdot60\cdot70+6\cdot60\cdot70$$

$$=25,200+25,200+25,200$$

$$=75,600.$$

The minimum total cost of the building is $75,600.

27. $C(x,y)=C(x)+C(y)$

$$C(x,y)=10+\frac{x^2}{6}+200+\frac{y^3}{9}$$

$$=210+\frac{x^2}{6}+\frac{y^3}{9}$$

We need to minimize

$$C(x,y)=210+\frac{x^2}{6}+\frac{y^3}{9}$$

subject to the constraint
$x+y=10,100.$
We first express $x+y=10,100$ as
$x+y-10,100=0.$

We form the new function F, given by:

$$F(x,y,\lambda)$$

$$=210+\frac{x^2}{6}+\frac{y^3}{9}-\lambda(x+y-10,100).$$

We find the first partial derivatives:

$$F_x=\frac{x}{3}-\lambda,$$

$$F_y=\frac{1}{3}y^2-\lambda,$$

$$F_\lambda=-(x+y-10,100).$$

We set each derivative equal to 0 and solve the resulting system

$$\frac{x}{3}-\lambda=0 \quad (1)$$

$$\frac{1}{3}y^2-\lambda=0 \quad (2)$$

$$x+y-10,100=0 \quad (3) \begin{bmatrix} -(x+y-10,100)=0, \text{ or} \\ x+y-10,100=0 \end{bmatrix}$$

From Eq. (1) and Eq. (2) we see:

$$x=3\lambda=y^2.$$

Thus, $x=y^2$.

Substituting y^2 for x in Eq. (3), we get:

$$y^2+y-10,100=0$$

$$(y+101)(y-100)=0$$

$$y+101=0 \quad \text{or} \quad y-100=0$$

$$y=-101 \quad \text{or} \quad y=100.$$

Since y cannot be -101 in the original problem, we only consider $y=100$. If $y=100$, then

$x=100^2-10,000$. To minimize total costs, 10,000 units should be made on machine A and 100 units should be made on machine B.

29. Placing the house on the point $(-2,1)$ on the xy-plane, we want to minimize the distance from the house and the water main. Since the water main is on the 45 degree line from the origin, the equation of the water main is $x+y=0$. Therefore, we want to minimize

$$D=\sqrt{(x-(-2))^2+(y-1)^2}$$

$$=\sqrt{(x+2)^2+(y-1)^2}$$

Subject to the constraint
$x+y=0.$

The solution is continued on the next page.

We form the new function F, given by:

$$F(x,y,\lambda)=\sqrt{(x+2)^2+(y-1)^2}-\lambda(x+y).$$

We find the first partial derivatives:

$$F(x,y,\lambda)=\sqrt{(\underline{x}+2)^2+(y-1)^2}-\lambda(\underline{x}+y)$$

$$F_x,=\frac{x+2}{\sqrt{(x+2)^2+(y-1)^2}}-\lambda$$

$$F(x,y,\lambda)=\sqrt{(x+2)^2+(\underline{y}-1)^2}-\lambda(x+\underline{y})$$

$$F_y=\frac{y-1}{\sqrt{(x+2)^2+(y-1)^2}}-\lambda,$$

$$F(x,y,\lambda)=\sqrt{(x+2)^2+(y-1)^2}-\underline{\lambda}(x+y)$$

$$F_\lambda=-(x+y).$$

We set each derivative equal to 0 and solve the resulting system:

$$\frac{x+2}{\sqrt{(x+2)^2+(y-1)^2}}-\lambda=0 \qquad (1)$$

$$\frac{y-1}{\sqrt{(x+2)^2+(y-1)^2}}-\lambda=0 \qquad (2)$$

$$x+y=0 \qquad (3)$$

Solving Eq. (2) for λ, we get:

$$\lambda=\frac{y-1}{\sqrt{(x+2)^2+(y-1)^2}}.$$

Substituting into Eq. (1) for λ, we get:

$$\frac{x+2}{\sqrt{(x+2)^2+(y-1)^2}}-\left(\frac{y-1}{\sqrt{(x+2)^2+(y-1)^2}}\right)=0$$

$$\frac{x-y+3}{\sqrt{(x+2)^2+(y-1)^2}}=0$$

$$x-y+3=0 \qquad \begin{array}{l}\text{Multiplying both sides}\\ \text{by }\sqrt{(x+2)^2+(y-1)^2}.\end{array}$$

$$x=y-3 \qquad (4)$$

Substituting $y-3$ for x in Eq. (3), we get:

$$y-3+y=0$$
$$2y=3$$
$$y=\frac{3}{2}=1.5$$

Then, using Eq. (4), we have:

$$x=\left(\frac{3}{2}\right)-3=-\frac{3}{2}=-1.5.$$

Therefore, the point on the water main that is closest to home is at $\left(-\frac{3}{2},\frac{3}{2}\right)$ on the water main. We determine the the shortest possible distance by substituting into the objective function.

$$D=\sqrt{(x+2)^2+(y-1)^2}$$

$$=\sqrt{\left(-\frac{3}{2}+2\right)^2+\left(\frac{3}{2}-1\right)^2}$$

$$=\sqrt{\left(\frac{1}{2}\right)^2+\left(\frac{1}{2}\right)^2}$$

$$=\sqrt{\frac{1}{4}+\frac{1}{4}}=\sqrt{\frac{2}{4}}=\frac{\sqrt{2}}{2}\approx0.7071$$

The shortest distance between the home and the water main is about 0.7071 miles. At a cost of $8,000 per mile, the minimum cost to connect the house to the water main would be $8000\times0.7071=\$5656.8\approx\$5657$.

31. Find the absolute minimum and maximum values of

$$g(x,y)=x^2+2y^2$$

Subject to the constraints of
$$-1\le x\le1$$
$$-1\le y\le2.$$

Using the constraints, we sketch the region of feasibility below.

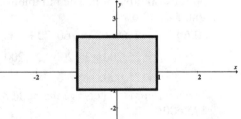

There are four corner points of the feasible region. They are:

$(-1,-1),(-1,2),(1,2),$ and $(1,-1)$. These are all critical points.

Next we check for critical points in the interior. We find the partial derivative of g with respect to x and with respect to y:

$$g_x(x,y)=2x$$
$$g_y(x,y)=4y$$

The solution is continued on the next page.

The derivatives on the previous page are set equal to 0 and we solve the system for x and y.

$2x = 0$

$4y = 0$

The solution to this system is

$x = 0$

$y = 0$.

The point $(0,0)$ is in the feasible region so it is a critical point.

Next we check the boundaries for possible critical points.

Along the boundary $x = -1$ we subsititute $x = -1$ into g.

$g(-1, y) = 2y^2 + 1$.

The derivative is $g_y(-1, y) = 2y$. Set equal to 0, we have $y = 0$. This results in the point $(-1, 0)$, which is in the feasible region.

Along the boundary $y = 2$ we substitute $y = 2$ into g.

$g(x, 2) = x^2 + 8$.

The derivative is $g_x(x, 2) = 2x$. Set equal to 0, we have $x = 0$. This results in the point $(0, 2)$, which is in the feasible region.

Along the boundary $x = 1$, we subsititute $x = 1$ into g.

$g(1, y) = 2y^2 + 1$.

The derivative is $g_y(1, y) = 4y$. Set equal to 0, we have $y = 0$. This results in the point $(1, 0)$, which is in the feasible region.

Along boundary $y = -1$ we substitute $y = -1$ into g.

$g(x, -1) = x^2 + 2$.

The derivative is $g_x(x, -1) = 2x$. Set equal to 0, we have $x = 0$. This results in the point $(0, -1)$, which is in the feasible region.

Therefore, the critical points are:

$(-1, -1), (-1, 2), (1, 2), (1, -1), (0, 0), (-1, 0),$

$(0, 2), (1, 0),$ and $(0, -1)$.

Evaluating each of the critical points we have:

$g(-1, -1) = (-1)^2 + 2(-1)^2 = 3$

$g(-1, 2) = (-1)^2 + 2(2)^2 = 9$

$g(1, 2) = (1)^2 + 2(2)^2 = 9$

$g(1, -1) = (1)^2 + 2(-1)^2 = 3$

$g(0, 0) = (0)^2 + 2(0)^2 = 0$

$g(-1, 0) = (-1)^2 + 2(0)^2 = 1$

$g(0, 2) = (0)^2 + 2(2)^2 = 8$

$g(1, 0) = (1)^2 + 2(0)^2 = 1$

$g(0, -1) = (0)^2 + 2(-1)^2 = 2$

From the function values, we determine that the absolute minimum is 0 and occurs at $(0,0)$. The absolute maximum is 9 and occurs at $(-1,2)$ and $(1,2)$.

33. Find the absolute minimum and maximum values of

$k(x, y) = -x^2 - y^2 + 4x + 4y$

subject to the constraints of

$0 \le x \le 3,$

$y \ge 0$

$x + y \le 6$.

Using the constraints, we sketch the region of feasibility below.

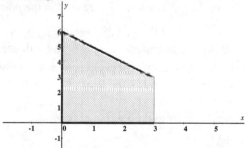

There are four corner points of the feasible region. They are:

$(0,0), (0,6), (3,3)$ and $(3,0)$. These are all critical points.

Next we check for critical points in the interior. We find the partial derivative of k with respect to x and with respect to y:

$k_x(x, y) = -2x + 4$

$k_y(x, y) = -2y + 4$

The solution is continued on the next page.

The derivatives on the previous page are set equal to 0 and we solve the system for x and y.

$$-2x+4=0$$

$$-2y+4=0$$

The solution to this system is

$$x=2$$

$$y=2.$$

The point $(2,2)$ is in the feasible region so it is a critical point.

Next we check the boundaries for possible critical points.

Along the y-axis, we substitute $x=0$ into k.

$$k(0,y)=-y^2+4y.$$

The derivative is $k_y(0,y)=-2y+4$. Set equal to 0, we have $y=2$. This results in the point $(0,2)$, which is in the feasible region.

Along the x-axis we substitute $y=0$ into k.

$$k(x,0)=-x^2+4x.$$

The derivative is $k_x(x,0)=-2x+4$. Set equal to 0, we have $x=2$. This results in the point $(2,0)$, which is in the feasible region.

Along the boundary $x=3$ we substitute $x=3$ into k.

$$k(3,y)=-(3)^2-y^2+4(3)+4y$$

$$=-y^2+4y+3$$

The derivative is $k_y(3,y)=-2y+4$. Set equal to 0, we have $y=2$. This results in the point $(3,2)$, which is in the feasible region.

Along the boundary $x+y\le 6$, we will use Lagrange multipliers to determine any critical points. We will rewrite the constraint to be:

$$x+y-6=0.$$

Next, we form the Lagrange function:

$$L(x,y,\lambda)$$

$$=-x^2-y^2+4x+4y-\lambda(x+y-6).$$

We determine the first partial derivatives:

$$L_x=-2x+4-\lambda$$

$$L_y=-2y+4-\lambda$$

$$L_\lambda=-x-y+6.$$

We set each partial derivative equal to 0.

$$-2x+4-\lambda=0 \qquad (1)$$

$$-2y+4-\lambda=0 \qquad (2)$$

$$-x-y+6=0 \qquad (3)$$

Solving equation (1) for λ we get:

$$\lambda=-2x+4.$$

Solving equation (2) for λ we get:

$$\lambda=-2y+4.$$

Equating the λ, we can simplify the two equations into a single equation involving x and y.

$$-2y+4=-2x+4$$

$$-2y=-2x$$

$$y=x$$

Substituting $y=x$ into equation (3) we get:

$$-x-(x)+6=0$$

$$-2x+6=0$$

$$-2x=-6$$

$$x=3.$$

Therefore,

$$y=x$$

$$y=3.$$

The critical point on the constraint $x+y=6$ is the point $(3,3)$, which we had already determined as a corner point of the region of feasibility.

Therefore, the critical points are:

$$(0,0),(0,6),(3,3),(3,0),(2,2),$$

$$(0,2),\text{ and }(2,0).$$

Evaluating each of the critical points we have:

$$k(0,0)=-(0)^2-(0)^2+4(0)+4(0)=0$$

$$k(0,6)=-(0)^2-(6)^2+4(0)+4(6)=-12$$

$$k(3,3)=-(3)^2-(3)^2+4(3)+4(3)=6$$

$$k(3,0)=-(3)^2-(0)^2+4(3)+4(0)=3$$

$$k(2,2)=-(2)^2-(2)^2+4(2)+4(2)=8$$

$$k(0,2)=-(0)^2-(2)^2+4(0)+4(2)=4$$

$$k(2,0)=-(2)^2-(0)^2+4(2)+4(0)=4.$$

From the function values, we determine that the absolute minimum is -12 and occurs at $(0,6)$.

The absolute maximum is 8 and occurs at $(2,2)$.

35. The warehouse can only store 90 tables. Therefore, the storage constraint is $x + y \le 90$.

Find the absolute maximum of
$$P(x, y) = -x^2 - 2y^2 - xy + 140x + 210y - 4300$$
subject to the constraints of
$$0 \le x, 0 \le y$$
$$x + y \le 90.$$
Using the constraints, we sketch the region of feasibility

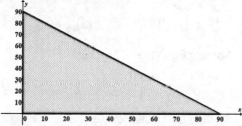

There are three vertices of the feasible region. They are:
$(0,0), (0,90)$ and $(90,0)$. These are all critical points.

Next we check for critical points in the interior. We find the partial derivative of P with respect to x and with respect to y:
$$P_x(x, y) = -2x - y + 140$$
$$P_y(x, y) = -4y - x + 210$$
The derivatives are set equal to 0 and we solve the system for x and y.
$$-2x - y + 140 = 0 \qquad (1)$$
$$-4y - x + 210 = 0 \qquad (2)$$
Multiplying Equation (2) by -2 and adding it to equation 1 we have:
$$7y - 280 = 0.$$
Solving for y we have:
$$7y = 280$$
$$y = 40.$$
Therefore, substituting $y = 40$ into Equation (1) we have:
$$-2x - (40) + 140 = 0$$
$$-2x + 100 = 0$$
$$-2x = -100$$
$$x = 50.$$
The solution to this system is
$$x = 50$$
$$y = 40.$$
The point $(50, 40)$ is in the feasible region so it is a critical point.

Next we check the boundaries for possible critical points.

Along the y-axis, we substitute $x = 0$ into P.
$$P(0, y) = -2y^2 + 210y - 4300.$$
The derivative is $P_y(0, y) = -4y + 210$. Set equal to 0, we have $y = 52.5$. This results in the point $(0, 52.5)$, which is in the feasible region. Realistically, it would not make sense to make 0.5 of a table.

Along the x-axis we substitute $y = 0$ into P.
$$P(x, 0) = -x^2 + 140x - 4300.$$
The derivative is $P_x(x, 0) = -2x + 140$. Set equal to 0, we have $x = 70$. This results in the point $(70, 0)$, which is in the feasible region.

Along the boundary $x + y \le 90$, we will use Lagrange multipliers to determine any critical points. We will rewrite the constraint to be:
$$x + y - 90 = 0.$$
Next, we form the Lagrange function:
$$L(x, y, \lambda)$$
$$= -x^2 - 2y^2 - xy + 140x + 210y - 4300$$
$$\qquad\qquad - \lambda(x + y - 90).$$
We determine the first partial derivatives:
$$L_x = -2x - y + 140 - \lambda$$
$$L_y = -4y - x + 210 - \lambda$$
$$L_\lambda = -x - y + 90.$$
We set each partial derivative equal to 0.
$$-2x - y + 140 - \lambda = 0 \qquad (1)$$
$$-4y - x + 210 - \lambda = 0 \qquad (2)$$
$$-x - y + 90 = 0 \qquad (3)$$
Solving equation (1) for λ we get:
$$\lambda = -2x - y + 140.$$
Solving equation (2) for λ we get:
$$\lambda = -4y - x + 210.$$
Equating the λ, we can simplify the two equations into a single equation involving x and y.
$$-4y - x + 210 = -2x - y + 140$$
$$-3y = -x - 70$$
$$y = \frac{1}{3}x + \frac{70}{3}$$

The solution is continued on the next page.

Substituting $y = \frac{1}{3}x + \frac{70}{3}$ into Eq. (3) on the previous page, we get:

$$-x - \left(\frac{1}{3}x + \frac{70}{3}\right) + 90 = 0$$

$$-\frac{4}{3}x + \frac{200}{3} = 0$$

$$-4x = -200$$

$$x = \frac{200}{4} = 50.$$

Therefore,

$$y = \frac{1}{3}x + \frac{70}{3} = \frac{1}{3}(50) + \frac{70}{3} = 40.$$

The critical point on the constraint $x + y = 90$ is the point $(50, 40)$, which we have already found.

Therefore, the critical points are:

$(0,0), (0,90), (90,0), (0,52.5), (70,0),$ and $(50,40)$.

Evaluating each of the critical points, we have:

$P(0,0) = 0$

$P(0,90) = -1300$

$P(90,0) = 200$

$P(0,52.5) = 1212.50$

$P(70,0) = 600$

$P(50,40) = 3400$

From the function values, we determine that the absolute maximum is 3400 and occurs at $(50, 40)$.

Therefore, the manufacture would maximize profit by producing 50 basic tables and 40 large model tables each week. The maximum profit would be $3400 per week.

37. Find the minimum value of

$$f(x, y) = 2x^2 + y^2 + 2xy + 3x + 2y$$

subject to the constraint

$$y^2 = x + 1.$$

We first express $y^2 = x + 1$ as $y^2 - x - 1 = 0$.
We form the new function F, given by:

$$F(x, y, \lambda)$$
$$= 2x^2 + y^2 + 2xy + 3x + 2y - \lambda\left(y^2 - x - 1\right).$$

We find the first partial derivatives:

$F_x = 4x + 2y + 3 + \lambda,$

$F_y = 2y + 2x + 2 - 2\lambda y,$

$F_\lambda = -\left(y^2 - x - 1\right).$

We set each derivative equal to 0 and solve the resulting system:

$$4x + 2y + 3 + \lambda = 0 \qquad (1)$$

$$2y + 2x + 2 - 2\lambda y = 0 \qquad (2)$$

$$y^2 - x - 1 = 0 \qquad (3) \begin{bmatrix} -\left(y^2 - x - 1\right) = 0, \text{ or} \\ y^2 - x - 1 = 0 \end{bmatrix}$$

Solving Eq. (1) for λ, we get:

$$4x + 2y + 3 + \lambda = 0$$

$$\lambda = -4x - 2y - 3. \qquad (4)$$

Solving Eq. (2) for λ, we get:

$$2y + 2x + 2 - 2\lambda y = 0$$

$$2\lambda y = 2y + 2x + 2$$

$$\lambda = \frac{2y + 2x + 2}{2y}$$

$$\lambda = \frac{y + x + 1}{y}. \qquad (5)$$

Setting Eq. (4) equal to Eq. (5) and solving for x we have:

$$-4x - 2y - 3 = \frac{y + x + 1}{y}$$

$$-4xy - 2y^2 - 3y = y + x + 1$$

$$-2y^2 - 3y - y - 1 = x + 4xy$$

$$-2y^2 - 4y - 1 = x(1 + 4y)$$

$$\frac{-2y^2 - 4y - 1}{1 + 4y} = x. \qquad (6)$$

Solving Eq. (3) for x we have:

$$y^2 - x - 1 = 0$$

$$y^2 - 1 = x \qquad (7)$$

The solution is continued on the next page.

Substituting Eq. (7) into Eq. (6) from the previous page, we have:

$$\frac{-2y^2 - 4y - 1}{1 + 4y} = y^2 - 1$$

$$-2y^2 - 4y - 1 = \left(y^2 - 1\right)\left(1 + 4y\right)$$

$$-2y^2 - 4y - 1 = y^2 + 4y^3 - 1 - 4y$$

$$0 = 4y^3 + 3y^2$$

$$0 = y^2\left(4y + 3\right)$$

$$y^2 = 0 \quad \text{or} \quad 4y + 3 = 0$$

$$y = 0 \quad \text{or} \quad y = -\frac{3}{4}$$

Using equation (7) when $y = 0$,

$$x = (0)^2 - 1$$

$$= -1$$

$$f(-1, 0) - 2(-1)^2 + (0)^2 + 2(-1)(0) +$$

$$3(-1) + 2(0)$$

$$= 2 - 3$$

$$= -1.$$

Using Eq. (7), when $y = -\dfrac{3}{4}$,

$$x = \left(\frac{-3}{4}\right)^2 - 1$$

$$= \frac{9}{16} - 1$$

$$= -\frac{7}{16}.$$

Evaluating the function we have:

$$f\left(-\frac{7}{16}, -\frac{3}{4}\right) = 2\left(-\frac{7}{16}\right)^2 + \left(-\frac{3}{4}\right)^2 +$$

$$2\left(-\frac{7}{16}\right)\left(-\frac{3}{4}\right) + 3\left(-\frac{7}{16}\right) +$$

$$2\left(-\frac{3}{4}\right)$$

$$= \frac{49}{128} + \frac{9}{16} + \frac{21}{32} - \frac{21}{16} - \frac{3}{2}$$

$$= -\frac{155}{128}.$$

The minimum value of f subject to the constraint occurs at $\left(-\dfrac{7}{16}, -\dfrac{3}{4}\right)$ and is

$$f\left(-\frac{7}{16}, -\frac{3}{4}\right) = -\frac{155}{128}.$$

39. Find the maximum value of

$$f(x, y, z) = x^2 y^2 z^2$$

subject to the constraint

$$x^2 + y^2 + z^2 = 2.$$

We first express $x^2 + y^2 + z^2 = 2$ as

$$x^2 + y^2 + z^2 - 2 = 0.$$

We form the new function F, given by:

$$F(x, y, z, \lambda)$$

$$= x^2 y^2 z^2 - \lambda\left(x^2 + y^2 + z^2 - 2\right).$$

We find the first partial derivatives:

$$F_x = 2xy^2 z^2 - 2\lambda x$$

$$F_y = 2x^2 yz^2 - 2\lambda y,$$

$$F_z = 2x^2 y^2 z - 2\lambda z,$$

$$F_\lambda = -\left(x^2 + y^2 + z^2 - 2\right).$$

We set each derivative equal to 0 and solve the resulting system:

$$2xy^2 z^2 - 2\lambda x = 0$$

$$2x^2 yz^2 - 2\lambda y = 0$$

$$2x^2 y^2 z - 2\lambda z = 0$$

$$x^2 + y^2 + z^2 - 2 = 0 \quad \begin{bmatrix} \left(x^2 + y^2 + z^2 - 2\right) = 0, \text{ or} \\ x^2 + y^2 + z^2 - 2 = 0 \end{bmatrix}$$

Rewriting the system we get:

$$2x\left(y^2 z^2 - \lambda\right) = 0 \quad (1)$$

$$2y\left(x^2 z^2 - \lambda\right) = 0 \quad (2)$$

$$2z\left(x^2 y^2 - \lambda\right) = 0 \quad (3)$$

$$x^2 + y^2 + z^2 - 2 = 0 \quad (4) \begin{bmatrix} -\left(x^2 + y^2 + z^2 - 2\right) = 0, \text{ or} \\ x^2 + y^2 + z^2 - 2 = 0 \end{bmatrix}$$

Note that for

$x = 0$, $y = 0$, or $z = 0$, $f(x, y, z) = 0$. For all values of x, y, and $z \neq 0$, $f(x, y, z) > 0$. Thus the maximum value of f cannot occur when any or all of the variables is 0. Thus we will only consider nonzero values of x, y, and z. Using the Principle of Zero Products, we get:

From Eq. (1)   From Eq. (2)   From Eq. (3)

$$y^2 z^2 - \lambda = 0 \quad x^2 z^2 - \lambda = 0 \quad x^2 y^2 - \lambda = 0$$

$$y^2 z^2 = \lambda \quad\quad x^2 z^2 = \lambda \quad\quad x^2 y^2 = \lambda$$

Thus, $y^2 z^2 = x^2 z^2 = x^2 y^2$ and $x^2 = y^2 = z^2$.

The solution is continued on the next page.

Substituting x^2 for y^2 and z^2 in Eq. (4) on the previous page, we have:

$$x^2 + x^2 + x^2 - 2 = 0$$
$$3x^2 = 2$$
$$x^2 = \frac{2}{3}$$
$$x = \pm\sqrt{\frac{2}{3}}$$

Since $x^2 = y^2 = z^2$ it follows that

$$y^2 = \frac{2}{3} \quad \text{and} \quad z^2 = \frac{2}{3}$$
$$y = \pm\sqrt{\frac{2}{3}} \quad \text{and} \quad z = \pm\sqrt{\frac{2}{3}}.$$

For $\left(\pm\sqrt{\frac{2}{3}}, \pm\sqrt{\frac{2}{3}}, \pm\sqrt{\frac{2}{3}}\right)$:

$$f(x,y,z) = \left(\pm\sqrt{\frac{2}{3}}\right)^2 \left(\pm\sqrt{\frac{2}{3}}\right)^2 \left(\pm\sqrt{\frac{2}{3}}\right)^2$$
$$= \frac{2}{3} \cdot \frac{2}{3} \cdot \frac{2}{3}$$
$$= \frac{8}{27}$$

Thus $f(x,y,z)$ has a maximum value of $\frac{8}{27}$ at

$$\left(\pm\sqrt{\frac{2}{3}}, \pm\sqrt{\frac{2}{3}}, \pm\sqrt{\frac{2}{3}}\right).$$

41. Find the maximum value of
$$f(x,y,z,t) = x+y+z+t$$
subject to the constraint
$$x^2+y^2+z^2+t^2 = 1.$$
We first express $x^2+y^2+z^2+t^2=1$ as
$$x^2+y^2+z^2+t^2-1=0.$$
We form the new function F, given by:
$$F(x,y,z,t,\lambda)$$
$$= x+y+z+t-\lambda(x^2+y^2+z^2+t^2-1)$$
We find the first partial derivatives:
$$F_x = 1-2\lambda x,$$
$$F_y = 1-2\lambda y,$$
$$F_z = 1-2\lambda z,$$
$$F_t = 1-2\lambda t,$$
$$F_\lambda = -(x^2+y^2+z^2+t^2-1).$$

We set each derivative equal to 0 and solve the resulting system:
$$1-2\lambda x = 0 \quad (1)$$
$$1-2\lambda y = 0 \quad (2)$$
$$1-2\lambda z = 0 \quad (3)$$
$$1-2\lambda t = 0 \quad (4)$$
$$x^2+y^2+z^2+t^2-1=0 \quad (5)$$
$$\left[\begin{array}{l}-(x^2+y^2+z^2+t^2-1)=0,\text{ or}\\ x^2+y^2+z^2+t^2-1=0\end{array}\right]$$

From Eq. (1), Eq. (2), Eq. (3), and Eq. (4), we see:
$$\frac{1}{2\lambda} = x = y = z = t.$$
Substituting x for y, z, and t in Eq. (5), we have:
$$x^2 + (x)^2 + (x)^2 + (x)^2 - 1 = 0$$
$$4x^2 = 1$$
$$x^2 = \frac{1}{4}$$
$$x = \pm\frac{1}{2}$$
Since $x = y = z = t$, it follows that

When $x = \frac{1}{2}$,
$$y = \frac{1}{2} \text{ and } z = \frac{1}{2} \text{ and } t = \frac{1}{2}.$$

When $x = -\frac{1}{2}$,
$$y = -\frac{1}{2} \text{ and } z = -\frac{1}{2} \text{ and } t = -\frac{1}{2}.$$
However, it is clear looking at the function that the point that will yield a maximum value is:
$$\left(\frac{1}{2}, \frac{1}{2}, \frac{1}{2}, \frac{1}{2}\right).$$
The maximum value is found as follows:
$$f\left(\frac{1}{2}, \frac{1}{2}, \frac{1}{2}, \frac{1}{2}\right) = \frac{1}{2}+\frac{1}{2}+\frac{1}{2}+\frac{1}{2} = 2.$$

43. We want to maximize
$$p(x,y)$$
subject to the constraint.
$$B = c_1 x + c_2 y.$$
We first express $B = c_1 x + c_2 y$ as
$$c_1 x + c_2 y - B = 0$$

The solution is continued on the next page.

Using the information on the previous page, we form the new function P, given by:

$$P(x, y, \lambda) = p(x, y) - \lambda(c_1 x + c_2 y - B).$$

We find the first partial derivatives.

$$P_x = p_x - \lambda c_1$$
$$P_y = p_y - \lambda c_2.$$

We set these derivatives equal to 0 and solve for λ.

$$p_x - \lambda c_1 = 0 \qquad p_y - \lambda c_2 = 0$$
$$p_x = \lambda c_1 \qquad p_y = \lambda c_2$$
$$\frac{p_x}{c_1} = \lambda \qquad \frac{p_y}{c_2} = \lambda$$

Thus, $\lambda = \dfrac{p_x}{c_1} = \dfrac{p_y}{c_2}$.

45. ✎

47. We want to minimize the distance

$$D = \sqrt{x^2 + y^2}$$

subject to the constraint

$$y = x^2 + 2x - 5$$

We rewrite the constraint to be:

$$-x^2 + y - 2x + 5 = 0$$

We form the new function F, given by:

$$F(x, y, \lambda) = \sqrt{x^2 + y^2} - \lambda\left(-x^2 + y - 2x + 5\right).$$

We find the first partial derivatives:

$$F(x, y, \lambda) = \sqrt{\underline{x}^2 + y^2} - \lambda\left(-\underline{x}^2 + y - 2\underline{x} + 5\right)$$

$$F_x = \frac{x}{\sqrt{x^2 + y^2}} + (2x + 2)\lambda$$

$$F(x, y, \lambda) = \sqrt{x^2 + \underline{y}^2} - \lambda\left(-x^2 + \underline{y} - 2x + 5\right)$$

$$F_y = \frac{y}{\sqrt{x^2 + y^2}} - \lambda,$$

$$F(x, y, \lambda) = \sqrt{x^2 + y^2} - \underline{\lambda}\left(-x^2 + y - 2x + 5\right)$$

$$F_\lambda = -\left(-x^2 + y - 2x + 5\right).$$

We set each derivative equal to 0 and solve the resulting system:

$$\frac{x}{\sqrt{x^2 + y^2}} + (2x + 2)\lambda = 0 \qquad (1)$$

$$\frac{y}{\sqrt{x^2 + y^2}} - \lambda = 0 \qquad (2)$$

$$-\left(-x^2 + y - 2x + 5\right) = 0 \qquad (3)$$

Solving Eq. (2) for λ, we get:

$$\lambda = \frac{y}{\sqrt{x^2 + y^2}}.$$

Substituting into Eq. (1) for λ, we get:

$$\frac{x}{\sqrt{x^2 + y^2}} + (2x + 2)\left(\frac{y}{\sqrt{x^2 + y^2}}\right) = 0$$

$$\frac{x + 2xy + 2y}{\sqrt{x^2 + y^2}} = 0$$

$$x + 2xy + 2y = 0$$

$$y = \frac{-x}{2x + 2} \qquad (4)$$

Substituting $-\dfrac{x}{2x + 2}$ for y in Eq. (3), we get:

$$x^2 - \left(\frac{-x}{2x + 2}\right) + 2x - 5 = 0$$

$$\frac{2x^3 + 2x^2 + x + 4x^2 + 4x - 10x - 10}{2x + 2} = 0$$

$$2x^3 + 6x^2 - 5x - 10 = 0$$

Using a graphing utility, we determine the solutions to this equation are:
$x \approx -3.298$ or $x \approx -1.091$ or $x = 1.389$.
Using equation (4) we find the values of y.

$$y = \frac{-(-3.298)}{2(-3.298) + 2} \approx -0.718$$

$$y = \frac{-(-1.091)}{2(-1.091) + 2} \approx -5.995$$

$$y = \frac{-(1.389)}{2(1.389) + 2} \approx -0.291$$

Substituting back into the objective function we have:

$$D(-3.298, -0.7176) = \sqrt{(-3.298)^2 + (-0.718)^2}$$
$$\approx 3.375$$

$$D(-1.091, -5.9945) = \sqrt{(-1.091)^2 + (-5.995)^2}$$
$$\approx 6.093$$

$$D(1.389, -0.291) = \sqrt{(1.389)^2 + (-0.291)^2}$$
$$\approx 1.419$$

Therefore, the point on the parabola that is closest to the origin is $(1.389, -0.291)$.

49 – 56. Left to the student.

Exercise Set 6.6

1. $\displaystyle\int_0^4\int_0^3 3x\,dx\,dy$

$=\displaystyle\int_0^4\left(\int_0^3 3x\,dx\right)dy$

We first evaluate the inside x-integral:

$\displaystyle\int_0^3 3x\,dx=\left[\frac{3x^2}{2}\right]_0^3$

$\qquad=\dfrac{3}{2}\left[3^2-0^2\right]$

$\qquad=\dfrac{3}{2}[9]=\dfrac{27}{2}.$

Then we evaluate the outside y-integral:

$\displaystyle\int_0^4\left(\int_0^3 3x\,dx\right)dy=\int_0^4\frac{27}{2}\,dy\quad\left(\int_0^3 3x\,dx=\frac{27}{2}\right)$

$\qquad=\left[\dfrac{27}{2}\,y\right]_0^4$

$\qquad=\dfrac{27}{2}\cdot4-\dfrac{27}{2}\cdot0$

$\qquad=54.$

3. $\displaystyle\int_{-1}^3\int_1^2 x^2y\,dy\,dx$

$=\displaystyle\int_{-1}^3\left(\int_1^2 x^2y\,dy\right)dx$

We first evaluate the inside y-integral, treating x as a constant:

$\displaystyle\int_1^2 x^2y\,dy=x^2\left[\frac{1}{2}y^2\right]_1^2$

$\qquad=x^2\left[\dfrac{1}{2}(2)^2-\dfrac{1}{2}(1)^2\right]$

$\qquad=x^2\left[2-\dfrac{1}{2}\right]$

$\qquad=\dfrac{3}{2}x^2.$

Then we evaluate the outside x-integral:

$\displaystyle\int_{-1}^3\left(\int_1^2 x^2y\,dy\right)dx=\int_{-1}^3\frac{3}{2}x^2\,dx\quad\left(\int_1^2 x^2y\,dy=\frac{3}{2}x^2\right)$

$\qquad=\left[\dfrac{1}{2}x^3\right]_{-1}^3$

$\qquad=\dfrac{1}{2}\left[3^3-(-1)^3\right]$

$\qquad=\dfrac{1}{2}\left[27-(-1)\right]$

$\qquad=\dfrac{1}{2}[28]$

$\qquad=14.$

5. $\displaystyle\int_{-4}^{-1}\int_1^3 (x+5y)\,dx\,dy$

$=\displaystyle\int_{-4}^{-1}\left(\int_1^3 (x+5y)\,dx\right)dy$

We first evaluate the inside x-integral, treating y as a constant:

$\displaystyle\int_1^3 (x+5y)\,dx$

$=\left[\dfrac{1}{2}x^2+5yx\right]_1^3$

$=\left[\dfrac{1}{2}(3)^2+5y(3)\right]-\left[\dfrac{1}{2}(1)^2+5y(1)\right]$

$=\dfrac{9}{2}+15y-\dfrac{1}{2}-5y$

$=10y+4.$

Then we evaluate the outside y-integral:

$\displaystyle\int_{-4}^{-1}\left(\int_1^3 (x+5y)\,dx\right)dy$

$=\displaystyle\int_{-4}^{-1}(10y+4)\,dy\quad\left(\int_1^3 (x+5y)\,dx=10y+4\right)$

$=\left[5y^2+4y\right]_{-4}^{-1}$

$=\left[5(-1)^2+4(-1)\right]-\left[5(-4)^2+4(-4)\right]$

$=(5-4)-\left[5(16)-16\right]$

$=1-80+16$

$=-63.$

7. $\displaystyle\int_0^1\int_x^1 xy\,dy\,dx$

$$=\int_0^1\left(\int_x^1 xy\,dy\right)dx$$

We first evaluate the inside y-integral, treating x as a constant:

$$\int_x^1 xy\,dy=x\left[\frac{1}{2}y^2\right]_x^1$$

$$=x\left[\frac{1}{2}(1)^2-\frac{1}{2}(x)^2\right]$$

$$=x\left[\frac{1}{2}-\frac{1}{2}x^2\right]$$

$$-\frac{1}{2}x-\frac{1}{2}x^3.$$

Then we evaluate the outside x-integral:

$$\int_0^1\left(\int_x^1 xy\,dy\right)dx$$

$$=\int_0^1\left(\frac{1}{2}x-\frac{1}{2}x^3\right)dx\quad\left(\int_x^1 xy\,dy=\tfrac{1}{2}x-\tfrac{1}{2}x^3\right)$$

$$=\left[\frac{1}{4}x^2-\frac{1}{8}x^4\right]_0^1$$

$$=\left[\frac{1}{4}(1)^2-\frac{1}{8}(1)^4\right]-\left[\frac{1}{4}(0)^2-\frac{1}{8}(0)^4\right]$$

$$=\left[\frac{1}{4}-\frac{1}{8}\right]-[0]$$

$$=\frac{1}{8}.$$

9. $\displaystyle\int_0^1\int_{x^2}^x (x+y)\,dy\,dx$

$$-\int_0^1\left(\int_{x^2}^x(x+y)\,dy\right)dx$$

We first evaluate the inside y-integral, treating x as a constant:

$$\int_{x^2}^x (x+y)\,dy$$

$$=\left[xy+\frac{1}{2}y^2\right]_{x^2}^x$$

$$=\left[x(x)+\frac{1}{2}(x)^2\right]-\left[x(x^2)+\frac{1}{2}(x^2)^2\right]$$

$$=\left[x^2+\frac{1}{2}x^2\right]-\left[x^3+\frac{1}{2}x^4\right]$$

$$=-\frac{1}{2}x^4-x^3+\frac{3}{2}x^2.$$

Then we evaluate the outside x-integral:

$$\int_0^1\left(\int_{x^2}^x(x+y)\,dy\right)dx$$

$$=\int_0^1\left(-\frac{1}{2}x^4-x^3+\frac{3}{2}x^2\right)dx$$

$$\left(\int_{x^2}^x(x+y)\,dy=-\tfrac{1}{2}x^4-x^3+\tfrac{3}{2}x^2\right)$$

$$=\left[-\frac{1}{10}x^5-\frac{1}{4}x^4+\frac{1}{2}x^3\right]_0^1$$

$$=\left[-\frac{1}{10}(1)^5-\frac{1}{4}(1)^4+\frac{1}{2}(1)^3\right]-$$

$$\qquad\left[-\frac{1}{10}(0)^5-\frac{1}{4}(0)^4+\frac{1}{2}(0)^3\right]$$

$$=\left[-\frac{1}{10}-\frac{1}{4}+\frac{1}{2}\right]-[0]$$

$$=\frac{3}{20}.$$

11. $\displaystyle\int_0^1\int_1^{e^x}\frac{1}{y}\,dy\,dx$

$$=\int_0^1\left(\int_1^{e^x}\frac{1}{y}\,dy\right)dx$$

We first evaluate the inside y-integral, treating x as a constant:

$$\int_1^{e^x}\frac{1}{y}\,dy=\left[\ln y\right]_1^{e^x}$$

$$=\ln e^x-\ln 1$$

$$=x.$$

Then we evaluate the outside x-integral.

$$\int_0^1\left(\int_1^{e^x}\frac{1}{y}\,dy\right)dx$$

$$=\int_0^1 x\,dx\qquad\left(\int_1^{e^x}\frac{1}{y}\,dy=x\right)$$

$$=\left[\frac{1}{2}x^2\right]_0^1$$

$$=\left[\frac{1}{2}(1)^2-\frac{1}{2}(0)^2\right]$$

$$=\frac{1}{2}.$$

13. $\int_0^2 \int_0^x (x+y^2)\, dy\, dx$

$= \int_0^2 \left(\int_0^x (x+y^2)\, dy \right) dx$

We first evaluate the inside y-integral, treating x as a constant:

$\int_0^x (x+y^2)\, dy$

$= \left[xy + \frac{1}{3}y^3 \right]_0^x$

$= \left[x(x) + \frac{1}{3}(x)^3 \right] - \left[x(0) + \frac{1}{3}(0)^3 \right]$

$= \left[x^2 + \frac{1}{3}x^3 \right] - [0]$

$= \frac{1}{3}x^3 + x^2.$

Then we evaluate the outside x-integral:

$\int_0^2 \left(\int_0^x (x+y^2)\, dy \right) dx$

$= \int_0^2 \left(\frac{1}{3}x^3 + x^2 \right) dx$

$\left(\int_0^x (x+y^2)\, dy = \frac{1}{3}x^3 + x^2 \right)$

$= \left[\frac{1}{12}x^4 + \frac{1}{3}x^3 \right]_0^2$

$= \left[\frac{1}{12}(2)^4 + \frac{1}{3}(2)^3 \right] - \left[\frac{1}{12}(0)^4 + \frac{1}{3}(0)^3 \right]$

$= \left[\frac{16}{12} + \frac{8}{3} \right] - [0]$

$= \frac{4}{3} + \frac{8}{3}$

$= 4.$

15. $\int_0^1 \int_0^{1-x^2} (1-y-x^2)\, dy\, dx$

$= \int_0^1 \left(\int_0^{1-x^2} (1-y-x^2)\, dy \right) dx$

We first evaluate the inside y-integral, treating x as a constant:

$\int_0^{1-x^2} (1-y-x^2)\, dy$

$= \left[y - \frac{1}{2}y^2 - x^2 y \right]_0^{1-x^2}$

$= \left[(1-x^2) - \frac{1}{2}(1-x^2)^2 - x^2(1-x^2) \right] -$

$\left[(0) - \frac{1}{2}(0)^2 - x^2(0) \right]$

$= 1 - x^2 - \frac{1}{2}(1-2x^2+x^4) - x^2 + x^4$

$= 1 - x^2 - \frac{1}{2} + x^2 - \frac{1}{2}x^4 - x^2 + x^4$

$= \frac{1}{2}x^4 - x^2 + \frac{1}{2}.$

Then we evaluate the outside x-integral:

$= \int_0^1 \left(\int_0^{1-x^2} (1-y-x^2)\, dy \right) dx$

$= \int_0^1 \left(\frac{1}{2}x^4 - x^2 + \frac{1}{2} \right) dx$

$= \left[\frac{1}{10}x^5 - \frac{1}{3}x^3 + \frac{1}{2}x \right]_0^1$

$= \left[\frac{1}{10}(1)^5 - \frac{1}{3}(1)^3 + \frac{1}{2}(1) \right] -$

$\left[\frac{1}{10}(0)^5 - \frac{1}{3}(0)^3 + \frac{1}{2}(0) \right]$

$= \frac{1}{10} - \frac{1}{3} + \frac{1}{2}$

$= \frac{4}{15}.$

The volume of the solid is $\frac{4}{15}$ units^3.

17. $f(x, y) = 2x - y$

$0 \le x \le 2$

$2 \le y \le 3$

Find

$$z_{av} = \frac{1}{A(R)} \int_2^3 \int_0^2 f(x, y) dx dy$$

The area of the rectangular region is

$A(R) = (2)(1) = 2.$

So we have:

$$z_{av} = \frac{1}{2} \int_2^3 \left(\int_0^2 (2x - y) dx \right) dy$$

We first evaluate the inside x-integral, treating y as a constant:

$$\int_0^2 (2x - y) dx$$

$$= \left[x^2 - yx \right]_0^2$$

$$= \left[(2)^2 - y(2) \right] - \left[(0)^2 + y(0) \right]$$

$$= 4 - 2y.$$

Then we evaluate the outside y-integral:

$$z_{av} = \frac{1}{2} \int_2^3 \left(\int_0^2 (2x - y) dx \right) dy$$

$$= \frac{1}{2} \int_2^3 (4 - 2y) dy$$

$$= \frac{1}{2} \left[4y - y^2 \right]_2^3$$

$$= \frac{1}{2} \left[\left[4(3) - (3)^2 \right] - \left[4(2) - (2)^2 \right] \right]$$

$$= \frac{1}{2} [3 - 2]$$

$$= \frac{1}{2}.$$

19. First we must determine the boundaries of the triangular region. We decided to integrate along y first. We find the equation of the line segment connecting the points $(0, 0)$ and $(6, 3)$. The slope of this line is given by:

$$m = \frac{3 - 0}{6 - 0} = \frac{1}{2}.$$

Therefore the equation is given by:

$$y = \frac{1}{2} x + 0$$

$$y = \frac{1}{2} x.$$

Therefore, the region is bounded vertically by the lines $y = 0$ and $y = \frac{1}{2} x$.

The region is bounded horizontally $x = 0$ and $x = 6.$

The area of this triangular region is given by

$$A(R) = \frac{1}{2}(6)(3) = 9$$

Therefore the average value of $f(x, y)$ over the region is given by :

$$z_{av} = \frac{1}{A(R)} \int_0^6 \int_0^{\frac{x}{2}} f(x, y) dy dx$$

$$= \frac{1}{9} \int_0^6 \left(\int_0^{\frac{x}{2}} (x^2 y) dy \right) dx.$$

We first evaluate the inside y-integral, treating x as a constant.

$$\int_0^{\frac{x}{2}} x^2 y \, dy = x^2 \left[\frac{1}{2} y^2 \right]_0^{\frac{x}{2}}$$

$$= x^2 \left[\frac{1}{2} \left(\frac{x}{2} \right)^2 - \frac{1}{2} (0)^2 \right]$$

$$= x^2 \left[\frac{x^2}{8} \right]$$

$$= \frac{1}{8} x^4.$$

Next we evaluate the outside x-integral:

$$z_{av} = \frac{1}{9} \int_0^6 \left(\int_0^{\frac{x}{2}} (x^2 y) dy \right) dx$$

$$= \frac{1}{9} \int_0^6 \frac{1}{8} x^4 dx$$

$$= \frac{1}{72} \left[\frac{1}{5} x^5 \right]_0^6$$

$$= \frac{1}{360} \left[x^5 \right]_0^6$$

$$= \frac{1}{360} \left[(6)^5 - (0)^5 \right]$$

$$= \frac{1}{360} (7776)$$

$$= \frac{108}{5}.$$

21. a) $\rho(x,y)=\dfrac{1}{100}x^2 y$

$0 \le x \le 30$

$0 \le y \le 20$

The population of fireflies in this field is given by:

$$\int_0^{30}\int_0^{20}\rho(x,y)\,dy\,dx$$

$$=\int_0^{30}\left(\int_0^{20}\left(\frac{1}{100}x^2 y\right)dy\right)dx$$

We first evaluate the inside y-integral, treating x as a constant:

$$\int_0^{20}\left(\frac{1}{100}x^2 y\right)dy$$

$$=\left[\frac{1}{100}x^2\left(\frac{1}{2}y^2\right)\right]_0^{20}$$

$$=\left[\frac{1}{200}x^2 y^2\right]_0^{20}$$

$$=\frac{1}{200}x^2(20)^2-\frac{1}{200}x^2(0)^2$$

$$=2x^2.$$

Then we evaluate the outside x-integral:

$$\int_0^{30}2x^2\,dx$$

$$=\left[2\left(\frac{1}{3}x^3\right)\right]_0^{30}$$

$$=\frac{2}{3}(30)^3-\frac{2}{3}(0)^3$$

$$=18,000.$$

There are 18,000 fireflies in the field.

b) The Average Value over the field is given by:

$$z_{av}=\frac{1}{A(R)}\int_0^{30}\int_0^{20}\rho(x,y)\,dy\,dx$$

$$=\frac{1}{A(R)}\int_0^{30}\left(\int_0^{20}\left(\frac{1}{100}x^2 y\right)dy\right)dx$$

The area of the rectangular region bounded by $0 \le x \le 30$ and $0 \le y \le 20$ is given by

$$A(R)=(30)(20)=600.$$

Therefore, substituting the value found in part a) we have

$$z_{av}=\frac{1}{A(R)}\int_0^{30}\left(\int_0^{20}\left(\frac{1}{100}x^2 y\right)dy\right)dx$$

$$=\frac{1}{600}[18,000]$$

$$=30$$

Ther are 30 fireflies per square yard.

23. $\displaystyle\int_0^1\int_1^3\int_{-1}^2(2x+3y-z)\,dx\,dy\,dz$

$$=\int_0^1\int_1^3\left(\int_{-1}^2(2x+3y-z)\,dx\right)dy\,dz$$

We first evaluate the inside x-integral, treating y and z as constants:

$$\int_{-1}^2(2x+3y-z)\,dx$$

$$=\left[x^2+3yx-zx\right]_{-1}^2$$

$$=\left[(2)^2+3y(2)-z(2)\right]-$$
$$\quad\left[(-1)^2+3y(-1)-z(-1)\right]$$

$$=[4+6y-2z]-[1-3y+z]$$

$$=3+9y-3z.$$

Then we evaluate the middle y-integral, treating z as a constant:

$$\int_1^3\left(\int_{-1}^2(2x+3y-z)\,dx\right)dy$$

$$=\int_1^3(3+9y-3z)\,dy$$

$$=\left[3y+\frac{9}{2}y^2-3zy\right]_1^3$$

$$=\left[3(3)+\frac{9}{2}(3)^2-3z(3)\right]-$$
$$\quad\left[3(1)+\frac{9}{2}(1)^2-3z(1)\right]$$

$$=\left[9+\frac{81}{2}-9z\right]-\left[3+\frac{9}{2}-3z\right]$$

$$=42-6z.$$

The solution is continued on the next page.

Finally, we evaluate the outside z integral:

$$\int_0^1 \left(\int_1^3 \left(\int_{-1}^2 (2x+3y-z)\,dx \right) dy \right) dz$$

$$= \int_0^1 (42-6z)\,dz$$

$$= \left[42z - 3z^2 \right]_0^1$$

$$= \left[42(1) - 3(1)^2 \right] - \left[42(0) - 3(0)^2 \right]$$

$$= 42 - 3$$

$$= 39.$$

25. $\displaystyle \int_0^1 \int_0^{1-x} \int_0^{2-x} (xyz)\,dz\,dy\,dx$

$$= \int_0^1 \int_0^{1-x} \left(\int_0^{2-x} (xyz)\,dz \right) dy\,dx$$

We first evaluate the inside z-integral, treating x and y as constants:

$$\int_0^{2-x} (xyz)\,dz$$

$$= \left[\frac{1}{2} xyz^2 \right]_0^{2-x}$$

$$= \left[\frac{1}{2} xy(2-x)^2 \right] - \left[\frac{1}{2} xy(0)^2 \right]$$

$$= \left[\frac{1}{2} xy(2-x)^2 \right] - [0]$$

$$= \frac{1}{2} x(2-x)^2 y.$$

Then we evaluate the middle y-integral, treating x as a constant:

$$\int_0^{1-x} \left(\int_0^{2-x} (xyz)\,dz \right) dy$$

$$= \int_0^{1-x} \left(\frac{1}{2} x(2-x)^2 y \right) dy$$

$$= \frac{1}{2} x(2-x)^2 \left[\frac{1}{2} y^2 \right]_0^{1-x}$$

$$= \frac{1}{2} x(2-x)^2 \left[\frac{1}{2}(1-x)^2 - \frac{1}{2}(0)^2 \right]$$

$$= \frac{1}{4} x(2-x)^2 \left[(1-x)^2 \right]$$

$$- \frac{1}{4} x(4-4x+x^2)\left[1-2x+x^2 \right]$$

$$= \left[x - x^2 + \frac{1}{4} x^3 \right]\left[1-2x+x^2 \right]$$

$$= x - 2x^2 + x^3 - x^2 + 2x^3 - x^4 +$$

$$\frac{1}{4} x^3 - \frac{1}{2} x^4 + \frac{1}{4} x^5$$

$$= \frac{1}{4} x^5 - \frac{3}{2} x^4 + \frac{13}{4} x^3 - 3x^2 + x.$$

Finally, we evaluate the outside x integral:

$$= \int_0^1 \int_0^{1-x} \left(\int_0^{2-x} (xyz)\,dz \right) dy\,dx$$

$$= \int_0^1 \left(\frac{1}{4} x^5 - \frac{3}{2} x^4 + \frac{13}{4} x^3 - 3x^2 + x \right) dx$$

$$= \left[\frac{1}{24} x^6 - \frac{3}{10} x^5 + \frac{13}{16} x^4 - x^3 + \frac{1}{2} x^2 \right]_0^1$$

$$= \left[\frac{1}{24}(1)^6 - \frac{3}{10}(1)^5 + \frac{13}{16}(1)^4 - (1)^3 + \frac{1}{2}(1)^2 \right] -$$

$$\left[\frac{1}{24}(0)^6 - \frac{3}{10}(0)^5 + \frac{13}{16}(0)^4 - (0)^3 + \frac{1}{2}(0)^2 \right]$$

$$= \left(\frac{1}{24} - \frac{3}{10} + \frac{13}{16} - 1 + \frac{1}{2} \right) - (0)$$

$$= \frac{10}{240} - \frac{72}{240} + \frac{195}{240} - \frac{240}{240} + \frac{120}{240}$$

$$= \frac{13}{240}.$$

27. ✎

29. $f(x,y) = x^2 - 3x + \dfrac{1}{3}xy - \dfrac{1}{3}y + 2$

On the region $1 \le x \le 2$ and $3 \le y \le 5$

a) We integrate the function over the region as follows:

$\displaystyle\int_3^5 \int_1^2 f(x,y)\,dx\,dy$

$= \displaystyle\int_3^5 \left(\int_1^2 \left(x^2 - 3x + \frac{1}{3}xy - \frac{1}{3}y + 2 \right) dx \right) dy$

We first evaluate the inside x-integral, treating y as a constant:

$\displaystyle\int_1^2 \left(x^2 - 3x + \frac{1}{3}xy - \frac{1}{3}y + 2 \right) dx$

$= \left[\dfrac{1}{3}x^3 - \dfrac{3}{2}x^2 + \dfrac{1}{6}x^2 y - \dfrac{1}{3}xy + 2x \right]_1^2$

$= \left[\dfrac{1}{3}(2)^3 - \dfrac{3}{2}(2)^2 + \dfrac{1}{6}(2)^2 y - \dfrac{1}{3}(2)y + 2(2) \right] -$

$\qquad \left[\dfrac{1}{3}(1)^3 - \dfrac{3}{2}(1)^2 + \dfrac{1}{6}(1)^2 y - \dfrac{1}{3}(1)y + 2(1) \right]$

$= \left[\dfrac{8}{3} - 6 + \dfrac{2}{3}y - \dfrac{2}{3}y + 4 \right] -$

$\qquad \left[\dfrac{1}{3} - \dfrac{3}{2} + \dfrac{1}{6}y - \dfrac{1}{3}y + 2 \right]$

$= \dfrac{2}{3} - \left[\dfrac{5}{6} - \dfrac{1}{6}y \right]$

$= \dfrac{1}{6}y - \dfrac{1}{6}.$

Then we evaluate the outside y-integral:

$= \displaystyle\int_3^5 \left(\int_1^2 \left(x^2 - 3x + \frac{1}{3}xy - \frac{1}{3}y + 2 \right) dx \right) dy$

$= \displaystyle\int_3^5 \left(\dfrac{1}{6}y - \dfrac{1}{6} \right) dy$

$= \left[\dfrac{1}{12}y^2 - \dfrac{1}{6}y \right]_3^5$

$= \left[\dfrac{1}{12}(5)^2 - \dfrac{1}{6}(5) \right] - \left[\dfrac{1}{12}(3)^2 - \dfrac{1}{6}(3) \right]$

$= \left[\dfrac{25}{12} - \dfrac{5}{6} \right] - \left[\dfrac{3}{4} - \dfrac{1}{2} \right]$

$= \dfrac{5}{4} - \dfrac{1}{4}$

$= 1.$

Therefore, $\displaystyle\int_3^5 \int_1^2 f(x,y)\,dx\,dy = 1$.

b) The probability that the dart lands at a point in R for which $1 \le x \le 2$ and $3 \le y \le 4$ is given by:

$\displaystyle\int_3^4 \int_1^2 f(x,y)\,dx\,dy$

$= \displaystyle\int_3^4 \left(\int_1^2 \left(x^2 - 3x + \frac{1}{3}xy - \frac{1}{3}y + 2 \right) dx \right) dy$

We first evaluate the inside x-integral, treating y as a constant, we know from part a) that

$\displaystyle\int_1^2 \left(x^2 - 3x + \frac{1}{3}xy - \frac{1}{3}y + 2 \right) dx = \dfrac{1}{6}y - \dfrac{1}{6}.$

Then we evaluate the outside y-integral:

$= \displaystyle\int_3^4 \left(\int_1^2 \left(x^2 - 3x + \frac{1}{3}xy - \frac{1}{3}y + 2 \right) dx \right) dy$

$= \displaystyle\int_3^4 \left(\dfrac{1}{6}y - \dfrac{1}{6} \right) dy$

$= \left[\dfrac{1}{12}y^2 - \dfrac{1}{6}y \right]_3^4$

$= \left[\dfrac{1}{12}(4)^2 - \dfrac{1}{6}(4) \right] - \left[\dfrac{1}{12}(3)^2 - \dfrac{1}{6}(3) \right]$

$= \left[\dfrac{16}{12} - \dfrac{4}{6} \right] - \left[\dfrac{3}{4} - \dfrac{1}{2} \right]$

$= \dfrac{2}{3} - \dfrac{1}{4} = \dfrac{5}{12}.$

The probabity the dart will land at a point in the region is $\dfrac{5}{12}$.

c) The probability that the dart lands at a point in R for which $1 \le x \le 2$ and $4 \le y \le 5$ is given by:

$\displaystyle\int_4^5 \int_1^2 f(x,y)\,dx\,dy$

$= \displaystyle\int_4^5 \left(\int_1^2 \left(x^2 - 3x + \frac{1}{3}xy - \frac{1}{3}y + 2 \right) dx \right) dy$

We first evaluate the inside x-integral, treating y as a constant, we know from part a) that

$\displaystyle\int_1^2 \left(x^2 - 3x + \frac{1}{3}xy - \frac{1}{3}y + 2 \right) dx = \dfrac{1}{6}y - \dfrac{1}{6}.$

The solution is continued on the next page.

Then we evaluate the outside y-integral:

$$= \int_4^5 \left(\int_1^2 \left(x^2 - 3x + \frac{1}{3}xy - \frac{1}{3}y + 2 \right) dx \right) dy$$

$$= \int_4^5 \left(\frac{1}{6}y - \frac{1}{6} \right) dy$$

$$= \left[\frac{1}{12}y^2 - \frac{1}{6}y \right]_4^5$$

$$= \left[\frac{1}{12}(5)^2 - \frac{1}{6}(5) \right] - \left[\frac{1}{12}(4)^2 - \frac{1}{6}(4) \right]$$

$$= \left[\frac{25}{12} - \frac{5}{6} \right] - \left[\frac{16}{12} - \frac{4}{6} \right]$$

$$= \frac{5}{4} - \frac{2}{3}$$

$$= \frac{7}{12}.$$

Therefore, The probabity the dart will land at a point in the region is $\frac{7}{12}$.

Note: We recognizing this is a probability distribution function, we could have used our answer from part b) to determine the probability that the dart would land in the region as follows

$$P(1 \le x \le 2, 4 \le y \le 5) = 1 - P(1 \le x \le 2, 3 \le y \le 4)$$

$$= 1 - \frac{5}{12}$$

$$= \frac{7}{12}.$$

d) The probability that the dart lands at a point in R for which $y \le x + 2$ is given by:

$$\int_1^2 \int_3^{x+2} f(x, y) \, dy \, dx$$

$$- \int_1^2 \left(\int_3^{x+2} \left(x^2 - 3x + \frac{1}{3}xy - \frac{1}{3}y + 2 \right) dy \right) dx$$

We first evaluate the inside y-integral, treating x as a constant:

$$\int_3^{x+2} \left(x^2 - 3x + \frac{1}{3}xy - \frac{1}{3}y + 2 \right) dy$$

$$= \left[x^2 y - 3xy + \frac{1}{6}xy^2 - \frac{1}{6}y^2 + 2y \right]_3^{x+2}$$

$$= \left[(x^2 - 3x + 2)(x+2) + \frac{1}{6}x(x+2)^2 - \frac{1}{6}(x+2)^2 \right] -$$

$$\left[(x^2 - 3x + 2)(3) + \frac{1}{6}x(3)^2 - \frac{1}{6}(3)^2 \right]$$

$$= \left[x^3 - x^2 - 4x + 4 + \frac{x^3 + 4x^2 + 4x}{6} - \frac{x^2 + 4x + 4}{6} \right] -$$

$$\left[3x^2 - 9x + 6 + \frac{3}{2}x - \frac{3}{2} \right]$$

$$- \left[\frac{7x^3}{6} - \frac{x^2}{2} - 4x + \frac{10}{3} \right] - \left[3x^2 - \frac{15x}{2} + \frac{9}{2} \right]$$

$$= \frac{7x^3}{6} - \frac{7x^2}{2} + \frac{7x}{2} - \frac{7}{6} = \frac{7}{6}(x^3 - 3x^2 + 3x - 1).$$

Then we evaluate the outside x-integral:

$$= \int_1^2 \left(\int_3^{x+2} \left(x^2 - 3x + \frac{1}{3}xy - \frac{1}{3}y + 2 \right) dy \right) dx$$

$$= \int_1^2 \frac{7}{6} (x^3 - 3x^2 + 3x - 1) \, dx$$

$$= \frac{7}{6} \left[\frac{1}{4}x^4 - x^3 + \frac{3}{2}x^2 - x \right]_1^2$$

$$= \frac{7}{6} \left[\frac{1}{4}(2)^4 - (2)^3 + \frac{3}{2}(2)^2 - (2) - \right.$$

$$\left. \frac{1}{4}(1)^4 - (1)^3 + \frac{3}{2}(1)^2 - (1) \right]$$

$$= \frac{7}{6} \left[[0] - \left[-\frac{1}{4} \right] \right]$$

$$= \frac{7}{24}.$$

The probability the dart will land in the region is $\frac{7}{24}$.

31. Left to the student.